Polymeric Materials and Their Application in 3D Printing

Polymeric Materials and Their Application in 3D Printing

Editors

Cristina-Elisabeta Pelin
Anton Ficai

MDPI

Basel • Beijing • Wuhan • Barcelona • Belgrade • Novi Sad • Cluj • Manchester

Cristina-Elisabeta Pelin
Materials and Tribology
National Institute for
Aerospace Research
"Elie Carafoli" - I.N.C.A.S.
Bucharest
Romania

Anton Ficai
Department of Science
and Engineering of Oxide
Materials and Nanomaterials
National University for
Science and Technology
Politehnica Bucharest
Bucharest
Romania

Editorial Office
MDPI AG
Grosspeteranlage 5
4052 Basel, Switzerland

This is a reprint of the Special Issue, published open access by the journal *Polymers* (ISSN 2073-4360), freely accessible at: www.mdpi.com/journal/polymers/special_issues/1SQ2S72IR0.

For citation purposes, cite each article independently as indicated on the article page online and using the guide below:

Lastname, A.A.; Lastname, B.B. Article Title. *Journal Name* **Year**, *Volume Number*, Page Range.

ISBN 978-3-7258-1406-0 (Hbk)
ISBN 978-3-7258-1405-3 (PDF)
https://doi.org/10.3390/books978-3-7258-1405-3

Contents

About the Editors

Cristina-Elisabeta Pelin

Dr. Cristina Elisabeta Pelin has been a Science Researcher at the National Institute for Aerospace Research "Elie Carafoli"—INCAS since 2011 and Head of the Materials Team since 2023. She received her BSc (2010) and MSc (2012) in Polymer Science and Engineering and a PhD (2015) in Chemical Engineering at the University Politehnica of Bucharest-Faculty of Applied Chemistry and Materials Science. Her thesis received 2nd place in the ICAS-IFAR Award in 2018. She is a member of the EREA Future Sky Circular Aviation Theme, representing INCAS. She is the guest editor for a Special Issue of MDPI's *Polymers*. Her research interests include polymeric composite materials, nanocomposites, fiber-reinforced composites, fibers, ablative materials, carbon-carbon composites, mechanical properties, and thermal properties of materials from aeronautics and aerospace.

Anton Ficai

Anton FICAI (born 1981) is full professor and PhD advisor in the Faculty of Chemical Engineering and Biotehnologies, University POLITEHNICA of Bucharest, who is actively involved in both academic and scientific life at the university. His major academic interests are related to the coordination of the teaching activities related to the following classes: Chemistry, Chemistry and Characterization of Materials, Composite Materials for Medicine, NanoBioMaterials for Tissue Engineering and Drug Delivery Systems. His research interests are much broader and cover the following topics: tissue engineering; drug delivery systems; multifunctional materials; composite materials; coatings, antimicrobial/antitumoral materials; nanoparticles synthesis and characterization; and surface modification. Till now, over 300 scientific papers, of which over 250 were ISI papers and 18 were books or chapters (including 2 edited books), were published along with 28 patent applications (10 of them being already released). The international recognition he has garnered for his R&D activity can be highlighted by the multiple invitations to speak at international conferences, to serve as a guest editor, and to serve as a member of the editorial boards of different national and international journals, as well as the Section Editor-in-Chief of *Coatings*. Valedictorian of UPB and former participant and laureate of the National Chemistry Olympiads, he was awarded with over 100 Gold Medals, Special Awards, or Best Paper Awards. He is also an active member of The Academy of Romanian Scientists and several professional societies.

polymers

MDPI

Review

The Use of Additive Manufacturing Techniques in the Development of Polymeric Molds: A Review

George Pelin [1], Maria Sonmez [2] and Cristina-Elisabeta Pelin [1,*]

1 INCAS—National Institute for Aerospace Research "Elie Carafoli", Bd. Iuliu Maniu 220, 061126 Bucharest, Romania; pelin.george@incas.ro
2 INCDTP-ICPI—National Research and Development Institute for Textile and Leather—Division Leather and Footwear Research Institute, Ion Minulescu St. 93, 031215 Bucharest, Romania; maria.sonmez@icpi.ro
* Correspondence: pelin.cristina@incas.ro; Tel.: +40-724-815-527

Abstract: The continuous growth of additive manufacturing in worldwide industrial and research fields is driven by its main feature which allows the customization of items according to the customers' requirements and limitations. There is an expanding competitiveness in the product development sector as well as applicative research that serves special-use domains. Besides the direct use of additive manufacturing in the production of final products, 3D printing is a viable solution that can help manufacturers and researchers produce their support tooling devices (such as molds and dies) more efficiently, in terms of design complexity and flexibility, timeframe, costs, and material consumption reduction as well as functionality and quality enhancements. The compatibility of the features of 3D printing of molds with the requirements of low-volume production and individual-use customized items development makes this class of techniques extremely attractive to a multitude of areas. This review paper presents a synthesis of the use of 3D-printed polymeric molds in the main applications where molds exhibit a major role, from industrially oriented ones (injection, casting, thermoforming, vacuum forming, composite fabrication) to research or single-use oriented ones (tissue engineering, biomedicine, soft lithography), with an emphasis on the benefits of using 3D-printed polymeric molds, compared to traditional tooling.

Keywords: additive manufacturing; 3D-printed molds; polymeric materials

check for updates

Citation: Pelin, G.; Sonmez, M.; Pelin, C.-E.. The Use of Additive Manufacturing Techniques in the Development of Polymeric Molds: A Review. *Polymers* **2024**, *16*, 1055. https://doi.org/10.3390/polym16081055

Academic Editor: Annalisa Chiappone

Received: 9 March 2024
Revised: 3 April 2024
Accepted: 8 April 2024
Published: 11 April 2024

1. Additive Manufacturing Introduction

Additive manufacturing (AM), commercially as known 3D printing, rapid prototyping, solid freeform fabrication, rapid manufacturing, desktop manufacturing, direct digital manufacturing, layered manufacturing, generative manufacturing, tool-less model making, etc., originates in the principles of topography and photo sculpture that uses a layered method to create 3D-shaped objects [1]. Additive manufacturing research studies were first conducted in the 1960s; techniques based on it were first commercialized around the 1980s by 3D Systems company [2], and since then, it is under constant growth and evolution. The layer-by-layer principle creates the most powerful advantage of AM, which is the ability to create almost any possible shape, while decreasing the time of product development, making it a solution to build complex and exotic structures that are difficult to achieve with conventional manufacturing strategies [2,3]. Besides the complexity of geometries, AM is promising due to the rapid production time, low to zero waste, and reduced labor costs, with high precision and accuracy [4,5].

Although the most popular term "3D printing" is often used to refer to additive manufacturing, in fact there are several individual layers processing manufacturing, depending on the materials, machining, and design used. Besides the technological process that develops with the aid of computer-assisted automated equipment, additive manufacturing has an entire engineering process preceding it consisting of model construction [3] generally using a virtual model built in a CAD software [6] or an acquisition of a physical model by a 3D scanner [7]. However, despite the complexity of the designing process, additive manufacturing techniques still have multiple advantages in several technical situations and applications in comparison to the multi-step conventional manufacturing methods [1], as synthesized in Figure 1.

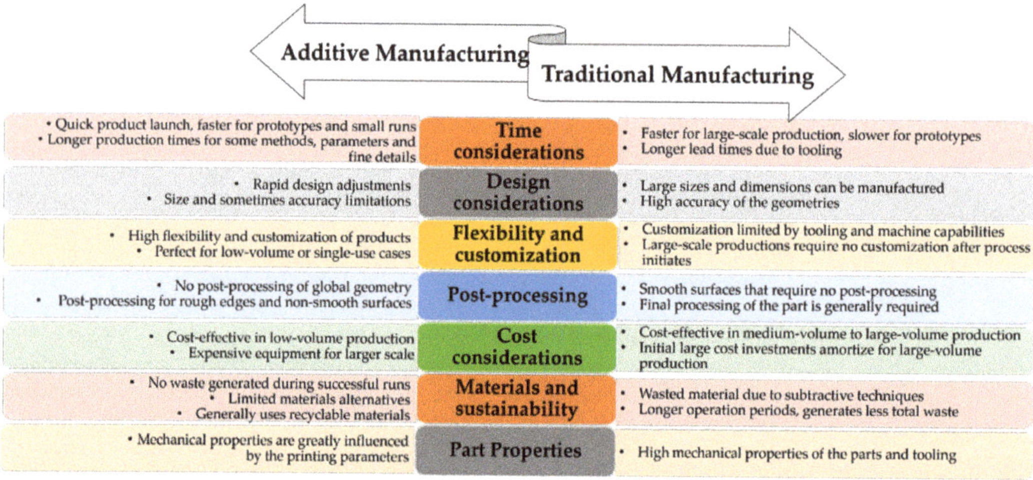

Figure 1. Main advantages versus main disadvantages of additive manufacturing compared to traditional manufacturing techniques.

It is important to mention that although AM is becoming more and more attractive and utilized in an expanding range of fields, there are still some areas in which traditional manufacturing exhibits major advantages compared to AM. Probably the most important consideration to be analyzed when choosing between these two routes is the production volume needed, which is a clear decisive factor, that consequently engages several other factors like additional time, total costs, and process global efficiency. The balance of all these factors in medium- to large-volume production is without a doubt ensured by the traditional manufacturing features, as the continuous production of a large number of parts will drastically diminish the cost per each part as well as amortize the initial high-cost investments. More than that, series production does not require design flexibility or product customization at any point after the production has started but rather requires a high resistance of the obtained products; therefore, the design benefits of AM are canceled in this situation. On the other hand, for low-volume production or customized products (such as individual-use cases that are crucial in patient-related medical areas), the issues associated with a large number of parts and less flexibility in the design become excrescent; therefore, AM features become major advantages in these situations. Besides the major issues related to time and costs, performance needs to be considered, in terms of materials and products' properties. For AM, there is a smaller range of polymeric materials available, and most of them are thermoplastics with medium-range mechanical and thermal performance especially over repeated cycles of production and use, while the metallic alternatives generally require expensive and high energetic consumption equipment. In terms of product performance, when fine details together with complex functional design are required, the increase in AM parameters demands greatly diminishes the time-reduction advantages, with the risk of eventual fatal errors generating waste, and consequently, additional and unforeseen costs as well as time delays [8].

For both additive and traditional manufacturing, the features that represent major advantages in some application domains become major disadvantages in other domains and vice versa. Therefore, although the innovative and rapid benefits of AM could create the belief that this technique could replace traditional manufacturing, in reality, these classes of manufacturing techniques complement each other, so they cover all necessary industrial fields and respond to the specific needs of a growing number of applications and newly appearing situations, such as the COVID-19 crisis. Thus, an effective approach would be to consider AM's unique features as an opportunity to cover areas where traditional manufacturing stumbles, or extend the use of traditional production processes, or combine AM with traditional techniques, where the product is manufactured using a hybrid technique (overprinting, over-molding).

Considering the great influence that the application specificity has on the selection of additive manufacturing as a production method, ASTM F42-Additive Manufacturing [9] introduced a classification of the AM processes into seven categories, according to the field that it is applied in, that are presented in Figure 2.

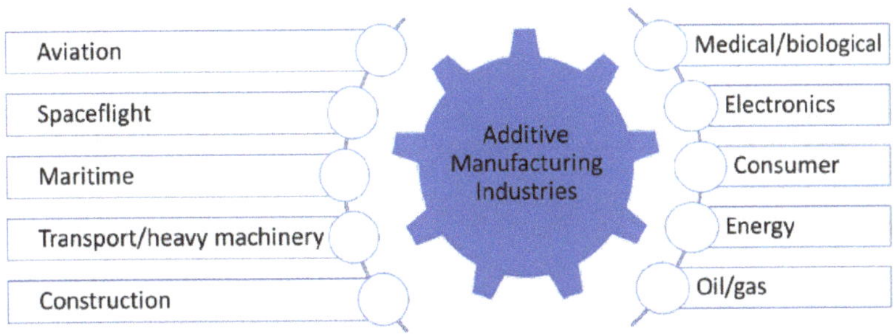

Figure 2. Main industry domains in which additive manufacturing is used.

In close connection to the categories of AM identified, a classification of the main AM methods is presented in Table 1.

Additive manufacturing usage is expanding continuously in the last decade; nowadays, these technologies are successfully implemented in a wide variety of industries that use concept models, functional models, patterns for investment and vacuum casting, medical models, and models for engineering analysis [6]. Therefore, besides the classification of AM techniques, the ASTM Committee F42 on Additive Manufacturing Technologies that is in charge of developing standards for additive manufacturing in a variety of industry-specific applications, settings, and conditions divided them by applications into 10 subsections: aviation, spaceflight, medical/biological, maritime, transport and heavy machinery, electronics, construction, oil/gas, consumer, and energy [10].

Nowadays, additive manufacturing still generates concerns about the quality of produced objects, high process failure rate, and/or higher associated cost and time of the production process as opposed to traditional manufacturing [8]. As Sztorch et al. [8] emphasized in their study concerning the production of personal protection products required in the COVID-19 crisis, there are situations in which the functionality of the product becomes a decisive factor and the product is required to enter the market in a short period of time and to be produced in large quantities, as extraordinary situations from the COVID-19 pandemic proved. They compared the launch of traditional injection molding together with FFF to produce face shields, considering unit costs and production possibilities at various timeframes, responding to the emerging immediate need for the quick provision of personal protective equipment for medical services. The comparison showed that FFF printing needs to be optimized by increasing the process speed by 6–10 times concomitantly with increasing reproducibility and part quality and mechanical strength caused by interlayer defects that need to be reduced. Polyamide 6 helmets proved to be a viable alternative for rapidly launching the production of products by mold injection, while for 3D printing to compete with this, using large groups of printers could be considered an option, but 3D printing can cover the buffer period until traditional injection molding enters into production, which responds to the immediate and urgent crisis requirements by some niche fields.

Table 1. Classification of the main AM methods [9,11–15].

AM Class	Materials Used	Principle	Techniques	Advantages	Disadvantages
Vat Photo Polymerization	Polymers (UV-curable photopolymer resins)	A liquid photopolymer in a vat is exposed to light source to be selectively cured into solid form	Stereolithography (SLA); Digital Light Processing (DLP); Continuous Liquid Interface Production (CLIP); Daylight Polymer Printing (DPP)	Rapid processing High quality finish of the part	High costs Extracting the 3D object from the mold generates issues
Material Jetting	Polymers (PP, HDPE, PS, PMMA), Waxes	Droplets of material are selectively deposited (jetted) on a substrate to build a 3D object	Material Jetting (MJ); Multi-jet Modeling (MJM); Nanoparticles Jetting (NPJ); Drop on Demand (DOD)	Less to zero waste	Difficult to apply in structural parts Post-processing required
Binder Jetting	Polymers (PA, ABS), Metals (stainless steel), Ceramics (Sand Glass)	Liquid bonding agent that acts as adhesive is selectively deposited to join materials in powder form	Powder Bed and Inkjet Head (PBIH); Plaster-based 3D Printing (PP)	Rapid processing No melting	Lower mechanical performance Post-processing required
Material Extrusion	Polymers (ABS, Polyamides, PC, PEI, PLA)	Thermoplastic polymer filament is extruded through a nozzle to build a 3D object	Fused Deposition Modeling (FDM); Fused Filament Fabrication (FFF)	Lower costs Good mechanical and structural properties High availability materials	Lower precision—many factors influence final model quality Accuracy and speed Nozzle requires technical attention
Sheet Lamination	Paper, Sheet Metals	Layers of material are joined together using an adhesive and printed one after the other (layer by layer) to build a 3D object	Laminated Object Manufacturing (LOM)	Low costs Acceptable accuracy	Limited material alternatives Post-processing required
Power Bed Fusion (PBF)	Metals (Stainless Steel, Aluminum, Titanium), Polymers (Polyamides)	Laser or electron beam melts or sinters the material in powder to build a 3D object	Selective Laser Sintering (SLS); Selective Laser Melting (SLM); Electron Beam Melting (EBM); Multi-Jet Fusion (MJF)	Suitable for prototyping Complex geometries	High costs Difficult to apply in structural parts Size limitations
Powder-fed Directed Energy Deposition (DED)	Metals (Stainless Steel, Aluminum, Titanium, etc.), Ceramics, Polymers	An electron beam, laser or arc energy source is directed toward a substrate material where it impinges with wire or powder feedstock material and melts, depositing the material on the substrate and building the part layer by layer	Wire Arc Additive Manufacturing (WAAM); Laser Metal Deposition (LMD); Laser Engineered Net Shaping (LENS); Laser Solid Forming (LSF); Directed Light Fabrication (DLF); 3D laser cladding	Suitable for repair/coat existing parts Machine large parts with high mechanical properties	Not suitable for small parts Lower detail accuracy and simple geometries

Where: PP—Polypropylene, HDPE—high density polyethylene, PS—polystyrene, PMMA—polymethyl methacrylate, PA—polyamide, ABS—acrylonitrile butadiene styrene, PC—polycarbonate, PEI—polyetherimides, PLA—polylactic acid.

Therefore, it becomes more and more clear that additive manufacturing and traditional manufacturing techniques as well as associated tooling are indeed not competitors, but rather complement each other in order to be able to respond successfully and efficiently to all the emerging requirements, technological evolution tendencies as well as exceptional situations, such as the ones generated by the pandemic years.

2. Additive Manufacturing Technologies That Use Polymers

The rapid evolution of additive manufacturing techniques adapted for polymeric composites development has evolved together with the circular economy growth and need for sustainability progress. As additive manufacturing of polymeric products and tools successfully supports the recycling and reusing of waste and/or used products to reintegrate them in a process chain, it greatly contributes to the circular economy concepts related to reducing raw-material consumption, waste, energetic consumption as well as costs related to manufacturing [16].

Since the birth of additive manufacturing, a multitude of methods have been introduced, customized, and personalized in several applications, from medicine, biomedicine, and tissue engineering to architectural design, automotive, aeronautics, and aerospace [17]. For plastic-based 3D-shaped products or tools, most studies and companies use material extrusion and vat photopolymerization, as they both allow the integration of reinforcing fibers into the polymer, and thus develop 3D-printed polymeric composites [3]. Material extrusion uses Fused Deposition Modeling and/or Fused Filament Fabrication techniques. The FDM additive manufacturing method was patented by Stratasys company in 1989 [18]; the term was trademarked in 1991 [19]. The term "Fused Filament Fabrication" began to be used when referring to other devices than that patented by Stratasys which used the same principles, in order to avoid litigation for copying their "FDM" trademark. Technically, both terms describe the same principle.

The three most used 3D printing techniques of plastic materials are fused deposition modeling, stereolithography, and selective laser sintering [20]. FFF/FDM and SLS use thermoplastics, while SLA uses thermosets; each of the techniques is presented below.

2.1. Fused Deposition Modeling

Fused filament fabrication is one of the most common techniques for polymer-based AM being widely used for printing components (from prototypes to functional end-use parts) manufactured from thermoplastic polymers. As described in Table 1, fused deposition modeling uses thermoplastic filaments as extrusion materials. As Figure 3 illustrates, the filament is subjected to heating until it reaches a molten state and extruded via the rollers rotating in opposite directions, through the nozzle of the printer, which moves in three degrees of freedom and deposits the polymer on a platform, building the part layer by layer according to the instructions and coordinates given through the design software-generated file [21].

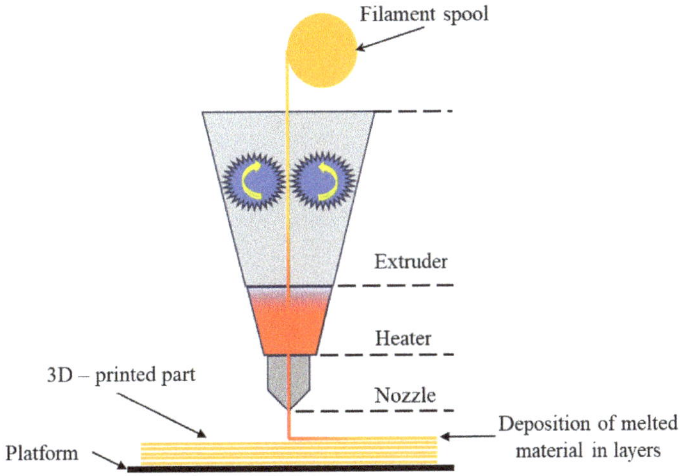

Figure 3. Working principle of the FFF/FDM technique.

In general, the consumer-level FFF/FDM technique ensures lower resolution and accuracy compared to other 3D printing processes using plastics, these two features being greatly influenced by the thermoplastic filament properties and the parts generally needing surface post-processing (i.e., chemical or mechanical polishing). Also, during the deposition of the layers, the formation of voids between them is a common problem which imprints a high closed porosity to the parts, influencing their capacity to bear mechanical loads. Therefore, this technique using consumer-level equipment is generally not suitable for complex designs or highly detailed parts, but it is a very attractive alternative for hobbies, DIY (Do-it-Yourself), and basic laboratory research studies helping students, researchers, engineers, and technicians. When using this technique on an industrial level, the available equipment provides some solutions to the drawbacks, and a larger variety of thermoplastics and even composites, but the price is commensurate with all the extra features [20,21].

The mechanical properties of components that are produced by FFF depend on the printing parameters, which are optimized to maximize the part quality, the microstructure, and the overall printing process economy [22,23]. FFF/FDM is currently confidently used in space hardware manufacturing applications for launch vehicles and spacecrafts [23]. FFF/FDM use a wide range of thermoplastics, from engineering nylons, ABS, and PLA to polyphenylene sulfide (PPS), polyetherimides reinforced with different fillers or blended with polycarbonate (known as ULTEM 9085 [24]), glycol-added polyethylene terephthalate (PET-G), thermoplastic polyurethanes (TPU), and high-tech thermoplastic consisting of polyetheretherketone (PEEK) and polyetherketoneketone (PEKK) [25]. The different classes of materials available to be 3D printed via FDM each possess specific advantages by their unique properties including transparency, biocompatibility, FST (flame–smoke–toxicity) certification, chemical resistance, heat resistance and strength, durability, etc., facilitating the material selection according to the target application [25].

Table 2 presents a summary of some of the most widely used polymers for 3D printing via FDM processing, showing their advantages and disadvantages when used in AM, together with their major fields of application.

Table 2. Main polymers used in FFF/FDM 3D printing—advantages, disadvantages, applications [20,26–37].

FFF Thermoplastics	Advantages	Disadvantages	Applications
PLA	Biodegradable, easy to print, cost-effective	Low strength, low durability, brittle	Consumer goods, toys, DYI, packaging, biomedical
ABS	More durable than PLA, impact-, heat-, chemical-, abrasion-resistant	More challenging to print, prone to warping	Consumer goods, tools, automotive, electrical enclosures
Polyamides	Durable, high strength, flexible	Water uptake, delamination, and poor adhesion when filled	Prosthetics, tools, encapsulations, working prototypes, mechanical components
PET-G	Versatile, flexible, mechanical strength, easy to print	Prone to dampness, easily scratched	Packaging, mechanical parts, printer parts, protective components
TPU	Rubber-like, flexible, durable	Challenging to print	Seals, gaskets, automotive, medical supplies
HIPS	Strength, flexible	Only compatible with ABS, easy to recycle, good support material	Protective packaging, containers
PVA	Biodegradable, cost-effective	Moisture uptake	Support in overhanging parts, sacrificial molds
PPS	Mechanical strength, thermally stable, chemically resistant	Low T_g, brittleness, low impact strength, prone to warping without fillers	Mechanical parts

Table 2. *Cont.*

FFF Thermoplastics	Advantages	Disadvantages	Applications
PEI	High T_g, flame retardant, mechanical strength	Expensive, susceptible to cracking	Automotive, aircraft parts
PEI/PC	High T_g, thermally stable, mechanical strength, chemically resistant	Water uptake	Transport, automotive, space applications
Carbon, glass, aramid fibers composites	Rigid, strong, tough	Compatibility limited to expensive industrial FDM 3D printers	Functional prototypes, jigs, fixtures, tooling

Where: HIPS—high-impact polystyrene, PVA—polyvinyl alcohol.

Besides the basic thermoplastic solutions for printing 3D parts, innovations in technology and materials have expanded their unique properties and usage by adding different compounds into the polymer and strongly enhancing the final products' performance and capacities [18]. There are applications that use FDM-printed parts from filaments infused with metallic, glass or ceramic compounds, in which the polymers are melted away by debinding and sintering to produce robust materials for electronics [18].

Amongst the tailored polymers for 3D printing are ULTEM materials, developed by Stratasys, ULTEM 9085 [24] being widely used for space applications as it offers high thermal stability, flame-retardant performance, chemical resistance, and high specific strength [38]. Tailoring of ULTEM properties for the improvement of its performance is presented in several research papers; most of them focus on improving its mechanical properties by the modification of printing parameters (i.e., layer thickness, raster angle and width, chamber temperature, print orientation, etc.) as well as the limitation of water uptake [39–41], the strength variation range being between 45–85% compared to injected parts.

It is clear that polymers can be successfully processed via AM methods, using them in a multitude of forms and compositions, from single polymer and polymeric blends, micro and nano composites, to short and long fiber-reinforced polymer composites, the used technique depending on the chosen compounds' processability features and target application. In the past decade, notable progress has been made in the field of 3D printing polymeric composites reinforced by fibers; considering the unique properties of polymers combined with the enhancements ensured by fiber-reinforcing agents, the immense benefits provided to the additive manufacturing sectors are of great value [42]. In the present, Stratasys manufactures FDM filaments composed of Nylon 12 and carbon fibers to produce parts as strong as aluminum, allowing the replacement of metal in different applications, exhibiting the highest flexural strength of any FDM thermoplastic, which leads to the highest stiffness-to-weight ratio [43]. However, for the development of molds, mechanical stresses are only some of the factors that influence their viability in different applications; therefore, strengthening with the aid of fibers is still limited to specific uses.

2.2. Stereolithography

Stereolithography was the first 3D printing technology and it remained until nowadays one of the most widely used for professional applications, due to the highest resolution and accuracy, high level of details, and high-quality surface finish that requires no further processing. Due to the high precision of the technology and chemical bonding formation between layers, the resulting parts are isotropic, and their mechanical performance is not influenced by process parameter variation. Given all these factors, the technique is optimum for highly detailed prototypes, such as molds, functional parts, patterns, jigs and fixtures, jewelry, dental implants, and end-use parts [20].

SLA belongs to the VAT polymerization class of AM techniques. According to Figure 4, a liquid, photosensitive thermoset resin is poured into a vat (tank) and interacts with a UV light for selective polymerization, the UV light curing the resin layer by layer until the final part is obtained. In SLA, layer thickness (or height) is generally approximately 50 μm but it can reach 10 μm, when extremely high quality is required, and time allows it.

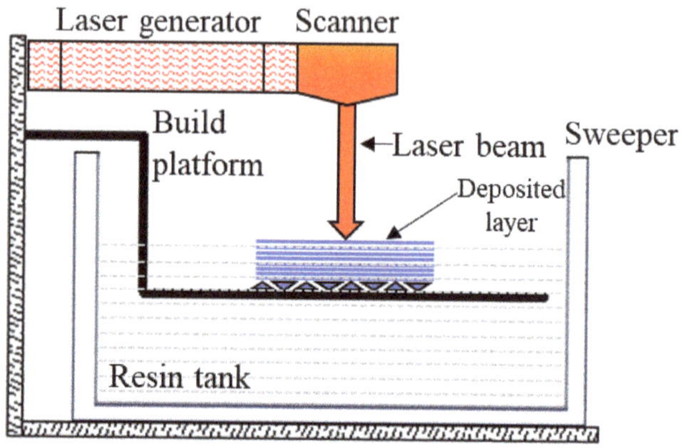

Figure 4. Working principle of the SLA technique.

Besides all these advantages, the SLA technique uses a wide and versatile resin formulation, covering several properties of tailoring (optical, mechanical, thermal, biocompatible). The materials' availability and properties are strongly dependent on the manufacturer and associated printer equipment.

The most important advantages of all SLA resins are high stiffness, highly smoothness of surface, and fine and high-level details, while the most important disadvantages are their low elongation at break that leads to brittle fracture, susceptibility to creep and UV radiation that affects their properties over time in outdoor applications [44]. Available SLA resins exhibit properties similar to some thermoplastics (i.e., ABS, PC, PP, etc.), being heat resistant by their high heat deflection temperatures (HDT), hard, flexible, impact-resistant, biocompatible or transparent, depending on their type [45]. Table 3 presents some of the main types of SLA resins, generally available at most manufacturers in different registered tradenames. The most commonly commercial SLA resins are manufactured by Formlabs, Protolabs, 3DLite, etc., each offering its own customized range of products for this application. Amongst them, Formlabs offers the most comprehensive resin library with over 40 SLA 3D printing material alternatives. In addition, the of the main types presented, Formlabs offers additional SLA resin alternatives such as flame-resistant resins (designed for indoor and industrial environments with high temperatures or ignition sources, like interior parts in aircrafts, cars, trains, protective and internal consumer/medical electronics components), Silicone 40A resins (first accessible 100% silicone 3D printing material with superior properties of cast silicone suitable for small batches of silicone parts, customized medical devices, flexible fixtures, masking tools, soft molds for casting urethane or resin), draft resins (up to 4 times faster than standard resins and 10 times faster than FDM), polyurethane resin (excellent long-term durability, stability to UV, temperature, humidity, flame retardancy, chemical and abrasion resistance, sterilizability), resins for medical and dental parts (biocompatible resins for producing medical and dental appliances), and jewelry (for easy investment casting and vulcanized rubber molding, with intricate details and strong shape retention) [20].

Table 3. Main polymers used in SLA 3D printing—advantages, disadvantages, applications [20,44,45].

SLA Resins	Advantages	Disadvantages	Applications
Standard	High tensile strength, high resolution, smooth surface finish	Very brittle (low elongation at break)	Visual prototypes, art models, concept models, looks-like prototypes
Tough (ABS-like)	High stiffness, excellent resistance to cyclic loads, compromise between properties of durable and standard resin	Not for parts with thin walls (minimum 1 mm), low HDT, brittle (low elongation at break)	Functional prototypes, mechanical assemblies, rigid parts that require high stiffness, housings and enclosures, jigs and fixtures, connectors, wear-and-tear prototypes
Durable (PP-like)	Highest impact strength and elongation at break, wear-resistant, flexible	Not for parts with thin walls (minimum 1 mm), low HDT, low tensile strength (lower than tough resin)	Prototyping parts with moving elements and snap-fits, consumer products, and low-friction and low-wear mechanical parts, housings and enclosures, jigs and fixtures, connectors, wear-and-tear prototypes
Heat-resistant	HDT between 200–300 °C, smooth surface finish	Poor impact strength, brittle, not for parts with thin walls (minimum 1 mm), temperature resistance increase decreases elongation	Heat-resistant fixtures, mold prototypes, hot air, gas and fluid flow equipment, and casting and thermoforming tooling, heat-resistant mounts, housings, and fixtures, molds and inserts
Ceramic-filled	Very stiff and rigid (high modulus and low creep), very smooth surface finish, good thermal stability and heat resistance)	More brittle than the tough and durable resins, brittle (low elongation at break), low impact strength	Molds and tooling, jigs, manifolds, fixtures, electrical application housings, and automotive parts
Flexible and elastic resin (rubber, TPU, silicone-like)	High flexibility (high elongation at break), low hardness (simulates an 80A durometer rubber), high impact resistance, flexibility of rubber, TPU, or silicone, bending, flexing, and compression resistance, repeated cycles without tearing	Lack the properties of true rubber, require extensive support structures, UV radiation sensibility, not for parts with thin walls (minimum 1 mm)	Objects that will be bent or compressed, wearables prototyping, multi-material assemblies, handles, grips, and overmolding, consumer goods prototyping, compliant features for robotics, medical devices, and anatomical models, special effects props and models
Clear resin	Polishes to near optical transparency, moisture-resistant, durable, large format available, stiff	Requires secondary operations for functional part clarity	Parts requiring optical transparency, millifluidics

SLA is known for creating high-resolution parts with good surface finish, but tensile strength can sometimes be affected; therefore, the careful choice of the material used is an important parameter for this technique as well [45].

2.3. Selective Laser Sintering

The SLS 3D printing technique belongs to the powder bed fusion AM class, which is generally applied for metals but can be applied for polyamides and a few other thermoplastics within the polymeric materials class. Its ability to produce strong functional parts at a high productivity rate generating low costs per part makes this technique trusted in a wide range of industries for applications such as rapid prototyping, manufacturing aids, low volume or custom production [20].

SLS 3D printing uses a high-power laser to sinter small particles of polymer powder into a solid structure based on a 3D model, as illustrated in Figure 5. The printing process develops over three main stages: (1) preheating—during which the powder bed is heated to a predefined temperature (bed temperature just below the softening temperature of the polymer that is used to minimize the laser energy and eliminate any distortion of the piece during cooling), held constant throughout the part-building process; (2) building phase—core phase of the fabrication process involving several operations (the lowering of the platform to receive the powder particles dragged by the roller or by the spreading blade, laser beam melting of the layer of particles along a computerized trajectory, gradually cooling down to the bed temperature for solidification); (3) cooling phase—during which the heat source is switched off and the powder bed cools, gradually cooling until it reaches the extraction temperature of the piece [46]. The unfused powder supports the printed part during the process, so it eliminates the need for dedicated support.

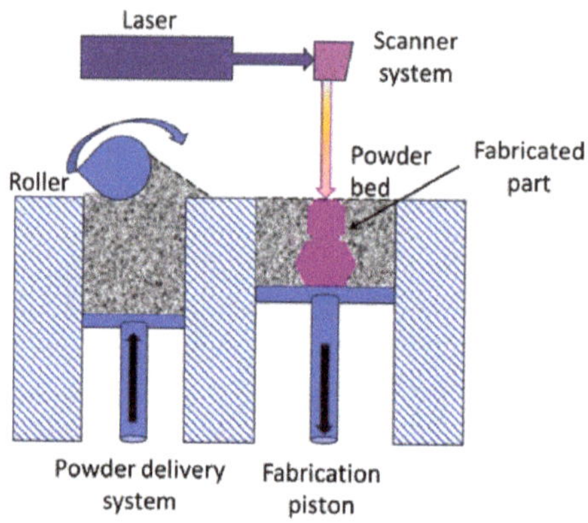

Figure 5. Working principle of the SLS technique.

SLS is an optimum choice for the printing of complex geometries, generating isotropic structures, and although surface finish is rather rough, post-processing is easy. As mentioned, compared to FFF/FDM and SLA, the available materials for SLS are very limited (mainly polyamides, sometimes PP, flexible TPU, TPE, and more recently, PEEK and PEKK), but the small class exhibits excellent mechanical performance, similar to injected parts [47], as presented in Table 4.

SLS is one of the 3D printing techniques that generates parts with one of the most isotropic compositions, and together with the use of high strength plastics, results in high-performance products.

Comparing the three 3D printing processes that use polymers, each of them can be a choice for different applications and requirements, as each of them has its own advantages as well as disadvantages. FF/FDM offers low-cost consumer equipment alternatives and uses widely available materials, SLA offers high accuracy, precision, and a smooth surface finish using a large variety of functional materials, while SLS ensures strong functional parts, without support structures during printing, and freedom of design. On the other hand, each of the three techniques exhibit drawbacks that could make them unsuitable for some applications. FDM only provides low accuracy and detail level with limited design when consumer equipment is used, with professional equipment mitigating some of the drawbacks but coming at a high cost. Parts that can be printed with SLA materials are often sensitive to long-term exposure to UV light, making them generally inaccessible for outdoor applications. SLS can be used with a limited range of materials and hardware equipment costs are higher. The selection between the three technologies needs to take into consideration all these aspects in the context of the cost investments, sustainability, application requirements, and available materials and equipment.

Table 4. Main polymers used in SLS 3D printing—advantages, disadvantages, applications [20,47–49].

SLS Resins	Advantages	Disadvantages	Applications
PA12	Strong, stiff, durable, impact-resistant and can endure repeated wear and tear; Resistant to UV, light, heat, moisture, solvents, temperature, and water	High porosity and low molecular weight deteriorate its mechanical properties, especially ductility and toughness	Functional and high-performance prototyping, end-use parts, medical devices, permanent jigs, fixtures, and tooling
PA11	Similar to PA12, but higher elasticity, elongation at break, and impact resistance	Lower stiffness than PA12	Functional prototyping, structural end-use parts, jigs, and fixtures, snaps, clips, and hinges, orthotics and prosthetics
Glass-filled PA12	Enhanced stiffness and thermal stability	More brittle, reduced impact resistance and flexibility	Robust jigs, fixtures, replacement parts, parts subjected to sustained loadings and high temperature, threads, and sockets
Carbon fiber-filled PA11	Highly stable, lightweight, high-performance material	More brittle, reduced impact resistance	Replacement for metal parts, tooling, jigs, fixtures, high-impact equipment, functional composite prototypes
Mineral-filled PA	Enhanced thermal properties, dimensional stability, rigidity, high HDT	Reduced impact resistance and flexibility, rougher surface than unfilled PA	Parts to withstand high temperatures or mechanical loads
Aluminum-filled PA	Dense, thermal, and conductive properties	Reduced impact resistance and flexibility	Parts with a metallic appearance, mechanical parts that do not experience high stress
Polypropylene	Ductile, durable, chemically resistant, watertight, weldable	Not as strong or rigid as other 3D-printed materials	Functional prototyping, end-use parts, watertight housings, cases, packaging prototypes, medical devices (orthotics and prosthetics), automotive interior components, strong and chemically resistant fixtures, tools, jigs
TPU	Flexible, elastic, rubbery, resilient to deformation, high UV stability, great shock absorption	Limited heat resistance, moisture sensitivity	Functional prototyping, flexible, rubber-like end-use parts, wearables and soft-touch elements, padding, dampers, cushions, grippers, gaskets, seals, masks, belts, plugs, tubes, medical devices (soles, splints, orthotics, prosthetics)
TPE	Elasticity, resistance to abrasion and good UV and ozone resistance	Temperature-sensitive, prone to shrinking	Seals, gaskets, plugs, grips, handles, over-molds, tubes, masks, and gloves
PEEK, PEKK	Excellent mechanical strength, stiffness, chemical resistance, wear resistance, thermal resistance	Low resistance to UV light, low flexibility, expensive	Components subject to friction or wear, surgical tools and implant, applications that require superior thermal resistance

3. Technologies That Use Molds

Molds represents one of the most used tools in the manufacturing industry with applications in several fields, from consumer goods to sports, medical, transport, and security. In the high competitiveness encountered in the mold and tooling industry nowadays, the time needed for a product to reach the market (time-to-market) represents an important factor to be considered by the companies, along with the quality of both the product and the mold, when building their tooling for development of products.

In today's competitive mold industry, a product's time-to-market plays an important role in the success of a company producing quality molds [50,51].

The molds and dies industry is the root of the manufacturing world, as they represent key elements in mass production. Molds find extended use in a wide range of technological processes, especially related to plastics (casting, injection, extrusion, compression, blow, rotational molding, resin transfer, etc.). Dies are mostly associated with metals, being implicated in technological processes like stamping, forming, metal injection, etc.

However, although tools are crucial for worldwide industrial fields, the molds industry faces some challenges as digital tooling expands. First, it is a capital-intensive industry in which the manufacturing costs are consistently increasing but the price of mold and die increase does not occur at the same rate, thus threatening the survival of competitors, automated shops, and factories which imply a decrease in human personnel and a lack of trained personnel to operate the machines [52]. Last but not least, the materials used for most of the industrial molds are expensive metallic ones, which require long timeframes and expensive manufacturing as well as secondary preparation stages; these factors represent unbalanced investment when low-volume production is needed or in application where the manufactured parts need constant tailoring and customization.

Therefore, considering all these aspects in the global economic and technological circumstances, it becomes urgent to direct the mold industry towards optimizing costs, improved efficiency, and advanced forecasting. There are a series of promoters that can significantly contribute to the aligning and allowing of the growth of the mold industry in the current economical/technological worldwide trend, some of them being 3D printing for prototyping, 5-Axis CNC precision machining, rapid tooling systems, and advanced CAM/CAD tools [52].

The main technologies that use molds as a main tooling method are injection molding, melt compounding, vacuum bagging liquid injection molding, casting, thermoforming, and composite fabrication, as well as different specific applications such as dedicated/customized biomedical devices.

These molding processes imply the use of different mold components, depending on the part targeted to be manufactured as well as the polymer type used. Thermoplastics can be molded by melting followed by cooling, while thermoset can be molded into different shapes by pouring/laying-up of resins in a liquid state and curing (at room or high temperatures). Some of the most important molding processes are listed below [53]:

- Casting—it is the simplest molding process, as it requires simple tooling and low costs, and can be performed at low pressures. The thermoplastic is heated until it reaches a molten state, poured into the mold, and allowed to cool before extraction from the mold. Although it allows the production of complex shapes, it can be used for parts with a thickness higher than 12–13 mm.
- Injection molding—it is one of the most extensively used techniques for molding plastics or metals as it allows the production of three-dimensional parts which can be easily reproduced. The material brought in liquid form is inserted/injected at a high pressure into a closed, cooled mold, filling it and taking its shape. The molded material is extracted after complete cooling and solidification. It is a process suitable for large quantity production (i.e., more than 30,000 parts per year). Despite the use of expensive tooling (i.e., expensive metallic molds), the large volume production ensures its cost-effectiveness; however, recent trends promote its use for smaller production volumes with the tooling adaption.
- Extrusion molding—it is similar to injection molding, but with the difference that the molten material is inserted/injected through a die and the obtained structure is linear and rod-like (not necessarily cylindrical). After cooling, the rod structure can be cut at different lengths depending on necessities.
- Compression molding—it is the most complicated molding process, in terms of labor, being used only for large-scale production (such as a higher number of small parts in boats, the automotive industry, etc.), and not for mass production. The liquid

molten material is poured into a lower mold and compressed with an upper mold into the desired shape and extracted after complete cooling and solidification. The high temperatures used ensure material strength.

- Blow molding—it is a process mainly used for pipes and milk bottle production, allowing the production of up to 1400 parts in a 12 h shift, with uniform wall thickness achievement. Although it uses the standard concept, it requires several different mold parts. The plastic in a melted state is injected into a cold mold, concomitant with air blowing into an attached tube, pressing the plastic against the walls of the mold so that it takes the shape of the mold. After complete cooling, the part is extracted.
- Rotational molding—it is an environmentally compatible process, as raw material does not go to waste. The process involves high-speed rotating using two mechanical arms, the mold that contains the hot liquid material, which uniformly coats the mold surface, and the final part has a uniform wall thickness and hollow shape. It is widely used for toys, tanks, and different other consumer goods.

The minimal requirements of 3D-printed molds come from the requirements imposed by the molding technology used together with the molded material properties. The most important requirements that a 3D-printed material has to respond to so that it can be used as mold tooling are referred to in [54]:

- Suitable mechanical properties, especially in terms of high stiffness—for example, injection molds must exhibit suitable mechanical performance to withstand the high pressure used during injection while maintaining a good dimensional stability (no deformation) and accuracy over multiple-use cycles.
- Suitable thermo-mechanical properties, in terms of resistance to high temperatures without showing deformation, meaning that the polymer used needs to exhibit a high value of heat deflection temperature in order to ensure a precise control of the process and the required dimensional stability.
- Dimensional accuracy is crucial for the production of parts with a high level of details.

Considering the limitations that polymers have by their own physico-chemical nature in the context of the materials requirements for molds and molded parts, using 3D-printed molds, especially for the injection molding technologies, narrows down their beneficial use to some technological situations such as referenced in [54]:

- When fast turnaround times are needed (1–2 weeks for 3D-printed molds as opposed to 5–7 weeks for traditional ones);
- Low-volume production (applications where a maximum number of 50–100 parts are needed);
- Small-size parts are to be produced (up to a maximum of 150 mm);
- Applications where design changes or iterations are foreseen.

The two 3D printing processes that can produce parts with high accuracy and smooth surfaces without requiring complex post-processing are material jetting and SLA. Materials jetting is used exclusively on an industrial scale, while SLA is available on both an industrial and a consumer level, although the materials and capabilities cannot be considered for high-end production [54].

Considering that in molding processes, the final part quality is greatly influenced by the mold features, there are a number of factors to be taken into consideration in terms of design to obtain the desired product quality via the desired process efficiency [55]:

- Selection of optimum material—the used materials need to withstand the parameters required to be implemented during the molding process (i.e., temperature, pressure) without melting, warping or deforming.
- Design considerations—the design of the mold needs to be optimized to build molds for any molding processes, especially injection molding, as design items (i.e., number of walls, wall thickness, draft angles, infill patterns, etc.) generate significant modifications to the quality and durability of the mold and consequently to the quality of the part and cost investments in the technology for the product.

- Testing trials and validation stages—as with any product or other processes, molds printed via 3D need to be tested in terms of resistance to the conditions required by the parameters used (thermal resistance, mechanical resistance and dimensional stability at the processing temperatures, pressures generating mechanical loads and during a required number of cycles), in order to establish the molds' limitations and perform adjustments if needed, before production starts.
- Production volume considerations—especially for injection molding that generally is suitable for thousands of cycles, 3D-printed molds cannot surpass traditional metallic tooling and can only be used when low-volume production (50–100) is possible due to the modification of their properties after a number of cycles; therefore, they can only be used in rapid prototyping, low-volume production, and other molding techniques that require single use or constant tailoring of the design.
- Size and shape of the molds—the selection of the mold type needs to take into consideration that it has to handle the size of the part to be manufactured, as generally mold machines by CNC are larger, and molds produced by 3D printing exhibit some size limitations compared to them.
- Surface finish—considering the high degree of surface finish offered by metallic molds (aluminum or steel), 3D-printed molds tend to exhibit generally rougher surfaces, decreasing the surface finish quality, without taking into consideration the degradation scenarios during injection molding, for example, rendering them the less suitable candidate in some applications.
- Draft angle—this factor needs to be considered especially for injection molding and composite fabrication, as its correct selection can contribute significantly to the facile extraction/demolding of the part at the end of the process.

As 3D printing technologies are in constant development and improving dynamics, the use of tooling produced by additive manufacturing techniques continues to expand. However, although AM appears to be replacing traditional tooling manufacturing, in reality, these two classes of techniques are partners rather than competitors at the risk of eliminating one another. Traditional manufacturing exhibits some clear benefits that could not be ensured (at least in the near future) by the AM alternatives, such as the fact that it allows high-volume production with lines that can run for 24 h daily, reduces cost-per-unit due to amortization of upfront tooling costs, and provides strong part consistency due to the possibility of repeating the same manufacturing cycles without deviation from the original design intended [56]. The major drawbacks of traditional manufacturing represented by the high waste of materials, inflexibility of the original design tailoring, and high costs for production quantity below a large volume (medium to low to single-use) are actually the major benefits of AM in their reversed form. Therefore, additive manufacturing emerges as a solution for the fields where the major drawbacks of traditional manufacturing generate a high level of impediments and disadvantages.

4. Applications That Use 3D-Printed Polymeric Molds

As already mentioned, there are sectors and applications in which traditional manufacturing techniques and their additional tooling cause significant technological and economic issues. Therefore, the use of AM tooling can optimize the supply chain and productivity by allowing the advanced and rapid customization of products, with improved functionality and weight implying reduced lead times and costs. AM tooling manufactured with polymeric materials is particularly useful in the low-volume production of high-complexity parts, where reiteration of design is a major requirement allowing the functionality improvement of the final product, and in the cases where weight reduction and fast lead times of the tooling are an advantage [57].

4.1. 3D Printing of Molds for Injection Techniques

Injection molding is one of the most established and important processes for mass production of objects and products from thermoplastics, usually without the need for additional finishing [58], being the second technology in plastic industry production, after extrusion technologies [59]. Injection molding is a mass production process as it allows the manufacturing of a large series of the same product with high quality [60,61].

One of the most important drawbacks of injection molding manufacturing is the high costs and extended lead time for designing and procuring the molds [62]. Industrially, the most commonly used materials for molds manufacturing are steels and aluminum, considering their machinability, variety in composition and properties, heat treatment possibility, higher thermal conductivity, and the ability to be coated for improved surface finish and good polishing ability [63,64]. Molds that are manufactured for high-volume production (up to millions of parts) require extreme durability and hardness/toughness and have to maintain their dimensional stability and special properties for thousands of thermal cycles [65]; therefore, steel alloys are the chosen material solution. However, although these harder materials exhibit all the positive properties, their requirement for special tooling imposed by the higher effort needed to mill them leads to a significant cost increase [66]. Therefore, for lower volume production, the higher costs exhibited by steel alloys machining into molds (generally achieved via CNC or electrical discharge machining techniques [62]) would increase the technology expenses beyond economic effectiveness. Additive manufacturing can bring important optimization advantages when combined with formative manufacturing such as injection molding, due to the advances in machines' design and materials [67]. Combining 3D printing for mold tooling seems to ensure a more cost-effective route compared to traditional metallic molds. Recently, plastics companies have shown an interest in using AM to manufacture molds for injection molding which can be used to produce end components in low-volume production; however, until now, there is no clear indication whether these parts are brought to market as independent products or components of a product, nor is there any indication of the cycle life of a mold produced using AM [68]. Rapid tooling is a term that describes the use of AM to achieve molds ensuring shorter lead times compared to conventional techniques [69]. Three-dimensional printing is a powerful solution for fabricating injection molds rapidly, with high flexibility and involving low costs, as it requires limited equipment, saving valuable CNC time and skilled operators for other high-value tasks. Molds manufactured via AM techniques can be obtained from industrial machines as well as from small-size laboratory equipment allowing design testing and iteration at a lower scale before investing in expensive tooling for mass production [62]; this diminishes raw material consumption during trials and non-profitable investment risks.

When choosing between 3D-printed and traditional tooling for injection molding processes, the volume of production is a crucial factor to be taken into consideration, as the features of AM can pass from advantages to disadvantages when large-volume production is used. Therefore, the additive manufacturing features of fast launching of the concept, high versatility and flexibility towards corrections required at almost any point during production, achieving high complexity geometries without significant increase in costs, and time and cost-effectiveness [64] represent major advantages when they are used as tooling methods instead of traditional ones, in low-volume, single-use or highly customizable case production.

Stratasys [70] summarizes a comparison between different methods to produce prototypes via injection molding, in terms of the number of parts, materials, average mold cost and average cost/past, which is presented in Table 5. However, Stratasys [70] mentions the important observation that with the use of FDM methods, the mechanical properties of the developed thermoplastic prototypes are not comparable with the ones obtained via traditional injection molding, as both processes and injected materials are different. Also, besides this, when producing a medium to a large volume of parts, although the time, cost, and post-processing required to produce the injection mold are significant, the long operation lifetime and large number of parts able to be manufactured compensate the investment to such an extent that it fully amortizes.

Table 5. Methods of producing polymer prototypes using different tooling methods [70–72].

Prototype Production Methods	Mold Durability	Average Mold Cost	Average Cost/Part	Production Average Cost	Lead Time	Design Flexibility
FDM direct 3D printing	N/A	N/A	Low to high	Low to high	Short to medium	High
Conventional Molds and Tooling	High (>10,000 parts)	High (2000 USD)	Low	High	Long	Low
3D-Printed Polymer Molds and Tooling	Low (1–10 parts)	Low (50–80 USD)	Low to medium	Low	Short	High
3D-Printed Metal Molds and Tooling	High (>10,000 parts)	High	Medium to high	Low to High	Short to long	Low

Considering all these aspects, polymeric materials are the most appropriate candidates for low-volume production of molds via additive manufacturing. Special care needs to be given to the properties of the polymer used for the mold versus the polymer to be injected into the mold, as the choice for the mold needs to present melting temperature above the one exhibited by the polymer used for injection. In this sense, there are companies that developed molds via 3D printing, as well as research studies investigating these technical alternatives. The company Formlabs manufactures polymeric molds via 3D printing stereolithography using their customized photo-curable resins, suitable to replace aluminum molds in injection molding applications for low-volume manufacturing, with cost reductions that could reach 80–90% and time reduction by 90% [73]. Depending on the necessities, Formlabs provides a large range of resins, each of them having one or more advantages such as high molding temperature and pressure properties/increased number of operation cycles/wall thickness/reduced costs [73]. However, their available classes of resins exhibit heat deflection temperature values up to a maximum of 238 °C [73], limiting their use with the injected materials to polymers with thermal resistance below this temperature, such as commodity thermoplastics (i.e., PLA, PE, PP, PS). Formlabs often applies encapsulation of the plastic molds into aluminum frames for better pressure withstanding and preventing warping and deformation after several thermal cycles [64]. Figure 6 illustrates the major steps in the workflow of the injection molding process when using 3D-printed molds.

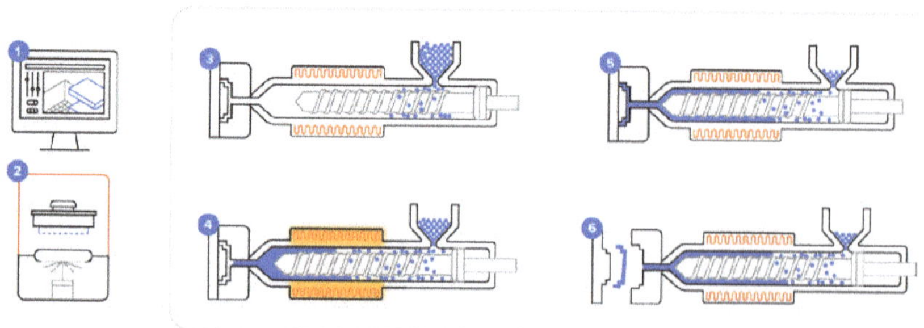

Figure 6. Workflow of injection molding process with 3D-printed molds: 1—mold design; 2—mold 3D printing; 3—mold clamping; 4—injection; 5—cooling; 6—demolding (image reproduced with Formlabs' permission [73]).

Stratasys also manufactures molds from their customized ABS photo-curable resin, but using the PolyJet technique, in which the resin is jetted and UV-cured [74], provided with additional cooling systems to maintain the temperature below 58 °C when injecting ABS.

Using Stratasys printers, the Canadian Javelin company provided 3D-printed ABS molds for injection applications, attesting that their products ensure a 50–70% cost reduction compared to aluminum tools, while maintaining the same advantages offered by the metal molds [75]. Meanwhile, Protolabs compared three manufacturing routes of a custom plastic fitting for a motor housing: injection molding ABS parts with SLA 3D-printed molds, industrial FDM ABS 3D printing, and traditional injection-molded ABS part [72], with their summary recording that the first method reduced the lead time by four times and the costs by almost 24 times compared to the traditional one. Their study concluded that 3D printing the injection molds is the most cost-effective way for low-run injection molding and that material jetting and SLA are the most suited technologies for 3D printing injection molds, while the lifetime of molds can be improved by using some technical issues (i.e., using wide draft angles, release compounds, keeping the part volume below 165 cm^3) [72]. They attested that depending on the material injected, 3D-printed molds could be used for cycles between 30–100 runs [72].

Dizon et al. [67] investigated the possibility of using different 3D printing methods and materials to manufacture polymeric molds for injection molding applications, starting from Formlabs and Stratasys materials and techniques. Injection molds having a chosen geometry (in their case, a cube) were printed via three different techniques: stereolithography—using the Formlabs printer and resin, PolyJet—using the Stratasys printer and materials, and Fused Filament Fabrication—using the Intamsys printer and Evonik PEEK material. For the molds manufactured by SLA and PolyJet made of photoreactive methacrylate-based resin and ABS-based material, respectively, excellent finishes were acquired, but for the ones manufactured via FFF from PEEK, the structure delaminated after the process. Good dimensional accuracy of injected parts manufactured from PLA material was achieved using molds manufactured by SLA and PolyJet printing.

As with the 3D printing market, stakeholders are facing more and more competition; several companies and research laboratories have extended their applications to 3D-printed molds for prototype part production, shortening research and development activities time to 35% and reducing costs up to 90% [64]. In addition to this, applicative research has attested that using 3D-printed molds allows an incomparable flexibility in terms of geometry and design.

Chung et al. [76] have conceived a method for rapid and low-cost production of liquid elastomer injection-molded devices that utilizes fused deposition modeling 3D printers for mold design, enabling rapid prototyping of elastomer devices with complex geometries and requirements, which is a hallmark of fields such as production of medical devices. The authors created the mold from ABS material via fused modeling deposition, taking into consideration surface smoothing for fine-tuning the mold by oversizing the mold (adding extra material) and sanding to desired dimensions or treating ABS with acetone for gradual dissolution. The low costs and reduced production time allow for several iterations of the design that allow corrections or modifications according to the device geometry. However, when using ABS material for molds, the low heat deflection temperature (90–100 °C) needs to be taken into consideration as it lowers the operating temperature down to 70 °C, limiting the range of polymers that can be injected. If higher curing temperatures are needed, other FDM/FFF suitable materials can be taken into consideration, such as polycarbonate. An ABS mold lasts around 20 uses before the ABS plastic wears out, cracks, or suffers damage due to compressional stresses and heating cycles [76], and generally FDM/FFF-printed molds can be used in direct rapid tooling for the limited number of shots in injection molding [64].

In 2018, Altaf et al. [77] conducted a study during which parts made by ABS and nylon mold inserts printed by the FDM technique showed a good performance, comparable to the machine metal mold, for a small number of metal injection molding cycles, concluding that enhanced polymer mold inserts could be a feasible choice in this process for low-volume part production, prototype manufacturing, design validation, form and fit analysis, and other upstream processes, prior to permanent mold manufacturing.

Depending on the material used, geometries and sizes of injection molded parts, as well as the additive manufacturing route used, Stratasys [70] attests that 3D-printed molds can withstand producing from a few dozen to dozens of thousands of parts. The ideal required molding temperature should not exceed 250–300 °C; therefore, polymers with melting/molding temperature higher than 250 °C or that exhibit high viscosity in the processing temperature domain will generate issues regarding the final products' quality, and they will shorten the mold life. Depending on the parts' geometries, size, and complexity, and most importantly, the class of material injected, molds' lifespan can vary from a few dozens to tens of thousands of cycles. Generally, traditional molds withstand more than 10,000 cycles with any polymeric-based material injected, while the metal laser-sintered ones can reach this number only when injecting standard polyolefins, PS, ABS or thermoplastic elastomers, the number decreasing below 100 parts when injecting fiber glass-reinforced PC or PA, PPS or PPO (polyphenylene oxide) polymers. When injecting products in cast resin manufactured molds, the standard thermoplastics can be formed using the same mold up to hundreds of cycles, and only a few dozen when injecting fiber glass-reinforced PC or PA, PPS or PPO. PolyJet molds can be used to produce standard thermoplastics (polyolefins, PS, ABS or thermoplastic elastomers) in an average number of 200–300 parts and a few dozen parts when injecting plastics like acetals, PC/ABS, and glass fiber-reinforced PP [70].

Godec et al. [78] studied the AM PolyJet process and its possible application for the production of bridge polymer molds for injection molding of a small quantity of the molded parts together with design rules for PolyJet bridge molds, dividing 3D-printed molds into three categories, depending on the durability [78]:

- soft (temporary) tool/molds (i.e., silicone molds)—as expected, they can be used for a very limited number of cycles before they reach their usage period.
- bridge tool/molds (i.e., plastic molds)—can be used for small-batch production (i.e., hundreds to thousands) and they generally require shorter manufacturing periods, their durability being strongly influenced by the material used for production within them.
- hard tool/molds (i.e., metallic molds)—can be used for large-batch production (i.e., hundred thousand), similar to the molds manufactured by conventional methods, but they require longer processing time and costs, compared to the other two categories.

In their studies, Godec et al. [78,79] attest that PolyJet molds are not intended to be designed to replace soft or hard tools used in medium- and high-volume production, their purpose being to fill the gap between them and sometimes act as substitutes for 3D-printed prototypes. Although the major advantages are a short time for manufacturing and printing at room temperature, successful injection molding using these molds requires taking into consideration additional factors such (such as design, manufacturing, and post-processing).

Another study [80] focused on comparing different AM technologies (SLA, Laser Sintering, and PolyJet) with different additive manufacturing polymers. The PolyJet resin had similar behavior and properties to ABS and the mold was tested to inject elastomeric polyethylene (injection temperature of 95 °C), polypropylene (injection temperature of 200 °C), and ABS (injection temperature of 270 °C). The mold withstood before cracking to a total of 20 parts (6 PE, 8 PP, and 6 ABS). For the LS method, the mold was produced from polyamide 12 filled with 50% Al, and for the SLA, the materials used were tough resins and high-temperature resins filled with 1/5% carbon nanotubes and graphene nanoplate. The SLA-produced mold with tough resin suffered from warping after printing and UV curing, producing a curved surface that makes this mold unusable for the injection of plastic materials. The laser-sintered mold with PA50Al had the lowest surface definition (detail finish), as expected, whereas the 3D-printed mold with ABS-like resin had the highest surface definition. The manufactured molds were successfully validated for short series productions and for obtaining final parts ready for product validation by using conventional polymers as PP and technical polymers as ABS. Also using UV-curable resin, but this time an acrylate-based one, Noble et al. [81] used an inkjet 3D printer to develop molds for the injection of parts for artificial photosynthesis device prototypes. The results were promising; although the directly 3D-printed parts did not have adequate surface finish for molding optical components, the surface finishing treatment (steel-shaft hot pressing, printer resin coating, scrapper and buffer polishing) tested afterwards added improvement to the final samples.

Also, in medical devices when higher resolutions are needed, FFF methods can be replaced by SLA, if costs justify it. SLA is commonly used for prototyping and low-volume runs of polyurethane devices by printing a mold master and casting a silicone mold around the mold master to create the mold for polyurethane injection [76].

Although considered in the last decade as problematic materials from a sustainability and circularity point of view, epoxy resins were also studied as candidates for injection molds manufacturing [82]. Rahmati and Dickens [83] produced SLA injection molds using SL epoxy that was successfully used to inject 500 PP and ABS parts. The molds' failure was caused by mechanical loadings in flexural or shear during the injection process, as the temperature of the epoxy molds was reduced to 45 °C before each new cycle.

The studies performed by researchers and small-scale manufacturing companies tend to attest that SLA-printed molds are generally feasible in replacing the expensive metallic molds needed when producing medium-scale quantities with low melting temperature thermoplastics such as PLA in a small-scale production facility.

Besides the widely known thermoplastic polymers processed by injection molding, depending on their customization requirements, rubber molded products are also processed via injection molding (organic rubber molding, Liquid Injection Molding or Thermoplastic Rubber Injection) [84]. Structur3d, a developer of soft materials for additive manufacturing builds, customized water-soluble PVA molds for use in custom-manufacturing rubber parts through liquid injection molding. Their solution consisting of sacrificial dissolvable 3D-printed molds allows the manufacturing of fine and complex design parts while removing the drawbacks associated with damaging the rubber parts during extraction from the molds. Moreover, sustainability is addressed, as the 3D-printed PVA molds exhibit suitable thermal stability to withstand rubber processing temperatures, while being 100% biodegradable, non-hazardous compounds that generate no hazardous by-products during removal by water dissolution [85].

Besides the mold material selection and surface finish treatments, the mold design greatly influences the time and costs invested in manufacturing and using the molds (whether polymeric or metallic) in injection molding. The cooling system choice of injection molding tools is an important factor that greatly influences the total production time, as the cooling stage represents about half of the overall production cycle [86] and cooling temperature, speed, and time generate strong effects on the injected polymer crystallization kinetics [87].

Some of the earliest research studies involving 3D-printed mold cooling systems design were performed in the early 2000s. Sachs et al. [88] compared surface temperature achieved using 3D-printed molds with conformal channels and machined molds made of stainless steel with straight channels, concluding that the printed ones exhibited a more uniform surface temperature. Xu et al. [89] demonstrated simultaneous improvements achieved with 3D-printed tools with conformal cooling channels in terms of production rate and part quality as compared with conventional production tools.

Since the inception of research on the topic, several studies have been conducted [90–93]; the subject still remains a challenge nowadays, as optimum configurations are still discussed. Injection molding tools with conformal cooling channels can only be achieved by additive manufactured molds, traditional die design being limited to straight drilled cooling channels. Jahan and El-Mounayri [94] recently proposed a methodology to determine the optimum design of conformal cooling channels in injection molds, their results showing that for different plastic part designs, different channel configurations provide optimum solutions when taking into consideration cross-section dimensions, section size, pitch distance, and mold wall to channel centerline distance. Their study provides a guideline for an easier selection of conformal channels' design parameters.

Besides the improvement of the thermo-mechanical performance of 3D-printed materials for injection molds requirements considering structural design and geometries, literature attests to significant opportunities in terms of the research of polymers, composites, and nanocomposites to enable rapid tooling with toughened materials via 3D printing techniques [64]. Considering injection molds, the need for toughened high-performance polymer-based materials in terms of thermo-mechanical properties and behavior lead to intense research on improving the fracture toughness, delamination, thermal properties, and heat transfer. All these items could be achieved to a high extent with the use of 3D printing methods to develop improved semi-crystalline thermoplastics as well as thermoset, in formulations that involve nanoparticles addition [95,96]. Addition of graphene oxide nanoparticles to 3D-printed TPU/PLA [95] led to high-quality complex shape nanocomposites parts with improved crystalline structure, 90 °C lower degradation temperature, and approximately 170% higher compression modulus and 75% higher tensile modulus. Besides graphene oxide, carbon nanotubes, nanoclays, nanosilica, and nanocellulose are the most commonly used nanofillers added to 3D-printed materials [96]. Still, the use of expensive nanofillers in applications destined for short lifecycle injection molds needs to be very well balanced in terms of performance versus costs evaluation.

Besides the growing application of 3D-printed molds for injection molding of polymers, research studies [97] extend injection molding to ceramic feedstocks from Al_2O_3 and $MoSi_2$ containing composite to produce a variety of parts with demanding geometries such as spirals, cages, and helices. Sacrificial molds from PLA were 3D printed via FDM and compared with DLP-printed ones, from water-soluble resin based on Polyvinylpyrrolidone, showing that the latter one is more suitable for the high resolution required by the products with small structural features. Although sacrificial, these molds imply costs smaller than 10 USD and production time in days of magnitude, compared to the traditional steel ones that can costs from 10,000 to 100,000 USD depending on their complexity and require production time from weeks to months.

4.2. 3D Printing of Molds for Casting Techniques

In casting, a hollow mold is created from a master, which can be hand-sculpted or more recently 3D printed, that is afterwards immersed in a casting material (i.e., sand, clay, concrete, epoxy, plaster, silicone) that hardens. Plastic or metal is poured into the mold, and the master is either removed or burnt out to create the final part [98]. Metal casting is widely used in jewelry, health care (especially dentistry), and engineering and manufacturing (especially aerospace and automotive) applications for parts with fine features or complex geometry [98]. Traditional molds designed for casting have a dense structure, which makes the cooling stage problematic due to uneven capability in this sense as the casting is wrapped inside a thick sand mold with low thermal conductivity [99]. Also, traditional casting techniques require very high up-front tooling costs together with slow, expensive, and laborious mold manufacturing [100]. In casting production techniques, additive manufacturing has been utilized for the manufacturing of prototypes, patterns (replicas of the final part), sand molds, cores and castings themselves, with an increasing interest in the molds and cores production using AM [99].

Replacing expendable wax patterns with 3D-printed patterns in the process of investment casting (lost-wax casting) can generate substantial cost reduction, even after adding printing equipment and material costs, by significant savings in terms of eliminating labor and materials for injection-molded master patterns, soft inner molds, and wax filling-associated expenses [100]. Literature attests to a wide range of studies for casting materials using 3D printing of inorganic sand molds [101–105], and recent interest is moving towards making the 3D-printed molds out of polymeric materials.

Photopolymerization technologies like SLA produce smooth and ultra-fine structure detailed parts and are consequently a compatible technology to manufacture smooth and detailed molds [100]. SLA materials are available as casting resins containing wax for direct investment (lost-wax) casting, which can be "burnt out" at the end of the process, ensuring intact molds. Formlabs offers a solution in this sense as well, through their low or high wax content resins for casting miniature parts design from ultra-fine structures (i.e., wire filigree) to wide range (stone holes or engravings) [106]. Long before the 3D printing era, vulcanized rubber molds were a major advancement in serial production, allowing investment casting at scale [107]. Depending on the requirements in terms of durability, three major rubber classes are available: organic rubber (destined for intense-use wax molds as it has the highest tear strength), heat vulcanized silicone rubber (can respond to a high level of detail, but has lower tear strength), and RTV silicon (destined for molding around delicate details, but has the lowest tear strength) [107]. Three-dimensional printing of vulcanized rubber molds for room temperature or even high temperature can be used for the production of wax models' quantity for investment casting wax of miniature-size metal parts [108].

Recently, Fraunhofer IPA researchers combined additive manufacturing and injection molding to create the Additive Freeform Casting process which benefits from the advantages of both technologies. They utilized the FDM process to print a mold (shell) using water-soluble polymer, polyvinyl acetate (PVAc), which was afterwards filled with polyurethane or epoxy resin and dried or cured, respectively. The shell was removed by water immersion [109,110]. This combined free-form casting was found to bring advantages when large, complex components are required in small quantities, while also saving weight.

Although casting using hard traditional molds ensures replication accuracy to the nanometric level, these methods have a major disadvantage when complex designs are needed, as they require the use of multipart or articulated molds and demolding becomes challenging [111]. Koivikko and Sariola [111] tested different sacrificial molds made of dissolvable materials (HIPS, ABS, polyvinyl butyral-PVB, PVA) to cast silicone elastomers. The 3D-printed molds fabricated by fused filament were subjected to dissolution in limonene, acetone, isopropanol/ethanol and water, applying different magnetic stirring and ultrasonication methods in order to evaluate their effect on dissolution time. ABS, PVB, and especially PVA exhibited successful behavior; however, PVA-water is the material-solvent team that is based on non-hazardous components and exhibits suitable dissolution rates, with no secondary effect on the casted elastomer (although HIPS exhibited the fasted dissolution time, limonene caused swelling and cracking in the elastomer during the drying stage). The proposed solutions allowed the manufacturing of overhangs and channels via single-step cast.

Polyvinyl alcohol, derived from the hydrolysis of polyvinyl acetate, is also one of the most accessible polymers from a technological and economical point of view for mold development in both business and the Do-It-Yourself sector. Three-dimensional-printed PVA molds allow the casting of highly detailed objects from metal fluid (mix of metal grit in a resin binder that resembles bronze-like metals perfectly) that could not be made with any other DIY or low-cost casting method, as it is incomparable, easier, and time-efficient compared to using mold making and metal casting [112]. Designer Eliza Wrobel made disposable 3D-printed PVA molds to cast a highly detailed figurine. The PVA molds, printed using a ZMorph 2.0 SX multitool 3D printer, have the advantage of being ready-to-use, not deforming once the material starts to give back heat, and dissolving completely after 24 h water immersion. The cast figurines only needed sanding to remove resin residues and 3D printing layers [112].

Polymer 3D printing of molds extended its use even in the more sensitive fields, like medical implants. In 2016, in a preclinical study conducted in Singapore, Tan et al. [113] obtained excellent cosmetic and cranial models results with patient-specific polymethylmethacrylate PMMA implants produced with low-cost 3D-printed PLA molds. In 2017, the subject was applied in a clinic study, when a team of medical doctors at Joseph University of Beirut [114] adopted a similar route by printing single-piece molds from low-cost PLA and using them to cast a customized PMMA cranioplasty implant, the applied work concluding that the technique is a cost-effective one for delayed reconstruction of various cranial defects. Three-dimensional prints of anatomical structures could be produced with sub-millimeter accuracy (<0.5 mm) compared to the original specimens. Although the low-cost desktop printers for PLA can facilitate the access to this rapid prototyping technology, the major disadvantage of applying this technique in medical fields and hospitals is the need to master software programs by which the digital model of the mold is designed. However, this drawback seems to become less and less major; considering the high demand for 3D-printed tooling, the programs are constantly improving into more user-friendly versions.

Still in the medical field, but towards pharmaceutical applications, Ajmal et al. [115] cast tablets of indomethacin in hydroxypropyl methylcellulose (HPMC) and polyethylene glycol (PEG) formulation using commercial PLA molds 3D printed via FDM with four different designs (with a designed disintegration functionality, composed mainly of two parts: a detachable cylinder and base/lid, which would separate into up to six sections) established by CAD software. The PLA molds' surfaces were lubricated with corn starch for easier tablet removal; after drying the tablets at room temperature for 24 h, the 3D-printed cylinder parts were removed and the tablets were detached from the molds (base/lid part) using a scalpel. The experiments showed that the resolution influences the ease of detachment in this method and proved in laboratory scale that fast customization of patient-oriented pharmaceutical products can be successfully achieved by means of rapid prototyping.

3D printing allows complex geometry and tailoring of different properties for the optimization of the casting process especially in terms of easier and damage-free demolding. Lv et al. [116] experimented with an innovative damage-free demolding method using a soft ultra-fine mold made of polycaprolactone deposited via electrohydrodynamic printing on a substrate in a predesigned printing path with high precision, used for the effective casting of bio-hydrogels and tested for potential applications in microfluids and cell patterns. The soft ultra-fine mold was framed and hydrogel precursor was poured into the frame and cured. After the mold was detached from the substrate, the fibers were softly peeled from the hydrogel with almost zero damage. The method allowed the damage-free detachment of the generally brittle bio-hydrogels by reducing the demolding stress, with the method showing potential to evolve as a general technique for micro/nanofabrication of brittle materials.

4.3. 3D Printing of Molds for Thermoforming

Thermoforming is a widely used technique in the processing of thermoplastics (generally ABS, PET, PETG, HIPS, PC, PP, PE) that involves heating of a plastic sheet over a specific design tool (mold) so that it takes the design of the tool, which is intensively utilized in packaging and consumer goods products, but also extended to automotive, transport or other high-tech industries.

For the manufacturing of parts needed in small quantities, tools made of hardwood are generally used and exhibit satisfactory behavior, while higher quantities, which implicate superior wear stresses, and metallic materials, such as aluminum, are used for tools. Traditional molds require additional processes such as drilling and milling, performed with the use of robust equipment with high investment; therefore, the process can become cost-effective when mass production of parts is needed. Small quantities require the use of molds that are easy, quick, and inexpensive to manufacture. Therefore, additive manufacturing seems to be the perfect solution in this sense as well.

Thermoforming can be performed using vacuum pressure (ideal to obtain parts precisely formed on one side), around 6.9 bar (for complex and intricate details, with surface finish similar to injection molding), and mechanical forming (negative and positive molds are pressed together, ideal for deep profiles). Thermoforming is mostly used for thermoplastics. Once again, Formlabs developed their own guidelines and cases for the optimization of thermoforming via 3D printing of molds or tooling, made of PS, PC, ABS, and HIPS, PETG,

PE, and PP, which were evaluated for the replacement of aluminum molds for low-volume manufacturing [117].

When designing a thermoforming tool for 3D printing (Figure 7), both the principles of thermoforming and the ones of additive manufacturing should be considered. Three-dimensional-printed molds can ensure the same features as metal molds, but allow increased design freedom with more intricate geometries [117]. Thermoforming tooling requirements are related to their successful resistance to assembly, forming, and demolding forces, temperatures, any coolants or mold release agents. Depending on the number of parts to be thermoformed, the design, and the product requirements, the Formlabs resin used to build the 3D-printed mold choice can be draft resin—for a quick simple design iteration of large parts and one or more pieces, lower resolution but up to four times faster than standard materials; grey resin—for high surface finish quality and detail parts in one or more pieces, better accuracy, consistency, simpler support removal; rigid 10 K resin—industrial-grade, highly glass-filled material capable of forming limited series of dozens of parts with close to production cycle times, high HDT values (up to 218 °C), and tensile modulus (10 GPa), it is suitable when conditions of forming are challenging [117]. Formlabs tested thermoforming of thick PS sheets for up to 50 cycles, using 3D-printed molds from Rigid 10 K Resin with cooling channels embedded, with execution times shorter by 3–7 times and costs reduced in half compared to traditional tooling, which exhibited quality similar to aluminum tooling [117]. For materials with stronger performance, consisting of PC, Formlabs tested molds 3D printed from draft resin and grey resin, exhibiting a production time of 1 day and production cost lower than USD 400. For the testing of ABS and HIPS molds, Grey Resin, Rigid 10 K Resin, and High-Temp Resin at 100 microns layer height were used, achieving quality similar to those achieved with traditional tooling. For the PETG, P,E and PP thermoforming, up to 20 parts of each were manufactured from Rigid 10 K and Grey Resin molds, without reaching mold degradation. For thinner sheets, after around 10 iterations of short cycle time, demolding issues started to appear, while with ticker sheets produced using longer cycle times, there were no demolding issues and quality was superior [117].

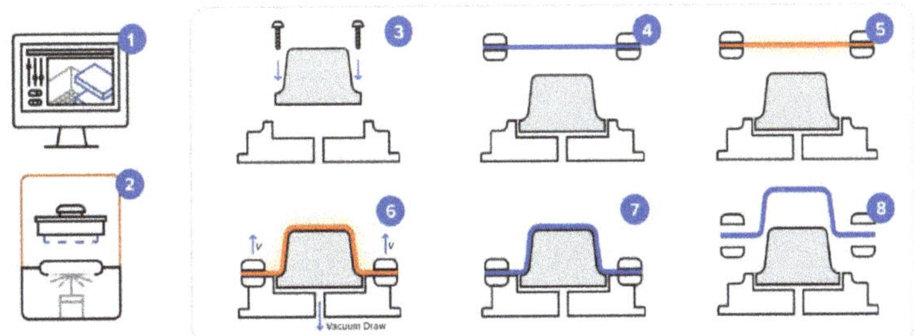

Figure 7. Thermoforming with 3D-printed molds (1—Mold design, 2—Mold 3D printing, 3—Mold assembly, 4—Sheet clamping, 5—Heating, 6—Forming, 7—Cooling, 8—Demolding and trimming) (image reproduced with Formlabs permission [117]).

Chimento et al. [118] have used 3D-printed molds manufactured from Zcorp 3DP Zp130 (mixture of plaster, vinyl polymer and sulphate salt [119]) that were subjected to post-processing using diluted cyanoacrylate (CA) and steam to increase strength while maintaining a porous surface suitable for thermoforming, and Zcorp 3DP Zp140 designed for water curing. The Zcorp-printed parts with different post-processing treatments were compared to the traditional mold material—plaster of Paris (calcium sulfate hemihydrate). Zp130 CA treated shower flexural strength comparable with 100% plaster samples, while exhibiting smaller wear areas. In addition, no differences in thermoforming performance were observed between the rapid prototyped specimen and traditional plaster specimens. All the results indicate that 3D-printed molds are feasible for thermoforming prosthetic and orthotic devices such as prosthetic sockets while providing new flexibility, confirming once again that high customizability prosthetic/orthotic devices can be easily fabricated by 3D-printed materials for rapid tooling.

Junk et al. [120] tested rigid PVC and PS sheets for thermoforming over an automotive shape mold produced via 3D printing, concluding that although materials-associated costs were higher than conventional aluminum or hardwood molds, the manufacturing process hourly rate decreased to 19% and process overall costs decreased to 14% of the metal mold-based process values. Besides economic considerations, the design can be easily modified, by adding channels, holes or other additional geometry (spacers), and additional operations related to the mold post-processing (such as drilling, CNC preparations) are completely removed.

Serrano-Mira et al. [121] analyzed the feasibility of using low-cost AM techniques as rapid tooling techniques to obtain thermoforming molds to quickly manufacture small production batches of tactile graphics. They compared two low-cost AM techniques, 3DP with cyanoacrylate infiltration and FDM with PLA, analyzing geometrical reproduction of the molds and their suitability for 0.2 mm thick PVC sheets thermoforming of tactile graphics. When printing small batches (tens of parts), 3DP appeared to be fast (approximately four times faster than PU prototypes) and economical, while FDM with low-cost equipment appeared to be slower, but implicated lower materials and operating costs. Also, compared to 3DP, FDM offers decreased results regarding details reproduction required in tactile graphics, although this issue can be improved by using smaller diameter nozzles and tailoring parameters.

Literature attests to a multitude of both research studies and small-scale production cases in which thermoforming and vacuum-forming methods are performed with the use of 3D-printed tools; this review points out some of the most diverse found.

4.4. 3D Printing of Molds for Composites Fabrication

Traditional manufacturing methods for fiber-reinforced polymer matrix composites (FRP) require hard tooling for the molds or mandrels shaping the obtained part. In the thermoset polymers area, one of the most important fields that uses molds extensively is the composite fabrication, which is applied for fiber-reinforced thermoset via vacuum-assisted transfer molding, resin transfer molding, prepreg processing, etc. These techniques can develop at the room temperature and vacuum pressure, or high temperature and supplemental pressure (in autoclave).

Traditional molds for composite fabrication are manufactured from metallic materials (generally aluminum, steel or different alloys) but also non-metallic (specialized tooling materials), but regardless of the raw material they are built from, they require significant labor and machining, and consequently high costs, material waste, and long lead times for even relatively simple part shapes. In this case, FDM printing demonstrated that it could ensure significant cost and time reduction, while allowing design flexibility as well as rapid and easy iteration even when complex geometries are required [122].

For the production of composite materials, different mold architectures are implemented to obtain different types of geometry of the composite parts [123]:

- one-part mold—used in vacuum bagging methods (i.e., for hand lay-up, resin infusion, prepregs, etc.) and generally for parts that need a glossy finish for one of the sides;
- two-parts mold—used in compression molding for parts that need both sides with a glossy finish;
- bladder mold—used in pressure molding where one side is the mold, the other is the bladder surface, for complex geometry that cannot be achieved via vacuum bagging or compression molding due to the impossibility of demolding the composite;
- mold pattern for negative mold—used when multiple molds are needed for production increase, a single pattern can be used to manufacture several molds.

Formlabs mentions some major factors to be considered in terms of the designing of the molds for composite fabrication such as draft angle, minimum radius, the inclusion of locating pins and indents, inclusion of surface overrun, adding trim lines, all intended to ease the process of technological challenges (i.e., ease of demolding, precise alignment, air entrapment avoidance, repeatable quality, etc.). After design fractures are established, there are also technology-related factors that need to be considered, such as the use of the smallest layer height to optimize the resolution and demolding, the use of a release agent for ease of demolding, avoiding the use of supports on molding faces not to interfere with surface finish, and allowing resin to degas to avoid air inclusion [123]. Formlabs presented three case studies using 3D-printed molds for composite fabrication [123]. The first one was the development of three-layered carbon fabric epoxy composite by hand lay-up and vacuum bagging using their Tough 1500 Resin to 3D print the mold via the SLA process; in the end, compared with outsourced CNC-machined molds, the 3D-printed mold took 2 days to be produced compared to 4–6 weeks. With CNC machined, the total cost of mold production was 310 USD compared to 900 USD for CNC machined. The second one was the development of bidimensional carbon fiber-reinforced epoxy composites prepregs in autoclave using their High Temp Resin on an SLA printer, estimating that the mold would withstand around 10–15 similar cycles, due to the high temperature and pressure in the autoclave. Although it is certainly not suitable for high-volume production, it can be used for dedicated high-performance applications such as dedicated sports equipment, customized tooling for aerospace or personalized prosthetics. The third case analyzed the 3D printing of patterns to cast molds for large series productions of prepreg composites, using their High-Temp Resin with an SLA printer. Comparing the costs to CNC machining, the 3D printing labor time extended over 1.5 h at a cost of USD 300 compared to 5.5 h at a cost of USD 1100 for CNC, mold materials cost USD 50 for 3D printing compared to USD 220 for CNC, while the total cost of the process cost USD 350 for printing compared to USD 1320 for CNC. The costs were reduced around four times on a basic part when using printed pattern for molds.

Stratasys successfully applied its FDM technology for tooling applications to manufacture and repair different composite lay-up configuration in low-volume quantity. However, the materials limitation delayed the progress of this application as the prepreg required temperature in the autoclave exceeding 180 °C was widely used in aircraft structures. Until they developed ULTEM 1010 resin, based on high-performance polyetherimide, able to withstand temperatures above 200 °C without deformation under mechanical loads [124], Stratasys offered ABS, PC, and ULTEM 9085 materials as alternatives for withstanding temperature values up to 85 °C, 135 °C, and 150 °C, respectively. Although PC and ULTEM 9035 HDT cover the 120–125 °C curing temperatures required by CFRP in the autoclave, the use of ULTEM 1010 manages to successfully minimize thermal expansion impact [122]. The guidelines of FDM-printed ULTEM 1010 tooling to build CFRP offered by Stratasys took into account some key considerations for design, material, and testing of the tooling characteristics. ULTEM 1010 performed successfully under harsher flexural loading conditions (using a lower threshold for acceptance) for the equivalent of dozens of high-temperature and -pressure autoclave cycles, anticipating they could exceed 100 cycles with successful behavior of the tooling, the use of lower pressure, and temperature conditions increasing the number of cycles even more [122]. In addition, the data presented suggest that for the vacuum bagging only small pressures (out-of-autoclave) method widely used in aerospace parts production, tool life ceases to be a major problem from creep-induced deformation perspectives [122].

Besides tool life and thermo-mechanical performance on several cycles, when composite materials are manufactured in molds, the materials compatibility is very important, so that debonding of the part from the mold does not generate issues. There are several polymer alternatives to be analyzed for 3D printing the molds for this application. For example, polyethylene terephthalate glycol (PET-G) is highly recommended due to its good capability to detach from the epoxy resin, while ABS molds should be avoided as detachment of the epoxy resin composite could be problematic [125].

In 2016, Oak Ridge National Laboratory collaborated with a team of industry partners to 3D print and machine several large molds and test them in Boeing's industrial autoclaves to produce carbon fiber composite. The thermoplastic molds survived the high-temperature, high-pressure conditions in the autoclave, which is used to cure aerospace-grade composite parts [126]. The successful testing resulted in high-quality composite parts that can be used in primary aircraft structures. Furthermore, the tools can be re-used to produce part replicates—resulting in further time and energy savings [127]. Different tools made from (PPS) with 50% by weight carbon fiber and Polyphenylsulfone (PPSU) with 25% carbon fiber were developed. The initial tests performed on the molds intended to stabilize the polymer system to withstand variable exposure to elevated temperature without substantial changes in the polymer viscosity. The printed tools were used to fabricate aerospace-grade epoxy reinforced by eight layers of carbon fiber. The tools were cleaned using isopropyl alcohol (IPA), and then three coats of mold release (Frekote 700NC) were applied to the mold surface. The pre-impregnated epoxy/carbon fibers were layed-up in the molds and vacuum bagged. The tools were placed in a production autoclave and exposed to a two-hour cure cycle of 176.6 °C and 620 kPa (90 psi). The tools were scanned after the autoclave process and dimensional analysis and deviation measurements were performed showing that deformations did not exceed 0.1 mm at the composite layup area. The project demonstrated the viability of using additively manufactured parts in the tooling industry to significantly reduce manufacturing costs and energy requirements by accelerating production times [128].

However, even considering all these advancements, 3D-printed tools are not yet common in serial production of high-temperature, autoclave-cured parts for aerospace, as there is still a need to expand the limited material alternatives and use certified properties and behavior of the tooling for these high-demanding applications. But, significant advancements are steadily developing, as in 2019, CEAD (Netherlands) together with partners produced 17 tools printed with short carbon fiber-reinforced polyethersulfone thermoplastic that have been used for more than two years by GKN Aerospace in Germany, for the serial production of CFRP landing flaps for Airbus (France) A350 aircraft, moving even further in 2023 by producing the tools using advanced tape layer additive manufacturing, that involve long fiber in tape form instead of previously used short carbon fibers [129].

On a more research- and education-oriented level, Dynamism, leading provider of professional 3D printing and Industry 4.0 solutions for enterprise, industrial, and education applications, describes the development of the bare-bones carbon fiber process without the specialized equipment needed for more technical processes and high-temperature epoxies. They recommend some major guiding steps: the mold needs to be prepared with a release agent (i.e., PVA helps to smooth out layer lines while providing a reliable release from the epoxies), as it allows its use with most conventional resin systems (epoxy, polyester, vinylester); this method is well compatible with hand layup with or without a vacuum bag, in case resin infusion is required; considering that 3D prints are not 100% airtight, the use of an envelope bagging method needs to be considered; if prepregs are used, the high temperature required for curing makes the PRT-G mold incompatible with the process, as the stress of the vacuum bag will lead to excessive warping and distortion [125].

Besides the widely known application in aerospace parts, CFRP can be used in other various fields (i.e., medical domain). Munoz-Guijosa et al. [130] presented their study on rapid printing of molds for lamination and autoclave curing of epoxy/carbon fibers composite based in prepregs for customized articular orthoses. The molds were manufactured via fused deposition modeling from PLA. In order to respond to the requirements of the epoxy prepregs curing and lamination, in accordance with final product properties related to ankle immobilizing, supporting, or protecting splint, the molds need to meet strict geometrical, mechanical, and thermal specifications. Therefore, the molds need to withstand the mechanical loads generated by the contraction of laminate during the curing process, and those related added by pressure during temperature curing (0.1–0.8 MPa in vacuum bagging or autoclave) maintain the required stiffness and strength at the curing temperature (that may be above 180 °C when an autoclave is used), and exhibit small mean surface roughness (0.5 μm order) to ensure ergonomic/esthetical properties. Considering these requirements, the design must compensate for the drawbacks related to rapid prototyping of molds. The authors created ABS outer shells of the mold, using a precise dual-extruder BCN3D Sigma machine, and in the case of vacuum/pressure curing, the shells were designed with flat areas for the attachment of the supplementary materials needed (release films, breathers, vacuum valves, sealing tape). After printing of the shells, the lamination surface is coated with epoxy resin to tailor the surface roughness. The printed mold shells are filled with a plaster slurry (hardened and dehydrated at 50 °C/2 h), having the role to improve mechanical endurance and heat absorption capacity. Therefore, the thickness of the ABS shell must be minimum (1 mm in this case) as it gives the desired geometry, but the thermal and mechanical properties are ensured by the plaster core. Compared with mold manufactured by machining aluminum, the proposed rapid tooling process ensures almost the same roughness (0.5 μm compared to 0.4 μm for Al), and costs reduced more than 30 times. Although the 3D-printed molds are estimated to withstand 5–10 cycles, considering the customization for each personalized orthoses of patients, the mold is not meant to be used more cycles than the maximum it withstands. The rapid tooling process presented in this paper innovates through the use of conventional FDM of basic thermoplastic polymers ensuring the improved mechanical and thermal properties of the final tooling by filling the 3D-printed shells with clay, making the mold suitable for epoxy CFRP lamination and autoclave curing.

Using 3D-printed molds and patterns in composite fabrication allows businesses to reduce workflow complexity, expand flexibility and design opportunities, and reduce costs and lead time.

4.5. 3D Printing of Molds for Tissue Engineering Scaffolds and Medical Applications

Tissue engineering seems to attract extensive research and medicine effort to develop off-the-shelf scaffolds, as they are able to provide a framework for cell proliferation, migration, and attachment, emerging as popular treatments for bone regeneration and wound healing, due to the mechanical properties that support tissue growth, and they also provide a temporary framework for regeneration [131]. Additive manufacturing is a new and emerging field in the tissue engineering sector in medicine. While 3D printing technologies (mainly fusion deposition modeling, stereolithography, laser sintering, inkjet printing) have been clinically deployed in cranio-maxillo-facial surgery, they are primarily used in the areas of models, guides, splints, and implants [131,132]. The three main approaches to 3DP in tissue engineering are bioprinting (printing live cells), printing acellular scaffolds, and printing molds to be filled with engineered tissue [131,133]. One of the first preliminary studies regarding the use of molds manufactured by 3D printing for scaffolds fabrication for bone regeneration [134] used cryogel together with 3D printing to create CT-derived, patient-tailored molds for scaffold fabrication. However, without sacrificial molds, the debonding resulted in the scaffold damage when the mold was opened. Therefore, the group of researchers advanced towards sacrificial (dissolvable) 3D-printed molds, manufactured from PVA, ABS, and HIPS, which dissolve in water, acetone, and d-limonene, to be used to manufacture tissue engineering scaffolds (cryogels, hydrogels) for cleft-craniofacial defects, which were characterized in terms of porosity, swelling kinetics, mechanical integrity, and cell compatibility. Cryogels were fabricated in PVA and ABS molds, while hydrogels were fabricated in PVA and HIPS molds having 1 mm thickness. HIPS molds required a long time to dissolve (5–8 h), making it difficult to remove the cryogels, being fully formed after 24 h. PVA and ABS dissolved in 2–4 h, but the hydrogels in ABS were very fragile and fractured during removal from the mold. All cryogels maintained accurate shape, and showed spongious morphostructure, mechanical durability with approximately 27 μm average pore size, and 80–87% porosity and good biocompatibility. The nanoporous and brittle structure of hydrogel scaffolds was somehow unsuitable for bone regeneration application, but further improvement studies could mitigate these drawbacks [131].

Sacrificial molds are a very attractive solution for tissue engineering scaffolds; the 3D printing of this type of molds has been studied more and more in the past decade. PVA can be readily printed using FDM printing devices both at the professional and DIY level [135], being intensively used in medical applications due to its cytocompatibility [136,137]. Mohanty et al. [138] studied the 3D printing of PVA via FDM for sacrificial molds to cast elastomeric polydimethylsiloxane (PDMS) polymer scaffolds with structured channels. Printing infill density was tailored between 20 and 80% during the process to obtain different porosities of scaffolds, achieving 81.2% porosity at 80% infill for 150 cm^2/cm^3 surface to volume ratio. This was the largest scaffold with so many channels fabricated at that time (75 cm^2 scaffold with 16,000 interconnected channels. The scaffolds were tested for in vitro hepatocytes cells culture for a 12-day period and the results indicated that the scaffolds produced in 3D-printed PVA molds led to a rapid, cheap, scalable, and compatible with cell culture process. PDMS microfluidic channels structures were fabricated in 3D-printed ABS molds, afterwards dissolved in acetone, resulting in channels down to 90 μm, with 500 μm diameter [139]. In a recent study, Brooks-Richard et al. [140] presented the design and fabrication of MEW (melt electro-writing) tubular scaffolds with complex geometry mimicking patient-specific vascular structures, on FDM 3D-printed PVA molds. The results showed that PVA was a more suitable material than metal mandrel due to its low insulative properties that improve the ability to produce highly ordered scaffolds, which were easy and fast to remove in water without affecting the MEW scaffold fibers' morphology and alignment.

PVA molds are used in medical applications not only as sacrificial molds. In 2014, a team of medical doctors fabricated an inverse replicate of the normal ear for a template in first-stage microtia surgery. A negative mold of the ear was fabricated using rapid prototyping with PLA, the printing process took 90 min, and required less than 1 USD total cost for disposal, and the mold was sterilized for intraoperative use as a template to create an autologous costochondral implant in its likeness [141].

The stomatology area researched the use of 3D printing of polymeric molds as a tool for their patients' customized needs. Yang et al. [142] fabricated novel TNZ dental fillers which were indirectly produced by thermal pressing using customized 3D-printed molds, manufactured from commercial filaments of PLA and ABS using a desktop printer.

Three-dimensional printers can produce anatomic models based on 3D ultrasound, magnetic resonance imaging (MRI), and computed tomography (CT) scans [143–145]; therefore, they can be successfully used to generate patient-specific molds. MRI investigations are of great use in the development of molds design and CAD architecture, which help by offering predictions for future cases. Pokorni and Tesarik [146] developed molds from PET-G polymer to produce phantoms of the human head tissues (skin, bone, cerebrospinal fluid, brain), to mimic head geometry and evaluate stroke detection mechanisms that can be further applied to patients. The design of the molds was developed from MRI-derived scans. Different shapes and sizes of head forms were 3D printed via FDM with Prusa i3 MK2, using a 0.35 mm layer height and 0% infill for a faster and more material-effective process. The printed molds were hollow, so basanite filler material was used to improve their mechanical strength.

MRI investigation information was used in a medical case presented by Costa et. al. [147], in which the anatomical registration of preoperative MRI and prostate whole-mount obtained with 3D-printed, patient-specific, MRI-derived molds was compared with conventional whole-mount sectioning, the study showing that 3D-printed molds for prostate specimen whole-mount sectioning provides significantly superior anatomical registration of in vivo multiparametric MRI and ex vivo prostate whole-mounts than conventional whole-mount sectioning. The design was composed of several stages, using multiple software (i.e., Matlab for volumetric reconstruction extract and conversion to STL file, Netfabb for molds generation based on a generic, SolidWorks for building of a parametrically controlled three-part slicing mold with holes for fixative perfusion and slots for slicing alignment). In the initial trial, the molds were printed on a commercial-grade ProJet 3510 Plus 3D Systems printer using a UV-curable resin (Visijet Crystals, 3D System), and after the parameters establishment, the MRI-derived molds were fabricated on a consumer-grade 3D printer (Leapfrog Creatr XL) using polylactic acid. Another study of medical cases presented the development of 3D-printed PLA patient-specific molds in a prostate phantom model which reduced the MRI-whole mount registration error relative to conventional sectioning. The 3D-printed molds showed the potential to improve prostate MRI-pathology correlations, with the potential to be applied to other organs [148].

A team of medical doctors from the USA presented an algorithm to automatically create 3D-printed molds guiding medial temporal lobe extraction for postmortem MRI, with interactively positioned cut planes used in four hemispheres, their method reducing errors and dependence on anatomical expertise while allowing more tissue to be spared from each brain donation and enabling postmortem imaging at a larger scale [149].

Still in the field of high-resolution imagining medical tooling, Weadock et al. [150] used 3D-printed molds for shaping bioabsorbable implants for customized surgical orbital repair, improving fit, reducing tissue handling and postoperative edema, and reducing surgical times. The orbital area images captured by computed tomography (CT) techniques were used to create STL models of the molds and were edited to create the mirror of the area and overlap it with the fractured side. Sterile or sterilizable molds printed using Formlabs Form 2 printer were fabricated using the images and taken to the operating rooms and used to shape the customized orbital implant for fracture repair in three patients, using bioabsorbable implants.

Three-dimensional-printed molds from PLA were used to fabricate replicas of uterine and fibroid elements, and a realistic model with silicone material uterus and fibroids was used to help resistant simulated laparoscopic myomectomy at low cost. Previously used molds can be repaired with silicone and reused by other residents [151]. Also, breast reconstructive surgery benefits from the use of 3D-printed molds. Patient-specific 3D-printed templates for intraoperative use were designed based on 3D stereophotogrammetry images. The molds were printed from PLA using an Ultimaker 2 printer and then placed in a sterile plastic sleeve to be used for the fitting of the free flap. Prior to anastomosis, the flap was positioned in this sterile covered template, where the contours of the free flap could be traced with a marker pen along the 3D-printed mold, and sutures can be placed to maintain the flap shape. During breast reconstruction, the autologous flap was placed inside the printed template to aid the surgeon in determining the shape and volume of the autologous flap creating the desired breast dimensions. Patients were 3D-photographed 6 to 9 months post-operatively. The study showed that for both unilateral and bilateral breast reconstructions, a mold can represent a useful, low cost, and fast processing tool added to the autologous reconstruction procedure [152].

4.6. 3D Printing f Molds for Soft Lithography

A special use of casting method involving molds is soft lithography, a technique used to create micro devices or three-dimensional structures by means of casting liquid polymer precursor against a topographically patterned mold. Although it involves casting of a polymer, it cannot be included in the general casting molding, as it is not an industrial-type technique, but rather a science-oriented one, as it is broadly used in bio-imprinting and micro/nanofabrication [153]. Soft lithography includes a cluster of methods that uses soft polymeric materials to fabricate small-size devices such as stamps, channels, or membranes with micro-sized features, being a reliable, easy, and low-cost process that allows replicating 3D structures from cm down to micrometric dimensions. The most common devices fabricated with this technique are microfluidics, intensively used in cell biology.

The most common elastomer used in this technique is PDMS, a soft bio-compatible elastomer that has high thermal and chemical stability, low toxicity, chemically inertness, insulating properties, gas permeability, excellent optical transparency to UV and visible light, low cost, mechanically flexible and durable, and last but not least, it is easy to mold [154,155].

Considering its unique properties, PDMS is of great interest in microfluidics applications (widely used in fluid mechanics, reagent mixture, cell biology, particle and cell separation, metabolomics and proteomics, forensic, genetic analysis) as microchips, using soft lithography. However, the costly and time-consuming master mold preparation, the silane surface treatment of the mold required to prevent PDMS detachment problems that can intervene in cell related studies, as well as different required designs of the structure that can be technologically complicated to obtain represent some major impediments. Therefore, 3D-printed molds stood out as an attractive alternative for molds fabrication in soft lithography, methods like stereolithography and digital light processing being some of the most suitable, especially for microfluidics and biomedical areas [156]. Resin or silicone are the generally used materials for PDMS molds fabrication, but beside the fact that they have higher costs than other materials available for 3D printing and require dedicated printers [157], they also generate an effect of inhibition of the curing process of the resin at the contact area of the PDMS with the mold [158], as full curing would be influenced by residual catalysts and monomers [159,160]; therefore, mold surface treatment before PDMS casting remains a challenge even for the 3D-printed ones. Studies attest to the use of different standard pre-treatments of 3D-printed molds via UV curing, ethanol-sonication surface cleaning, preheating, and silanization [156,161] while other studies use alternative treatments such as ink airbrushing [162], a multiple-step procedure including UV treatment, ethanol immersion, air plasma, and perfluorooctyl triethoxysilane treatment [163]. However, the protocols adopted in different research studies seem to be influenced by a variety of factors; therefore, a standard protocol could not be established so far.

Bazaz et al. [156] proposed a method of casting PDMS directly over a 3D-printed mold fabricated directly by the DLP method using a resin based on methacrylated oligomers and monomers, without any pretreatment/surface treatment of the mold, reducing the timeframe for mold fabrication to less than 5 h, compared to several days (for standard soft lithography). Using this resin allowed the removal of mold treatment, as the methacrylated monomers in the resin composition do not react with the casted PDMS, as there are no residual monomer units on the mold surface to impede PDMS polymerization. The PDMS detached from the molds without difficulties. Four microfluidic devices were designed for separation, micro-mixing, concentration gradient generation, and cell culturing applications, the results indicating the biocompatibility of the resin and stable gradient indicating the potential to be used in drug delivery systems.

An Australian research study [164] experimented with a simple fabrication technique of lung-on-a-chip devices using surface-treated DLP 3D-printed molds using photopolymerizable resins based on acrylate polymers for the casting of PDMS parts. The use of acrylate polymer-printed molds allowed a multiple step treatment of their surface (isopropanol washing, UV curing, ethanol immersion, plasma treatment, silanization) in order to prevent PDMS from sticking to the molds and consequently making them suitable for repeated long-term PDMS casting. The approached simple, robust, and cost-effective method allows fabrication of the chip in less than a day, and the use of re-usable molds. In the field of PDMS casting, more advanced studies have developed recently. Yasuda et al. [165] presented the manufacturing of a shark skin-like silicone rubber film that mimics the simplified 2D surface of a shark's skin. The study developed and optimized 3D-printed molds for silicone rubber casting, choosing a 2D-surface version as first prototype. The 3D printing of the full 3D shape remains challenging as supporters are required for 3D printing overhangs of 30° or smaller relative to the horizontal plane, and these supporters would need to be removed during post-processing. The 3D-printed mold proposed by the authors allows for re-use of the molds to increase the manufacturing output. The mold was printed using PLA 2.82 mm filament on an Ultimaker 3 printer. PDMS silicone was casted into the printed molds, the method enabling production of large surfaces of orientable micropatterned repetitive structures at a very reasonable cost performance.

Once again, one of the alternative materials that belongs to the more accessible class is polylactic acid; besides the cost effectiveness, it exhibits biocompatibility and biodegradability features, which are crucial for the PDMS molds applications. However, for the use in PDMS casting for cellular applications, PLA molds need to be subjected to a further step after printing, for the surface fine details adjustment in order to smooth the rough edges. Van der Borg et al. [166] used 1.75 mm diameter PLA filaments to print molds, using a commercial 3D filament printer for the use in casting of PDMS to study biological samples by light microscopy. Printing parameters used were 190 °C, on a tape-covered metal phase heated at 60 °C, 0.1 mm layer height, 10 mm/s print speed, without supports. After printing, the mold surface and edges were smoothed by heated chloroform vapors treatment and afterwards left suspended in the fume hood for 1 h and placed in a vacuum desiccator for 12 h. PDMS was casted into the assembled molds, desiccated and cured at 60 °C/4 h. After detachment from the mold, PDSM excess was removed with a scalpel, obtaining 3 mm height rings. The results indicated that PLA 3D printing of molds represents a promising alternative to be used as molds for cellular studies. Others developed a modular microfluidic system for PDMS casting in PLA 3D-printed molds for high-resolution imaging and analyses of leukocyte adherence to differentially treated endothelial cultures. The molds for PDMS casting were printed with a Form 2 printer using black resin and layers of 25 μm thickness, and the alignment tool was printed using an Ultimaker 3 Extended printer using black PLA filament in a 0.4 mm nozzle and 150 μm thick layers. PDMS modules casted for microfluidic chips were bonded to glass slides by connection to vacuum. The 3D printing of tools in this study contributed to the optimization of the functionality of modular microfluidic systems, by using customizable, user-designed devices [167]. The soft lithography technique implies a very diverse scientific and technological set-up, being greatly influenced by the specificity of each of the study features (used geometries, materials, and target applications); therefore, it still remains a sector in which trying to identify a generally applicable design and parameters set-up is a challenge.

4.7. 3D Printing of Sacrificial Molds

Sacrificial molds are a class of non-reusable molds that can be destroyed after the part has been produced. They can be made of low melting point materials such as wax that are typically destroyed by heating, or by dissolvable materials that can be washed in water or other solvents. Unlike reusable molds for which disassembly and demolding considerations drive the mold decomposition, in the case of sacrificial molds primary considerations for decomposition are manufacturability of individual mold components [168]. When using sacrificial molds, rather than mimic the conventional functionality of a tool, the soluble/meltable tooling uses the same technologies and equipment, but the material that creates the mold is changed. Soluble tooling allows for a flexible workflow from geometry to molds to parts [169]. Besides the sacrificial molds cases already mentioned in the previous section dedicated to injection molding [85,97], casting [111], and tissue engineering [131,136–138,140] molding using sacrificial molds does not constitute a standalone technology, but it is often used as an alternative to build parts via different customized technological routes, where reusable tools and tool life are not issues to be considered.

Some of the most important motivations when choosing sacrificial molds are encountered in situations like the following:

- when small size features complex geometries like the ones provided with microchannels or overhangs, seamless or hollow areas are needed;
- when removing/debonding the part from a fix mold is technologically challenging or generates significant damage to the formed part;
- when complex geometry requires the use of multipart or articulated molds and demolding becomes challenging;
- when the volume of production allows the use of molds that become waste once a part is produced.

Sacrificial molds can be used in individual or combined situations as mentioned above. Sacrificial tooling allows designers, engineers, and researchers to create hollow, seamless, and complex structures with smooth internal surfaces and simplified tool removal [170]. Some of the traditional sacrificial molds are made of eutectic salts, ceramics, cast urethanes, or other similar materials, but they are generally difficult to handle due to brittleness, require additional tooling, or are limited in terms of geometries flexibility due to production or removal challenges [170].

Three-dimensional-printed sacrificial molds have been widely used for manufacturing microfluidic channels, polymeric scaffolds, engineering vasculatures, inorganic 3D matrix materials, and microneedles [138,171,172]. In terms of 3D-printed sacrificial molds materials, the alternatives are still even more limited than the ones for reusable 3D-printed tooling; however, the research conducted so far is promising in this sense.

One of the most intensively used polymers for development of sacrificial molds is PVA, as it is a hydrophilic and therefore a water-soluble, biocompatible, mechanically stable with low toxicity compound that can be easily processed as it can be printed at around 180 °C. PVA was often used to produce sacrificial molds for scaffolds made of PDMS [138,171,173], gelatin [174], fibrin or other materials used to produce different small-scale detailed patterns needed in engineering vasculature or other channel networks applications. The fabrication of sacrificial PLA templates or molds is generally performed via FDM printing [175].

Nagarajan et al. [176] presented the use of FDM-printed sacrificial PVA molds to fabricate self-standing water-insoluble gelatin scaffolds with tunable pore size and porosity. Varying the PVA infill density, they obtained porosity values between 400 and 1200 µm, and that proved to be stable in a phosphate-buffered saline swelling agent. Their results show that the sacrificial mold approach allows the fabrication of gelatin scaffolds with tunable pore size and architecture suitable for tissue engineering applications, which could be further extended to customized scaffolds using various other biopolymers or synthetic polymers. Zou et al. [177] used PVA sacrificial molds to fabricate a pre-vascularized face-like construction based on a 3D tai-chi pattern. The PVA mold scaffold was printed by FDM and filled by printing with hydrogel composites (nanocellulose, agarose, and sodium alginate with HUVECs and human fibroblasts), and removed with PBS solution after crosslinking with $CaCl_2$. PVA 3D-printed sacrificial templates were also used by Park et al. [178] to produce customized ultrathin tubes with adequate mechanical flexibility to mimic bile ducts. The PVA templates were printed at high temperatures and the surface was smoothed by ultrasonication at 50 °C; they were coated with polycaprolactone (PCL) by immersion and removed by water dissolution and ultrasonication. Another study [179] presented the coating of 3D-printed PVA sacrificial templates, with PCL and TPU for tailored porous surfaces with flexibility compatible with soft tissues. Hu et al. [180] used sacrificial PVA molds printed by FDM for microchannels development in tissue engineering applications, which were embedded into three different matrix materials (matrigel, fibrin, gelatin) and removed afterwards by perfusing.

Like previously mentioned, there are studies that introduced the use of other polymers as sacrificial AM molds, such as ABS, HIPS, and PVB, that can be dissolved in different chemical solvents (acetone, limonene, isopropanol/ethanol), that sometimes generate environmental issues, and might as well affect the produced part if the proper compatibility between mold/part is not taken into consideration [111,131]. Besides these, PLA is another attractive polymer, suitable to be used as sacrificial template. PLA was 3D printed, immersed in a gelatin solution at 4 °C, and dissolved with the use of dichloromethane solvent to form a gelatin template, and gelatin methacrylate solution with cells was used to cast the template, which was subsequently removed as well at 37 °C, resulting in a gelatin methacrylate human tissue model with a microchannel network [181]. Montazerian et al. [182] developed 3D-printed PLA shell molds with superior structural integrity to fabricate porous channel network PDMS scaffolds that were removed in dichloromethane solvent.

Poly(N-isopropylacrylamide) (PNIPAM) is another attractive polymer for biomanufacturing applications, often used in drug delivery, soft robotics and engineered vasculature, due to its biocompatibility, ease of processing, and solubility in water at low temperatures [183]. Lee et al. conducted research studies using thermosensitive PNIPAM as a sacrificial template to fabricate microvascular networks within gelatin scaffolds, removing the mold/template by the solvent-spinning method [184], and further comparing the effects of PNIPAM-fabricated microchannels and macrochannels on the formation of normal functional vessels [185]. PNIPAM sacrificial molds are generally produced by electrospinning with microfibers, increasing the scalability of the 3D-printed sacrificial template [175].

In the same category of thermo-responsive polymers for sacrificial molds, other studies used Poloxamer 407 (also known by the trademark Pluronic F127), a triblock copolymer consisting of a central hydrophobic block of polypropylene glycol flanked by two hydrophilic blocks of polyethylene glycol, a water-soluble polymer, that displays a reversible thermal characteristic, as it is liquid at room temperature exhibiting good printability, liquefies at 4 °C, and takes a gel form when administered at body temperature, which makes them attractive candidates as pharmaceutical drug carriers or complex vascular network sacrificial templates [186,187]. Nothdurfter et al. [188] printed Pluronic F127 as a sacrificial mold, on a layer of crosslinked cell-laden hydrogel and fabricated hollow channels in a micro-jetted cell-laden hydrogel chip, having a PMMA rigid shell to mimic a neuroblastoma tumor-environment model. The Pluronic F127 mold was removed by liquefying below 15 °C. While other studies used liquefication at 4 °C to remove Pluronic F127 sacrificial molds [189,190], some used a Pluronic F127 3D-printed sacrificial mold to fabricate photocurable hydrogel scaffolds with customized channels by printing the photocured matrix and removed the mold by immersion in PBS [191]. Others studied improved the mechanical properties and fidelity of the Pluronic F127 3D-printed mold by adding nanoclays into the composition, followed by encapsulation in PDMS and curing and removal by liquification in water at 4 °C [192]. However, although promising and easy to remove, the weak mechanical properties of Pluronic F127 need to be considered when casting in situ scaffold matrix [193].

Another AM polymer that can be used for sacrificial molds is polycaprolactone, synthetic, semi-crystalline, biodegradable polyester with a melting temperature of 60 °C, which can be dissolved in chloroform, dichloromethane, and dioxane [194]. PCL sacrificial molds were used to produce vascular niches and sweat gland interactive models and were removed by incubation with chloroform after dehydration, leaving behind porous constructs [195]. For sacrificial PCL templates with small-size features, electrospinning and electrohydrodynamic jet printing are often used, being extremely useful for engineering vasculature [175].

Other dedicated polymers can be implemented as sacrificial molds via 3D printing and removing, such as water-soluble Poly(2-cyclopropyl-2-oxazoline) [196], potassium bromide soluble polyelectrolyte complex [197], water-soluble butanediol vinyl alcohol copolymer [198], PDMS [199], petroleum jelly–liquid paraffin [200], and water-soluble thermo-responsive polyisocyanide [201].

In terms of meltable sacrificial molds, wax is one of the most used materials. Three-dimensional microvascular networks within an epoxy polymer matrix were fabricated by casting into 3D printing molds made of sacrificial wax, which were subsequently removed by heating above the melting temperature of 60 °C [202]. However, the melting temperature value of most wax limits the material casted for scaffold formation, as polymers that requires higher curing temperatures than the mold can resist are not an option.

When selecting the material of the sacrificial mold or template for biomedical applications, it is very important to consider the compound and/or temperature for mold removal, as some solvents can damage the material of the scaffold, while high temperatures needed to melt the mold material could exceed the thermal resistance of the scaffold material.

Three-dimensional printing sacrificial templates have shown remarkable potential for fabricating intricate structured engineered vasculatures due to their feasibility and versatility, but there are still studies needed to overcome the challenges in producing biomimetic vasculature, related to building of hierarchical vasculature within tissue engineering scaffolds [175].

5. Conclusions

The outbreak of additive manufacturing use in almost all industries worldwide changes not only the form, functionality, and pathway of products to the market, but also the methods and routes that are followed to build the products, revolutionizing the way the products are created. Additive manufacturing application in mold development ensures an important mitigation of the area in which traditional manufacturing exhibits limitations, allowing the development of unique designs and tools that can be continuously customized and adapted according to the application requirements and customer's needs while maintaining the reduced costs and time provided through its characteristics. Three-dimensional-printed molds bring significant benefits in industrial and business fields, as they contribute to the optimization of supply chains and business strategies in small- to medium-scale production in industries like automotive, aerospace and transport, electronics and construction. In the special use applications from tissue engineering and biomedicine, the use of 3D-printed molds allows high quality and detailed customization of dedicated or individual-use products that would not be achievable by traditional techniques. From complex geometries to mass customization, 3D-printed molds can provide significant technical and financial advantages for the manufacturing process and quality of obtained products. Three-dimensional printing of molds is encountered in laboratory research studies, small- to even large-size (in some situations that allow it) industries as well as companies that developed offering 3D printing services for other beneficiaries. The increasing availability of 3D printing services allows researchers without expertise in design or manufacturing to acquire molds already customized to their required characteristics and produce their own devices at low cost, while experienced researchers in the field can fabricate and customize the molds and continuously adapt them for their specific applications.

Author Contributions: Conceptualization, C.-E.P., M.S. and G.P.; methodology, C.-E.P.; software, G.P.; validation, M.S. and G.P.; formal analysis, G.P.; investigation, C.-E.P.; resources, C.-E.P., M.S. and G.P.; data curation, G.P. and M.S.; writing—original draft preparation, C.-E.P.; writing—review and editing, C.-E.P., M.S. and G.P.; visualization, C.-E.P., M.S. and G.P.; supervision, M.S.; project administration, C.-E.P.; funding acquisition, C.-E.P. and M.S. All authors have read and agreed to the published version of the manuscript.

Funding: This work was supported by a grant of the Romanian Ministry of Research, Innovation and Digitization, CNCS/CCCDI—UEFISCDI, project number 601PED/2022 entitled "Lightweight reinforced thermoplastic materials for vacuum thermoformed encapsulation applications in Unmanned Aerial Vehicles", within PNCDI III.

Institutional Review Board Statement: Not applicable.

Data Availability Statement: Data are contained within the article.

Conflicts of Interest: The authors declare no conflicts of interest. The funders had no role in the design of this study; in the collection, analyses, or interpretation of data; in the writing of the manuscript; or in the decision to publish the results.

References

1. Balasubramanian, K.R.; Senthilkumar, V. (Eds.) *Additive Manufacturing Applications for Metals and Composites*; Advances in Civil and Industrial Engineering; IGI Global: Hershey, PA, USA, 2020; ISBN 978-1-79984-054-1.
2. Gardan, J. Additive manufacturing technologies: State of the art and trends. *Int. J. Prod. Res.* **2016**, *54*, 3118–3132. [CrossRef]
3. Gardan, J. Additive manufacturing technologies for polymer composites: State-of-the-art and future trends. In *Structure and Properties of Additive Manufactured Polymer Components*; Woodhead Publishing: Cambridge, UK, 2020.
4. Rane, K.; Strano, M. A comprehensive review of extrusion-based additive manufacturing processes for rapid production of metallic and ceramic parts. *Adv. Manuf.* **2019**, *7*, 155–173. [CrossRef]

5. Al Rashid, A.; Khan, S.A.; Al-Ghamdi, S.G.; Koç, M. Additive manufacturing: Technology, applications, markets, and opportunities for the built environment. *Autom. Constr.* **2020**, *118*, 103268. [CrossRef]
6. Živanović, S.; Popović, M.; Vorkapić, N.; Pjević, M.; Slavković, N. An overview of rapid prototyping technologies using subtractive, additive and formative processes. *FME Trans.* **2020**, *48*, 246–253. [CrossRef]
7. Chua, C.K.; Leong, K.F.; Lim, C.S. *Rapid Prototyping: Principles and Applications*, 3rd ed.; World Scientific: River Edge, NJ, USA, 2010; ISBN 978-981-277-897-0.
8. Sztorch, B.; Brząkalski, D.; Jałbrzykowski, M.; Przekop, R.E. Processing Technologies for Crisis Response on the Example of COVID-19 Pandemic—Injection Molding and FFF Case Study. *Processes* **2021**, *9*, 791. [CrossRef]
9. Picariello, P. *Committee F42 on Additive Manufacturing Technologies, ASTM F42-Additive Manufacturing Technologies*; ASTM International: West Conshohocken, PA, USA, 2017.
10. Collins, S. What's New with ASTM F42 On Additive Manufacturing—ASTM Center of Excellence—Research to Standards—Additive Manufacturing. 2022. Available online: https://gain.inl.gov/AMMQualificationWorkshopPresentations/08-Collins-ASTM_AMMQualWkshp.Rev1_2021.08.24.pdf (accessed on 12 February 2024).
11. Reddy, K.S.; Dufera, S. Additive manufacturing technologies. *BEST Int. J. Manag. Inf. Technol. Eng.* **2016**, *4*, 89–112.
12. Srivastava, M.; Rathee, S.; Patel, V.; Kumar, A.; Koppad, P.G. A review of various materials for additive manufacturing: Recent trends and processing issues. *J. Mater. Res. Technol.* **2022**, *21*, 2612–2641. [CrossRef]
13. Sefene, E.M. State-of-the-art of selective laser melting process: A comprehensive review. *J. Manuf. Syst.* **2022**, *63*, 250–274. [CrossRef]
14. Vartanian, K.; Brewer, L.; Manley, K.; Cobbs, T. *Powder Bed Fusion vs. Directed Energy Deposition Benchmark Study*; Optomec Corporation: Albuquerque, NM, USA, 2016.
15. 3Dnatives. PBF vs. DED: Which Metal 3D Printing Process Should You Choose? Available online: https://www.3dnatives.com/en/pbf-vs-ded-which-metal-3d-printing-process-should-you-choose-140320234/ (accessed on 26 February 2024).
16. EREA Future Sky Circular Aviation White Paper—A Research Programme of the Future Sky Joint Research Initiative 2019. Available online: https://futuresky.eu/themes/circular-aviation/ (accessed on 13 January 2024).
17. Guo, N.; Leu, M.C. Additive manufacturing: Technology, applications and research needs. *Front. Mech. Eng.* **2013**, *8*, 215–243. [CrossRef]
18. 3DEXPERIENCE Make Dasault Systems—FDM—Fused Deposition Modeling. Available online: https://www.3ds.com/make/service/3d-printing-service/fdm-fused-deposition-modeling (accessed on 14 February 2024).
19. Registered Trademark: Stratasys Inc FDM—Trademark Details—Serial Number 74133656, Registration Number 1663961 1991. Available online: https://trademarks.justia.com/741/33/fdm-74133656.html (accessed on 23 February 2024).
20. Formlabs. Guide to 3D Printing Materials: Types, Applications, and Properties. Available online: https://formlabs.com/blog/3d-printing-materials/ (accessed on 28 January 2024).
21. Mwema, F.M.; Akinlabi, E.T. Basics of Fused Deposition Modelling (FDM). In *Fused Deposition Modeling*; Springer Briefs in Applied Sciences and Technology; Springer International Publishing: Cham, Switzerland, 2020; pp. 1–15, ISBN 978-3-030-48258-9.
22. Głowacki, M.; Mazurkiewicz, A.; Słomion, M.; Skórczewska, K. Resistance of 3D-Printed Components, Test Specimens and Products to Work under Environmental Conditions—Review. *Materials* **2022**, *15*, 6162. [CrossRef]
23. Zaldivar, R.J.; Mclouth, T.D.; Ferrelli, G.L.; Patel, D.N.; Hopkins, A.R.; Witkin, D. Effect of initial filament moisture content on the microstructure and mechanical performance of ULTEM® 9085 3D printed parts. *Addit. Manuf.* **2018**, *24*, 457–466. [CrossRef]
24. Stratasys. ULTEMTM 9085 Resin—High-Performance FDM PEI Thermoplastic—Stratasys Catalog. Available online: https://www.stratasys.com/en/materials/materials-catalog/fdm-materials/ultem-9085/ (accessed on 28 February 2024).
25. Intamsys Technology Co., Ltd. 3D Printing in Aerospace. Available online: https://www.intamsys.com/aerospace/ (accessed on 29 February 2024).
26. Iftekar, S.F.; Aabid, A.; Amir, A.; Baig, M. Advancements and Limitations in 3D Printing Materials and Technologies: A Critical Review. *Polymers* **2023**, *15*, 2519. [CrossRef] [PubMed]
27. Liu, Z.; Wang, Y.; Wu, B.; Cui, C.; Guo, Y.; Yan, C. A critical review of fused deposition modeling 3D printing technology in manufacturing polylactic acid parts. *Int. J. Adv. Manuf. Technol.* **2019**, *102*, 2877–2889. [CrossRef]
28. Moradi, M.; Beygi, R.; Yusof, N.M.; Amiri, A.; Da Silva, L.F.M.; Sharif, S. 3D Printing of Acrylonitrile Butadiene Styrene by Fused Deposition Modeling: Artificial Neural Network and Response Surface Method Analyses. *J. Mater. Eng. Perform.* **2023**, *32*, 2016–2028. [CrossRef]
29. Kotomin, S.V.; Kramarev, D.V.; Obidin, I.M.; Polunin, S.V. Influence of 3D Printing Conditions of Polyethylene Terephthalate Glycol on the Mechanical Properties of Products Based on It. *Polym. Sci. Ser. A* **2022**, *64*, 617–623. [CrossRef]
30. León-Calero, M.; Reyburn Valés, S.C.; Marcos-Fernández, Á.; Rodríguez-Hernandez, J. 3D Printing of Thermoplastic Elastomers: Role of the Chemical Composition and Printing Parameters in the Production of Parts with Controlled Energy Absorption and Damping Capacity. *Polymers* **2021**, *13*, 3551. [CrossRef] [PubMed]
31. Mohd Radzuan, N.A.; Khalid, N.N.; Foudzi, F.M.; Rajendran Royan, N.R.; Sulong, A.B. Mechanical Analysis of 3D Printed Polyamide Composites under Different Filler Loadings. *Polymers* **2023**, *15*, 1846. [CrossRef] [PubMed]
32. Hanitio, E.W.; Lutfhyansyah, N.R.; Efendi, B.M.; Mardiyati, Y.; Steven, S. From Electronic Waste to 3D-Printed Product, How Multiple Recycling Affects High-Impact Polystyrene (HIPS) Filament Performances. *Materials* **2023**, *16*, 3412. [CrossRef]

33. Mallakpour, S.; Tabesh, F.; Hussain, C.M. A new trend of using poly(vinyl alcohol) in 3D and 4D printing technologies: Process and applications. *Adv. Colloid Interface Sci.* **2022**, *301*, 102605. [CrossRef] [PubMed]
34. Geng, P.; Zhao, J.; Gao, Z.; Wu, W.; Ye, W.; Li, G.; Qu, H. Effects of Printing Parameters on the Mechanical Properties of High-Performance Polyphenylene Sulfide Three-Dimensional Printing. *3D Print. Addit. Manuf.* **2021**, *8*, 33–41. [CrossRef]
35. Geng, P.; Zhao, J.; Wu, W.; Wang, Y.; Wang, B.; Wang, S.; Li, G. Effect of Thermal Processing and Heat Treatment Condition on 3D Printing PPS Properties. *Polymers* **2018**, *10*, 875. [CrossRef] [PubMed]
36. Jiang, S.; Liao, G.; Xu, D.; Liu, F.; Li, W.; Cheng, Y.; Li, Z.; Xu, G. Mechanical properties analysis of polyetherimide parts fabricated by fused deposition modeling. *High Perform. Polym.* **2019**, *31*, 97–106. [CrossRef]
37. SyBridge Technologies. 5 of the Highest-Performing Industrial FDM Plastics. Available online: https://sybridge.com/highest-performing-industrial-fdm-plastics/ (accessed on 20 January 2024).
38. McLouth, T.; Severino, J.; Adams, P.; Patel, P.; Zaldivar, D. The impact of print orientation and raster pattern on fracture toughness in additively manufactured ABS. *Addit. Manuf.* **2017**, *18*, 103–109. [CrossRef]
39. Ahn, S.H.; Baek, C.; Lee, S.; Ahn, I.S. Anisotropic Tensile Failure Model of Rapid Prototyping Parts—Fused Deposition Modeling (FDM). *Int. J. Mod. Phys. B* **2003**, *17*, 1510–1516. [CrossRef]
40. Bellehumeur, C.; Li, L.; Sun, Q.; Gu, P. Modeling of Bond Formation Between Polymer Filaments in the Fused Deposition Modeling Process. *J. Manuf. Process.* **2004**, *6*, 170–178. [CrossRef]
41. Zaldivar, R.J.; Witkin, D.B.; McLouth, T.; Patel, D.N.; Schmitt, K.; Nokes, J.P. Influence of processing and orientation print effects on the mechanical and thermal behavior of 3D-Printed ULTEM® 9085 Material. *Addit. Manuf.* **2017**, *13*, 71–80. [CrossRef]
42. Parandoush, P.; Lin, D. A review on additive manufacturing of polymer-fiber composites. *Compos. Struct.* **2017**, *182*, 36–53. [CrossRef]
43. Stratasys Inc. FDM Nylon 12CF—FDM Thermoplastic Filament—Datasheet. Available online: https://www.stratasys.com/en/stratasysdirect/materials/thermoplastics/nylon/nylon-12-carbon-fiber/ (accessed on 21 January 2024).
44. Protolabs Network. What's the Right Resin for SLA? 3D Printing Materials Compared. Available online: https://www.hubs.com/knowledge-base/sla-3d-printing-materials-compared/ (accessed on 1 April 2024).
45. Protolabs. Selecting a Material for Stereolithography (SLA) 3D Printing. Available online: https://www.protolabs.com/resources/design-tips/a-guide-to-stereolithography-3d-printing-materials/ (accessed on 1 April 2024).
46. Morano, C.; Pagnotta, L. Additive Manufactured Parts Produced Using Selective Laser Sintering Technology: Comparison between Porosity of Pure and Blended Polymers. *Polymers* **2023**, *15*, 4446. [CrossRef] [PubMed]
47. Formlabs. Guide to Selective Laser Sintering (SLS) 3D Printing. Available online: https://formlabs.com/eu/blog/what-is-selective-laser-sintering/ (accessed on 1 April 2024).
48. Aurum3D. SLS 3D Printing Materials. Available online: https://www.aurum3d.com/blog/sls-3d-printing-materials/ (accessed on 1 April 2024).
49. 3DSPRO. Selective Laser Sintering (SLS) Materials Guide. Available online: https://3dspro.com/resources/blog/selective-laser-sintering-materials-guide (accessed on 3 April 2024).
50. Bobby, S.; Singamneni, S. Conformal cooling through thin shell moulds produced by 3D printing. *Aust. J. Multi-Discip. Eng.* **2013**, *9*, 155–163. [CrossRef]
51. Masato, D.; Kim, S.K. Global Workforce Challenges for the Mold Making and Engineering Industry. *Sustainability* **2023**, *16*, 346. [CrossRef]
52. Epictool. The Mold and Die Tooling Industry. Available online: https://epictool.ca/the-mold-and-die-tooling-industry/ (accessed on 1 April 2024).
53. Prototool. The Complete Introduction of Mold Parts. Available online: https://prototool.com/mold-parts/ (accessed on 30 March 2024).
54. Protolabs Network. 3D Printed Injection Molds: Materials Compared. Available online: https://www.hubs.com/knowledge-base/3d-printed-injection-molds-materials-compared/#requirements (accessed on 28 February 2024).
55. Xometry. How to Create 3D-Printed Molds for Injection Molding. Available online: https://www.xometry.com/resources/3d-printing/3d-printed-molds-for-injection-molding/ (accessed on 27 February 2024).
56. Engineering.com. Friends or Foes: Additive and Traditional Manufacturing. Available online: https://www.engineering.com/story/friends-or-foes-additive-and-traditional-manufacturing (accessed on 21 January 2024).
57. The CASTOR Team. Additive Manufacturing Tooling—Why 3D Printing Technology Is Ideal for Jigs, Tools & Fixtures? Available online: https://www.3dcastor.com/post/why-3d-printing-is-the-best-solution-for-tooling (accessed on 23 January 2024).
58. Ebnesajjad, S. Injection Molding. In *Fluoroplastics*; Elsevier: Amsterdam, The Netherlands, 2015; pp. 236–281, ISBN 978-1-4557-3197-8.
59. Rosato, D.V.; Rosato, M.G. 4—INJECTION MOLDING. In *Plastic Product Material and Process Selection Handbook*; Elsevier Science: Amsterdam, The Netherlands, 2004. [CrossRef]
60. Khan, R.M.; Acharya, G. Plastic Injection Molding Process and Its Aspects for Quality: A Review. *Eur. J. Adv. Eng. Technol.* **2016**, *3*, 66–70.
61. Kausalyah, V.; Nik Maisara, E.N.M.; Shasthri, S. Experimental analysis on the effects of pigmentation in the defects formation of the polypropylene plastic injection moulding. *IOP Conf. Ser. Mater. Sci. Eng.* **2020**, *834*, 012038. [CrossRef]

62. Formlabs. How to Estimate Injection Molding Cost? Available online: https://formlabs.com/blog/injection-molding-cost/ (accessed on 19 January 2024).
63. DesignWorld. Mold Design and Tooling for Injection Molding. 2013. Available online: https://www.designworldonline.com/mold-design-tooling-for-injection-molding/ (accessed on 20 January 2024).
64. Dizon, J.R.C.; Valino, A.D.; Souza, L.R.; Espera, A.H.; Chen, Q.; Advincula, R.C. Three-dimensional-printed molds and materials for injection molding and rapid tooling applications. *MRS Commun.* **2019**, *9*, 1267–1283. [CrossRef]
65. Wang, H.S.; Wang, Y.N.; Wang, Y.C. Cost estimation of plastic injection molding parts through integration of PSO and BP neural network. *Expert Syst. Appl.* **2013**, *40*, 418–428. [CrossRef]
66. Kerkstra, R. How to Select the Right Tool Steel for Mold Cavities. Available online: https://www.ptonline.com/articles/tooling-how-to-select-the-right-tool-steel-for-mold-cavities (accessed on 20 January 2024).
67. Dizon, J.R.C.; Valino, A.D.; Souza, L.R.; Espera, A.H.; Chen, Q.; Advincula, R.C. 3D Printed Injection Molds Using Various 3D Printing Technologies. *Mater. Sci. Forum* **2020**, *1005*, 150–156. [CrossRef]
68. Whlean, C.; Sheahan, C. Using Additive Manufacturing to Produce Injection Moulds Suitable for Short Series Production. *Procedia Manuf.* **2019**, *38*, 60–68. [CrossRef]
69. Afonso, D.; Pires, L.; De Sousa, R.A.; Torcato, R. Direct rapid tooling for polymer processing using sheet metal tools. *Procedia Manuf.* **2017**, *13*, 102–108. [CrossRef]
70. Stratasys. Precision Prototyping: The Role of 3D Printed Molds in the Injection Molding Industry, White Paper—Stratasys. 2014. Available online: https://www.stratasys.com/en/resources/whitepapers/precision-prototyping/ (accessed on 20 January 2024).
71. Schwaar, C. 3D Printing Injection Molds—The Ultimate Guide (All3DPro). Available online: https://all3dp.com/1/3d-printing-injection-molds-the-ultimate-guide/ (accessed on 3 February 2024).
72. Protolabs. 3D Printing Low-Run Injection Molds. Available online: https://www.hubs.com/knowledge-base/3d-printing-low-run-injection-molds/ (accessed on 3 February 2024).
73. Formlabs. *Formlabs White Paper—Low-Volume Rapid Injection Molding with 3D Printed Molds*; Formlabs: Somerville, MA, USA, 2020.
74. Stratasys. Whie Paper—White PaperTop Five Reasons to Integrate PolyJet Technology into your Product Development Lifecycle. 2020. Available online: https://www.stratasys.com/en/guide-to-3d-printing/technologies-and-materials/polyjet-technology/ (accessed on 8 March 2024).
75. Javelin. 3D Print Production Molds. Available online: https://www.javelin-tech.com/3d/manufacture/molds/ (accessed on 8 March 2024).
76. Chung, P.; Heller, J.A.; Etemadi, M.; Ottoson, P.E.; Liu, J.A.; Rand, L.; Roy, S. Rapid and Low-cost Prototyping of Medical Devices Using 3D Printed Molds for Liquid Injection Molding. *J. Vis. Exp.* **2014**, *88*, e51745. [CrossRef]
77. Altaf, K.; Qayyum, J.; Rani, A.; Ahmad, F.; Megat-Yusoff, P.; Baharom, M.; Aziz, A.; Jahanzaib, M.; German, R. Performance Analysis of Enhanced 3D Printed Polymer Molds for Metal Injection Molding Process. *Metals* **2018**, *8*, 433. [CrossRef]
78. Godec, D.; Breški, T.; Katalenić, M. Additive Manufacturing of Polymer Moulds for Small-Batch Injection Moulding. *Teh. Glas.* **2020**, *14*, 218–223. [CrossRef]
79. Godec, D.; Gabrić, I.; PilipoviC, A.; Smoljkić, A.; Surma, R. Tensile properties of multi-material 3D-printed parts. In Proceedings of the MOTSP 2018/10th International scientific conference on Management of Technology Step to Sustainable Production, Primošten, Croatia, 6–8 June 2018.
80. León-Cabezas, M.A.; Martínez-García, A.; Varela-Gandía, F.J. Innovative advances in additive manufactured moulds for short plastic injection series. *Procedia Manuf.* **2017**, *13*, 732–737. [CrossRef]
81. Noble, J.; Walczak, K.; Dornfeld, D. Rapid Tooling Injection Molded Prototypes: A Case Study in Artificial Photosynthesis Technology. *Procedia CIRP* **2014**, *14*, 251–256. [CrossRef]
82. Ma, S.; Gibson, I.; Balaji, G.; Hu, Q.J. Development of epoxy matrix composites for rapid tooling applications. *J. Mater. Process. Technol.* **2007**, *192–193*, 75–82. [CrossRef]
83. Rahmati, S.; Dickens, P. Rapid tooling analysis of Stereolithography injection mould tooling. *Int. J. Mach. Tools Manuf.* **2007**, *47*, 740–747. [CrossRef]
84. RPM Industrial Rubber Parts. 3 Types of Rubber Injection Molding | RPM Rubber Parts. Available online: https://www.rpmrubberparts.com/blog/3-types-of-rubber-injection-molding (accessed on 2 March 2024).
85. Langnau, L.; 3D Print Molds for Custom Manufacture of Rubber Parts. MakePartsFast 21 February 2018. Available online: https://www.makepartsfast.com/3d-print-molds-for-custom-manufacture-of-rubber-parts/ (accessed on 3 February 2024).
86. Dimla, D.E.; Camilotto, M.; Miani, F. Design and optimisation of conformal cooling channels in injection moulding tools. *J. Mater. Process. Technol.* **2005**, *164*, 1294–1300. [CrossRef]
87. Spina, R.; Spekowius, M.; Hopmann, C. Multiphysics simulation of thermoplastic polymer crystallization. *Mater. Des.* **2016**, *95*, 455–469. [CrossRef]
88. Sachs, E.; Wylonis, E.; Allen, S.; Cima, M.; Guo, H. Production of injection molding tooling with conformal cooling channels using the three dimensional printing process. *Polym. Eng. Sci.* **2000**, *40*, 1232–1247. [CrossRef]
89. Xu, X.; Sachs, E.; Allen, S. The design of conformal cooling channels in injection molding tooling. *Polym. Eng. Sci.* **2001**, *41*, 1265–1279. [CrossRef]
90. Hsu, F.H.; Wang, K.; Huang, C.T.; Chang, R.Y. Investigation on conformal cooling system design in injection molding. *Adv. Prod. Eng. Manag.* **2013**, *8*, 107–115. [CrossRef]

91. Khan, M.; Afaq, S.K.; Khan, N.U.; Ahmad, S. Cycle Time Reduction in Injection Molding Process by Selection of Robust Cooling Channel Design. *ISRN Mech. Eng.* **2014**, *2014*, 1–8. [CrossRef]
92. Saifullah, A.B.M.; Masood, S.H.; Sbarski, I. New Cooling Channel Design for Injection Moulding. In Proceedings of the World Congress on Engineering, London, UK, 1–3 July 2009.
93. Feng, S.; Kamat, A.M.; Pei, Y. Design and fabrication of conformal cooling channels in molds: Review and progress updates. *Int. J. Heat Mass Transf.* **2021**, *171*, 121082. [CrossRef]
94. Jahan, S.; El-Mounayri, H. A Thermomechanical Analysis of Conformal Cooling Channels in 3D Printed Plastic Injection Molds. *Appl. Sci.* **2018**, *8*, 2567. [CrossRef]
95. Chen, Q.; Mangadlao, J.D.; Wallat, J.; De Leon, A.; Pokorski, J.K.; Advincula, R.C. 3D Printing Biocompatible Polyurethane/Poly(lactic acid)/Graphene Oxide Nanocomposites: Anisotropic Properties. *ACS Appl. Mater. Interfaces* **2017**, *9*, 4015–4023. [CrossRef]
96. De Leon, A.C.; Chen, Q.; Palaganas, N.B.; Palaganas, J.O.; Manapat, J.; Advincula, R.C. High performance polymer nanocomposites for additive manufacturing applications. *React. Funct. Polym.* **2016**, *103*, 141–155. [CrossRef]
97. Wick-Joliat, R.; Tschamper, M.; Kontic, R.; Penner, D. Water-soluble sacrificial 3D printed molds for fast prototyping in ceramic injection molding. *Addit. Manuf.* **2021**, *48*, 102408. [CrossRef]
98. An Introduction to Injection Molding, Casting, and Thermoforming. Available online: https://formlabs.com/eu/blog/low-volume-production-injection-molding-casting-thermoforming/ (accessed on 2 March 2024).
99. Kang, J.; Shangguan, H.; Deng, C.; Hu, Y.; Yi, J.; Wang, X.; Zhang, X.; Huang, T. Additive manufacturing-driven mold design for castings. *Addit. Manuf.* **2018**, *22*, 472–478. [CrossRef]
100. O'Neill, B. Anniwa—Guide to 3D Printing Molds for Metal Casting. Available online: https://www.aniwaa.com/guide/3d-printers/guide-3d-printing-molds-for-metal-casting/ (accessed on 2 March 2024).
101. Martinez, D.; Bate, C.; Manogharan, G. Towards Functionally Graded Sand Molds for Metal Casting: Engineering Thermo-mechanical Properties Using 3D Sand Printing. *JOM* **2020**, *72*, 1340–1354. [CrossRef]
102. Mitra, S.; Rodríguez De Castro, A.; El Mansori, M. The effect of ageing process on three-point bending strength and permeability of 3D printed sand molds. *Int. J. Adv. Manuf. Technol.* **2018**, *97*, 1241–1251. [CrossRef]
103. Sivarupan, T.; El Mansori, M.; Coniglio, N.; Dargusch, M. Effect of process parameters on flexure strength and gas permeability of 3D printed sand molds. *J. Manuf. Process.* **2020**, *54*, 420–437. [CrossRef]
104. Sivarupan, T.; Upadhyay, M.; Ali, Y.; El Mansori, M.; Dargusch, M.S. Reduced consumption of materials and hazardous chemicals for energy efficient production of metal parts through 3D printing of sand molds. *J. Clean. Prod.* **2019**, *224*, 411–420. [CrossRef]
105. Snelling, D.; Li, Q.; Meisel, N.; Williams, C.B.; Batra, R.C.; Druschitz, A.P. Lightweight Metal Cellular Structures Fabricated via 3D Printing of Sand Cast Molds. *Adv. Eng. Mater.* **2015**, *17*, 923–932. [CrossRef]
106. Formlabs. Formlabs White Paper—Introduction to Casting for 3D Printed Jewelry Patterns. 2021. Available online: https://3d.formlabs.com/white-paper-introduction-casting-3d-printed-jewelry-patterns/ (accessed on 2 April 2024).
107. Formlabs. FormLabs White Paper—Vulcanized Rubber Molding with 3D Printed Masters. 2020. Available online: https://3d.formlabs.com/white-paper-vulcanized-rubber-molding-3d-printed-masters/ (accessed on 2 April 2024).
108. Formlabs. Injection Molding, Casting, and Thermoforming with 3D Printed Molds. Available online: https://formlabs.com/blog/low-volume-production-injection-molding-casting-thermoforming/ (accessed on 2 April 2024).
109. Fraunhofer Institute for Production Engineering and Automation IPA. Combined Advantages of 3D Printing and Injection Molding. 14 March 2018. Available online: https://www.ipa.fraunhofer.de/de/presse/presseinformationen/vorzuege-von-3d-druck-und-spritzguss-kombiniert.html (accessed on 2 March 2024).
110. Watkin, H. Fraunhofer IPA Combine 3D Printing and Injection Molding Advantages. 15 March 2018. Available online: https://all3dp.com/fraunhofer-ipa-combine-3d-printing-injection-molding-advantages/ (accessed on 2 March 2024).
111. Koivikko, A.; Sariola, V. Fabrication of Soft Devices with Buried Fluid Channels by Using Sacrificial 3D Printed Molds. In Proceedings of the 2019 2nd IEEE International Conference on Soft Robotics (RoboSoft), Seoul, Republic of Korea, 14–18 April 2019; IEEE: Seoul, Republic of Korea, 2019; pp. 509–513.
112. Traczyk, M. Zmorph3D—How to Create 3D Printed PVA Molds Ready for Metal Fluid Casting. Available online: https://zmorph3d.com/blog/3d-printed-pva-molds-metal-fluid/ (accessed on 2 March 2024).
113. Tan, E.T.W.; Ling, J.M.; Dinesh, S.K. The feasibility of producing patient-specific acrylic cranioplasty implants with a low-cost 3D printer. *J. Neurosurg.* **2016**, *124*, 1531–1537. [CrossRef] [PubMed]
114. Abdel Hay, J.; Smayra, T.; Moussa, R. Customized Polymethylmethacrylate Cranioplasty Implants Using 3-Dimensional Printed Polylactic Acid Molds: Technical Note with 2 Illustrative Cases. *World Neurosurg.* **2017**, *105*, 971–979. [CrossRef]
115. Ajmal, A.; Meskarzadeh, A.; Genina, N.; Hirschberg, C.; Boetker, J.P.; Rantanen, J. The Use of 3D Printed Molds to Cast Tablets with a Designed Disintegration Profile. *AAPS PharmSciTech* **2019**, *20*, 127. [CrossRef]
116. Lv, S.; Nie, J.; Gao, Q.; Xie, C.; Zhou, L.; Qiu, J.; Fu, J.; Zhao, X.; He, Y. Micro/nanofabrication of brittle hydrogels using 3D printed soft ultrafine fiber molds for damage-free demolding. *Biofabrication* **2020**, *12*, 025015. [CrossRef] [PubMed]
117. Formlabs. FormLabs White Paper—Low-Volume Rapid Thermoforming with 3D Printed Molds. 2022. Available online: https://3d.formlabs.com/thermoforming/#form (accessed on 2 March 2024).
118. Chimento, J.; Jason Highsmith, M.; Crane, N. 3D printed tooling for thermoforming of medical devices. *Rapid Prototyp. J.* **2011**, *17*, 387–392. [CrossRef]

119. Z Corp. ZP—Material Safety Data Sheet zp130 Powder 2007. Available online: http://ytec3d.com/wp-content/uploads/2014/08/zp130.pdf (accessed on 2 February 2024).
120. Junk, S.; Sämann-Sun, J.; Niederhofer, M. Application of 3D Printing for the Rapid Tooling of Thermoforming Moulds. In *Proceedings of the 36th International MATADOR Conference*; Hinduja, S., Li, L., Eds.; Springer: London, UK, 2010; pp. 369–372, ISBN 978-1-84996-431-9.
121. Serrano-Mira, J.; Gual-Ortí, J.; Bruscas-Bellido, G.; Abellán-Nebot, J.V. Use of additive manufacturing to obtain moulds to thermoform tactile graphics for people with visual impairment. *Procedia Manuf.* **2017**, *13*, 810–817. [CrossRef]
122. Stratasys. Stratasys White Paper—Additively Manufactured Composite Tooling. 2020. Available online: https://www.stratasys.com/en/resources/whitepapers/additively-manufactured-composite-tooling/ (accessed on 2 March 2024).
123. Formlabs. *WHITE PAPER Carbon Fiber Parts Manufacturing with 3D Printed Molds*; Formlabs: Somerville, MA, USA, 2021.
124. Stratasys. ULTEMTM 1010 Resin Data Sheet. 2024. Available online: https://www.stratasys.com/siteassets/materials/materials-catalog/fdm-materials/ultem1010/mds_fdm_ultem-1010-resin_0224a.pdf?v=49221c (accessed on 2 March 2024).
125. Dynamism. How to 3D Print Molds for Carbon Fiber Parts. Available online: https://www.dynamism.com/learn/automotive/how-to-3d-print-carbon-fiber-molds.html (accessed on 2 March 2024).
126. Manufacturing—3D-Printed Tools. 2016. Available online: https://www.ornl.gov/news/manufacturing-3d-printed-tools (accessed on 2 March 2024).
127. ORNL and Boeing Perform First Successful Autoclave Testing of 100% 3D Printed Tools. Available online: https://www.3ders.org/articles/20160506-ornl-and-boeing-perform-first-successful-autoclave-testing-of-3d-printed-tools.html (accessed on 1 March 2024).
128. Kunc, V.; Lindahl, J.; Dinwiddie, R.; Post, B.; Love, L.; Matlack, M.; Fahey, R.L., Jr.; Hassen, A.A. *Investigation of In-Autoclave Additive Manufacturing Composite Tooling*; CAMX: Anaheim, CA, USA, 2016.
129. Gardiner, G. 3D-Printed CFRP Tools for Serial Production of Composite Landing Flaps. 28 June 2023. Available online: https://www.compositesworld.com/articles/3d-printed-cfrp-tools-for-serial-production-of-composite-landing-flaps (accessed on 1 March 2024).
130. Munoz-Guijosa, J.M.; Zapata Martínez, R.; Martínez Cendrero, A.; Díaz Lantada, A. Rapid Prototyping of Personalized Articular Orthoses by Lamination of Composite Fibers upon 3D-Printed Molds. *Materials* **2020**, *13*, 939. [CrossRef]
131. De La Lastra, A.; Hixon, K.; Aryan, L.; Banks, A.; Lin, A.; Hall, A.; Sell, S. Tissue Engineering Scaffolds Fabricated in Dissolvable 3D-Printed Molds for Patient-Specific Craniofacial Bone Regeneration. *J. Funct. Biomater.* **2018**, *9*, 46. [CrossRef]
132. Jacobs, C.; Lin, A. A New Classification of Three-Dimensional Printing Technologies: Systematic Review of Three-Dimensional Printing for Patient-Specific Craniomaxillofacial Surgery. *Plast. Reconstr. Surg.* **2017**, *139*, 1211–1220. [CrossRef] [PubMed]
133. Mani, M.; Sadia, M.; Jaganathan, S.; Khudzari, A.; Supriyanto, E.; Saidin, S.; Ramakrishna, S.; Ismail, A.; Mohd Faudzi, A.A. A review on 3D printing in tissue engineering applications. *J. Polym. Eng.* **2022**, *42*, 243–265. [CrossRef]
134. Hixon, K.; Melvin, A.; Lin, A.; Hall, A.; Sell, S. Cryogel scaffolds from patient-specific 3D-printed molds for personalized tissue-engineered bone regeneration in pediatric cleft-craniofacial defects. *J. Biomater. Appl.* **2017**, *32*, 088532821773482. [CrossRef]
135. Duran, C.; Subbian, V.; Giovanetti, M.; Simkins, J.; Beyette, F. Experimental desktop 3D printing using dual extrusion and water-soluble polyvinyl alcohol. *Rapid Prototyp. J.* **2015**, *21*, 528–534. [CrossRef]
136. Paradossi, G.; Cavalieri, F.; Chiessi, E.; Spagnoli, C.; Cowman, M.K. Poly(vinyl alcohol) as versatile biomaterial for potential biomedical applications. *J. Mater. Sci. Mater. Med.* **2003**, *14*, 687–691. [CrossRef]
137. Baker, M.; Walsh, S.; Schwartz, Z.; Boyan, B. A review of polyvinyl alcohol and its uses in cartilage and orthopedic applications. *J. Biomed. Mater. Res. B Appl. Biomater.* **2012**, *100*, 1451–1457. [CrossRef]
138. Mohanty, S.; Larsen, L.B.; Trifol, J.; Szabo, P.; Burri, H.V.R.; Canali, C.; Dufva, M.; Emnéus, J.; Wolff, A. Fabrication of scalable and structured tissue engineering scaffolds using water dissolvable sacrificial 3D printed moulds. *Mater. Sci. Eng. C* **2015**, *55*, 569–578. [CrossRef]
139. Saggiomo, V.; Velders, A.H. Simple 3D Printed Scaffold-Removal Method for the Fabrication of Intricate Microfluidic Devices. *Adv. Sci.* **2015**, *2*, 1500125. [CrossRef]
140. Brooks-Richards, T.L.; Paxton, N.C.; Allenby, M.C.; Woodruff, M.A. Dissolvable 3D printed PVA moulds for melt electrowriting tubular scaffolds with patient-specific geometry. *Mater. Des.* **2022**, *215*, 110466. [CrossRef]
141. Rankin, T.; Mailey, B.; Cucher, D.; Giovinco, N.; Armstrong, D.; Gosman, A. Use of 3D Printing for Auricular Template Molds in First Stage Microtia. *Plast. Reconstr. Surg.* **2014**, *134*, 16–17. [CrossRef]
142. Yang, Y.; Li, H.; Xu, Y.; Dong, Y.; Shan, W.; Shen, J. Fabrication and evaluation of dental fillers using customized molds via 3D printing technology. *Int. J. Pharm.* **2019**, *562*, 66–75. [CrossRef] [PubMed]
143. Miller, M.; Hutchins, G. Development of anatomically realistic PET and PET/CT phantoms with rapid prototyping technology. In Proceedings of the 2007 IEEE Nuclear Science Symposium Conference Record, Honolulu, HI, USA, 26 October–3 November 2007; IEEE: Piscataway, NJ, USA, 2007; Volume 2007, p. 4257, ISBN 978-1-4244-0922-8.
144. Heron, W.; dos Santos, J.; Fontes, R.; Daltro, P.; Gasparetto, E.; Marchiori, E.; Campbell, S. Additive manufacturing models of fetuses built from three-dimensional ultrasound, magnetic resonance imaging and computed tomography scan data. *Ultrasound Obstet. Gynecol. Off. J. Int. Soc. Ultrasound Obstet. Gynecol.* **2010**, *36*, 355–361. [CrossRef]
145. Samuel, B.; Pinto, C.; Pietila, T.; Vettukattil, J. Ultrasound-derived three-dimensional printing in congenital heart disease. *J. Digit. Imaging* **2015**, *28*, 459–461. [CrossRef] [PubMed]

146. Pokorny, T.; Tesarik, J. 3D Printed Multi-layer Molds of Human Head Tissue Phantoms for Microwave Stroke Detection. In Proceedings of the 2019 PhotonIcs & Electromagnetics Research Symposium—Spring (PIERS-Spring), Rome, Italy, 17–20 June 2019; IEEE: Rome, Italy, 2019; pp. 1424–1427.
147. Costa, D.N.; Chatzinoff, Y.; Passoni, N.M.; Kapur, P.; Roehrborn, C.G.; Xi, Y.; Rofsky, N.M.; Torrealba, J.; Francis, F.; Futch, C.; et al. Improved Magnetic Resonance Imaging-Pathology Correlation with Imaging-Derived, 3D-Printed, Patient-Specific Whole-Mount Molds of the Prostate. *Investig. Radiol.* **2017**, *52*, 507–513. [CrossRef]
148. Priester, A.; Wu, H.; Khoshnoodi, P.; Schneider, D.; Zhang, Z.; Asvadi, N.H.; Sisk, A.; Raman, S.; Reiter, R.; Grundfest, W.; et al. Registration Accuracy of Patient-Specific, Three-Dimensional-Printed Prostate Molds for Correlating Pathology with Magnetic Resonance Imaging. *IEEE Trans. Biomed. Eng.* **2019**, *66*, 14–22. [CrossRef]
149. Lasserve, J.; Lim, S.A.; Wisse, L.; Ittyerah, R.; Ravikumar, S.; Lavery, M.; Robinson, J.L.; Schuck, T.; Grossman, M.; Lee, E.B.; et al. Optimized extraction of the medial temporal lobe for postmortem MRI based on custom 3D printed molds: Neuroimaging/New imaging methods. *Alzheimers Dement.* **2020**, *16*, e043254. [CrossRef]
150. Weadock, W.J.; Heisel, C.J.; Kahana, A.; Kim, J. Use of 3D Printed Models to Create Molds for Shaping Implants for Surgical Repair of Orbital Fractures. *Acad. Radiol.* **2020**, *27*, 536–542. [CrossRef]
151. Winkel, A.; Honart, A.; Robinson, A.; Jones, A.-A.; Squires, A.; Royce, C.; McKeon, B.A.; Bookman, L.; Vicari, L.; Dalrymple, J.; et al. Validation of a Simulation Model for Laparoscopic Myomectomy Developed with 3D-Printed Molds. *Obstet. Gynecol.* **2018**, *132*, 33S. [CrossRef]
152. Hummelink, S.; Verhulst, A.C.; Maal, T.J.J.; Ulrich, D.J.O. Applications and limitations of using patient-specific 3D printed molds in autologous breast reconstruction. *Eur. J. Plast. Surg.* **2018**, *41*, 571–576. [CrossRef] [PubMed]
153. Moreira, F.T.C.; Guerreiro, J.R.L.; Brandão, L.; Sales, M.G.F. 1—Synthesis of molecular biomimetics. In *Biomimetic Technologies*; Ngo, T.D., Ed.; Woodhead Publishing: Sawston, UK, 2015; pp. 3–31, ISBN 978-0-08-100249-0.
154. Chow, W.W.Y.; Lei, K.F.; Shi, G.; Li, W.J.; Huang, Q. Microfluidic channel fabrication by PDMS-interface bonding. *Smart Mater. Struct.* **2006**, *15*, S112–S116. [CrossRef]
155. Miranda, I.; Souza, A.; Sousa, P.; Ribeiro, J.; Castanheira, E.M.S.; Lima, R.; Minas, G. Properties and Applications of PDMS for Biomedical Engineering: A Review. *J. Funct. Biomater.* **2021**, *13*, 2. [CrossRef]
156. Razavi Bazaz, S.; Kashaninejad, N.; Azadi, S.; Patel, K.; Asadnia, M.; Jin, D.-Y.; Ebrahimi Warkiani, M. Rapid Softlithography Using 3D-Printed Molds. *Adv. Mater.* **2019**, *4*, 1900425. [CrossRef]
157. Mi, S.; Du, Z.; Xu, Y.; Sun, W. The crossing and integration between microfluidic technology and 3D printing for organ-on-chips. *J. Mater. Chem. B* **2018**, *6*, 6191–6206. [CrossRef]
158. Hwang, Y.; Paydar, O.H.; Candler, R.N. 3D printed molds for non-planar PDMS microfluidic channels. *Sens. Actuators Phys.* **2015**, *226*, 137–142. [CrossRef]
159. Dinh, T.; Phan, H.-P.; Kashaninejad, N.; Nguyen, T.-K.; Dao, D.V.; Nguyen, N.-T. An On-Chip SiC MEMS Device with Integrated Heating, Sensing, and Microfluidic Cooling Systems. *Adv. Mater. Interfaces* **2018**, *5*, 1800764. [CrossRef]
160. Chan, H.N.; Chen, Y.; Shu, Y.; Chen, Y.; Tian, Q.; Wu, H. Direct, one-step molding of 3D-printed structures for convenient fabrication of truly 3D PDMS microfluidic chips. *Microfluid. Nanofluidics* **2015**, *19*, 9–18. [CrossRef]
161. Villegas, M.; Cetinic, Z.; Shakeri, A.; Didar, T.F. Fabricating smooth PDMS microfluidic channels from low-resolution 3D printed molds using an omniphobic lubricant-infused coating. *Anal. Chim. Acta* **2018**, *1000*, 248–255. [CrossRef]
162. Comina, G.; Suska, A.; Filippini, D. Low cost lab-on-a-chip prototyping with a consumer grade 3D printer. *Lab. Chip* **2014**, *14*, 2978–2982. [CrossRef] [PubMed]
163. Waheed, S.; Cabot, J.M.; Macdonald, N.P.; Kalsoom, U.; Farajikhah, S.; Innis, P.C.; Nesterenko, P.N.; Lewis, T.W.; Breadmore, M.C.; Paull, B. Enhanced physicochemical properties of polydimethylsiloxane based microfluidic devices and thin films by incorporating synthetic micro-diamond. *Sci. Rep.* **2017**, *7*, 15109. [CrossRef]
164. Shrestha, J.; Ghadiri, M.; Shanmugavel, M.; Razavi Bazaz, S.; Vasilescu, S.; Ding, L.; Ebrahimi Warkiani, M. A rapidly prototyped lung-on-a-chip model using 3D-printed molds. *Organs-on-a-Chip* **2019**, *1*, 100001. [CrossRef]
165. Yasuda, Y.; Zhang, K.; Sasaki, O.; Tomita, M.; Rival, D.; Galipon, J. Manufacturing of Biomimetic Silicone Rubber Films for Experimental Fluid Mechanics: 3D Printed Shark Skin Molds. *J. Electrochem. Soc.* **2019**, *166*, B3302–B3308. [CrossRef]
166. Van Der Borg, G.; Warner, H.; Ioannidis, M.; Van Den Bogaart, G.; Roos, W.H. PLA 3D Printing as a Straightforward and Versatile Fabrication Method for PDMS Molds. *Polymers* **2023**, *15*, 1498. [CrossRef]
167. Hernández Vera, R.; O'Callaghan, P.; Fatsis-Kavalopoulos, N.; Kreuger, J. Modular microfluidic systems cast from 3D-printed molds for imaging leukocyte adherence to differentially treated endothelial cultures. *Sci. Rep.* **2019**, *9*, 11321. [CrossRef] [PubMed]
168. Dhaliwal, S.; Gupta, S.K.; Huang, J.; Kumar, M. A Feature-Based Approach to Automated Design of Multi-Piece Sacrificial Molds. *J. Comput. Inf. Sci. Eng.* **2001**, *1*, 225–234. [CrossRef]
169. Spielman, S. MachineDesign: Dissolving Molds: A New Way to Think About Injection Molding. Available online: https://www.machinedesign.com/3d-printing-cad/video/21271382/nexa3d-dissolving-molds-a-new-way-to-think-about-injection-molding (accessed on 1 March 2024).
170. Erickson, A. The Value and Benefits of FDM Sacrificial Tooling. Available online: https://www.cati.com/blog/value-benefits-fdm-sacrificial-tooling/ (accessed on 1 March 2024).
171. Goh, W.; Hashimoto, M. Fabrication of 3D Microfluidic Channels and In-Channel Features Using 3D Printed, Water-Soluble Sacrificial Mold. *Macromol. Mater. Eng.* **2018**, *303*, 1700484. [CrossRef]

172. Lu, R.; Chandrasekaran, S.; Du Frane, W.L.; Landingham, R.L.; Worsley, M.A.; Kuntz, J.D. Complex shaped boron carbides from negative additive manufacturing. *Mater. Des.* **2018**, *148*, 8–16. [CrossRef]
173. Xiao, S.; Zhao, T.; Wang, J.; Wang, C.; Du, J.; Ying, L.; Lin, J.; Zhang, C.; Hu, W.; Wang, L.; et al. Gelatin Methacrylate (GelMA)-Based Hydrogels for Cell Transplantation: An Effective Strategy for Tissue Engineering. *Stem Cell Rev. Rep.* **2019**, *15*, 664–679. [CrossRef] [PubMed]
174. Mohanty, S.; Sanger, K.; Heiskanen, A.; Trifol, J.; Szabo, P.; Dufva, M.; Emnéus, J.; Wolff, A. Fabrication of scalable tissue engineering scaffolds with dual-pore microarchitecture by combining 3D printing and particle leaching. *Mater. Sci. Eng. C* **2016**, *61*, 180–189. [CrossRef]
175. Li, S.; Li, H.; Shang, X.; He, J.; Hu, Y. Recent advances in 3D printing sacrificial templates for fabricating engineered vasculature. *MedComm—Biomater. Appl.* **2023**, *2*, e46. [CrossRef]
176. Nagarajan, S.; Belaid, H.; Radhakrishnan, S.; Teyssier, C.; Balme, S.; Miele, P.; Cornu, D.; Subbaraya, N.K.; Cavaillès, V.; Bechelany, M. Sacrificial mold-assisted 3D printing of stable biocompatible gelatin scaffolds. *Bioprinting* **2021**, *22*, e00140. [CrossRef]
177. Zou, Q.; Tian, X.; Luo, S.; Yuan, D.; Xu, S.; Yang, L.; Ma, M.; Ye, C. Agarose composite hydrogel and PVA sacrificial materials for bioprinting large-scale, personalized face-like with nutrient networks. *Carbohydr. Polym.* **2021**, *269*, 118222. [CrossRef]
178. Park, S.-H.; Kang, B.-K.; Lee, J.E.; Chun, S.W.; Jang, K.; Kim, Y.H.; Jeong, M.A.; Kim, Y.; Kang, K.; Lee, N.K.; et al. Design and Fabrication of a Thin-Walled Free-Form Scaffold on the Basis of Medical Image Data and a 3D Printed Template: Its Potential Use in Bile Duct Regeneration. *ACS Appl. Mater. Interfaces* **2017**, *9*, 12290–12298. [CrossRef]
179. Lee, J.E.; Park, S.J.; Yoon, Y.; Son, Y.; Park, S.-H. Fabrication of 3D freeform porous tubular constructs with mechanical flexibility mimicking that of soft vascular tissue. *J. Mech. Behav. Biomed. Mater.* **2018**, *91*, 193–201. [CrossRef]
180. Hu, M.; Dailamy, A.; Lei, X.; Parekh, U.; McDonald, D.; Kumar, A.; Mali, P. Facile Engineering of Long-Term Culturable Ex Vivo Vascularized Tissues Using Biologically Derived Matrices. *Adv. Healthc. Mater.* **2018**, *7*, 1800845. [CrossRef]
181. Davoodi, E.; Montazerian, H.; Zhianmanesh, M.; Abbasgholizadeh, R.; Haghniaz, R.; Baidya, A.; Pourmohammadali, H.; Annabi, N.; Weiss, P.; Toyserkani, E.; et al. Template-Enabled Biofabrication of Thick 3D Tissues with Patterned Perfusable Macrochannels (Adv. Healthcare Mater. 7/2022). *Adv. Healthc. Mater.* **2022**, *11*, 2270038. [CrossRef]
182. Montazerian, H.; Mohamed, M.G.A.; Montazeri, M.M.; Kheiri, S.; Milani, A.S.; Kim, K.; Hoorfar, M. Permeability and mechanical properties of gradient porous PDMS scaffolds fabricated by 3D-printed sacrificial templates designed with minimal surfaces. *Acta Biomater.* **2019**, *96*, 149–160. [CrossRef]
183. Haq, M.A.; Su, Y.; Wang, D. Mechanical properties of PNIPAM based hydrogels: A review. *Mater. Sci. Eng. C* **2016**, *70*, 842–855. [CrossRef] [PubMed]
184. Lee, J.; Wang, X.; Faley, S.; Baer, B.; Balikov, D.; Sung, H.-J.; Bellan, L. Development of 3D Microvascular Networks within Gelatin Hydrogels Using Thermoresponsive Sacrificial Microfibers. *Adv. Healthc. Mater.* **2016**, *5*, 781. [CrossRef]
185. Lee, J.; Kim, D.-H.; Yoon, J.-K.; Park, D.; Kim, H.-S.; Shin, Y.M.; Baek, W.; Kang, M.-L.; Sung, H.-J. Microchannel network hydrogel induced ischemic blood perfusion connection. *Nat. Commun.* **2020**, *11*, 615. [CrossRef] [PubMed]
186. Giuliano, E.; Paolino, D.; Fresta, M.; Cosco, D. Mucosal Applications of Poloxamer 407-Based Hydrogels: An Overview. *Pharmaceutics* **2018**, *10*, 159. [CrossRef] [PubMed]
187. Kolesky, D.B.; Homan, K.A.; Skylar-Scott, M.A.; Lewis, J.A. Three-dimensional bioprinting of thick vascularized tissues. *Proc. Natl. Acad. Sci. USA* **2016**, *113*, 3179–3184. [CrossRef] [PubMed]
188. Nothdurfter, D.; Ploner, C.; Coraça-Huber, D.; Wilflingseder, D.; Müller, T.; Hermann, M.; Hagenbuchner, J.; Ausserlechner, M. 3D bioprinted, vascularized neuroblastoma tumor environment in fluidic chip devices for precision medicine drug testing. *Biofabrication* **2022**, *14*, 035002. [CrossRef]
189. Neufeld, L.; Yeini, E.; Reisman, N.; Shtilerman, Y.; Ben-Shushan, D.; Pozzi, S.; Madi, A.; Tiram, G.; Eldar-Boock, A.; Ferber, S.; et al. Microengineered perfusable 3D-bioprinted glioblastoma model for in vivo mimicry of tumor microenvironment. *Sci. Adv.* **2021**, *7*, eabi9119. [CrossRef]
190. Hynes, W.; Pepona, M.; Robertson, C.; Alvarado, J.; Dubbin, K.; Triplett, M.; Adorno, J.; Randles, A.; Moya, M. Examining metastatic behavior within 3D bioprinted vasculature for the validation of a 3D computational flow model. *Sci. Adv.* **2020**, *6*, eabb3308. [CrossRef]
191. Ji, S.; Almeida, E.; Guvendiren, M. 3D Bioprinting of Complex Channels within Cell-Laden Hydrogels. *Acta Biomater.* **2019**, *95*, 214–224. [CrossRef]
192. Zhou, K.; Dey, M.; Ayan, B.; Zhang, Z.; Ozbolat, V.; Kim, M.H.; Khristov, V.; Ozbolat, I. Fabrication of PDMS microfluidic devices using nanoclay-reinforced Pluronic F-127 as a sacrificial ink. *Biomed. Mater.* **2021**, *16*, 045005. [CrossRef]
193. Wang, A.; Dong, L.; Guo, Z.; Sun, W.; Mi, S. A methacrylated hyaluronic acid network reinforced Pluronic F-127 gel for treatment of bacterial keratitis. *Biomed. Mater.* **2022**, *17*, 045017. [CrossRef]
194. Ludueña, L.; Alvarez, V.; Vázquez, A. Processing and microstructure of PCL/clay nanocomposites. *Mater. Sci. Eng. A* **2007**, *460*, 121–129. [CrossRef]
195. Yuan, X.; Duan, X.; Enhejirigala; Li, Z.; Yao, B.; Song, W.; Wang, Y.; Kong, Y.; Zhu, S.; Zhang, F.; et al. Reciprocal interaction between vascular niche and sweat gland promotes sweat gland regeneration. *Bioact. Mater.* **2023**, *21*, 340–357. [CrossRef]
196. Ryma, M.; Genç, H.; Nadernezhad, A.; Paulus, I.; Schneidereit, D.; Friedrich, O.; Andelovic, K.; Lyer, S.; Alexiou, C.; Cicha, I.; et al. A Print-and-Fuse Strategy for Sacrificial Filaments Enables Biomimetically Structured Perfusable Microvascular Networks with Functional Endothelium Inside 3D Hydrogels. *Adv. Mater.* **2022**, *34*, 2200653. [CrossRef] [PubMed]

197. Wang, H.; Zhou, X.; Wang, J.; Zhang, X.; Zhu, M.; Wang, H. Fabrication of channeled scaffolds through polyelectrolyte complex (PEC) printed sacrificial templates for tissue formation. *Bioact. Mater.* **2022**, *17*, 261–275. [CrossRef]
198. Szklanny, A.; Machour, M.; Redenski, I.; Chochola, V.; Goldfracht, I.; Kaplan, B.; Epshtein, M.; Yameen, H.; Merdler, U.; Feinberg, A.; et al. 3D Bioprinting of Engineered Tissue Flaps with Hierarchical Vessel Networks (VesselNet) for Direct Host-To-Implant Perfusion. *Adv. Mater.* **2021**, *33*, 2102661. [CrossRef] [PubMed]
199. Shakeri, A.; Khan, S.; Didar, T. Conventional and emerging strategies for the fabrication and functionalization of PDMS-based microfluidic devices. *Lab Chip* **2021**, *21*, 3053–3075. [CrossRef] [PubMed]
200. Cheng, F.; Cao, X.; Li, H.; Liu, T.; Xie, X.; Huang, D.; Maharjan, S.; Bei, H.P.; Gomez Espinosa, A.; Li, J.; et al. Generation of Cost-Effective Paper-Based Tissue Models through Matrix-Assisted Sacrificial 3D Printing. *Nano Lett.* **2019**, *19*, 3603–3611. [CrossRef] [PubMed]
201. Celikkin, N.; Simó Padial, J.; Costantini, M.; Hendrikse, H.; Cohn, R.; Wilson, C.; Rowan, A.; Święszkowski, W. 3D Printing of Thermoresponsive Polyisocyanide (PIC) Hydrogels as Bioink and Fugitive Material for Tissue Engineering. *Polymers* **2018**, *10*, 555. [CrossRef]
202. Therriault, D.; White, S.R.; Lewis, J.A. Chaotic mixing in three-dimensional microvascular networks fabricated by direct-write assembly. *Nat. Mater.* **2003**, *2*, 265–271. [CrossRef] [PubMed]

polymers

Review

Additive Manufactured Parts Produced Using Selective Laser Sintering Technology: Comparison between Porosity of Pure and Blended Polymers

Chiara Morano and Leonardo Pagnotta *

Department of Mechanical, Energy and Management Engineering, University of Calabria, 87036 Rende, CS, Italy; chiara.morano@unical.it
* Correspondence: leonardo.pagnotta@unical.it

Abstract: For different manufacturing processes, porosity occurs in parts made using selective laser sintering (SLS) technology, representing one of the weakest points of materials produced with these processes. Even though there are different studies involving many polymeric materials employed via SLS, and different manuscripts in the literature that discuss the porosity occurrence in pure or blended polymers, to date, no researcher has reported a systematic and exhaustive comparison of the porosity percentage. A direct comparison of the available data may prove pivotal in advancing our understanding within the field of additively manufactured polymers. This work aims to collect and compare the results obtained by researchers who have studied SLS's applicability to different amorphous or semi-crystalline polymers and pure or blended materials. In particular, the porosity values obtained by different researchers are compared, and tables are provided that show, for each material, the process parameters and the measured porosity values.

Keywords: additive manufacturing; 3D printing; selective laser sintering; porosity; pure polymers; blended polymers

Citation: Morano, C.; Pagnotta, L. Additive Manufactured Parts Produced Using Selective Laser Sintering Technology: Comparison between Porosity of Pure and Blended Polymers. *Polymers* **2023**, *15*, 4446. https://doi.org/10.3390/polym15224446

Academic Editors: Anton Ficai and Cristina-Elisabeta Pelin

Received: 30 September 2023
Revised: 10 November 2023
Accepted: 14 November 2023
Published: 17 November 2023

1. Introduction

Although the use of metals is constantly growing [1], polymers are still the most used materials today [2] in additive manufacturing (AM) [3].

Currently, most components made with polymeric materials are manufactured using the selective laser sintering (SLS) process [4]. This process belongs to one of the first-born families of AM processes, called Powder Bed Fusion (identified with the acronym PBF), which is based on the fusion of layers of powdered material. The SLS process, in particular, uses thermoplastic polymeric powders, and their fusion is obtained, layer by layer, using a laser beam that acts along directions selected using a computerized system. SLS is one of the most widespread AM processes. In principle, any polymer that is available in powder form, which can be melted and bonded without decomposition via heating, would appear to be processable using selective laser sintering. In practice, however, today, due to the very complicated and difficult-to-control physical phenomena involved in the process [5,6], there are only a few polymers that are suitable for SLS [7–10]. Both amorphous and semi-crystalline polymers have been studied and employed in SLS processes, with the latter being the most popular [11,12].

In the market of materials that are available for SLS processes, polyamide-based powders 11 (PA11) and 12 (PA12) dominate, followed by other polymeric powders such as BPT, PC, PE, PEBA, PEEK, PET, PMMA, PP, PS, SEBS, TPE, TPU [7], and very few other types. More than 90% of the industrial consumption of polymers for SLS comprises pure Polyamide 12 (PA12) or reinforced blends, such as dry blends of glass-, aluminum-, and carbon-fiber-filled polyamides [10].

The quality of the parts produced using SLS could greatly be affected by the fabrication process, i.e., the powder state, powder particle size, and shape [13,14], and the process parameters [15–17]. It is also well known that, as for different manufacturing processes, porosity occurs in parts made using SLS technology. Porosity occurs due to the intrinsic phenomena involved during the melting, sintering, and consolidation processes of powders [18] and represents one of the weakest points of materials produced with these processes. Due to that, the key point behind the widespread use of 3D-printed parts for structural application in different industrial fields is the improvement in product reliability, e.g., defect and porosity reduction. In fact, porosity dramatically affects the quality and reliability of additively manufactured parts and, therefore, deserves great consideration. In recent years, additively manufactured materials have been studied through both numerical and experimental methods to try to understand the effect of porosity (shape, size, number, and position of pores) on critical mechanical properties, such as stiffness, strength, and toughness, and to establish the correlations between these [9,19,20]. Many efforts were made to study how the process parameters affect the porosity level. It was demonstrated that, by optimizing these parameters in the best possible way, the porosity level of the manufactured parts is reduced and becomes dependent only on the type of material used. Overall, the porosity of the parts fabricated with amorphous materials is higher than that of the parts made with semi-crystalline polymers [18].

The common goal of these studies is to mitigate the effects of porosity by developing methodologies that are capable of reducing or, more ambitiously, controlling the generation of pores during the process and introducing post-processing techniques for their elimination.

Evidently, in this context, porosity measurements play a role of primary importance. The scientific literature boasts a large number of articles that study the various polymers that can be processed with SLS, reporting data on their porosity. Measurements of the porosity and density were carried out with different investigation methodologies, passing from traditional measurement techniques to more modern and sophisticated ones [21]. The easiest technique that could be employed for porosity quantification is Archimedes' method, which allows for porosity evaluation through a density measurement. However, this technique offers some difficulties related to theoretical density knowledge and does not give any information about the pores' characteristics [21]. The distributions and shapes of the pores can be observed directly using the microscopy analysis technique [22]. The latter, however, has the disadvantages of being a destructive technique and only allowing for the observation of small sections of the sample. Among other techniques for measuring porosity, microcomputed tomography (μ-CT) is certainly the most powerful. This methodology offers the advantage of providing, in addition to the porosity value, the spatial distribution, shape, and size of the pores, in a non-destructive way [23,24]. Its main drawback is the high cost of the equipment.

The purpose of this paper is to collect and compare the porosity percentage measured by researchers on 3D-printed parts. In particular, works that evaluated the applicability of different polymers to SLS were taken into account. Moreover, both amorphous or semi-crystalline as well as pure or blended polymers were considered, even if, as it stands, the highest amount of research is focused on PA12 parts. As assessed before, the importance of having the correct knowledge of porosity in SLS fabricated parts led to the possibility of improving the mechanical properties of 3D-printed parts and, consequently, their reliability.

Despite the intense research carried out and the numerous papers published, to the knowledge of the authors of this work, to date, no investigator has endeavored to juxtapose the assessed levels of porosity. It is believed that a comparison of the available data may be pivotal for augmenting knowledge in the field.

In this paper, after an overview of the various materials processed using SLS, reported in Section 2, the data that are present in the literature relating to the porosity measured on polymers are collected and compared. In particular, in the Section 3, the data related to parts fabricated by polymers belonging to the polyamide family are first discussed, starting from PA12 and moving on to PA6, PA1010, and PA11 and their blends.

Additionally, the porosity measurements of other pure polymers and polymer blends are compared. Tables are provided, which show, for each material, the process parameters and the measured porosity values. Finally, Section 4 is also included for the critical analysis of the main results found in the literature.

2. Brief Outline of the SLS Process and Porosity

A typical system scheme used for 3D printing parts fabricated through SLS technology is shown in Figure 1. For a detailed description of this system, the SLS process, and the influence of the various process parameters on the formation of porosity in the particles produced, please refer to [18].

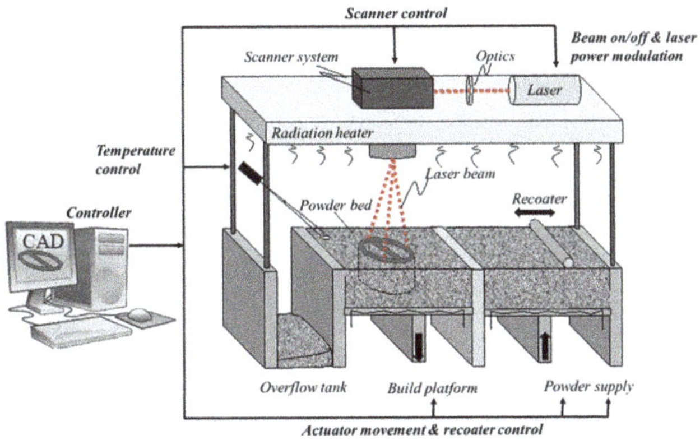

Figure 1. Schematic of an SLS machine's main components (reprinted from [18]).

In short, the manufacturing process involves three stages:

(i) Preheating phase. In this phase, the powder bed is heated to a predefined temperature (bed temperature, T_b), which is held constant throughout the part-building process. The T_b is kept just below the softening temperature of the polymer that is used to minimize the laser energy and eliminate any distortion of the piece during cooling.

(ii) Building phase. This is the core phase of the fabrication process that involves different operations. First of all, the platform is lowered to receive the powder particles dragged by the roller or by the spreading blade. After that, the laser beam melts the layer of particles along the computerized trajectory. Finally, the piece is gradually cooled down to the T_b value for solidification.

(iii) Cooling phase. In this phase, the heat source is switched off with the consequent gradual cooling of the powder bed until it reaches the extraction temperature of the piece.

For the numerous parameters involved in the process, refer to Table 1 [18].

Table 1. SLS process parameters (reprinted from [18]).

SLS Process Parameters			
Powder-Based	**Laser-Based**	**Temperature-Based**	**Scan-Based**
Particle shape, size, number, and spatial distribution	Laser power	Powder bed temperature	Scan speed
Powder flowability	Spot size	Powder feeder temperature	Hatching distance
Recoating speed, layer thickness, and powder density	Pulse duration	Temperature distribution	Scanning pattern
Material Properties	Pulse frequency		

In addition to the parameters reported in Table 1, it is important to introduce the Energy Density (*ED*) supplied by the laser to the powder bed. The *ED* stands as an exceptional metric employed by numerous researchers to assess the impact of process parameters on the final part's quality and porosity. Termed as Andrew's number, the *ED* quantifies the energy dispensed to particles per unit area of the powder bed surface. Its computation is expressed by the following equation:

$$ED = P/(v{\cdot}s) \quad (1)$$

where P represents the laser power (in W), v is the scan speed of the laser beam (in mm/s), and s is the laser scan spacing, i.e., the distance between two consecutive laser tracks (in mm). The supplied energy density *ED* is then usually given in J/cm^2, and its value could affect the porosity percentage in the SLS parts. In particular, if the *ED* received by the powder layer is too low or too high, this could lead to an increase in the porosity measured in the parts. Moreover, each of the parameters included in the equation has been found to affect significantly the porosity percentage in the SLS parts [18]. Besides the *ED* parameter, several factors that are not included in the equation could affect the porosity percentage. In particular, among all, the powder bed temperature and the layer thickness are those of the greatest importance. The powder bed temperature influences the cooling rate and viscosity of the polymer during the fabrication process. On the other hand, the layer thickness influences the adhesion characteristics between two consecutive printing layers. To add to these parameters, other factors influence the porosity development mechanisms such as the powder particle sizes, powder re-usage, laser spot diameters, laser scanning strategy, and material properties.

A schematic of the different kinds of porosity that could be found in SLS parts in conjunction with the processing parameters that contribute to their development is reported in Figure 2. In general, porosity is an intrinsic phenomenon of the SLS process. During the melting process, the air could remain entrapped between two adjacent particles, leading to the development of an intra-layer porosity. The amount of these voids is affected by different processing parameters, i.e., the laser power and speed, particle shapes and re-usage, and material properties (viscosity). Beyond that, the porosity may arise due to inconsistent powder deposition as well as an inconsistent energy density received by the deposited powder layer. If the laser power or scan speed is too high or too low, the material layer is too thick, or the hatch distance determining the overlap area and therefore the connection between two hatch lines is too short or too long, this will cause the incomplete melting of the particles by promoting the formation of pores [18,25,26], i.e., a lack of fusion porosity. Finally, porosity could develop between two consecutive layers, i.e., inter-layer porosity. For an updated overview of the nomenclature and measurement methods, refer to [21]. Porosity can be defined through the following ratio:

$$\varepsilon = V_p/V \quad (2)$$

where V is the part volume and V_p is the pores' volume. The V_p value could be calculated using different approaches depending on the pore classification. Pore classification could be conducted according to different characteristics. First of all, it is possible to distinguish between open pores and closed pores, in function to the capability to intercept external fluid. A second classification could be conducted based on the pores' geometry, e.g., cylinders, prisms, spherical cavities, and windows. However, for 3D-printed parts, the occurrence of irregular pores is very high and, consequently, it is not possible to employ this classification method. A third classification is conducted based on the pore size, identified as the smallest pore dimension, i.e., pore width. In this case, it is possible to distinguish between micropores (i.e., pore width < 2 nm), mesopores (i.e., pore width > 2 and <50 nm), and macropores (i.e., pore width > 50 nm) according to the IUPAC classification. However, the one-dimension classification is sometimes not exhaustive, and very often, other 2D and 3D parameters are involved, e.g., areas or volumes. Finally, a fourth classification could be conducted based on the pores' origins. In this case, it is possible to distinguish between intrinsic pores, i.e., unintentional pores, and extrinsic pores, i.e., pores that are intentionally introduced for a specific application. For a more detailed pore classification, please refer to Morano and Pagnotta's work [18].

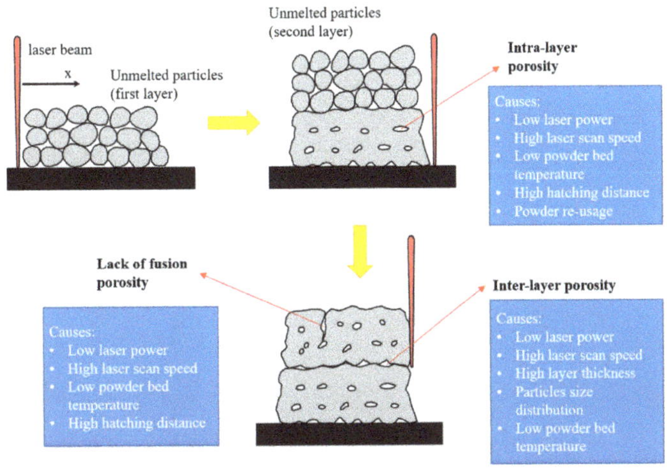

Figure 2. Porosity in SLS parts with respect to processing parameters.

3. Polymers for the SLS Process

The SLS process exhibits characteristics that make it suitable for processing different kinds of materials and/or blends. However, thermoplastic polymers (both amorphous and semi-crystalline) are the most widely applied materials in SLS since they require low processing temperatures and, consequently, low laser powers.

A rather comprehensive overview of the SLS polymer powders that are commercially available or reported in the scientific literature has been presented by Tan et al. [10]. The authors classified the different thermoplastic polymers using the pyramidal scheme reported in Figure 2. The first classification is between amorphous (on the left side) and (partially) crystalline (on the right side) polymers. Beyond that, moving upwards, polymers are distinguished based on their mechanical properties, operating temperatures, and costs. In particular, on the bottom, we find the so-called "commodity" polymers, i.e., low-cost polymers for high-consumption applications. In the middle area, it is possible to find the "engineering" polymers, i.e., materials for applications requiring few advanced characteristics, such as moderate temperature resistance and good mechanical properties. Finally, on the top of the pyramid, we find the "high-performance" polymers, i.e., polymers with high costs and high mechanical properties and/or service temperatures. Moreover, the red boxes identify polymers that are commercially available, and the yellow boxes are for polymers that were studied in the laboratory and reported in the scientific literature, while the white boxes are for polymers that are not available for SLS (refer to the abbreviation listed at the end of the paper for the meaning of acronyms).

By analyzing the data reported in Figure 2, it can be seen that, among the thermoplastic polymers available, approximately 25% are not suitable for SLS, 40% are not commercially available even if they have already been tested, while only the remaining part (corresponding to approximately 35%) is currently used in the additive manufacturing industry. It should also be noted that the latter is represented by approximately 85% of semi-crystalline polymers, including polyamides. The latter, even if numerically few by type, are polymers that quantitatively represent almost all of the production of powders. Due to that, these polymers have been and are still the most studied, both in their pure and blended forms.

In the following sections, the polymers of the polyamide family and their compounds will be first discussed. After that, the other semi-crystalline polymers will be considered, starting from the base of the pyramid and ending with the most performing polymer shown on the top of the pyramid. The same methodological sequence will be used to describe amorphous polymers and elastomers.

It should be noted that not all of the polymers that have been investigated up to now are shown in Figure 3. The missing ones, found in the literature by the authors of this paper, will also be reported and discussed, as far as possible.

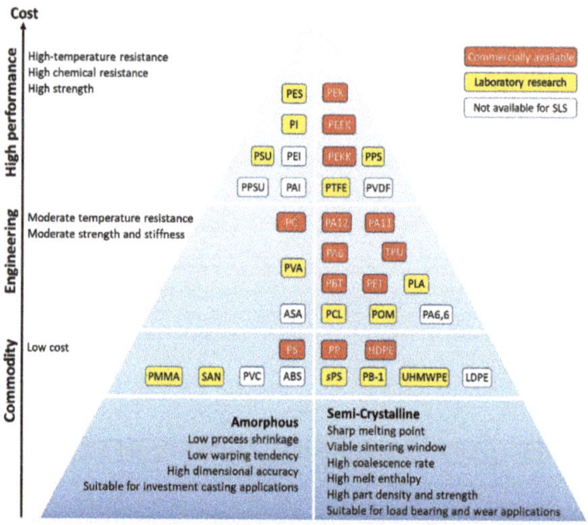

Figure 3. Polymer powders commercially available for SLS. Reprinted from [10].

3.1. *The Polyamide Family*

Polyamide is the most popular polymer in the SLS application since it allows for the creation of parts that have good mechanical properties, good finishes, low costs, and recyclability [27]. In addition, they are also suitable for the production of composite materials and medical applications [28,29].

In addition to the more widespread Polyamide 12, the polyamide family also includes Polyamide 6 (PA6), Polyamide 1010 (PA1010), Polyamide 11 (PA11), and the blends obtained by mixing different polyamide powders. The main results are discussed in the following sections.

3.1.1. Polyamide 12

Actually, polyamide 12 is the most applied and studied material for the SLS process [30,31]. Despite this, PA12 parts are still characterized by the occurrence of fabrication voids and defects. Due to that, achieving low levels of porosity in manufactured parts remains a major challenge, and multiple and accurate studies have been conducted to try to identify and explain the mechanisms of pore formation [18,32]. Several researchers have studied how the SLS processing parameters influence the total contents of pores and their distributions within the polymer parts.

Dupin et al. [33] compared the closed porosity, shapes, dimensions, and positions of pores of SLS parts produced from two different PA12 powders (Duraform and InnovPA) by varying the energy density value. In particular, the authors decided to modify only the laser power value by keeping the other parameters fixed. Seven different laser power levels were selected and, consequently, seven ED values. The porosity characteristics were evaluated using both Archimedes' principle and X-ray tomography. They found that the quantities of open and closed porosities decrease as the ED increases. These results can be explained by taking into account that increasing the ED induces more particles to melt, so the amount of molten polymer increases too. This molten matter promotes the coalescence process between adjacent particles and therefore enhances the densification of the parts. Overall, it is possible to assess that the energy density has a great impact on the residual porosity. The authors also show that the particle size distribution and the crystallization temperature of the powder material are the key parameters in pore formation. It has been observed that the presence of small particles affects the density of the final part as it promotes the adhesion between the layers producing a lower interlayer porosity. This feature also influences the fusion phase since it affects the coalescence process of the particles. On the other hand, the crystallization temperature affects the porosity during the last stage of the process, i.e., the cooling stage. In fact, lower crystallization temperatures imply an increase in the time that the materials spent in the molten stage, i.e., lower porosity [33].

Tontowi and Childs [34] investigated the effect of the powder bed temperature (ambient build powder surface temperature) on the part density. The effect was evaluated both experimentally and numerically by developing a 2D model. The authors showed that small temperature variations have a marked effect on the part density. This result was observed both numerically and experimentally. In particular, the lower the powder bed temperature, the lower the sintered part density. The effect of the bed temperature could be mitigated by varying the energy density value according to temperature fluctuation.

Gomes et al. [35] analyzed the influence of the dust lap on the quality of the PA12 printed parts using a CT analysis. The authors found an increase in the porosity percentage by increasing the recycling cycles. In particular, a very low porosity percentage was measured for the parts fabricated with virgin powder, i.e., around 1.5%. By increasing the number of printing cycles, this value increased up to 9%. The porosity increase was further accompanied by geometrical errors. Powder recycling is a key point for the SLS process since it allows for the reduction in production costs as well as process waste.

Dewulf et al. [36] investigated the influence of laser power, hatch spacing, and scan speed on porosity development. Each parameter influence was evaluated separately by keeping the other values constant. With this approach, it was possible to obtain samples fabricated with the same ED value but with different processing parameter values. It was shown that an increase in the energy density leads to different porosity contents depending on the varied parameters. Moreover, it was found that by reducing the hatching distance, it was possible to reduce the porosity value. Conversely, the minimum porosity value does not correspond to the maximum laser power or the minimum scanning speed. This research demonstrated that the ED value alone is not enough to predict the microstructure of 3D-printed parts.

Pavan et al. [37] analyzed the part density as a function of both the intra-layer time and energy density values. The intra-layer time, i.e., the time between the scanning of a certain point of the layer and the recoating operation, is responsible for the temperature that is locally reached by the powder during the printing process and, consequently, it could significantly affect the morphology of the 3D-printed parts. The authors revealed that the porosity is significantly affected by the combination of the inter-layer time and ED used during the printing process. Even if it is well known that the ED value is a crucial factor in the part density, the authors demonstrated that the intra-layer time has a similar effect. Ensuring a more uniform inter-layer time during the process would allow for a significant reduction in the variation of the product quality.

Stichel et al. [25,38] presented the results of a Round Robin study involving mechanical tensile tests and a microstructural pore morphology analysis of various samples fabricated using different manufacturing machines. The pore morphologies, as assessed through X-ray computed tomography, were juxtaposed and examined in relation to the process parameters utilized and their resultant mechanical properties. Their investigation revealed that the laser energy input parameters exhibited a limited impact on the porosity, in contrast to the prevailing literature, which suggests that a reduction in porosity can be achieved by increasing the laser energy. Conversely, the process temperature, specifically the powder bed temperature, appeared to exert influence over the pore density, with higher temperatures correlating with lower pore densities.

Rüsenberg et al. [39] investigated the porosity at different regions of SLS PA12 cubes, realized by modifying the laser power value, and evaluated the correlation with the main mechanical properties. The authors demonstrated that a higher part density was obtained with a higher energy density, and the mechanical properties were improved. Moreover, the authors found a skin that appears to be significantly denser compared to the internal region. Similar results were obtained by Ajoku et al. [40], by Rouholamin and Hopkinson [41], and by Morano et al. [42].

Liebrich et al. [43] evaluated the occurrence of porosity on thin-walled structures. The measurements were carried out using X-ray microtomography. The authors proved that the porosity within thin-walled structures produced by SLS strongly depends on the wall thickness as well as on the orientation in the building chamber. Overall, the measured porosity values were significantly lower compared to the overall porosity levels reported for laser-sintered parts of greater dimensions.

Morano et al. [42] analyzed the change in the shape and distribution of the pores during quasi-static loading conditions, inducing plastic strain, by employing X-ray micro tomography. The authors found a significant variation in the porosity percentage by increasing the residual deformation. This result was accompanied by a variation of pore shapes and dimensions. The analysis made it possible to follow the main mechanism that contributes to sample failure, e.g., pores' coalescence.

3.1.2. Porosity and Pore Size Distribution of PA12

The porosity values of PA, measured over the past decade by some of the researchers cited in the previous section, are summarized in Table 2. Alongside the porosity percentage ranges, the table also provides the ranges of values of the process parameters used by various authors to produce the analyzed parts. The measurements were taken at different times and places on parts made with different process parameters and, in some cases, using different techniques. Nonetheless, important considerations can be drawn from their analysis.

First of all, it can be verified that all of the percentage porosities measured are included in the wide range from 0.7% to 16%. The differences between these values can be mainly attributable to the processing parameters used for SLS printing. The latter, as discussed in the previous section, has a strong impact on the structure and on the distribution of pores inside of a finished product and, consequentially, on its mechanical properties [44,45].

However, it should be noted that the maximum values of porosity, reported by Dupin et al. [33], were obtained at lower energy density values, while the minimum values, measured by Liebric et al. [43], were obtained for the case of thin-walled structures. If these particular data are not considered, the variability of porosity can be considered restricted to the range of 2.5–4.8%.

Another important consideration is that the porosity measurements reported by various researchers confirm the correlation between the ED value end porosity percentage, as already observed by Caulfield et al. [26] in 2007. The increase in the ED value, almost always, leads to an increase in the density of the material (or, equivalently, a decrease in its porosity). This correlation could be demonstrated by individually plotting the values provided by various authors in their papers. For the sake of brevity, these data are not all reported in this work. However, trends can be verified by analyzing the data summarized in Table 2, which, for each author, reports only the extremes of the variation intervals. Note that, for each group of data, the ED values increase from left to right, while, on the contrary, the porosity values decrease.

It should be noted that it is not possible to observe a direct correspondence between the ED value and the measured porosity. That could be explained considering that the ED value depends on different factors, i.e., the laser power, hatching distance, and scanning speed (see Equation (1)) [25,46]. Furthermore, porosity also depends on all of the other processing parameters (e.g., bed temperature, layer thickness, etc.; see Table 1), as well as on the measurement method employed for its quantification.

As an example, Figure 4a reports a comparison between the porosity values obtained by various authors (Stichel et al. [25], Dewulf et al. [36], Morano et al. [42], and Pavan et al. [47]) with approximately equal ED values (ED = 3.4 $\pm$ 0.1 J/cm^2), while, in Figure 4b, the contribution of pores with a specific diameter on the total porosity can be observed. The average value of the measured porosity is equal to 3.5 $\pm$ 0.3%. Note that the differences between the measured porosity values are not directly related to the changes in the ED. They are probably attributable to the different process parameters used.

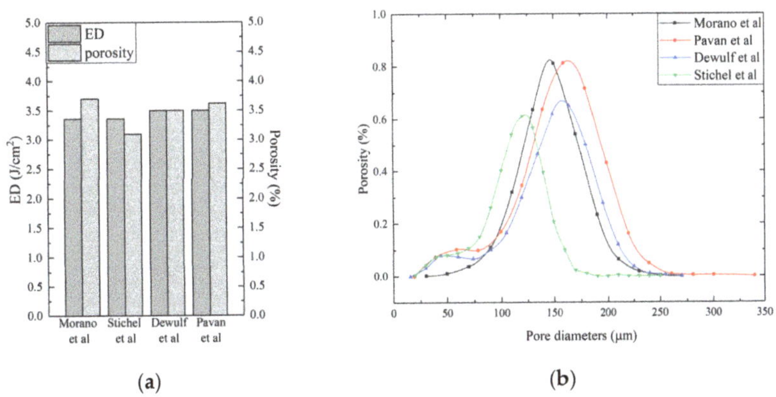

Figure 4. Comparison between (**a**) porosity and (**b**) equivalent pore diameter distribution available in the literature [25,36,42,47].

Another general conclusion is that, in all cases, the pore diameter distributions are slightly different. The greatest contribution is given by the pores with an average size that is typically contained between 120 μm and 180 μm. The remaining pores of smaller or larger dimensions, while contributing in a limited way to the overall porosity, can have a great influence on the mechanical properties of the material.

Returning to the correlation between the ED and porosity, it is important to highlight that Erdal et al. [48], as well as Rouholamin et al. [41] and Stichel et al. [25], reported the existence of a maximum optimal energy density. In fact, the authors found that by increasing the energy density beyond this maximum, the level of porosity of the part can remain unchanged or even deteriorate because of thermal degradation.

The sources of variability mentioned above uniquely affect the porosity values obtained by each researcher. This, unfortunately, does not allow for a direct comparison of all the data that are available in the literature in order to extract more general information.

To confirm this, Figure 5 shows the curve obtained using all the data available in the literature (approximately fifty porosity values measured for different ED levels).

The average porosity obtained is 4.6%, with a large standard deviation of approximately 2.6%. However, when the data are filtered by eliminating the most unlikely values, the average value drops to approximately 4.0% with a standard deviation of 0.94%. However, these values are very far from those presented previously.

It is difficult to draw other conclusions from the data that are available in the literature on the porosity of PA12. Generally, each researcher has developed their studies by keeping some parameters constant (very often without indicating their values in their published papers) and varying only those of interest, and, except for Stichel et al. [25], everyone used their equipment. It is therefore impossible to try to determine any correlations with the degree of porosity from the published data to understand what method to use to further decrease the porosity level.

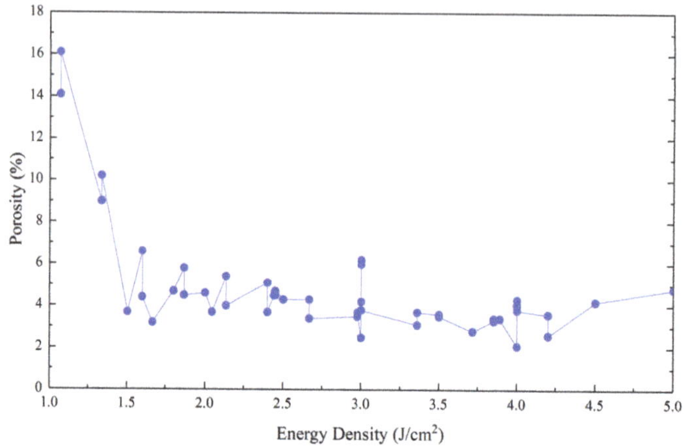

Figure 5. Porosity as a function of energy density. The graph was obtained by grouping all the data available in the literature for PA12.

Table 2. Summary of measured PA12 porosity values available in the literature.

Material	Powder Size (μm)	Layer Thickness (μm)	Powder Bed Temperature °C	Energy Density J/cm²	Porosity %	Measurement Technique
PA2200 [25]	60	100–150	160–178	1.67–3.72	3.20–2.80	μ-CT
PA2200 [36]	60	120	-	2.44–4.20	4.70–2.60	μ-CT
PA2200 [42]	56	100	168	3.36	3.70	μ-CT
PA2200 [43]	56	100	-	-	2.60–0.70	μ-CT
PA2200 [47]	56	120	-	2.00–5.00	4.80–3.60	μ-CT
PA2200 [46]	60–80	–	173	3.00–4.00	6.50–2.50	μ-CT
Duraform [25]	58	101–120	165.5–182	1.50–2.04	3.80–3.60	μ-CT
Duraform [33]	60	100	150	1.07–2.67	16.10–4.30	Archimedes μ-CT
Duraform [49]	58	100	175	1.80	4.70	μ-CT
InnovPA [33]	43	100	150	1.07–2.67	14.10–3.40	Archimedes μ-CT

It would be desirable for researchers to follow a single direction not only thematically but also for the presentation of the results. Everyone could thus proceed independently and obtain and present results that could be useful to the entire scientific community that studies the porosity of polymers. Currently, many works deal with the influence of the ED, but few investigate the influences of other parameters, so researchers should work in this direction in the future.

3.2. Polyamide 6 (PA6), 1010 (PA1010), and 11 (PA11) and Polyamide Blends

Despite the commercial availability of different polyamide powders for SLS printing, the literature on polymer provides, in general, only little information on their processing, since the vast majority of published results are focused on PA12.

Nevertheless, the data in the literature about the porosity percentage measured on SLS parts fabricated using different polyamide powders are summarized in Table 3.

Table 3. Summary of measured porosity values available in the literature for different polyamide parts.

Material	Powder Size (μm)	Layer Thickness (μm)	Powder Bed Temperature °C	Energy Density J/cm^2	Porosity %	Measurement Technique
PA6 [50]	180	500	25–150	-	1.9–3.9	Micrographs
PA6 [51]	100	100	25–180	2–36.6	39.7–60	Archimedes
PA1010 [52]	20–110	100–250	90–102	-	-	-
PA11 [53]	-	80–120	187	1.83–5.40	-	-
PA6/PA12 [54]	150–160	150	120	-	-	-
PA4,6/PA12 10/90 [55]	-	100	162	2.5–8	3.5–4.5	Archimedes
PA4,6/PA12 50/50 [55]	-	100	162	2.5–8	4–5	Archimedes
PA4,6/PA12 90/10 [55]	-	100	162	2.5–8	4.5–6.5	Archimedes

One of the other polyamide powders available for 3D printing is PA6. Zhou et al. [50] investigated Polyamide 6 single-layer specimens. The hatch spacing and the processing temperature were varied to evaluate their influences on the sample characteristics and porosity development. The authors demonstrated that the hatch spacing significantly affects the occurrence of layer porosity. In particular, the porosity ratio increases drastically as the scan spaces enlarge, passing from 1.95% to 3.89% when the scanning space increases from 0.25 to 0.45 mm, with an increment of about 50%. On the other hand, the processing temperature affects the mechanical properties.

Ling et al. [51], instead, analyzed, among other things, the porosity ratio of sintered specimens with different ambient temperatures and layer thicknesses. The experimental results demonstrated a decrease in the porosity ratio by increasing the processing temperature, i.e., from 60% measured at 25 °C to approximately 39% measured for temperatures up to 180 °C. These measured values are significantly higher than those reported by Zhou et al. [50]. This difference could be attributed to different printing parameters and sample characteristics. In particular, such high porosity values were measured on one-layer samples. This aspect demonstrated that the layer-by-layer process helps to reduce porosity since some void could be closed during the second layer melting. In fact, the porosity percentage measured on the samples realized with different layers, from 3 to 10, showed a decrease of up to 21% by increasing the number of layers. By further increasing the number of layers, the porosity decrease was slow, and for a number of layers that exceeded 14, it was negligible.

Another commercial powder that is available is PA1010. Liu-Ian et al. [52] investigated the morphology changes of a modified PA1010 by varying different processing parameters, i.e., the laser power, powder bed temperature, and layer thickness. The author found that by increasing the laser power, i.e., by more than 8 W, it is possible to obtain a well-defined morphology. However, for a laser power greater than 15 W, polymer degradation was observed. Similarly, increasing the bed temperature allows for a reduction in the dimensions of the detected pores. Finally, by reducing the layer thickness, it is possible to improve the sample morphology, i.e., lower the porosity, even if, for a thickness lower than 0.05 mm, the roller compromises the sample surface. Overall, even if authors analyze the porosity morphology, they do not quantify the amount.

Finally, PA11 has seen increased interest in general use due to its sustainable nature, since it is unique among other polyamides, as it is non-petroleum sourced. To fill the gap left by the literature on how to achieve optimal processing conditions, Wegner et al. [53] studied the correlations between the process parameters and part properties using a Design Of Experiments (DOE) approach. In particular, the main processing parameters, i.e., the scan speed, the laser power, hatch distance, and layer thickness, were modified, and their influence on the sample characteristics was analyzed. In particular, the authors evaluated the part density, surface roughness, and the final mechanical properties. Also, in this case, the authors did not furnish data about the porosity percentage. However, the experimental results demonstrated that energy density values that are significantly higher than those employed for PA12 fabrication are requested to achieve dense parts.

For the sake of completeness, this section closes by highlighting that, recently, some studies have been carried out to verify the possibility of using mixtures of polyamide powders in the SLS process. In particular, the works of Salmoria et al. [54] and Strobbe et al. [55] examined the properties of PA12 blends with PA6 and PA4,6 powders, respectively.

3.3. Other Pure Polymers and Polymer Blends

Among the AM techniques, SLS gives the possibility to process a wider range of polymeric powders [11], including a variety of pure-polymer-based powders. Comprehensive reviews on materials and process development are given by Kruth et al. [56], Schmid et al. [6,57], and, more recently, by Tan et al. [10]. Schmid et al. [57], in particular, discussed why several approaches adopted for new types of polymers failed and the reasons for the difficulties in developing new SLS powders.

Typically, polymer powders that are employed for the SLS process are semi-crystalline thermoplastic, even if it is also possible to find amorphous polymeric powder as well as elastomers. Thermoplastic polymer materials are well suited for laser sintering because of their relatively low melting temperatures.

In fact, if we exclude the polyamide varieties that were already examined in the previous section, a very limited variety of other kinds of polymers has been the subject of scientific publication. In this section, studies in which the porosity has been investigated are discussed, and the main results are summarized in Table 4.

Schmid et al. [58,59] presented a process chain for the production of spherical polybutylene terephthalate (PBT) microparticles. Their PBT powder, having a melting point of 223 °C, could be processed using a building temperature of 210 °C. Overall, by carrying out a rounding and a drying process on powder particles, it was possible to obtain a material suitable for 3D printing. However, further optimization is needed to improve the density of bulk parts.

Arai et al. [22] proposed to use a copolymer PBT (cPBT) for the fabrication of polymeric powder for the SLS process. The authors employed a cryomilling process for powder fabrication. It was found that the employed methods led to the occurrence of some metallic particle contaminations. These particles are responsible for an increased crystallization temperature that reduces the process windows. Nevertheless, the so-obtained cPBT powder was successfully employed for part fabrication with the SLS process. Table 4 summarizes the results in terms of porosity for different layer thicknesses and energy densities.

Table 4. Summary of measured porosity values available in the literature for different polymers or blends.

Material	Powder Size (μm)	Layer Thickness (μm)	Powder Bed Temperature °C	Energy Density J/cm^2	Porosity %	Measurement Technique
PBT [58]	25	295–450	210	8.4–12.6	6.3–14.1	Archimedes
PBT [22]	76	100	190–193	6.7–40	1.7–20.8	Micrographs
HDPE [60]	150–212	200	95	44	35	Archimedes
UHMPE [61]	125	100	142	1.6–3.2	60–65	Archimedes and μ-CT
PP [9]	45	150	150	1.8–1.9	8.4–10.1	μ-CT
PET [62]	59	100	200–240	2–5	2	Micrographs
PEEK 150PF [63]	56	100–200	345–357	1–3.6	0.2–15	Archimedes
PEEK 450PF [64]	50	120	-	1.47–3.24	0.35–17	μ-CT

Table 4. *Cont.*

Material	Powder Size (μm)	Layer Thickness (μm)	Powder Bed Temperature °C	Energy Density J/cm^2	Porosity %	Measurement Technique
PEEK HP3 [65]	60	100	-	-	4.36	Mercury intrusion
PEK HP3 [66]	70	120	368	-	0.3–10.4	μ-CT
PEK HP3 [67]	37–63	120	340	-	-	-
POM [68]	87–146	200	154–159	-	-	-
			BLENDS			
PA12/PEEK [69]	80	100	-	4.5	-	-
SEBS/PP [70]	85–107	100	100–160	7.9	-	-
PBT/PC [71]	161/218	-	205	5–10	10–40	Micrographs
PP/PA12 [72]	-	100	158	-	-	-
PA12/PBT [73]	60/200	-	140	38	-	-
PA12/HDPE [74]	60/120	150	60	-	-	-

Other semi-crystalline polymers such as polyethylene, polypropylene, polyoxymethylene, poly(ether ketone), and poly(ether ether ketone) are being actively researched, and some have been commercialized [7].

Bai et al. [75] explored, for the first time, the processability of polyethylene via selective laser sintering. The authors evaluated the influence of the processing parameters, e.g., the powder bed temperature, and laser power on the mechanical properties of 3D-printed parts. Moreover, the effect of the thermal history during the laser sintering process has also been evaluated. Unfortunately, the authors did not report any data about the porosity but evaluated only the quality of the fabricated parts.

Salmoria et al. [60] investigated the fabrication of HDPE specimens via SLS, employing particles with different sizes to control the porosity variation. The authors showed that the pore dimension depends on the sintering degree as well as on the particle size. In particular, the dimension of closed pores increases by increasing the dimension of the particles employed for 3D printing.

Khali et al. [61] carried out a mechanical and morphological characterization of porous Ultra-High-Molecular-Weight Polyethylene (UHMWPE) laser-sintered samples realized by using different processing parameters. Different works in the literature demonstrate that is difficult to fabricate UHMWPE parts via additive manufacturing. Due to that, the authors evaluated the influence of laser power variation on 3D-printed UHMWPE parts. The results demonstrated that the porosity level remains high (ranging between 60% and 65%) with no significant variation by modifying the laser power value and, consequently, the flexural properties are compromised.

Poly(ethylene terephthalate) (PET) was found to be suitable for application in SLS. Bashir et al. [62] analyzed the feasibility of processing highly crystalline PET for SLS 3D printing. It was found that the material exhibits a wide operating window and the recyclability of unmelted powder for new cycles is good. Overall, it seems that the printability of PET is similar to that of PA12. Moreover, the authors measured a 2% residual porosity.

In recent years, high interest was given to a new type of polymers that are suitable for high temperatures, i.e., Poly Aryl Ether Ketones (PAEKs), for the SLS process. Examples are Poly Ether Ketone (PEK) and Poly Ether Ether Ketone (PEEK), which could be successfully employed in different industrial fields, thanks to their high melting temperatures, chemical and wear resistance, and biocompatibility [63]. Even if PA and PS polymer families are widespread in different industrial processes, the processability of PEEK through 3D printing is currently a challenge [66]. However, it is necessary to include some printer variation to increase the process temperature up to 350 °C, i.e., the melting temperature of this kind of polymer. Moreover, it is also necessary to improve the materials' flowability. It was also shown that it is possible to reduce the porosity from 15% to a nearly zero value by properly selecting the process parameters. In this way, it is also possible to improve the mechanical properties that appear to be interconnected to the porosity percentage. In the literature, different studies were carried out on the feasibility of PEK, PEEK, and EOS PEEK HP3 parts using the SLS process [64,65,67,76].

The SLS technology could be employed for PP powder processing. However, it is necessary to deeply analyze the process parameters' influence on PP 3D-printed parts for reliable manufacturing. Flores Ituarte et al. [9] evaluated the influence of the main processing parameters' variation on the porosity percentage. In particular, a DOE was developed to investigate the influence of both the laser power and scanning speed. The porosity was measured through computed tomography. It was found that the occurrence of a high porous structure, i.e., a porosity percentage ranging between 8.46% and 10.08%, and, moreover, the highest porosity appears to be located in the interlayer planes.

Other polymers that exhibit good mechanical properties such as high stiffness, high wear, and creep resistance are PBT and POM. However, studies on this kind of polymer are scarce. Recently, Wegner [76] analyzed the percentage of bulk density of different polymeric parts fabricated through two different laser sintering machines. A porosity percentage between 1% and 2% was found, which is lower than the typical values measured on PA 12. Dechet et al. [68] employed a non-mechanical method, based on the solution–dissolution process, for the fabrication of POM powders that are suitable for PBF. The quality of the as-manufactured powder was demonstrated through the manufacturing of multi-layered samples.

Finally, polymer blends were developed and analyzed for the fabrication of parts with improved properties. This kind of material offers an alternative approach for obtaining parts with specific characteristics, thus allowing for the development of new applications. Nonetheless, polymer blends have received considerably less attention in research compared to pure polymers. This disparity arises from the necessity for chemical compatibility between the constituent materials in the blend and the thermal limitations that make the sintering of such blends more challenging. Additionally, the temperature ranges within which the sintering process must occur tend to be narrower for polymer blends than for their pure polymer constituents. This implies that polymer blends are more susceptible to variations in the bed temperature of the part, underscoring the critical importance of precise temperature control. The utilization of SLS for polymer blends is contingent upon a broad selection of compatible blend constituents. Nevertheless, there have been noteworthy developments in the application of SLS to various polymer blends including PA12/PEEK [69], PA12/HDPE [74], PA12/PBT [73], PA12/PP [72], PBT/PC [71], PMMA/PS [77], PP/POM [76], and SEBS/PP [70]. The data are reported in Table 4.

3.4. *Amorphous Polymers and Elastomers*

Amorphous polymers were the first kind of polymers employed for SLS. The main results regarding amorphous polymers and elastomers are summarized in Table 5.

Table 5. Summary of measured porosity values available in the literature for different amorphous polymers or elastomers.

Material	Powder Size (μm)	Layer Thickness (μm)	Powder Bed Temperature °C	Energy Density J/cm^2	Porosity %	Measurement Technique
PC [78]	30–180	130	145	3–12	10–45	Archimedes
PS [79]	25–106	150	85	2–12	12–60	Archimedes

Table 5. *Cont.*

Material	Powder Size (μm)	Layer Thickness (μm)	Powder Bed Temperature °C	Energy Density J/cm^2	Porosity %	Measurement Technique
PS [80]	75–100	150	90–95	6–7	-	-
PS [81]	-	100	90–100	4–14	5–25	Archimedes
PMMA [82]	75	-	100	15–40	50–61	Archimedes
TPU [83]	63–75	100	70–125	5–14	10–21	Archimedes
TPU [84]	45.7–62.8	100	125	25	0–0.2	Archimedes and μ-CT
SAN [85]	59.08	100	99	2–12	30–55	Archimedes

Among amorphous polymers, polycarbonate is widespread, and it includes bisphenol-A PC and aliphatic PC. Bisphenol-A PC exhibits good mechanical properties, and due to that, it is possible to find different studies [78,86,87]. However, bisphenol-A is classified as a low-poison chemical material, and due to that, in different countries, its use is forbidden for applications in food and medical fields. Consequently, aliphatic PC is subjected to increasing interest. The influence of the main processing parameters and, in particular, of the laser power energy on the aliphatic polycarbonate porosity was analyzed by Song et al. [87]. The experimental results show the occurrence of high porosity, between 50% and 70%. Overall, the process allows for the feasibility of the fabrication of aliphatic PC samples via 3D printing, even if it is still necessary to reduce the porosity.

As regards polystyrene (PS) and high-impact polystyrene (HIPS), the research in this field is limited. Only very few works have been published in the open literature [79–81]. For example, Shi et al. [80] evaluated the printability of high-impact polystyrene and found good dimensional accuracy as well as mechanical properties. Similarly, Strobbe et al. [81] analyzed the same material using a single-layer approach and evaluated the printing parameters' influence. A good consolidation of 3D-printed parts and a small amount of porosity was found by properly selecting 3D printing parameters.

Polymethylmethacrylate (PMMA), a synthetic resin [77,82], is employed in medical fields. Typically, this material is mixed with other additives to obtain a soft substrate that can harden gradually. In fact, this material is processed using two different approaches: (i) molding, which is employed when the PMMA is in the soft condition, or (ii) machining after the reaching of the hard form. However, these technologies are accompanied by some limitations such as the possibility to fabricate complex parts. Additive manufacturing could overcome this issue. Velu et al. [82] evaluated the influence of processing parameters on PMMA parts fabricated through SLS. The authors found a correlation between porosity and processing parameters as well as a correlation to mechanical properties. In particular, through the proper selection of process parameters, it is possible to reduce the porosity (52% instead of 61%) and to improve the mechanical properties (tensile strength is two times higher).

As regards the elastomers, among them, one of the most used for SLS is TPU, even if its applicability fields are limited. Currently, there are limited studies on the correlation between the processing parameters and TPU sample quality, e.g., porosity occurrence. Verbelen et al. [83] carried out an experimental analysis of different TPU grades characterized by very distinct characteristics. The authors demonstrated that it is possible to employ the SLS process for the fabrication of TPU parts. However, the final parts are characterized by high porosity and degradation. Due to that, further studies are requested. Ziegelmeier et al. [84,88] carried out different studies on TPU processability using SLS. Firstly, the authors evaluated the correlation between the behavior of the powder and the final properties of the 3D-printed parts. With this aim, the authors selected two different kinds of elastomers, i.e., TPU and a commercial thermoplastic elastomer (i.e., the Duraform

Flex-DF). The authors demonstrated that improved packing and flow capability of the powder particles could lead to bulk parts characterized by lower porosity.

Yan et al. [85] studied a styrene–acrylonitrile copolymer (SAN), another kind of amorphous polymer, as an SLS material to make parts with good dimensional accuracy and sintering properties and, consequently, better mechanical properties. In particular, the authors evaluated the influence of processing parameters, e.g., the ED value, on the quality of 3D-printed parts, e.g., the porosity and dimensional accuracy. Moreover, the results were compared with the data on the parts fabricated using PS powder. It was found that SAN could be successfully employed for the fabrication of parts with complex shapes and good dimensional accuracy. However, a high occurrence of voids was detected, and that issue was overcome by employing a post-processing treatment, i.e., infiltrating epoxy.

4. Discussion

For the convenience of treatment, all of the porosity values found in the literature are summarized in the graphs of Figure 6. In particular, Figure 6a reports the data relating to semi-crystalline polymers, while Figure 6b shows the porosities of the amorphous polymers. The values presented are the average of the data reported in Tables 2–5, neglecting out-of-range values and blends.

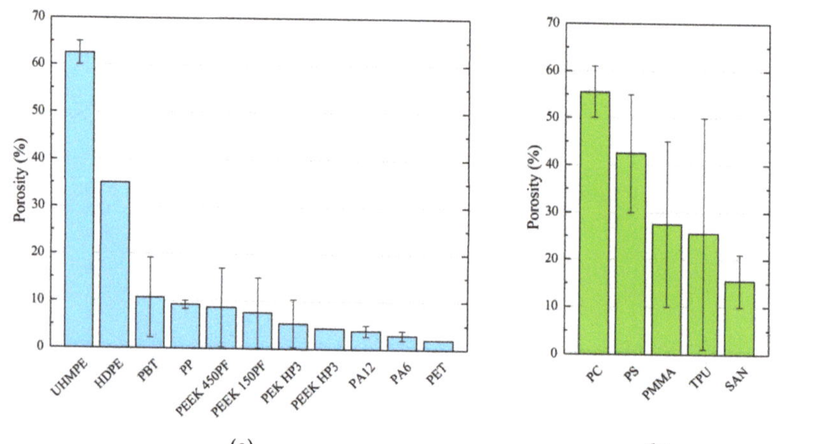

(a) (b)

Figure 6. Porosity average value found in the literature for (**a**) semi-crystalline polymers and (**b**) amorphous polymers and elastomers.

The comparison between the porosities of semi-crystalline polymers (Figure 6a) and the porosities of amorphous polymers and elastomers (Figure 6b) clearly confirms what has already been highlighted previously. Semi-crystalline polymers are characterized by lower porosity values. Except for UHMPE and HDPE, which represent two particular cases, the ratio between the porosity values of the two polymer families is greater than 5/1. UHMPE and HDPE are not generally used for the SLS process, and their printability is still a challenge. Porosity values will certainly reduce drastically in the future. Excluding these two polymers, the porosity in semi-crystalline polymers remains less than 10%, and it should be highlighted that an appropriate choice of processing parameters could allow for values close to zero to be reached. The porosities of amorphous and elastomeric polymers are, however, higher and vary between 15% and 55%.

By considering each polymer family, the first outcome that emerged from the literature analysis is that the polyamide family represents the highest percentage of the polymers employed for SLS. In particular, PA12 is a widespread material, and several works have been carried out on this material. The experimental evaluation of porosity on that material demonstrates a high variability, from a minimum of 2.5% to a maximum of 17% [33,36,46]. These values were measured on the samples that were obtained with different processing parameters, thus demonstrating that the fabrication process significantly affects the morphology of 3D-printed parts. Overall, a reduction in the measured porosity was observed by increasing the ED value. Higher supplied energies give the possibility of increasing the dimension of the melting pool and reducing the material viscosity, thus obtaining a denser part. However, it is not possible to limit the correlation of the porosity value to the ED value. In fact, in some cases, with similar ED values, different porosity values were measured (see Figure 4 and Table 2). These discrepancies could be attributed to the other processing parameters that are not involved in the ED equation. Moreover, the same ED values could be obtained, starting from different processing parameters that could influence the quality of 3D-printed parts in different ways [36]. Two main considerations follow from these results. First of all, it is clear that it is possible to reduce the porosity by acting on fabrication parameters; secondly, the complete elimination of the porosity could not be obtained just by acting on process parameters. Moreover, it is also important to consider that different techniques employed for porosity evaluation could lead to different measuring errors, thus contributing to the variability. All of these considerations suggest the need to develop standardized strategies and protocols for the fabrication of 3D-printed parts and their quality assessment.

Similar considerations could be made on the other polyamide raw powders as well as for the other polymers or blends. However, for these materials, the data in the literature about porosity are limited and reveal a very high variability (see Tables 3–5). Moreover, in some cases, the information about fabrication parameters does not give the possibility to make specific comparisons between different analyses carried out by different researchers. Overall, the higher porosity values observed for some SLS polymers could be attributed to two main factors. On one hand, some of the considered materials have been employed in the SLS process in the past few years. Due to that, the process is still under development, and further optimization is needed. On the other hand, some of these materials are not suitable for traditional SLS processes, such as those materials that require high melting temperatures. In this case, the SLS process and machines should be optimized for this kind of material. Overall, it is possible to conclude that porosity is still an open issue for SLS-printed polymers and that further analysis is required, especially for new emerging polymers with high mechanical properties, e.g., PEEK.

5. Conclusions and Future Remarks

In this review paper, a wide description of the main polymeric materials employed for the SLS process is reported. An analysis of the literature demonstrated that the widespread material for SLS is PA12. However, the analysis of the process parameters and their influences on the mechanical properties and part density is actually a crucial aspect, and different researchers are focusing their efforts on investigating this point. In fact, a deep understanding of this correlation is a key point for improving the reliability of SLS parts. The experimental results obtained by many researchers on PA12 allowed us to observe a correlation between the ED value and the porosity percentage that decreases by increasing the ED. However, the ED value alone is not enough to predict the porosity percentage in 3D-printed parts, since other processing parameters significantly affect the porosity development, e.g., the powder bed temperature or layer thickness. It is also important to highlight that the porosity values refer only to the percentage of porosity without giving information about the pores' shapes. However, it was demonstrated that the pores' dimensions and shapes could affect the mechanical properties as well as the porosity percentage itself. For this reason, further investigation is needed to analyze the influence of the processing parameters on the pores' geometrical characteristics.

An analysis of the literature also demonstrated the increasing research around other kinds of polymers for structural application, e.g., PEEK, or medical application, e.g., PMMA. The data about porosity are reported in specific tables, and the results demonstrated that the high product quality of 3D-printed parts is still an open issue, and further investigation is requested. Overall, the way to achieve the deployment of SLS technology in different industrial fields lies in the possibility of developing standard strategies for part fabrication and, as a final goal, fabrication protocols.

Author Contributions: Conceptualization, L.P. and C.M.; data curation, L.P. and C.M.; writing—original draft preparation, C.M.; writing—review and editing, L.P.; supervision, L.P.; project administration, L.P. All authors have read and agreed to the published version of the manuscript.

Funding: This research received no external funding.

Institutional Review Board Statement: Not applicable.

Data Availability Statement: The data reported in this work are available upon request.

Conflicts of Interest: The authors declare no conflict of interest.

Abbreviations

ABS	Acrylonitrile Butadiene Styrene
AM	Additive Manufacturing
ASA	Acrylonitrile Styrene Acrylate
ASTM	American Society for Testing and Materials
CAD	Computer-Aided Design
CT	Computed Tomography
DOE	Design Of Experiments
HDPE	High-Density Polyethylene
PA	PolyAmide
PAEK	PolyArylEtherKetones
PAI	PolyAmideImide
PBF	Powder Bed Fusion
PB	PolyButylene
PBT	PolyButylene Terephthalate
PC	PolyCarbonate
PCL	PolyCaproLactone
PEEK	PolyEtherEtherKetone
PEI	PolyEtherImide;
PES	PolyEtherSulfone
PET	PolyEthylene Terephthalate
PI	PolyImide
PLA	PolyLacticAcid
PMMA	Poly(Methyl MethAcrylate)
POM	PolyOxyMethylene
PP	PolyPropylene
PPF	Poly(PropyleneFumarate)
PPSF	PolyPhenyl Sulfone
PPSU	Poly(Phenyl Sulfone)
PS	PolyStyrene
PSU	PolySulfone
PTFE	PolyTetraFluoroEthylene
PU	PolyUrethane
PVA	Poly(VinylAlcohol)
PVAc	PolyVinylAcetate
PVB	PolyVinylButyral
PVC	Poly(VinylChloride)
PVDF	PolyVinyliDene Fluoride
RP	Rapid Prototyping
SAN	Styrene-AcryloNitrile

SLS	Selective Laser Sintering
sPS	Syndiotactic PolyStyrene
TPE	ThermoPlastic Elastomers
TPO	Phosphineoxide
TPU	Thermoplastic PolyUrethane elastomer
UHMWPE	Ultra-High-Molecular-Weight PolyEthylene

References

1. Yadroitsev, I.; Yadroitsava, I.; Du Plessis, A.; MacDonald, E. Additive Manufacturing Materials and Technologies. In *Fundamentals of Laser Powder Bed Fusion of Metals*; Elsevier: Amsterdam, The Netherlands, 2021; ISBN 978-0-12-824090-8.
2. Wong, K. *Wohlers 2017 Report on 3D Printing Industry Points to Softened Growth*; Wohlers Associates: Fort Collins, CO, USA, 2017; p. 1.
3. *ASTM 52900*; Additive Manufacturing—General Principles—Fundamentals and Vocabulary. ASTM International: Conshohocken, PA, USA, 2022.
4. Yan, C.; Shi, Y.; Li, Z.; Wen, S.; Wei, Q. *Selective Laser Sintering Additive Manufacturing Technology*; Elsevier: Amsterdam, The Netherlands, 2021; ISBN 978-0-08-102993-0.
5. Goodridge, R.D.; Tuck, C.J.; Hague, R.J.M. Laser Sintering of Polyamides and Other Polymers. *Prog. Mater. Sci.* **2012**, *57*, 229–267. [CrossRef]
6. Schmid, M.; Amado, A.; Wegener, K. Materials Perspective of Polymers for Additive Manufacturing with Selective Laser Sintering. *J. Mater. Res.* **2014**, *29*, 1824–1832. [CrossRef]
7. Schmid, M.; Wegener, K. Additive Manufacturing: Polymers Applicable for Laser Sintering (LS). *Procedia Eng.* **2016**, *149*, 457–464. [CrossRef]
8. Schmid, M.; Wegener, K. Thermal and Molecular Properties of Polymer Powders for Selective Laser Sintering (SLS). In Proceedings of the Regional Conference Graz 2015—Polymer Processing Society, Graz, Austria, 21–25 September 2015; p. 100003.
9. Flores Ituarte, I.; Wiikinkoski, O.; Jansson, A. Additive Manufacturing of Polypropylene: A Screening Design of Experiment Using Laser-Based Powder Bed Fusion. *Polymers* **2018**, *10*, 1293. [CrossRef]
10. Tan, L.J.; Zhu, W.; Zhou, K. Recent Progress on Polymer Materials for Additive Manufacturing. *Adv. Funct. Mater.* **2020**, *30*, 2003062. [CrossRef]
11. Bourell, D.; Kruth, J.P.; Leu, M.; Levy, G.; Rosen, D.; Beese, A.M.; Clare, A. Materials for Additive Manufacturing. *CIRP Ann.* **2017**, *66*, 659–681. [CrossRef]
12. Brighenti, R.; Cosma, M.P.; Marsavina, L.; Spagnoli, A.; Terzano, M. Laser-Based Additively Manufactured Polymers: A Review on Processes and Mechanical Models. *J. Mater. Sci.* **2021**, *56*, 961–998. [CrossRef]
13. Zhang, Z.; Yao, X.X.; Ge, P. Phase-Field-Model-Based Analysis of the Effects of Powder Particle on Porosities and Densities in Selective Laser Sintering Additive Manufacturing. *Int. J. Mech. Sci.* **2020**, *166*, 105230. [CrossRef]
14. Vock, S.; Klöden, B.; Kirchner, A.; Weißgärber, T.; Kieback, B. Powders for Powder Bed Fusion: A Review. *Prog. Addit. Manuf.* **2019**, *4*, 383–397. [CrossRef]
15. Gibson, I.; Rosen, D.; Stucker, B.; Khorasani, M. *Additive Manufacturing Technologies*, 3rd ed.; Springer International Publishing: Cham, Switzerland, 2021; ISBN 978-3-030-56126-0.
16. Beal, V.E.; Paggi, R.A.; Salmoria, G.V.; Lago, A. Statistical Evaluation of Laser Energy Density Effect on Mechanical Properties of Polyamide Parts Manufactured by Selective Laser Sintering. *J. Appl. Polym. Sci.* **2009**, *113*, 2910–2919. [CrossRef]
17. El Magri, A.; Bencaid, S.E.; Vanaei, H.R.; Vaudreuil, S. Effects of Laser Power and Hatch Orientation on Final Properties of PA12 Parts Produced by Selective Laser Sintering. *Polymers* **2022**, *14*, 3674. [CrossRef] [PubMed]
18. Morano, C.; Pagnotta, L. Additive Manufactured Parts Produced by Selective Laser Sintering Technology: Porosity Formation Mechanisms. *J. Polym. Eng.* **2023**, *43*, 537–555. [CrossRef]
19. Zhu, W.; Yan, C.; Shi, Y.; Wen, S.; Liu, J.; Shi, Y. Investigation into Mechanical and Microstructural Properties of Polypropylene Manufactured by Selective Laser Sintering in Comparison with Injection Molding Counterparts. *Mater. Des.* **2015**, *82*, 37–45. [CrossRef]
20. Yan, C.; Shi, Y.; Hao, L. Investigation into the Differences in the Selective Laser Sintering between Amorphous and Semi-Crystalline Polymers. *Int. Polym. Process.* **2011**, *26*, 416–423. [CrossRef]
21. Morano, C.; Pagnotta, L. On Powder Bed Fusion Manufactured Parts: Porosity and Its Measurement. *Curr. Mater. Sci.* **2023**, *17*, 185–197. [CrossRef]
22. Arai, S.; Tsunoda, S.; Kawamura, R.; Kuboyama, K.; Ougizawa, T. Comparison of Crystallization Characteristics and Mechanical Properties of Poly(Butylene Terephthalate) Processed by Laser Sintering and Injection Molding. *Mater. Des.* **2017**, *113*, 214–222. [CrossRef]
23. Vidakis, N.; David, C.; Petousis, M.; Sagris, D.; Mountakis, N.; Moutsopoulou, A. The Effect of Six Key Process Control Parameters on the Surface Roughness, Dimensional Accuracy, and Porosity in Material Extrusion 3D Printing of Polylactic Acid: Prediction Models and Optimization Supported by Robust Design Analysis. *Adv. Ind. Manuf. Eng.* **2022**, *5*, 100104. [CrossRef]

24. Vidakis, N.; David, C.; Petousis, M.; Sagris, D.; Mountakis, N. Optimization of Key Quality Indicators in Material Extrusion 3D Printing of Acrylonitrile Butadiene Styrene: The Impact of Critical Process Control Parameters on the Surface Roughness, Dimensional Accuracy, and Porosity. *Mater. Today Commun.* **2023**, *34*, 105171. [CrossRef]
25. Stichel, T.; Frick, T.; Laumer, T.; Tenner, F.; Hausotte, T.; Merklein, M.; Schmidt, M. A Round Robin Study for Selective Laser Sintering of Polymers: Back Tracing of the Pore Morphology to the Process Parameters. *J. Mater. Process. Technol.* **2018**, *252*, 537–545. [CrossRef]
26. Caulfield, B.; McHugh, P.E.; Lohfeld, S. Dependence of Mechanical Properties of Polyamide Components on Build Parameters in the SLS Process. *J. Mater. Process. Technol.* **2007**, *182*, 477–488. [CrossRef]
27. Morano, C.; Alfano, M.; Pagnotta, L. Effect of Strain Rates and Heat Exposure on Polyamide (PA12) Processed via Selective Laser Sintering. *Materials* **2023**, *16*, 4654. [CrossRef] [PubMed]
28. David, C.; Sagris, D.; Petousis, M.; Nasikas, N.K.; Moutsopoulou, A.; Sfakiotakis, E.; Mountakis, N.; Charou, C.; Vidakis, N. Operational Performance and Energy Efficiency of MEX 3D Printing with Polyamide 6 (PA6): Multi-Objective Optimization of Seven Control Settings Supported by L27 Robust Design. *Appl. Sci.* **2023**, *13*, 8819. [CrossRef]
29. Vidakis, N.; Petousis, M.; Michailidis, N.; Mountakis, N.; Papadakis, V.; Argyros, A.; Charou, C. Medical Grade Polyamide 12 Silver Nanoparticle Filaments Fabricated with In-Situ Reactive Reduction Melt-Extrusion: Rheological, Thermomechanical, and Bactericidal Performance in MEX 3D Printing. *Appl. Nanosci.* **2023**. [CrossRef]
30. Cai, C.; Tey, W.S.; Chen, J.; Zhu, W.; Liu, X.; Liu, T.; Zhao, L.; Zhou, K. Comparative Study on 3D Printing of Polyamide 12 by Selective Laser Sintering and Multi Jet Fusion. *J. Mater. Process. Technol.* **2021**, *288*, 116882. [CrossRef]
31. Simha Martynková, G.; Slíva, A.; Kratošová, G.; Čech Barabaszová, K.; Študentová, S.; Klusák, J.; Brožová, S.; Dokoupil, T.; Holešová, S. Polyamide 12 Materials Study of Morpho-Structural Changes during Laser Sintering of 3D Printing. *Polymers* **2021**, *13*, 810. [CrossRef] [PubMed]
32. Ly, H.-B.; Monteiro, E.; Dal, M.; Regnier, G. On the Factors Affecting Porosity Dissolution in Selective Laser Sintering Process. In Proceedings of the 21st International Esaform Conference on Material Forming: ESAFORM 2018, Palermo, Italy, 23–25 April 2018; p. 120014.
33. Dupin, S.; Lame, O.; Barrès, C.; Charmeau, J.-Y. Microstructural Origin of Physical and Mechanical Properties of Polyamide 12 Processed by Laser Sintering. *Eur. Polym. J.* **2012**, *48*, 1611–1621. [CrossRef]
34. Tontowi, A.E.; Childs, T.H.C. Density Prediction of Crystalline Polymer Sintered Parts at Various Powder Bed Temperatures. *Rapid Prototyp. J.* **2001**, *7*, 180–184. [CrossRef]
35. Gomes, P.C.; Piñeiro, O.G.; Alves, A.C.; Carneiro, O.S. On the Reuse of SLS Polyamide 12 Powder. *Materials* **2022**, *15*, 5486. [CrossRef]
36. Dewulf, W.; Pavan, M.; Craeghs, T.; Kruth, J.-P. Using X-ray Computed Tomography to Improve the Porosity Level of Polyamide-12 Laser Sintered Parts. *CIRP Ann.* **2016**, *65*, 205–208. [CrossRef]
37. Pavan, M.; Faes, M.; Strobbe, D.; Van Hooreweder, B.; Craeghs, T.; Moens, D.; Dewulf, W. On the Influence of Inter-Layer Time and Energy Density on Selected Critical-to-Quality Properties of PA12 Parts Produced via Laser Sintering. *Polym. Test.* **2017**, *61*, 386–395. [CrossRef]
38. Stichel, T.; Frick, T.; Laumer, T.; Tenner, F.; Hausotte, T.; Merklein, M.; Schmidt, M. A Round Robin Study for Selective Laser Sintering of Polyamide 12: Microstructural Origin of the Mechanical Properties. *Opt. Laser Technol.* **2017**, *89*, 31–40. [CrossRef]
39. Rüsenberg, S.; Schimdt, L.; Schmid, H.-J. *Mechanical and Physical Properties—A Way to Assess Quality of Laser Sintered Parts*; The University of Texas in Austin: Austin, TX, USA, 2011.
40. Ajoku, U.; Saleh, N.; Hopkinson, N.; Hague, R.; Erasenthiran, P. Investigating Mechanical Anisotropy and End-of-Vector Effect in Laser-Sintered Nylon Parts. *Proc. IMechE* **2006**, *220*, 1077–1086. [CrossRef]
41. Rouholamin, D.; Hopkinson, N. An Investigation on the Suitability of Micro-Computed Tomography as a Non-Destructive Technique to Assess the Morphology of Laser Sintered Nylon 12 Parts. *Proc. Inst. Mech. Eng. Part B J. Eng. Manuf.* **2014**, *228*, 1529–1542. [CrossRef]
42. Morano, C.; Crocco, M.C.; Formoso, V.; Pagnotta, L. Effect of Induced Plastic Strain on the Porosity of PA12 Printed through Selective Laser Sintering Studied by X-ray Computed Micro-Tomography. *Int. J. Adv. Manuf. Technol.* **2023**, *125*, 3229–3240. [CrossRef]
43. Liebrich, A.; Langowski, H.C.; Schreiber, R.; Pinzer, B.R. Porosity Distribution in Laser-Sintered Polymeric Thin Sheets as Revealed by X-ray Micro Tomography. *Polym. Test.* **2019**, *76*, 286–297. [CrossRef]
44. Wörz, A.; Wudy, K.; Drummer, D. *Understanding the Influence of Energy-Density of the Layer Dependent Part Properties in Laser-Sintering of PA12*; The University of Texas in Austin: Austin, TX, USA, 2019. [CrossRef]
45. Hofland, E.C.; Baran, I.; Wismeijer, D.A. Correlation of Process Parameters with Mechanical Properties of Laser Sintered PA12 Parts. *Adv. Mater. Sci. Eng.* **2017**, *2017*, 4953173. [CrossRef]
46. Lexow, M.M.; Drexler, M.; Drummer, D. Fundamental Investigation of Part Properties at Accelerated Beam Speeds in the Selective Laser Sintering Process. *Rapid Prototyp. J.* **2017**, *23*, 1099–1106. [CrossRef]
47. Pavan, M.; Craeghs, T.; Puyvelde, P.V.; Kruth, J.-P.; Dewulf, W. Understanding The Link Between Process Parameters, Microstructure And Mechanical Properties Of Laser Sintered PA12 Parts Through X-ray Computed Tomography. In Proceedings of the 2nd International Conference on Progress in Additive Manufacturing, Singapore, 16–19 May 2016; p. 9.

48. Erdal, M.; Dag, S.; Jande, Y.A.C.; Tekin, C.M. Production and Characterization of Uniform and Graded Porous Polyamide Structures Using Selective Laser Sintering. *Workshop Rapid Technol.* **2009**, *24*, 43–50.
49. Flodberg, G.; Pettersson, H.; Yang, L. Pore Analysis and Mechanical Performance of Selective Laser Sintered Objects. *Addit. Manuf.* **2018**, *24*, 307–315. [CrossRef]
50. Zhou, W.; Wang, X.; Hu, J.; Zhu, X. Melting Process and Mechanics on Laser Sintering of Single Layer Polyamide 6 Powder. *Int. J. Adv. Manuf. Technol.* **2013**, *69*, 901–908. [CrossRef]
51. Ling, Z.; Wu, J.; Wang, X.; Li, X.; Zheng, J. Experimental Study on the Variance of Mechanical Properties of Polyamide 6 during Multi-Layer Sintering Process in Selective Laser Sintering. *Int. J. Adv. Manuf. Technol.* **2019**, *101*, 1227–1234. [CrossRef]
52. Lin, L.-L.; Shi, Y.-S.; Zeng, F.-D.; Huang, S.-H. Microstructure of Selective Laser Sintered Polyamide. *J. Wuhan Univ. Technol. Mat. Sci. Edit.* **2003**, *18*, 60–63. [CrossRef]
53. Wegner, A.; Harder, R.; Witt, G.; Drummer, D. Determination of Optimal Processing Conditions for the Production of Polyamide 11 Parts Using the Laser Sintering Process. *Int. J. Recent. Contrib. Eng. Sci. IT* **2015**, *3*, 5. [CrossRef]
54. Salmoria, G.V.; Leite, J.L.; Paggi, R.A. The Microstructural Characterization of PA6/PA12 Blend Specimens Fabricated by Selective Laser Sintering. *Polym. Test.* **2009**, *28*, 746–751. [CrossRef]
55. Strobbe, D.; Van Puyvelde, P.; Kruth, J.-P.; Van Hooreweder, B. *Laser Sintering of PA12/PA4,6 Polymer Composites*; The University of Texas in Austin: Austin, TX, USA, 2018; p. 10.
56. Kruth, J.-P.; Levy, G.; Klocke, F.; Childs, T.H.C. Consolidation Phenomena in Laser and Powder-Bed Based Layered Manufacturing. *CIRP Ann.* **2007**, *56*, 730–759. [CrossRef]
57. Schmid, M.; Amado, A.; Wegener, K. Polymer Powders for Selective Laser Sintering (SLS). In Proceedings of the PPS-30: The 30th International Conference of the Polymer Processing Society, Cleveland, OH, USA, 6–12 June 2014; p. 160009.
58. Schmidt, J.; Sachs, M.; Fanselow, S.; Zhao, M.; Romeis, S.; Drummer, D.; Wirth, K.-E.; Peukert, W. Optimized Polybutylene Terephthalate Powders for Selective Laser Beam Melting. *Chem. Eng. Sci.* **2016**, *156*, 1–10. [CrossRef]
59. Schmidt, J.; Sachs, M.; Zhao, M.; Fanselow, S.; Wudy, K.; Drexler, M.; Drummer, D.; Wirth, K.-E.; Peukert, W. A Novel Process for Production of Spherical PBT Powders and Their Processing Behavior during Laser Beam Melting. In Proceedings of the PPS-31: The 31st International Conference of the Polymer Processing Society, Jeju Island, Republic of Korea, 7–11 June 2015; p. 140008.
60. Salmoria, G.V.; Ahrens, C.H.; Klauss, P.; Paggi, R.A.; Oliveira, R.G.; Lago, A. Rapid Manufacturing of Polyethylene Parts with Controlled Pore Size Gradients Using Selective Laser Sintering. *Mat. Res.* **2007**, *10*, 211–214. [CrossRef]
61. Khalil, Y.; Hopkinson, N.; Kowalski, A.; Fairclough, J.P.A. Influence of Laser Power on Morphology and Properties of Laser-Sintered UHMWPE. In Proceedings of the 27th Annual International Solid Freeform Fabrication Symposium, Austin, TX, USA, 8 August 2016; p. 21.
62. Bashir, Z.; Gu, H.; Yang, L. Evaluation of Poly(Ethylene Terephthalate) Powder as a Material for Selective Laser Sintering, and Characterization of Printed Part. *Polym. Eng. Sci.* **2018**, *58*, 1888–1900. [CrossRef]
63. Schmidt, M.; Pohle, D.; Rechtenwald, T. Selective Laser Sintering of PEEK. *CIRP Ann.* **2007**, *56*, 205–208. [CrossRef]
64. Berretta, S.; Evans, K.E.; Ghita, O. Processability of PEEK, a New Polymer for High Temperature Laser Sintering (HT-LS). *Eur. Polym. J.* **2015**, *68*, 243–266. [CrossRef]
65. Hoskins, T.J.; Dearn, K.D.; Kukureka, S.N. Mechanical Performance of PEEK Produced by Additive Manufacturing. *Polym. Test.* **2018**, *70*, 511–519. [CrossRef]
66. Ghita, O.R.; James, E.; Trimble, R.; Evans, K.E. Physico-Chemical Behaviour of Poly (Ether Ketone) (PEK) in High Temperature Laser Sintering (HT-LS). *J. Mater. Process. Technol.* **2014**, *214*, 969–978. [CrossRef]
67. Ghita, O.; James, E.; Davies, R.; Berretta, S.; Singh, B.; Flint, S.; Evans, K.E. High Temperature Laser Sintering (HT-LS): An Investigation into Mechanical Properties and Shrinkage Characteristics of Poly (Ether Ketone) (PEK) Structures. *Mater. Des.* **2014**, *61*, 124–132. [CrossRef]
68. Dechet, M.A.; Baumeister, I.; Schmidt, J. Development of Polyoxymethylene Particles via the Solution-Dissolution Process and Application to the Powder Bed Fusion of Polymers. *Materials* **2020**, *13*, 1535. [CrossRef] [PubMed]
69. Schultz, J.P.; Martin, J.P.; Kander, R.G.; Suchicital, C.T.A. *Selective Laser Sintering of Nylon 12-PEEK Blends Formed by Cryogenic Mechanical Alloying*; The University of Texas in Austin: Austin, TX, USA, 2000; p. 6.
70. He, Z.; Ren, C.; Zhang, A.; Bao, J. Preparation and Properties of Styrene Ethylene Butylene Styrene/Polypropylene Thermoplastic Elastomer Powder for Selective Laser Sintering 3D Printing. *J. Appl. Polym. Sci.* **2021**, *138*, 50908. [CrossRef]
71. Greiner, S.; Wudy, K.; Lanzl, L.; Drummer, D. Selective Laser Sintering of Polymer Blends: Bulk Properties and Process Behavior. *Polym. Test.* **2017**, *64*, 136–144. [CrossRef]
72. Drummer, D.; Wudy, K.; Kühnlein, F.; Drexler, M. Polymer Blends for Selective Laser Sintering: Material and Process Requirements. *Phys. Procedia* **2012**, *39*, 509–517. [CrossRef]
73. Salmoria, G.V.; Lauth, V.R.; Cardenuto, M.R.; Magnago, R.F. Characterization of PA12/PBT Specimens Prepared by Selective Laser Sintering. *Opt. Laser Technol.* **2018**, *98*, 92–96. [CrossRef]
74. Salmoria, G.V.; Leite, J.L.; Ahrens, C.H.; Lago, A.; Pires, A.T.N. Rapid Manufacturing of PA/HDPE Blend Specimens by Selective Laser Sintering: Microstructural Characterization. *Polym. Test.* **2007**, *26*, 361–368. [CrossRef]
75. Bai, J.; Zhang, B.; Song, J.; Bi, G.; Wang, P.; Wei, J. The Effect of Processing Conditions on the Mechanical Properties of Polyethylene Produced by Selective Laser Sintering. *Polym. Test.* **2016**, *52*, 89–93. [CrossRef]

76. Wegner, A. New Polymer Materials for the Laser Sintering Process: Polypropylene and Others. *Phys. Procedia* **2016**, *83*, 1003–1012. [CrossRef]
77. Leite, J.L.; Salmoria, G.V.; Paggi, R.A.; Ahrens, C.H.; Pouzada, A.S. A Study on Morphological Properties of Laser Sintered Functionally Graded Blends of Amorphous Thermoplastics. *Int. J. Mater. Prod. Technol.* **2010**, *39*, 205. [CrossRef]
78. Ho, H.C.H.; Gibson, I.; Cheung, W.L. Effects of Energy Density on Morphology and Properties of Selective Laser Sintered Polycarbonate. *J. Mater. Process. Technol.* **1999**, *89–90*, 204–210. [CrossRef]
79. Ku, C.W.; Gibson, I.; Cheung, W.L. *Selective Laser Sintered CastForm^{TM} Polystyrene with Controlled Porosity and Its Infiltration Characteristics by Red Wax*; The University of Texas in Austin: Austin, TX, USA, 2002.
80. Shi, Y.; Wang, Y.; Chen, J.; Huang, S. Experimental Investigation into the Selective Laser Sintering of High-Impact Polystyrene. *J. Appl. Polym. Sci.* **2008**, *108*, 535–540. [CrossRef]
81. Strobbe, D.; Dadbakhsh, S.; Verbelen, L.; Van Puyvelde, P.; Kruth, J.-P. Selective Laser Sintering of Polystyrene: A Single-Layer Approach. *Plast. Rubber Compos.* **2018**, *47*, 2–8. [CrossRef]
82. Velu, R.; Singamneni, S. Evaluation of the Influences of Process Parameters While Selective Laser Sintering PMMA Powders. *Proc. Inst. Mech. Eng. Part C J. Mech. Eng. Sci.* **2015**, *229*, 603–613. [CrossRef]
83. Verbelen, L.; Dadbakhsh, S.; Van den Eynde, M.; Strobbe, D.; Kruth, J.-P.; Goderis, B.; Van Puyvelde, P. Analysis of the Material Properties Involved in Laser Sintering of Thermoplastic Polyurethane. *Addit. Manuf.* **2017**, *15*, 12–19. [CrossRef]
84. Ziegelmeier, S.; Christou, P.; Wöllecke, F.; Tuck, C.; Goodridge, R.; Hague, R.; Krampe, E.; Wintermantel, E. An Experimental Study into the Effects of Bulk and Flow Behaviour of Laser Sintering Polymer Powders on Resulting Part Properties. *J. Mater. Process. Technol.* **2015**, *215*, 239–250. [CrossRef]
85. Yan, C.; Shi, Y.; Yang, J.; Liu, J. Investigation into the Selective Laser Sintering of Styrene–Acrylonitrile Copolymer and Postprocessing. *Int. J. Adv. Manuf. Technol.* **2010**, *51*, 973–982. [CrossRef]
86. Ho, H.C.H.; Cheung, W.L.; Gibson, I. Morphology and Properties of Selective Laser Sintered Bisphenol A Polycarbonate. *Ind. Eng. Chem. Res.* **2003**, *42*, 1850–1862. [CrossRef]
87. Song, X.H.; Shi, Y.S.; Song, P.H.; Wei, Q.S.; Li, W. Effects of the Processing Parameters on Porosity of Selective Laser Sintered Aliphatic Polycarbonate. *Adv. Mater. Res.* **2014**, *915–916*, 1000–1004. [CrossRef]
88. Ziegelmeier, S.; Wöllecke, F.; Tuck, C.; Goodridge, R. *Characterizing the Bulk & Flow Behaviour of LS Polymer Powders*; The University of Texas in Austin: Austin, TX, USA, 2013.

polymers

Review

Self-Healing Polymeric Materials and Composites for Additive Manufacturing

Yixue Jiang [1,2], Evelyn Ling Ling Ng [1], Danielle Xinyun Han [1], Yinjia Yan [1,3], Siew Yin Chan [1], John Wang [1,2,*] and Benjamin Qi Yu Chan [1,*]

1 Institute of Materials Research and Engineering (IMRE), Agency for Science, Technology and Research (A*STAR), 2 Fusionopolis Way, Innovis #08-03, Singapore 138634, Singapore
2 Department of Materials Science and Engineering, College of Design and Engineering, National University of Singapore, 9 Engineering Drive 1, Singapore 117575, Singapore
3 Frontiers Science Center for Flexible Electronics (FSCFE), Xi'an Institute of Flexible Electronics (IFE), Xi'an Institute of Biomedical Materials and Engineering (IBME), and Ningbo Institute, Northwestern Polytechnical University, 127 West Youyi Road, Xi'an 710072, China
* Correspondence: msewangj@nus.edu.sg (J.W.); benjamin_chan@imre.a-star.edu.sg (B.Q.Y.C.)

Abstract: Self-healing polymers have received widespread attention due to their ability to repair damage autonomously and increase material stability, reliability, and economy. However, the processability of self-healing materials has yet to be studied, limiting the application of rich self-healing mechanisms. Additive manufacturing effectively improves the shortcomings of conventional processing while increasing production speed, accuracy, and complexity, offering great promise for self-healing polymer applications. This article summarizes the current self-healing mechanisms of self-healing polymers and their corresponding additive manufacturing methods, and provides an outlook on future developments in the field.

Keywords: 3D printing; self-healing; polymers and composites; additive manufacturing; smart polymers

check for updates

Citation: Jiang, Y.; Ng, E.L.L.; Han, D.X.; Yan, Y.; Chan, S.Y.; Wang, J.; Chan, B.Q.Y. Self-Healing Polymeric Materials and Composites for Additive Manufacturing. *Polymers* **2023**, *15*, 4206. https://doi.org/10.3390/polym15214206

Academic Editors: Anton Ficai and Cristina-Elisabeta Pelin

Received: 21 September 2023
Revised: 17 October 2023
Accepted: 23 October 2023
Published: 24 October 2023

1. Introduction

Inspired by nature, researchers have introduced self-healing mechanisms into materials, enabling them to repair internal or external damage, regain functionality, and extend their lifetime. These are called self-healing materials [1–5]. Self-healing materials are regarded as a new generation of advanced materials, and have been applied in many fields due to their advantages [6–8]. It has attracted the attention of modern science, materials engineering, and commodity research over the last decades [7,9,10]. Self-healing polymers endow fabricated products with product durability, mechanical stability, and reliability, as well as having a low environmental impact [6,9]. For example, soft robots, which are susceptible to damage, commonly require damaged components to be replaced entirely. This is costly, time consuming, and waste generating. Developing soft robots using self-healing polymers as an alternative presents a more economical and sustainable solution [11–13].

Generally, self-healing mechanisms of self-healing materials fall into two categories: extrinsic healing and intrinsic healing. Extrinsic healing includes microcapsules, microvascular networks, and nanoparticles. In contrast, intrinsic healing is divided into dynamic covalent bonding types, including Diels–Alder reaction, disulfide, and dynamic non-covalent bonding, involving hydrogen bonds, ionic interactions, and coordination.

Although the development of self-healing materials has attracted increasing interest, reports have focused on studies of healing ability and underlying mechanisms, with very little attention paid to their processability [14,15]. In addition, traditional processing techniques such as casting, electrospinning, or extrusion limit the complexity, heterogeneous material integration, precision, and yield of objects, while manufacturing with three-dimensional printing (3D printing) can effectively improve these deficiencies [12,16–19].

Three-dimensional printing, also referred to as additive manufacturing (AM), fabricates 3D physical objects consistent with corresponding digital models by stacking materials layer-by-layer, without the need for molds or machining [15,20,21].

Three-dimensional printing has many advantages, such as fast production speeds and the ability to precisely control structures' dimensions, shape, and density, allowing for the mass customization of complex equipment [12,22]. At the same time, because no additional tools or secondary processing is required, the effect of reducing manufacturing waste, and saving production time and energy consumption can be achieved [22,23]. However, the evaluation of self-healing polymers in the context of additive manufacturing (AM) necessitates a thorough assessment of various performance metrics. Of greatest importance is the healing efficiency, which serves as a measure of the material's ability to restore its properties after sustaining damage. In order to ensure structural integrity, it becomes imperative to maintain mechanical properties such as strength and stiffness. However, striking a delicate balance between maintaining a reasonable viscosity—an essential requirement for resin-based 3D printing—and preserving optimal healing efficiency proves to be quite challenging.

Manufacturing self-healing materials with 3D printing can combine the advantages of both, facilitating the manufacture of smart devices and enabling a more comprehensive range of applications for self-healing materials. These applications include, for example, aerospace applications where they hold promise in reducing maintenance requirements for critical components; biomedical applications where they enhance device lifespan leading to improved patient outcomes; automotive manufacturing, which stands to gain from impact-resistant parts; consumer electronics benefiting from durable self-healing coatings; and infrastructure sector enjoying reduced maintenance costs along with enhanced durability. The advent of additive manufacturing has revolutionized these domains by enabling customization options, facilitating production complexities inherent in intricate geometries while ensuring on-demand fabrication capabilities resulting in heightened operational efficiency and resilience [16,24,25].

So far, the 3D printing methods reported for the manufacturing of self-healing materials have been divided into extrusion-based AM processes, including mainly fused deposition modelling (FDM) and direct-ink writing (DIW), and vat photopolymerized AM processes, primarily including stereolithography (SLA) and digital light processing (DLP). With research in this field becoming more extensive and comprehensive, there is a need now to review and summarize existing AM methods of self-healing materials, as well as its projected applications, to ensure its long-term relevance. This review focuses on various self-healing mechanisms and 3D printing techniques, as well as an analysis of the compatibility between them. Readers are directed to other detailed reviews on self-healing materials in related fields such as sensors [26], soft robotics [11], material design [27,28], etc. The review is organized into three parts: firstly, on main self-healing mechanisms; secondly, an overview of the leading 3D printing methods used at this stage for self-healing polymers; thirdly, proposed applications for future additive manufacturing of self-healing polymers. A brief summary of existing work on 3D-printable self-healing materials classified into their various printing and self-healing mechanisms are consolidated in Table 1.

Table 1. Brief summary of the existing literature on printable self-healing materials.

	Extrinsic Healing			Intrinsic Healing				
	Microcapsule	**Microvascular**	**Nanoparticles**	**Diels–Alder Reaction**	**Disulfide**	**Hydrogen Bonds**	**Ionic Interactions**	**Coordination**
Fused deposition modeling (FDM)		O'Harra et al. (2020) [29]	Peng et al. (2021) [30]	Bi et al. (2022) [31] Zhou et al. (2020) [32]		O'Harra et al. (2020) [29] Bi et al. (2022) [31] Zuo et al. (2021) [33]	O'Harra et al. (2020) [29]	Zuo et al. (2021) [33] Lai et al. (2018) [34]
Direct-ink writing (DIW)			Kuang et al. (2018) [35]	Zhang et al. (2019) [16] Yuan et al. (2020) [36]		Wang et al. (2021) [21]	Wang et al. (2021) [21]	
Stereolithography (SLA)	Sanders et al. (2019) [37]			Durand-Silva et al. (2021) [38]			Liu et al. (2020) [39]	
Digital light processing (DLP)		Kuang et al. (2018) [35]	Zhang et al. (2019) [40]		Li et al. (2019) [41] Miao et al. (2022) [42] Zhang et al. (2022) [43]	Zhang et al. (2022) [43] Caprioli et al. (2021) [44]	Zhang et al. (2022) [43]	

2. Self-Healing Mechanisms

A wide range of self-healing mechanisms have been developed since the 1990s when Dry et al. researched the development of self-healing fibre-reinforced composites and polymeric smart materials [45]. In 2001, White et al. investigated the microencapsulation of healing agents embedded in polymeric matrices [10,45,46]. Self-healing mechanisms can be classified as extrinsic or intrinsic. The healing ability of extrinsic self-healing materials arise from integrating self-healing properties into the original material system. In contrast, intrinsic self-healing materials depend on the material's inherent chemical groups and properties [11].

2.1. Extrinsic Healing

2.1.1. Capsule-Based Healing Systems

The most common extrinsic self-healing methods utilize micro containers made from brittle polymers, such as microcapsules containing a healing agent, appropriate catalyst, curing agent, and reaction initiator embedded into the polymer matrix [9,10]. During object fracture, the micro containers are broken, automatically releasing healing agents into the crack site and filling the damaged area (Figure 1a) [10,37]. This method is often used to enable materials with stiff polymer matrices to achieve self-healing. The damaging force (trigger) is required to activate the healing action rather than the molecular diffusion of the matrix [11,37]. One advantage of microencapsulation technology is that the many different healing chemistries and encapsulation techniques allow the healing mechanism to be adapted to different matrices. However, the volume of healing agent is limited, limiting healing to microscopic damage, and/or it can only occur once [9,11].

Recently, Ma et al. embedded melamine–formaldehyde (MF) microcapsules, which wrap the epoxy oxide as a repairing agent, and $Cu(MI)_4Br_2$ as a curing agent in an epoxy oxide-based self-healing system, and printed out samples using a 3D printer (Figure 1c) [47]. The curing agent decomposed above the decomposition temperature and cured the damaged area for repair. It was found that adding 10 wt% microcapsules to the matrix could improve the tensile strength of the 3D-printed samples. After heating the scratches, the mechanical strength of the fractured material could be restored to 44.42 MPa with a high self-healing efficiency of 89.98% [47]. The 4D printing microencapsulated epoxy oxide self-healing system in the study was shown to form self-healing materials with high tensile strength and stability properties. This is beneficial to various practical 4D printing-related applications and shows excellent potential biological, medical, and bionic applications [47].

2.1.2. Vascular-Based Healing Systems

Microvascular healing systems store healing agents in hollow channels or interconnected networks and heal on a similar principle to microcapsules, i.e., that injury triggers self-healing (Figure 1b). The difference is that they can store more healing agent and transport it over greater distances. Hence, more significant injuries may be repaired, and reparation can occur multiple times within a particular area [9].

However, due to the solid nature of the vascular system, it cannot be used directly in resins or inks, and cannot be directly 3D-printed to obtain the self-healing product. More experiments were performed using 3D-printed vascular systems in combination with other methods, such as casting to obtain self-healing objects (Figure 1d,e) [48–50].

2.1.3. Phase-Separated Additives Healing Systems

Self-healing can also be achieved by embedding solid thermoplastic particles with a low melting temperature in the matrix. At a temperature above its T_g, thermoplastic additives melt and diffuse into the interfaces of the crack, and achieve self-healing by adhering the two surfaces together as they recrystallizes [30,51]. Unlike the encapsulation approach, this healing process is non-autonomous because thermal activation is needed, and can perform several times on the same fracture surface [9,11]. The disadvantage of nanoparticles is similar to microcapsules in that the thermoplastic particles do not entirely fill the damaged area, limiting the healing effect [9] (Figure 1f).

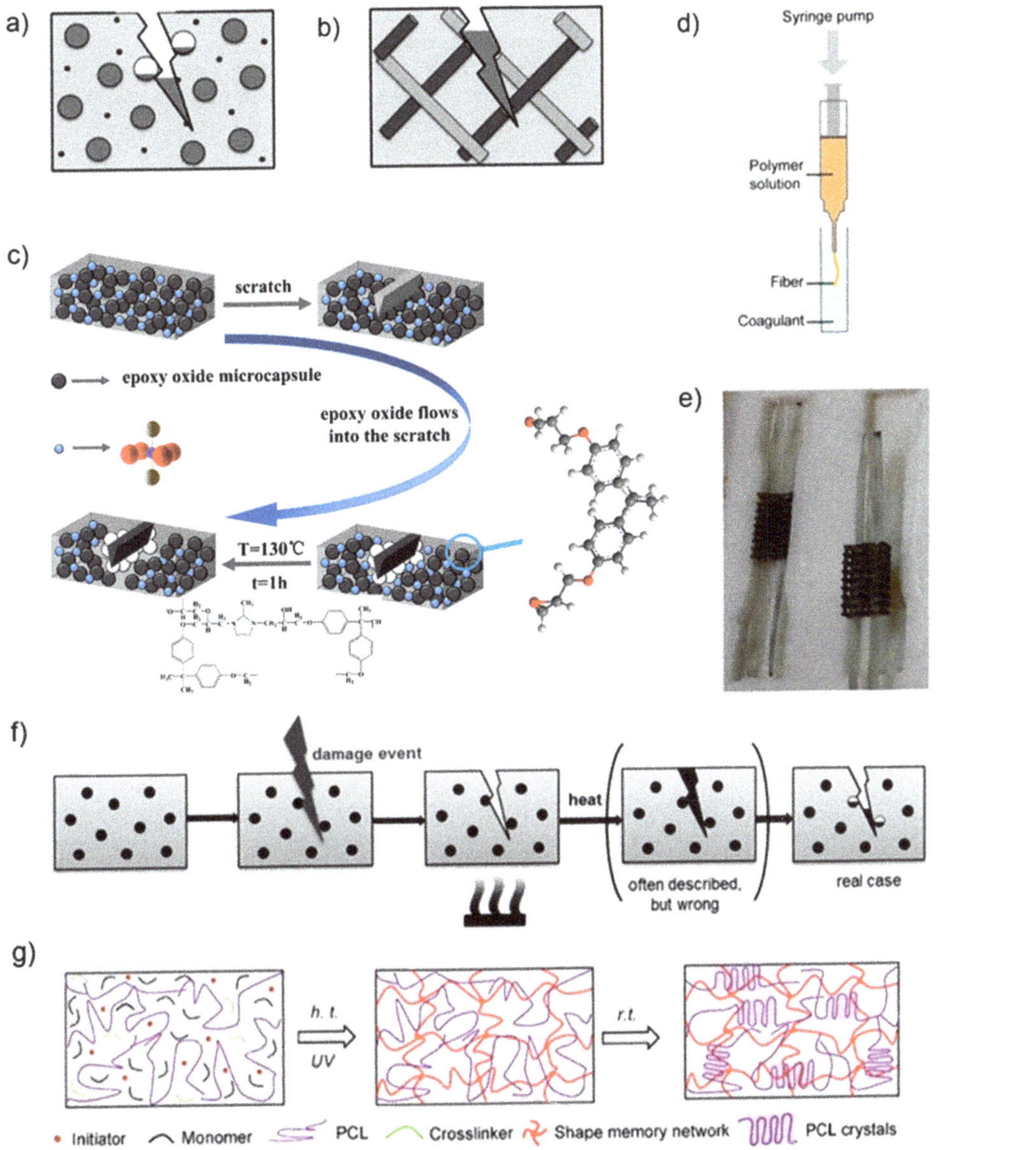

Figure 1. (**a**) Capsule-based healing systems, the gray ball is the healing agent, and the black ball is the catalyst or curing agent [9]. (**b**) Vascular-based healing systems, the gray ball is the healing agent, and the dark gray ball is the catalyst or curing agent [9]. (**c**) Schematic diagram of scratch repair principle with MF microcapsules embedded in epoxy oxides [47]. (**d**) Schematic diagram of 3D-printed vascularized fibres [52]. (**e**) Microvascular networks infused with self-healing and curing agents are embedded in silicon moulds [50]. (**f**) Phase-separated additives healing systems, thermoplastic particles (black balls) are dispersed in the matrix [9]. (**g**) Chemical structure evolution of the repair principle of the SH-SMP solution during UV-based 3D printing at high temperature (h.t.) and cooling down to room temperature (r.t.) [40].

Zhang et al. successfully incorporated the semicrystalline linear polymer polycaprolactone (PCL) into a methacrylate-based shape memory polymer (SMP) system. They printed complex structures with self-healing properties with a DLP 3D printer [40]. PCL acts as a self-healing agent to provide self-healing properties through the above principles and can recover the mechanical properties of damaged structures to over 90% [40] (Figure 1g). Similar to Zhang et al.'s experimental approach, Peng et al. printed a highly extensible, self-healing shape memory elastomer by a desktop FFF 3D printer, and the self-healing properties were achieved by semicrystalline thermoplastic PCL, which was incorporated into the thermoplastic elastomer [30].

2.2. *Intrinsic Self-Healing*

More researchers are combining intrinsic self-healing with 3D printing because of its theoretically infinite healing capacity [53]. Self-healing behavior is triggered by damage, driven by surface tension and elastic energy of stress sources or by appropriate external stimuli such as temperature, heat, electromagnetic radiation, pH, light. or ionic strength changes, and can be permanently fractured and reorganized [9,10]. Due to the different types of interactions used to achieve healing, intrinsic healing mechanisms can be further subdivided into those based on covalent interactions, such as the Diels–Alder reaction and disulfide; or non-covalent interactions, such as hydrogen bonding, ionic interactions, and coordination [1,9]. Due to the high bond strength of 150–550 kJ mol^{-1}, the breakage and reformation of covalent bonds is usually non-autonomous and requires relatively large amounts of energy from external stimuli. Still, they typically exhibit better elasticity and durability. Non-covalent bonds usually result from intermolecular interactions, and the many non-covalent bonds that crosslink polymers can lead to the formation of supramolecular networks [11]. Compared to dynamic covalent chemistry, non-covalent bond strengths are lower. This weaker crosslink means that less energy is required to break these bonds, being more susceptible to the external environments, thus allowing materials to exhibit extraordinary healing and stimuli-responsive properties [1,43].

2.2.1. Dynamic Covalent Bonds

Diels–Alder Reactions

The most common dynamic covalent bonds are thermoreversible Diels–Alder (DA) reactions, whose bond strength is relatively high, so its healing process usually occurs at high temperatures [11,54]. DA reactions are usually formed by equilibrium reactions between dienes and dienophiles, both present as functional groups on the constituting monomers (or prepolymers), the most common of which is the reversible DA cycloaddition between furans and maleimides (Figure 2a) [36]. DA reactions require high temperatures to activate healing. On one hand, this shifts the equilibrium of the exothermic DA reaction from the major part of the DA bonds formed at ambient temperature to the breaking of these bonds, leading to higher concentrations of reactive dienes and dienophilic functional groups. On the other hand, this increases molecular mobility, further facilitating contact between the fracture surfaces and facilitating self-healing [11]. For example, the furan/maleimide DA polymer network requires thermal activation at 80–130 °C, including decoupling of the furan/maleimide pair when temperatures above the dissociation temperature (TD), molecular movement in the polymer network, and subsequent furan/maleimide coupling at the condensation temperature [32].

It has been shown that macroscopic damage can heal with very high healing efficiency by this mechanism, and that multiple damage–healing cycles can be performed without a significant decrease in healing performance (Figure 2b) [16]. In addition to self-healing through high-temperature activation of molecular segments, Yang et al. achieved the process of self-healing triggered by near-infrared (NIR) light through photothermal conversion and the method can be remotely controlled to irradiate the self-healing area precisely without damaging their original 3D structures (Figure 2c) [16]. The mechanism can also be introduced into 3D printing, and it has been reported that by introducing the DA response into the new 3DP ink, DIW prints maintain their excellent self-healing properties [16,36].

Diels–Alder reactions, while promising for intrinsic self-healing systems in polymers and materials, have some limitations. These include temperature sensitivity, reversibility issues, slow kinetics at lower temperatures, stoichiometric requirements, limited chemical compatibility, environmental sensitivity, potential brittleness, and limited reusability [55]. These challenges can impact the efficiency, speed, and applicability of Diels–Alder-based self-healing materials in various real-world applications. Ongoing research aims to address these limitations by developing innovative catalysts, more reversible reaction mechanisms, and improved material designs [56].

Disulfide Bonds

Among the intrinsic methods, the self-healing effect of disulfide bonds has attracted much attention due to its ability to allow the full recovery of mechanical properties under mild healing conditions [41,54]. The disulfide groups can be cleaved by external forces, reduction reaction, etc., to form two thiol groups and can be reformed under stimulation for disulfide metathesis, which is responsible for the self-healing properties (Figure 2e) [54]. Disulphide metathesis reactions can be accelerated by catalysts. For example, Qureshi et al. managed to successfully reduce the reaction time by adding tributylphosphine (TBP) to a self-healing UV-cured ink [57]. In this experiment, the nucleophilic attack of TBP on the disulfide bond produced a TBP cationic intermediate and a thiolate anion [57,58]. The cross-nucleophilic attack of the thiolate anions on the other sulfur atoms occurs via the intermediate, leading to the exchange of network chains and the return of the catalyst to its original state, where the disulfide metathesis reaction is repeated (Figure 2d) [57,58]. Due to the limited mobility of chains in glassy polymers, the self-healing process needs to take place in the mobile chain segments above the T_g for the interchange reaction to take place. Hence, this method is widely used for low T_g materials such as polyurethanes and polyesters [54].

Li et al. introduced disulfide bonds into polyurethane acrylates and printed them by digital light processing 3D printing technology. The resulting polyurethane elastomer showed good self-healing ability, with the healed samples recovering to 95% of their original strength after 12 h of healing at 80 °C, and could be healed multiple times [41]. In the same vein, Yu et al. designed a photo elastomer ink containing disulfide groups that can achieve self-healing properties by disulfide metathesis reaction, and elastomers obtained by 3D printing can heal themselves wholly and quickly [59]. Similar to the Diels–Alder reaction, dynamic disulfide bonds can be used to exploit the principle of photothermal conversion through a self-healing process triggered by near-infrared (NIR) light (Figure 2f) [42].

However, it is noteworthy that disulfide-based reactions still exhibit certain limitations. These include relatively slow kinetics, temperature sensitivity, potential catalyst dependency, stoichiometry requirements, sensitivity to environmental factors, limited chemical compatibility, and issues related to durability and reusability. These limitations can affect the speed, efficiency, and applicability of disulfide-based self-healing, especially in environments with temperature fluctuations or when rapid repairs are essential [11]. Nevertheless, ongoing research seeks to overcome these challenges by developing more effective catalysts, optimized reactant designs, and more resilient materials, enhancing the potential of disulfide reactions for self-healing applications [60].

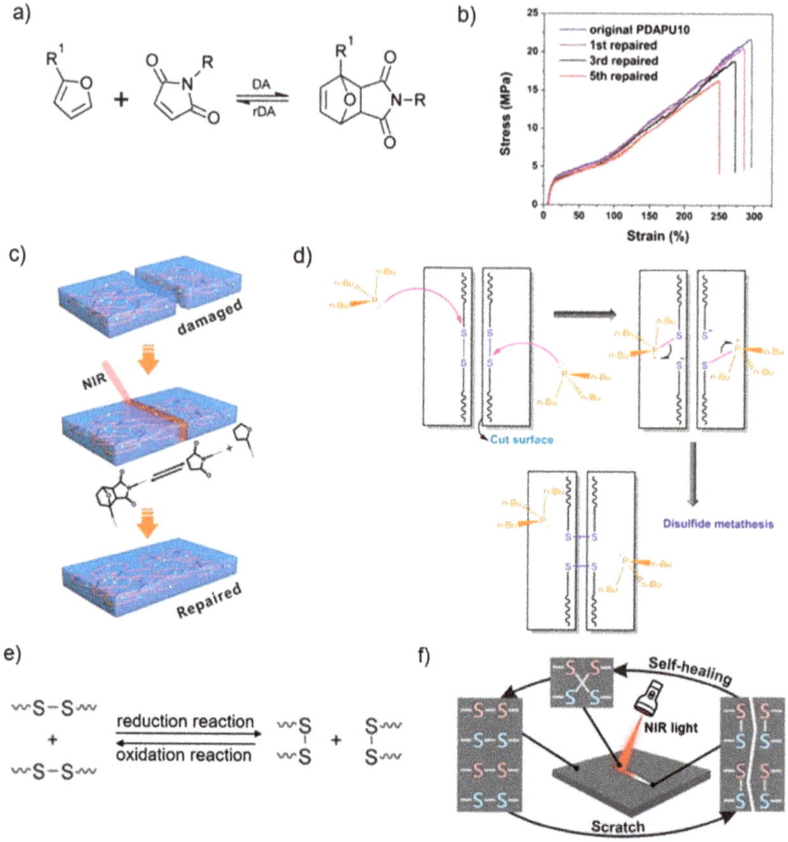

Figure 2. (**a**) Schematic diagram of reversible DA reaction between furan and maleimide groups [9]. (**b**) Stress–strain curve of the self-healing material (named PDAPU10) designed by Yang et al. for five stretch–recovery cycles with a self-healing efficiency of 92% [16]. (**c**) Schematic representation of DA-reaction-based precise self-healing triggered by NIR [16]. (**d**) Schematic of the disulfide metathesis reaction with catalyst TBP [57]. (**e**) Schematic of disulfide interchange reaction [54]. (**f**) Schematic of disulfide-bonds-based self-healing response activated by NIR [42].

2.2.2. Non-Covalent Interactions

Hydrogen Bonds

The strength of hydrogen bonds varies from 2 to 40 kcal mol^{-1}, depending on the hydrogen donor and acceptor [61,62]. Due to its low bond strength, hydrogen-bonded SH polymers can easily undergo the self-healing process by increasing temperatures, as high temperatures result in the breakage of hydrogen bond crosslinks and lead to viscous flow, thereby contributing to self-healing. This is in addition to the relatively fast reformation of hydrogen bonds due to their affinity, offering the possibility of developing autonomous self-healing polymers (Figure 3a) [7,11,63]. Some substances can be repaired after macroscopic damage, even for non-functionalized substances, due to the presence of their hydrogen bonds, e.g., polyurethanes, where interfacial healing is efficiently achieved mainly through hydrogen bonding among urethane units at the broken interface and the highly efficient healing properties are also retained even after 3D printing [10,14,64].

However, low-bond-energy crosslinks are detrimental to mechanical strength, strain recovery, and creep behavior. To improve mechanical strength and structural stability, multivalent hydrogen bonds can be utilized, increasing the number of hydrogen bonds formed per crosslink, and then increasing the strength of the crosslinks [11,65]. This theory is used in many 3D printing materials to provide self-healing properties. For example, Caprioli et al. mixed unmodified non-crosslinked poly (vinyl alcohol) (PVA) with acrylic acid (AAc), the cross-linker poly (ethylene glycol) diacrylate (PEGDA), and photoinitiator diphenyl (2,4,6-trimethylbenzoyl)phosphine oxide (TPO) to make a photocurable ink [44]. PVA has the inherent property of chain interdiffusion and the presence of hydroxyl groups formed by hydrogen bonds, and the carboxylic groups of AAc can form multiple hydrogen bonds with PVA chains. These give the 3D-printed hydrogel its high self-healing property (Figure 3f) [44,66].

Ureidopyrimidinone (UPy) methacrylate monomers are also often used in 3D-printed polymer matrices because of their ability to form four hydrogen bonds with a self-complementary unit, responsible for the self-healing behavior [13]. Alternatively, a healing agent can be added to the matrix to exploit hydrogen bonding interactions. For example, Triton X-100 can be used as a healing agent because of the hydrogen bonding of its hydrophilic polyethene oxide parts [67]. For example, Kee et al. added Triton X-100 to 3D-printable thermoelectric composites, allowing the fractured material to retain over 85% of its properties after healing [67].

Ionic Interactions

Self-healing polymers based on supramolecular dynamic networks can also be made using ionic interactions between ionic polymers [11]. Typically, ionomers are polymer chains partially modified with ionic side groups and the corresponding counter ions (Figure 3b) [9]. The ionomers contain between 1 and 15% of charged or ionic species. These ionic groups tend to aggregate, forming ionic clusters due to physical crosslinking, and allowing the reversible formation and reformation of the network structure to provide self-healing properties via reversible breakage and reformation of ionic bonds [9,10]. Compared to other types of non-covalent bonds, ionic interactions have higher aggregate strength, which facilitates increased tensile strength, toughness, and fracture resistance. But, they also require more external energy to break [11]. Like hydrogen bonding, ionomers effectively behave as an autonomous self-healing material in some applications. An example of this is in a ballistic impact healing application where the damage event provides enough energy in the form of heat generated by friction to complete healing [11,68,69]. Dynamic ionic crosslinking has proven to be an efficient way to heal damages in 3D-printed polymers [39,70]. Liu et al. exploited the dynamic ionic crosslinked network formed between the carboxyl and amino-functional polysiloxanes to ensure the self-healing and reprocessing capabilities of 3D-printed silicone elastomers. This showed excellent healing efficiency of 97% and a healing process that could be repeated multiple times, in addition to repeatedly reprocessing these elastomers and still repairing the damage with over 90% efficiency (Figure 3d) [39].

Polyacrylic acid is often used to achieve autonomous intrinsic self-healing properties in hydrogels that can be used as 3D printing materials, and self-healing hydrogels allow for ion transport while maintaining excellent mechanical stability [21,71,72]. Darabi et al. utilized dynamic ionic interactions between the carboxylic groups of poly(acrylic acid), NH groups of polypyrrole and ferric ions, and a combination of both physical and chemical crosslinking to enable the hydrogel to meet the highest self-healing efficiency while maintaining mechanical stability and electrical conductivity [71].

Coordination Interactions

Among various non-covalent interactions, metal–ligand interactions can also form supramolecular networks and have unique properties [7,11]. Coordination complexes are formed between positively charged metal ions and partially negatively charged groups on ligand molecules. There are a large number of accessible metal ions and ligand molecules that make coordination chemistry particularly attractive by careful selection of the combination of ligands and metal ions (Figure 3c) [1,11]. It is possible to tune the bond strength into the desired weak and dynamic, or, with the presence of functional metal ions or ligands and dynamic metal–ligand bonds, polymers can display a variety of advanced functions such as dielectric, magnetic, luminescent, catalytic, and stimulating reactivity [1,7]. For example, by replacing the metal ion with which the pyridine interacts, the Zn–pyridine interaction is strong but not dynamic compared to the Fe–pyridine and Tb–pyridine interactions used for self-healing [10]. The charge on the ligand molecule is usually much smaller than that on the metal ion. Since dipole–ion interactions are weaker than ion interactions, the strength of coordination bonds is weaker even if they can be adjusted within a specific range. These weaker crosslinks allow for the healing of macroscopic damage at mild temperatures without the need for external stimulus [1,11].

Similar to the two non-covalent bonds mentioned above, coordination bonds can also be used to achieve self-healing of 3D-printed materials. Lai et al. used the weak but abundant Zn(II)–carboxylate coordination bond to design polydimethylsiloxane (PDMS) polymers and achieve rigid and healable materials through 3D printing. The coordination equilibrium is temperature sensitive, with the mutual crosslinking of Zn(II) and carboxylate at room temperature. The equilibrium shifts toward the dissociated state when temperature is increased, producing an increasing number of non-cross-linked chains, which can resume crosslinking when the temperature is cooled down again (Figure 3g) [34]. Shi et al. utilized dynamic coordination bonds between bisphosphonate (BP) ligands and calcium ions based on the polysaccharide hyaluronic acid (HA) backbone to design a new hydrogel ink that can be used for extrusion-based 3D printing. Similarly, Wang et al. used ionic interactions between poly(acrylic acid) acid and calcium ions combined with dual-crosslinked networks to develop a multifunctional hydrogel with extraordinary mechanical strength and self-healing efficiency (Figure 3j) [72,73].

2.3. Development Status

One strategy for developing more advanced self-healing materials is to combine multiple self-healing mechanisms. This approach is also applicable to developing self-healing materials for 3D printing, as reported in several research papers [10,29,31,33,43,74].

Wu et al. developed a Cu(II)–dimethylglyoxime–urethan-complex-based polyurethane polymer link (Cu–DOU–CPU) with a synergistic triple dynamic bond, including the presence of dynamic covalent bonds (oxime amine bonds) and dynamic non-covalent bonds (metal–ligand bonds and hydrogen bonds), which can enhance both the self-healing properties and the mechanical properties of the material [33]. The relative recovery rate of 3D-printed objects reached 94% in the absence of any external stimuli (Figure 3h) [33].

Xu et al. designed a double-network hydrogel consisting of a chitosan–citrate (CS) network crosslinked by citrate ions via electrostatic interaction and poly(sulfobetaine-co-acrylic acid) (P(SBMA-co-Ac)) network crosslinked by hydrogen bonding between carboxyl groups and ionic interaction between zwitterionic moieties (Figure 3i). This had a self-healing-property of 95.4% [74]. The hydrogel was 3D-printed with good electrical conductivity and sensitivity. It can be used as a strain sensor for detecting human motion, retaining good sensitivity after fracture, and self-healing [74].

Similarly, Chen et al. have produced ionic gels with remarkable self-healing properties by combining dynamic disulfide, hydrogen bonds, and ionic interactions by 3D printing, achieving healing efficiencies of over 95% under heating, and over 99% under UV irradiation, while offering advantages of high elasticity and durability (Figure 3e) [43]. Xu et al. prepared 3D-printed excellent performance composites with self-healing properties that are attributed by a combination of Diels–Alder (DA) reversible covalent bonding and hydrogen bonding [31].

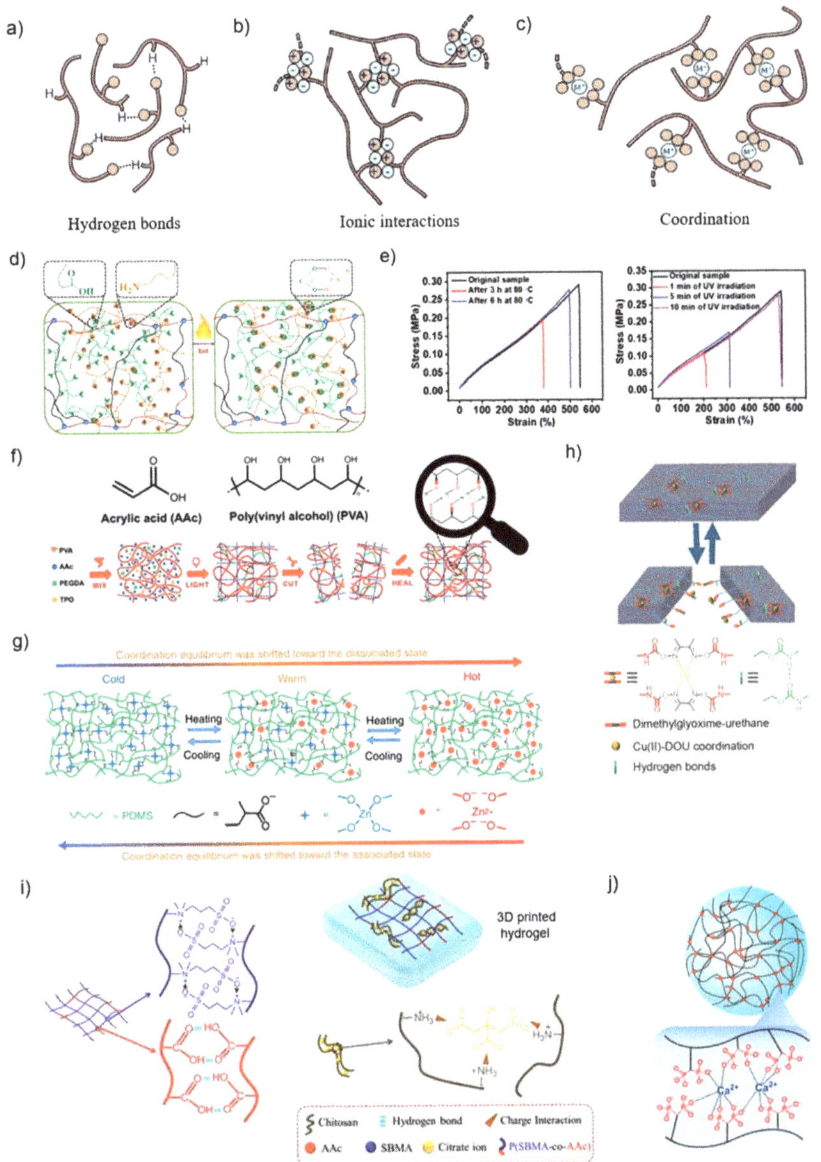

Figure 3. (**a–c**) Non-covalent interactions, including hydrogen bonding, ionic interactions, or coordination metal–ligand complexes [11]. (**d**) Schematic illustration of the dynamic ionic crosslinked network between the carboxyl and amino-functional polysiloxanes [39]. (**e**) Stress–strain curves of original and healed samples at 80 °C for 3 and 6 h, with healing efficiency over 95%. Stress–strain curves of original and healed samples at 1, 5, and 10 min of UV irradiation, with healing efficiency over 99% [43]. (**f**) Schematic of the chemical structure of AAc and PVA in a light-curing resin, the multiple hydrogen bonds formed, and the self-healing process [44]. (**g**) Schematic structure of the polymer network for self-healing through Zn(II)–carboxylate coordination bonding [34]. (**h**) Schematic of triple dynamic self-healing of Cu–DOU–CPU polymers [33]. (**i**) Schematic diagram of the self-healing principle of P(SBMA-co-Ac)/CS–Cit DN hydrogels [74]. (**j**) Schematic presentation of a hydrogel network formed by HA–BP macromolecules upon coordination bonding with Ca^{2+} ions [45].

3. Additive Manufacturing of Self-Healing Materials

3.1. Extrusion-Based AM Processes

Among several 3D printing techniques, extrusion-based AM processes have great potential as multi-material or multifunctional fabrication methods. They are widespread at amateur and professional levels because of their low cost, simple operation, low environmental impact, and relatively high prototyping speed [75–79]. Two of the most common extrusion-based AM processes are FDM and DIW [80].

3.1.1. Fused Deposition Modelling (FDM)

FDM, also called fused filament fabrication (FFF), is an extrusion-based method in which a continuous filament of polymer melts from a heated nozzle, depositing the molten material in a consecutive, layer-by-layer fashion on a print bed that may also be heated (Figure 4a) [14,81]. The material cools and solidifies upon deposition. Then, the next layer is deposited similarly to build a digitally modelled part iteratively [80].

Polymers used for FDM are typically thermoplastic or nanocomposites, and there are many advantages to adding self-healing properties to the material. FDM 3D-printed polymeric parts have been shown to completely retain their self-healing ability as the bulk self-healing polymer while showing the potential to obtain improved (3D-printed polymeric components) properties such as mechanical properties, durability, damage tolerance, and extensibility [14,30]. For example, Ritzen et al. found through a compression cut test that self-healing thermoplastic polyurethanes (SH-TPU) printed by FDM had the same mechanical and healing behavior as a bulk self-healing polymer, implying that the self-healing property of the polymer was unaffected by processing steps and printing (Figure 4c) [14]. Peng et al. reported that the mechanical properties and extensibility of the material were considerably enhanced when the morphology was preferentially oriented along the printing direction [30].

Nevertheless, the FDM method results in weak interlayer adhesion during layer-by-layer deposition, possibly due to the high viscosity of the solvent-free resin. This disadvantage may tend to cause cracks and deformations during use, thus, significantly reducing the service life of the manufactured object. This problem can be solved using functional materials with a higher affinity for fusion and layer adhesion, such as self-healing materials [29,32,33]. Zhou et al. used DA-based polymers (DAPs) that dissociate into low-viscosity liquids when their temperature exceeds that of its dissociation temperature (T_D), successfully addressing current significant challenges in FDM (Figure 4i) [32]. Similarly, in another report, a polymer ink with synergetic multiple dynamic bonds, including reversible coordination and hydrogen bonds used by Zuo et al., was also developed to optimize FDM printing conditions [33]. It reported that the method increases interlayer adhesion and reduces the number of support structures available by taking advantage of room-temperature self-healing property to assemble 3D-printed blocks into large complex objects, with the added benefit of decreasing the complexity and cost (Figure 4d) [33]. O'Harra et al. introduced ionic interactions and H-bonding into elastomeric materials printed by FDM, which provided the opportunity to not only eliminate the inherent weaknesses of FDM 3D printing but also to retain the homogeneity and desirable thermophysical properties of the material [29].

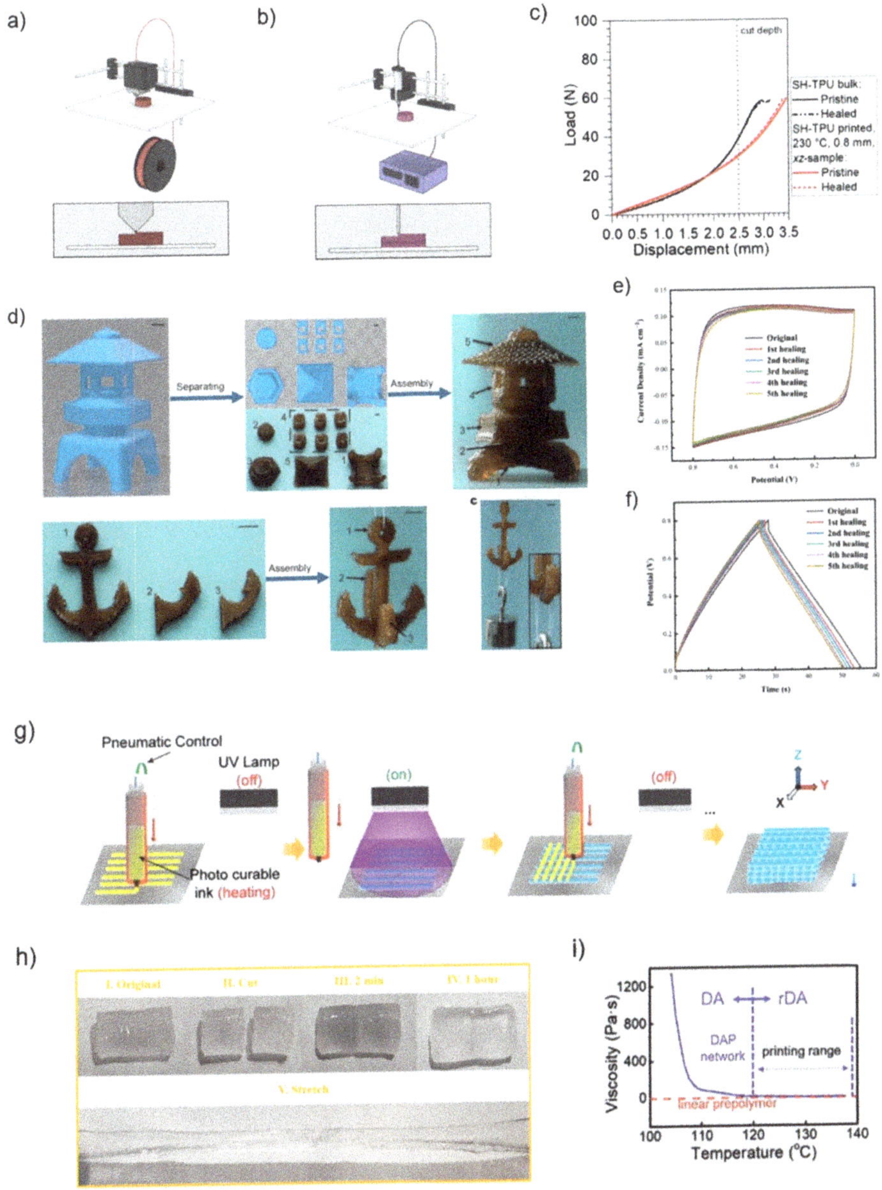

Figure 4. (**a**,**b**) Schematic diagram of fused deposition modelling (FDM) and direct-ink writing (DIW) [80]. (**c**) Compression-cut mechanical characterization of healed FDM-printed and bulk SH-TPU polymers [14]. (**d**) Example of free assembly of 3D-printed parts into complex objects. The numbers in the photographs represent the order of assembly [33]. (**e**) CV curves of the MSC at different damage/self-healing cycles. (**f**) GCD curves of the MSC at different damage/self-healing cycles [21]. (**g**) DIW-based 3D printer equipped with heating elements prints each filament layer followed by shining UV light to cure the resin [35]. (**h**) Digital photographs demonstrating the self-healing process of the hydrogel designed by Wang et al.: I original state; II cut into two pieces; III self-healing in 2 min; IV self-healing in 1 h; V stretching after healing [21]. (**i**) Viscosity of the DAPs and the linear prepolymer as a function of temperature [32].

3.1.2. Direct-Ink Writing (DIW)

DIW, also called liquid deposition modelling (LDM), utilizes a pneumatic, piston, or screw to extrude the print material through a nozzle or syringe needle (Figure 4b) [36,82]. Some curing processes can be performed by thermal and photopolymerization curing or dispensing two reactive components using mixing nozzles to facilitate the deposition and subsequent stabilization of the material (Figure 4g) [80,83,84].

Like FDM, the nozzle or syringe needle can be moved over the build surface in three dimensions at a constant height. The extruded materials are joined together layer-by-layer to form the final 3D construct [80,84]. DIW printing inks require unique rheological properties, such as low viscosity and shear-thinning, to ensure continuity of printing without resorting to excessively high pneumatic or mechanical pressure, and to stop the flow or restore mechanical integrity after extrusion [80,83]. Compared to thermoplastic printing in FDM, DIW may be utilized to print thermoset materials that are inherently less processable. Introducing self-healing properties to thermoset materials printed via DIW allows for the retention of mechanical strength and durability of traditional thermosets, and endows the materials with additional self-healing properties and processability [16,36]. In Yuan's report, thermally reversible Diels–Alder (DA) was used in a thermoset to develop a self-healing ink for DIW. The self-healing properties in materials enable high strength recovery of up to 85%, and repeatable healing without a significant decrease in healing performance. The material can also be reprocessed and remolded. In addition, the relatively low viscosity of the ink eliminates voids between the printed filaments, resulting in high tensile strength [36]. Despite the many advantages of DIW for manufacturing thermoset materials, challenges remain in developing printing inks with rheological optimization [36].

Extant research has shown that results have been achieved with DIW-printed self-healing materials. Wang et al. developed a self-healing and highly stretchable hydrogel achieved by dynamic borate ester and multi-network hydrogen bonds (Figure 4h). After direct DIW printing, the micro-supercapacitor (MSC) showed initial structural self-healing properties, and enabled rapid electrochemical restoration with little change in cyclic voltammetry (CV) curves and galvanostatic charge/discharge (GCD) curves after multiple physical damage/healing cycles. This is beneficial for the study of self-healing hydrogel systems in portable wearable electronic devices (Figure 4e,f) [21].

Since Kuang et al. first combined the attributes of shape memory and self-healing with 3D printing in 2018, shape-memory-assisted self-healing has received increasing attention [35]. This makes sense, as the introduction of self-healing improves the damage accumulated during the repeated deformation–recovery process of the printed part without external healing agents [16,85]. In Kuang's report, the embedded semicrystalline thermoplastic plays a dual role as a switching phase for shape memory and a healing agent for self-healing behavior, and complex structures with functional properties such as shape memory and self-healing were printed by DIW [35]. Similar to Kuang's experiment, Zhang et al. introduced the DA reaction into SMPs, and obtained final printed objects with designed shapes, higher quality, and precise self-healing properties by DIW [16].

3.2. *Vat Photopolymerization*

Vat photopolymerization, such as SLA and DLP 3D printing, avoid the problems of extrusion printing with extensive rheological optimization of inks, and the possibility of displaying warped and slightly distorted planes. They also have better print resolution and higher efficiency [25,37,44].

Vat photopolymerization is a continuous, layered technique for fabricating a part by using UV or visible light to project a 2D pattern onto a liquid photopolymer resin, with the first layer of the part being cured directly onto the build surface and each subsequent layer being cured onto the previous layer, repeating the process until the manufacturing is complete. SLA and DLP are subdivided by light source configuration—in SLA 3D printing, the laser scans point-by-point and cures the photopolymer resin to complete each layer of the pattern; while in DLP systems, UV light can cure the entire patterned area of each layer in a single exposure (Figure 5a,b) [44,80].

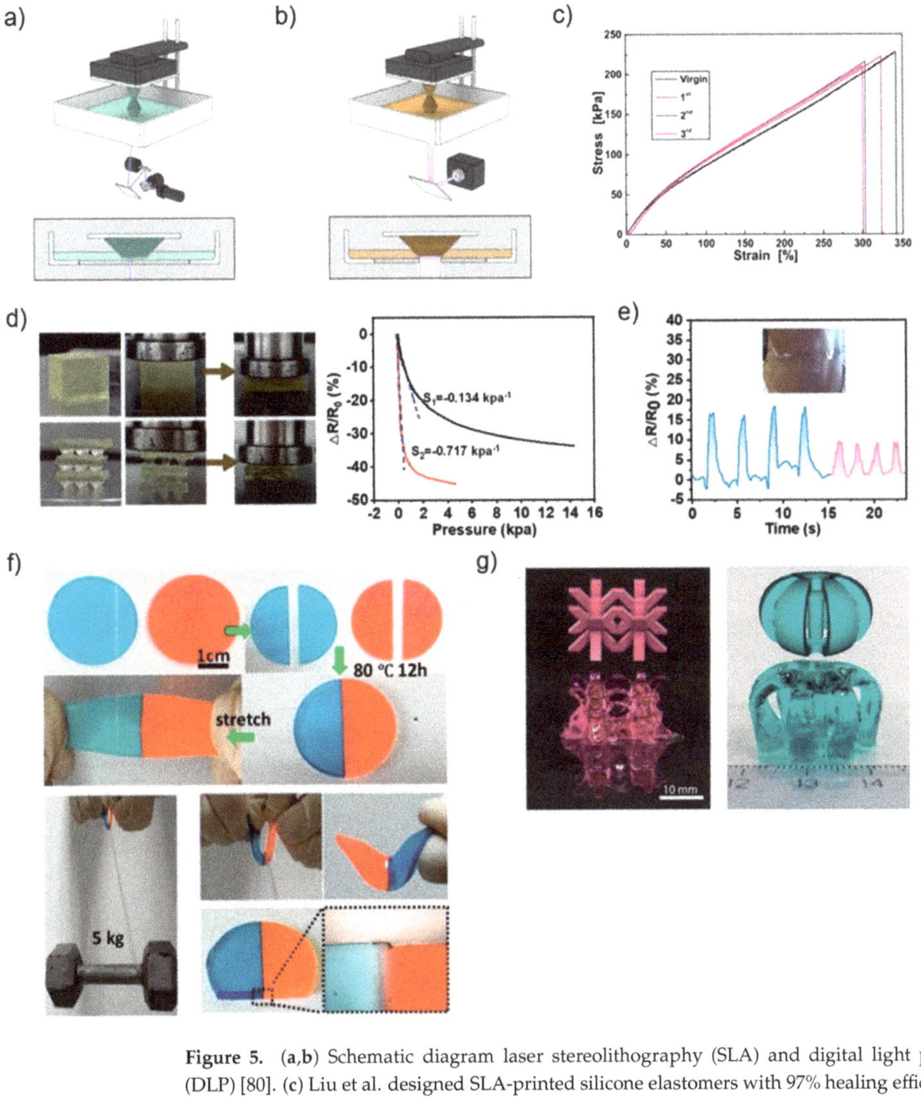

Figure 5. (**a**,**b**) Schematic diagram laser stereolithography (SLA) and digital light processing (DLP) [80]. (**c**) Liu et al. designed SLA-printed silicone elastomers with 97% healing efficiency [39]. (**d**) Comparison of the pressure sensitivity of the planar sensors (S_1) and micro-structured (S_2). Pressure sensitivity (S) is expressed as the relative resistance change ratio over pressure $(\Delta R/R_0)/P$ [43]. (**e**) The micro-structured pressure sensor senses the subtle muscle movements of the throat during the swallowing of water (the first four cycles) and saliva (the last four cycles) [43]. (**f**) Image of the polyurethane elastomer designed by Li et al. cut into two pieces, connected, healed for 12 h at 80 °C and, finally, subjected to first stretching manually to a large deformation and then to a 5 kg weight lifting test; optical microscope image of the healing interface of the healed elastomer [41]. (**g**) Complex structures with flat surfaces and sharp edges printed by DLP. Body-centered cubic lattice-like structure printed with methyl red sodium salt dye (**left**); axisymmetric structure with central pillar printed with brilliant green dye (**right**).

Resins for vat photopolymerization usually consist of monomers, oligomers, functionalized polymers, and photoinitiators, and the resulting resin must be translucent to the light source used for curing to allow UV light to penetrate and cure the designed layer height. At the same time, the viscosity must be manageable to ensure fluidity [80].

Three-dimensional photopolymerization is a free radical polymerization reaction that occurs in the presence of a photoinitiator using light as the stimulus, and (meth)acrylates and (meth)acrylamides are usually used because of their fast reaction rate, which allows for faster conversion of reactive liquid resins into solid materials [86].

3.2.1. Stereolithography (SLA)

Photopolymerization is widely used because of some excellent characteristics. For example, the leading additive manufacturing methods for silicone elastomers are extrusion 3D printing and photopolymerization. These are popular because they break some limitations of extrusion 3D printing, and also have high efficiency, higher resolution, and better surface quality [39,87]. Liu et al. have fabricated self-healable and reprocessed silicone elastomers by SLA 3D printing and achieved a healing efficiency of 97% (Figure 5c) [39]. The first 3D-printed structure with extrinsic self-healing properties was also achieved by SLA, where Sanders et al. successfully printed 3D self-healing composites by adding self-healing capsules to SLA resin to provide a solvent-welding effect [37].

Additionally, 3D printing techniques based on vat photopolymerization have been widely used to produce high-resolution thermoset materials. Objects created with vat photopolymerization techniques offer superior thermal stability and quality due to the formation of covalent crosslinks between the layers [38]. Zhang et al. successfully printed thermoset materials with self-healing properties, excellent mechanical properties, and high modulus by incorporating a reversible addition–fragmentation chain transfer agent into the SLA resin [86]. Similarly, Durand-Silva et al. successfully used SLA to print thermoset polymers with self-healing properties enabled by Diels–Alder reactions [38].

3.2.2. Digital Light Processing (DLP)

Because DLP can irradiate and cure one layer at a time, it has a shorter print time than SLA, and a more comprehensive range of applications. For example, in 2022, Zhang et al. took advantage of the high printing accuracy of DLP 3D printing technology and applied it to the fabrication of micro-structured pressure sensors that meet the requirements of ultra-high elasticity and durability and remarkable self-healing properties, while significantly improving the sensitivity of the sensors to detect complex muscle movements and subtle motions (Figure 5d,e) [43]. In addition, the printable ionic gel possesses tunable mechanical properties due to the easy modification of the ratio of the individual components of the DLP resin [43].

For the first time in 2019, Li et al. fabricated a self-healing polyurethane elastomer by DLP 3D printing, which can heal up to 95% efficiency and multiple times because of hydrogen and disulfide bond metathesis. The polyurethane elastomer has a good prospect for application due to ease of manufacture, excellent performance, high precision, and complex structure (Figure 5f) [41]. Additionally, there are also improvements in the additive manufacturing of hydrogels, which were previously only processed by extrusion-based additive manufacturing techniques with limited freedom in terms of design and resolution.

However with DLP 3D printing, it is possible to print bespoke structures with overhanging, hollow features, and high precision without the need for support materials [44]. For example, Caprioli et al. in 2021 used the dispersive forces between materials to effectively interact to overcome the inherent incompatibility between vat photopolymerization and self-healing properties, printing complex structures with sharp edges and a self-healing efficiency of 72% via DLP (Figure 5g) [44].

4. Applications and Perspectives

In recent years, the field of additive manufacturing has witnessed remarkable achievements in the printing of self-repairing polymers and composites. These advancements have paved the way for groundbreaking applications spanning a wide range of industries. Notably, aerospace engineering has been revolutionized by the emergence of 3D-printed polymer matrices capable of autonomously mending microcracks and damages found in structural components [88,89]. This development has significantly impacted maintenance strategies while ensuring enhanced longevity and safety for critical aircraft parts. Similarly, the biomedical industry has experienced a paradigm shift with the fabrication of customized biocompatible implants possessing inherent self-healing properties [90,91]. The introduction of these innovative medical devices minimizes invasive replacement procedures while maximizing patient comfort levels. By harnessing this technology, patients can benefit from prolonged use without compromising their quality of life. The automotive industry is yet another beneficiary where significant advantages have emerged through the utilization of self-repairing polymer composites for impact-resistant car parts, enabling vehicles to be more resilient to external forces encountered on roads or during accidents [92]. These developments highlight not only the immense potential but also underscore key benefits such as increased sustainability, cost-effectiveness, and resilience across diverse sectors when combining additive manufacturing techniques with self-healing polymers and composites [93].

Although breakthroughs have been made in the field over recent years, its potential is hindered by several limitations. For example, the selection of appropriate self-healing mechanisms, which involves careful consideration of trade-offs. Chemical mechanisms exhibit commendable healing efficiency but may necessitate specific activation conditions that can potentially impact AM processes adversely. On the other hand, physical mechanisms like shape-memory polymers offer greater compatibility with AM techniques. The choice between microcapsules and vascular networks depends on the unique demands posed by each application. Microcapsules enable localized healing capabilities while vascular networks provide broader avenues for restoration. Material development poses significant challenges due to the need for harmonizing printability with mechanical performance, thereby demanding innovative formulations. Achieving material compatibility between the healing agent and polymer matrix emerges as a critical factor in this pursuit; furthermore, scaling up 3D printing for large-scale applications remains an obstacle that needs to be overcome. While there are many advantages to manufacturing self-healing polymers by AM, the inherent material properties present unique processing challenges for the field. For extrusion-based AM processes, the printing ink has to meet the rheological requirements while ensuring that the printed objects are sufficiently rigid. In addition, 3D-printed self-healing polymers with a high crosslinking density can reduce the molecular motion of the polymer chains, which can affect the self-healing effect. These require both polymer chemistry innovations and printer design to achieve a balance between performance. Until this can be achieved, the benefits of manufacturing self-healing polymers by AM cannot be fully taken advantage of and will remain theoretical.

The integration of multiple self-healing mechanisms with hybrid 3D printing represents another highly promising avenue for further advancing the utilization of self-healing polymers and composites. By introducing redundancy in the healing process, this approach not only enhances healing efficiency but also ensures greater reliability in critical applications. Additionally, it enables the tailored integration of diverse mechanisms, allowing for optimized material performance based on specific conditions and requirements. The versatility offered by hybrid 3D printing allows for the precise incorporation of these mechanisms into intricate structures, pushing the boundaries of what can be achieved with self-healing materials. This advancement has profound implications across a myriad of industries that demand durable and resilient materials. Notably, it offers sustainability benefits through waste reduction, on-demand production capabilities, and extended product lifecycles. While challenges such as material compatibility and design complexities must be addressed, the synergistic combination of multiple self-healing mechanisms with hybrid 3D printing holds potential to revolutionize material design and application.

This paper summarizes the status quo of developments in self-healing mechanisms for self-healing polymers, as well as its advantages and its disadvantages. It also highlights examples of applications of the main 3D printing methods used to manufacture polymers. However, research into the field is far from maturity—damaged parts are repaired but not fully restored to their original surface, and damage cure has only been demonstrated at the laboratory level. While there appears to be great potential for real-world applications of self-healing polymers, it will be some time before its projected economic and environmental impact materializes [94,95]. Functional self-healing polymers are a growing trend—combining self-healing materials with additional functions such as shape memory property [96–102], electrical conductivity [103–105], etc., could further expand the market for applications.

Author Contributions: Writing—original draft preparation: Y.J.; Writing—review and editing: E.L.L.N., D.X.H., Y.Y., S.Y.C. and B.Q.Y.C.; Supervision: J.W. and B.Q.Y.C. All authors have read and agreed to the published version of the manuscript.

Funding: This research is supported by the Agency for Science, Technology and Research (A*STAR) under its Career Development Award scheme (C210112028).

Conflicts of Interest: The authors declare no conflict of interest.

References

1. Li, C.H.; Zuo, J.L. Self-Healing Polymers Based on Coordination Bonds. *Adv. Mater.* **2020**, *32*, e1903762. [CrossRef] [PubMed]
2. Rekondo, A.; Martin, R.; Ruiz de Luzuriaga, A.; Cabañero, G.; Grande, H.J.; Odriozola, I. Catalyst-free room-temperature self-healing elastomers based on aromatic disulfide metathesis. *Mater. Horiz.* **2014**, *1*, 237–240. [CrossRef]
3. Rodriguez, E.D.; Luo, X.; Mather, P.T. Linear/network poly(epsilon-caprolactone) blends exhibiting shape memory assisted self-healing (SMASH). *ACS Appl. Mater. Interfaces* **2011**, *3*, 152–161. [CrossRef] [PubMed]
4. Wang, S.; Urban, M.W. Self-healing polymers. *Nat. Rev. Mater.* **2020**, *5*, 562–583. [CrossRef]
5. Utrera-Barrios, S.; Verdejo, R.; López-Manchado, M.Á.; Hernández Santana, M. Self-Healing Elastomers: A sustainable solution for automotive applications. *Eur. Polym. J.* **2023**, *190*, 112023. [CrossRef]
6. Zhang, W.; Jiang, H.; Chang, Z.; Wu, W.; Wu, G.; Wu, R.; Li, J. Recent achievements in self-healing materials based on ionic liquids: A review. *J. Mater. Sci.* **2020**, *55*, 13543–13558. [CrossRef]
7. Yang, Y.; Urban, M.W. Self-healing polymeric materials. *Chem. Soc. Rev.* **2013**, *42*, 7446–7467. [CrossRef]
8. Chan, B.Q.Y.; Low, Z.W.K.; Heng, S.J.W.; Chan, S.Y.; Owh, C.; Loh, X.J. Recent Advances in Shape Memory Soft Materials for Biomedical Applications. *ACS Appl. Mater. Interfaces* **2016**, *8*, 10070–10087. [CrossRef]
9. Urdl, K.; Kandelbauer, A.; Kern, W.; Müller, U.; Thebault, M.; Zikulnig-Rusch, E. Self-healing of densely crosslinked thermoset polymers—A critical review. *Prog. Org. Coat.* **2017**, *104*, 232–249. [CrossRef]
10. Willocq, B.; Odent, J.; Dubois, P.; Raquez, J.M. Advances in intrinsic self-healing polyurethanes and related composites. *RSC Adv.* **2020**, *10*, 13766–13782. [CrossRef]
11. Terryn, S.; Langenbach, J.; Roels, E.; Brancart, J.; Bakkali-Hassani, C.; Poutrel, Q.-A.; Georgopoulou, A.; George Thuruthel, T.; Safaei, A.; Ferrentino, P.; et al. A review on self-healing polymers for soft robotics. *Mater. Today* **2021**, *47*, 187–205. [CrossRef]
12. Nadgorny, M.; Xiao, Z.; Connal, L.A. 2D and 3D-printing of self-healing gels: Design and extrusion of self-rolling objects. *Mol. Syst. Des. Eng.* **2017**, *2*, 283–292. [CrossRef]
13. Suriano, R.; Bernasconi, R.; Magagnin, L.; Levi, M. 4D Printing of Smart Stimuli-Responsive Polymers. *J. Electrochem. Soc.* **2019**, *166*, B3274–B3281. [CrossRef]
14. Ritzen, L.; Montano, V.; Garcia, S.J. 3D Printing of a Self-Healing Thermoplastic Polyurethane through FDM: From Polymer Slab to Mechanical Assessment. *Polymers* **2021**, *13*, 305. [CrossRef]
15. Nadgorny, M.; Collins, J.; Xiao, Z.; Scales, P.J.; Connal, L.A. 3D-printing of dynamic self-healing cryogels with tuneable properties. *Polym. Chem.* **2018**, *9*, 1684–1692. [CrossRef]
16. Zhang, Y.; Yin, X.-Y.; Zheng, M.; Moorlag, C.; Yang, J.; Wang, Z.L. 3D printing of thermoreversible polyurethanes with targeted shape memory and precise in situ self-healing properties. *J. Mater. Chem. A* **2019**, *7*, 6972–6984. [CrossRef]
17. Gomez, E.F.; Wanasinghe, S.V.; Flynn, A.E.; Dodo, O.J.; Sparks, J.L.; Baldwin, L.A.; Tabor, C.E.; Durstock, M.F.; Konkolewicz, D.; Thrasher, C.J. 3D-Printed Self-Healing Elastomers for Modular Soft Robotics. *ACS Appl. Mater. Interfaces* **2021**, *13*, 28870–28877. [CrossRef]
18. Jiang, T.; Kai, D.; Liu, S.; Huang, X.; Heng, S.; Zhao, J.; Chan, B.Q.Y.; Loh, X.J.; Zhu, Y.; Mao, C.; et al. Mechanically cartilage-mimicking poly(PCL-PTHF urethane)/collagen nanofibers induce chondrogenesis by blocking NF–kappa B signaling pathway. *Biomaterials* **2018**, *178*, 281–292. [CrossRef]

19. Chan, S.Y.; Chan, B.Q.Y.; Liu, Z.; Parikh, B.H.; Zhang, K.; Lin, Q.; Su, X.; Kai, D.; Choo, W.S.; Young, D.J.; et al. Electrospun Pectin-Polyhydroxybutyrate Nanofibers for Retinal Tissue Engineering. *ACS Omega* **2017**, *2*, 8959–8968. [CrossRef]
20. Jiang, Y.; Leng, Q.Y.; Yan, Y.; Ng, E.L.L.; Chee, H.L.; Wang, F.; Chan, S.Y.; Loh, X.J.; Wang, J.; Chan, B.Q.Y. 4D Printing of Single-Network Shape Memory Polyurethanes with Two-Way Actuation Properties. *ACS Appl. Polym. Mater.* **2022**, *4*, 8574–8583. [CrossRef]
21. Wang, Y.; Shi, Y.; Gu, Y.; Xue, P.; Xu, X. Self-Healing and Highly Stretchable Hydrogel for Interfacial Compatible Flexible Paper-Based Micro-Supercapacitor. *Materials* **2021**, *14*, 1852. [CrossRef] [PubMed]
22. Yang, G.; Sun, Y.; Limin, q.; Li, M.; Ou, K.; Fang, J.; Fu, Q. Direct-ink-writing (DIW) 3D printing functional composite materials based on supra-molecular interaction. *Compos. Sci. Technol.* **2021**, *215*, 109013. [CrossRef]
23. Ouyang, L.; Highley, C.B.; Rodell, C.B.; Sun, W.; Burdick, J.A. 3D Printing of Shear-Thinning Hyaluronic Acid Hydrogels with Secondary Cross-Linking. *ACS Biomater. Sci. Eng.* **2016**, *2*, 1743–1751. [CrossRef] [PubMed]
24. He, M.; Zhao, Y.; Liu, Y.; Wei, D. A 3D printable self-healing composite conductive polymer for sensitive temperature detection. *Chin. Chem. Lett.* **2020**, *31*, 826–830. [CrossRef]
25. Guo, B.; Ji, X.; Chen, X.; Li, G.; Lu, Y.; Bai, J. A highly stretchable and intrinsically self-healing strain sensor produced by 3D printing. *Virtual Phys. Prototyp.* **2020**, *15*, 520–531. [CrossRef]
26. Khatib, M.; Zohar, O.; Haick, H. Self-Healing Soft Sensors: From Material Design to Implementation. *Adv. Mater.* **2021**, *33*, 2004190. [CrossRef]
27. Jayabalakrishnan, D.; Naga Muruga, D.B.; Bhaskar, K.; Pavan, P.; Balaji, K.; Rajakumar, P.S.; Priya, C.; Deepa, R.A.B.; Sendilvelan, S.; Prabhahar, M. Self-Healing materials—A review. *Mater. Today Proc.* **2021**, *45*, 7195–7199. [CrossRef]
28. Ikura, R.; Park, J.; Osaki, M.; Yamaguchi, H.; Harada, A.; Takashima, Y. Design of self-healing and self-restoring materials utilizing reversible and movable crosslinks. *NPG Asia Mater.* **2022**, *14*, 10. [CrossRef]
29. O'Harra, K.; Sadaba, N.; Irigoyen, M.; Ruipérez, F.; Aguirresarobe, R.; Sardon, H.; Bara, J. Nearly Perfect 3D Structures Obtained by Assembly of Printed Parts of Polyamide Ionene Self-Healing Elastomer. *ACS Appl. Polym. Mater.* **2020**, *2*, 4352–4359. [CrossRef]
30. Peng, B.; Yang, Y.; Ju, T.; Cavicchi, K.A. Fused Filament Fabrication 4D Printing of a Highly Extensible, Self-Healing, Shape Memory Elastomer Based on Thermoplastic Polymer Blends. *ACS Appl. Mater. Interfaces* **2021**, *13*, 12777–12788. [CrossRef]
31. Bi, H.; Ye, G.; Sun, H.; Ren, Z.; Gu, T.; Xu, M. Mechanically robust, shape memory, self-healing and 3D printable thermoreversible cross-linked polymer composites toward conductive and biomimetic skin devices applications. *Addit. Manuf.* **2022**, *49*, 102487. [CrossRef]
32. Zhou, Q.; Gardea, F.; Sang, Z.; Lee, S.; Pharr, M.; Sukhishvili, S.A. A Tailorable Family of Elastomeric-to-Rigid, 3D Printable, Interbonding Polymer Networks. *Adv. Funct. Mater.* **2020**, *30*, 2002374. [CrossRef]
33. Zuo, H.; Liu, Z.; Zhang, L.; Liu, G.; Ouyang, X.; Guan, Q.; Wu, Q.; You, Z. Self-healing materials enable free-standing seamless large-scale 3D printing. *Sci. China Mater.* **2021**, *64*, 1791–1800. [CrossRef]
34. Lai, J.C.; Li, L.; Wang, D.P.; Zhang, M.H.; Mo, S.R.; Wang, X.; Zeng, K.Y.; Li, C.H.; Jiang, Q.; You, X.Z.; et al. A rigid and healable polymer cross-linked by weak but abundant Zn(II)-carboxylate interactions. *Nat. Commun.* **2018**, *9*, 2725. [CrossRef] [PubMed]
35. Kuang, X.; Chen, K.; Dunn, C.K.; Wu, J.; Li, V.C.F.; Qi, H.J. 3D Printing of Highly Stretchable, Shape-Memory, and Self-Healing Elastomer toward Novel 4D Printing. *ACS Appl. Mater. Interfaces* **2018**, *10*, 7381–7388. [CrossRef] [PubMed]
36. Yuan, T.; Zhang, L.; Li, T.; Tu, R.; Sodano, H.A. 3D Printing of a self-healing, high strength, and reprocessable thermoset. *Polym. Chem.* **2020**, *11*, 6441–6452. [CrossRef]
37. Sanders, P.; Young, A.J.; Qin, Y.; Fancey, K.S.; Reithofer, M.R.; Guillet-Nicolas, R.; Kleitz, F.; Pamme, N.; Chin, J.M. Stereolithographic 3D printing of extrinsically self-healing composites. *Sci. Rep.* **2019**, *9*, 388. [CrossRef] [PubMed]
38. Durand-Silva, A.; Cortes-Guzman, K.P.; Johnson, R.M.; Perera, S.D.; Diwakara, S.D.; Smaldone, R.A. Balancing Self-Healing and Shape Stability in Dynamic Covalent Photoresins for Stereolithography 3D Printing. *ACS Macro Lett.* **2021**, *10*, 486–491. [CrossRef]
39. Liu, Z.; Hong, P.; Huang, Z.; Zhang, T.; Xu, R.; Chen, L.; Xiang, H.; Liu, X. Self-healing, reprocessing and 3D printing of transparent and hydrolysis-resistant silicone elastomers. *Chem. Eng. J.* **2020**, *387*, 124142. [CrossRef]
40. Zhang, B.; Zhang, W.; Zhang, Z.; Zhang, Y.F.; Hingorani, H.; Liu, Z.; Liu, J.; Ge, Q. Self-Healing Four-Dimensional Printing with an Ultraviolet Curable Double-Network Shape Memory Polymer System. *ACS Appl. Mater. Interfaces* **2019**, *11*, 10328–10336. [CrossRef]
41. Li, X.; Yu, R.; He, Y.; Zhang, Y.; Yang, X.; Zhao, X.; Huang, W. Self-Healing Polyurethane Elastomers Based on a Disulfide Bond by Digital Light Processing 3D Printing. *ACS Macro Lett.* **2019**, *8*, 1511–1516. [CrossRef] [PubMed]
42. Miao, J.-T.; Ge, M.; Wu, Y.; Peng, S.; Zheng, L.; Chou, T.Y.; Wu, L. 3D printing of sacrificial thermosetting mold for building near-infrared irradiation induced self-healable 3D smart structures. *Chem. Eng. J.* **2022**, *427*, 131580. [CrossRef]
43. Zhang, M.; Tao, X.; Yu, R.; He, Y.; Li, X.; Chen, X.; Huang, W. Self-healing, mechanically robust, 3D printable ionogel for highly sensitive and long-term reliable ionotronics. *J. Mater. Chem. A* **2022**, *10*, 12005–12015. [CrossRef]
44. Caprioli, M.; Roppolo, I.; Chiappone, A.; Larush, L.; Pirri, C.F.; Magdassi, S. 3D-printed self-healing hydrogels via Digital Light Processing. *Nat. Commun.* **2021**, *12*, 2462. [CrossRef]
45. Dry, C. Matrix cracking repair and filling using active and passive modes for smart timed release of chemicals from fibers into cement matrices. *Smart Mater. Struct.* **1994**, *3*, 118–123. [CrossRef]

46. White, S.R.; Sottos, N.R.; Geubelle, P.H.; Moore, J.S.; Kessler, M.R.; Sriram, S.R.; Brown, E.N.; Viswanathan, S. Autonomic healing of polymer composites. *Nature* **2001**, *409*, 794–797. [CrossRef]
47. Ma, B.; Zhang, Y.; Wei, Y.; Li, M.; Li, D. Graphene Oxide-Modified Microcapsule Self-Healing System for 4D Printing. *Front. Mater.* **2021**, *8*, 657777. [CrossRef]
48. Postiglione, G.; Alberini, M.; Leigh, S.; Levi, M.; Turri, S. Effect of 3D-Printed Microvascular Network Design on the Self-Healing Behavior of Cross-Linked Polymers. *ACS Appl. Mater. Interfaces* **2017**, *9*, 14371–14378. [CrossRef]
49. Zhang, Y.; Li, W.; Pan, P.; Tang, J.; Dong, B.; Xing, F.; Zhu, G. Programmable construction of vasculature by printing in cementitious materials for self-healing application. *Compos. Part B Eng.* **2022**, *242*, 110056. [CrossRef]
50. Eslami-Farsani, R.; Khalili, S.M.R.; Khademoltoliati, A.; Saeedi, A. Tensile and creep behavior of microvascular based self-healing composites: Experimental study. *Mech. Adv. Mater. Struct.* **2019**, *28*, 384–390. [CrossRef]
51. Zhang, P.F.; Li, G.Q. Healing-on-demand composites based on polymer artificial muscle. *Polymer* **2015**, *64*, 29–38. [CrossRef]
52. Garg, M.; Ladd, A.C.; Aw, J.E.; Zhang, X.; Sottos, N.R. Sacrificial Cyclic Poly(phthalaldehyde) Templates for Low-Temperature Vascularization of Polymer Matrices. *ACS Appl. Polym. Mater.* **2021**, *4*, 479–487. [CrossRef]
53. Garcia, S.J. Effect of polymer architecture on the intrinsic self-healing character of polymers. *Eur. Polym. J.* **2014**, *53*, 118–125. [CrossRef]
54. Canadell, J.; Goossens, H.; Klumperman, B. Self-Healing Materials Based on Disulfide Links. *Macromolecules* **2011**, *44*, 2536–2541. [CrossRef]
55. Ratwani, C.R.; Kamali, A.R.; Abdelkader, A.M. Self-healing by Diels-Alder cycloaddition in advanced functional polymers: A review. *Prog. Mater. Sci.* **2023**, *131*, 101001. [CrossRef]
56. Goyal, M.; Agarwal, S.N.; Bhatnagar, N. A review on self-healing polymers for applications in spacecraft and construction of roads. *J. Appl. Polym. Sci.* **2022**, *139*, e52816. [CrossRef]
57. Rahman, S.S.; Arshad, M.; Qureshi, A.; Ullah, A. Fabrication of a Self-Healing, 3D Printable, and Reprocessable Biobased Elastomer. *ACS Appl. Mater. Interfaces* **2020**, *12*, 51927–51939. [CrossRef]
58. Lei, Z.Q.; Xiang, H.P.; Yuan, Y.J.; Rong, M.Z.; Zhang, M.Q. Room-Temperature Self-Healable and Remoldable Cross-linked Polymer Based on the Dynamic Exchange of Disulfide Bonds. *Chem. Mater.* **2014**, *26*, 2038–2046. [CrossRef]
59. Yu, K.; Xin, A.; Du, H.; Li, Y.; Wang, Q. Additive manufacturing of self-healing elastomers. *NPG Asia Mater.* **2019**, *11*, 7. [CrossRef]
60. Sanka, R.V.S.P.; Krishnakumar, B.; Leterrier, Y.; Pandey, S.; Rana, S.; Michaud, V. Soft Self-Healing Nanocomposites. *Front. Mater.* **2019**, *6*, 137. [CrossRef]
61. Larson, J.W.; McMahon, T.B. Hydrogen-bonding in gas-phase anions—An experimental investigation of the interaction between chloride-ion and bronsted acids from ion-cyclotron resonance chloride exchange equilibria. *J. Am. Chem. Soc.* **1984**, *106*, 517–521. [CrossRef]
62. Emsley, J. Very strong hydrogen-bonding. *Chem. Soc. Rev.* **1980**, *9*, 91–124. [CrossRef]
63. Cordier, P.; Tournilhac, F.; Soulie-Ziakovic, C.; Leibler, L. Self-healing and thermoreversible rubber from supramolecular assembly. *Nature* **2008**, *451*, 977–980. [CrossRef] [PubMed]
64. Kim, Y.J.; Huh, P.H.; Kim, B.K. Synthesis of Self-Healing Polyurethane Urea-Based Supramolecular Materials. *J. Polym. Sci. Part B Polym. Phys.* **2015**, *53*, 468–474. [CrossRef]
65. Cui, J.X.; del Campo, A. Multivalent H-bonds for self-healing hydrogels. *Chem. Commun.* **2012**, *48*, 9302–9304. [CrossRef] [PubMed]
66. Liu, T.Q.; Jiao, C.; Peng, X.; Chen, Y.N.; Chen, Y.Y.; He, C.C.; Liu, R.G.; Wang, H.L. Super-strong and tough poly(vinyl alcohol)/poly(acrylic acid) hydrogels reinforced by hydrogen bonding. *J. Mater. Chem. B* **2018**, *6*, 8105–8114. [CrossRef]
67. Kee, S.; Haque, M.A.; Corzo, D.; Alshareef, H.N.; Baran, D. Self-Healing and Stretchable 3D-Printed Organic Thermoelectrics. *Adv. Funct. Mater.* **2019**, *29*, 1905426. [CrossRef]
68. Varley, R.J.; van der Zwaag, S. Towards an understanding of thermally activated self-healing of an ionomer system during ballistic penetration. *Acta Mater.* **2008**, *56*, 5737–5750. [CrossRef]
69. Varley, R.J.; van der Zwaag, S. Development of a quasi-static test method to investigate the origin of self-healing in ionomers under ballistic conditions. *Polym. Test.* **2008**, *27*, 11–19. [CrossRef]
70. Li, X.P.; Yu, R.; Zhao, T.T.; Zhang, Y.; Yang, X.; Zhao, X.J.; Huang, W. A self-healing polysiloxane elastomer based on siloxane equilibration synthesized through amino-ene Michael addition reaction. *Eur. Polym. J.* **2018**, *108*, 399–405. [CrossRef]
71. Darabi, M.A.; Khosrozadeh, A.; Mbeleck, R.; Liu, Y.; Chang, Q.; Jiang, J.; Cai, J.; Wang, Q.; Luo, G.; Xing, M. Skin-Inspired Multi-functional Autonomic-Intrinsic Conductive Self-Healing Hydrogels with Pressure Sensitivity, Stretchability, and 3D Printability. *Adv. Mater.* **2017**, *29*, 1700533. [CrossRef]
72. Wang, Y.; Chang, Q.; Zhan, R.; Xu, K.; Wang, Y.; Zhang, X.; Li, B.; Luo, G.; Xing, M.; Zhong, W. Tough but self-healing and 3D printable hydrogels for E-skin, E-noses and laser controlled actuators. *J. Mater. Chem. A* **2019**, *7*, 24814–24829. [CrossRef]
73. Shi, L.; Carstensen, H.; Hölzl, K.; Lunzer, M.; Li, H.; Hilborn, J.; Ovsianikov, A.; Ossipov, D.A. Dynamic Coordination Chemistry Enables Free Directional Printing of Biopolymer Hydrogel. *Chem. Mater.* **2017**, *29*, 5816–5823. [CrossRef]
74. Zhang, J.; Chen, L.; Shen, B.; Wang, Y.; Peng, P.; Tang, F.; Feng, J. Highly transparent, self-healing, injectable and self-adhesive chitosan/polyzwitterion-based double network hydrogel for potential 3D printing wearable strain sensor. *Mater. Sci. Eng. C Mater. Biol. Appl.* **2020**, *117*, 111298. [CrossRef] [PubMed]

75. Rim, Y.S.; Bae, S.H.; Chen, H.J.; De Marco, N.; Yang, Y. Recent Progress in Materials and Devices toward Printable and Flexible Sensors. *Adv. Mater.* **2016**, *28*, 4415–4440. [CrossRef]
76. Mannoor, M.S.; Jiang, Z.; James, T.; Kong, Y.L.; Malatesta, K.A.; Soboyejo, W.O.; Verma, N.; Gracias, D.H.; McAlpine, M.C. 3D Printed Bionic Ears. *Nano Lett.* **2013**, *13*, 2634–2639. [CrossRef] [PubMed]
77. Truby, R.L.; Lewis, J.A. Printing soft matter in three dimensions. *Nature* **2016**, *540*, 371–378. [CrossRef]
78. Nadgorny, M.; Ameli, A. Functional Polymers and Nanocomposites for 3D Printing of Smart Structures and Devices. *ACS Appl. Mater. Interfaces* **2018**, *10*, 17489–17507. [CrossRef]
79. Yang, K.; Grant, J.C.; Lamey, P.; Joshi-Imre, A.; Lund, B.R.; Smaldone, R.A.; Voit, W. Diels-Alder Reversible Thermoset 3D Printing: Isotropic Thermoset Polymers via Fused Filament Fabrication. *Adv. Funct. Mater.* **2017**, *27*, 1700318. [CrossRef]
80. Shafranek, R.T.; Millik, S.C.; Smith, P.T.; Lee, C.-U.; Boydston, A.J.; Nelson, A. Stimuli-responsive materials in additive manufacturing. *Prog. Polym. Sci.* **2019**, *93*, 36–67. [CrossRef]
81. Zein, I.; Hutmacher, D.W.; Tan, K.C.; Teoh, S.H. Fused deposition modeling of novel scaffold architectures for tissue engineering applications. *Biomaterials* **2002**, *23*, 1169–1185. [CrossRef] [PubMed]
82. Guzzi, E.A.; Bovone, G.; Tibbitt, M.W. Universal Nanocarrier Ink Platform for Biomaterials Additive Manufacturing. *Small* **2019**, *15*, e1905421. [CrossRef]
83. Gul, J.Z.; Sajid, M.; Rehman, M.M.; Siddiqui, G.U.; Shah, I.; Kim, K.H.; Lee, J.W.; Choi, K.H. 3D printing for soft robotics—A review. *Sci. Technol. Adv. Mater.* **2018**, *19*, 243–262. [CrossRef] [PubMed]
84. Wang, X.; Jiang, M.; Zhou, Z.; Gou, J.; Hui, D. 3D printing of polymer matrix composites: A review and prospective. *Compos. Part B Eng.* **2017**, *110*, 442–458. [CrossRef]
85. Luo, X.; Mather, P.T. Shape Memory Assisted Self-Healing Coating. *ACS Macro Lett.* **2013**, *2*, 152–156. [CrossRef] [PubMed]
86. Zhang, Z.; Corrigan, N.; Boyer, C. A Photoinduced Dual-Wavelength Approach for 3D Printing and Self-Healing of Thermosetting Materials. *Angew. Chem. Int. Ed.* **2022**, *61*, e202114111. [CrossRef] [PubMed]
87. Wu, A.S.; Small Iv, W.; Bryson, T.M.; Cheng, E.; Metz, T.R.; Schulze, S.E.; Duoss, E.B.; Wilson, T.S. 3D Printed Silicones with Shape Memory. *Sci. Rep.* **2017**, *7*, 4664. [CrossRef]
88. Paladugu, S.R.; Sreekanth, P.S.R.; Sahu, S.K.; Naresh, K.; Karthick, S.A.; Venkateshwaran, N.; Ramoni, M.; Mensah, R.A.; Das, O.; Shanmugam, R. A Comprehensive Review of Self-Healing Polymer, Metal, and Ceramic Matrix Composites and Their Modeling Aspects for Aerospace Applications. *Materials* **2022**, *15*, 8521. [CrossRef]
89. Das, R.; Melchior, C.; Karumbaiah, K.M. Self-healing composites for aerospace applications. In *Advanced Composite Materials for Aerospace Engineering*; Rana, S., Fangueiro, R., Eds.; Woodhead Publishing: Sawston, UK, 2016; pp. 333–364. [CrossRef]
90. Brochu, A.B.W.; Craig, S.L.; Reichert, W.M. Self-healing biomaterials. *J. Biomed. Mater. Res. Part A* **2011**, *96A*, 492–506. [CrossRef]
91. Menikheim, S.; Leckron, J.; Duffy, M.; Zupan, M.; Mallory, A.; Lien, W.; Lavik, E. Biocompatible Nanocapsules for Self-Healing Dental Resins and Bone Cements. *ACS Omega* **2022**, *7*, 31726–31735. [CrossRef]
92. Priyadarsini, M.; Rekha Sahoo, D.; Biswal, T. A new generation self-healing composite materials. *Mater. Today Proc.* **2021**, *47*, 1229–1233. [CrossRef]
93. Shinde, V.V.; Wang, Y.; Salek, M.F.; Auad, M.L.; Beckingham, L.E.; Beckingham, B.S. Material Design for Enhancing Properties of 3D Printed Polymer Composites for Target Applications. *Technologies* **2022**, *10*, 45. [CrossRef]
94. Khan, A.; Ahmed, N.; Rabnawaz, M. Covalent Adaptable Network and Self-Healing Materials: Current Trends and Future Prospects in Sustainability. *Polymers* **2020**, *12*, 2027. [CrossRef] [PubMed]
95. Haines-Gadd, M.; Charnley, F.; Encinas-Oropesa, A. Self-healing materials: A pathway to immortal products or a risk to circular economy systems? *J. Clean. Prod.* **2021**, *315*, 128193. [CrossRef]
96. Li, J.; Chee, H.L.; Chong, Y.T.; Chan, B.Q.Y.; Xue, K.; Lim, P.C.; Loh, X.J.; Wang, F. Hofmeister Effect Mediated Strong PHEMA-Gelatin Hydrogel Actuator. *ACS Appl. Mater. Interfaces* **2022**, *14*, 23826–23838. [CrossRef] [PubMed]
97. Wong, J.H.M.; Tan, R.P.T.; Chang, J.J.; Chan, B.Q.Y.; Zhao, X.; Cheng, J.J.W.; Yu, Y.; Boo, Y.J.; Lin, Q.; Ow, V.; et al. Injectable Hybrid-Crosslinked Hydrogels as Fatigue-Resistant and Shape-Stable Skin Depots. *Biomacromolecules* **2022**, *23*, 3698–3712. [CrossRef] [PubMed]
98. Chan, S.Y.; Goh, S.S.; Dou, Q.; Chan, B.Q.Y.; Choo, W.S.; Young, D.J.; Loh, X.J. Unprecedented Acid-Promoted Polymerization and Gelation of Acrylamide: A Serendipitous Discovery. *Chem. Asian J.* **2018**, *13*, 1797–1804. [CrossRef]
99. Chan, B.Q.Y.; Heng, S.J.W.; Liow, S.S.; Zhang, K.; Loh, X.J. Dual-responsive hybrid thermoplastic shape memory polyurethane. *Mater. Chem. Front.* **2017**, *1*, 767–779. [CrossRef]
100. Ang, J.Y.; Chan, B.Q.Y.; Kai, D.; Loh, X.J. Engineering Porous Water-Responsive Poly(PEG/PCL/PDMS Urethane) Shape Memory Polymers. *Macromol. Mater. Eng.* **2017**, *302*, 1700174. [CrossRef]
101. Kai, D.; Prabhakaran, M.P.; Chan, B.Q.Y.; Liow, S.S.; Ramakrishna, S.; Xu, F.; Loh, X.J. Elastic poly(ε-caprolactone)-polydimethylsiloxane copolymer fibers with shape memory effect for bone tissue engineering. *Biomed. Mater.* **2016**, *11*, 015007. [CrossRef]
102. Chan, B.Q.Y.; Liow, S.S.; Loh, X.J. Organic–inorganic shape memory thermoplastic polyurethane based on polycaprolactone and polydimethylsiloxane. *RSC Adv.* **2016**, *6*, 34946–34954. [CrossRef]
103. Yan, Y.; Jiang, Y.; Ng, E.L.L.; Zhang, Y.; Owh, C.; Wang, F.; Song, Q.; Feng, T.; Zhang, B.; Li, P.; et al. Progress and opportunities in additive manufacturing of electrically conductive polymer composites. *Mater. Today Adv.* **2023**, *17*, 100333. [CrossRef]

104. Chan, B.Q.Y.; Chong, Y.T.; Wang, S.; Lee, C.J.J.; Owh, C.; Wang, F.; Wang, F. Synergistic combination of 4D printing and electroless metallic plating for the fabrication of a highly conductive electrical device. *Chem. Eng. J.* **2022**, *430*, 132513. [CrossRef]
105. Kai, D.; Tan, M.J.; Prabhakaran, M.P.; Chan, B.Q.Y.; Liow, S.S.; Ramakrishna, S.; Loh, X.J. Biocompatible electrically conductive nanofibers from inorganic-organic shape memory polymers. *Colloids Surf. B Biointerfaces* **2016**, *148*, 557–565. [CrossRef] [PubMed]

polymers

MDPI

Article

Influence of 3D Printing Conditions on Some Physical–Mechanical and Technological Properties of PCL Wood-Based Polymer Parts Manufactured by FDM

Irina Beşliu-Băncescu [1,*], Ioan Tamaşag [1,*] and Laurenţiu Slătineanu [2]

1 Faculty of Mechanical Engineering, Automotive and Robotics, "Stefan cel Mare" University, 720229 Suceava, Romania
2 Faculty of Machine Manufacturing and Industrial Management, "Gheorghe Asachi" Technical University of Iasi, 700050 Iaşi, Romania; lslati@yahoo.com
* Correspondence: irina.besliu@usm.ro (I.B.-B.); ioan.tamasag@usm.ro (I.T.); Tel.: +40-744322449 (I.B.-B.); +40-741564971 (I.T.)

Abstract: The paper investigates the influence of some 3D printing conditions on some physical–mechanical and technological properties of polycaprolactone (PCL) wood-based biopolymer parts manufactured by FDM. Parts with 100% infill and the geometry according to ISO 527 Type 1B were printed on a semiprofessional desktop FDM printer. A full factorial design with three independent variables at three levels was considered. Some physical–mechanical properties (weight error, fracture temperature, ultimate tensile strength) and technological properties (top and lateral surface roughness, cutting machinability) were experimentally assessed. For the surface texture analysis, a white light interferometer was used. Regression equations for some of the investigated parameters were obtained and analysed. Higher printing speeds than those usually reported in the existing literature dealing with wood-based polymers' 3D printing had been tested. Overall, the highest level chosen for the printing speed positively influenced the surface roughness and the ultimate tensile strength of the 3D-printed parts. The cutting machinability of the printed parts was investigated by means of cutting force criteria. The results showed that the PCL wood-based polymer analysed in this study had lower machinability than natural wood.

Keywords: wood-based biopolymer; surface quality; tensile strength; machinability

Citation: Beşliu-Băncescu, I.; Tamaşag, I.; Slătineanu, L. Influence of 3D Printing Conditions on Some Physical–Mechanical and Technological Properties of PCL Wood-Based Polymer Parts Manufactured by FDM. *Polymers* **2023**, *15*, 2305. https://doi.org/10.3390/polym15102305

Academic Editors: Cristina-Elisabeta Pelin and Anton Ficai

Received: 29 April 2023
Revised: 10 May 2023
Accepted: 12 May 2023
Published: 14 May 2023

1. Introduction

The ability to quickly generate complex surfaces and structures at lower costs and significantly lower material losses in the case of traditional mechanical processing technologies recommend 3D printing technologies for many industrial applications. There are several types of 3D printing processes, such as selective laser sintering (SLS), stereolithography (SLA), multi-jet fusion (MJF), digital light processing (DLP), digital light processing (DLP), fused deposition modelling (FDM), etc. FDM, also known as MEX (Material Extrusion) [1], is one of the most commonly used 3D printing processes because of the wide range of materials that can be processed/manufactured. The FDM process input parameters, such as the layer thickness, wall shell thickness, printing temperature, infill structure, infill density percentage, and printing speeds, strongly influence the mechanical proprieties of the printed products.

FDM is an emerging technology implemented in sectors such as the automotive, aerospace, medical, architecture, fashion, and food industries [2]. The main drawbacks reported for these technologies are the anisotropic nature and poor mechanical properties of the 3D-printed parts [2]. The principle of this manufacturing technique is that the wire material is heated and deposited layer by layer into the desired part shape. The part material must be pre-processed by hot melt extrusion to be transformed into filaments.

The literature provides multiple studies that analyse the influence of printing conditions and parameters and post-processing methods on the mechanical properties of 3D-printed parts, especially FDM [3–11]. Several researchers have contributed with comprehensive reviews on these issues [12–14]. Furthermore, the subject continuously develops due to the increasing interest in different industries, requiring more attention from the scientific community.

Voids usually appear between the deposited filament layers in the FDM printing process. These voids are believed to be one of the main causes of low tensile strength and anisotropy [15] and may also affect the 3D-printed parts' cutting machinability. In the case of WPC (wood-based composite polymer) 3D-printed parts, it had been considered that wood fibres might encourage void formation. Comparing unfilled printed specimens with reinforced ones with natural fibres has shown a negative influence of the fibres on strength, while stiffness either increases slightly or remains constant [16].

The use of wood is increasing due to the growth of the world population, the development of new wood products, and the identification of new applications in various fields. Wood is a renewable and carbon-storing resource [16] with excellent properties but is limited to forest land. In recent decades, the wood demand increased significantly and overcame disposable supplies. Sustainability targets and growing environmental concerns have increased the demand for renewable and recyclable materials with compatible proprieties and behaviour/performance. In recent years, wood-based composite polymers (WPCs) have been gaining popularity [15,17–19]. These materials are composed of one or more natural wood chips, fibres, or flours and one or a mixture of polymers, most commonly thermoplastic polymers such as polyethene (PE), polylactide (PLA), or polypropylene (PP). Compared to natural timber products, WPCs present higher resistance to weathering and biological deterioration, thermal resistance, and expose sufficient strength for structural applications [20]. WPCs are mainly used for outdoor and indoor furniture, window and door frames, moulding, different construction purposes, and the automotive and marine industries [15,18,21]. The main drawbacks are the slightly higher prices and lower thermal resistance compared to natural wood.

The mechanical performance of WPCs is the main objective addressed by research in this field. Most of the research dealing with wood-based polymers analyses some mechanical proprieties for commercially available filaments [22,23] or develops and tests new wood composite filaments by mixing different amounts and types of wood fibres, polymers, additives, and fillers [15,16,24–29].

The most popular wood-based polymer type is obtained with a polylactic acid (PLA) polymer matrix and different percentages of wood fibres, dust, or chips. The performance of WPC material can be enhanced by using a proper combination of polymers and providing different fillers and additives. The research carried out in this field showed that beech sawdust can contribute to the reinforcement of flexural stress and tensile strength and that sawdust also helps reduce WPC density [21]. Additionally, WPCs are often brittle. Styrene and butadiene rubber (SBR), ethylene-propylene monomers leather (EPDM), or plastic elastomers can be used for toughening purposes [18]. Because of their strong flammability level, flame retardants, usually polyphosphate (APP), must be provided in their composition [19]. The addition of lignocellulosic fibres to WPC filaments was reported to lower the mechanical proprieties of the 3D-printed composites [27].

Hydrothermal degradation tests were performed [27] to establish its effect on the mechanical properties. The results showed that adding natural fillers and different levels of infilling resulted in a similar level of reduction in the properties. Additionally, the addition of natural fillers resulted in a slightly lower drop than the lowered infilling rate for tensile strength [27].

Results from another study [30] indicated that thickness swell, water uptake, mechanical strength, and stiffness increased, and elongation at break and impact energy decreased with an increasing wood fibre proportion.

The influence of the shape of wood particles on the mechanical proprieties of WPCs was also investigated. Huang et al. [29] showed that the shape and surface roughness of the wood particles, rather than the wood species, play an essential role in determining the properties of 3D-printed WPC products. Additionally, it was reported that wood particles with more rounded shapes and smoother surfaces are more suitable for obtaining a denser and stronger 3D-printed WPC product [29].

The machinability of wood–plastic composites has been approached by a relatively small number of studies. Most of these studies were carried on parts generated by other machining processes than 3D printing. Zhu et al. [31] explored the cutting performance of wood–plastic composites based on cutting forces, cutting temperature, surface quality, chip formation, and tool wear during peripheral milling experiments using cemented carbide cutters. The wood–plastic composites tested were processed by extrusion, moulding, and injection moulding. WPPC exhibited the highest cutting forces and cutting temperatures under the same cutting conditions, followed by WPEC and WPVCC. Wu et al. [32] had studied the helical milling performance of the WPC obtained by mixing poplar flour and polyethylene followed by extrusion at high temperatures. They reported that in WPC helical milling, the cutting force increases with increased spindle speed, cutting depth, and tool helical angle.

Biopolymers have attracted increased attention in recent years mainly because of their abundant and sustainable sources and versatile properties [2]. Biowood, produced by Rosa3D, is a wood-based composite biopolymer. The main components of Biowood filament are polycaprolactone (PCL), polyester, starch, lignin, natural resins, waxes and oils, natural fatty acids, cellulose, and natural fibres [33]. Polycaprolactone (PCL) is a biodegradable polyester with a low melting point of 60 °C [25] that is usually blended with other polymers. Studies [25] showed that natural fibres generally enhance polycaprolactone's biodegradability and mechanical proprieties. Combining cellulose with polycaprolactone increased the tensile modulus but decreased the tensile strength of the composites [25]. Lignin is a natural polymer that binds cellulose fibres together, assuring stiffness for the wood-based polymer composites. Starch is not only used for binding and as a glue agent. The blending of starch with plastics has been reported to improve water resistance, processing properties, and mechanical properties [27].

Zgodavová K. et al. [23] have tested different thermoplastic materials for printing shield frames in terms of mechanical properties, geometric accuracy, weight, printing time, filament price, and environmental sustainability. Among them, they tested PHABiowood Rosa3D. The input parameters considered were the layer thickness, number of perimeters, extrusion width, infill density, and nozzle temperature. The tensile stress of the PHA Biowood varied from 10.8 MPa to 21.8 MPa, and the factors with significant influence over the mechanical properties were the infill and the interaction between the layer height and printing infill.

The aim of this paper was to highlight the results of some experimental studies dealing with the influence of some specific factors that characterise the 3D printing conditions of PCL wood-based polymer parts on some physical–mechanical and technological properties of the material incorporated in those parts. As input factors of the 3D-printing process, printing temperature, layer height, and printing speed were considered. Some physical–mechanical properties (weight error, fracture temperature, ultimate tensile strength) and technological properties (top and lateral surface roughness, cutting machinability) constituted output parameters that were subjected to the analysis. This study's novelty consists in analysing the influence of some printing parameters (printing temperature, layer height, and printing speed) on some of the qualitative aspects and mechanical proprieties of the Biowood Rosa3D wood-based biopolymer parts generated by FDM. The values selected for the printing speed parameter were significantly superior to those usually tested in previous research in this field or those recommended by the filament producer. Another novelty aspect of this study is the slot milling machinability analysis by means of cutting force levels of the FDM-printed parts.

Even if complex shape parts can be generated by additive manufacturing, there are several situations where 3D-printed parts may require future processing.

The use of cutting as a secondary processing operation for 3D-printed parts can address for parts with high functional and tolerance requirements. Usually, the FDM parts achieved accuracies of ±0.5 mm for desktop printers and ±0.2 mm for industrial printers, and with CNC machining, accuracies of ±0.05 mm can be obtained.

Another reason for combining the two technologies is productivity. Even with the recent advancements in 3D-printing technology, printing speed is still a major drawback for considering these technologies for industrial applications. By considering cutting technologies for some of the part features, the machining time of the parts can be significantly improved.

This study can be a starting point for other researchers that aim to establish the proper printing conditions for PCL wood-based polymers and for industry agents interested in developing biodegradable wood-like products.

2. Experimental Setup

In Figure 1, a schematic representation of the experimental program used in the study is presented. The model offers information about the input parameters, equipment, procedures, and the investigated parameters considered in the study.

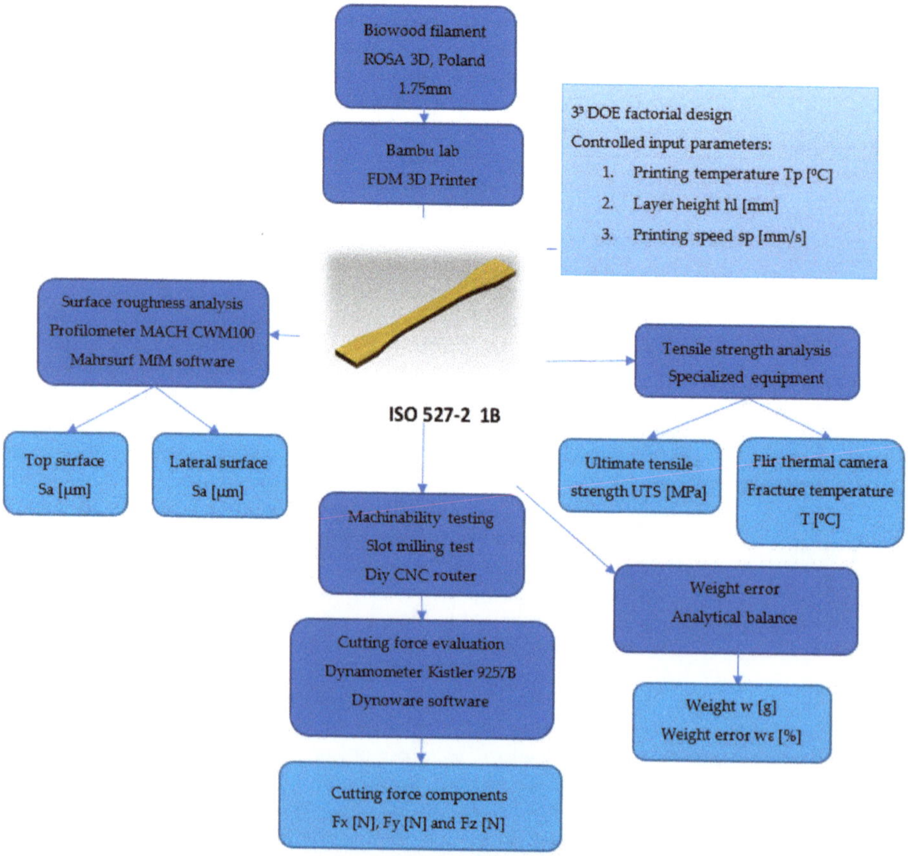

Figure 1. Schematic representation of the experimental program.

The parts, in the form of test specimens with dimensions according to ISO 527-2 1B, were manufactured using a BambuLab X1C 3D printer and then tested and analysed from four perspectives:

1. Surface quality by obtaining values for surface roughness (Sq) using a Mahr CWM 100 profilometer;
2. Tensile strength, obtaining values for UTS but also for the temperature at the time of specimen rupture;
3. Analysis of the density variation of the resulting parts in terms of weight;
4. Machinability of the parts, where the values for the components Fx, Fy, and Fz of the cutting force were obtained using a Kistler type 9257B dynamometer.

A full factorial experiment was considered to achieve the desired research objectives. The independent variable factors that were changed in the experimental procedure were the following: the printing temperature Tp (°C), the layer height, hl (mm), and printing speed, sp (mm/s). The values of the input factor levels selected in this study for each of them are presented in Table 1.

Table 1. Values of the input factors corresponding to the full factorial design.

Level	Input Parameters		
	Printing Temperature Tp (°C)	Layer Height hl (mm)	Printing Speed sp (mm/s)
1	175	0.2	150
2	190	0.28	200
3	220	0.4	300

The results obtained under experimental conditions according to the DOE 3^3 factorial design, were analysed to obtain variation plots, and then ANOVA was applied to determine the factors with statistically significant influence.

As output interest parameters of the proposed study, the following parameters were considered:

- Weight error (%);
- Arithmetical mean height, Sa (μm) of the top and lateral surfaces of the specimens;
- Ultimate tensile strength UTS (MPa);
- Fracture temperature T (°C);
- Cutting force components Fx (N), Fy (N), and Fz (N).

The advancements in 3D printing equipment have opened new opportunities in terms of reducing the printing time. The producers of 3D printers have focused on addressing one of the main drawbacks of additive manufacturing technologies, which is the printing time. Printing time is directly proportional to the printing speed that can be achieved. Therefore, in recent years, new 3D printers with higher printing speed facilities were produced. Even if high printing speeds can be achieved by using these new 3D printers available in the market, the testing of these capabilities is still limited. In this study, significantly higher printing speeds than those usually reported in the scientific literature were considered.

The factor levels were chosen to preserve the randomness of the results. In the case of temperature, the minimum level was chosen to be 175 °C, the second level 190 °C, which is most often used in FDM 3D printing especially for biopolymers (such as PLA—polylactic acid), and 220 °C, 10 °C more than the manufacturer's recommendation.

In terms of layer height, level 1 of 0.2 mm was chosen because it is the most common in the literature, 0.4 mm because the nozzle used has a diameter of 0.6 mm (dimensions suggested by the filament manufacturer), and the layer height represents under 75% of the nozzle diameter. The middle value of 0.28 mm was chosen to be able to observe inter-layer overlap and part density variation when the levels did not have a multiple character.

In terms of printing speed, high random speeds in the range 150–300 mm/s were chosen. These values were chosen because this range is less studied and the printer used, being core XY, allows printing at high speeds obtaining high qualities, comparable to 3D printing at low speeds.

The choice of level values was made in such a way that the midpoint was not close to the end values, thus avoiding the possibility of intercalation of mean effects.

2.1. Materials

Biowood is a raw polymer consisting of only renewable resources. The test samples used in the experiments were produced by Rosa3D Filaments (Poland). The main components of this wooden thermoplastic polymer filament are the following: polycaprolactone (PCL), polyester, starch, lignin, natural resins, waxes and oils, natural fatty acids, cellulose, and natural fibres [33]. The wood fibre content is considered to facilitate mechanical processing. In Table 2, the main physical properties of biowood polymer are presented. According to the filament producer, biowood filaments require low extrusion temperatures, between 170 and 210 °C. Moreover, printing speeds in the range 60–80 mm/s and build platform temperatures of 30–50 °C are recommended [33].

Table 2. Physical properties of Biowood [34].

Softening point (°C)	50
Density (kg/m^3)	1260
Elastic modulus (MPa)	3200
Tensile strength (MPa)	36

2.2. Sample Preparation and Equipment

Experimental tests were conducted considering standardised tensile test specimens ISO 527 Type 1B. The probes had the geometry and dimensions presented in Figure 2. For these studies, 100% infill specimens were considered.

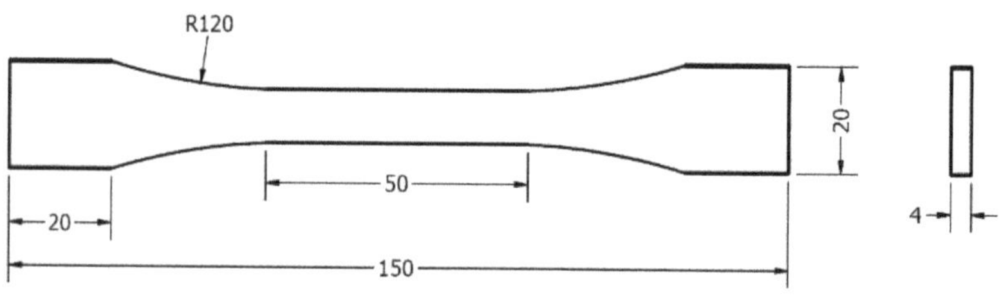

Figure 2. Tensile testing specimen ISO 527 Type 1B.

Specimens were manufactured on an FDM Desktop enclosed printer type X1-Carbon Combo produced by Bambu Lab (Austin, TX, USA) (Figure 3). The printer has a lidar resolution of 7 µm, 20 m/s^2 acceleration, and a maximum speed of 500 m/s, and it works with a Prusa-type slicer. A hardened steel nozzle of 50 HRC with a diameter of 0.5 mm was used. The weight of the specimens was determined using an analytical balance produced by Kern (Balingen, Germany) type ADB 200-4 with a resolution of 0.0001 g. A 100% infill for all the tested samples was considered. The build platform temperature was set to 35 °C.

The theoretical part weight was determined by calculating the theoretical volume based on the nominal dimensions of the ISO 527-2 1B specimen and after multiplying it with the density provided by the producers of Rosa Biowood filaments in the technical data sheet. The estimated theoretical weight was used to determine the weight error for the 3D-printed parts. The weight error was calculated as the difference between the theoretical and the measured weight and divided by the theoretical weight as follows

$$\varepsilon_w = \frac{(w_t - w)}{w_t} \cdot 100\ [\%]; \quad (1)$$

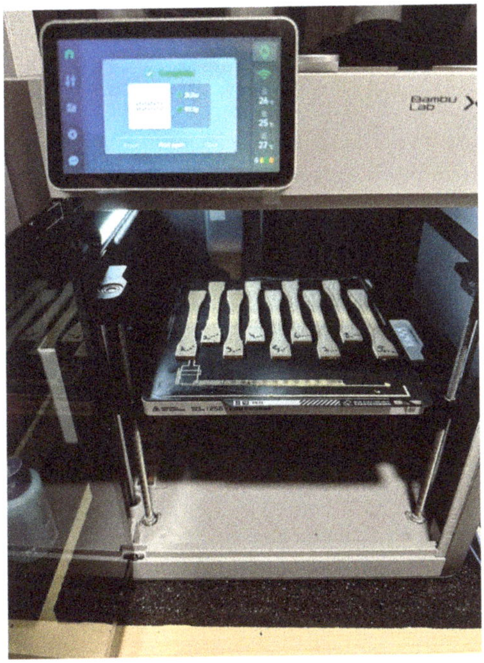

Figure 3. The Bambu Lab 3D FDM printer used in the experiments.

The tensile strength of the specimens was measured using experimental equipment (Figure 4) previously designed and executed within the Faculty of Mechanical Engineering, Automotive, and Robotics at the "Stefan cel Mare" University of Suceava, Romania. The tensile strength measuring device comprises specimen grips mounted on the crossheads of the tensile testing device body. The drive system controls the up or down motion of the moving crosshead. Sensors measure the specific elongation and traction force. After the amplifier amplifies the signal, the measuring results are introduced to a computer via a data acquisition device and processed by specialised software.

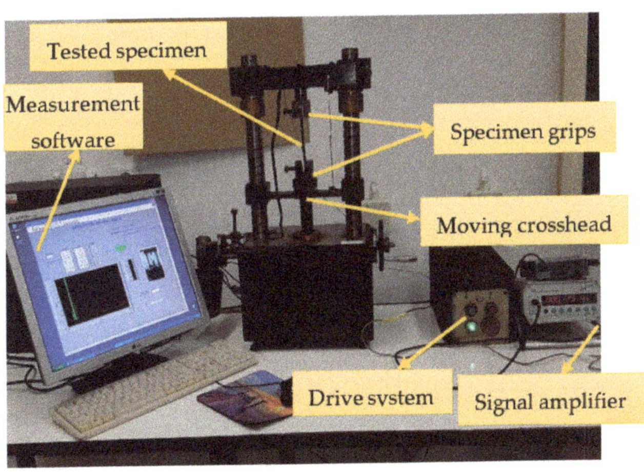

Figure 4. The ultimate tensile strength measuring device.

Fracture temperatures can be used to analyse the energy levels absorbed by the specimen material and the strain developed in the material before the rupture. The fracture temperature was measured using a high-speed thermal camera produced by Flir type X6540sc, produced by Teledyne FLIR (Wilsonville, OR, USA) (Figure 5), with an accuracy of ±1 °C/1%. The data provided by the camera were analysed and processed using Research IR specialized software. The maximum temperature before the rapture of the samples was retained and analysed in this study.

(**a**)

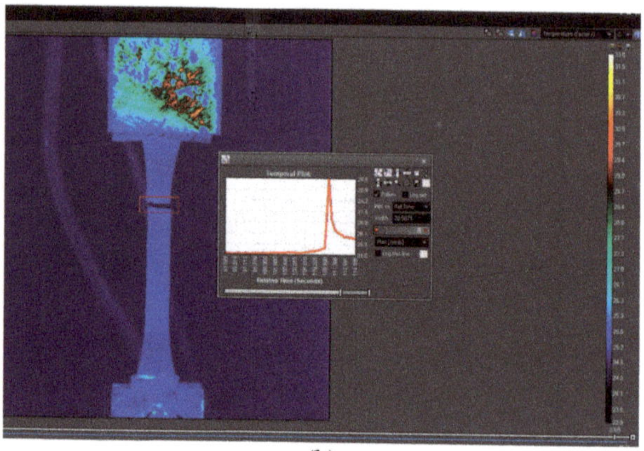

(**b**)

Figure 5. Fracture temperature measurement: (**a**) Flir X6540sc thermal camera; (**b**) fracture temperature analysis.

Surface quality was investigated by the surface area roughness parameter Sa (arithmetical mean height). According to ISO 25178, this parameter expresses the arithmetic mean of the height's absolute value from the surface's mean plane [35]. It is known that the most frequently used parameter for characterising the surface texture in a section through the machined surface of a part is the average arithmetic deviation Ra of the evaluated profile. In many situations, only values for the roughness parameter Ra are prescribed in part technical drawings. It is appreciated that, in relation to other roughness parameters, the Ra parameter provides the most information regarding the future operating behaviour of the surface it characterises. When the question arises of evaluating the roughness of a specific surface, the roughness parameter Sa has a similar meaning and importance to that of the roughness parameter Ra in the case of the profile of a surface in a certain section through the workpiece.

Because of the specific way the FDM printing processes are carried out, the printed parts' top surface and lateral surfaces will expose different surface textures. These textures are a result of how the melted material layers are deposited. That is why both surfaces were considered. The measurements were carried out on three different surface areas, and the average value was determined and used in the study.

Sa surface roughness values were obtained using the Mahr CWM 100 confocal microscope and white light interferometer, produced by Mahr GmbH, Gottingen, Germany (Figure 6a), and surface topography (Figure 6b) was analysed using the related Mahrsurf MfM software Version 7.4.8676.

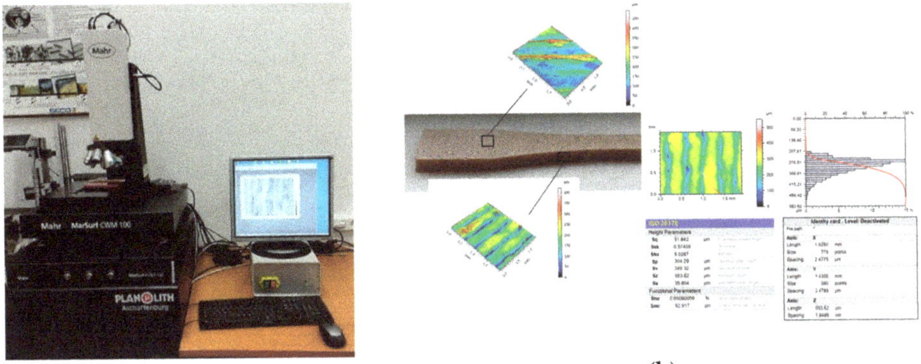

(a) (b)

Figure 6. Surface roughness measuring procedure: (**a**) Mahr CWM 100 interferometer; (**b**) top and lateral surface analysis.

In Figure 7, the end milling setup for the cutting machinability testing is presented. The machining tests were carried out on a Diy CNC router. The cutting forces' magnitude was measured using a Kistler dynamometer (produced by Kistler Group, Wien, Austria) type 9257B. The cutting parameters used for the machining tests were the following: cutting speed—150 m/min, cutting feed—800 mm/min, and depth of cut—ap = 1.5 mm. The cutting tool used was a two-flute end mill with a diameter of 3.17 mm made of ultrafine carbide Co10%, produced by Jiangsu Weixiang Tools Manufacturing Co., Ltd., Zhenjiang, China (Figure 7b). The obtained graphs for the cutting forces were processed and analysed in the related specialised software Dynoware version 3.3.1.0.

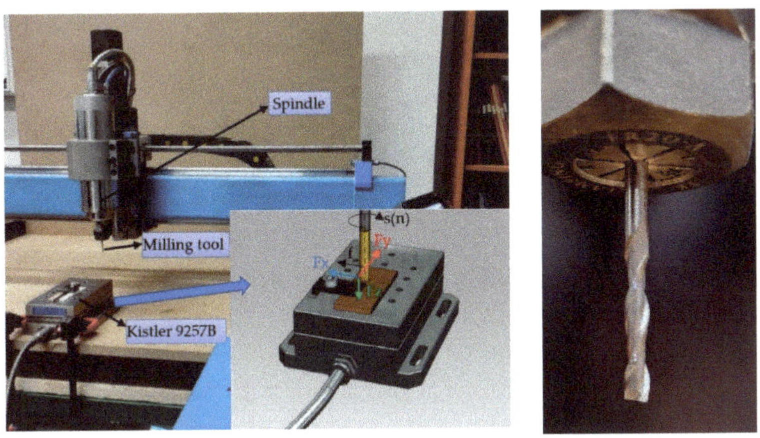

(a) (b)

Figure 7. Slot milling tests setup: (**a**) cutting force measurement; (**b**) the end-milling tool.

The experimental data obtained were analysed using a trial version of the DOE statistical software Minitab.

3. Results

The experimental results of the main output parameters investigated in this study are presented in Table 3.

Table 3. Testing conditions and experimental results.

	Printing Temperature Tp (°C)	Layer Height hl (mm)	Printing Speed sp (mm/s)	Weight Error ε_w (%)	Top Surface Roughness, Sa (μm)	Lateral Surface Roughness Sal (μm)	Ultimate Tensile Strength, UTM (MPa)	Fracture Temperature T (°C)
1	175	0.2	150	7.100%	14.995	38.161	7.5000	28.1
2			200	6.971%	14.549	35.6655	13.1250	26.7
3			300	6.948%	14.441	35.343	13.2787	29.6
4		0.28	150	6.071%	13.845	40.418	10.0000	27.3
5			200	6.108%	16.238	33.848	11.4062	27.5
6			300	6.056%	14.034	37.807	12.4992	28.1
7		0.4	150	7.328%	55.260	46.174	12.5000	27.2
8			200	7.403%	59.768	48.912	12.7840	28.7
9			300	7.509%	58.225	37.840	11.7187	26.9
10	190	0.2	150	7.262%	14.9585	45.719	9.2160	26.7
11			200	7.286%	16.599	36.305	12.3437	29.2
12			300	7.130%	15.560	36.400	15.3125	27.1
13		0.28	150	6.379%	62.972	44.391	13.1250	28.2
14			200	6.413%	67.723	35.031	15.7824	27.9
15			300	6.388%	67.476	31.971	15.4688	28.9
16		0.4	150	7.619%	113.845	36.211	9.0624	27.3
17			200	7.608%	118.236	43.638	11.5625	27.6
18			300	7.605%	109.852	52.905	11.8750	29.3
19	220	0.2	150	7.260%	64.635	38.328	12.0313	28
20			200	7.117%	64.556	42.606	11.4062	28.4
21			300	7.119%	82.446	43.400	10.7812	29.4
22		0.28	150	5.906%	106.59	36.131	12.0313	27.7
23			200	6.004%	112.97	54.213	13.2800	28.5
24			300	6.049%	72.425	36.364	13.4375	29
25		0.4	300	7.098%	131.640	44.051	14.2188	27.8
26			150	7.045%	122.390	57.922	17.1872	28.8
27			200	7.164%	149.720	39.632	19.0624	27.9

3.1. Part Weight and Weight Error

Figure 8 presents the main effects and interaction plots obtained for the part weight.

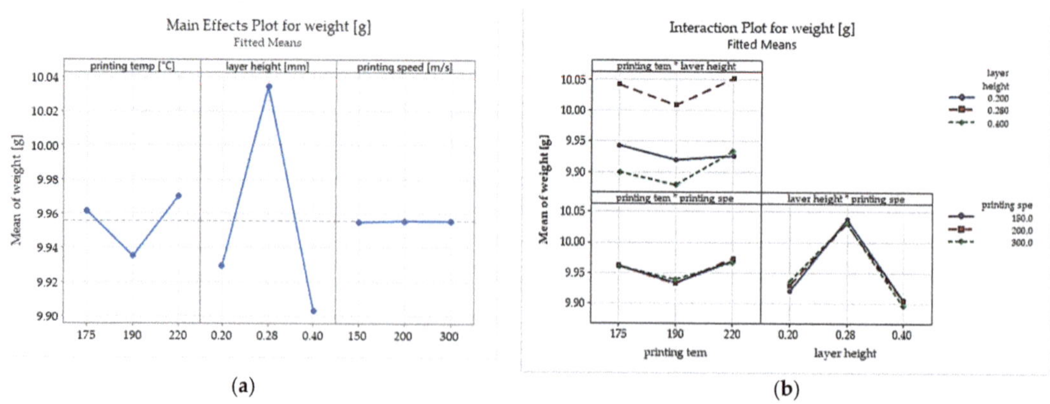

Figure 8. The influence of the selected input parameters on specimen weight: (a) the main effects plots; (b) interaction plots.

The results show that the specimens printed with layer heights of 0.28 mm have a significantly higher weight and a smaller weight error than those with layer heights of 0.2 mm and 0.4 mm.

An explanation could be that this is caused by how the Prusa slicer determines the extrusion width and, more precisely, the overlapping between the extrusion lines when the height of the part is not an integer multiple of the layer height value. The overlap factor greatly impacts the FDM parts' voids' volume, conducting denser structures and lower weight error for the FDM-printed parts. Besides the layer height, the printing temperature is another important factor strongly influencing the part weight and weight error. This parameter influences the printed material's thermal expansion, fluidity, layer adhesion, hardness, and tensile properties. Experimental results show a minimum weight error for the parts printed with a temperature of 220 °C which exceeds the range recommended by the Biowood filament producers. At lower printing temperatures, the material does reach the proper fluidity and causes bad adhesion and voids between the extrusion lines and layers.

Table 4 presents the analysis of variance (ANOVA) carried out to analyse how the selected input factors affect the experimental values obtained for the part weights. The test shows that with a 95% confidence interval, none of the inputs are statistically significant.

Table 4. Analysis of Variance for weight best fit regression.

Source	DF	Adj SS	Adj MS	F-Value	*p*-Value
Regression	3	0.008754	0.002918	0.77	0.521
printing temp (°C)	1	0.001036	0.001036	0.27	0.605
layer height (mm)	1	0.007717	0.007717	2.04	0.166
printing speed (m/s)	1	0.000001	0.000001	0.00	0.986
Error	23	0.086840	0.003776		
Total	26	0.095594			

3.2. *Sa Surface Roughness Parameter*

3.2.1. Roughness of the Top Surface of the Specimens

Figure 9 presents the influence exerted by the selected input parameters on Sa surface roughness of the top surface of the specimens. As it can be observed the printing temperature and layer height have a strong influence on the Sa surface roughness parameter variations.

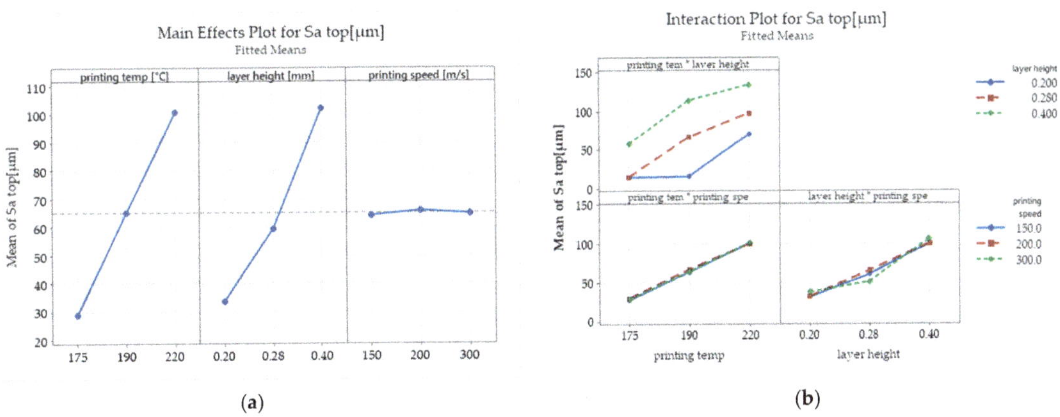

Figure 9. The influence exerted by the selected input parameters on Sa surface roughness of the top surface of the specimens: (**a**) the main effects plots; (**b**) interaction plots.

Figure 10 presents the isometric images of the surface topography of the top surfaces of the 3D-printed parts obtained using the Mahr CWM 100 white light interferometer and confocal microscope. It can be seen that higher printing temperatures result in better layer adhesion and fewer pores. Additionally, when higher temperatures and higher layer heights are used, the upper top surfaces of the specimen expose significantly higher surface asperities that result from over-extrusion and signalise a bad material flow.

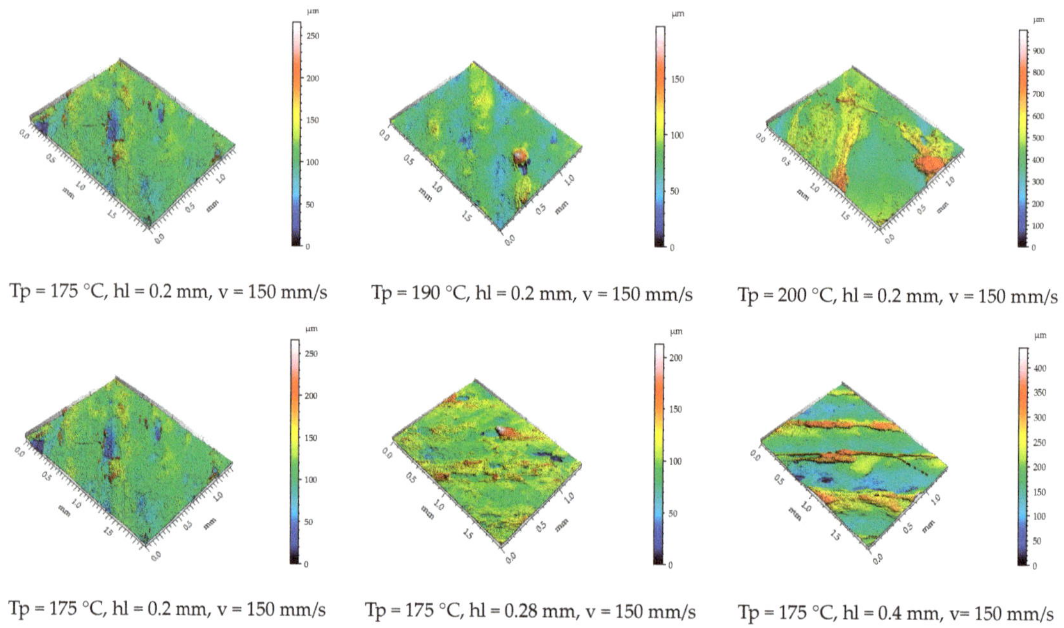

Figure 10. Isometric images top surface topography of the printed specimens.

The ANOVA test carried out for the top surface Sa roughness parameter is presented in Table 5. The test result indicates that all the input parameters investigated are statistically significant, with a reliability coefficient of 0.95.

Table 5. Analysis of Variance for Sa top best fit regression.

Source	DF	Adj SS	Adj MS	F-Value	*p*-Value
Regression	3	43,847.9	14,616.0	62.55	0.000
printing temp (°C)	1	22,324.0	22,324.0	95.53	0.000
layer height (mm)	1	21,523.4	21,523.4	92.11	0.000
printing speed (m/s)	1	0.5	0.5	0.00	0.964
Error	23	5374.7	233.7		
Total	26	49,222.6			

3.2.2. Roughness of the Lateral Surface of the Specimens

Even if in the scientific literature [16] it is stipulated that increasing the printing speed is chosen at the expense of lower surface quality, in this study, for the PCL wood-based biopolymer investigated, the results show contrary aspects (Figures 9 and 11). The printing speed exhibits a relatively low influence over the Sa surface roughness parameter measured for the top surface of the tested samples. Lower surface roughness values for the lateral surfaces of the samples were obtained for the parts printed with the highest level chosen for the printing speed sp = 300 mm/s. The arithmetical mean height roughness parameter increases with the increase in printing speed but tends to decrease after a certain value.

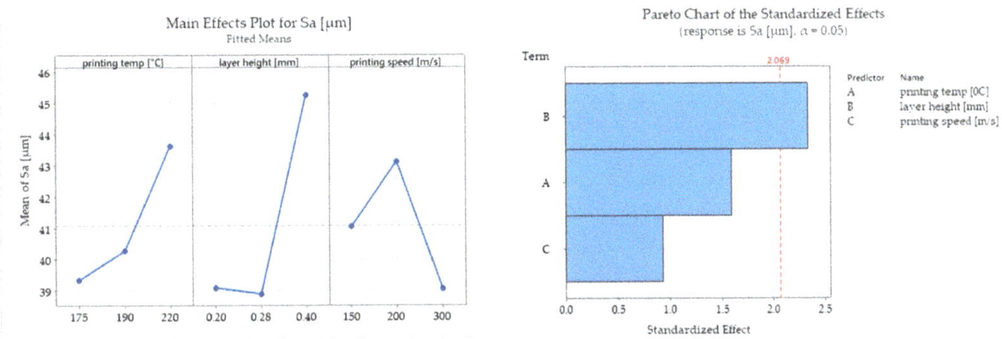

(**a**) (**b**)

Figure 11. Main effects plots for Sa roughness parameter measured on the lateral surface of the specimens: (**a**) main effects plots for Sa; (**b**) Pareto chart.

Figure 12 presents isometric images of the lateral surface texture of the specimens printed with a printing temperature of Tp = 190 °C and with a layer height of 0.4 mm at different printing speeds. The arithmetic means indicate that the height of the asperities is significantly lower when the highest level of the printing speed is adopted.

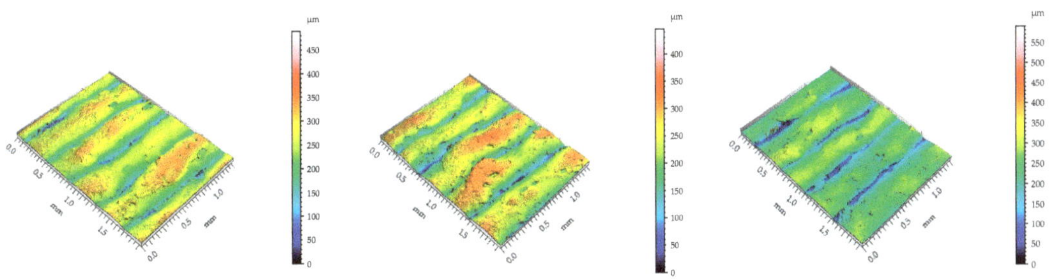

Tp = 190 °C, hl = 0.4 mm, sp = 150 m/s Tp = 190 °C, hl = 0.4 mm, sp = 200 m/s Tp = 190 °C, hl = 0.4 mm, sp = 300 m/s

Figure 12. Isometric images of lateral surface topography of the printed specimens.

These could be a result of the rapid cooling of the melted deposit layers due to the ventilation effect associated with the high velocity of the nozzle.

The ANOVA test carried out for the Sa roughness parameter measured for the lateral surfaces (Table 6) of the 3D-printed parts indicates that the layer height is statistically significant with a reliability coefficient of 0.95 (α = 0.05). This is also sustained by the Pareto graph (Figure 11b), which indicates that the interest parameter Sa variation is likely attributable to the layer height parameter variation.

Table 6. Analysis of Variance for Sa.

Source	DF	Adj SS	Adj MS	F-Value	*p*-Value
Regression	3	313.20	104.40	2.95	0.054
printing temp (°C)	1	89.57	89.57	2.53	0.125
layer height (mm)	1	192.69	192.69	5.44	0.029
printing speed (m/s)	1	30.95	30.95	0.87	0.360
Error	23	814.71	35.42		
Total	26	1127.91			

3.3. Ultimate Tensile Strength

Figure 13 presents images of the fracture surfaces obtained in the tensile strength tests. The fracture appearance presents different proportions of brittle or ductile failure modes. It could be observed that the specimens obtained at higher printing temperatures exposed higher percentages of ductile fracture. This means that by using higher printing temperatures, the parts will have more toughness.

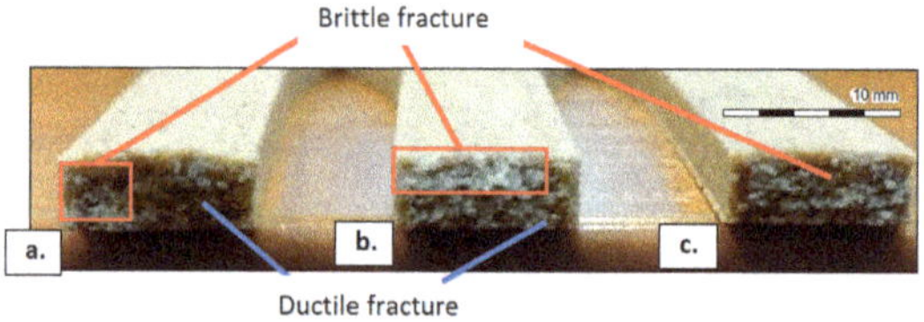

Figure 13. Fracture surfaces of the specimens after the tensile strength tests: (**a**) specimen printed at Tp = 175 °C with hl = 0.4 mm and sp = 200 mm/s; (**b**) specimen printed at Tp = 190 °C with hl = 0.4 mm and sp = 200 mm/s; (**c**) specimen printed at Tp = 220 °C with hl = 0.4 mm and sp = 200 mm/s.

This was also reflected in the ultimate tensile strength values obtained in the tensile strength tests that were carried out. The printing temperature exposed a significant influence on ultimate tensile strength values. In the Pareto chart (Figure 14b), the level of significance of each input factor chosen for this study can be analysed. The results show that among the studied factors, the printing speed and the interaction between the printing temperature and layer height are statistically significant at the 0.05 level.

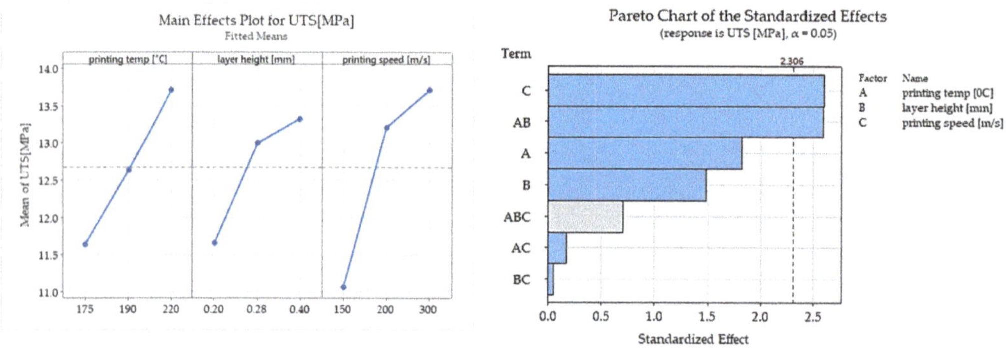

Figure 14. The influence of the selected input parameters over the ultimate tensile strength: (**a**) the main effects plots; (**b**) Pareto chart for the significance of the studied input parameters.

This result is also sustained by the analysis of variance carried out (Table 7). According to the ANOVA test, the printing speed is statistically significant for the ultimate tensile strength variation with a reliability coefficient of 0.95.

In this study, significantly higher printing speeds than those usually reported as being studied in the scientific literature (range 15–170 m/s) [1,3] were used. Higher printing speeds can prevent the alteration of biocomponents of the filaments due to intense exposure to high temperatures. Even if it is a general belief that higher printing speeds conduct weaker structures due to insufficient cooling time between layers and bad layer adhesion, the results obtained in this study indicate that a higher printing speed significantly increases the ultimate tensile strength and the fracture temperature of the printed parts (Figures 14 and 15).

Table 7. Analysis of Variance for UTS (MPa).

Source	DF	Adj SS	Adj MS	F-Value	p-Value
Regression	3	56.29	18.763	4.19	0.017
printing temp (°C)	1	18.75	18.748	4.19	0.052
layer height (mm)	1	11.32	11.321	2.53	0.125
printing speed (m/s)	1	26.22	26.220	5.86	0.024
Error	23	102.95	4.476		
Total	26	159.24			

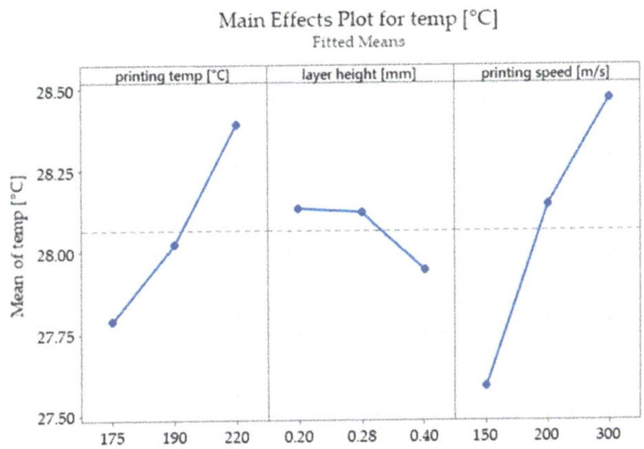

Figure 15. The main effects plots for the fracture temperature.

3.4. 3D-printed Parts Machinability

The cutting machinability of Biowood Rosa3D printed parts was also investigated. Slot milling tests were carried out, and the cutting force values obtained were compared with the ones achieved by machining in identical condition pinewood and beech wood samples. Pine and beech wood were selected as representatives for the soft and hard wood categories. A measurement of the cutting force components for slot milling operations of some samples from three distinct wooden materials was carried out, one of which was the biowood. The tests were carried out on a three-axis DIY milling router-type machine tool, using a two-flute tungsten carbide end mill type 10113117 produced by Weix tools, China. The geometry of the active zone of the end mill is typical for wood bits. As for cutting conditions, the following values were chosen: ap = 1.5 mm for depth of cut, f = 800 mm/min for cutting feed, and vc = 150 m/min for cutting speed.

By machining the samples obtained by FDM 3D printing of Biowood Rosa filaments, significantly higher cutting forces were obtained (Figure 16). The average cutting forces generated by machining Biowood Rosa samples were up to 10× higher than those obtained by end-milling softwood samples and up to 2.5× higher than those obtained for the hardwood samples.

Figure 17 presents the main influence of the 3D printing input parameters analysed in the study over the cutting force components. The printing temperature and layer height positively affect the cutting force components' magnitude. Printing speed negatively influences the machinability of Biowood Rosa parts according to the force-cutting criteria. Even if the material becomes more ductile because of the exposure to high temperatures and therefore requires higher efforts to be machined, at high printing temperatures, over-extrusion phenomena could be observed by analysing the top surface topography of the specimens. This phenomenon can lead to weak structural bonds and is conducive to lower values for the cutting forces.

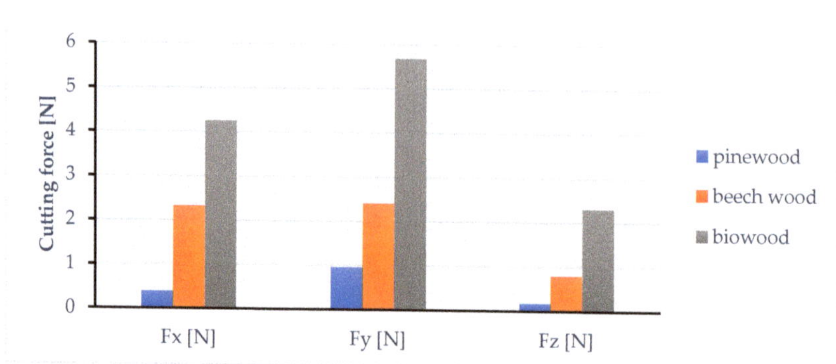

Figure 16. Cutting force components comparison between pinewood, beech wood, and Biowood parts generated during slot milling with cutting speeds of 150 m/min, cutting feeds of 800 mm/min, and cutting depth of 1.5 mm.

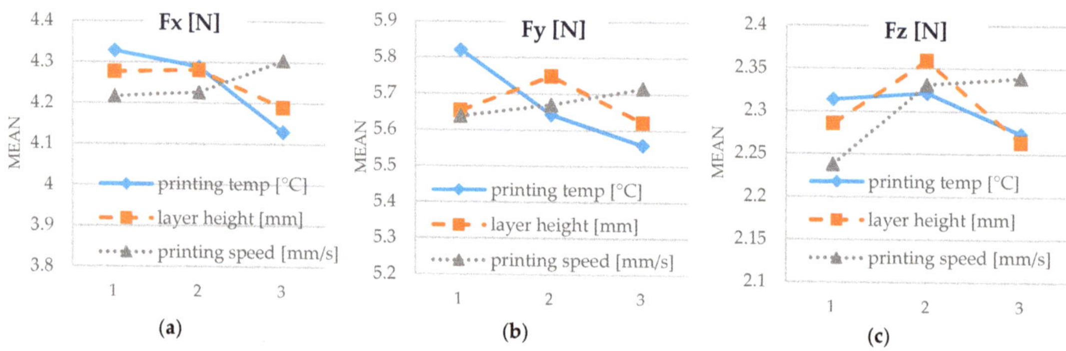

Figure 17. Mean effects plots for the cutting forces' components. (**a**) main effect plot for Fx [N]; (**b**) main effect plot for Fy[N]; (**c**) main effect plot for Fz[N].

3.5. Nonlinear Regression Analysis

Through the mathematical processing of the experimental results, it became possible to determine some empirical power, function-type mathematical models. With these empirical mathematical models, additional information was obtained regarding the order of influence and the intensity of the influence exerted by some factors on the output parameters of the investigated process. Microsoft Excel software was used for the mathematical processing of the experimental results. In this way, the following empirical mathematical models were obtained:

- For the lateral surface roughness (standard error of the regression S = 6.0487, correlation coefficient R = 0.5039):

$$Sa = 7.3154 \cdot Tp^{0.4558} hl^{0.2210} sp^{-0.0747} \ [\mu m]; \quad (2)$$

- For the ultimate tensile strength (standard error of the regression S = 2.076, correlation coefficient R = 0.6140):

$$UTS = 0.08885 \ T^{0.7137} hl^{0.1974} sp^{0.2708} \ [MPa]; \quad (3)$$

- For the part weight (standard error of the regression S = 0.0629, correlation coefficient R = 0.2167):

$$W = 9.5844 \cdot T^{0.06157} hl-^{0.004079} sp^{0.000083}\ [g]; \quad (4)$$

Figure 18 shows normally distributed data for the regression equations determined for the lateral surface roughness of the printed parts and the ultimate tensile strength interest parameters. Additionally, the distances between the residuals versus their expected values for the regressions are relatively small. The parameter estimation errors are presented in Tables 8–10. The small values of the coefficient standard error (SE) indicate a precise estimation.

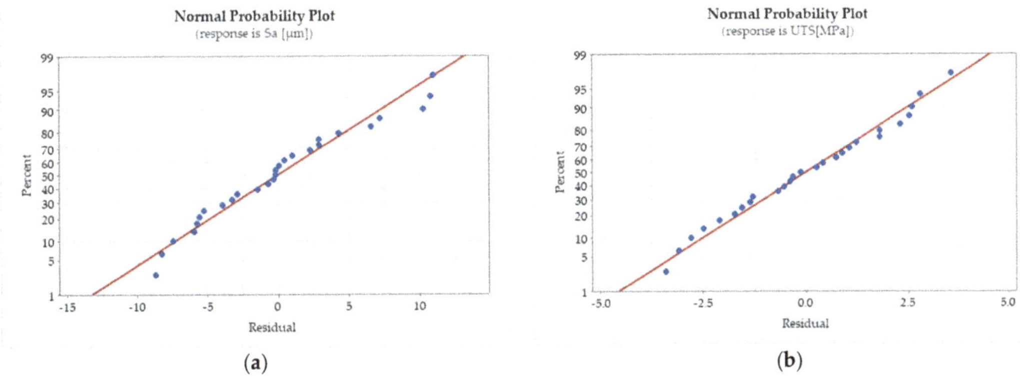

Figure 18. Normal probability plots for the nonlinear regressions (**a**) for the Sa roughness parameter measured on the lateral surfaces of the printed part and (**b**) for the ultimate tensile strength.

Table 8. The parameter estimation errors for Equation (2).

Parameter	Estimate	SE Estimate
×1	7.31547	12.0722
×2	0.45580	0.2951
×3	0.22107	0.1002
×4	−0.07472	0.0998

Table 9. The parameter estimation errors for Equation (3).

Parameter	Estimate	SE Estimate
×1	0.088580	0.161801
×2	0.713767	0.326272
×3	0.197467	0.111027
×4	0.270825	0.109504

Table 10. The parameter estimation errors for Equation (4).

Parameter	Estimate	SE Estimate
×1	9.58440	0.686965
×2	0.00615	0.012852
×3	−0.00408	0.004299
×4	0.00008	0.004279

By examining the mathematical model corresponding to the lateral surface, Sa roughness parameter, it could be seen that these parameters will register an increase when the printing temperature TP and layer height hl increase and decrease with the increase in the printing speed sp. The printing temperature Tp exerts the most substantial influence on the Sa parameter, which, in the empirical mathematical model, corresponds to the highest value of the corresponding exponent compared to the values of the exponents attached to the rest of the analysed process input factors.

It should be noted that increasing the value of any of the three factors considered will increase the ultimate tensile strength UTS because the values of all exponents are positive. The printing temperature Tp is also the input factor with the strongest influence in the ultimate tensile strength UTS because, in this case, the value of the exponent attached to this factor also has the maximum value of the values of the exponents of the other input factors studied. An explanation of the increase in the value of the UTS parameter when increasing the printing temperature Tp could result from better adhesion of the deposited layers due to the higher values of the printing temperature.

The three input factors have a relatively small influence on the weight output parameter. This finding is based on the very low values of the exponents obtained for the input factors in the empirical mathematical model corresponding to the parameter W. However, it can be observed that in this case, the strongest influence also seems to be exerted by the printing temperature Tp, whose exponent has the maximum value.

4. Discussion

Even when 100% infill was set as the printing condition for the specimen manufacturing process, the resulting parts' weight was smaller than the theoretical weight (determined by the theoretical volume and the material density value provided by the filament producers). The weight error calculated for the specimens ranged between 5.9–7.6%.

The surface roughness parameter Sa measured on the top surfaces of the 3D-printed samples ranged between 13.8 and 149.7 (μm). For this output parameter, all of the considered input factors were reported as statistically significant according to the ANOVA test carried out. Even if the printing temperature levels tested in this study did not exceed the temperature range recommended by the filament producer, the area surface roughness parameter Sa of the top surfaces of the samples printed at 190 °C recorded an average increase of 152%, and those printed at 220 °C showed an average increase of 347% compared with those printed with a temperature of 175 °C. The layer height parameter also exposed a similar influence on the Sa roughness of the top surfaces of the printed parts. The roughness parameter had an average increase of 176% when the layer height was set at 0.28 mm and an average increase of 303% when a layer height of 0.4 mm was adopted compared with that resulting from parts printed with a layer height of 0.2 mm. These variations result due to over-extrusion caused by inefficient flow rates.

The surface roughness parameter Sa values measured on the lateral surfaces of the printed specimens ranged between 35.34 and 57.92 (μm). According to the main effect plots (Figure 11a), the printing conditions that assured a better surface roughness were the lowest value of the printing temperature (175°), the layer height of 0.28 mm, and the maximum value of the printing speed (300 m/s). Among the input factors investigated, only the layer height tested as statistically significant according to the ANOVA test (Table 6).

The ultimate tensile strength values obtained were in the range of 7.5–19.06 MPa. Significant correlations were found between printing speed, mechanical strength (ultimate tensile strength), printing temperature–layer height interaction, and mechanical strength.

The machinability was investigated using cutting forces criteria. Machinability is rated relative to the results achieved for a representative/reference material. To evaluate the machinability of Biowood printed parts, machining tests were carried out in similar conditions for pinewood and beech wood as representatives of softwood and hardwood materials. The average cutting forces generated by machining Biowood Rosa samples were up to 10× higher than those obtained by end-milling softwood samples and up to 2.5× higher than those obtained for the hardwood samples. The printing temperature and layer height tend to positively affect the cutting force components' magnitude, while printing speed negatively influences the machinability of Biowood Rosa printed parts.

5. Conclusions

Biopolymers are a natural alternative to synthetic polymers that exhibit reduced carbon dioxide emissions in their synthesis. In recent decades, more and more emphasis has been placed on using biopolymers for various medical, food, and industrial applications.

Few studies have been carried out on testing the capabilities of wood biopolymer composites. Most of these studies usually address only the mechanical properties of WPC. The present research explores the effect of printing temperature, layer height, and printing speed on surface quality, tensile performance, and cutting machinability of parts obtained by FDM printing of Rosa3D Biowood filament. Biowood produced by Rosa3D is a wood-based composite biopolymer obtained by amalgamating wood fibres in a polycaprolactone PCL and polyester polymeric matrix and adding fillers (starch, lignin) and additives (natural resins, waxes, and oils, natural fatty acids) to the mix.

The novelty of this study consists in exploring some of the qualitative aspects and mechanical proprieties of the PCL wood-based biopolymer parts generated by FDM with different printing conditions. The printing parameters varied in the experimental study were the printing temperature, the layer height, and the printing speed. The printing speed levels selected for the experiment were significantly superior to those usually tested in previous research in this field or those recommended by the filament producer.

The surface roughness of the parts was investigated. Higher Sa (arithmetical mean height) values were obtained when high printing temperatures and layer height were used. The surface texture obtained for these specific printing conditions exhibits signs of over-extrusion. Overall, the highest level chosen for the printing speed positively influenced the surface roughness.

Another novelty aspect of this study is the cutting machinability as a secondary machining operation of the FDM-printed wood-based composite biopolymer. Machinability is the property that characterizes the ease with which a material can be machined with a cutting tool. The machinability testing criteria used in this study was the cutting force components' magnitude. The Biowood Rosa3D printed parts exhibit poor machinability in reference to natural wood parts (pinewood and beech wood). Therefore, lower cutting forces and better cutting machinability were obtained for the parts printed at higher printing temperatures and layer heights and with lower printing speeds.

Testing the capabilities of newly developed polymer composites, especially biopolymer composites, should be a constant concern for researchers to achieve competitive products for the industry. Many drawbacks of FDM 3D printing of wood-based biopolymers could be overcome by carefully choosing the processing parameters.

Author Contributions: Conceptualization I.B.-B.; methodology, I.B.-B. and I.T.; software I.B.-B.; formal analysis, I.B.-B., I.T. and L.S.; investigation, I.B.-B. and I.T.; writing—original draft preparation, I.B.-B., I.T. and L.S.; writing—review and editing, I.B.-B. and L.S.; supervision, L.S. All authors have read and agreed to the published version of the manuscript.

Funding: This research received no external funding.

Institutional Review Board Statement: Not applicable.

Data Availability Statement: Some or all data, models, or code generated or used during the study are available from the corresponding author by request.

Acknowledgments: The work of the author Ioan Tamașag was supported by the project "PROIN-VENT", Contract no. 62487/03.06.2022-POCU/993/6/13-Code 153299, financed by The Human Capital Operational Programme 2014–2020 (POCU), Romania.

Conflicts of Interest: The authors declare no conflict of interest.

References

1. Wasti, S.; Adhikari, S. Use of biomaterials for 3D printing by fused deposition modeling technique: A review. *Front. Chem.* **2020**, *8*, 315. [CrossRef]
2. *ISO/ASTM 52900:2021*; Additive Manufacturing—General Principles—Fundamentals and Vocabulary. ISO/ASTM International: Geneva, Switzerland, 2021.
3. Rahmatabadi, D.; Ghasemi, I.; Baniassadi, M.; Abrinia, K.; Baghani, M. 3D Printing of PLA-TPU with Different Component Ratios: Fracture Toughness, Mechanical Properties, and Morphology. *J. Mater. Res. Technol.* **2022**, *21*, 3970–3981. [CrossRef]
4. Trzaskowski, M.; Mańka-Malara, K.; Szczesio-Włodarczyk, A.; Sokołowski, J.; Kostrzewa-Janicka, J.; Mierzwińska-Nastalska, E. Evaluation of Mechanical Properties of 3D-Printed Polymeric Materials for Possible Application in Mouthguards. *Polymers* **2023**, *15*, 898. [CrossRef]
5. Moradi, M.; Aminzadeh, A.; Rahmatabadi, D.; Hakimi, A. Experimental Investigation on Mechanical Characterization of 3D Printed PLA Produced by Fused Deposition Modeling (FDM). *Mater. Res. Express* **2021**, *8*, 035304. [CrossRef]
6. Butt, J.; Bhaskar, R. Investigating the Effects of Annealing on the Mechanical Properties of FFF-Printed Thermoplastics. *JMMP* **2020**, *4*, 38. [CrossRef]
7. Rahmatabadi, D.; Aberoumand, M.; Soltanmohammadi, K.; Soleyman, E.; Ghasemi, I.; Baniassadi, M.; Abrinia, K.; Bodaghi, M.; Baghani, M. 4D Printing-Encapsulated Polycaprolactone–Thermoplastic Polyurethane with High Shape Memory Performances. *Adv. Eng. Mater.* **2023**, *25*, 2201309. [CrossRef]
8. Beniak, J.; Holdy, M.; Križan, P.; Matúš, M. Research on Parameters Optimization for the Additive Manufacturing Process. *Transp. Res. Procedia* **2019**, *40*, 144–149. [CrossRef]
9. Shbanah, M.; Jordanov, M.; Nyikes, Z.; Tóth, L.; Kovács, T.A. The Effect of Heat Treatment on a 3D-Printed PLA Polymer's Mechanical Properties. *Polymers* **2023**, *15*, 1587. [CrossRef]
10. Tamașag, I.; Suciu, C.; Beșliu-Băncescu, I.; Dulucheanu, C.; Cerlincă, D.-A. Experimental Study on the Possibilities of FDM Direct Colour Printing and Its Implications on Mechanical Properties and Surface Quality of the Resulting Parts. *Polymers* **2022**, *14*, 5173. [CrossRef] [PubMed]
11. Sedlak, J.; Joska, Z.; Hrbackova, L.; Jurickova, E.; Hrusecka, D.; Horak, O. Determination of Mechanical Properties of Plastic Components Made by 3D Printing. *Manuf. Technol.* **2023**, *22*, 733–746. [CrossRef]
12. Bakhtiari, H.; Aamir, M.; Tolouei-Rad, M. Effect of 3D Printing Parameters on the Fatigue Properties of Parts Manufactured by Fused Filament Fabrication: A Review. *Appl. Sci.* **2023**, *13*, 904. [CrossRef]
13. Mazzanti, V.; Malagutti, L.; Mollica, F. FDM 3D Printing of Polymers Containing Natural Fillers: A Review of Their Mechanical Properties. *Polymers* **2019**, *11*, 1094. [CrossRef] [PubMed]
14. Wickramasinghe, S.; Do, T.; Tran, P. FDM-Based 3D Printing of Polymer and Associated Composite: A Review on Mechanical Properties, Defects and Treatments. *Polymers* **2020**, *12*, 1529. [CrossRef]
15. Zarna, C.; Chinga-Carrasco, G.; Echtermeyer, A.T. Bending properties and numerical modelling of cellular panels manufactured from wood fibre/PLA biocomposite by 3D printing. *Compos. Part A Appl.* **2023**, *165*, 107368. [CrossRef]
16. Burgert, I.; Keplinger, T.; Cabane, E.; Merk, V.; Rüggeberg, M. Chapter 13—Biomaterial Wood: Wood-Based and Bioinspired Materials. In *Secondary Xylem Biology*; Kim, Y.S., Funada, R., Singh, A.P., Eds.; Academic Press: Cambridge, MA, USA, 2016; pp. 259–281. ISBN 9780128021859. [CrossRef]
17. Babu, R.P.; O'Connor, K.; Seeram, R. Current progress on bio-based polymers and their future trends. *Prog. Biomater.* **2013**, *2*, 8. [CrossRef] [PubMed]
18. Ramesh, M.; Rajeshkumar, L.; Sasikala, G.; Balaji, D.; Saravanakumar, A.; Bhuvaneswari, V.; Bhoopathi, R. A critical review on wood-based polymer composites: Processing, properties, and prospects. *Polymers* **2022**, *14*, 589. [CrossRef]
19. Mandala, R.; Bannoth, A.P.; Akella, S.; Rangari, V.K.; Kodali, D. A short review on fused deposition modeling 3D printing of bio-based polymer nanocomposites. *J. Appl. Polym. Sci.* **2022**, *139*, 51904. [CrossRef]
20. Herrera, N.; Olsén, P.; Berglund, L.A. Strongly improved mechanical properties of thermoplastic biocomposites by PCL grafting inside holocellulose wood fibers. *ACS Sustain. Chem. Eng.* **2020**, *8*, 11977–11985. [CrossRef]
21. Kariz, M.; Sernek, M.; Obućina, M.; Kuzman, M.K. Effect of wood content in FDM filament on properties of 3D printed parts. *Mater. Today Commun.* **2018**, *14*, 135–140. [CrossRef]
22. Le Duigou, A.; Castro, M.; Bevan, R.; Martin, N. 3D printing of wood fibre biocomposites: From mechanical to actuation functionality. *Mater. Des.* **2016**, *96*, 106–114. [CrossRef]
23. Zgodavová, K.; Lengyelová, K.; Bober, P.; Eguren, J.A.; Moreno, A. 3D printing optimization for environmental sustainability: Experimenting with materials of protective face shield frames. *Materials* **2021**, *14*, 6595. [CrossRef]
24. Sabo, R.; Jin, L.; Stark, N.; Ibach, R.E. Effect of environmental conditions on the mechanical properties and fungal degradation of polycaprolactone/ microcrystalline cellulose/wood flour composites. *BioResources* **2013**, *8*, 3322–3335. [CrossRef]
25. Jo, J.; Kim, H.; Jeong, S.-Y.; Park, C.; Hwang, H.S.; Koo, B. Changes in mechanical properties of polyhydroxyalkanoate with double silanized cellulose nanocrystals using different organosiloxanes. *Nanomaterials* **2021**, *11*, 1542. [CrossRef]
26. Bouafif, H.; Koubaa, A.; Perré, P.; Cloutier, A. Effects of fiber characteristics on the physical and mechanical properties of wood plastic composites. *Compos. Part A Appl.* **2009**, *40*, 1975–1981. [CrossRef]
27. Gadhave, R.V.; Das, A.; Mahanwar, P.A.; Gadekar, P.T. Starch Based Bio-Plastics: The Future of Sustainable Packaging. *Open J. Polym. Chem.* **2018**, *8*, 21–33. [CrossRef]

28. Mazur, K.E.; Borucka, A.; Kaczor, P.; Gądek, S.; Bogucki, R.; Mirzewiński, D.; Kuciel, S. Mechanical, Thermal and Microstructural Characteristic of 3D Printed Polylactide Composites with Natural Fibers: Wood, Bamboo and Cork. *J. Polym. Environ.* **2022**, *30*, 2341–2354. [CrossRef]
29. Migneault, S.; Koubaa, A.; Perré, P. Effect of Fiber Origin, Proportion, and Chemical Composition on the Mechanical and Physical Properties of Wood-Plastic Composites. *J. Wood Chem. Technol.* **2014**, *34*, 241–261. [CrossRef]
30. Huang, Y.; Löschke, S.; Proust, G. In the mix: The effect of wood composition on the 3D printability and mechanical performance of wood-plastic composites. *Compos. Part C Open Access* **2021**, *5*, 100140. [CrossRef]
31. Zhu, Z.; Buck, D.; Wang, J.; Wu, Z.; Xu, W.; Guo, X. Machinability of different wood-plastic composites during peripheral milling. *Materials* **2022**, *15*, 1303. [CrossRef] [PubMed]
32. Wu, Z.; Zhang, F.; Hu, Y.; Zhu, Z.; Guo, X. Study on helical milling performance of wood-plastic composites. *CIRP J. Manuf. Sci. Technol.* **2022**, *37*, 143–154. [CrossRef]
33. BioWOOD—Compostable Filament with Wood Fibers. Available online: https://rosa3d.pl/en/biowood-compostable-filament-with-wood-fibers/ (accessed on 20 March 2023).
34. 3D Filament BIOWood Technical Data Sheet. Available online: https://sernia.ru/upload/pdf_files/Introduction%20to%20surface%20roughness%20measurement.pdf (accessed on 20 March 2023).
35. *ISO 25178-2:2021*; Geometrical Product Specifications (GPS)—Surface Texture: Areal—Part 2: Terms, Definitions and Surface Texture Parameters. International Organization for Standardization: Geneva, Switzerland, 2021.

polymers MDPI

Article

Influence of Three-Dimensional Printing Parameters on Compressive Properties and Surface Smoothness of Polylactic Acid Specimens

Hamed Bakhtiari [1,*], Mostafa Nikzad [2] and Majid Tolouei-Rad [1,*]

[1] School of Engineering, Edith Cowan University, Joondalup, WA 6027, Australia
[2] Department of Mechanical and Product Design Engineering, School of Engineering, Swinburne University of Technology, Hawthorn, VIC 3122, Australia; mnikzad@swin.edu.au
* Correspondence: h.bakhtiari@ecu.edu.au (H.B.); m.rad@ecu.edu.au (M.T.-R.); Tel.: +61-8-6304-5883 (M.T.-R.)

Abstract: While the mechanical performance of fused filament fabrication (FFF) parts has been extensively studied in terms of the tensile and bending strength, limited research accounts for their compressive performance. This study investigates the effect of four process parameters (layer height, extrusion width, nozzle temperature, and printing speed) on the compressive properties and surface smoothness of FFF parts made of Polylactic Acid (PLA). The orthogonal Taguchi method was employed for designing the experiments. The surface roughness and compressive properties of the specimens were then measured and optimized using the analysis of variance (ANOVA). A microscopic analysis was also performed to identify the failure mechanism under static compression. The results indicated that the layer height had the most significant influence on all studied properties, followed by the print speed in the case of compressive modulus, hysteresis loss, and residual strain; extrusion width in the case of compressive strength and specific strength; and nozzle temperature in the case of toughness and failure strain. The optimal design for both high compressive properties and surface smoothness were determined as a 0.05 mm layer height, 0.65 mm extrusion width, 205 °C nozzle temperature, and 70 mm/s print speed. The main failure mechanism observed by SEM analysis was delamination between layers, occurring at highly stressed points near the stitch line of the PLA prints.

Keywords: additive manufacturing; process optimization; compressive properties; surface smoothness; fused filament fabrication

Citation: Bakhtiari, H.; Nikzad, M.; Tolouei-Rad, M. Influence of Three-Dimensional Printing Parameters on Compressive Properties and Surface Smoothness of Polylactic Acid Specimens. *Polymers* **2023**, *15*, 3827. https://doi.org/10.3390/polym15183827

Academic Editors: Cristina-Elisabeta Pelin and Anton Ficai

Received: 18 August 2023
Revised: 14 September 2023
Accepted: 16 September 2023
Published: 19 September 2023

1. Introduction

Material extrusion (ME) refers to additive manufacturing techniques that involve the layer-by-layer extrusion of molten or semi-liquid materials to create plastic, metal, or composite parts [1]. FFF (fused filament fabrication), also known as FDM (Fused Deposition Modeling), is an ME technique that utilizes a heated nozzle to melt and deposit materials onto a build platform. FFF-printed parts have found extensive applications as biomedical scaffolds [2], composite structures [3], shape memory components [3], and functional prototypes in various industries, such as medical, automotive, and electronics [4–6]. A schematic of an FFF machine and the resulting printed part is depicted in Figure 1. The print head can move freely in the XY plane while the height is adjusted by the movement of the build platform in the Z direction. In the deposition stage, the new layer is bonded to the successive layer through the fusion mechanism [7].

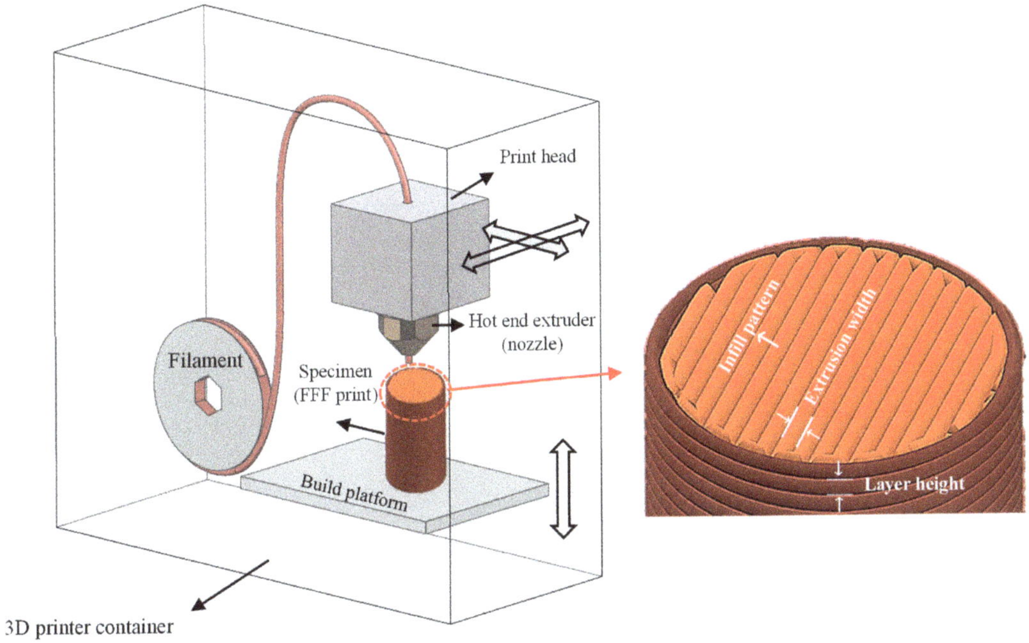

Figure 1. Schematic illustration of the FFF process, and some geometrical features of an FFF print.

The compressive strength and surface finish in fused filament fabrication (FFF) parts play critical roles in various demanding applications. Notably, FFF finds application in metal forming and molding tools [8–10], jigs and fixtures [11,12], full-scale molds for wind turbine blades [13], packaging products [14,15], bone scaffolds [2], and medical implants [16], where compressive strength is essential to withstand compressive stresses. The surface finish is equally crucial in such applications. For instance, in 3D printing die tools for sheet metal drawing, the surface roughness influences the friction between the die and sheet metal, affecting the quality of the drawn metal. Similarly, in the automotive industry, strict dimensional requirements necessitate a smooth surface finish for body panels to ensure a proper fit and alignment, highlighting the significance of surface quality in precise part manufacturing.

The mechanical strength of the FFF parts (also referred to as "print" in this paper) is greatly influenced by the process and geometrical parameters [17,18]. Several studies in the literature have investigated the effect of process parameters on the mechanical performance of FFF parts. Cojocaru et al. [19] conducted a review study on the effect of process parameters on the mechanical properties of Polylactic Acid (PLA) prints, concluding that thinner layers resulted in smaller internal cavities and improved surface quality, layer bonding, and mechanical properties. It was also shown that mechanical properties can be improved when a grid infill pattern and horizontal build direction were used in the 3D printing of specimens. However, high printing speeds can negatively impact surface quality due to incomplete solidification, and low or high nozzle temperatures can cause incomplete melting or material flow issues. Process parameters such as the nozzle temperature and print speed impact the temperature profile (heating and cooling cycles), subsequently affecting the bonding strength between layers [20,21].

The effect of the process and geometrical parameters on the mechanical behavior of prints can vary based on the type of loading. For example, research has shown that decreasing the raster angle improves the tensile strength of a print [22–25] but decreases its compressive strength [25,26]. This is while no clear correlation has been found between the raster angle and fatigue life of prints [27]. This discrepancy stems from the different failure modes under different loads. In tension, raster breaking is the main failure mode, so the highest load-carrying capacity occurs when the extruded filaments are aligned with the loading direction. On the other hand, parts are more likely to buckle under compressive force, so having the filaments arranged perpendicular to the applied load leads to the best compressive strength results [26]. Gordelier et al. [28] found that the optimal tensile strength in ABS and PLA prints was attained with a raster angle of 0°, minimum layer height, and horizontal build direction. Bakhtiari et al. [27] studied the influence of printing parameters on the fatigue performance of FFF prints. They found that cross-over infill patterns like the grid and honeycomb outperformed unidirectional raster patterns in tensile fatigue strength. Notably, the 0° and 45° raster angles exhibited the highest fatigue strength, while the printing speed (25–70 mm/s) had a comparatively weaker influence, inversely affecting the tensile fatigue life.

Although the tensile and bending strength of FFF prints has been extensively studied in the literature, there is a limited account of their compressive performance. Gabor et al. [29] investigated the compressive strength of PLA prints at various build orientations (0°, 15°, 30°, 45°, 60°, and 90°) and found that a 0° orientation yielded the highest strength. In another study [30], the build orientation was found to have the biggest influence on the compressive strength of ABS prints followed by the layer height, raster angle, and air gap, while the extrusion width exhibited an insignificant impact. Compressive strength was directly correlated with layer height and extrusion width but inversely with build orientation and air gap. Dave et al. [31] examined the compressive strength of PLA prints at different layer heights (0.1–0.3 mm), infill densities (60–80%), and print speeds (30–50 mm/min). The findings indicated that the compressive strength peaked at a layer height of 0.2 mm before declining. They also demonstrated that compressive strength directly correlated with infill density due to the reduced interlayer cavities and greater material volume at higher infill densities, which enhanced the mechanical support. Print speed had a minor effect, showing a slight increase up to 40 mm/min.

Infill patterns significantly impact print strength by influencing the contact area and layer bonding. Figure 2 depicts various infill patterns studied for the compressive behavior of FFF prints. Prajapati et al. [32] examined PLA prints with varying infill patterns (rectilinear, concentric, and Hilbert curve) with the highest compression strength achieved using the rectilinear pattern. In [33], PLA prints with a triangular infill design demonstrated a superior compressive strength than the grid, quarter cubic, and tri-hexagon patterns due to having a larger contact area between the layers. Similarly, the Hilbert curve pattern exhibited a superior compressive strength than the other designs (honeycomb, line, rectilinear, Archimedean curve, and Octagram spiral) due to the strong bonding between the rasters and layers [34].

Material choice significantly affects the compressive strength in FFF prints due to its rheological and microstructural characteristics. ABS and nylon, for example, exhibit increased compressive strength as the layer height decreases [35,36]. Reducing the layer height leads to an increased shear rate within the extruding polymer, causing a viscosity drop in materials with shear-thinning properties [37], such as ABS and nylon [38,39]. This, in turn, enhances the layer fusion and overall strength. In contrast, PEEK is a non-shear-thinning polymer, resulting in the layer height having minimal influence on the compressive strength [40]. In addition, ABS is an amorphous polymer, while PEEK is semi-crystalline. The microstructural differences between these materials can also affect their fusion and recrystallization behavior during the 3D printing process. Other studies have looked at the effect of the printing speed on the strength of PLA [31] and PEEK [41] prints, with little to no significant changes observed. Table 1 presents the results found in the literature regarding the influence of the FFF process parameters on the compressive properties and surface quality of different prints.

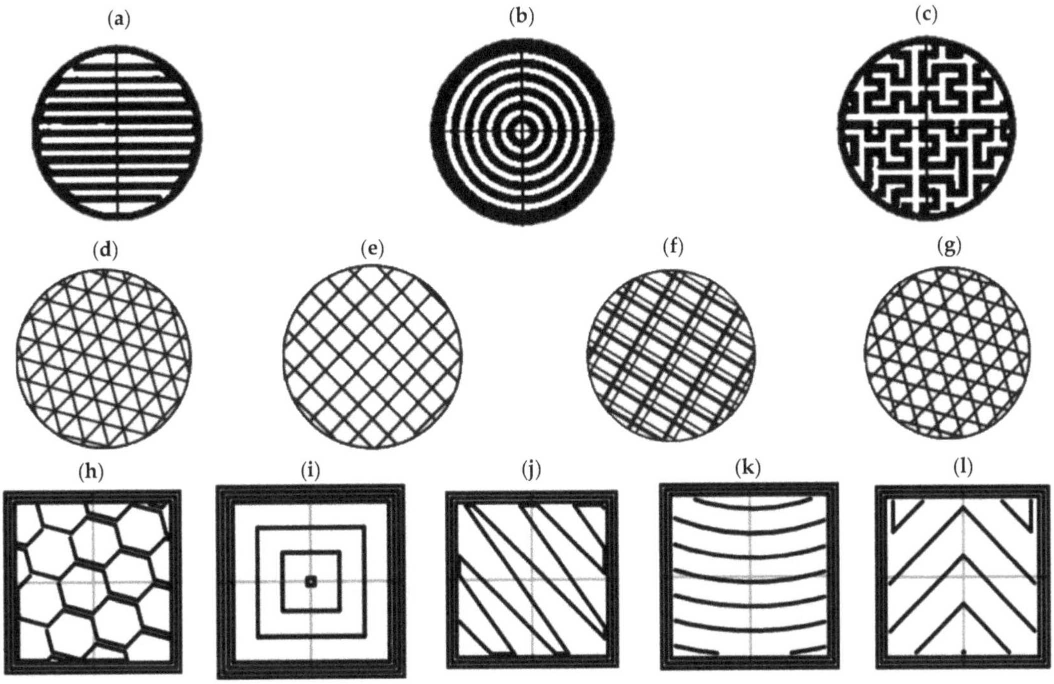

Figure 2. Infill patterns investigated in compression tests: (**a**) rectilinear, (**b**) concentric-circle, (**c**) Hilbert curve, (**d**) triangle, (**e**) grid, (**f**) quarter cubic, (**g**) tri-hexagon, (**h**) honeycomb, (**i**) concentric-rectangle, (**j**) line, (**k**) Archimedean, and (**l**) Octagram spiral (**a–c**) reproduced with permission from Springer [32], (**d–g**) reproduced with permission from Elsevier [33], (**h–l**) reproduced with permission from Springer [34]).

The surface smoothness of 3D-printed PLA parts is also influenced by various printing parameters as summarized in Table 1. The surface roughness is a measure of surface smoothness, meaning that lower surface roughness is an indicator of a smoother surface. While the raster angle [42] has been shown to have a negligible effect on the surface finish in some studies, the layer height [42–44], extrusion width [42,45,46], and print speed [46] were found to enhance the surface finish when reduced. Higher infill densities also contribute to smoother surfaces [34,47]. The impact of the air gap and nozzle temperature varies in different studies, with some suggesting them influential [48,49] and others finding them negligible [42,43,46]. Similarly, the build orientation had varying effects on surface roughness, exhibiting either a decreasing trend [43] or an oscillating pattern [44].

While the tensile and flexural strengths of FFF prints have received extensive attention in the literature, their compressive performance has been relatively underexplored. The aim of this study was to examine the effect of the layer height, extrusion width, nozzle temperature, and printing speed on the surface roughness and compressive properties of PLA specimens. The Taguchi method was employed to design experiments and optimize the printing parameters for maximum compressive strength and surface smoothness. Physical and mechanical tests were conducted to determine the density, surface quality, and compressive properties of the specimens. Furthermore, a microscopic analysis was carried out to identify the underlying failure mechanism under static loading. The findings of this study provide further insights into optimizing the 3D printing process for applications requiring strong compressive properties and high surface quality.

Table 1. Influence of FFF process parameters on the surface and compressive properties of FFF prints.

		Influence on Mechanical and Surface Properties	
		Compressive Properties	**Surface Quality**
3D printing parameters	**Nozzle temperature**	• High nozzle temperature can result in hardness increase and thermos-oxidative degradation of polymers at the same time, leading to increase in strength [35]	• Changing the nozzle temperature shows no significant effect on the surface roughness of FFF prints [43,46] • The surface roughness decreases with increasing nozzle temperature [49]
	Raster angle	• Rasters perpendicular to the loading direction yield the highest compressive strength [26,30]	• The highest surface finish is attained at zero raster angle; however, its effect is negligible [42]
	Infill pattern	• Compressive strength of rectilinear pattern > concentric and Hilbert curve pattern [32] • Compressive strength and specific strength of hexagonal pattern > linear > triangular pattern [50] • Compressive strength of 0/90° and 45/−45° infill patterns are almost equal in horizontal prints made of Ultem 9085 [51] • Compressive strength of grid pattern > triangle > tri-hexagon > quarter cubic [33] • Compressive strength of Hilbert curve pattern (121.35 MPa) > rectilinear > line > Archimedean > honeycomb > Octagram [34]	• Rectilinear pattern exhibits the lowest surface roughness as compared to Hilbert curve and line patterns [34]
	Infill density	• Compressive strength and modulus increase by increasing the infill density [31,32,34,50,52]	• Surface roughness decreases significantly by increasing the infill density [47]
	Layer height	• Within 0.1–0.3 mm range, 0.2 mm layer height produces the highest compressive strength in PLA print [31] • Compressive strength of ABS prints exhibits both inverse [35] and direct relationship [30] with the layer height • Compressive modulus increases with decreasing layer height • Layer height shows little impact on compressive strength of PEEK prints [40]	• Decreasing the layer height improves the surface quality [42,43,49] • Reducing the layer height results in a decrease in surface roughness in 20 and 45° build orientations, but it increases at 70° build orientation [44]

Table 1. *Cont.*

		Influence on Mechanical and Surface Properties	
		Compressive Properties	**Surface Quality**
3D printing parameters	**Printing speed**	• No significant effect [31,41]	• Decreasing the print speed enhances the surface finish of PLA prints [46]
	Build orientation	• Compressive strength decreases by increasing the build orientation • [29,30,32,51] • Horizontal build direction exhibits 15–40% higher compressive strength than that of vertical build direction in Ultem 9085 prints [51] • Compressive strength of ABS prints is not impacted by build direction [53]	• Increasing the build orientation angle decreases the surface roughness in [43] • and shows an increasing–decreasing trend in [44]
	Air gap	• Compressive strength exhibits inverse relationship with the air gap [30]	• The lowest surface finish is attained at zero air gap [48] • Negative air gap degrades the surface quality [42] • Changing the air gap shows no significant effect on the surface roughness of FFF prints [43]
	Extrusion width	• Extrusion width exhibits insignificant impact on compressive strength of ABS prints [30]	• The best surface finish is attained at the highest extrusion width [42,45] • Decreasing the extrusion width enhances the surface roughness [46] in the extrusion width direction • Changing the extrusion width shows no significant effect on the surface roughness of FFF prints [43]

2. Materials and Methods

2.1. Sample Preparation

In the present study, PLA filament (X3D pro PLA), with a diameter of 1.75 mm, was used to fabricate samples. Cylindrical samples (ϕ12.7 × 25.4 mm) were designed and layered using ideaMaker 4.4.0 software and then 3D printed using a Raise3D Pro3 Plus printer. Figure 3 shows a typical printed sample. To help improve the adhesion of the sample to the build plate, a raft platform was added to the base of each specimen and the build plate was pre-heated to a 60 °C. 0°/45° grid pattern with the infill density of 100%, and one solid shell layer was used to build the specimen.

2.2. Design of Experiments

The layer height, extrusion width, nozzle temperature, and print speed were considered as the variable parameters. The design of the experiments was performed by employing the orthogonal array Taguchi method in Minitab 21.4.0.0 software. Table 2 shows the variable with different levels (low, medium, and high) used as well as the response parameters and their definitions.

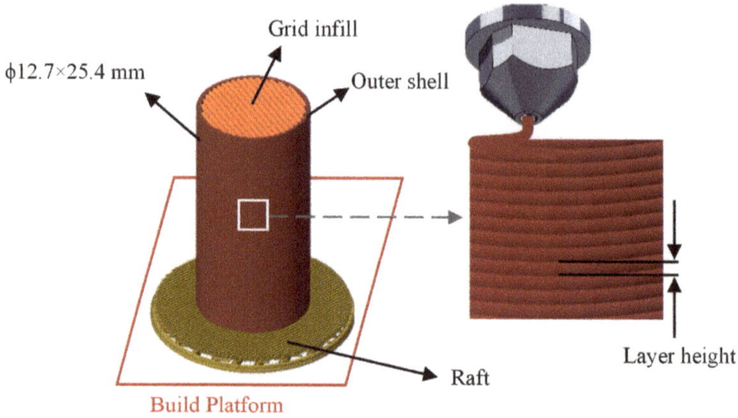

Figure 3. Schematic illustration of a sample printed in this study.

Table 2. Variables, responses, and levels specified for the design of experiments.

Variables	Unit	Low (−1)	Medium (0)	High (+1)
Layer height (LH)	mm	0.05	0.15	0.25
Extrusion width (EW)	mm	0.45	0.55	0.65
Nozzle temperature (T)	°C	190	205	220
Print speed (V)	mm/s	30	50	70
Responses				
Relative density	$\rho_{rel} = \rho_{sample}/\rho_{Filament} \times 100$			
Surface roughness	$R_a = \int_0^1 \lvert Z(x)\rvert dx/1$			
Compressive strength	$\sigma_y = 0.2\%\text{proofstrength}$			
Compressive modulus	$E = \text{Slopeofthelinearpartof}(\sigma - \varepsilon)\text{curve}$			
Specific strength	$\sigma_s = \sigma_y/\rho_{sample}$			
Failure strain	$\varepsilon_f = \text{Strainatfailurepoint}$			
Hysteresis loss (Area)	$A_{hys} = \text{Areaconfinedbytheloading} - \text{unloadingcurves}$			
Residual strain	$\varepsilon_r = \text{Residualstrainafterunloading}$			

The configuration of the designed experiments is shown in Table 3. The minimum and maximum level of the factors were selected according to the range given by the manufacturer as well as some trial and error. The relative density, surface roughness, compressive strength, compressive modulus, specific strength, failure strain, hysteresis loss, and residual strain were set as the response parameters.

Table 3. Design of experiment (orthogonal array Taguchi method).

Sample No.	Layer Height (mm)	Extrusion Width (mm)	Nozzle Temperature (°C)	Print Speed (mm/s)
S1	0.25	0.55	190	70
S2	0.15	0.55	220	30
S3	0.15	0.45	205	70
S4	0.05	0.65	220	70
S5	0.05	0.55	205	50
S6	0.15	0.65	190	50
S7	0.25	0.45	220	50
S8	0.25	0.65	205	30
S9	0.05	0.45	190	30

2.3. Density Measurement

The density measurements were conducted using the Archimedes method according to ASTM D792 [54]. A digital balance with a precision of 0.0001 gr was used for weighing the samples in air and in submerged states, and distilled water was used for submerging the samples (Figure 4). Each sample was weighed in air and then submerged in distilled water using a sinker. The test was carried out in the standard laboratory temperature and the water temperature of 28 °C. The apparent density of each sample was calculated according to Equation (1).

$$\rho_{\text{sample}}\left(\text{g}/\text{cm}^3\right) = \frac{m_a}{(m_a - m_w)} \times \rho_w \tag{1}$$

where ρ_w is the density of water at the test temperature (0.996 gr/cm^3), and m_a and m_b are the apparent mass of the sample in air and in water, respectively. To calculate the relative density of each sample, its apparent density was divided by the density of the filament (1.26 gr/cm^3) which represents the density of a fully dense sample (Equation (2)).

$$\rho_{\text{rel}} = \frac{\rho_{\text{sample}}}{\rho_{\text{Filament}}} \times 100 \tag{2}$$

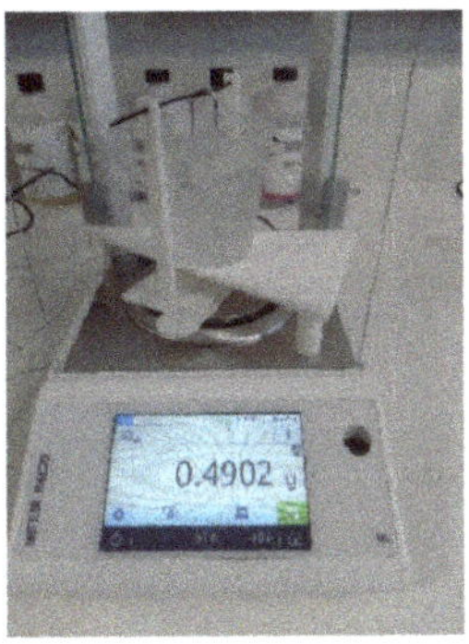

Figure 4. Digital balance for density measurements.

To make sure about the validity of the results, three specimens were tested for each group, making a total of 27 experiments. Then, the arithmetic average of the measured densities in each group was reported.

Following the density measurement, the samples underwent drying via a fan-forced flow of air at a temperature of 35 °C for a duration of 1 h, utilizing a sunbeam electronic dehydrator machine. Subsequently, the mass of each sample was determined and compared to its initial mass to confirm the absence of any water within the samples. The findings revealed a maximum mass change of 0.0005 g among the 27 specimens tested, indicating the complete evaporation of all the water present within the samples.

2.4. Surface Roughness

Having a high-quality surface finish is of paramount importance for improving both the functionality and appearance of 3D-printed parts. In addition, it can also help reduce costs by minimizing the amount of post-processing needed and speeding up the overall prototyping process. In the present study, the TMR200 surface roughness tester, manufactured by PCWI Co., was employed for the surface measurements. The instrument featured a 5µm tip, which was employed for all surface roughness measurements. The cutoff length was set at 2.5 mm as recommended by the manufacturer, ensuring coverage of the entire length of at least five consecutive layers in the 3D-printed samples. A stainless-steel V-block was used to mount the specimens in the lateral direction. Figure 5 shows the configuration of the surface roughness setup.

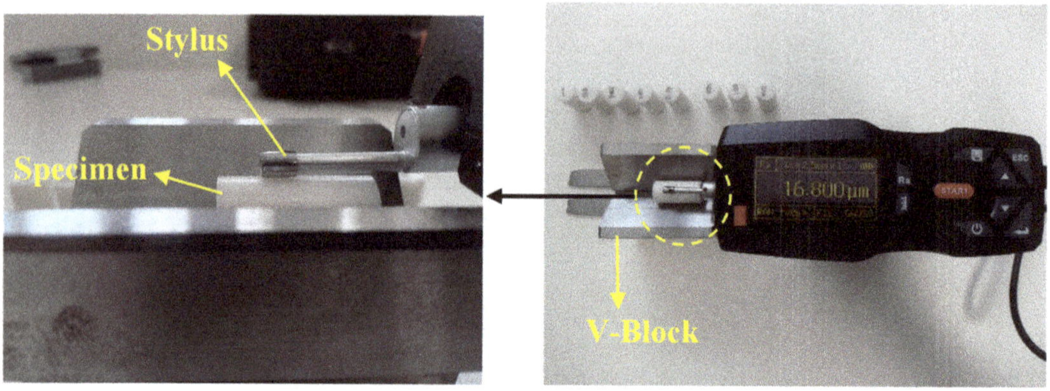

Figure 5. Setup for surface roughness measurement.

The roughness tester was calibrated before testing using a calibration block provided by the manufacturer. The stylus was brought into contact with the surface of interest while maintaining consistent pressure. The stylus was then moved along the sample's length in a straight line, and its displacement was measured and converted into an electrical signal. The processed signal was then used to calculate the Ra using Equation (3).

$$R_a = \frac{1}{L}\int_0^l |Z(x)|dx \tag{3}$$

where R_a is the average surface roughness, L is the sampling length along the x direction, and Z (x) is the height of the surface profile relative to the mean line at the distance x. Figure 6 depicts the configuration of the FFF prints and the surface roughness measurement.

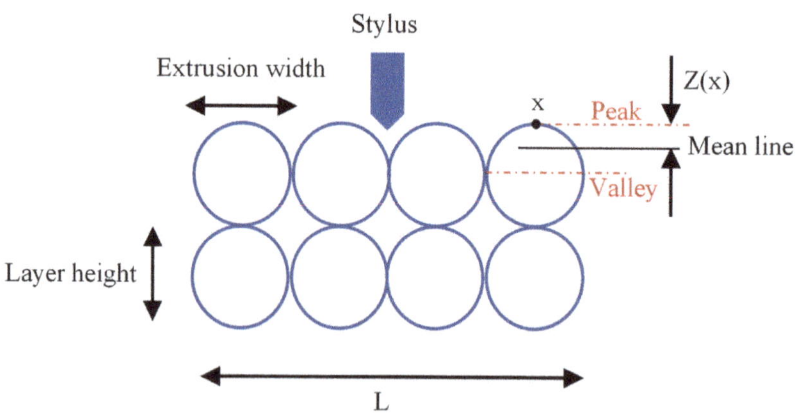

Figure 6. FFF prints configuration and surface roughness measurement.

Each test was repeated three times at different locations, and the average of three recorded R_a was reported as the surface roughness.

2.5. Compression and Hysteresis Compression

The quasi-static compression tests were conducted using an Instron 8801 device. The device was equipped with a double-acting servo hydraulic actuator, capable of applying forces up to ±100 kN, within a 150 mm stroke range, and at loading rates between 0.1 mm·min^{-1} and 240 mm·s^{-1}. All the compression tests were carried out in accordance with the ASTM D695 standard [55], which covers the determination of the compressive properties of rigid plastics. The compression tests were conducted at a temperature of 23 ± 2 °C and relative humidity of ~6%. The compression speed was set at 1.3 mm/min and each test was repeated five times to guarantee the statistical accuracy and reliability of the results.

Prior to testing, the diameter and height of the fabricated specimens were measured to the nearest 0.01 mm at several points for the stress and strain measurements, respectively. The minimum diameter was used to calculate the cross-sectional area. The load–deformation curve was recorded at a data recording frequency of 20 Hz, and the resulting stress–strain curve was obtained by dividing the load and deformation values to the minimum cross-sectional area and initial sample's length, respectively. The zero-stress portions of the stress–strain curves were then ditched from the graphs, as shown in Figure 7a. All the samples were subjected to both a uniaxial destructive compression test and hysteresis (loading–unloading) compression, and each test was repeated three times for statistical validation. Figure 7 illustrates a typical stress–strain curve of a specimen under compression and hysteresis compression.

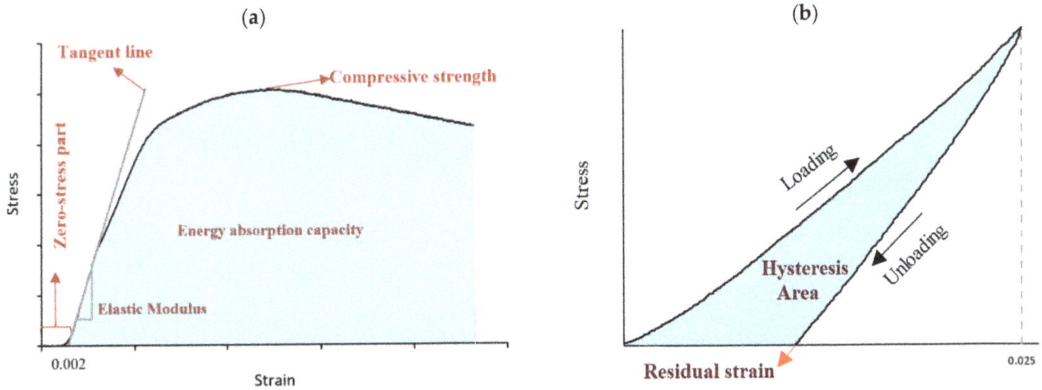

Figure 7. Typical stress–strain curve of a specimen under (**a**) compression and (**b**) hysteresis compression.

The compressive modulus was calculated by drawing a tangent line to the initial linear portion of the stress–strain curve, selecting a point on this line, and dividing the compressive stress represented by this point by the corresponding strain. The compressive strength and specific strength of each sample were then obtained by identifying the maximum compressive stress and dividing it by the sample's density, respectively. The specific strength of the FFF prints is studied in this paper because it is an important mechanical property that indicates the strength-to-weight ratio of the printed parts, which is crucial in many engineering applications where lightweight and high-strength components are required. Finally, the toughness (energy absorption capacity) of each sample was acquired by measuring the area under the stress–strain curve up to the failure point in Figure 7a.

The area between the loading and unloading curves in Figure 7b indicates the amount of energy that is lost due to the material's viscoelastic nature. In fact, when plastics are deformed, a phase shift between the stresses and strains occurs, resulting in the hysteresis loop [56]. The energy that is dissipated in one full cycle is represented by the area of this hysteresis loop. Like the compression test, the loading and unloading in the hysteresis compression were performed at the rate of 1.3 mm/min. The samples were kept in the unloading state for 10 min for static recovery. To make sure that the loading does not surpass the elastic region, the yield strain of each sample was determined from the compression tests, and hysteresis loading was performed up to the 0.025 strain, which is in the elastic region of all the samples. After loading achieved the pre-defined stroke, the unloading stage followed until the complete release of the load. The remaining deformation after unloading was measured and reported as the residual strain. Also, the area confined between the loading and unloading curves was calculated as the hysteresis loss.

2.6. Scanning Electron Microscopy (SEM)

Scanning electron microscopy was carried out using a JEOL benchtop SEM machine to investigate the fracture characteristics of the specimens. Three-dimensional printed samples were mounted on a holder and imaging was conducted around the fractured areas with an accelerating voltage of 15 kV.

3. Results and Discussion

Table 4 provides the results obtained from the physical and mechanical experiments for the 3D-printed samples.

Table 4. Compressive properties and surface roughness of the PLA prints.

No.	T (min)	ρ_{rel} (%)	Ra (μm)	σ_{max}(MPa)	E (GPa)	σ/ρ (kN.m/kg)	T_C (MJ.m^{-3})	ε_f (%)	A_{Hys} (MJ.m^{-3})	ε_r (%)
S1	14	92.50	17.35	61.30	1.42	52.53	13.21	0.29	0.39	1.07
S2	32	91.80	10.67	62.78	1.84	54.17	14.05	0.31	0.50	0.57
S3	25	91.80	10.40	62.55	1.71	54.01	12.56	0.28	0.45	0.78
S4	60	93.20	6.59	71.49	2.04	60.79	12.15	0.25	0.52	0.64
S5	70	92.30	6.22	66.38	2.02	56.98	11.57	0.24	0.59	0.43
S6	23	93.20	14.32	65.50	1.75	55.65	12.79	0.27	0.44	0.80
S7	17	93.50	16.83	55.46	1.52	46.96	14.20	0.32	0.40	0.82
S8	18	93.00	17.95	61.22	1.74	52.15	14.81	0.32	0.46	0.71
S9	106	90.30	7.30	62.23	2.02	54.59	10.68	0.24	0.57	0.39

3.1. Analysis of Variance (ANOVA)

The analysis of variance (ANOVA) determines the statistical significance of various factors on different responses. Each response is fitted using the linear regression as Equation (4).

$$y = \beta_0 + (\beta_k \times x_k) \quad (4)$$

where y, β_k, and x_k represent the response, coefficients, and factors, respectively. Minitab 21.4.0.0 software was used in this study for the statistical analyses. Depending on the regression data, the transformation of responses might be necessary to meet the assumptions of normality and homogeneity of variance [57]. In the present study, the Box–Cox equation ($y' = y^{\lambda}$) was utilized to apply transformation to the surface roughness, compressive strength, compressive modulus, specific strength, toughness, failure strain, hysteresis loss, and residual strain using λ values of 0.5, 3, 3, 4, 4, 7, −1, and 0.5, respectively. Table 5 provides the results obtained by the ANOVA for each response, including the developed regression equations. The significance of each factor was investigated by a *t*-test at 95% confidence, and the results were indicated by the *p*-value, with *p*-values less than 0.05 showing a significant influence on the response. To avoid overfitting in the regression models, the reduced models were constructed using only the significant factors. As presented in Table 6, the reduced regression equations demonstrate higher predicted R-squared values, signifying the generalization capability and improved representation of the data relationships.

Table 5. Analysis of variance (results) *.

	DF	Adj SS	Adj MS	F-Value	*p*-Value	Remarks
Surface roughness	$= (2.208 + 7.903\text{LH})^2$					
Regression	4.000	178.390	44.598	52.060	0.001	Significant
Layer height (mm)	1.000	170.880	170.880	199.480	<0.001	Significant
Extrusion width (mm)	1.000	3.125	3.125	3.650	0.129	Insignificant
Nozzle temperature (C)	1.000	3.969	3.969	4.630	0.098	Insignificant
Print speed (mm/s)	1.000	0.416	0.416	0.490	0.524	Insignificant
Error	4.000	3.426	0.857			
Total	8.000	181.816				
Compressive strength	$= 109.25\sqrt{5.92 - 37.49\,LH + 30.65\,EW}$					
Regression	4.000	167.300	41.825	94.000	<0.001	Significant
Layer height (mm)	1.000	86.762	86.762	195.000	<0.001	Significant
Extrusion width (mm)	1.000	58.500	58.500	131.480	<0.001	Significant
Nozzle temperature (C)	1.000	2.134	2.134	4.800	0.094	Insignificant
Print speed (mm/s)	1.000	19.904	19.904	44.730	0.003	Significant
Error	4.000	1.780	0.445			
Total	8.000	169.079				
Compressive modulus	$= 3.07 \times \sqrt[3]{1.0252 - 2.359\,LH - 0.003\,V}$					
Regression	4.000	4.664	1.166	29.950	0.003	Significant
Layer height (mm)	1.000	4.114	4.114	105.650	0.001	Significant
Extrusion width (mm)	1.000	0.146	0.146	3.760	0.125	Insignificant
Nozzle temperature (C)	1.000	0.081	0.081	2.090	0.222	Insignificant
Print speed (mm/s)	1.000	0.323	0.323	8.290	0.045	Significant
Error	4.000	0.156	0.039			
Total	8.000	4.819				
Specific strength	$= 85.85\sqrt[3]{3.83 - 34.77\,LH + 22.1\,EW + 0.064\,V}$					
Regression	4.000	116.649	29.162	41.060	0.002	Significant
Layer height (mm)	1.000	73.994	73.994	104.170	0.001	Significant
Extrusion width (mm)	1.000	30.294	30.294	42.650	0.003	Significant
Nozzle temperature (C)	1.000	1.032	1.032	1.450	0.295	Insignificant
Print speed (mm/s)	1.000	11.330	11.330	15.950	0.016	Significant
Error	4.000	2.841	0.710			
Total	8.000	119.491				
Toughness	$= 20.36 \times \sqrt[3]{-7.76 + 13.122\,LH + 3.573\,EW + 0.04104\,T - 0.02265\,V}$					
Regression	4.000	14.602	3.651	80.440	<0.001	Significant
Layer height (mm)	1.000	10.331	10.331	227.640	<0.001	Significant
Extrusion width (mm)	1.000	0.766	0.766	16.880	0.015	Significant
Nozzle temperature (C)	1.000	2.274	2.274	50.100	0.002	Significant
Print speed (mm/s)	1.000	1.231	1.231	27.130	0.006	Significant
Error	4.000	0.182	0.045			
Total	8.000	14.784				
Failure strain	$= 0.385 \times \sqrt[6]{-307.54 + 0.3621\,LH + 0.001218\,T - 0.000758\,V}$					
Regression	4.000	0.011	0.003	325.020	<0.001	Significant
Layer height (mm)	1.000	0.008	0.008	908.720	<0.001	Significant
Extrusion width (mm)	1.000	0.000	0.000	0.420	0.553	Insignificant
Nozzle temperature (C)	1.000	0.002	0.002	231.480	<0.001	Significant
Print speed (mm/s)	1.000	0.001	0.001	159.450	<0.001	Significant
Error	4.000	0.000	0.000			
Total	8.000	0.011				
Hysteresis loss	$= (3.109\,LH + 0.00651\,V - 1.331)^{-1}$					
Regression	4.000	12.544	3.136	12.540	0.016	Significant
Layer height (mm)	1.000	10.406	10.406	41.610	0.003	Significant
Extrusion width (mm)	1.000	0.076	0.076	0.310	0.610	Insignificant
Nozzle temperature (C)	1.000	0.126	0.126	0.500	0.517	Insignificant
Print speed (mm/s)	1.000	1.935	1.935	7.740	0.050	Significant
Error	4.000	1.001	0.250			
Total	8.000	13.544				

Table 5. *Cont.*

	DF	Adj SS	Adj MS	F-Value	*p*-Value	Remarks
Residual strain	$= (0.4397 + 1.171\ LH + 0.004129\ V)^2$					
Regression	4.000	0.342	0.085	23.330	0.005	Significant
Layer height (mm)	1.000	0.217	0.217	59.140	0.002	Significant
Extrusion width (mm)	1.000	0.004	0.004	1.160	0.341	Insignificant
Nozzle temperature (C)	1.000	0.009	0.009	2.410	0.196	Insignificant
Print speed (mm/s)	1.000	0.112	0.112	30.600	0.005	Significant
Error	4.000	0.015	0.004			
Total	8.000	0.356				

* DF: degrees of freedom; Adj SS: adjusted sum of squares; Adj MS: Adj SS/DF; F-value: the ratio of the mean square for each factor to the mean square for error; *p*-value: the probability of obtaining a test statistic as extreme or more extreme than the observed value, assuming the null hypothesis is true.

Table 6. Predictive performance of regression models.

	Predicted R-sq (%) (Full Model)	Predicted R-sq (%) (Reduced Model)	Improvement (%)
Surface roughness	89.28	92.15	3.21
Compressive strength	93.53	94.5	1.04
Compressive modulus	82.25	83.24	1.2
Specific strength	84.87	91.54	7.86
Failure strain	98.53	99.05	0.53
Hysteresis loss	64.48	81.53	26.44
Residual strain	82.71	87.66	5.98

To gain a better understanding of the influence of each factor, the signal-to-noise (S/N) ratio was calculated with the help of Equation (5).

$$\mathrm{S/N} = -10log_{10}\left[\frac{1}{n}\sum_{\mathrm{i}=1}^{\mathrm{n}}\frac{1}{\mathrm{y_i}^2}\right] \quad (5)$$

where n is the total number of experiments, and y_i is the response value for the i_{th} experiment. A higher S/N ratio indicates that the factor has a stronger influence on the response variable. Figures 8 and 9 illustrate the main effect plots of the means and S/N ratios for each response, respectively. "Main effect" refers to the impact of a single factor on the response variable, irrespective of the influence of other factors. The main effect of the means (or S/N ratios) for factor x is calculated by subtracting the overall mean of y (or S/N ratios) from the mean of y (or S/N ratios) at each level of x.

From Figure 8 it can be seen that changing the layer height leads to the highest variation in all the responses as further supported by Figure 9, which displays the magnitude of the S/N ratios for each factor. Figure 9 demonstrates that for all the investigated responses, the S/N ratio is maximized at the 0.05 mm layer height, while the minimal S/N value is attained at 0.25 mm, suggesting that the layer height has a predominant influence on the studied properties.

Figure 8 also provides correlations between the responses and factors. While the surface roughness, failure strain, toughness, and residual strain show a positive correlation with the layer height, the compressive strength, compressive modulus, specific strength, and hysteresis loss exhibit a negative correlation. The layer height was found to have the most significant influence on all the studied properties, followed by the print speed (for compressive modulus, hysteresis loss, and residual strain), extrusion width (for compressive strength and specific strength), and nozzle temperature (for toughness and failure strain) as the second most influential factor.

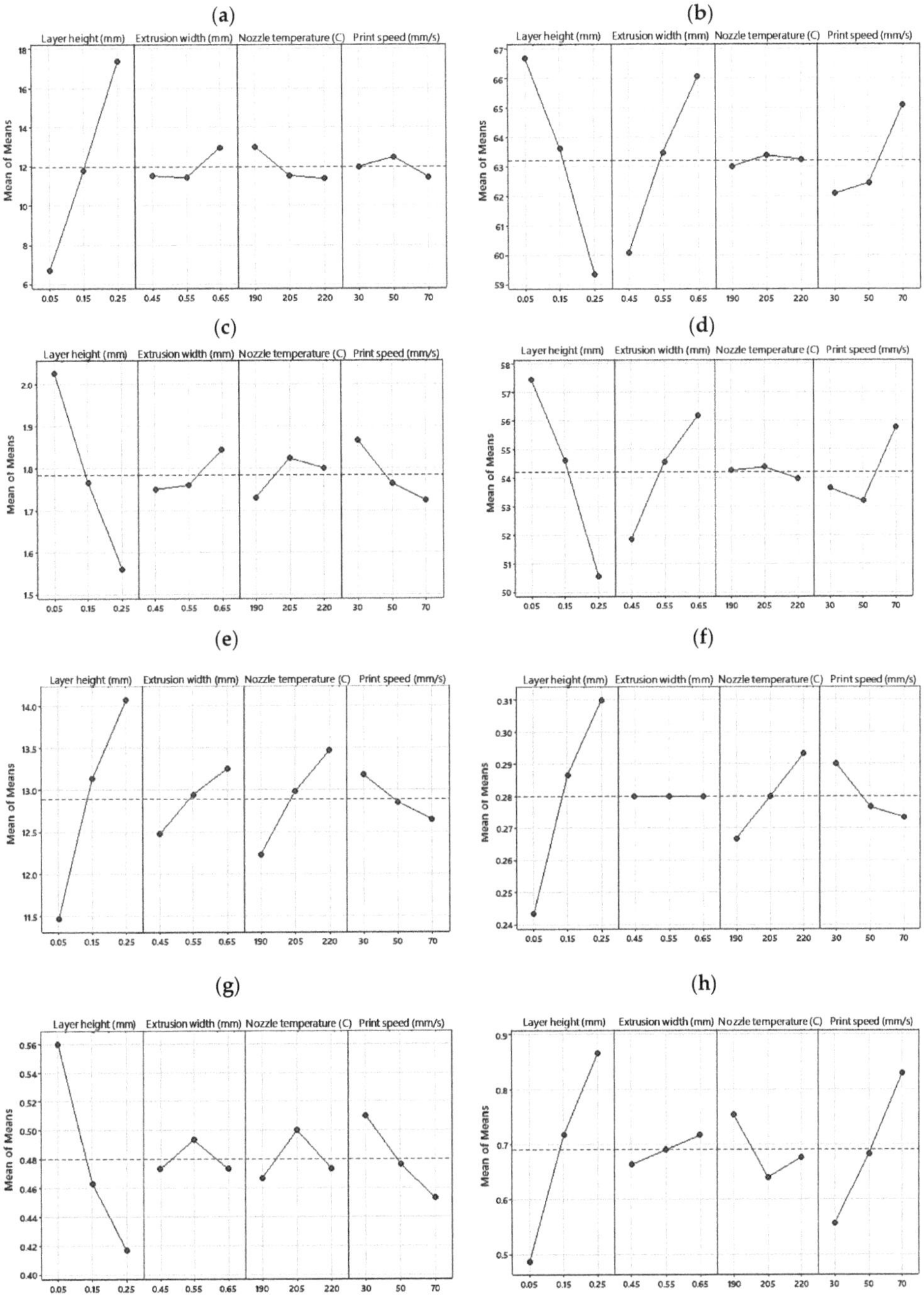

Figure 8. Main effects plot for means of (**a**) surface roughness, (**b**) compressive strength, (**c**) compressive modulus, (**d**) specific strength, (**e**) toughness, (**f**) failure strain, (**g**) hysteresis loss, and (**h**) residual strain.

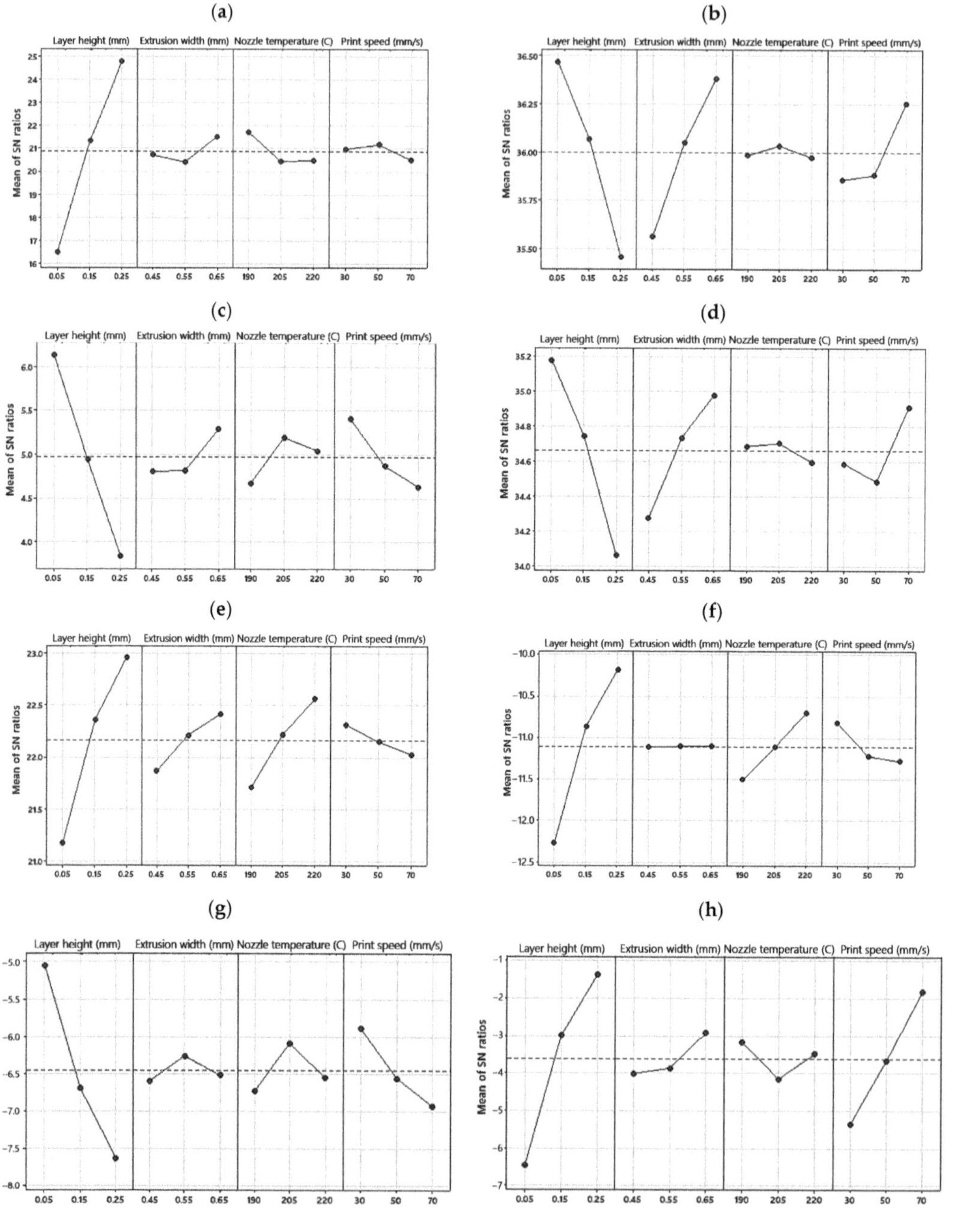

Figure 9. Main effects plot for S/N ratios of (**a**) surface roughness, (**b**) compressive strength, (**c**) compressive modulus, (**d**) specific strength, (**e**) toughness, (**f**) failure strain, (**g**) hysteresis loss, and (**h**) residual strain.

In the following, the impact of the process parameters on the studied responses will be discussed in detail.

3.2. Relative Density

The density of the 3D-printed specimens is depicted in Figure 10.

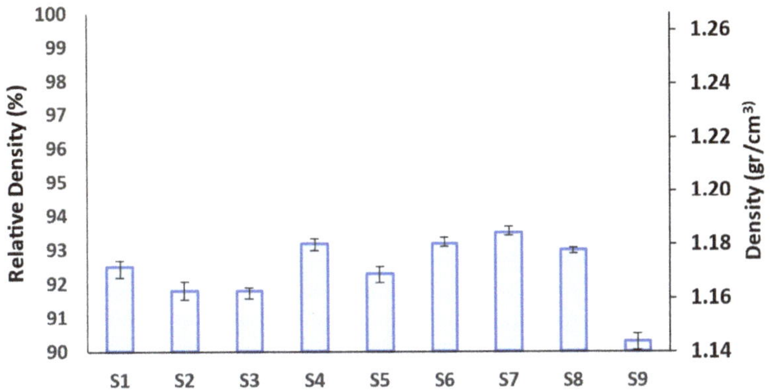

Figure 10. Relative density of 3D-printed PLA specimens (error bars: standard deviations).

As can be seen in Figure 10, the relative density of the samples is almost identical, ranging between 90.3% and 93.5%. This shows that altering the design parameters does not significantly affect the density of the samples. The remaining porosity within the samples is attributed to the incomplete bonding between the layers as well as the round shape of the extruded filaments as reported elsewhere [58,59]. Nevertheless, some reports suggest that increasing the nozzle temperature [60] and reducing the print speed [61] can greatly enhance the relative density of FFF prints.

3.3. Surface Roughness

Figure 11a illustrates the surface roughness of the PLA prints. As can be seen, the surface texture can vary greatly depending on the process parameters, with the S5 and S8 samples having the lowest (~6μm) and the highest (~18μm) surface roughness, respectively. To simplify the significance of each factor and its correlation to the surface roughness, the Pareto chart of the standardized effects is shown in Figure 11b. In this chart, the factors were ranked in order of their impact on the properties studied. Standardized effects larger than the vertical dotted line show a significant effect on the response, with the factors having a higher standardized effect showing a higher influence. The black arrows indicate the correlation/trend of each factor with the response.

It has been shown that parameters such as layer height, extrusion width, raster angle, and nozzle temperature are influential on the surface quality of FFF prints [62]. In the present study, the layer height was identified as the only influential factor on the surface quality of the prints and other parameters, i.e., the extrusion width, print speed, and nozzle temperature exhibited a nonsignificant impact on the results as demonstrated in Figure 11b. In the literature, the layer height was shown to be the most influential factor among other parameters when the surface quality is measured in the height direction [17,43]. This is because the surface roughness is highly affected by the peaks and valleys between the deposited layers in a print, as depicted by Z(x) in Figure 6. According to Equation (1), decreasing the layer height reduces the Z distance and thus decreases the surface roughness. Figure 12 illustrates the surface profiles of the FFF samples along the height dimension. As can be seen, the samples with the lowest layer height (S4, S5, and S9) exhibit the smoothest surfaces with the minimal distance between the peaks and valleys.

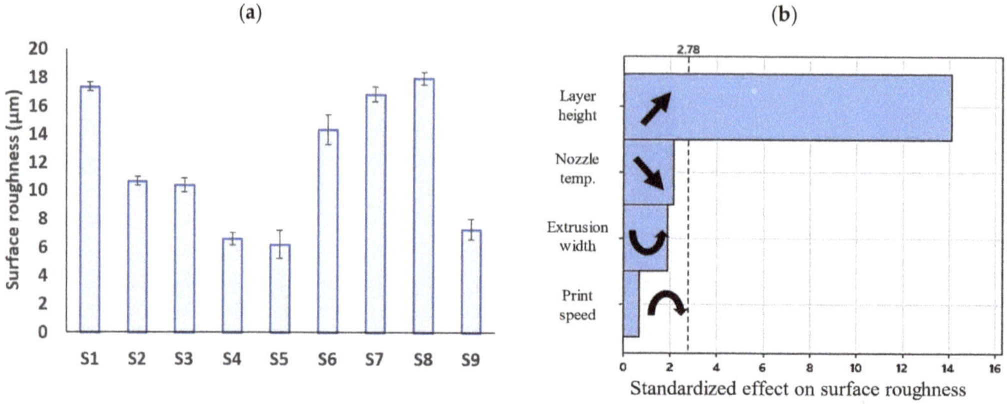

Figure 11. (**a**) Surface roughness of the fabricated PLA prints (error bars: standard deviations), and (**b**) Pareto charts of standardized effects of 3D printing parameters on surface roughness (black arrows in the figure indicate the correlation/trend between the process parameters and the response).

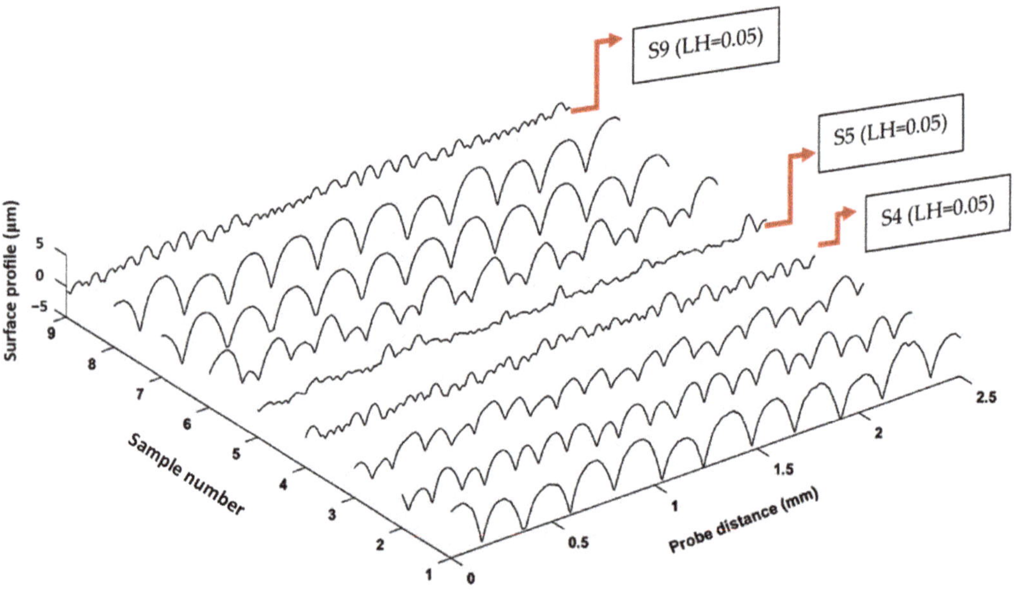

Figure 12. Surface profile of FFF samples along the height direction.

A similar result was reported in [46] where the extrusion width was identified as the most influential factor in the surface roughness of the PLA prints, while the nozzle temperature showed little influence. In the mentioned study, the surface roughness was measured at the top surface of the prints where the stylus sweeps the extruded filaments from the width direction.

3.4. Compressive Properties

Figure 13 demonstrates the compressive properties extracted from the compression tests as well as the Pareto charts for each property. As can be seen, the compressive modulus, compressive strength, and specific strength fall within the range of 1.42–2.04 GPa,

55.5–71.5 MPa, and 46.96–60.79 kN.m/kg, respectively. The S4 sample exhibited the highest compressive strength, compressive modulus, and specific strength among the samples.

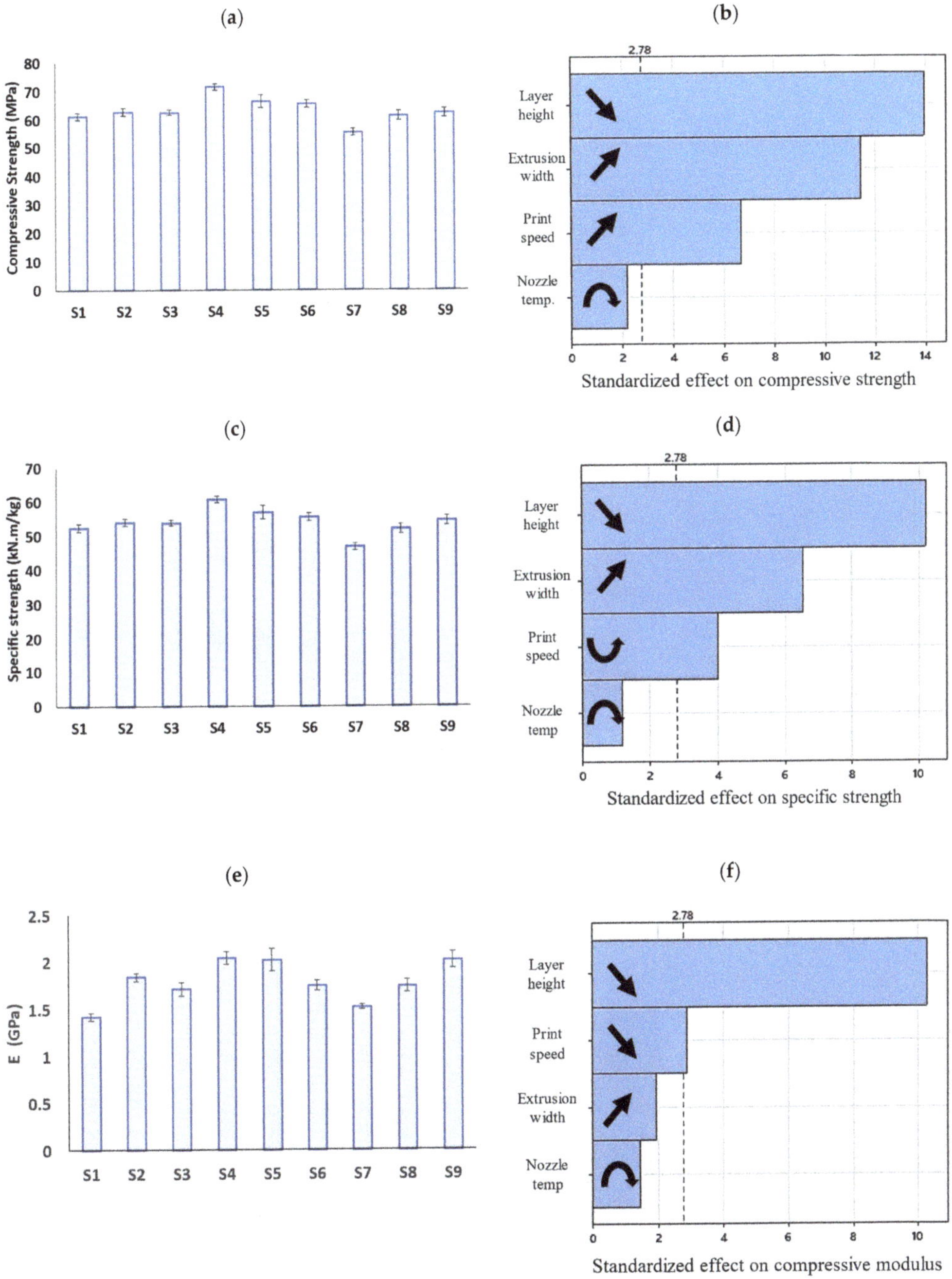

Figure 13. Bar graphs of (**a**) compressive strength, (**c**) specific strength, and (**e**) compressive modulus of PLA prints (error bars: standard deviations) and (**b**,**d**,**f**) their corresponding Pareto charts.

According to the Pareto charts in Figure 13, the compressive strength, modulus, and specific strength are inversely proportional to the layer height. However, there has been contradicting reports in the literature, exhibiting direct [30], inverse [35], and insignificant [40] effects of the layer height on the compressive strength of FFF prints. Wu et al. [40] reported that while the thickness of the layers greatly affected the tensile strength, it had minimal impact on the bending and compressive strengths of PEEK prints. Sood et al. [30] examined the effect of different layer heights (0.127, 0.178, and 0.254 mm) on the compressive strength of ABS prints and found that a decrease in the layer height causes a decrease in compressive stress. They concluded that reducing the layer height results in a higher number of layers and this leads to higher distortions arising from the thermal stress accumulating between the layers. Conversely, Nomani et al. [35] achieved contradicting results which are in agreement with the present study. Based on their results, the compression testing of ABS prints at different layer heights ranging from 0.2 mm to 0.8 mm revealed samples printed at the smallest examined layer height of 0.2 mm attained the highest compressive strength and modulus. Their results showed that a higher layer height led to a greater residual porosity and lower hardness in the bulk material, which could explain the observed decrease in mechanical strength. In fact, due to the circular cross section of the filaments, some internal cavities are formed between the rasters (Figure 14), resulting in anisotropic properties and decreased strength in FFF prints under compression [58,63]. Reducing the layer height leads to smaller internal cavities (porosity) between layers, resulting in enhanced mechanical properties. A porosity increase with the layer height has been reported in the literature [27,59]. Furthermore, decreasing the layer height has been shown to increase the contact area between extruded filaments. This increases the heat transmission between layers and promotes layer adhesion [26].

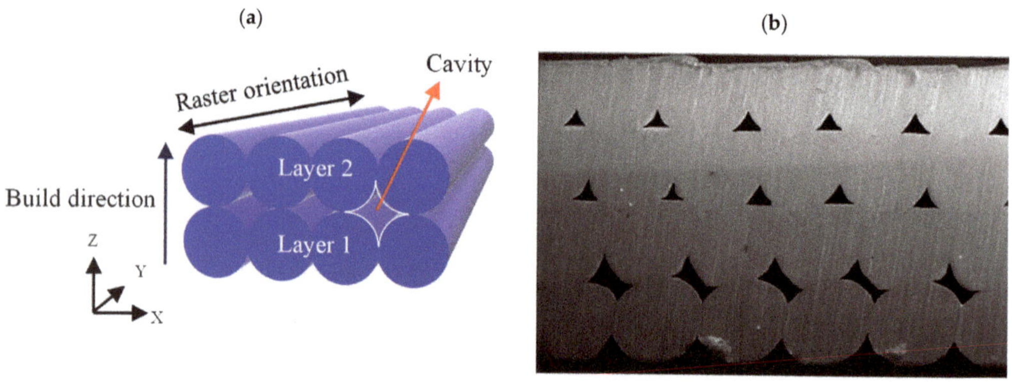

Figure 14. (**a**) Anisotropy of FFF prints caused by build orientation, raster orientation, and cavities, and (**b**) internal structure of an ABS print [64] (CC BY 4.0).

The results obtained by the ANOVA analysis also revealed a direct relationship between the compressive properties of the PLA prints and the extrusion width. While it has been reported that the extrusion width has an insignificant impact on the compressive strength of ABS prints [14], the results of the present study, in Figure 8b–d, showed that altering the extrusion width from 0.45 mm to 0.65 mm led to an increase in the compressive strength and specific strength by 10% and 8.5%, respectively. However, the impact of the extrusion width on the compressive modulus was not significant. Altering the extrusion width can affect the compressive strength of FFF prints as it influences the bonding characteristics between layers. By increasing the extrusion width, the contact area between layers is increased, leading to stronger bonding and greater strength in the printed part.

Finally, the print speed exhibited a different impact on the compressive properties of the PLA prints. While increasing the print speed resulted in a higher compressive strength, it had an inverse effect on the compressive modulus. When polymers are printed at lower speeds, the filaments remain in contact with the heated nozzle for a longer period of time, increasing the temperature within the filament. High-temperature exposure affects the crystallization of PLA and induces thermal degradation, resulting in a loss in mechanical characteristics [65]. In addition, at lower print speeds, the time between depositing a layer and the hot nozzle touching it again to deposit another layer on top (known as the thermal cycle) increases, leading to a higher temperature gradient between adjacent layers. Because the preceding layer has undergone significant cooling, the overall energy of the polymer at the interface may fall short of facilitating adequate molecular chain fusion and coalescence, ultimately leading to suboptimal interlayer bonding. This results in weaker bonds and lower compressive strength. If the temperature between these layers is insufficient, it can hinder molecular chain fusion, resulting in weak interlayer bonding. Higher print speeds reduce the time between layers, minimizing cooling and maintaining a higher temperature, and thus enhancing mechanical performance. Zhang et al. [66] predicted a positive link between the print speed and FDM component strength due to better thermal coalescence. Likewise, Samy et al. [66] found that higher nozzle speeds in FFF printing prevented significant cooling between layers, reducing the residual stress and strengthening the bonds. However, there are some reports suggesting that print speed has no significant effect on compressive strength [15,21].

As shown in Figure 15, the layer height and nozzle temperature exhibited the highest impact with a direct correlation to the toughness and failure strain of the PLA prints. Although the data on the compressive toughness and failure strain of FFF prints are lacking in the literature, some data show that these properties in tensile loading are not remarkably affected by 3D printing parameters [67]. It is believed that when the layer height is increased, each layer has more material deposited, resulting in a larger surface area for the next layer to bond to. Similarly, when the nozzle temperature is increased, the PLA material is heated to a higher temperature, which makes it more malleable and easier to bond to adjacent layers. Enhanced interlayer bonding leads to a higher toughness in PLA prints, as the layers are less likely to separate under stress or impact. The print speed was also shown to have a direct correlation with the toughness and an inverse relation with the failure strain.

3.5. Hysteresis Properties

Figure 16 illustrates the hysteresis loss and residual strains extracted from the hysteresis curves of the PLA prints. The S5 sample exhibited the biggest hysteresis loss, showing a higher damping capacity than the other samples. The S9 sample also had the lowest residual strain after unloading.

According to the Pareto charts in Figure 16b,d, the hysteresis properties of the PLA prints are affected the most by the layer height, followed by the print speed, while the nozzle temperature and extrusion width showed a little impact on the results. The layer height and print speed exhibited a direct correlation with the residual strain and an inverse relation with the hysteresis loss. That is, by decreasing the layer height or print speed, the residual strain of a PLA print decreases while its hysteresis loss increases, which is a favorable outcome. Because the compression hysteresis of FFF prints has not been studied thoroughly in the literature, the underlying mechanism of these observations is not fully understood. It is believed that by decreasing the layer height, the porosity of the structure will decrease, resulting in a higher hysteresis loss. The direct relation between the density of polymers and their hysteresis loss has been reported for polymeric foams [68] and FFF prints [56]. In addition, samples with a higher relative density possess more rigid structures which allow them to recover their shape after unloading, resulting in lower residual strain. Decreasing the print speed has also resulted in an increase in the hysteresis loss. This is mainly due to the higher residual stresses within the prints induced by higher magnitudes of temperature gradients at low printing speeds. It has been shown that the induction of compressive residual stresses can increase the damping capacity of materials [69].

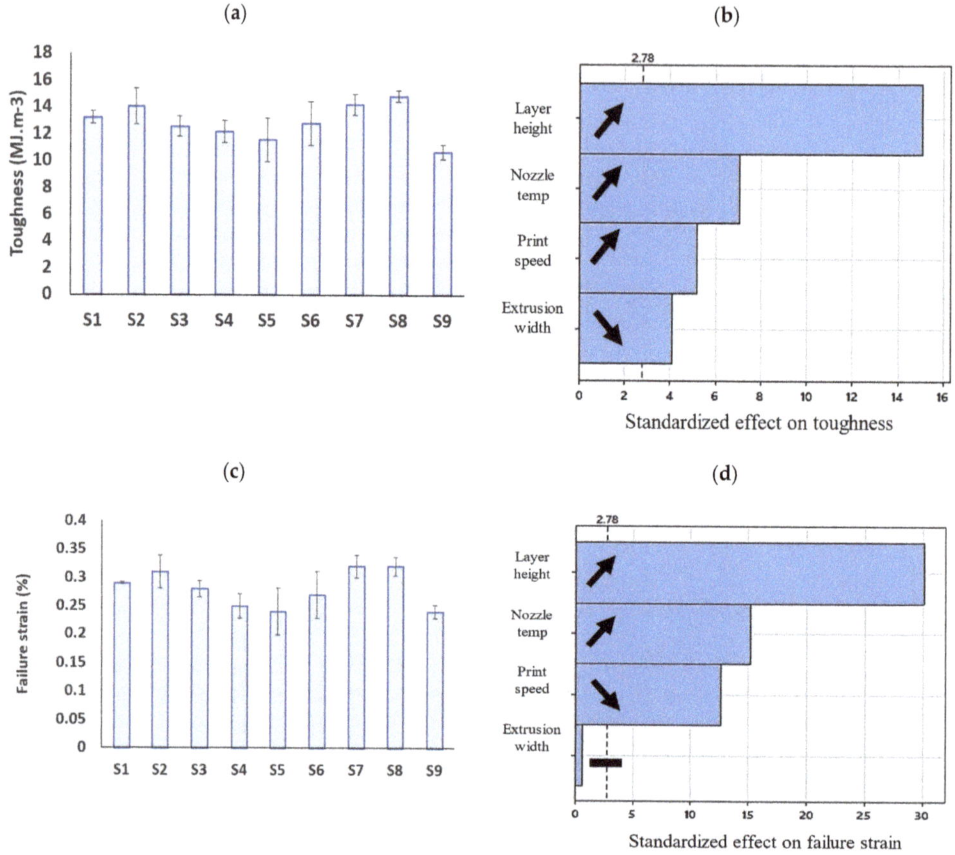

Figure 15. Bar graphs of (**a**) toughness and (**c**) failure strain of PLA prints (error bars: standard deviations) and (**b**,**d**) their corresponding Pareto charts.

3.6. Optimization

Two optimization schemes were followed in this study. In the first scheme (A), the goal was to achieve the highest compressive strength, regardless of the other characteristics. In the second scheme (B), the optimization problem was solved to maximize the compressive strength, compressive modulus, specific strength, and toughness while minimizing the surface roughness. The optimization goals for each scheme were chosen based on the potential industrial and biomedical applications of PLA prints. For example, in plastic enclosures for electronic equipment, plastic tools handles, automotive interior components, or some surgical instruments, the main focus is on the static strength of the prints. However, in various biomedical applications such as bone scaffolds [70], a combination of factors including the mechanical strength, compressive modulus, toughness, and surface roughness of the prints are all taken into consideration.

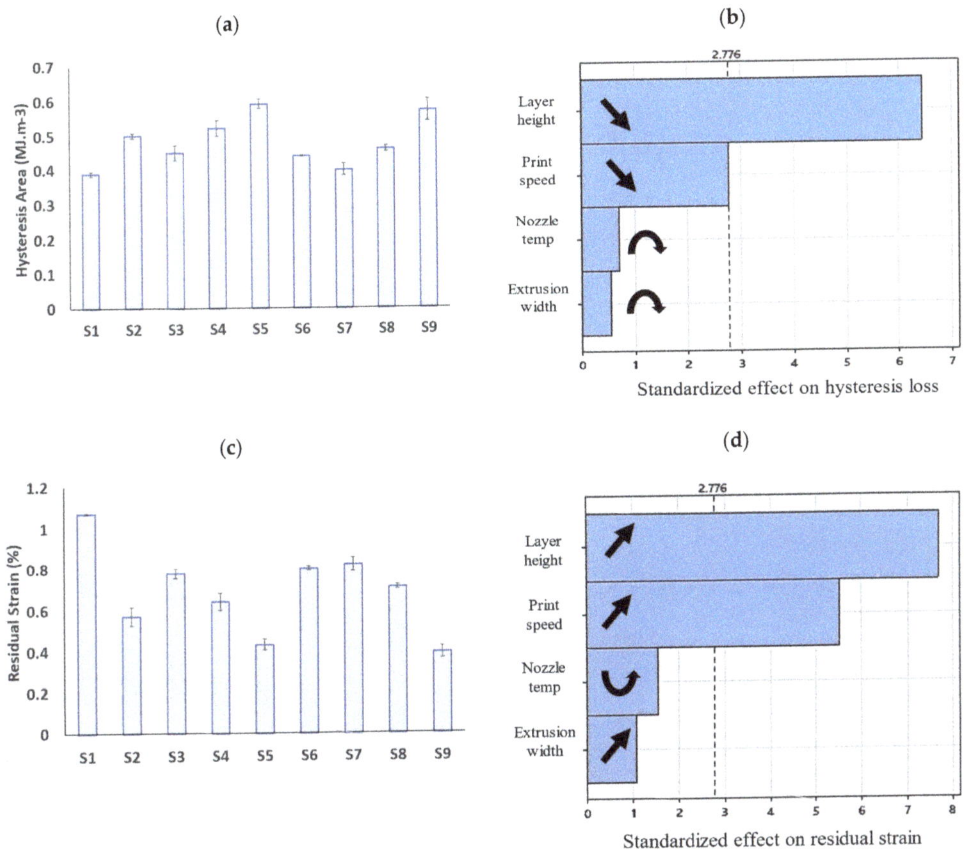

Figure 16. Bar graphs of (**a**) hysteresis area (or hysteresis loss) and (**c**) residual strain of PLA prints (error bars: standard deviations) and (**b**,**d**) their corresponding Pareto charts.

Table 7 provides five candidate designs for each scheme with the predicted responses and their corresponding desirability index (DI). Letters A and B denote the optimum designs for schemes A and B, respectively. The first two optimum designs with the highest desirability index were chosen for experimental validation, making a total of four designs (A1, A2, B1, and B2). To validate the optimization results, three samples were fabricated for each design and then were subjected to mechanical and physical tests as described earlier. As can be seen in Table 7, A2 and B1 represent the same design, implying that this design is optimum for both schemes. In addition, when only the highest compressive strength is desired (scheme A), the first optimum design (A1) is the same as sample S4 which was fabricated and tested earlier. So, there was no need to test the A1 sample again as its properties already existed. As shown in Table 8, the average of the measured values for each characteristic was computed and compared to the corresponding predicted values. The A1 design possesses the highest compressive strength (71.49 MPa) followed by A2 and B2, respectively. The highest performance was achieved in the B1 (=A2) sample when the compressive properties and surface quality are desired. Compared to B2, the B1 sample showed a higher compressive strength, compressive modulus, failure strain, toughness, and surface quality, while the B2 sample marginally possessed a higher relative density, specific strength, and hysteresis loss.

Table 7. Optimum designs for schemes A and B (selected designs with the highest DI indexes are shown in bold format).

No.	LH	EW	T	V	ρ_{rel}	Ra	σ_{max}	E	σ/ρ	T_C	ε_f	A_{Hys}	ε_r (%)	DI
A1	**0.05**	**0.65**	**220**	**70**	**92.47**	**6.77**	**71.46**	**2.02**	**61.02**	**12.28**	**0.25**	**0.52**	**0.61**	**1.00**
A2	**0.05**	**0.65**	**205**	**70**	**92.01**	**6.66**	**71.46**	**2.02**	**61.02**	**11.82**	**0.24**	**0.55**	**0.58**	**1.00**
A3	0.05	0.65	190	70	91.63	7.45	71.46	2.02	61.02	11.20	0.22	0.51	0.69	0.97
A4	0.05	0.55	190	70	91.63	6.86	68.88	1.98	59.38	10.98	0.22	0.51	0.69	0.84
A5	0.05	0.55	220	70	92.47	6.28	68.88	1.98	59.38	12.02	0.25	0.52	0.61	0.84
B1	**0.05**	**0.65**	**205**	**70**	**92.01**	**6.66**	**71.46**	**2.02**	**61.02**	**11.82**	**0.24**	**0.55**	**0.58**	**0.85**
B2	**0.05**	**0.65**	**220**	**30**	**91.66**	**6.77**	**68.42**	**2.11**	**58.87**	**12.28**	**0.26**	**0.60**	**0.34**	**0.84**
B3	0.05	0.65	220	70	92.47	6.77	71.46	2.02	61.02	12.28	0.25	0.52	0.61	0.84
B4	0.05	0.65	205	30	91.21	6.66	68.42	2.11	58.87	11.82	0.25	0.63	0.30	0.82
B5	0.05	0.65	205	50	92.51	6.66	68.79	2.05	58.44	11.82	0.24	0.58	0.43	0.81

Table 8. Mechanical and physical properties of optimum samples.

	ρ_{rel}	σ_{max}	E	σ/ρ	A_{Hys}	T	Ra	ε_f
A1 (Experimental)	**93.2**	**71.49**	**2.04**	**60.79**	**0.52**	**12.15**	**6.59**	**0.25**
A1 (Predicted)	92.47	71.46	2.02	61.02	0.52	12.28	6.77	0.25
A2 (=B1) (Experimental)	**93.6**	**67.79**	**2.29**	**56.36**	**0.56**	**12.38**	**4.92**	**0.26**
A2 (=B1) (Predicted)	92	71.46	2.02	61.02	0.55	11.82	6.66	0.24
B2 (Experimental)	**93.7**	**67.15**	**2.26**	**56.77**	**0.65**	**9.82**	**5.86**	**0.21**
B2 (Predicted)	91.66	68.42	2.11	58.87	0.6	12.28	6.77	0.26

Figure 17 shows the compressive stress–strain and hysteresis curves of the optimum samples.

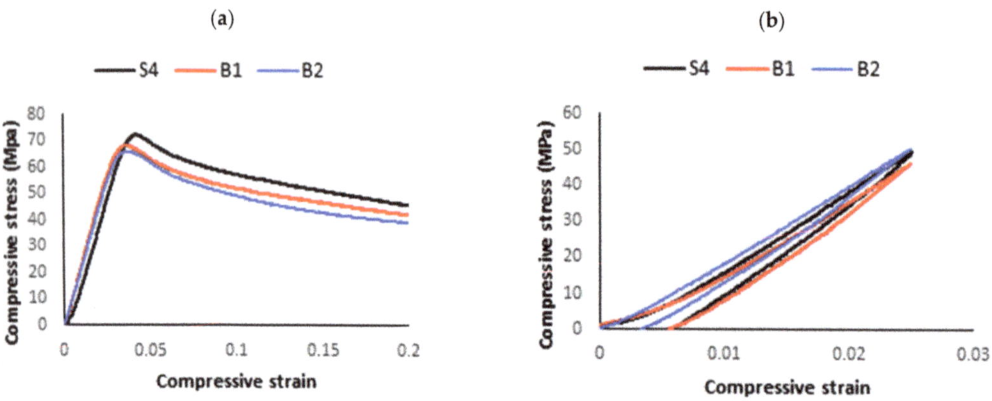

Figure 17. (**a**) Compressive stress–strain and (**b**) hysteresis–compression curves of optimum samples.

3.7. Failure Analysis

A failure analysis was performed visually and using scanning electron microscopy. The observations clearly showed that the main failure mechanism in all the tested samples was sliding thc layers in the middle section of the specimens. As evidenced in Figure 18a, further compressing the samples resulted in a rupture caused by the delamination between the layers. The same observation has been reported in the literature for FFF prints under compression [30,71]. The SEM images revealed that delamination occurs at the highly stressed points along the height of the samples. In Figure 18b,c, the failure images of the A1 and S1 samples are shown in two different magnifications, highlighting the delamination characteristics at the interfaces. Clean delamination suggests the possible existence of defects or the lack of interlayer adhesion as the layers were deposited. In contrast, microfibrils suggest strong interlayer bonding due to a high degree of fusion between the layers upon deposition, which is similar to cohesive failures in the fibers of perfectly fused (bonded) layers. The symmetry of these fractures was also noteworthy in the observations. It is believed that the interlaminar shearing, which seems to primarily drive the failure of the core, caused a stress concentration symmetrically around the axis. A stress concentration drives the existing defects into debonding the printed layers, resulting in the fracture of specimens.

(**a**)

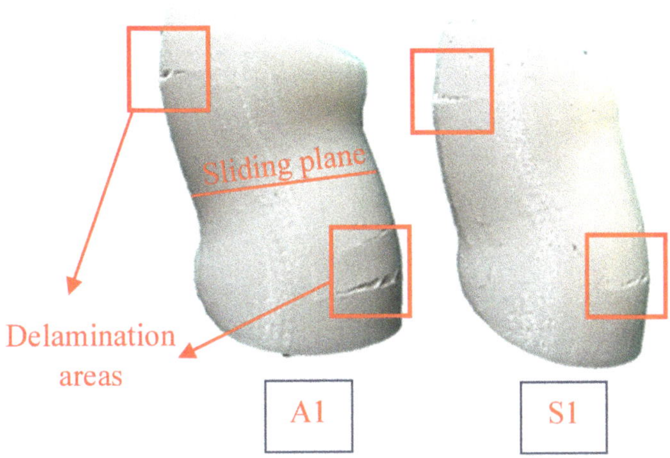

(**b**)

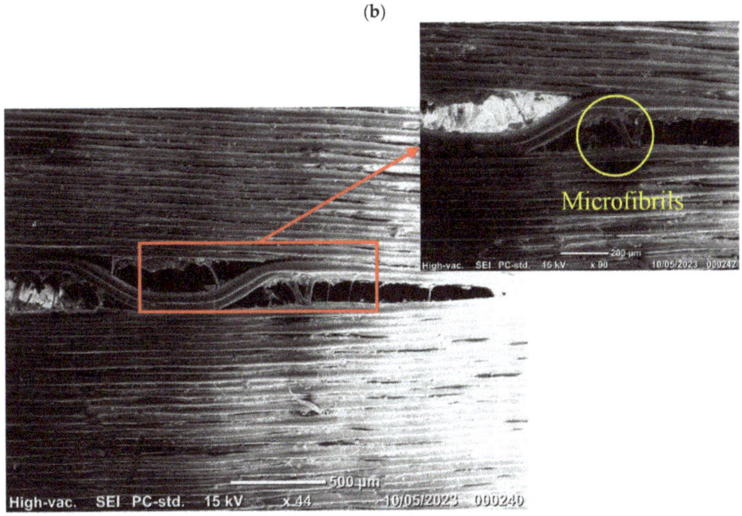

Figure 18. *Cont.*

(c)

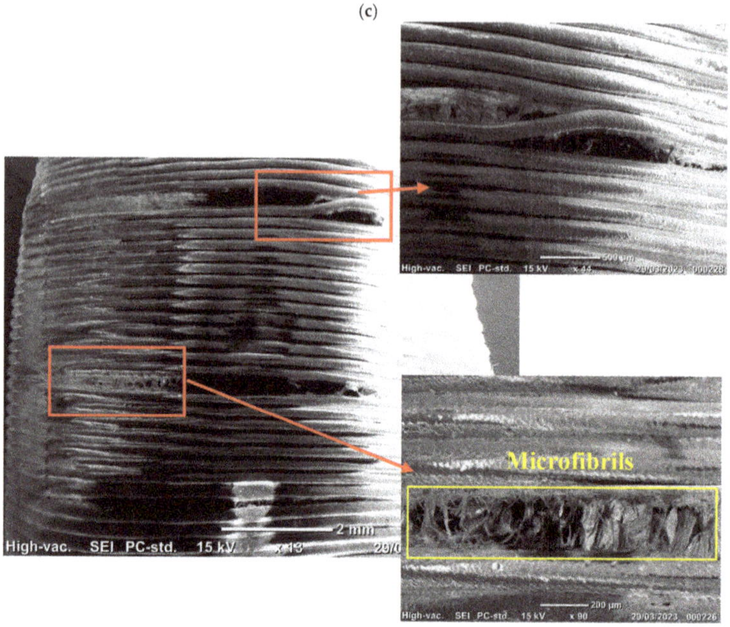

Figure 18. (**a**) Sliding and delamination of layers under compression, and SEM images of the delamination area in samples (**b**) A1 and (**c**) S1.

One common observation among all the samples that has received less attention in the literature is that all the delamination points were close to the so-called stitch line, as depicted by the yellow box in Figure 19. As shown in this figure, the bulged area in each layer is the starting point where the hot nozzle starts depositing the filament and finishes the layer at the same point. The stich line is a line connecting the starting points of the layers. The shape discontinuity and gaps in the stich line cause the stress concentration. The proximity of the interlayer gaps (as shown by the red box) to the stitch line has resulted in the failures starting in this stress-concentrated region. One solution to this might be changing the starting point in each layer, using a different path planning strategy.

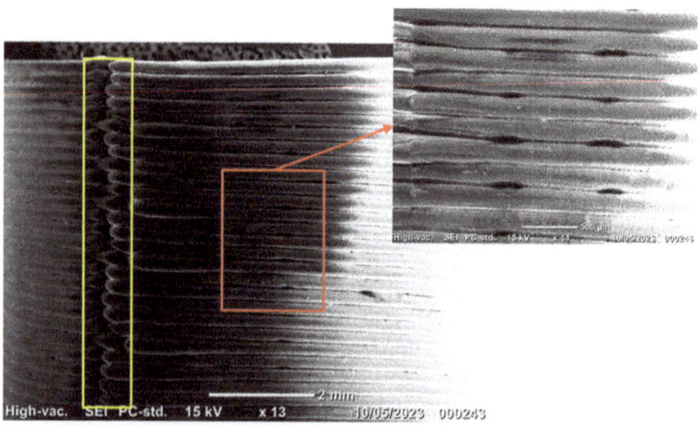

Figure 19. Gaps and discontinuities at the interface areas (red box) and stich line (yellow box) in as-fabricated S1 sample.

4. Conclusions

In this study, the effect of the layer height, extrusion width, nozzle temperature, and print speed on the compressive and surface properties of PLA specimens was investigated. The experiments were designed using the orthogonal array Taguchi method, and the 3D-printed specimens were subjected to uniaxial compression and hysteresis compression testing. The relative density and surface roughness of the specimens were also extracted. The most significant impact on the studied properties was found to be caused by changes in the layer height, which resulted in the highest variation in all the responses. Reducing the layer height was found to increase the compressive strength, compressive modulus, and hysteresis loss while reducing the surface roughness and residual strain. However, a decrease in the toughness and failure strain was observed for the PLA prints with a reduced layer height. Additionally, print speed was identified as the second most influential factor affecting the compressive modulus, hysteresis loss, and residual strain. The relative density of the specimens was found to be insensitive to the studied parameters, ranging from 90.3% to 93.5%. According to the optimization results, the highest compressive strength was obtained at a layer height of 0.05, an extrusion width of 0.65, a nozzle temperature of 220 °C, and a print speed of 70 mm/s. The failure analysis revealed that interlayer sliding and layer debonding were the main failure mechanisms for the PLA prints under compression. It was also noted that the interlayer gaps and shape discontinuity in the stich line in the FFF prints caused the formation of highly stress-concentrated areas, resulting in the subsequent layer debonding and rupture of the specimens.

One potential avenue for future research is the exploration of post-processing techniques to enhance print quality and compressive properties. Investigating various methods such as thermal and chemical treatments can yield valuable insights into improving the final product. Furthermore, it would be valuable to delve deeper into the impact of environmental factors, such as temperature and humidity, on the compressive properties of 3D prints. A comprehensive study could help establish guidelines for better design practices, including optimal printing conditions under different environmental conditions. Another area ripe for exploration is the detailed investigation on the effect of process parameters on the anisotropic behavior of prints under compression. By fine-tuning these parameters, researchers can optimize FFF settings for specific applications, potentially expanding the range of industries where FFF printing can be effectively utilized. Lastly, understanding the intricate relationship between the 3D printing parameters and microstructural properties of produced parts is of paramount importance for enhancing the compressive strength of prints. A thorough analysis in this regard can lead to innovative solutions for creating stronger and more resilient components.

Author Contributions: Conceptualization, M.T.-R. and H.B.; methodology, M.T.-R. and H.B.; software, H.B.; validation, H.B. and M.N.; formal analysis, H.B. and M.N.; investigation, H.B., M.T.-R. and M.N.; resources, H.B.; data curation, H.B.; writing—original draft preparation, H.B.; writing—review and editing, M.T.-R. and M.N.; visualization, H.B.; supervision, M.T.-R.; project administration, M.T.-R.; funding acquisition, M.T.-R. All authors have read and agreed to the published version of the manuscript.

Funding: This research received no external funding.

Institutional Review Board Statement: Not applicable.

Data Availability Statement: Not applicable.

Acknowledgments: The first author would like to thank Edith Cowan University for the awarded (ECU-HDR) higher degree research scholarship.

Conflicts of Interest: The authors declare no conflict of interest.

Nomenclature

LH	Layer height	**E**	Compressive modulus
EW	Extrusion width	σ/ρ	Specific strength
T	Nozzle temperature	T_C	Toughness
V	Print speed	ε_f	Failure strain
ρ_{rel}	Relative density	A_{Hys}	Hysteresis area (hysteresis loss)
Ra	Surface roughness	ε_r	Residual strain
σ_{max}	Compressive strength	**DI**	Desirability index
PLA	Polylactic Acid	**FFF**	Fused filament fabrication
ANOVA	Analysis of variance	**FDM**	Fused Deposition Modeling
PEEK	Polyether ether ketone	**ABS**	Acrylonitrile butadiene styrene

References

1. Altıparmak, S.C.; Yardley, V.A.; Shi, Z.; Lin, J. Extrusion-Based Additive Manufacturing Technologies: State of the Art and Future Perspectives. *J. Manuf. Process.* **2022**, *83*, 607–636. [CrossRef]
2. Zhang, Q.; Zhou, J.; Zhi, P.; Liu, L.; Liu, C.; Fang, A.; Zhang, Q. 3d Printing Method for Bone Tissue Engineering Scaffold. *Med. Nov. Technol. Devices* **2023**, *17*, 100205. [CrossRef]
3. Soleyman, E.; Aberoumand, M.; Soltanmohammadi, K.; Rahmatabadi, D.; Ghasemi, I.; Baniassadi, M.; Abrinia, K.; Baghani, M. 4d Printing of Pet-G Via Fdm Including Tailormade Excess Third Shape. *Manuf. Lett.* **2022**, *33*, 1–4. [CrossRef]
4. Sun, Z.P.; Guo, Y.B.; Shim, V.P. Characterisation and Modeling of Additively-Manufactured Polymeric Hybrid Lattice Structures for Energy Absorption. *Int. J. Mech. Sci.* **2021**, *191*, 106101. [CrossRef]
5. Liang, X.; Gao, J.; Xu, W.; Wang, X.; Shen, Y.; Tang, J.; Cui, S.; Yang, X.; Liu, Q.; Yu, L. Structural Mechanics of 3d-Printed Poly (Lactic Acid) Scaffolds with Tetragonal, Hexagonal and Wheel-Like Designs. *Biofabrication* **2019**, *11*, 035009. [CrossRef]
6. Palmić, T.B.; Slavič, J. Single-Process 3d-Printed Stacked Dielectric Actuator. *Int. J. Mech. Sci.* **2022**, *230*, 107555. [CrossRef]
7. Mwema, F.M.; Akinlabi, E.T.; Mwema, F.M.; Akinlabi, E.T. Basics of Fused Deposition Modelling (Fdm). In *Fused Deposition Modeling: Strategies for Quality Enhancement*; Springer International Publishing: Cham, Switzerland, 2020; pp. 1–15.
8. Giorleo, L.; Ceretti, E. Deep Drawing Punches Produced Using Fused Filament Fabrication Technology: Performance Evaluation. *J. Manuf. Process.* **2022**, *84*, 1–9. [CrossRef]
9. Strano, M.; Rane, K.; Farid, M.A.; Mussi, V.; Zaragoza, V.; Monno, M. Extrusion-Based Additive Manufacturing of Forming and Molding Tools. *Int. J. Adv. Manuf. Technol.* **2021**, *117*, 2059–2071. [CrossRef]
10. Schuh, G.; Bergweiler, G.; Fiedler, F.; Bickendorf, P.; Schumacher, P. Small Series Production and Geometric Analysis of Sheet Metal Car Body Parts Using Forming Tools Made of Fused Filament Fabricated Pla. In Proceedings of the 2020 IEEE International Conference on Industrial Engineering and Engineering Management (IEEM), Singapore, Singapore, 14–17 December 2020.
11. Kampker, A.; Bergweiler, G.; Hollah, A.; Lichtenthäler, K.; Leimbrink, S. Design and Testing of the Different Interfaces in a 3d Printed Welding Jig. *Procedia CIRP* **2019**, *81*, 45–50. [CrossRef]
12. Krznar, N.; Pilipović, A.; Šercer, M. Additive Manufacturing of Fixture for Automated 3d Scanning–Case Study. *Procedia Eng.* **2016**, *149*, 197–202. [CrossRef]
13. Post, B.; Richardson, B.; Lloyd, P.; Love, L.; Nolet, S.; Hannan, J. *Additive Manufacturing of Wind Turbine Molds*; Oak Ridge National Lab. (ORNL): Oak Ridge, TN, USA, 2017.
14. Sánchez-Belenguer, C.; Vendrell-Vidal, E.; Sánchez-López, M.; Díaz-Marín, C.; Aura-Castro, E. Automatic Production of Tailored Packaging for Fragile Archaeological Artifacts. *J. Comput. Cult. Heritage* **2015**, *8*, 1–11. [CrossRef]
15. Pal, A.K.; Mohanty, A.K.; Misra, M. Additive Manufacturing Technology of Polymeric Materials for Customized Products: Recent Developments and Future Prospective. *RSC Adv.* **2021**, *11*, 36398–36438. [CrossRef] [PubMed]
16. Nadagouda, M.N.; Rastogi, V.; Ginn, M. A Review on 3d Printing Techniques for Medical Applications. *Curr. Opin. Chem. Eng.* **2020**, *28*, 152–157. [CrossRef]
17. Mohamed, O.A.; Masood, S.H.; Bhowmik, J.L. Optimization of Fused Deposition Modeling Process Parameters: A Review of Current Research and Future Prospects. *Adv. Manuf.* **2015**, *3*, 42–53. [CrossRef]
18. Bian, Y.-H.; Yu, G.; Zhao, X.; Li, S.-X.; He, X.-L.; Tian, C.-X.; Li, Z.-Y. Exit Morphology and Mechanical Property of Fdm Printed Pla: Influence of Hot Melt Extrusion Process. *Adv. Manuf.* **2023**, *11*, 56–74. [CrossRef]
19. Cojocaru, V.; Frunzaverde, D.; Miclosina, C.-O.; Marginean, G. The Influence of the Process Parameters on the Mechanical Properties of Pla Specimens Produced by Fused Filament Fabrication—A Review. *Polymers* **2022**, *14*, 886. [CrossRef]
20. Vanaei, H.R.; Raissi, K.; Deligant, M.; Shirinbayan, M.; Fitoussi, J.; Khelladi, S.; Tcharkhtchi, A. Toward the Understanding of Temperature Effect on Bonding Strength, Dimensions and Geometry of 3d-Printed Parts. *J. Mater. Sci.* **2020**, *55*, 14677–14689. [CrossRef]
21. Kuznetsov, V.E.; Solonin, A.N.; Tavitov, A.; Urzhumtsev, O.; Vakulik, A. Increasing Strength of Fff Three-Dimensional Printed Parts by Influencing on Temperature-Related Parameters of the Process. *Rapid Prototyp. J.* **2020**, *26*, 107–121. [CrossRef]

22. Gurrala, P.K.; Regalla, S.P. Part Strength Evolution with Bonding between Filaments in Fused Deposition Modelling: This Paper Studies How Coalescence of Filaments Contributes to the Strength of Final Fdm Part. *Virtual Phys. Prototyp.* **2014**, *9*, 141–149. [CrossRef]
23. Gray, R.W.; Baird, D.G.; Bøhn, J.H. Effects of Processing Conditions on Short Tlcp Fiber Reinforced Fdm Parts. *Rapid Prototyp. J.* **1998**, *4*, 14–25. [CrossRef]
24. Zhong, W.; Li, F.; Zhang, Z.; Song, L.; Li, Z. Short Fiber Reinforced Composites for Fused Deposition Modeling. *Mater. Sci. Eng. A* **2001**, *301*, 125–130. [CrossRef]
25. Rahmatabadi, D.; Soltanmohammadi, K.; Aberoumand, M.; Soleyman, E.; Ghasemi, I.; Baniassadi, M.; Abrinia, K.; Bodaghi, M.; Baghani, M. Development of Pure Poly Vinyl Chloride (Pvc) with Excellent 3d Printability and Macro-and Micro-Structural Properties. *Macromol. Mater. Eng.* **2023**, *308*, 2200568. [CrossRef]
26. Sood, A.K.; Ohdar, R.K.; Mahapatra, S.S. Parametric Appraisal of Mechanical Property of Fused Deposition Modelling Processed Parts. *Mater. Des.* **2010**, *31*, 287–295. [CrossRef]
27. Bakhtiari, H.; Aamir, M.; Tolouei-Rad, M. Effect of 3d Printing Parameters on the Fatigue Properties of Parts Manufactured by Fused Filament Fabrication: A Review. *Appl. Sci.* **2023**, *13*, 904. [CrossRef]
28. Gordelier, T.J.; Thies, P.R.; Turner, L.; Johanning, L. Optimising the Fdm Additive Manufacturing Process to Achieve Maximum Tensile Strength: A State-of-the-Art Review. *Rapid Prototyp. J.* **2019**, *25*, 953–971. [CrossRef]
29. Gabor, C.; Pop, M.A.; Magli, D.; Bedo, T.; Munteanu, S.I.; Munteanu, D. The Optimization of the Production Procedure in Relation to the Mechanical Properties of Additively Manufactured Parts. *Mater. Today Proc.* **2019**, *19*, 1008–1013. [CrossRef]
30. Sood, A.K.; Ohdar, R.K.; Mahapatra, S.S. Experimental Investigation and Empirical Modelling of Fdm Process for Compressive Strength Improvement. *J. Adv. Res.* **2012**, *3*, 81–90. [CrossRef]
31. Dave, H.K.; Rajpurohit, S.R.; Patadiya, N.H.; Dave, S.J.; Sharma, K.S.; Thambad, S.S.; Srinivasn, V.P.; Sheth, K.V. Compressive Strength of Pla Based Scaffolds: Effect of Layer Height, Infill Density and Print Speed. *Int. J. Mod. Manuf. Technol.* **2019**, *11*, 21–27.
32. Prajapati, A.R.; Rajpurohit, S.R.; Patadiya, N.H.; Dave, H.K. Analysis of Compressive Strength of 3d Printed Pla Part. In *Advances in Manufacturing Processes: Select Proceedings of RAM 2020*; Springer Singapore: Singapore, 2021.
33. Aloyaydi, B.; Sivasankaran, S.; Mustafa, A. Investigation of Infill-Patterns on Mechanical Response of 3d Printed Poly-Lactic-Acid. *Polym. Test.* **2020**, *87*, 106557. [CrossRef]
34. Yadav, P.; Sahai, A.; Sharma, R.S. Strength and Surface Characteristics of Fdm-Based 3d Printed Pla Parts for Multiple Infill Design Patterns. *J. Inst. Eng. (India) Ser. C* **2021**, *102*, 197–207. [CrossRef]
35. Nomani, J.; Wilson, D.; Paulino, M.; Mohammed, M.I. Effect of Layer Thickness and Cross-Section Geometry on the Tensile and Compression Properties of 3d Printed Abs. *Mater. Today Commun.* **2020**, *22*, 100626. [CrossRef]
36. Moradi, M.; Aminzadeh, A.; Rahmatabadi, D.; Rasouli, S.A. Statistical and Experimental Analysis of Process Parameters of 3d Nylon Printed Parts by Fused Deposition Modeling: Response Surface Modeling and Optimization. *J. Mater. Eng. Perform.* **2021**, *30*, 5441–5454. [CrossRef]
37. Shaqour, B.; Abuabiah, M.; Abdel-Fattah, S.; Juaidi, A.; Abdallah, R.; Abuzaina, W.; Qarout, M.; Verleije, B.; Cos, P. Gaining a Better Understanding of the Extrusion Process in Fused Filament Fabrication 3d Printing: A Review. *Int. J. Adv. Manuf. Technol.* **2021**, *114*, 1279–1291. [CrossRef]
38. Khan, M.M.K.; Liang, R.F.; Gupta, R.K.; Agarwal, S. Rheological and Mechanical Properties of Abs/Pc Blends. *Korea-Aust. Rheol. J.* **2005**, *17*, 1–7.
39. Hu, G.-S.; Wang, B.-B.; Gao, F.-Z. Investigation on the Rheological Behavior of Nylon 6/11. *Mater. Sci. Eng. A* **2006**, *426*, 263–265. [CrossRef]
40. Wu, W.; Geng, P.; Li, G.; Zhao, D.; Zhang, H.; Zhao, J. Influence of Layer Thickness and Raster Angle on the Mechanical Properties of 3d-Printed Peek and a Comparative Mechanical Study between Peek and Abs. *Materials* **2015**, *8*, 5834–5846. [CrossRef]
41. Basgul, C.; Yu, T.; MacDonald, D.W.; Siskey, R.; Marcolongo, M.; Kurtz, S.M. Does Annealing Improve the Interlayer Adhesion and Structural Integrity of Fff 3d Printed Peek Lumbar Spinal Cages? *J. Mech. Behav. Biomed. Mater.* **2020**, *102*, 103455. [CrossRef]
42. Nancharaiah, T.R.D.R.V.; Raju, D.R.; Raju, V.R. An Experimental Investigation on Surface Quality and Dimensional Accuracy of Fdm Components. *Int. J. Emerg. Technol.* **2010**, *1*, 106–111.
43. Vasudevarao, B.; Natarajan, D.P.; Henderson, M.; Razdan, A. Sensitivity of Rp Surface Finish to Process Parameter Variation 251. In Proceedings of the 2000 International Solid Freeform Fabrication Symposium, Austin, TX, USA, 8–10 August 2000.
44. Vijay, P.; Danaiah, P.; Rajesh, K.V.D. Critical Parameters Effecting the Rapid Prototyping Surface Finish. *J. Mech. Eng. Autom.* **2011**, *1*, 17–20. [CrossRef]
45. Peng, A.H. Methods of Improving Part Accuracy During Rapid Prototyping. *Adv. Mater. Res.* **2012**, *430–432*, 760–763. [CrossRef]
46. Lužanin, O.; Movrin, D.; Plančak, M. Experimental Investigation of Extrusion Speed and Temperature Effects on Arithmetic Mean Surface Roughness in Fdm-Built Specimens. *J. Technol. Plast.* **2013**, *38*, 179–190.
47. Patel, D.M. Effects of Infill Patterns on Time, Surface Roughness and Tensile Strength in 3d Printing. *Int. J. Eng. Dev. Res.* **2017**, *5*, 566–569.
48. Sood, A.K. Study on Parametric Optimization of Fused Deposition Modelling (FDM) Process. Ph.D. Thesis, National Institute of Technology, Rourkela, India, 2011.
49. Wang, P.; Zou, B.; Ding, S. Modeling of Surface Roughness Based on Heat Transfer Considering Diffusion among Deposition Filaments for Fdm 3d Printing Heat-Resistant Resin. *Appl. Therm. Eng.* **2019**, *161*, 114064. [CrossRef]

50. Baig, A.M.; Moeed, K.M.; Haque, M.S. A Study on the Effect of Infill Percentage and Infill Pattern on the Compressive Behaviour of the Fdm Printed Polylactic Acid (Pla) Polymer. *GRD J. Glob. Res. Dev. J. Eng.* **2019**, *9*, 5–8.
51. Motaparti, K.P.; Taylor, G.; Leu, M.-C.; Chandrashekhara, K.; Castle, J.; Matlack, M. Effects of Build Parameters on Compression Properties for Ultem 9085 Parts by Fused Deposition Modeling. In Proceedings of the 27th Annual International Solid Freeform Fabrication Symposium, Austin, TX, USA, 8–10 August 2016.
52. Abbas, T.; Othman, F.M.; Ali, H.B. Effect of Infill Parameter on Compression Property in Fdm Process. *Int. J. Eng. Res. Appl.* **2017**, *7*, 16–19.
53. Ahn, S.-H.; Montero, M.; Odell, D.; Roundy, S.; Wright, P.K. Anisotropic Material Properties of Fused Deposition Modeling Abs. *Rapid Prototyp. J.* **2002**, *8*, 248–257. [CrossRef]
54. *ASTM D792-20*; Standard Test Methods for Density and Specific Gravity (Relative Density) of Plastics by Displacement. American Standard Methods. The American Society for Testing and Materials (ASTM): West Conshohocken, PA, USA, 2008.
55. Forster, A.M. *Materials Testing Standards for Additive Manufacturing of Polymer Materials: State of the Art and Standards Applicability*; National Institute of Standards and Technology (NIST): Gaithersburg, MD, USA, 2015.
56. Gaweł, A.; Kuciel, S. The Study of Physico-Mechanical Properties of Polylactide Composites with Different Level of Infill Produced by the Fdm Method. *Polymers* **2020**, *12*, 3056. [CrossRef]
57. Neter, J.; Kutner, M.H.; Nachtsheim, C.J.; Wasserman, W. *Applied Linear Statistical Models*; Marshall University: Huntington, WV, USA, 1996.
58. He, F.; Khan, M. Effects of Printing Parameters on the Fatigue Behaviour of 3d-Printed Abs under Dynamic Thermo-Mechanical Loads. *Polymers* **2021**, *13*, 2362. [CrossRef]
59. Kuznetsov, V.E.; Solonin, A.N.; Urzhumtsev, O.D.; Schilling, R.; Tavitov, A.G. Strength of Pla Components Fabricated with Fused Deposition Technology Using a Desktop 3d Printer as a Function of Geometrical Parameters of the Process. *Polymers* **2018**, *10*, 313. [CrossRef]
60. Ding, S.; Zou, B.; Wang, P.; Ding, H. Effects of Nozzle Temperature and Building Orientation on Mechanical Properties and Microstructure of Peek and Pei Printed by 3d-Fdm. *Polym. Test.* **2019**, *78*, 105948. [CrossRef]
61. Basgul, C.; Yu, T.; MacDonald, D.W.; Siskey, R.; Marcolongo, M.; Kurtz, S.M. Structure–Property Relationships for 3d-Printed Peek Intervertebral Lumbar Cages Produced Using Fused Filament Fabrication. *J. Mater. Res.* **2018**, *33*, 2040–2051. [CrossRef] [PubMed]
62. Hashmi, A.W.; Mali, H.S.; Meena, A. The Surface Quality Improvement Methods for Fdm Printed Parts: A Review. In *Fused Deposition Modeling Based 3D Printing*; Springer: Cham, Switzerland, 2021; pp. 167–194.
63. Guessasma, S.; Belhabib, S.; Nouri, H.; Ben Hassana, O. Anisotropic Damage Inferred to 3d Printed Polymers Using Fused Deposition Modelling and Subject to Severe Compression. *Eur. Polym. J.* **2016**, *85*, 324–340. [CrossRef]
64. Garzon-Hernandez, S.; Garcia-Gonzalez, D.; Jérusalem, A.; Arias, A. Design of Fdm 3d Printed Polymers: An Experimental-Modelling Methodology for the Prediction of Mechanical Properties. *Mater. Des.* **2020**, *188*, 108414. [CrossRef]
65. Calafel, I.; Aguirresarobe, R.H.; Peñas, M.I.; Santamaria, A.; Tierno, M.; Conde, J.I.; Pascual, B. Searching for Rheological Conditions for Fff 3d Printing with Pvc Based Flexible Compounds. *Materials* **2020**, *13*, 178. [CrossRef] [PubMed]
66. Zhang, J.; Wang, X.Z.; Yu, W.W.; Deng, Y.H. Numerical Investigation of the Influence of Process Conditions on the Temperature Variation in Fused Deposition Modeling. *Mater. Des.* **2017**, *130*, 59–68. [CrossRef]
67. Samykano, M.; Selvamani, S.K.; Kadirgama, K.; Ngui, W.K.; Kanagaraj, G.; Sudhakar, K. Mechanical Property of Fdm Printed Abs: Influence of Printing Parameters. *Int. J. Adv. Manuf. Technol.* **2019**, *102*, 2779–2796. [CrossRef]
68. Alzoubi, M.F.; Tanbour, E.Y.; Al-Waked, R. Compression and Hysteresis Curves of Nonlinear Polyurethane Foams under Different Densities, Strain Rates and Different Environmental Conditions. In Proceedings of the ASME International Mechanical Engineering Congress and Exposition, Denver, CO, USA, 11–17 November 2011.
69. Singh, L.; Khan, R.A.; Aggarwal, M.L. Relationship between Damping Factor and Compressive Residual Stress for Shot-Peened Austenitic Stainless Steel. *Int. Sch. Res. Not.* **2011**, *2011*, 867484. [CrossRef]
70. Prasadh, S.; Wong, R.C.W. Unraveling the Mechanical Strength of Biomaterials Used as a Bone Scaffold in Oral and Maxillofacial Defects. *Oral Sci. Int.* **2018**, *15*, 48–55. [CrossRef]
71. Ravandi, M.R.M.; Dezianian, S.; Ahmad, M.T.; Ghoddosian, A.; Azadi, M. Compressive Strength of Metamaterial Bones Fabricated by 3d Printing with Different Porosities in Cubic Cells. *Mater. Chem. Phys.* **2023**, *299*, 127515. [CrossRef]

polymers

MDPI

Article

Thermo-Mechanical Behavior and Strain Rate Sensitivity of 3D-Printed Polylactic Acid (PLA) below Glass Transition Temperature (T_g)

Vukašin Slavković [1,*], Blaž Hanželič [2], Vasja Plesec [2], Strahinja Milenković [1] and Gregor Harih [2]

[1] Faculty of Engineering, University of Kragujevac, Sestre Janjic 6, 34000 Kragujevac, Serbia; strahinja.milenkovic@fink.rs
[2] Laboratory for Integrated Product Development and CAD, Faculty of Mechanical Engineering, University of Maribor, Smetanova ulica 17, SI-2000 Maribor, Slovenia; blaz.hanzelic1@um.si (B.H.); vasja.plesec@um.si (V.P.); gregor.harih@um.si (G.H.)
* Correspondence: vukasin@fink.rs

Abstract: This study investigated the thermomechanical behavior of 4D-printed polylactic acid (PLA), focusing on its response to varying temperatures and strain rates in a wide range below the glass transition temperature (T_g). The material was characterized using tension, compression, and dynamic mechanical thermal analysis (DMTA), confirming PLA's strong dependency on strain rate and temperature. The glass transition temperature of 4D-printed PLA was determined to be 65 °C using a thermal analysis (DMTA). The elastic modulus changed from 1045.7 MPa in the glassy phase to 1.2 MPa in the rubber phase, showing the great shape memory potential of 4D-printed PLA. The filament tension tests revealed that the material's yield stress strongly depended on the strain rate at room temperature, with values ranging from 56 MPa to 43 MPA as the strain rate decreased. Using a commercial FDM Ultimaker printer, cylindrical compression samples were 3D-printed and then characterized under thermo-mechanical conditions. Thermo-mechanical compression tests were conducted at strain rates ranging from 0.0001 s^{-1} to 0.1 s^{-1} and at temperatures below the glass transition temperature (T_g) at 25, 37, and 50 °C. The conducted experimental tests showed that the material had distinct yield stress, strain softening, and strain hardening at very large deformations. Clear strain rate dependence was observed, particularly at quasi-static rates, with the temperature and strain rate significantly influencing PLA's mechanical properties, including yield stress. Yield stress values varied from 110 MPa at room temperature with a strain rate of 0.1 s^{-1} to 42 MPa at 50 °C with a strain rate of 0.0001 s^{-1}. This study also included thermo-mechanical adiabatic tests, which revealed that higher strain rates of 0.01 s^{-1} and 0.1 s^{-1} led to self-heating due to non-dissipated generated heat. This internal heating caused additional softening at higher strain rates and lower stress values. Thermal imaging revealed temperature increases of 15 °C and 18 °C for strain rates of 0.01 s^{-1} and 0.1 s^{-1}, respectively.

Keywords: smart materials; shape memory polymer; 3D printing; 4D printing; thermo-mechanical experiments

check for updates

Citation: Slavković, V.; Hanželič, B.; Plesec, V.; Milenković, S.; Harih, G. Thermo-Mechanical Behavior and Strain Rate Sensitivity of 3D-Printed Polylactic Acid (PLA) below Glass Transition Temperature (T_g). *Polymers* **2024**, *16*, 1526. https://doi.org/10.3390/polym16111526

Academic Editors: Anton Ficai and Cristina-Elisabeta Pelin

Received: 17 April 2024
Revised: 23 May 2024
Accepted: 24 May 2024
Published: 29 May 2024

1. Introduction

Smart materials are a prominent class of materials that have revolutionized both research and engineering. In general, materials with shape memory, usually named shape memory materials (SMMs), are characterized by the shape memory effect (SME). SMMs are divided into several groups: shape memory polymers (SMPs), shape memory alloys (SMAs), shape memory hydrogels (SMHs), and shape memory ceramics (SMCs). SMPs can respond to a various external stimulus and can recover their deformed shape and return to their permanent shape from a programmed (temporary) shape under the influence of light [1,2], heat [3], magnetic [4], electricity [5], moisture [6], and water [7,8]. For a long time, SMAs have been very prevalent, especially in human medicine [9], aerospace [10], and robotics [11]. However, today, SMPs and SMHs are slowly taking the lead among other SMMs due to their broad applicability and the relatively low cost of the raw material and manufacturing. The advantages of SMPs and SMHs over, primarily SMAs, are that the stiffness can be adjusted in a wide glass transition temperature range T_g (55–100 °C) [12]. Besides that, SMPs are characterized by low density (≈1.2 g/cm^3), large deformations, biodegradability, biocompatibility, as well as low thermal conductivity [13,14]. SMPs can also restore shape after being exposed to very large plastic deformations of ≈500%, while for SMAs, it is ≈6–7%. In addition to mechanical factors, technological factors such as cost, fabrication, toxicity, or recycling potential significantly affect the predominance of SMPs over SMAs [15] in the era of green technologies and green polymers [16].

The emergence of 4D printing represents an innovative fusion of smart materials and additive manufacturing techniques, propelling scientific exploration into material responsiveness to external stimuli and the development of intelligent structures for various applications. Smart materials in 4D printing adapt their properties or shapes in response to external stimuli. These materials can also harness energy, typically thermal, to perform mechanical tasks [17]. Four-dimensional printing technologies have facilitated scientific exploration into material research, stimulus responsiveness, mathematical modeling, and the subsequent development of intelligent structures. Four-dimensional printing has garnered increased interest lately, notably through the pioneering work of Professor Tibbits' research group at the Massachusetts Institute of Technology (MIT) [18]. Like most rapidly growing technologies, 4D printing relies on the rapid development of smart materials, 3D printers, mathematical modeling, and design [19]. Figure 1 shows that, in contrast to 3D printing, the output product of 4D printing is an active or dynamic structure that can be activated with appropriate external stimulus or energy input. The development of the new 4D printing industry is directly dependent on material science and the development of new materials. In addition to the materials growing and advancing in technologies such as fused deposition modeling (FDM) [20] or fused filament fabrication (FFF), digital light processing (DLP) [21], stereolithography (SLA) [22], selective laser melting (SLM), and inkjet [23,24], it is also a condition for further progress in this field. Various materials such as PVC [25,26], PETG [27], and photopolymers [28] are used in 4D printing, and even blends [29] and multimaterials for 3D-printed auxetic structures [30] are used in 4D printing. This variety of materials and printing technologies, and even the creation of composites, opens completely new perspectives and possibilities for the use of 4D printing in various fields.

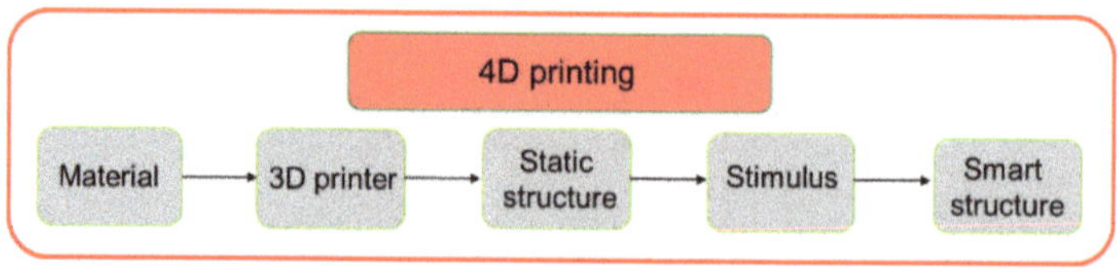

Figure 1. The 4D printing concept with PLA material.

PLA is a material that has many applications, both in medicine and non-medical fields. One of its key features is its biocompatibility, which makes it safe for use in medical treatments. As it is a product of the human body and is obtained from natural sources, it is also biodegradable. This is especially important for medical applications, where the device needs to be absorbed by the body after it has served its function.

There is a growing trend of replacing devices composed of metal or alloys with polymers to allow for the gradual healing of diseased tissue through the mechanical weakening of the polymer devices. Additionally, as biodegradation occurs over time, there is no need for additional procedures to remove the device [31]. Due to the ability to customize the chemical structure and mechanical characteristics to the biochemical environment, PLA is widely used in biomedicine. It is used in various applications, including stents [32], orthopedic screws [33], supports for growing various cells, muscle tissue, bone and cartilage regeneration, planting osteogenic stem cells and implantation into bone defects [34], and drug delivery and delivery devices [35]. The use of PLA in additive manufacturing enables the production of complex biomedical devices based on computer-aided design and construction (CAD); in particular, with the use of patient-specific anatomical data, it leads to the creation of one-of-a-kind implants [36] and prosthesis sockets [37]. A new challenge in the field of additive technologies is the application of 3D printing in the production of PLA composites, with or without reinforcement [38], scaffolds [39], biodegradable stents [40] and, lately, in auxetic energy absorption structures [41–47]. PLA can also be blended with other materials such as TPU in order to show that, by changing the composition and programming temperature, the desired properties for different applications can be achieved so that the highest fixity, recovery, and stress recovery are obtained in hot-, cold-, and warm-programmed samples by manipulating the input energy and temperature [48]. Besides other thermoplastics used in FFF, PLA also shows potential for blending with natural materials such as wood [49].

In the last decade, the number of papers with mechanical tests of FFF samples has increased. In ref. [50], the authors compared the mechanical characteristics of the unidirectional 3D-printed material with that of homogeneous injection-molded PLA, showing that manufacturing by 3D printing and annealing improves the toughness of samples. One of the latest research studies dealt with the influence of strain rate and temperature on the mechanical behavior of a PLA printed structure in tension [51]. The study aimed to analyze the effect of the infill line distance of 3D-printed circular samples on their compressive elastic behavior during cyclic compressive loading [52]. In the paper presented in [53], uniaxial tensile responses of 3D-printed polylactic acid (PLA) samples following standard ASTM-D412 have been studied to characterize the mechanical properties at three temperatures: 30 °C, 40 °C, and 50 °C. Also, this study includes quasi-static compressive experiments performed on polymetric tubes with different temperatures. In ref. [54], the authors conducted experimental testing to determine the compression performance and deformation behavior of 3D-printed PLA lattice structures.

In order to determine the influence of anisotropy and infill on the SME effect in printed materials, the authors in [55] examined the samples using uniaxial tensile tests and compressive tests to study the effect of infill patterns on mechanical properties. Paper [56] presents an experimental study on the compression of uniaxial properties of a PLA material manufactured with FFF in accordance with the requirements and conditions established in the ISO 604 standard, characterizing the compression stiffness, the compression yield stress, the field of displacements, and stress along its elastic area until it reaches the compression yield stress and ultimate yield stress data; the results showed that PLA material is promising for the manufacture of low-volume industrial components that are subject to compression.

The authors in [57] introduced a novel honeycomb structure that can enhance the compression property and energy absorption 4D printing with PLA materials, showing that the novel honeycomb had a high compression property and had high energy absorption capacity. In this work [58], the influence of several factors such as printing temperature, bed temperature, printing speed, fan speed, and flow was studied, showing that the parameters of extrusion-based 3D printing influence the transformability of PLA-based materials. In ref. [59], PLA was used in the 4D printing process for the manufacturing of complex geometry absorber components produced by FFF with varying printing parameters (temperature at the nozzle, the deposition speed, the layer thickness) and activation temperatures. The experiments showed that the components had good shape memory properties that were mostly influenced by activation temperature. Experimental tensile and compression tests were conducted in [60] on FFF PLA parts to evaluate the difference of main mechanical properties in the tensile and compressive state.

In paper [61], the monotonic, fatigue, and creep behavior of PLA under compression was studied using cylindrical specimens that were tested according to ASTM D695 to identify and quantify the effects of printing parameters on the compression behavior of these specimens and failure mechanisms, finding that compressive strength is linearly dependent with the density of the samples. In paper [62], the authors examined PLA and PLA-Cu samples under both static and dynamic loading using a universal testing machine and a split Hopkinson pressure bar apparatus, showing that the addition of copper powder increased the yield strength of the composite material significantly compared with pure PLA, with both materials being strain rate-sensitive. Also, study [63] examined the strain rate sensitivity of five thermoplastic materials (PLA, ABS, PC, CPE+, and nylon) under various tensile test speeds to study strain rate influence on the mechanical characteristics of FFF 3D-printed materials. The influence of strain rate on tensile strength and yield strength in dynamic conditions was examined.

The compression behavior of 4D-printed metamaterials with various Poisson ratios in [64] showed that cellular metamaterials with zero Poisson ratios possessed superior vibration isolation capability compared with negative or positive Poisson ratio cellular metamaterials at different deformation stages by using a comprehensive analysis. A very detailed study presented in [65] described the influence of printing parameters on the mechanical response of polylactic acid (PLA), high-impact polystyrene (HIPS), and acrylonitrile–butadiene–styrene (ABS), with special reference to shape memory in a 4D print while stretching at different speeds and at different temperatures. In order to examine the tensile strain rate performance of 3D-printed PLA with various printing orientations in paper, ref. [66] conducted a study using different strain rates ranging from the slowest to medium speed. The study, like most of the previous ones, showed different responses when the rate of deformation increased using an additional analysis of elongation and bending.

Even though the thermo-mechanical behavior of 4D-printed PLA has been studied for years, comprehensive stress–strain data regarding a strain of ≈50%, including loading and unloading, a variety of strain rates in the range of 0.0001 to 0.1 s^{-1}, and temperature ranges of 23 to 50 °C are not available. This paper aims to extensively and experimentally investigate the dependence of FDM 4D-printed PLA on a wide range of strain rates and temperatures in compression scenarios with large deformations. Due to high strain rates, test conditions can occur that are almost adiabatic. Determining the existence of self-heating in FDM PLA and the consequent additional softening is a special challenge. All tests were carried out in coupled thermo-mechanical conditions so that the research results contribute to the expansion of knowledge in the field and provide new insights into the behavior of 4D-printed PLA. The main motive is to determine all parameters related to the macro-mechanical characteristics of the material, which will assist the development of a coupled thermo-mechanical constitutive model for accurately modeling the behavior of the material using the finite element method (FEM). The most important motive is the possibility of expanding knowledge and further research in the field of auxetic structures, whose primary mode of use and exploitation is radial and uniaxial compression at various strain rates.

In the Section 2, a comprehensive outline of the materials and methods utilized is presented, encompassing details such as the filament used for the 3D printing of samples, sample annealing procedure, uniaxial isothermal tensile filament, compression of cylindrical printed samples, and DMTA; the findings obtained through DMTA analysis and the uniaxial tensile and compression testing are eloquently presented and thoroughly discussed within the context of the paper's primary objectives. The paper is concluded with a summary of the most critical findings and thoughtful suggestions for future research.

2. Materials and Methods

2.1. Three-Dimensional Printing of PLA Samples

The required PLA samples for all thermo-mechanical uniaxial compression tests were printed by 3D printing a PLA filament with a diameter of 1.75 mm, manufactured by Ultimaker (Utrecht, The Netherlands). Based on the data sheet of the manufacturer, the material has a density of 1.24 $\frac{g}{cm^3}$, a melting rate during printing (MFR) of 6.09 $\frac{g}{10\ min}$, and a melting temperature of 145–160 °C. The samples used in this study were manufactured using a UM2 + FDM 3D printer (Ultimaker, Utrecht, The Netherlands) equipped with a 0.4 mm nozzle. The 3D printing settings were chosen based on the manufacturer's recommendation and experience with a 100% infill, nozzle temperature of 210 °C, working plate temperature of 60 °C, layer deposition height of 0.1 mm, and printing speed of 40 $\frac{mm}{s}$. Cura 5.3.0 (Ultimaker, Utrecht, The Netherlands) slicer software was used for preparing the G-code for the 3D printer. Barreling and buckling of the samples were avoided with the model's orientation as observed in previous studies, where it shown that these phenomena can be avoided at a ratio of height and diameter of below 2 [67–69]. The geometry of the samples and the later test procedure have been defined according to the standard for compression tests ASTM D695 [70], as seen in Figure 2. All tests were performed to obtain stress–strain curves at the maximum safe strain value, e.g., before crack or fracture initiated in the samples. It should be noted that although one of the main advantages of additive manufacturing is the production speed, when considering smart materials in additive technologies, especially within 4D printing, the printing speed must be significantly reduced. As shown in [71], the production of samples in 4D printing at high printing speeds causes extremely high anisotropy due to residual thermal deformations. The basic parameters used in FDM printing for these samples are outlined in Table 1.

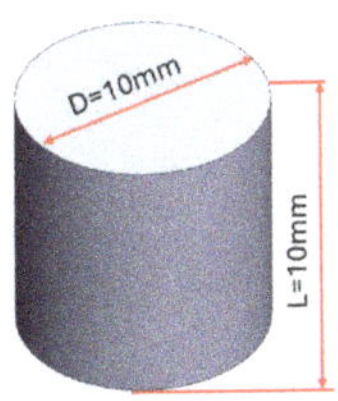

(a)

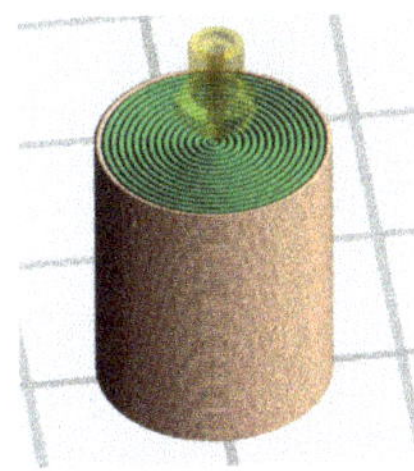

(b)

Figure 2. (**a**) geometry of 4D-printed sample and (**b**) 3D printing preview of slicing patterns.

Table 1. Printing parameters of testing samples.

Parameter	Value
Nozzle diameter	0.4 mm
Layer height	0.1 mm
Infill	100%
Printing speed	40 mm/s
Printing bed temperature	60 °C
Production time	35 min

2.2. *Dynamic Mechanical Thermal Analysis (DMTA)*

The dynamic viscoelastic properties and thermal behavior of 4D-printed PLA were investigated in this study. The experiments were conducted on a solid clamping tool for measurements according to DIN/ISO 6721-1 with the use of Thermo Scientific™ HAAKE™ MARS™ Rheometers (Thermo Fisher Scientific Inc., Waltham, MA, USA) in combination with a controlled test chamber (CTC). The measurements were taken over a temperature range of 40–80 °C, with a strain amplitude of 0.01% and deformation frequency of 1 Hz on standard test prismatic PLA samples; their dimensions were 40 × 10 × 1 mm, and they were fabricated using the same printing parameters that are given in Table 1. The heating rate was set to 2 °C/min, and samples were preheated from room temperature to 40 °C and then kept for 5 min at that temperature. This controlled setup adhered to the guidelines outlined in DIN/ISO 6721-1, ensuring accurate determination of the dynamic mechanical properties of the PLA material under investigation. A DMTA was performed in torsion with a rotational rheometer, where the material was subjected to oscillatory shear while undergoing continuous temperature variation. The geometry of DMTA test samples and the 3D printing preview are shown in Figure 3.

(a)

(b)

Figure 3. (**a**) Geometry of DMTA samples and (**b**) 3D printing preview of slicing patterns.

2.3. Annealing Printed Specimen

After production, the samples were annealed in an oven at a temperature that was ≈20 °C higher than T_g and kept at that temperature for two hours. After that, the samples were slowly cooled in the oven to room temperature over several hours. The annealing procedure after sample fabrication provides several improvements, including reducing imperfections in the samples, reducing porosity, causing better adhesion of material layers, and improving the quality of surface layers. All samples used in the experiments were left in the oven environment (room temperature and humidity of 50%) to eliminate possible external influences and material aging before testing.

2.4. Uniaxial Tests

In order to determine the large deformation behavior of 4D-printed PLA polymer at various strain rates and temperatures lower then T_g, a Shimadzu EHF-EV101K3-070-0A (Kyoto, Japan) universal testing machine equipped with a 100 kN calibrated load cell and temperature chamber was used. Displacement control during the test was performed via the RS485 controller (Schneider Electric, Regensburg, Germany), while the temperature in the temperature chamber was controlled via the EUROTHERM 2408 controller (Worthing, West Sussex, UK) and the iTOOLS 9.87 software package. As the diameter of the filament at 1.75 mm was not adapted to the tension grips, the adaptive tool shown in Figure 4 was used during testing. Tensile tests were performed at room temperature for three strain rates, 0.01, 0.001, and 0.0001 s^{-1}.

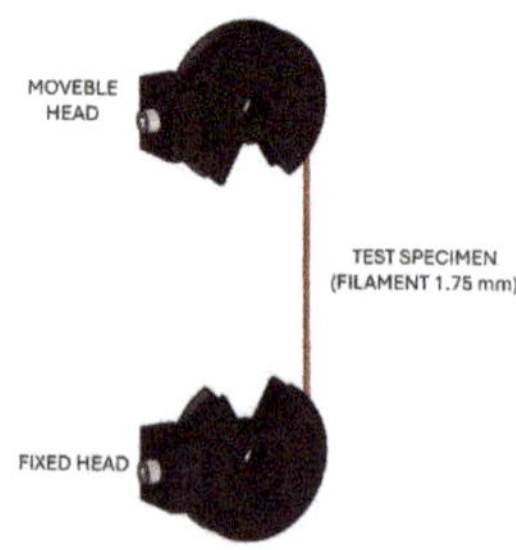

Figure 4. Uniaxial tensile testing of filament procedure: (**left**) equipment for uniaxial filament testing and (**right**) sketch of equipment.

Thermo-mechanical properties in uniaxial compression were measured using a Shimadzu EHF-EV101K3-070-0A universal testing machine (Kyoto, Japan) equipped with a 100 kN calibrated load cell and temperature chamber. Thermocouples were placed close to the surface and at the height of the sample, while the temperature of the chamber in the sample zone was maintained. Cylindrical samples with a height and diameter of 10 mm were used; the ratio of height and diameter was chosen in order to avoid the occurrence of barreling and buckling of the samples as observed in previous studies, where it was shown that these phenomena occur at a ratio of height and diameter exceeding 2. To reduce the friction, Teflon strips were placed between the sample and the surface of the compression platens. The procedure of isothermal tests was defined as follows: the specimen was placed on a previously applied Teflon (PTFE) strip on the bottom compression platen; then the chamber was heated to the desired test temperature. In order to achieve temperature equilibrium, the specimen was kept in the heated chamber ≈30 min before starting the uniaxial compression test. The upper moving platen moved freely for a given displacement at a constant strain rate for the compression tests. All successful experiments were repeated three times for each strain rate and temperature (36 specimens in total) to ensure the repeatability of results and exclude potential mistakes. Uniaxial compression experiments were conducted at four different strain rates, 0.1, 0.01, 0.001, and 0.0001 s^{-1}, and at three temperatures, 25, 37, and 50 °C. The mechanical test results, including the upper and lower limits of measured stresses, can be found in Supplementary Material Figure S1.

2.5. Measurement of Temperature Change at High Strain Rates

This study examined PLA material with a shape memory effect and 4D printing properties. The measurement of temperature increase in the samples was performed at room temperature and at strain rates of 0.01 s^{-1} and 0.1 s^{-1}. Even before the initial tests, these strain rates were identified as those for which an almost adiabatic scenario is established for thermoplastics [72–75]. A Flir I7 infrared camera was used for the measurement of temperature evolution during compression tests.

3. Results and Discussion

3.1. DMTA

Figure 5 allows for the identification of three characteristic zones of PLA: glass transition (blue area in the figure), a solid or glassy phase with high values of elastic moduli (left of the glass transition zone), and a rubbery or soft phase of low values of elastic moduli (right of the transition zone). Figure 5 shows changes in the storage modulus and tan delta in the DMTA results. According to Figure 5, the glass-to-rubber transition zone for PLA starts at 57 °C and continues to 73 °C. Also, the middle value of the storage modulus is detected at 65 °C, which represents the glass transition temperature. The extreme peak of the storage modulus drop occurs in a narrow temperature range. The glass transition temperature of PLA, whose position is defined by the peak of the loss tangent (tan delta), is approximately at T_g = 65 °C, at which a significant drop in the elastic modulus occurs. In the given temperature range, the storage modulus decreased from 1045.7 MPa (glassy phase) to 1.2 MPa (rubber phase), and their high ratio (more than two orders of magnitude) shows the great shape memory potential of 4D-printed PLA [76]. Both the drop in the elastic modulus and peak of the tan delta are clear in the marked blue area of the glass transition zone. The results are in agreement with previous studies [77,78] for 4D-printed PLA, Table 2 shows the summary of the results of measuring the transition temperature T_g and the changes in the storage modulus G', loss modulus G'' and loss tangent tan delta with the temperature obtained by the DMTA tests.

Table 2. DMTA results.

G' (MPa)	**G'' (MPa)**	**G'/G'' (–)**	**T_g (°C)**
1045	1.2	>100	65

3.2. Uniaxial Tensile Tests of Filament

This section provides confirmation of the dependence of the base material PLA, used for printing the samples, on the strain rate. As expected, Figure 6 shows a typical distribution of curves for thermoplastic materials that depend on the rate of deformation. At all three strain rates, the elastic range up to the point of over-yielding is expressed. The non-linear increase in stress follows up to the yield point, which is also the point of the highest stress, after which deformational softening of the material follows. The yield stresses are 55 MPa, 48 MPa, and 42 MPa for strain rates of 0.01 s^{-1}, 0.001 s^{-1}, and 0.0001 s^{-1}, respectively. Also, all yield points lie in the range of 6–8% of deformation. The results of the tensile filament test are summarized in Table 3.

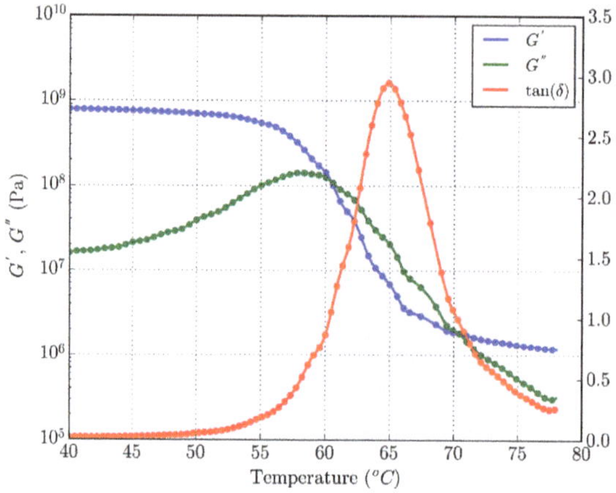

Figure 5. DMTA results for 4D-printed PLA.

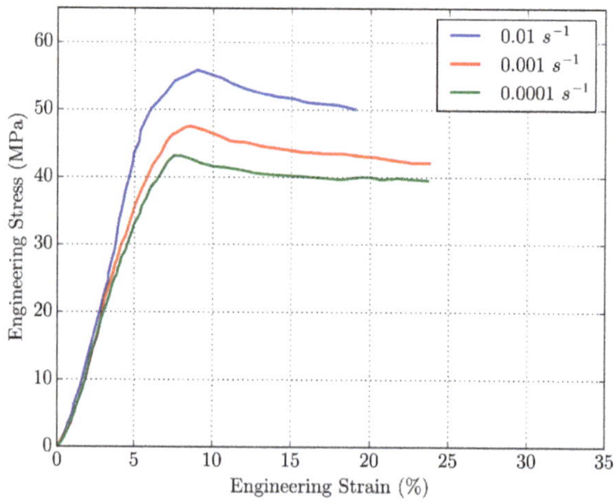

Figure 6. Stress–strain curves for PLA filament at various strain rates.

Table 3. Results of the filament tensile test.

Strain Rate	Temperature (°C)	Tensile Strength (MPa)	Elongation at Break (%)
0.01 s^{-1}	23	55.83 ± 1.54	21.83 ± 3.66
0.001 s^{-1}	23	47.83 ± 1.31	30 ± 4.9
0.0001 s^{-1}	23	43 ± 0.816	41.33 ± 3.09

3.3. *Uniaxial Compression Testing*

A series of uniaxial compression experiments have been conducted on the 4D-printed PLA. All tests were conducted for temperatures below the determined glass transition temperature of T_g = 65 °C. The cylindrical compression test samples were 10 mm in diameter and 10 mm tall, created according to previous studies of uniaxial compression.

Samples were annealed right after 3D printing by heating in an oven at a temperature about 20 °C above the determined glass transition temperature for two hours before slow cooling to room temperature. The experiments were conducted using a servo-hydraulic Shimadzu testing machine equipped with a thermal chamber. In order to heat the compression steel platens uniformly, sample and steel platens were allowed to heat at the testing temperature for about 30 min prior to testing. To reduce friction at the contact surfaces, Teflon (PTFE) films were applied. Figure 7 shows the as-printed cylindrical samples and samples after the compression test. It can be observed that although friction is present during the test, visual inspection of the edges of the sample shows a reduced effect of friction due to the use of Teflon strips. The uniaxial compression experiments on PLA were conducted for the temperatures of 25, 37, and 50 °C at four strain rates, 0.0001, 0.001, 0.01, and 0.1 s^{-1}. The compression tests were conducted at a strain level of ≈50% (0.68 true strain). Because the drop in the mechanical response is observed for the higher strain rates, the analysis of results was separated into two groups: isothermal testing at the strain rates of 0.0001 and 0.001 s^{-1} and adiabatic testing at the strain rates of 0.01 and 0.1 s^{-1}.

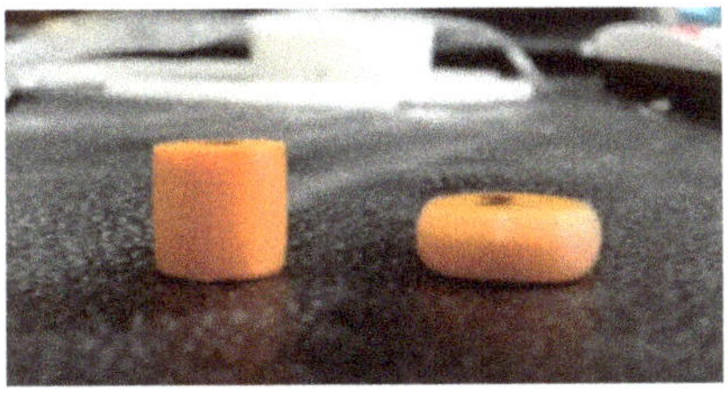

Figure 7. PLA 4D-printed cylindrical samples—as printed (**left**) and after compression (**right**).

Figures 8 and 9 show that 4D-printed PLA polymer exhibits a strain rate and temperature-dependent response typical for solid or glassy phases under isothermal experimental conditions. The material has shown a tendency of glassy polymer behavior under a T_g with an initial elastic region and rate-dependent yield point, followed by strain softening and strain hardening at larger strains. Figure 8 shows representative stress–strain curves for PLA at strain rates of 0.0001 and 0.001 s^{-1} at temperatures of 25, 37, and 50 °C. Referring to Figure 8, it can be observed that as the temperature increases from 25 to 50 °C, the yield stress decreases from ≈80 MPa to ≈40 MPa for the strain rate of 0.0001 s^{-1} and from ≈90 MPa to ≈60 MPa for the strain rate of 0.001 s^{-1}. In the case of both strain rates, strain hardening at large strains is present. Figure 9 shows a set of stress–strain curves for the strain rates of 0.0001 and 0.001 s^{-1} at temperatures of 25, 37, and 50 °C. Referring to Figure 9, which shows stress–strain curves at various fixed temperatures and at two different strain rates, a clear strain rate dependence is observed in the material. In this case, the yield stress of the material decreases by ≈10 MPa for each decade decrease in strain rate at given temperatures, resulting in ≈80 MPa and ≈70 MPa, ≈70 MPa and ≈60 MPa, and ≈60 MPa and ≈50 MPa for the temperatures of 25, 37, and 50 °C, respectively. Upon unloading, about a 4% strain is reversible when the temperature is held constant. Increasing the temperature at the same strain rate leads to a drop in stress during strain hardening, but the amount of hardening at larger strains is slightly affected for this temperature range. Even for the very large strains, there is no intersection of stress–strain curves, which indicates that no additional softening caused by the strain rate occurred.

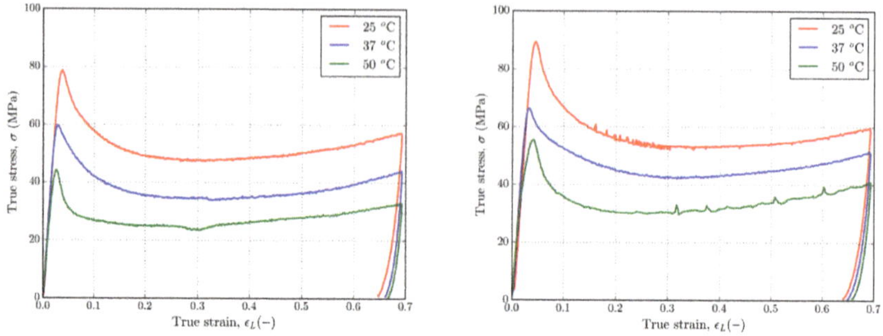

(**a**) 0.0001 s-1 (**b**) 0.001 s-1

Figure 8. Stress–strain curves in uniaxial compression for PLA at strain rates of (**a**) 0.0001 s^{-1} and (**b**) 0.001 s^{-1} and temperatures of 25, 37, and 50 °C.

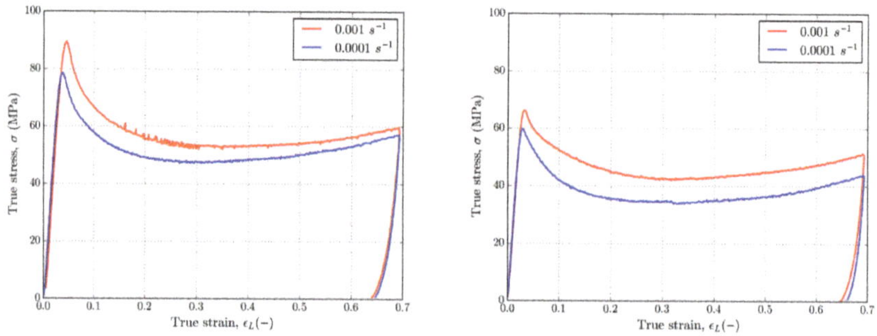

(**a**) 25 °C (**b**) 37 °C

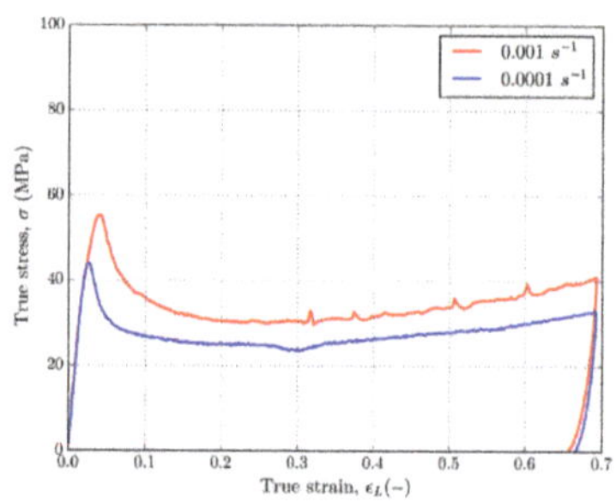

(**c**) 50 °C

Figure 9. Stress–strain curves in uniaxial compression for PLA at temperatures of (**a**) 25 °C, (**b**) 37 °C, and (**c**) 50 °C at strain rates of 0.0001 and 0.001 s^{-1}.

Figures 10 and 11 show that 4D-printed PLA polymer exhibits a strain rate and temperature-dependent response typical for solid or glassy phases at adiabatic experimental conditions. Figure 10 shows representative stress–strain curves for PLA at strain rates of 0.01 and 0.1 s^{-1} at temperatures of 25 °C, 37 °C, and 50 °C. Referring to Figure 10, it can be observed that as temperature increases from 25 to 50 °C, the yield stress decreased from ≈100 MPa to ≈60 MPa for the strain rate of 0.01 s^{-1} and from ≈110 MPa to ≈80 MPa for the strain rate of 0.1 s^{-1}. Slight strain hardening at large strains is present only at a strain rate of 0.01 s^{-1}. Figure 11 shows a set of stress–strain curves for the strain rates 0.001, 0.01, and 0.1 s^{-1} and temperatures 25 °C, 37 °C, and 50 °C. It should be noted that a 0.001 s^{-1}

strain rate curve has been added to this figure in order to show the amount of softening and make a clear distinction between isothermal and adiabatic strain rates. Referring to Figure 11, which shows stress–strain curves at various fixed temperatures and at the three different strain rates, clear strain rate dependence is observed in the material. In these cases, the yield stress of material decreased by ≈10 MPa for each decade decrease in strain rate at given temperatures, resulting in ≈100 MPa, ≈90 MPa, and ≈80 MPa; ≈90 MPa, ≈80 MPa, and ≈70 MPa; and ≈80 MPa, ≈60 MPa, and ≈50 MPa for the temperatures 25 °C, 37 °C, and 50 °C, respectively. In this case, another very important strain rate-dependent feature of PLA is observed for the higher strain rates 0.01 and 0.1 s^{-1} at all temperatures. At higher values of strain, in the case of both higher strain rates, heat generated during plastic deformation could not be dissipated to the surrounding area, which is a clear explanation for the crossing of curves at higher strains, which is in correlation with the tensile [79] and compression [73,74] results of previous studies. Upon unloading, about ≈5% strain is reversible when the temperature is held constant.

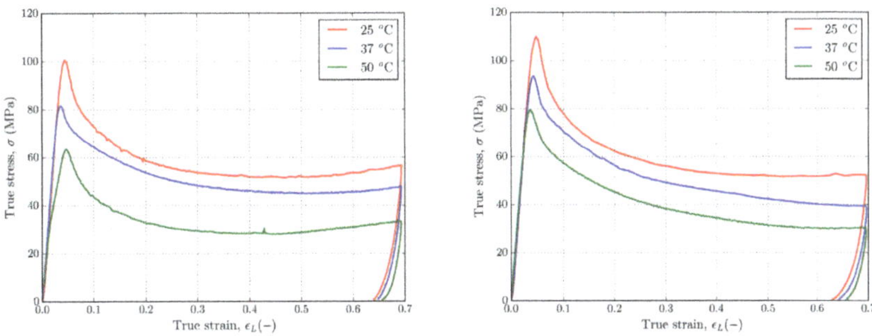

(**a**) 0.01 s^{-1} (**b**) 0.1 s^{-1}

Figure 10. Stress–strain curves in uniaxial compression for PLA at strain rates of (**a**) 0.01 s^{-1} and (**b**) 0.1 s^{-1} and temperatures of 25, 37, and 50 °C.

The characteristics that can be observed in the stress–strain curves are yield curves that have a clearly defined yield point, like the curves grouped by strain rate, and after unloading, ≈5% of strain is reversible. There is an intersection of the strain–stress curves due to the stress drop at strain rates of 0.1 s^{-1} and 0.01 s^{-1} at deformation ≈0.5, which is in agreement with the results for thermoplastics [75,80,81]. The stress drop is a consequence of self-heating in the case of both strain rates; heat cannot be dissipated from the surrounding area because of the speed of the process, which consequently leads to further softening of the thermo-sensitive material.

Figure 12 summarizes the dependence of yield stress on temperatures and strain rates. The yield stress lies in a wide range of values, ranging from 110 MPa for the most extreme case of room temperature and the highest strain rate to 42 MPa at a temperature of 50 °C and the lowest applied strain rate. It is important to emphasize that a logarithmic scale was utilized to illustrate the correlation between yield stress and strain rate. This method of organizing the data offers a clearer comprehension of the significance of yield stress in the thermo-mechanical uniaxial compression of PLA. As anticipated, the yield stress decreases as temperatures rise, and strain rates remain constant. On the other hand, increasing the strain rate at constant temperatures leads to a higher yield stress response of the material. These variations are nearly linear across the board, which can be beneficial in establishing the pattern of change when implementing the constitutive model for 4D-printed PLA.

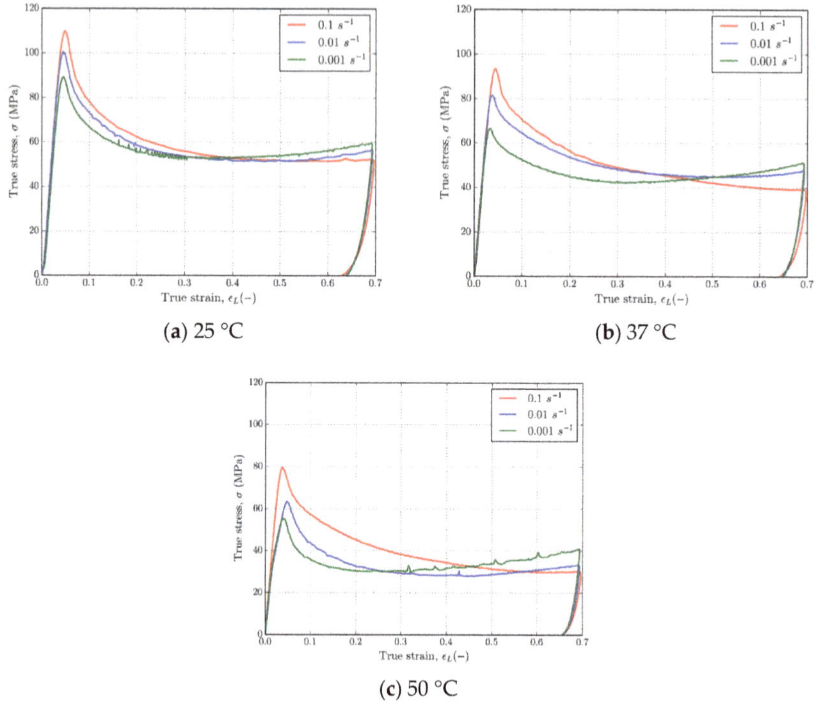

Figure 11. Stress–strain curves for PLA at temperatures of (**a**) 25 °C, (**b**) 37 °C, and (**c**) 50 °C at strain rates of 0.01 and 0.1 s^{-1}.

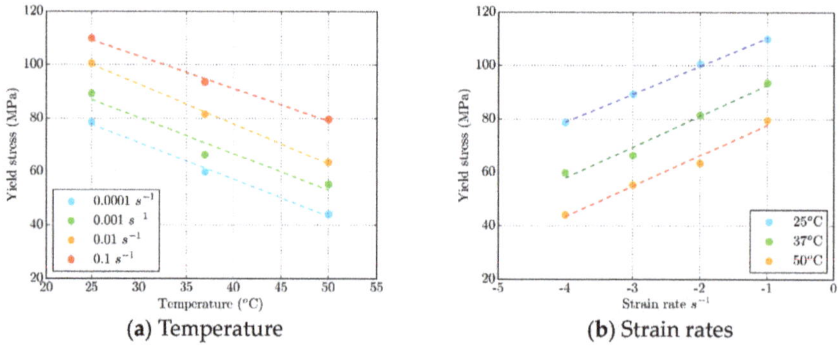

Figure 12. Yield stress value dependence on (**a**) temperature and (**b**) strain rate.

3.4. Adiabatic Tests with Self-Heating and Strain Softening

In this section, the results of self-heating in the material at strain rates of greater than 0.01 s^{-1} are presented, indicating that the mentioned processes can be considered almost adiabatic. A typical thermo-mechanical coupled curve is shown in Figure 13. The temperature was recorded with a thermal imaging camera for the PLA sample at strain rates of 0.01 and 0.1 s^{-1}. The recording shows that the temperature on the surface of the sample increases monotonously, with the temperature rising from room temperature to ≈40 °C. Thermal imaging reveals an observed temperature increase of 15 °C and 18 °C on the sample's surface for the strain rates of 0.01 s^{-1} and 0.1 s^{-1}, respectively. Although this temperature is below the determined Tg for PLA, as shown in the stress–strain curves, there is a drop in mechanical characteristics that can be attributed to the internal heating of the extremely thermo-sensitive material. The maximum temperature was consistently observed in the middle of the sample. It is also shown that the increase in temperature is insignificant before the flow in material and that the temperature continues to rise constantly. Immediately after the start of unloading, a temperature drop of ≈2 °C is observed. These findings will serve as the foundation for upcoming research, which will center on employing advanced material modeling methods, including thermo-mechanical coupling, to address self-heating effects.

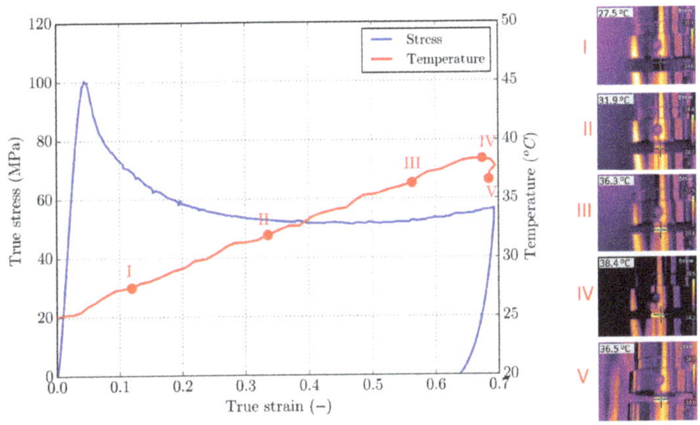

(**a**) 0.01 s-1

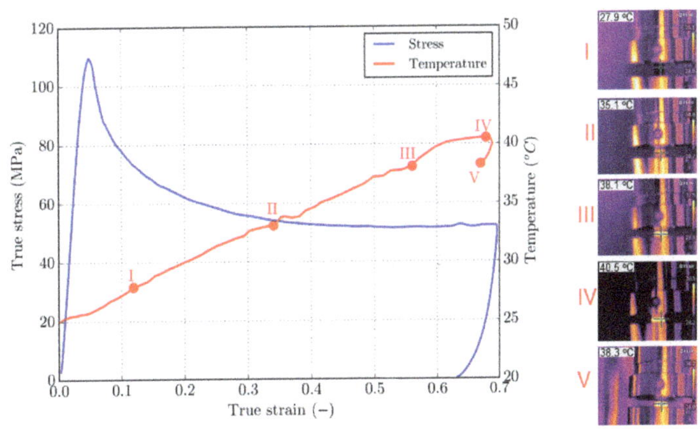

(**b**) 0.1 s-1

Figure 13. Temperature evolution images in 5 characteristic points (I-V) of 4D-printed PLA at strain rates of (**a**) 0.01 s^{-1} and (**b**) 0.1 s^{-1}.

4. Conclusions

This study confirmed PLA material's dependency on strain rate, with the stress–strain curves displaying typical thermoplastic behavior. Yield stresses varied with strain rates, underscoring the material's sensitivity to strain rates. The samples exhibited clear strain rate dependence, particularly at quasi-static rates, with temperature and strain rate variations significantly impacting mechanical properties, including yield stress and deformation behavior. Isothermal compression tests showed predictable stress–strain curves with distinct yield points, while adiabatic tests revealed additional complexities, such as heat accumulation leading to further softening. Observations at higher strain rates indicated self-heating phenomena in PLA, resembling adiabatic conditions. Thermal imaging revealed temperature increases during deformation, with maximum temperatures occurring at the sample's center. The drop in mechanical characteristics attributed to internal heating highlighted the material's thermo-sensitive nature. These findings deepen the understanding of PLA behavior and hold significant implications for practical applications, especially in 3D and 4D printing and manufacturing. Future research should focus on advanced modeling techniques to predict material behavior and explore mitigation strategies for self-heating effects, enhancing PLA-based product reliability and performance in applications with deformations at higher strain rates. The observed adiabatic processes that take place in the material during deformation at high strain rates require the development of an FEM coupled thermo-mechanical constitutive model to simulate self-heating processes with sufficient accuracy. In addition, the ultimate goal of following research is to expand the FEM model with the ability to simulate shape recovery in 4D-printed PLA samples and structures, both at cold (temperatures below T_g) and hot programming (temperatures over T_g). Cold programming is essential because most auxetic structures and metamaterials undergo deformation at temperatures lower than T_g, and the simulation of shape recovery occurs by heating above T_g. The 4D-printed PLA's remarkable stability at lower temperatures and ability to undergo significant deformations at higher temperatures make it an ideal candidate for shape recovery research, particularly in auxetics. Its capacity for precise shape retention and adaptive behavior offers innovative applications in biomedical devices, aerospace, and soft robotics, where dynamic responses to external stimuli, like temperature changes, are essential. These findings will serve as the foundation for upcoming research, which will center on employing advanced material modeling methods, including thermo-mechanical coupling, to address self-heating effects. The goal is to improve the reliability and performance of PLA-based products in applications at higher strain rates, especially auxetic and metamaterial structures, and to create models for simulating shape recovery in 4D-printed PLA structures at cold and hot programming temperatures. A study set up this way could be the basis for the successful and precise modeling of auxetics and metamaterials in cold and hot programming in consecutive research. The lower temperatures used in this research should serve to further focus on cold programming auxetics and research related to shape recovery by heating. Although the printing speeds and directions can also affect 4D printing properties, this study focused on a fully thermo-mechanical coupled characterization of PLA to determine characteristics for further developing the constitutive model.

Supplementary Materials: The following supporting information can be downloaded at: https://www.mdpi.com/article/10.3390/polym16111526/s1, Figure S1: True stress—true strain diagram with upper and lower limits of stresses for all strain rates and temperatures.

Author Contributions: Conceptualization, V.S. and G.H.; methodology, V.S., B.H. and G.H.; software, V.S. and B.H.; validation, V.S., B.H. and S.M.; formal analysis, V.S. and G.H.; investigation, V.S., B.H. and G.H.; resources, V.S.; data curation, V.S., V.P. and B.H.; writing—original draft preparation, V.S., G.H., B.H., V.P. and S.M.; writing—review and editing, V.S., G.H., B.H., V.P. and S.M.; visualization, V.S., B.H. and S.M.; supervision, G.H.; funding acquisition, G.H. All authors have read and agreed to the published version of the manuscript.

Funding: The authors acknowledge the financial support from the Slovenian Research and Innovation Agency (Research Core Funding No. P2-0063).

Institutional Review Board Statement: Not applicable.

Data Availability Statement: The original contributions presented in the study are included in the article and Supplementary Material, further inquiries can be directed to the corresponding author.

Conflicts of Interest: The authors declare no conflicts of interest.

References

1. Lendlein, A.; Jiang, H.; Jünger, O.; Langer, R. Light-Induced Shape-Memory Polymers. *Nature* **2005**, *434*, 879–882. [CrossRef] [PubMed]
2. Lee, K.M.; Koerner, H.; Vaia, R.A.; Bunning, T.J.; White, T.J. Light-Activated Shape Memory of Glassy, Azobenzene Liquid Crystalline Polymer Networks. *Soft Matter* **2011**, *7*, 4318–4324. [CrossRef]
3. Heuchel, M.; Sauter, T.; Kratz, K.; Lendlein, A. Thermally Induced Shape-Memory Effects in Polymers: Quantification and Related Modeling Approaches. *J. Polym. Sci. Part B Polym. Phys.* **2013**, *51*, 621–637. [CrossRef]
4. Razzaq, M.Y.; Behl, M.; Nöchel, U.; Lendlein, A. Magnetically Controlled Shape-Memory Effects of Hybrid Nanocomposites from Oligo(Omega-Pentadecalactone) and Covalently Integrated Magnetite Nanoparticles. *Polymer* **2014**, *55*, 5953–5960. [CrossRef]
5. Schmidt, A.M. Electromagnetic Activation of Shape Memory Polymer Networks Containing Magnetic Nanoparticles. *Macromol. Rapid Commun.* **2006**, *27*, 1168–1172. [CrossRef]
6. Garces, I.T.; Aslanzadeh, S.; Boluk, Y.; Ayranci, C. Effect of Moisture on Shape Memory Polyurethane Polymers for Extrusion-Based Additive Manufacturing. *Materials* **2019**, *12*, 244. [CrossRef] [PubMed]
7. Liu, Y.; Li, Y.; Chen, H.; Yang, G.; Zheng, X.; Zhou, S. Water-Induced Shape-Memory Poly(D,L-Lactide)/Microcrystalline Cellulose Composites. *Carbohydr. Polym.* **2014**, *104*, 101–108. [CrossRef]
8. Bai, Y.; Chen, X. A Fast Water-Induced Shape Memory Polymer Based on Hydroxyethyl Cellulose/Graphene Oxide Composites. *Compos. Part A Appl. Sci. Manuf.* **2017**, *103*, 9–16. [CrossRef]
9. Petrini, L.; Migliavacca, F. Biomedical Applications of Shape Memory Alloys. *J. Metall.* **2011**, *2011*, 501483. [CrossRef]
10. Hartl, D.J.; Lagoudas, D.C. Aerospace Applications of Shape Memory Alloys. *Proc. Inst. Mech. Eng. Part G J. Aerosp. Eng.* **2007**, *221*, 535–552. [CrossRef]
11. Kheirikhah, M.M.; Rabiee, S.; Edalat, M.E. A Review of Shape Memory Alloy Actuators in Robotics. In *RoboCup 2010: Robot Soccer World Cup XIV*; Ruiz-del-Solar, J., Chown, E., Plöger, P.G., Eds.; Springer: Berlin/Heidelberg, Germany, 2011; pp. 206–217.
12. Erkeçoğlu, S.; Sezer, A.D.; Bucak, S. Smart Delivery Systems with Shape Memory and Self-Folding Polymers. In *Smart Drug Delivery System*; Sezer, A.D., Ed.; IntechOpen: Rijeka, Croatia, 2016.
13. Korde, J.M.; Kandasubramanian, B. Naturally Biomimicked Smart Shape Memory Hydrogels for Biomedical Functions. *Chem. Eng. J.* **2020**, *379*, 122430. [CrossRef]
14. Huang, W.M.; Ding, Z.; Wang, C.C.; Wei, J.; Zhao, Y.; Purnawali, H. Shape Memory Materials. *Mater. Today* **2010**, *13*, 54–61. [CrossRef]
15. Liu, C.; Qin, H.; Mather, P.T. Review of Progress in Shape-Memory Polymers. *J. Mater. Chem.* **2007**, *17*, 1543–1558. [CrossRef]
16. Hasan, M.R.; Davies, I.J.; Pramanik, A.; John, M.; Biswas, W.K. Potential of Recycled PLA in 3D Printing: A Review. *Sustain. Manuf. Serv. Econ.* **2024**, *3*, 100020. [CrossRef]
17. Mehrpouya, M.; Vahabi, H.; Janbaz, S.; Darafsheh, A.; Mazur, T.R.; Ramakrishna, S. 4D Printing of Shape Memory Polylactic Acid (PLA). *Polymer* **2021**, *230*, 124080. [CrossRef]
18. Tibbits, S. The Emergence of "4D Printing". *TED Conf.* **2013**.
19. Liu, Y.; Zhang, W.; Zhang, F.; Lan, X.; Leng, J.; Liu, S.; Jia, X.; Cotton, C.; Sun, B.; Gu, B.; et al. Shape Memory Behavior and Recovery Force of 4D Printed Laminated Miura-Origami Structures Subjected to Compressive Loading. *Compos. Part B Eng.* **2018**, *153*, 233–242. [CrossRef]
20. Dudek, P. FDM 3D Printing Technology in Manufacturing Composite Elements. *Arch. Metall. Mater.* **2013**, *58*, 1415–1418. [CrossRef]
21. Jasveer, S.; Jian-bin, X. Comparison of Different Types of 3 D Printing Technologies. *Int. J. Sci. Res. Publ.* **2018**, *8*, 1–9. [CrossRef]
22. Kuang, X.; Zhao, Z.; Chen, K.; Fang, D.; Kang, G.; Qi, H.J. High-Speed 3D Printing of High-Performance Thermosetting Polymers via Two-Stage Curing. *Macromol. Rapid Commun.* **2018**, *39*, 1700809. [CrossRef]
23. Soleimani-Gorgani, A. 14—Inkjet Printing. In *Printing on Polymers*; Izdebska, J., Thomas, S., Eds.; William Andrew Publishing: Norwich, NY, USA, 2016; pp. 231–246. ISBN 978-0-323-37468-2.
24. Riheen, M.A.; Saha, T.K.; Sekhar, P.K. Inkjet Printing on PET Substrate. *J. Electrochem. Soc.* **2019**, *166*, B3036–B3039. [CrossRef]
25. Aberoumand, M.; Soltanmohammadi, K.; Rahmatabadi, D.; Soleyman, E.; Ghasemi, I.; Baniassadi, M.; Abrinia, K.; Bodaghi, M.; Baghani, M. 4D Printing of Polyvinyl Chloride (PVC): A Detailed Analysis of Microstructure, Programming, and Shape Memory Performance. *Macromol. Mater. Eng.* **2023**, *308*, 2200677. [CrossRef]
26. Aberoumand, M.; Rahmatabadi, D.; Soltanmohammadi, K.; Soleyman, E.; Ghasemi, I.; Baniassadi, M.; Abrinia, K.; Bodaghi, M.; Baghani, M. Stress Recovery and Stress Relaxation Behaviors of PVC 4D Printed by FDM Technology for High-Performance Actuation Applications. *Sens. Actuators A Phys.* **2023**, *361*, 114572. [CrossRef]
27. Soleyman, E.; Rahmatabadi, D.; Soltanmohammadi, K.; Aberoumand, M.; Ghasemi, I.; Abrinia, K.; Baniassadi, M.; Wang, K.; Baghani, M. Shape Memory Performance of PETG 4D Printed Parts under Compression in Cold, Warm, and Hot Programming. *Smart Mater. Struct.* **2022**, *31*, 085002. [CrossRef]
28. Sælen, R.L.; Hopperstad, O.S.; Clausen, A.H. Mechanical Behaviour and Constitutive Modelling of an Additively Manufactured Stereolithography Polymer. *Mech. Mater.* **2023**, *185*, 104777. [CrossRef]
29. Mirasadi, K.; Rahmatabadi, D.; Ghasemi, I.; Khodaei, M.; Baniassadi, M.; Baghani, M. Investigating the Effect of ABS on the Mechanical Properties, Morphology, Printability, and 4D Printing of PETG-ABS Blends. *Macromol. Mater. Eng.* **2024**, 2400038. [CrossRef]

30. Mehrpouya, M.; Ghalayaniesfahani, A.; Postmes, J.F.; Gibson, I. Tailoring Mechanical Properties in 3D Printed Multimaterial Architected Structures. *J. Mech. Behav. Biomed. Mater.* **2024**, *152*, 106431. [CrossRef]
31. Hamad, K.; Kaseem, M.; Yang, H.W.; Deri, F.; Ko, Y.G. Properties and Medical Applications of Polylactic Acid: A Review. *Express Polym. Lett.* **2015**, *9*, 435–455. [CrossRef]
32. Soares, J.S.; Moore, J.E.; Rajagopal, K.R. Constitutive Framework for Biodegradable Polymers with Applications to Biodegradable Stents. *ASAIO J.* **2008**, *54*, 295–301. [CrossRef]
33. Haers, P.E.; Suuronen, R.; Lindqvist, C.; Sailer, H. Biodegradable Polylactide Plates and Screws in Orthognathic Surgery: Technical Note. *J. Cranio-Maxillo-Facial Surg.* **2010**, *26*, 87–91. [CrossRef]
34. Wiebe, J.; Nef, H.M.; Hamm, C.W. Current Status of Bioresorbable Scaffolds in the Treatment of Coronary Artery Disease. *J. Am. Coll. Cardiol.* **2014**, *64*, 415–424. [CrossRef]
35. Lasprilla, A.J.R.; Martinez, G.A.R.; Lunelli, B.H.; Jardini, A.L.; Filho, R.M. Poly Lactic Acid Synthesis for Application in Biomedical Devices—A Review. *Biotechnol. Adv.* **2012**, *30*, 321–328. [CrossRef] [PubMed]
36. Gross, B.C.; Erkal, J.L.; Lockwood, S.Y.; Chen, C.; Spence, D.M. Evaluation of 3D Printing and Its Potential Impact on Biotechnology and the Chemical Sciences. *Anal. Chem.* **2014**, *86*, 3240–3253. [CrossRef] [PubMed]
37. Plesec, V.; Humar, J.; Dobnik-Dubrovski, P.; Harih, G. Numerical Analysis of a Transtibial Prosthesis Socket Using 3D-Printed Bio-Based PLA. *Materials* **2023**, *16*, 1985. [CrossRef] [PubMed]
38. Milenkovic, S.; Slavkovic, V.; Fragassa, C.; Grujovic, N.; Palic, N.; Zivic, F. Effect of the Raster Orientation on Strength of the Continuous Fiber Reinforced PVDF/PLA Composites, Fabricated by Hand-Layup and Fused Deposition Modeling. *Compos. Struct.* **2021**, *270*, 114063. [CrossRef]
39. Senatov, F.S.; Niaza, K.V.; Zadorozhnyy, M.Y.; Maksimkin, A.V.; Kaloshkin, S.D.; Estrin, Y.Z. Mechanical Properties and Shape Memory Effect of 3D-Printed PLA-Based Porous Scaffolds. *J. Mech. Behav. Biomed. Mater.* **2016**, *57*, 139–148. [CrossRef]
40. Slavkovic, V.; Palic, N.; Milenkovic, S.; Zivic, F.; Grujovic, N. Thermo-Mechanical Characterization of 4D-Printed Biodegradable Shape-Memory Scaffolds Using Four-Axis 3D-Printing System. *Materials* **2023**, *16*, 5186. [CrossRef] [PubMed]
41. Bodaghi, M.; Namvar, N.; Yousefi, A.; Teymouri, H.; Demoly, F.; Zolfagharian, A. Metamaterial Boat Fenders with Supreme Shape Recovery and Energy Absorption/Dissipation via FFF 4D Printing. *Smart Mater. Struct.* **2023**, *32*, 095028. [CrossRef]
42. Pham, D.B.; Huang, S.-C. A Novel Bio-Inspired Hierarchical Tetrachiral Structure That Enhances Energy Absorption Capacity. *J. Mech. Sci. Technol.* **2023**, *37*, 3229–3237. [CrossRef]
43. Choudhry, N.K.; Panda, B.; Dixit, U.S. Energy Absorption Characteristics of Fused Deposition Modeling 3D Printed Auxetic Re-Entrant Structures: A Review. *J. Mater. Eng. Perform.* **2023**, *32*, 8981–8999. [CrossRef]
44. Gisario, A.; Desole, M.P.; Mehrpouya, M.; Barletta, M. Energy Absorbing 4D Printed Meta-Sandwich Structures: Load Cycles and Shape Recovery. *Int. J. Adv. Manuf. Technol.* **2023**, *127*, 1779–1795. [CrossRef]
45. Novak, N.; Plesec, V.; Harih, G.; Cupar, A.; Kaljun, J.; Vesenjak, M. Development, Fabrication and Mechanical Characterisation of Auxetic Bicycle Handlebar Grip. *Sci. Rep.* **2023**, *13*, 8158. [CrossRef] [PubMed]
46. Zhao, W.; Yue, C.; Liu, L.; Leng, J.; Liu, Y. Mechanical Behavior Analyses of 4D Printed Metamaterials Structures with Excellent Energy Absorption Ability. *Compos. Struct.* **2023**, *304*, 116360. [CrossRef]
47. Bodaghi, M.; Serjouei, A.; Zolfagharian, A.; Fotouhi, M.; Rahman, H.; Durand, D. Reversible Energy Absorbing Meta-Sandwiches by FDM 4D Printing. *Int. J. Mech. Sci.* **2020**, *173*, 105451. [CrossRef]
48. Rahmatabadi, D.; Ghasemi, I.; Baniassadi, M.; Abrinia, K.; Baghani, M. 4D Printing of PLA-TPU Blends: Effect of PLA Concentration, Loading Mode, and Programming Temperature on the Shape Memory Effect. *J. Mater. Sci.* **2023**, *58*, 7227–7243. [CrossRef]
49. Morvayova, A.; Contuzzi, N.; Fabbiano, L.; Casalino, G. Multi-Attribute Decision Making: Parametric Optimization and Modeling of the FDM Manufacturing Process Using PLA/Wood Biocomposites. *Materials* **2024**, *17*, 924. [CrossRef] [PubMed]
50. Song, Y.; Li, Y.; Song, W.; Yee, K.; Lee, K.-Y.; Tagarielli, V.L. Measurements of the Mechanical Response of Unidirectional 3D-Printed PLA. *Mater. Des.* **2017**, *123*, 154–164. [CrossRef]
51. Luo, J.; Luo, Q.; Zhang, G.; Li, Q.; Sun, G. On Strain Rate and Temperature Dependent Mechanical Properties and Constitutive Models for Additively Manufactured Polylactic Acid (PLA) Materials. *Thin-Walled Struct.* **2022**, *179*, 109624. [CrossRef]
52. Pepelnjak, T.; Karimi, A.; Maček, A.; Mole, N. Altering the Elastic Properties of 3D Printed Poly-Lactic Acid (PLA) Parts by Compressive Cyclic Loading. *Materials* **2020**, *13*, 4456. [CrossRef]
53. Chen, W.; Guo, C.; Zuo, X.; Zhao, J.; Peng, Y.; Wang, Y. Experimental and Numerical Investigation of 3D Printing PLA Origami Tubes under Quasi-Static Uniaxial Compression. *Polymers* **2022**, *14*, 4135. [CrossRef]
54. Qin, D.; Sang, L.; Zhang, Z.; Lai, S.; Zhao, Y. Compression Performance and Deformation Behavior of 3D-Printed PLA-Based Lattice Structures. *Polymers* **2022**, *14*, 1062. [CrossRef] [PubMed]
55. Liu, T.; Liu, L.; Zeng, C.; Liu, Y.; Leng, J. 4D Printed Anisotropic Structures with Tailored Mechanical Behaviors and Shape Memory Effects. *Compos. Sci. Technol.* **2020**, *186*, 107935. [CrossRef]
56. Mercado-Colmenero, J.M.; Rubio-Paramio, M.A.; la Rubia-Garcia, M.D.; Lozano-Arjona, D.; Martin-Doñate, C. A Numerical and Experimental Study of the Compression Uniaxial Properties of PLA Manufactured with FDM Technology Based on Product Specifications. *Int. J. Adv. Manuf. Technol.* **2019**, *103*, 1893–1909. [CrossRef]
57. Peng, X.; Liu, G.; Li, J.; Wu, H.; Jia, W.; Jiang, S. Compression Property and Energy Absorption Capacity of 4D-Printed Deformable Honeycomb Structure. *Compos. Struct.* **2023**, *325*, 117591. [CrossRef]

58. Cadete, M.S.; Gomes, T.E.P.; Gonçalves, I.; Neto, V. Influence of 3D-Printing Deposition Parameters on Crystallinity and Morphing Properties of PLA-Based Materials. *Prog. Addit. Manuf.* **2024**. [CrossRef]
59. Barletta, M.; Gisario, A.; Mehrpouya, M. 4D Printing of Shape Memory Polylactic Acid (PLA) Components: Investigating the Role of the Operational Parameters in Fused Deposition Modelling (FDM). *J. Manuf. Process.* **2021**, *61*, 473–480. [CrossRef]
60. Brischetto, S.; Torre, R. Tensile and Compressive Behavior in the Experimental Tests for PLA Specimens Produced via Fused Deposition Modelling Technique. *J. Compos. Sci.* **2020**, *4*, 140. [CrossRef]
61. Cláudio, R.A.; Dupont, J.; Baptista, R.; Leite, M.; Reis, L. Behaviour Evaluation of 3D Printed Polylactic Acid under Compression. *J. Mater. Res. Technol.* **2022**, *21*, 4052–4066. [CrossRef]
62. JI, Q.; Wei, J.; YI, J.; Zhang, L.; Ma, J.; Wang, Z. Study on the Static and Dynamic Mechanical Properties and Constitutive Models of 3D Printed PLA and PLA-Cu Materials. *Mater. Today Commun.* **2024**, *39*, 108690. [CrossRef]
63. Hosseini, S.A.; Torabizadeh, M.; Eisazadeh, H. Experimental Study of the Effect of Strain Rate on the Mechanical Behavior of Assorted Thermoplastic Polymers. *J. Mater. Eng. Perform.* **2023**. [CrossRef]
64. Xu, P.; Lan, X.; Zeng, C.; Zhang, X.; Zhao, H.; Leng, J.; Liu, Y. Compression Behavior of 4D Printed Metamaterials with Various Poisson's Ratios. *Int. J. Mech. Sci.* **2024**, *264*, 108819. [CrossRef]
65. Rajkumar, A.R.; Shanmugam, K. Additive Manufacturing-Enabled Shape Transformations via FFF 4D Printing. *J. Mater. Res.* **2018**, *33*, 4362–4376. [CrossRef]
66. Balasubramanian, M.; Saravanan, R.; Shanmugam, V. Impact of Strain Rate on Mechanical Properties of Polylatic Acid Fabricated by Fusion Deposition Modeling. *Polym. Adv. Technol.* **2024**, *35*, e6335. [CrossRef]
67. Bergström, J.S.; Boyce, M.C. Constitutive Modeling of the Large Strain Time-Dependent Behavior of Elastomers. *J. Mech. Phys. Solids* **1998**, *46*, 931–954. [CrossRef]
68. Qi, H.J.; Boyce, M.C. Constitutive Model for Stretch-Induced Softening of the Stress–Stretch Behavior of Elastomeric Materials. *J. Mech. Phys. Solids* **2004**, *52*, 2187–2205. [CrossRef]
69. Qi, H.J.; Nguyen, T.D.; Castro, F.; Yakacki, C.M.; Shandas, R. Finite Deformation Thermo-Mechanical Behavior of Thermally Induced Shape Memory Polymers. *J. Mech. Phys. Solids* **2008**, *56*, 1730–1751. [CrossRef]
70. *ASTM D695-23*; Standard Test Method for Compressive Properties of Rigid Plastics. ASTM International: West Conshohocken, PA, USA, 2023.
71. Bodaghi, M.; Damanpack, A.R.; Liao, W.H. Triple Shape Memory Polymers by 4D Printing. *Smart Mater. Struct.* **2018**, *27*, 065010. [CrossRef]
72. Garg, M.; Mulliken, A.D.; Boyce, M.C. Temperature Rise in Polymeric Materials During High Rate Deformation. *J. Appl. Mech.* **2008**, *75*, 011009. [CrossRef]
73. Ames, N.M.; Srivastava, V.; Chester, S.A.; Anand, L. A Thermo-Mechanically Coupled Theory for Large Deformations of Amorphous Polymers. Part II: Applications. *Int. J. Plast.* **2009**, *25*, 1495–1539. [CrossRef]
74. Okereke, M.I.; Buckley, C.P.; Siviour, C.R. Compression of Polypropylene across a Wide Range of Strain Rates. *Mech. Time-Depend. Mater.* **2012**, *16*, 361–379. [CrossRef]
75. Hao, P.; Spronk, S.W.F.; Paepegem, W.V.; Gilabert, F.A. Hydraulic-Based Testing and Material Modelling to Investigate Uniaxial Compression of Thermoset and Thermoplastic Polymers in Quasistatic-to-Dynamic Regime. *Mater. Des.* **2022**, *224*, 111367. [CrossRef]
76. Staszczak, M.; Nabavian Kalat, M.; Golasiński, K.M.; Urbański, L.; Takeda, K.; Matsui, R.; Pieczyska, E.A. Characterization of Polyurethane Shape Memory Polymer and Determination of Shape Fixity and Shape Recovery in Subsequent Thermomechanical Cycles. *Polymers* **2022**, *14*, 4775. [CrossRef] [PubMed]
77. Bodaghi, M.; Damanpack, A.R.; Liao, W.H. Self-Expanding/Shrinking Structures by 4D Printing. *Smart Mater. Struct.* **2016**, *25*, 105034. [CrossRef]
78. Van Manen, T.; Janbaz, S.; Jansen, K.M.B.; Zadpoor, A.A. 4D Printing of Reconfigurable Metamaterials and Devices. *Commun. Mater.* **2021**, *2*, 56. [CrossRef]
79. Miehe, C.; Göktepe, S.; Méndez Diez, J. Finite Viscoplasticity of Amorphous Glassy Polymers in the Logarithmic Strain Space. *Int. J. Solids Struct.* **2009**, *46*, 181–202. [CrossRef]
80. Boyce, M.C.; Arruda, E.M.; Jayachandran, R. The Large Strain Compression, Tension, and Simple Shear of Polycarbonate. *Polym. Eng. Sci.* **1994**, *34*, 716–725. [CrossRef]
81. Wiersma, J.; Sain, T. A Coupled Viscoplastic-Damage Constitutive Model for Semicrystalline Polymers. *Mech. Mater.* **2023**, *176*, 104527. [CrossRef]

polymers

MDPI

Review

Status of Polymer Fused Deposition Modeling (FDM)-Based Three-Dimensional Printing (3DP) in the Pharmaceutical Industry

Heba Iqbal [1], Queenie Fernandes [2], Sourour Idoudi [1], Renuka Basineni [1] and Nashiru Billa [1,*]

[1] Pharmaceutical Sciences Department, College of Pharmacy, QU Health, Qatar University, Doha P.O. Box 2713, Qatar; hi1309307@qu.edu.qa (H.I.); si1602796@student.qu.edu.qa (S.I.); renuka.basineni@qu.edu.qa (R.B.)

[2] College of Medicine, QU Health, Qatar University, Doha P.O. Box 2713, Qatar

* Correspondence: nbilla@qu.edu.qa

Abstract: Additive manufacturing (AM) or 3D printing (3DP) is arguably a versatile and more efficient way for the production of solid dosage forms such as tablets. Of the various 3DP technologies currently available, fused deposition modeling (FDM) includes unique characteristics that offer a range of options in the production of various types of tablets. For example, amorphous solid dispersions (ASDs), enteric-coated tablets or poly pills can be produced using an appropriate drug/polymer combination during FDM 3DP. The technology offers the possibility of evolving personalized medicines into cost-effective production schemes at pharmacies and hospital dispensaries. In this review, we highlight key FDM features that may be exploited for the production of tablets and improvement of therapy, with emphasis on gastrointestinal delivery. We also highlight current constraints that must be surmounted to visualize the deployment of this technology in the pharmaceutical and healthcare industries.

Keywords: amorphous solid dispersion; fused deposition modeling; solubility; polymer; pharmaceutical; three-dimensional printing

Citation: Iqbal, H.; Fernandes, Q.; Idoudi, S.; Basineni, R.; Billa, N. Status of Polymer Fused Deposition Modeling (FDM)-Based Three-Dimensional Printing (3DP) in the Pharmaceutical Industry. *Polymers* **2024**, *16*, 386. https://doi.org/10.3390/polym16030386

Academic Editors: Cristina-Elisabeta Pelin and Anton Ficai

Received: 17 December 2023
Revised: 22 January 2024
Accepted: 29 January 2024
Published: 30 January 2024

1. Introduction

Three-dimensional printing (3DP) is an additive manufacturing (AM) technique whereby material is gradually added up, layer by layer, to construct a 3D geometric representation of a digitized imagery [1]. Since its inception in the 1980s, AM technology has evolved exponentially due to its unique features, such as an amenability for constructing complicated geometries with composite materials mimicking body parts, organs or pharmaceutical dosage forms, therefore eliminating traditional manufacturing processes that are time-consuming and convolute [2]. Thus, 3DP has an untapped potential in biomedical, pharmaceutical and industrial applications [2]. The 3DP market has boomed over the last decade, largely buoyed by cost-effectiveness, increased printing speed and precision on printed prototypes [3]. Therefore, for industrial production processes aimed at increasing industrial efficiency, 3DP techniques offer a formidable scope for product design and expansion. Three-dimensional printing technology is utilized in the automotive, aerospace, medical, food, electrical and construction industries [4]. Its application in the pharmaceutical industry has only begun to gain traction within the last decade due to the attributes presented above and, especially, the possibility for customization and affordability [5–9]. Various dosage forms, such as orally administered tablets [10], transdermal patches and microneedles have been successfully fabricated using AMTs [11]. Furthermore, AMTs have also been implicated for clinical use, including cardiology [12], neurosurgery [13], otolaryngology [14], pulmonology [15], podiatry [16], gastroenterology [17] and radiotherapy [18]. In light of its potential applications within the pharmaceutical industry, 3DP technology is likely to advance the sector's scope of products output and applications, especially in the realm of personalized medicines [19].

Pharmaceutical production methods like capsule filling and tableting have advanced in the last few decades, especially in the provision of modified drug-release profiles through novel drug-delivery formulations, including polymeric matrices, nanoparticles, functionalized liposomes and biomimetic particles [20]. However, these formulations go as far as to provide doses to patients based on label claims. They do not account for the variations in dosage requirements amongst patients imposed by genetic or metabolic predispositions and, therefore, cannot be tuned for the requirements of personalized medicines. This is mainly due to process restrictions within conventional production modalities. Absorption of active pharmaceutical ingredients (APIs) following oral administration of conventional dosage forms is associated with a variability in the release of APIs from the dosage form and unpredictable pharmacokinetics. AMT provides scope for the design of patient-centered dosage forms with programmable release capable of minimizing unpredictability in the absorption and maximizing therapy [20].

The first and only 3D-printed tablet currently on the market (Spritam® by Aprecia Pharmaceuticals, Blue Ash, OH, USA) received Food and Drug Administration (FDA) approval in 2015. The tablet is administered via the oral route and is used in patients with dysphagia [21]. Liquid dosage forms may also be used in dysphagic patients; however, these offer diluted dosing and promote instability of the API [22,23]. Interestingly, T19 is another 3D-printed drug produced by Triastek, a Chinese pharmaceuticals and 3D printing technology firm, that has received an investigational new drug (IND) approval from the FDA. T19 is been designed for treatment in rheumatoid arthritis [24].

Several approaches have currently been applied to improve the solubility of drugs, including particle size reduction, nanosuspension, salt formation, pH adjustment, use of surfactants and use of amorphous solid dispersions (ASDs) [25–30]; however, solubility enhancement techniques tag along peculiar constraints [31,32], as presented in Figure 1.

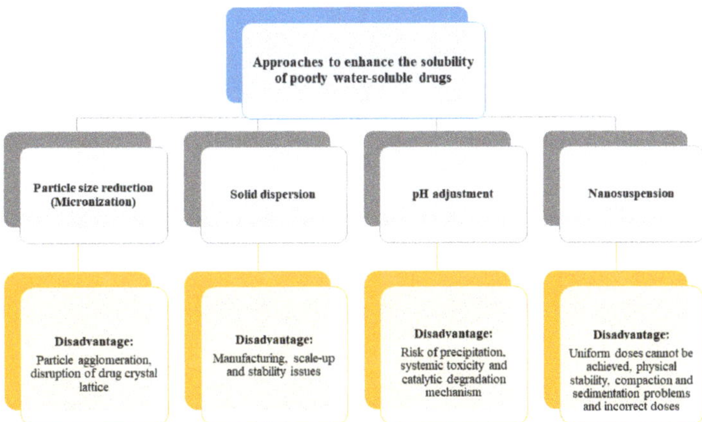

Figure 1. Solubility-enhancing approaches and disadvantages.

Three-dimensional printing technology has emerged in recent years as a possible tool for the production of ASDs, with the inclusion of variable doses of APIs in solid dosage forms [33]. In the context of customizing 3DP technology for individual patient needs, this dual approach of 3DP and ASD may be useful at dispensaries and pharmacies, with scope for the improvement in solubility of APIs and the production of variable doses of medications [33]. Furthermore, the approval of Spritam® has set the precedence for the utilization of 3DP in the manufacture of newer drug-delivery systems [34].

Three-dimensional printing offers different types of feed mechanisms (technologies) such as material jetting, powder bed fusion, direct energy deposition, binder jetting, light photopolymerization and extrusion. These types of 3DP technologies enable the 3D printing of materials such as fluids, waxes, powder and solids [35]. Thus, there is flexibility in the choice of the appropriate material as per requirement.

Aptly, AM techniques offer robust and potent platforms that can be largely employed by pharmaceutical companies for the manufacture of various doses and dosage forms as alluded to above. However, fused deposition modeling (FDM) is an extensively used extrusion-based approach, with consequential outcomes in the pharmaceutical industry due to the similarity to other extrusion techniques already utilized in the industry, i.e., hot melt extrusion (HME) [36]. Moreover, FDM offers the possibility of utilizing biodegradable thermoplastic polymers commonly employed in the formulation of drug products [37,38]. FDM technology is affordable, amenable to modifications, and is simple and may be adapted for desktop usage [39]. Therefore, there is potential for its adoption and evolution in pharmacies and dispensaries toward the provision of personalized medicines [40,41]. In this review, we expound the potential of the FDM 3D printing technique in the pharmaceutical industry along with the challenges it faces.

2. Key Elements of Fused Deposition Modeling (FDM)

In FDM, the print material, usually a thermoplastic filament, is deposited selectively onto a build platform as it melts and oozes out of a nozzle or orifice. The platform and nozzle move in synchrony to allow the layer-by-layer construction of a 3D model [42]. Materials including plastic prototypes and low-volume functionality components can be fabricated using FDM extrusion-based approaches. For instance, FDM is the most extensively used extrusion-based approach for modeling, prototyping and fabrication [43,44]. FDM 3D printing utilizes a variety of thermoplastic polymers including polylactic acid (PLA), acetonitrile butadiene styrene (ABS), polypropylene (PP) and polyvinyl alcohol (PVA). These polymers facilitate fabrication processes and the fabricated object may provide feedback on the efficiency of the filament for the printing [45].

The production of a filament for the FDM process can be achieved through HME, whereby the polymer is heated and squeezed through an extruder to produce the filament. Once the filament is formed, a 3D-printed object may be fabricated using an FDM 3D printer [36,46].

With regard to the potential applications of FDM by the pharmaceutical industry and pharmacies, the need for incorporation of an API in the filament is very crucial. An API may be incorporated in the filament via impregnation (IMP) after the filament is formed or during HME of the filament [47]. However, API loading via IMP after formation of the filament is inefficient because it results in very low drug loads. Maximum drug loading is achievable through incorporation during HME [47,48]. This approach also allows the incorporation of other additives, including plasticizers or polymers (as in the formation of ASDs) [48]. As previously reported, HME may also be used to produce ASDs of poorly water-soluble drugs, whereby the API is presented in an amorphous configuration [48]. This amorphous configuration of the API ensures improved solubility.

The HME mixer also ensures that the API is homogenous within the extruded filament. Mixing can be carried out on a double or single screw mixer [46]. Subsequently, the fabricated filament is fed into the FDM printing machine to print a tablet [48], as shown in Figure 2.

The coupling of HME with FDM-based 3DP allows for the production of pharmaceutical-grade filaments for the printing of medicines [49]. In this light, HME is favored for obtaining drug-loaded filaments, which are used as starting materials for the 3D printing of tablets and other dosage forms [50]. The use of HME to produce filaments with defined shapes and properties is crucial for the success of 3D printing in pharmaceutical applications [51]. The continuous and cost-efficient nature of HME makes it an attractive manufacturing process for drug-delivery systems [52]. Moreover, the use of HME in combination with 3D printing technology also supports the development of personalized medicine and targeted drug-delivery systems [53]. The production of filaments through HME is a critical step in the fabrication of 3D-printed dosage forms, enabling the incorporation of drugs into the matrix and ensuring the smooth structure of the filaments for successful 3DP [53].

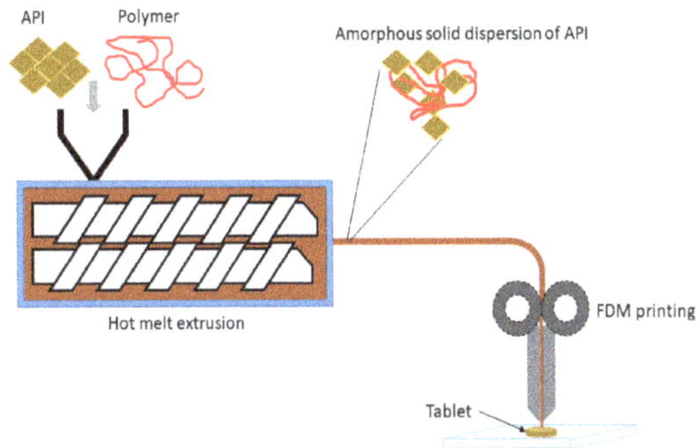

Figure 2. Polymer fused deposition modeling (FDM)-linked hot melt extrusion (HME).

However, the potential applications of HME coupled with FDM-based 3DP may be limited, especially in the bioprinting of thermolabile drugs. However, alternative strategies have been suggested in order to overcome such limitations. One approach involves reducing the FDM printing temperature to accommodate low-melting and thermolabile drugs, as demonstrated by a certain study [54]. Furthermore, the use of natural products in the preparation of 3D-printed drug-delivery systems has been investigated, providing a valuable potential for the fabrication of thermolabile drug-containing tablets via FDM [55]. In addition, inkjet printing of thermolabile model drugs onto FDM-printed substrates has also been explored, offering a potential alternative strategy for drug formulation and evaluation [56]. Moreover, the use of polymer blends to improve the printability and to regulate drug release from pharmaceutical solid dispersions prepared via FDM 3D printing has been investigated, indicating a promising approach to address the challenges associated with printing thermolabile drugs [57]. These alternative strategies demonstrate the ongoing efforts to overcome the limitations of FDM-based printing of thermolabile drug substances, offering potential solutions for the fabrication of patient-tailored dosage forms and drug-delivery systems.

Similarly, other 3D printing techniques such as selective laser sintering (SLS) and stereolithography (SLA) have also been evaluated as suitable alternatives to FDM-based printing. These techniques involve the loose packaging of polymers, ultimately giving rise to porous structures that serve as enhanced drug-delivery systems. However, due to the use of high-energy lasers in such manufacturing processes, damage to especially sensitive drugs is a huge possibility [58]. Therefore, these techniques may not be frequently employed in the manufacturing of drug-loaded formulations. Moreover, in comparison to FDM-based 3DP, they may prove to be less cost-efficient. For example, FDM processes require lower initial investment and operational costs, as compared to SLA. FDM printers are generally more affordable, and the materials used in FDM are often cheaper than SLA resins. In addition, the binder-jetting 3D printing technique can be used to produce amorphous dosage forms for heat-sensitive drugs with high speed using a liquid binding agent to bond thin layers of solid powder [58].

3. Polymers Utilized in FDM-Based 3DP

Commonly used filaments in FDM 3DP techniques include acrylonitrile butadiene styrene (ABS) and polylactic acid (PLA) [4]. ABS is a thermoplastic polymer manufactured from petroleum, through combining acrylonitrile, butadiene and styrene. Owing to its toughness and impact resistance, it has found numerous applications in the automotive industry and in the production of marine components and toys [5]. However, due to its nonbiodegradability and mild toxicity, it is not the preferred choice in pharmaceutical industries [3].

In comparison, polylactic acid (PLA) is another thermoplastic polymer that has an advantage over ABS, due to it being biodegradable [5]. Accordingly, it has received approvals from both the FDA and the European regulatory authorities for its application in the medical and food industries [59]. The utilization of PLA is linked to several advantages, including biocompatibility [60].

Likewise, owing to the growing interest in FDM-based 3DP, many studies have investigated the use of various other polymers in the fabrication of filaments for use in FDM-based 3D printing. Consequently, a particular study demonstrated the potential of pharmaceutical-grade polymers such as polyvinylpyrrolidone (PVP), polyvinyl alcohol (PVA), Kollicoat® IR (KIR), Soluplus® (SLP), polyethylene oxide (PEO), hydroxypropyl methyl cellulose (HPMC), Eudragit® L (ERD L), Eudragit® RL (ERD RL) and ethyl cellulose (EC) as filaments for 3DP [61]. Interestingly, this study reported that drug release is largely dependent on the type of polymers used in the manufacture of the filament and differs based on their water solubility. For example, polymers such as PEO and KIR offer an almost immediate drug release, while on the other hand, HPMC, PVA and SLP are used for sustained or extended drug release [61]. Conversely, the use of poorly permeable hydrophobic polymers such as EDR RL and EC often results in an extremely slow release of the drug [61]. Similarly, other agents, like "Ticagrelor", a blood-thinning drug, and "Tacrolimus", a macrolide inhibitor of calcineurin used in organ transplantations, have also been developed using 3DP techniques [62,63]. In addition, another slow-release drug, in the form of pH-responsive tablets for colon drug-delivery applications, have been developed through the use of 3DP technology [64]. Evidently, an array of polymers is currently available for use in the application of FDM-based 3DP in pharmaceutical industries. However, the choice of polymer is tightly governed by the specification of the type of drug, its biological target and the duration of the therapy [61].

4. FDM 3D Printing and Amorphous Solid Dispersions (ASDs)

ASDs may be defined as a solid dispersion that involves the melting of a solid mixture of API and a suitable vehicle, usually polymers that form eutectic mixtures [51,65]. The polymer or "solvent", interferes with the ordered arrangement of the crystalline API and thereby transforms the mixture into an amorphous solution [51,52,65,66]. The use of FDM in conjunction with ASDs of APIs with low solubility has shown that 3D-printed tablets derived as such show considerable improvement in the solubility of the API [67]. However, the type of polymer used in the printing is very crucial. In a particular study [57] on the formation of felodipine ASDs using FDM 3DP, they observed that the drug-release rate can be altered by varying its miscibility in the polymer blend. The polymer blending technique is an efficient formulation strategy widely used in the plastic and polymer industries to improve the processibility of the material [57]. In addition to the miscibility of the API in the polymer blend, the printability of the blend using FDM is equally crucial. A separate mixture of Eudragit EPO and Soluplus with PEG (polyethylene glycol), PEO (polyethylene oxide) and/or Tween 80 resulted in excellent printability of the blends, as opposed to only Eudragit EPO or Soluplus used alone, which exhibit poor fluidity as well as a high-melt viscosity. This is despite the fact that Eudragit EPO and Soluplus are extensively used in the HME, but are clearly not suitable for FDM 3DP [57]. Furthermore, the ratio of the API to the excipient in the filament can also affect the printability of the filament using FDM. In a study by Kissi et al., HME-extruded filaments containing naproxen were discovered to be amorphous and by increasing the API ratio from 0–10% to 10–20%, the filament brittleness was reduced and printability improved without a compromise on ASD stability. This was due to the plasticizing effect imparted to the filament by the API [68]. In another study, Tan et al. also showed that the configuration of the API in the filament and FDM-printed tablets remained amorphous and the polymeric solvent controls the ductility and flexibility of the formed filament, which, in turn, affects the efficiency of printing using FDM [69]. In the same study, the dissolution behavior of theophylline from 3DP tablets using FDM using hydroxyl propyl cellulose (HPC), polyethylene glycol (PEG) and Eudragit® RL/PO showed that both HPC and PEG tablets were fully disintegrated/dissolved, but not the Eudragit® RL/PO tablets, retaining its shape because of the insoluble nature of the Eudragit matrix [69]. Consequently, the rates of theophylline release from the HPC and

PEG tablets were higher than from the Eudragit® RL/PO [69]. The retention of tablet shape or disintegration of matrices during API release will also dictate the release mechanism. In the case of the Eudragit® RL PO, the matrices are permeable but not soluble; thus, release is likely a diffusion-mediated profile. Furthermore, biodegradability of a polymer in physiologic media will also impact the rate of the release mechanism of the API from the printed tablets; for example, polylactic acid (PLA) has a degradation half-life of approximately 210 days, hence the biodegradation-mediated release of API will be a slow process [70,71]. Tagami et al. showed that by increasing the hydrophobic filler component consisting of PLA and decreasing the hydrophilic component containing a hydrophilic API, e.g., calcein, a slow release profile was obtained and vice versa from the FDM polymer composite [72], as illustrated in Figure 3a–c (reproduced with permission). Thus, through careful choice of polymer blends, and printing geometries, we can modulate the rates of API release from FDM-printed ASD. In a related study, Jamróz et al., 2018 [73] separately utilized Kollicoat® IR (water-soluble polymer) or PLA (water insoluble) to construct tablets using FDM, whereby the tablets formulated with the PLA polymer presented a prolonged release profile of up to 70% after 6 h. On the other hand, the tablet constructed with Kollicoat® IR showed 90% of drug release within 45 min [73]. This shows that polymer combinations can be very useful for tailoring the drug release from the FDM 3D-printed tablets [73,74].

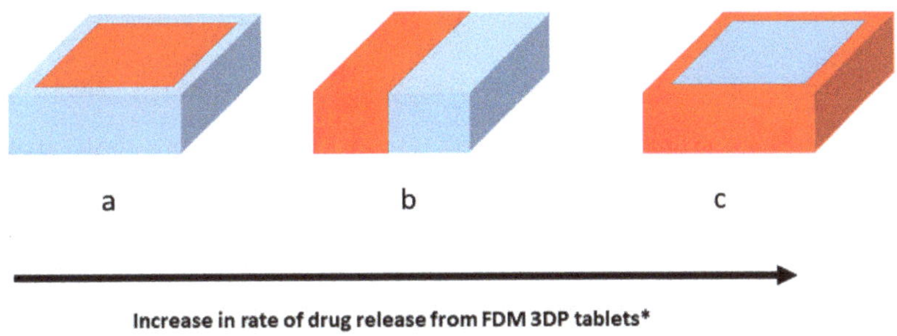

Figure 3. Red shade: drug in soluble polymer (PVA); blue shade: water-insoluble PLA filler only. (**a**) ASD surrounded by insoluble PLA in dosage form; (**b**) ASD adjacent to insoluble PLA in dosage form; (**c**) Insoluble PLA surrounded by ASD. * (Reproduced with permission from reference [59]).

However, some single water-soluble polymers are not printable; for example, Kollidon® VA64 (PVP-VA) due to brittleness [75]. Furthermore, some hydrophobic polymers such as PLA impede API release from the FDM 3D-printed tablets, which is exacerbated when the API is completely miscible in the polymer [76].

In addition, alternate strategies to FDM-based 3DP have been developed in order to eliminate the need to produce filaments for subsequent fabrication into 3DP tablets using FDM; for example, direct powder extrusion (DPE) may be used to prepare ASDs [77], whereby ASD/3D-printed tablets are produced in a single step, with effective improvement in the solubility of a poorly soluble API [77]. However, when compared to FDM, DPE is less reliable in the formation of ASDs due to the possibilities of recrystallization of API from the ASD. On the other hand, DPE is useful in the formation of ASDs of thermolabile APIs [39].

5. FDM-Printed Tablets for Targeted Gastrointestinal Drug Delivery

Since FDM-printed tablets are almost certainly destined for oral administration, this section is dedicated to relevant pharmaceutical technologies that may be employed in conjunction with FDM designs for effective gastrointestinal drug deployment. In addition to the possibilities of forming ASDs, the FDM technique may be used to produce individualized doses to match the severity of a disease or patient predispositions [78]. Combined ASD/FDM technologies can potentially be used for the production of poly pills in chronically ill patients, with the added provision of improving the solubility of APIs [78]. With regards to the oral administration of 3DP tablets, it is important to recognize the physiological and anatomical constraints along the gastrointestinal tract as well as features that can be exploited to maximize absorption. The enzymatic milieu, acidic environment of the stomach and neutral to alkaline pH of the small intestine and colon all provide challenges and opportunities for successful deployment of APIs from FDM-printed tablets [79]. In this regard, the careful selection of polymer/excipients in FDM design and production can be tuned to achieve targeted drug delivery.

For example, despite the low absorptive surface area of the colon, it may be optimal for the absorption of many API drugs due to low enzyme/proteolytic activity and extended transit of dosage forms [80]. Thus, colon-targeted delivery of APIs is recognized as a potential strategy for systemic deployment of protein and gene products [80]. FDM design may incorporate enteric polymers such as Eudragit® FS30 D or cellulose acetate phthalate as a printed coating, which dissolves at a pH of about 7, and is therefore ideal for shielding acid-labile drugs from the acidic pH of the stomach or for a colon-targeted delivery of therapeutics. The use of these polymers as coatings in conventional tablet dosage forms is well established. However, FDM printing technologies provide scope for the incorporation of additional polymers for pharmaceutical or therapeutic applications in a relatively simpler manner. Finally, there is more flexibility for the evolution of FDM tablet designs aimed at achieving targeted gastrointestinal delivery [81,82].

6. FDM-Based Bioprinting of Implant Transplantation Devices and Prosthetics

The applications of AMT in the biomedical sector are diverse. In this light, the role of FDM-based 3DP in personalized medicine is fundamental because customized patient therapy is growing into the future of targeted medicine.

The bioprinting of tissues and organs for suitable transplantation into the host is one such application with a high demand. This process utilizes a digitally guided pipette in order to layer living cells to artificially engineer living tissue according to a prespecified blueprint [83]. Currently, bioprinting has been used to develop tissues of the bone such as cartilaginous structures, and heart tissues such as vascular grafts, in addition to multilayered skin grafts and tracheal splints [84]. Interestingly, some studies [34,85] have also explored the embedding of drug-delivery systems inside the generated implants in order to facilitate enhanced surgical recovery and complete restoration of intended activity. In recent times, implant dosage forms engineered through the utilization of 3DP platforms is indeed gaining momentum for its crucial application in the pharmaceutical industry. Interestingly, this technology is able to produce implants in both micro- and macro-architecture settings. Accordingly, a particular study demonstrated that drug implants fabricated through 3DP have more clear advantages, as compared to implants engineered through conventional compressing-based methodologies [86]. This is particularly because 3DP techniques facilitate the generation of implants with a higher porosity, thereby offering a complex and sophisticated release profile that is difficult to achieve through other conventional methods. In this light, a variety of implants have been generated using 3DP technologies, namely, implants with the pulsed, bimodal, immediate, sustained, delayed and complex drug-release profiles [86,87].

Similarly, artificially generated organoids and tissues generated through the use of 3DP platforms have also found their application in medical research, due to their ability to mimic organs in vitro, thus offering an economical and sustainable experimental platform [83]. The use of medical models is fundamental in medical training, pre- and postoperative planning as well as in patient education [84,88]. However, more recently, through the use of AMTs, it is now also possible to engineer life-sized organ implants and anatomical models that have found their application in surgical practice and training, as well as in assisting medical diagnosis [89–93]. These models are often created from two-dimensional images like X-rays, CT scans or MRIs that are generally patient-specific, thereby offering ideal and realistic structures of intricate anatomical parts of the human body. In addition, 3DP has also found its application in the generation of tissue scaffolding and mechanical bone replicas [94,95]. Distinct techniques like electro-spinning, freeze-drying, emulsification and solvent casting are often employed in the development of tissue scaffolds [83]. Biomanufacturing is yet another technology that utilizes AMTs in combination with tissue engineering [96]. This process is often used to generate biocompatible structures to compensate for osteo defects [97].

FDM-based 3DP has also found its application in the development of medical devices that significantly enhance surgical and clinical procedures. It is now utilized to construct orthopedic instruments, and dental and surgical guides that can follow the patient's unique anatomy with high levels of precision [98]. Such tools are purposed for application in complex and intricate surgeries, thus providing enhanced safety to the patient [99]. These are often engineered according to patient-customized dimensions, such as in drilling guides [100]. Also, in the area of dental practice, FDM-based 3DP has been exploited in the development of dental pieces like bridges and crowns [101,102]. Currently, the generation of customized instruments for use in dental surgery are commonly preferred [103,104].

Another stratum of biomedicine where the use of AMTs is imperative is in the development of prosthetics. Prosthetic limbs can be engineered and personalized to the fit of the patient. Three-dimensional printing is extremely efficient in the generation of customized prostheses like cranial, maxillofacial and mandible implants, thus helping in the resolution of various orthopedic impediments [105]. For example, a particular study [106] used 3DP to design and develop joint prostheses that have been engineered on the basis of surgically resected tibial osteosarcomas. Similarly, another study [107] discussed the implantation of femur modular prosthesis in a patient with osteosarcoma that was generated through the use of AMT. The study stated that the patient was able to attain complete painless recovery with perfect weight-bearing capacity. The process of development of prosthetics usually begins with medical imaging and segmentation, followed by three-dimensional scanning and finally 3D modeling [108]. In certain cases, manufacturing techniques involving AMTs can also be computer numerical control technologies [109].

7. Challenges

There are some significant challenges associated with the use of FDM. This technique may be best suited for small-scale prototyping, although larger printers are in use in various industries. In this light, although FDM-based 3D printing may identify as an easily operable technique with less postprocessing requirements, it is also commonly associated with certain limitations [110]. First, it is a relatively slow and expensive method of production of final articles owing to the limited availability and high cost of raw materials [111]. Second, it often requires skilled labor in order to achieve precise and accurate end-products. Consequently, despite the fact that materials used in 3DP may last longer as compared to the conventional pharmaceutical and industrial manufacturing procedures, a higher level of precision is often demanded in order to achieve desirable results, thereby slowing down the duration of the process [112]. However, in foresight, a gradual decline in the costs of the materials and machinery is expected in the near future.

Moreover, in comparison to traditional processes of drug manufacture, FDM-based 3DP is led by a three-dimensional modeling of drugs that is largely software based [113]. As a result, an infinite amount of product variability is achievable without an additional cost, as compared to conventional technologies of drug manufacture that demand a detailed architecture of products and multistep processes for customized end products [113]. Thus, FDM-based 3DP may be the best-suited option for personalized therapies, where drugs are often produced/dispensed in relatively smaller quantities.

Further, under certain circumstances, such product schemes identify with a tradeoff for lack of precision. There is also inaccuracy in the nozzle temperature, in effective solidification and in poor layer adhesion [114]. FDM produces low-resolution finishing, and twisting/wrapping problems [115]. Thus, there is the need for post-manufacture processing, or worse, structural demolition to achieve the desired configuration of the prototype [114]. Moreover, the volume capacity of the printer sets a restriction on the size of the printed object [114].

With regards to FDM-based 3DP for pharmaceutical dosage forms, thermolabile drugs are likely to be degraded during printing [116]. Inadequate API loading in filaments and, subsequently, in printed tablets is a key constraint in the use of FDM leading slow rates of API release [116]. On the other hand, an increased API content modifies the crystallographic, thermal and rheological properties of the filament, with the possibility of rendering them unprintable [116]. Since FDM is a mechanical process, failure within moving parts can affect the printing process [117]. Crucially, there is only a handful of suitable thermoplastic polymers suited for use in FDM 3DP pharmaceutical dosage forms [72]. Use of a hydrophobic polymer such as PLA may cause a slow drug release from printed tablets. Innovative designs aimed at increasing the surface area of printed tablets exposing the drug to the media only marginally improved the rate of release [118]. These constraints have negatively impacted the utilization of FDM in the pharmaceutical industry [118]. However, with key advances in polymer chemistry and pharmaceutical technology, it is the view of the authors that these constraints are not insurmountable.

8. Conclusions

We conclude that FDM-based 3DP offers a huge potential to the pharmaceutical industry for the fabrication of a variety of solid dosage forms, particularly those intended for gastrointestinal delivery. It is versatile and permits the incorporation of several pharmaceutical excipients in relatively fewer production steps. The formation of ASDs is noteworthy, where improvement in the solubility of APIs is an added output. Furthermore, the technique provides a framework for polymer combinations in tablet designs, including gastrointestinal targeted. The FDM 3D printing technique also provides insights into the production of personalized medicines due to the efficiency in production. Notwithstanding, there are regulatory and production constraints, but these are not insurmountable. It is our view that research should focus on the production of biocompatible/biodegradable thermoplastic polymers that also promote API release from printed tablets, because this may shorten the trajectory to the realization of FDM 3DP tablets on the market.

Author Contributions: Conceptualization: N.B.; initial draft: H.I., Q.F., S.I. and R.B.; subsequent draft: H.I.; preparation of figures: H.I.; formatting: Q.F. and H.I.; final review: N.B. All authors have read and agreed to the published version of the manuscript.

Funding: Open access publication fee was funded by Qatar National Library.

Institutional Review Board Statement: Not applicable.

Data Availability Statement: Data are contained within the article.

Conflicts of Interest: The authors declare no conflict of interest.

References

1. Mwema, F.M.; Akinlabi, E.T. Basics of Fused Deposition Modelling (FDM). In *Fused Deposition Modeling*; Springer Nature: Berlin/Heidelberg, Germany, 2020; pp. 1–15. [CrossRef]
2. Gibson, I.; Rosen, D.; Stucker, B. *Additive Manufacturing Technologies*; Springer: New York, NY, USA, 2015. [CrossRef]
3. Yan, Q.; Dong, H.; Su, J.; Han, J.; Song, B.; Wei, Q.; Shi, Y. A Review of 3D Printing Technology for Medical Applications. *Engineering* **2018**, *4*, 729–742. [CrossRef]
4. Shahrubudin, N.; Lee, T.C.; Ramlan, R. An Overview on 3D Printing Technology: Technological, Materials, and Applications. *Procedia Manuf.* **2019**, *35*, 1286–1296. [CrossRef]
5. Banks, J. Adding Value in Additive Manufacturing: Researchers in the United Kingdom and Europe Look to 3D Printing for Customization. *IEEE Pulse* **2013**, *4*, 22–26. [CrossRef] [PubMed]
6. Mertz, L. Dream It, Design It, Print It in 3-D: What Can 3-D Printing Do for You? *IEEE Pulse* **2013**, *4*, 15–21. [CrossRef] [PubMed]
7. Kim, K.; Ratri, M.C.; Choe, G.; Nam, M.; Cho, D.; Shin, K. Three-Dimensional, Printed Water Filtration System for Economical, on-Site Arsenic Removal. *PLoS ONE* **2020**, *15*, e0231475. [CrossRef] [PubMed]
8. Rankin, T.M.; Giovinco, N.A.; Cucher, D.J.; Watts, G.; Hurwitz, B.; Armstrong, D.G. Dimensional Printing Surgical Instruments: Are We There Yet? *J. Surg. Res.* **2014**, *189*, 193–197. [CrossRef] [PubMed]
9. Ursan, I.D.; Chiu, L.; Pierce, A. Three-Dimensional Drug Printing: A Structured Review. *J. Am. Pharm. Assoc.* **2013**, *53*, 136–144. [CrossRef] [PubMed]

10. Sun, Y.; Soh, S. Printing Tablets with Fully Customizable Release Profiles for Personalized Medicine. *Adv. Mater.* **2015**, *27*, 7847–7853. [CrossRef] [PubMed]
11. Ye, Y.; Yu, J.; Wang, C.; Nguyen, N.; Walker, G.M.; Buse, J.B.; Gu, Z. Microneedles Integrated with Pancreatic Cells and Synthetic Glucose-Signal Amplifiers for Smart Insulin Delivery. *Adv. Mater.* **2016**, *28*, 3115–3121. [CrossRef] [PubMed]
12. Vukicevic, M.; Mosadegh, B.; Min, J.K.; Little, S.H. Cardiac 3D Printing and its Future Directions. *JACC Cardiovasc. Imaging* **2017**, *10*, 171–184. [CrossRef]
13. Thawani, J.; Randazzo, M.; Pisapia, J.; Singh, N. 3D printing in neurosurgery: A systematic review. *Surg. Neurol. Int.* **2016**, *7*, 801–809. [CrossRef]
14. Crafts, T.D.; Ellsperman, S.E.; Wannemuehler, T.J.; Bellicchi, T.D.; Shipchandler, T.Z.; Mantravadi, A.V. Three-Dimensional Printing and Its Applications in Otorhinolaryngology–Head and Neck Surgery. *Otolaryngol. Neck Surg.* **2016**, *156*, 999–1010. [CrossRef] [PubMed]
15. Guibert, N.; Mhanna, L.; Didier, A.; Moreno, B.; Leyx, P.; Plat, G.; Mazieres, J.; Hermant, C. Integration of 3D Printing and Additive Manufacturing in the Interventional Pulmonologist's Toolbox. *Respir. Med.* **2018**, *134*, 139–142. [CrossRef]
16. Williams, C.; James, A.; Chae, M.P.; Hunter-Smith, D.J. 3D Printing in Clinical Podiatry: A Pilot Study and Review. *J. Foot Ankle Res.* **2015**, *8*, O41. [CrossRef]
17. Jeon, H.; Kang, K.; Park, S.; Kim, W.D.; Paik, S.S.; Lee, S.-H.; Jeong, J.; Choi, D. Generation of Multilayered 3D Structures of HepG2 Cells Using a Bio-printing Technique. *Gut Liver* **2017**, *11*, 121–128. [CrossRef]
18. Su, S.; Moran, K.; Robar, J.L. Design and Production of 3D Printed Bolus for Electron Radiation Therapy. *J. Appl. Clin. Med. Phys.* **2014**, *15*, 194–211. [CrossRef] [PubMed]
19. Chen, G.; Xu, Y.; Kwok, P.C.L.; Kang, L. Pharmaceutical Applications of 3D Printing. *Addit. Manuf.* **2020**, *34*, 101209. [CrossRef]
20. Mohammed, A.; Elshaer, A.; Sareh, P.; Elsayed, M.; Hassanin, H. Additive Manufacturing Technologies for Drug Delivery Applications. *Int. J. Pharm.* **2020**, *580*, 119245. [CrossRef]
21. Reddy, C.V.; Balamuralidhara, V.; Venkatesh, M.; Kumar, T.P. First FDA Approved 3D Printed Drug Paved New Path for Increased Precision in Patient Care. *Appl. Clin. Res. Clin. Trials Regul. Aff.* **2020**, *7*, 93–103. [CrossRef]
22. Wright, D.; Begent, D.; Crawford, H. Medication Management of Adults with Swallowing Difficulties. In *Consensus Guideline on the Gedication Ganagement of Adults with Swallowing Difficulties*; Mebendium Gr. Publication Ltd.: Buckinghamshire, UK, 2017.
23. Pamudji, J.S.; Mauludin, R.; Nurhabibah. Influence of β-Cyclodextrin on Cefixime Stability in Liquid Suspension Dosage Form. *Procedia Chem.* **2014**, *13*, 119–127. [CrossRef]
24. Everett, H. Triastek Receives FDA IND Clearance for 3D Printed Drug to Treat Rheumatoid Arthritis. Available online: https://3dprintingindustry.com/news/triastek-receives-fda-ind-clearance-for-3d-printed-drug-to-treat-rheumatoid-arthritis-184159/ (accessed on 22 January 2024).
25. Sigfridsson, K.; Lundqvist, A.J.; Strimfors, M. Particle Size Reduction for Improvement of Oral Absorption of the Poorly Soluble Drug UG558 in Rats during Early Development. *Drug Dev. Ind. Pharm.* **2009**, *35*, 1479–1486. [CrossRef] [PubMed]
26. Doke, V.V.; Khutle, N.M.; Sharma, M.; Gupta, K. Solubility Enhancement of Poorly Soluble Drug Ezetimibe by Developing Self Nano Emulsifying Drug Delivery System. *Indian J. Sci. Technol.* **2022**, *15*, 1504–1516. [CrossRef]
27. Putra, O.D.; Umeda, D.; Fujita, E.; Haraguchi, T.; Uchida, T.; Yonemochi, E.; Uekusa, H. Solubility Improvement of Benexate through Salt Formation Using Artificial Sweetener. *Pharmaceutics* **2018**, *10*, 64. [CrossRef] [PubMed]
28. Choudhary, A.N.; Nayal, S. A Review: Hydrotropy a Solubility Enhancing Technique. *Pharma Innov. J.* **2019**, *8*, 1149–1153.
29. Hart, M.L. Brief Overview of Various Approaches to Enhance Drug Solubility. *J. Dev. Drugs* **2013**, *2*, 1149–1153. [CrossRef]
30. Dalvi, P.B.; Gerange, A.B.; Ingale, P.R. Solid dispersion: Strategy to enhance solubility. *J. Drug Deliv. Ther.* **2015**, *5*, 20–28. [CrossRef]
31. Vemula, V.R.; Lagishetty, V.; Lingala, S. ChemInform Abstract: Solubility Enhancement Techniques. *Int. J. Pharm. Sci. Rev. Res.* **2010**, *5*, 41–51. [CrossRef]
32. Pawar, S.S.; Dahifale, B.R.; Nagargoje, S.P.; Shendge, R.S. Nanosuspension Technologies for Delivery of Drugs. *Nanosci. Nanotech. Res.* **2017**, *4*, 59–66.
33. Trenfield, S.J.; Januskaite, P.; Goyanes, A.; Wilsdon, D.; Rowland, M.; Gaisford, S.; Basit, A.W. Prediction of Solid-State Form of SLS 3D Printed Medicines Using NIR and Raman Spectroscopy. *Pharmaceutics* **2022**, *14*, 589. [CrossRef] [PubMed]
34. Prasad, L.K.; Smyth, H. 3D Printing Technologies for Drug Delivery: A Review. *Drug Dev. Ind. Pharm.* **2015**, *42*, 1019–1031. [CrossRef] [PubMed]
35. Kamran, M.; Saxena, A. A Comprehensive Study on 3D Printing Technology. *MIT Int. J. Mech. Eng.* **2016**, *6*, 63–69.
36. Patil, H.; Tiwari, R.V.; Repka, M.A. Hot-Melt Extrusion: From Theory to Application in Pharmaceutical Formulation. *AAPS PharmSciTech* **2016**, *17*, 20–42. [CrossRef]
37. Zema, L.; Melocchi, A.; Maroni, A.; Gazzaniga, A. Three-Dimensional Printing of Medicinal Products and the Challenge of Personalized Therapy. *J. Pharm. Sci.* **2017**, *106*, 1697–1705. [CrossRef] [PubMed]
38. Thakkar, R.; Thakkar, R.; Pillai, A.; Ashour, E.A.; Repka, M.A. Systematic Screening of Pharmaceutical Polymers for Hot Melt Extrusion Processing: A Comprehensive Review. *Int. J. Pharm.* **2020**, *576*, 118989. [CrossRef] [PubMed]
39. Melocchi, A.; Briatico-Vangosa, F.; Uboldi, M.; Parietti, F.; Turchi, M.; von Zeppelin, D.; Maroni, A.; Zema, L.; Gazzaniga, A.; Zidan, A. Quality Considerations on the Pharmaceutical Applications of Fused Deposition Modeling 3D Printing. *Int. J. Pharm.* **2020**, *592*, 119901. [CrossRef] [PubMed]

40. Skowyra, J.; Pietrzak, K.; Alhnan, M.A. Fabrication of Extended-Release Patient-Tailored Prednisolone Tablets via Fused Deposition Modelling (FDM) 3D Printing. *Eur. J. Pharm. Sci.* **2015**, *68*, 11–17. [CrossRef] [PubMed]
41. Pietrzak, K.; Isreb, A.; Alhnan, M.A. A Flexible-Dose Dispenser for Immediate and Extended Release 3D Printed Tablets. *Eur. J. Pharm. Biopharm.* **2015**, *96*, 380–387. [CrossRef] [PubMed]
42. Friel, R.J. Power Ultrasonics for Additive Manufacturing and Consolidating of Materials. In *Power Ultrasonics*; Elsevier: Amsterdam, The Netherlands, 2015; pp. 313–335. [CrossRef]
43. Praveena, B.A.; Lokesh, N.; Buradi, A.; Santhosh, N.; Praveena, B.L.; Vignesh, R. Comprehensive Review of Emerging Additive Manufacturing (3D Printing Technology): Methods, Materials, Applications, Challenges, Trends and Future Potential. *Mater. Today Proc.* **2021**, *52*, 1309–1313. [CrossRef]
44. Zarek, M.; Layani, M.; Cooperstein, I.; Sachyani, E.; Cohn, D.; Magdassi, S. 3D Printing of Shape Memory Polymers for Flexible Electronic Devices. *Adv. Mater.* **2015**, *28*, 4449–4454. [CrossRef] [PubMed]
45. Kristiawan, R.B.; Imaduddin, F.; Ariawan, D.; Ubaidillah; Arifin, Z. A Review on the Fused Deposition Modeling (FDM) 3D Printing: Filament Processing, Materials, and Printing Parameters. *Open Eng.* **2021**, *11*, 639–649. [CrossRef]
46. Maniruzzaman, M.; Boateng, J.S.; Snowden, M.J.; Douroumis, D. A Review of Hot-Melt Extrusion: Process Technology to Pharmaceutical Products. *ISRN Pharm.* **2012**, *2012*, 436763. [CrossRef]
47. Thanawuth, K.; Sutthapitaksakul, L.; Konthong, S.; Suttiruengwong, S.; Huanbutta, K.; Dass, C.R.; Sriamornsak, P. Impact of Drug Loading Method on Drug Release from 3D-Printed Tablets Made from Filaments Fabricated by Hot-Melt Extrusion and Impregnation Processes. *Pharmaceutics* **2021**, *13*, 1607. [CrossRef]
48. Long, J.; Gholizadeh, H.; Lu, J.; Bunt, C.; Seyfoddin, A. Application of Fused Deposition Modelling (FDM) Method of 3D Printing in Drug Delivery. *Curr. Pharm. Des.* **2017**, *23*, 433–439. [CrossRef]
49. Chai, X.; Chai, H.; Wang, X.; Yang, J.; Li, J.; Zhao, Y.; Cai, W.; Tao, T.; Xiang, X. Fused Deposition Modeling (FDM) 3D Printed Tablets for Intragastric Floating Delivery of Domperidone. *Sci. Rep.* **2017**, *7*, 2829. [CrossRef]
50. Ponsar, H.; Wiedey, R.; Quodbach, J. Hot-Melt Extrusion Process Fluctuations and Their Impact on Critical Quality Attributes of Filaments and 3D-Printed Dosage Forms. *Pharmaceutics* **2020**, *12*, 511. [CrossRef]
51. Venâncio, N.; Pereira, G.G.; Pinto, J.F.; Fernandes, A.I. Influence of the Infill Geometry of 3D-Printed Tablets on Drug Dissolution. In Proceedings of the CiiEM 2021, Bucharest, Romania, 14–15 October 2021; p. 15.
52. Gallas, M.; Boulet, P.; de Margerie, V. Extrusion for pharma applications: An update. *SPE Polym.* **2023**, *4*, 16–23. [CrossRef]
53. Gadgeppa, B.O.; Tukaram, M.S.; Baburo, G.; Sidram, G.P.; Gaurav, A.; Agarwal, S.; Jyoti, G.; Singal, R. Formulation and Evaluation of 3D Printed Pregabalin Tablets Targeted for Neuropathic Pain by Qbd Approach for Personalized Medicine. *Int. J. Pharma Bio Sci.* **2021**, *11*, P1–P13. [CrossRef]
54. Kollamaram, G.; Croker, D.M.; Walker, G.M.; Goyanes, A.; Basit, A.W.; Gaisford, S. Low Temperature Fused Deposition Modeling (FDM) 3D Printing of Thermolabile Drugs. *Int. J. Pharm.* **2018**, *545*, 144–152. [CrossRef]
55. Aguilar-De-Leyva, Á.; Linares, V.; Casas, M.; Caraballo, I. 3D Printed Drug Delivery Systems Based on Natural Products. *Pharmaceutics* **2020**, *12*, 620. [CrossRef] [PubMed]
56. Eleftheriadis, G.K.; Katsiotis, C.S.; Andreadis, D.A.; Tzetzis, D.; Ritzoulis, C.; Bouropoulos, N.; Kanellopoulou, D.; Andriotis, E.G.; Tsibouklis, J.; Fatouros, D.G. Inkjet Printing of a Thermolabile Model Drug onto FDM-Printed Substrates: Formulation and Evaluation. *Drug Dev. Ind. Pharm.* **2020**, *46*, 1253–1264. [CrossRef] [PubMed]
57. Alhijjaj, M.; Belton, P.; Qi, S. An investigation into the use of polymer blends to improve the printability of and regulate drug release from pharmaceutical solid dispersions prepared via fused deposition modeling (FDM) 3D printing. *Eur. J. Pharm. Biopharm.* **2016**, *108*, 111–125. [CrossRef]
58. Mohapatra, S.; Kar, R.K.; Biswal, P.K.; Bindhani, S. Approaches of 3D Printing in Current Drug Delivery. *Sens. Int.* **2021**, *3*, 100146. [CrossRef]
59. Paul, A.; Sreedevi, K.; Sharma, S.S.; Anjana, V.N. *Polylactic Acid (PLA) BT—Handbook of Biopolymers*; Thomas, S., Ajitha, A.R., Jose Chirayil, C., Thomas, B., Eds.; Springer Nature: Singapore, 2022; pp. 1–33. [CrossRef]
60. Durand, B.; Marchand, C. Smart Features in Fibrous Implantable Medical Devices. In *Smart Textiles and Their Applications*; Elsevier: Amsterdam, The Netherlands, 2016. [CrossRef]
61. Melocchi, A.; Parietti, F.; Maroni, A.; Foppoli, A.; Gazzaniga, A.; Zema, L. Hot-Melt Extruded Filaments Based on Pharmaceutical Grade Polymers for 3D Printing by Fused Deposition Modeling. *Int. J. Pharm.* **2016**, *509*, 255–263. [CrossRef] [PubMed]
62. Rastpeiman, S.; Panahi, Z.; Akrami, M.; Haririan, I.; Asadi, M. Facile Fabrication of an Extended-Release Tablet of Ticagrelor Using Three Dimensional Printing Technology. *J. Biomed. Mater. Res. Part A* **2024**, *112*, 20–30. [CrossRef]
63. Abdollahi, A.; Ansari, Z.; Akrami, M.; Haririan, I.; Dashti-Khavidaki, S.; Irani, M.; Kamankesh, M.; Ghobadi, E. Additive Manufacturing of an Extended-Release Tablet of Tacrolimus. *Materials* **2023**, *16*, 4927. [CrossRef]
64. Asadi, M.; Salehi, Z.; Akrami, M.; Hosseinpour, M.; Jockenhövel, S.; Ghazanfari, S. 3D Printed PH-Responsive Tablets Containing N-Acetylglucosamine-Loaded Methylcellulose Hydrogel for Colon Drug Delivery Applications. *Int. J. Pharm.* **2023**, *645*, 123366. [CrossRef] [PubMed]
65. Walden, D.M.; Bundey, Y.; Jagarapu, A.; Antontsev, V.; Chakravarty, K.; Varshney, J. Molecular Simulation and Statistical Learning Methods toward Predicting Drug–Polymer Amorphous Solid Dispersion Miscibility, Stability, and Formulation Design. *Molecules* **2021**, *26*, 182. [CrossRef]
66. Chiou, W.L.; Riegelman, S. Pharmaceutical Applications of Solid Dispersion Systems. *J. Pharm. Sci.* **1971**, *60*, 1281–1302. [CrossRef]

67. Parulski, C.; Gresse, E.; Jennotte, O.; Felten, A.; Ziemons, E.; Lechanteur, A.; Evrard, B. Fused Deposition Modeling 3D Printing of Solid Oral Dosage Forms Containing Amorphous Solid Dispersions: How to Elucidate Drug Dissolution Mechanisms through Surface Spectral Analysis Techniques? *Int. J. Pharm.* **2022**, *626*, 122157. [CrossRef]
68. Kissi, E.O.; Nilsson, R.; Nogueira, L.P.; Larsson, A.; Tho, I. Influence of Drug Load on the Printability and Solid-State Properties of 3D-Printed Naproxen-Based Amorphous Solid Dispersion. *Molecules* **2021**, *26*, 4492. [CrossRef] [PubMed]
69. Tan, D.K.; Maniruzzaman, M.; Nokhodchi, A. Development and Optimisation of Novel Polymeric Compositions for Sustained Release Theophylline Caplets (PrintCap) via FDM 3D Printing. *Polymers* **2019**, *12*, 27. [CrossRef] [PubMed]
70. Lasprilla, A.J.R.; Martinez, G.A.R.; Lunelli, B.H.; Jardini, A.L.; Filho, R.M. Poly-Lactic Acid Synthesis for Application in Biomedical Devices—A review. *Biotechnol. Adv.* **2012**, *30*, 321–328. [CrossRef]
71. Da Silva, D.; Kaduri, M.; Poley, M.; Adir, O.; Krinsky, N.; Shainsky-Roitman, J.; Schroeder, A. Biocompatibility, Biodegradation and Excretion of Polylactic Acid (PLA) in Medical Implants and Theranostic Systems. *Chem. Eng. J.* **2018**, *340*, 9–14. [CrossRef] [PubMed]
72. Tagami, T.; Nagata, N.; Hayashi, N.; Ogawa, E.; Fukushige, K.; Sakai, N.; Ozeki, T. Defined Drug Release from 3D-Printed Composite Tablets Consisting of Drug-Loaded Polyvinylalcohol and a Water-Soluble or Water-Insoluble Polymer Filler. *Int. J. Pharm.* **2018**, *543*, 361–367. [CrossRef] [PubMed]
73. Jamróz, W.; Kurek, M.; Czech, A.; Szafraniec, J.; Gawlak, K.; Jachowicz, R. 3D Printing of Tablets Containing Amorphous Aripiprazole by Filaments Co-Extrusion. *Eur. J. Pharm. Biopharm.* **2018**, *131*, 44–47. [CrossRef] [PubMed]
74. Fuenmayor, E.; Forde, M.; Healy, A.V.; Devine, D.M.; Lyons, J.G.; McConville, C.; Major, I. Material Considerations for Fused-Filament Fabrication of Solid Dosage Forms. *Pharmaceutics* **2018**, *10*, 44. [CrossRef] [PubMed]
75. Kukkonen, J.; Ervasti, T.; Laitinen, R. Production and Characterization of Glibenclamide Incorporated PLA Filaments for 3D Printing by Fused Deposition Modeling. *J. Drug Deliv. Sci. Technol.* **2022**, *77*, 103843. [CrossRef]
76. Goyanes, A.; Allahham, N.; Trenfield, S.J.; Stoyanov, E.; Gaisford, S.; Basit, A.W. Direct Powder Extrusion 3D Printing: Fabrication of Drug Products Using a Novel Single-Step Process. *Int. J. Pharm.* **2019**, *567*, 118471. [CrossRef]
77. Gottschalk, N.; Bogdahn, M.; Quodbach, J. 3D Printing of Amorphous Solid Dispersions: A Comparison of Fused Deposition Modeling and Drop-on-Powder Printing. *Int. J. Pharm. X* **2023**, *5*, 100179. [CrossRef]
78. Lee, V.H.; Yamamoto, A. Penetration and Enzymatic Barriers to Peptide and Protein Absorption. *Adv. Drug Deliv. Rev.* **1989**, *4*, 171–207. [CrossRef]
79. Krishna, R.; Garg, A.; Jin, B.; Keshavarz, S.S.; Bieberdorf, F.A.; Chodakewitz, J.; Wagner, J.A. Assessment of a Pharmacokinetic and Pharmacodynamic Interaction between Simvastatin and Anacetrapib, a Potent Cholesteryl Ester Transfer Protein (CETP) Inhibitor, in Healthy Subjects. *Br. J. Clin. Pharmacol.* **2009**, *67*, 520–526. [CrossRef]
80. Nikam, V.K.; Kotade, K.B.; Gaware, V.M.; Dolas, R.T.; Dhamak, K.; Somwanshi, S.; Khadse, A.; Kashid, V. Eudragit a Versatile Polymer: A Review. *Pharmacologyonline* **2011**, *1*, 152–164.
81. Tambuwala, M.M.; Charbe, N.B.; McCarron, P.; Lane, M. Application of Three-Dimensional Printing for Colon Targeted Drug Delivery Systems. *Int. J. Pharm. Investig.* **2017**, *7*, 47–59. [CrossRef]
82. Ali, H.; Kurokawa, S.; Shehab, E.; Mukhtarkhanov, M. Development of a Large-Scale Multi-Extrusion FDM Printer, and Its Challenges. *Int. J. Lightweight Mater. Manuf.* **2023**, *6*, 198–213. [CrossRef]
83. Kumar, R.; Kumar, M.; Chohan, J.S. The Role of Additive Manufacturing for Biomedical Applications: A Critical Review. *J. Manuf. Process.* **2021**, *64*, 828–850. [CrossRef]
84. Ventola, C.L. Medical Applications for 3D Printing: Current and Projected Uses. *Pharm. Ther.* **2014**, *39*, 704–711.
85. Akmal, J.S.; Salmi, M.; Mäkitie, A.; Björkstrand, R.; Partanen, J. Implementation of Industrial Additive Manufacturing: Intelligent Implants and Drug Delivery Systems. *J. Funct. Biomater.* **2018**, *9*, 41. [CrossRef] [PubMed]
86. Huang, W.; Zheng, Q.; Sun, W.; Xu, H.; Yang, X. Levofloxacin Implants with Predefined Microstructure Fabricated by Three-Dimensional Printing Technique. *Int. J. Pharm.* **2007**, *339*, 33–38. [CrossRef] [PubMed]
87. Halliday, A.J.; Moulton, S.E.; Wallace, G.G.; Cook, M.J. Novel Methods of Antiepileptic Drug Delivery—Polymer-Based Implants. *Adv. Drug Deliv. Rev.* **2012**, *64*, 953–964. [CrossRef] [PubMed]
88. Tack, P.; Victor, J.; Gemmel, P.; Annemans, L. 3D-Printing Techniques in a Medical Setting: A Systematic Literature Review. *Biomed. Eng. Online* **2016**, *15*, 115. [CrossRef] [PubMed]
89. Wake, N.; Alexander, A.E.; Christensen, A.M.; Liacouras, P.C.; Schickel, M.; Pietila, T.; Matsumoto, J. Creating Patient-Specific Anatomical Models for 3D Printing and AR/VR: A Supplement for the 2018 Radiological Society of North America (RSNA) Hands-on Course. *3D Print. Med.* **2019**, *5*, 17. [CrossRef] [PubMed]
90. Pati, F.; Song, T.-H.; Rijal, G.; Jang, J.; Kim, S.W.; Cho, D.-W. Ornamenting 3D Ornamenting 3D Printed Scaffolds with Cell-Laid Extracellular Matrix for Bone Tissue Regeneration. *Biomaterials* **2015**, *37*, 230–241. [CrossRef]
91. Tam, M.D.; Laycock, S.D.; Bell, D.G.; Chojnowski, A. 3-D Printout of a DICOM File to Aid Surgical Planning in a 6 Year Old Patient with a Large Scapular Osteochondroma Complicating Congenital Diaphyseal Aclasia. *J. Radiol. Case Rep.* **2012**, *6*, 31–37. [CrossRef] [PubMed]
92. Goyanes, A.; Det-Amornrat, U.; Wang, J.; Basit, A.W.; Gaisford, S. 3D Scanning and 3D Printing as Innovative Technologies for Fabricating Personalized Topical Drug Delivery Systems. *J. Control. Release* **2016**, *234*, 41–48. [CrossRef]
93. Kwak, M.K.; Jeong, H.; Suh, K.Y. Rational Design and Enhanced Biocompatibility of a Dry Adhesive Medical Skin Patch. *Adv. Mater.* **2011**, *23*, 3949–3953. [CrossRef] [PubMed]

However, the mechanical and physical properties of thermoplastic polymers are heavily influenced by the crystallisation process and crystalline morphology [13]. This is of particular importance for critical end-use parts, such as aerospace components, where dimensional precision and mechanical performance are vital. During LS, crystallisation is associated with a volume reduction, which can affect the geometrical accuracy of the final part; non-uniform crystallisation within parts can result in warpage, rendering the component unserviceable [14]. Polyamide-12 (PA-12) is the most common commercial polymer within laser sintering, and a key characteristic of this material is a large processing window [2,4–6]. This ensures that there is a substantial difference between the crystallisation temperature (T_c) and the melting temperature (T_m), preventing crystallisation occurring during the build and reducing the shrinkage process. However, the crystallisation kinetics of PA-12 are also heavily dependent on multiple processing parameters, such as temperature, time, cooling rate, pressure, and melt orientation [7,15]. Furthermore, PA-12 is metastable and polymorphic, which can result in a multi-phase crystallisation process whereby, depending on temperature, PA-12 may transform into various different crystal phases [7]. From an industrial standpoint, understanding this behaviour is crucial because changes in the extent of crystallinity development, and type of crystalline morphology, can significantly affect the thermal and mechanical properties of a fabricated part [15].

Polymer crystallisation involves primary and secondary processes, which are often treated in isolation. Primary crystallisation is initiated when the polymer melt is cooled below T_m, via a process known as nucleation. Homogenous nucleation is where disordered molecules, within entangled amorphous polymer chains, are able to rotate, allowing small nuclei to form spontaneously throughout the polymer melt [16]. However, nucleation can also occur heterogeneously, whereby impurities present in the polymer system act as artificial primary nuclei [16]. Following the formation of stable nuclei, free rotation of flexible polymer chains allows chain folding and alignment into a lamellar structure, whereby adjacent crystalline blocks are separated by a mobile, inter-lamellar amorphous phase [17]. This results in the formation of spherulites, which grow radially from the nuclei until spherulite impingement, indicating completion of the primary process. The rate of primary crystallisation is related to the degree of supercooling from the melt. At temperatures near the T_m, crystal growth is thermodynamically favoured so nucleation is limited, resulting in a relatively smaller number of nuclei that grow into large spherulites. The presence of large spherulites increases the strength of the polymer but also renders the material brittle, emphasising how an understanding of crystallisation kinetics can allow close control of material morphology and, subsequently, the mechanical properties of LS parts.

Traditionally, it was considered that secondary crystallisation describes any further developments of crystallinity, which occur following impingement of spherulites, permitting previously "non-crystallisable units", which were unable to contribute to primary crystallisation, to integrate into the growing lamellae structure [18]. This can refer to lateral thickening of the crystalline lamella formed during primary crystallisation [19–21] and the growth of new, smaller crystal structures within the amorphous fraction through "infilling" [15,19,20] (Figure S1). During primary crystallisation, radial growth of lamellae generally occurs linearly with time until spherulite impingement [22]; however, an alternative kinetic mechanism is required to define the secondary process. Conventionally, a logarithmic increase in secondary crystallisation with time was reported [23–25], but more recently, it has been suggested that secondary crystallisation adheres to a root time dependence, which relates to a diffusion-controlled process [18,20–22]. This theory was proposed by Hay, who also presented evidence for primary and secondary crystallisation occurring simultaneously, contradicting previous suggestions that the secondary process can only initiate following termination of primary crystallisation through spherulite impingement [26]. Lamellae produced in the early stages of crystallisation, at the centre of the spherulite, were observed to be thicker than at the outer boundary, indicating that secondary crystallisation must occur within the spherulite as soon as lamellae are formed [22,27].

Another important consideration of LS is the requirement for significant amounts of support material, resulting in 80–90% of feedstock powder remaining un-sintered during each build. Therefore, to ensure the economic and environmental sustainability of the process, un-sintered powder remaining in the build chamber must be re-used in future builds [11,28]. However, during each build, PA-12 is exposed to temperatures close to the material melting point for prolonged time periods, leading to aging and degradation, which affect the quality and re-usability of the material. At the powder bed temperature of ~170 °C, multiple aging processes, namely, polycondensation, secondary crystallisation, cross-linking, and chain scission, often occur concurrently [29–31]. To counteract the effect of these aging phenomena, it is common for the un-sintered ("used") powder to be "refreshed" with virgin material before re-use in subsequent builds. As such, a refresh ratio is commonly used, which refers to the proportion of recycled to virgin powder, and typically ranges from 30–50% virgin powder [29,32]. Nonetheless, within the LS industry, there is currently a limited understanding regarding the effect of re-using PA-12 powder across multiple build cycles on the crystallisation kinetics of PA-12 powder.

Previous studies have shown that aging and degradation have the potential to alter the crystallisation behaviour, melt flowability, and particle coalescence of PA-12 powder. When polyamides are exposed to elevated temperatures, increased mobility of amorphous regions can result in increased crystallinity via secondary crystallisation [28,30,33]. Multiple papers have reported that successive powder re-use, or simulated oven-conditioning, causes a reduction in melt flow rate; this can be attributed to an increased cross-link density [9,30,33], or a rise in molecular weight (M_W), as a result of polycondensation [4,5,7,9,11]. In the context of LS, reductions in melt flowability can limit full coalescence of PA-12 particles. As LS occurs under atmospheric pressure, coalescence is particularly important because particle viscosity and surface tension become the driving force for complete consolidation and compaction of the powder layer [34]. Therefore, reductions in material coalescence can hinder the surface finish of LS parts [34] and cause increased part porosity, which often results in embrittlement of fabricated components [32,35,36]. However, there have been no previous investigations into the effect of powder re-use, aging, and virgin refresh rates on the isothermal crystallisation kinetics of PA-12 powder. As crystallisation is so crucial to fabricating LS parts with sufficient mechanical and structural properties, this research is necessary to help reveal more information about the behaviour of un-sintered powder and its suitability for re-use in future builds.

There are multiple models that aim to best illustrate the isothermal crystallisation kinetics of polymers. The Avrami model undoubtedly remains the most popular isothermal crystallisation model and has been used frequently to explain the crystallisation kinetics of various polyamides [37–39], including LS grade PA-12 powder [7,14,31,40]. However, despite widespread use, the Avrami model has been criticised due to a number of limitations [41]. In practice, the Avrami model is only suitable for modelling primary crystallisation and is unable to accurately describe the latter stages of crystallisation that follow spherulite impingement. As such, Avrami often returns non-integer values for the Avrami exponent, which provides limited information regarding the mechanism for nucleation and growth [22,26,42]. The Avrami model also fails to account for an induction time, and differential scanning calorimetry (DSC) often returns incomplete exotherms, leading to integration errors and inaccurate data [41]. These limitations have led various authors to propose alternative models that aim to resolve these problems by modifying the existing Avrami equation. These include, but are not limited to, the simplified Hillier [43], Tobin [44], Malkin [45], and Hay models [20,21,46,47].

Previous work has begun to explore the isothermal crystallisation kinetics of PA-12 powder in the context of laser sintering [14,40,48]. A study from Neugebauer et al. explored both the isothermal and non-isothermal crystallisation kinetics of polyamide-12 used within LS [14]. This study intended to emulate the laser exposure found within the LS procedure. However, they used a cooling rate of only 40 °C/min to reach the target isothermal temperature, which is likely slower than the rate of cool-down that the powder experiences after laser scanning. Similarly, this rate may not be quick enough to prevent crystallisation initiating during cooling prior to the isothermal step. Zhao, Wudy, and Drummer explored the isothermal crystallisation kinetics of PA-12 between 160–168 °C, using a faster cooling rate of 60 °C/min. They observed that the Avrami model can only accurately fit 70% of the full crystallisation process; beyond this point, the experimental data deviate away from the modelled curve. However, their work was limited to the Avrami model alone, and the various alternative models were not considered [40]. Furthermore, these papers do not explore the effect of powder re-use, and subsequent material aging, on the isothermal crystallisation kinetics of PA-12.

Paolucci et al. investigated the influence of aging on the crystallisation kinetics of PA-12 powder [7]. Virgin PA-12 powder was stored at a range of temperatures for extended time periods, in a nitrogen flushed oven, to replicate the conditions of an LS build chamber. Thermal treatment resulted in polycondensation, indicated by increases in M_w and melt viscosity. Fast scanning calorimetry experiments on the "annealed" PA-12 samples suggested that powder aging does not affect crystal nucleation but causes a reduction in the crystal growth rate of PA-12. Nevertheless, it was reported that this reduction in crystallisation kinetics only occurred in a specific temperature range between 100–150 °C, which is below the expected powder bed temperature during LS. Also, oven storage cannot directly replicate complex powder re-use procedures, such as the use of refresh ratios, which restricts the industrial relevance of their work.

This paper investigates the coalescence behaviour, melt flowability, and sintering kinetics of PA-12 powder in order to identify the key aging processes occurring when feedstock material is re-used across multiple LS build cycles with a 70:30 refresh ratio. Via differential scanning calorimetry experiments, the effect of powder re-use, and subsequent aging processes, on the isothermal crystallisation kinetics of PA-12 is quantified and modelled using the traditional Avrami model. Furthermore, to the best of the authors knowledge, the applicability of alternative models, namely, simplified Hillier, Tobin, Malkin, and Hay, to describe the crystallisation behaviour of LS grade PA-12 powder is yet to be determined. As such, this work also uses non-linear regression analysis to determine the suitability of each of these models for describing the isothermal crystallisation behaviour of virgin and re-used (aged) PA-12 powder. Finally, the most appropriate model for analysing the crystallisation kinetics of virgin PA-12 powders, and PA-12 feedstock re-used in up to six LS build cycles, is proposed.

2. Experimental Method

Prototal UK Ltd. (Newbury, UK) provided industrial grade virgin PA-12 (PA2200) powder (EOS GmbH, Krailling, Germany) and refreshed PA-12 powder samples that had been re-used in up to 7 laser sintering builds, using a refresh ratio of 70:30 (used: virgin) for each build cycle. In this study, virgin powder and powder re-used in 2, 4, and 6 build cycles were analysed. These are referred to as used 2, used 4, and used 6, respectively.

2.1. *Powder Characterisation*

2.1.1. Hot-Stage Microscopy (HSM)

The effect of powder re-use on the melting and coalescence behaviour of PA-12 powder was visualised using a Keyence 4K digital microscope (Keyence UK Ltd., Milton Keynes, UK), equipped with a Linkam THMS600 heating stage and a Linkam TMS 94 temperature controller (Linkam Scientific Instruments Ltd., Redhill, UK). To investigate the coalescence behaviour, virgin and re-used powders were heated from 25 °C to 220 °C at 10 °C/min, and samples were viewed using a microscope magnification of 300×. Alternatively, for the isothermal particle "sintering" experiment, samples were heated from 25 °C to the sintering temperature (195 °C, 200 °C, 205 °C) at 150 °C/min and held at that temperature for 5 min. In this case, to focus on two individual powder particles, the magnification was 1000×.

2.1.2. Differential Scanning Calorimetry (DSC)—Isothermal Crystallisation

A Mettler Toledo differential scanning calorimeter, DSC 1 (Mettler Toledo, Schwerzenbach, Switzerland) calibrated with zinc (T_m 419.5 °C, ΔH_f 107.5 Jg^{-1}) and indium (T_m 156.6 °C, ΔH_f 28.45 Jg^{-1}), was used to determine the isothermal crystallisation kinetics of the PA-12 powder. Powder samples (6 ± 0.5 mg), were placed into Mettler Toledo 40 μL aluminium DSC pans, capped with aluminium lids, and sealed with a press. All experiments were conducted under a nitrogen flow rate of 100 cm^3/min, and a pinhole was placed in the top of each DSC pan. A Huber TC100 immersion cooler (Huber Kaltemaschinenbau AG, Offenburg, Germany) provided temperature control for extended isothermal segments.

The DSC protocol involved isothermal experiments from the molten state, also known as melt-crystallisation. In each DSC experiment, samples were heated at 30 °C min^{-1} to 220 °C and held for 3 min to eliminate any residual crystals. They were then cooled at 70 °C/min to the designated isothermal T_c (162–169 °C) and held for up to 420 min. The cooling rate of 70 °C/min is the maximum capability of the DSC-1 system and was selected to try and prevent any crystallisation initiating on the cool-down. Following the isothermal hold, samples were immediately re-heated to 220 °C at 10 °C/min to analyse the effect of isothermal temperature on the polymers melt behaviour. The heat of fusion (ΔH_f) was determined from the melting peak and used to calculate percentage crystallinity (X_c) using Equation (1), whereby ΔH_f^0 = 209.3 Jg^{-1} (100% crystalline PA-12) [2].

$$X_c(\%) = \frac{\Delta H_f}{\Delta H_f^0} \times 100 \tag{1}$$

To investigate the isothermal crystallisation kinetics of virgin powder and material re-used in 2, 4, and 6 LS build cycles, raw data from the isothermal DSC segment was extracted and the change in heat flow, as a function of time, measured. These raw data, after being subjected to baseline and induction time corrections, were used to analyse and compare the Avrami [49], simplified Hillier [43,50], Tobin [44], and Malkin [45] models (Equations S1–S5). To enable visual comparison with the Hay model [22,51], cumulative crystallinity data were converted to true fractional crystallinity using Equation (2). For a detailed explanation of each model, refer to previous work from Kelly and Jenkins [15,47].

$$\text{Fractional crystallinity} = \frac{\text{Cumulative area under the exotherm } \left(\text{Wsg}^{-1}\right)}{\text{Heat of fusion } \left(\text{Jg}^{-1}\right)} \tag{2}$$

Following the isothermal crystallisation experiments, the crystallisation kinetic parameters were estimated from traditional double log plots for the Avrami, simplified Hillier, Tobin, and Hay models, whilst the Malkin parameters were established from the Avrami results. Using SPSS software (IBM, Portsmouth, UK), the calculated parameters were then employed with each model to produce fractional crystallinity versus time plots. The fit of these plots was compared with experimental data and analysed using visual evaluation, as well as values for the coefficient of determination (R^2) and standard error of regression, s (Equation (3)):

$$s = \sqrt{\frac{\Sigma\left(X_t - X'_t\right)^2}{n-2}}, \tag{3}$$

where s is the standard error of regression, X_t is experimental fractional crystallinity, X'_t is the modelled fractional crystallinity, and *n* is the number of data points. Additionally, SPSS was utilised to generate curve fittings for each model using non-linear regression. This produced multivariable kinetic parameters, which were also used to produce fractional crystallinity curves and analysed via the same methods. These various statistical analysis techniques were applied to understand the applicability of each model as a function of aging state; as such, the most suitable model for both virgin and re-used PA-12 powder was determined.

3. Results and Discussion

3.1. The Effect of Powder Re-Use on Sintering and Coalescence Behaviour of PA-12

The melt flowability and coalescence behaviour of PA-12 powder is a useful indicator of powder quality because it can have a significant effect on the structural and mechanical properties of final LS parts. Nonetheless, the use of hot-stage microscopy for characterising recycled polymeric powder is less common than other techniques such as DSC or melt-flow index. In this study, HSM showed a reduction in melt flowability and material coalescence as a function of powder re-use (Figure 1). Upon heating, virgin powder particles remain solid until at least 180 °C; however by 190 °C, melting has already begun, and at 195 °C, the material is in an entirely molten state. Therefore, melting of virgin material occurs sharply with a moderate melting point range. The molten polymer can flow easily without restriction; by 210 °C, there is minimal free volume remaining within the sample, which represents an increased coalescence rate and good melt flowability. However, with increased powder re-use, melting initiates at progressively higher temperatures (indicated by the theoretical yellow arrow in Figure 1) and over a notably broader temperature range. This is further shown by DSC experiments, where an increased peak T_m and T_m range is observed as a function of build number (Figure S2). In used powder samples, consolidation of the molten material is limited, even at temperatures as high as 210 °C, suggesting a significant decrease in powder coalescence and melt flowability. Similarly, in used 4 and 6 powder samples, un-molten particle cores are present at 200 °C and 210 °C, respectively, indicating that increased powder re-use could lead to incomplete melting during the sintering process.

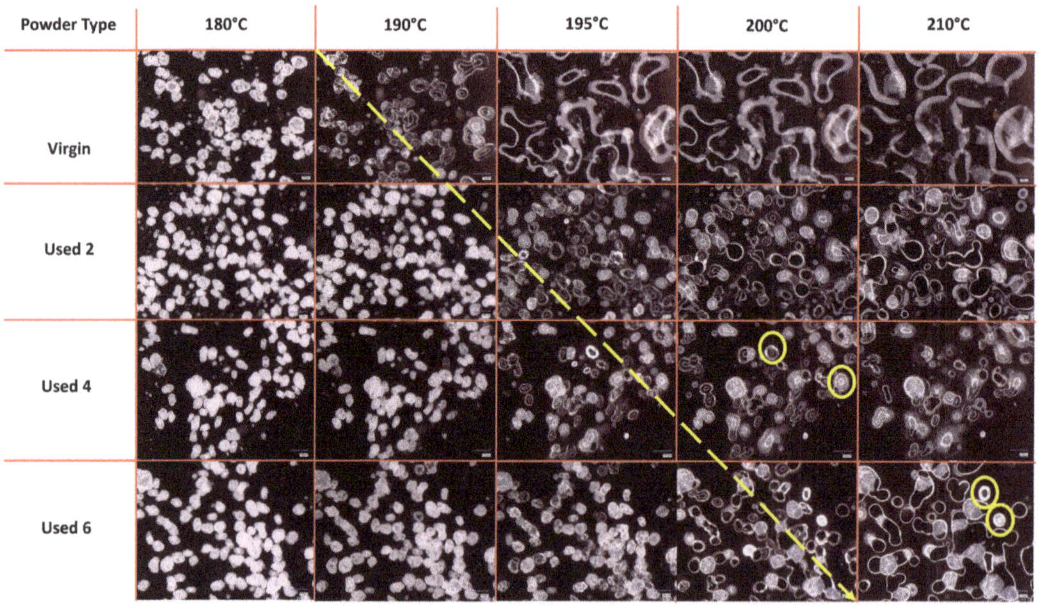

Figure 1. Powder melting and coalescence behaviour of virgin and used powder, observed using hot stage microscopy by heating samples at a constant heating rate of 10 °C/min. The theoretical yellow arrow indicates the change from particle softening to melting and coalescence during heating; yellow circles provide examples of incomplete melting and the presence of un-molten particle cores.

Successive re-use of PA-12 powder in LS builds could result in polycondensation, an aging process that causes macromolecular chain growth via reactions of free-chain end groups (Figure S3). Ultimately, this often results in a significant increase in M_w [2,6,8,28,52] and melt viscosity, which hinders the flowability of the molten phase [4,7,10,30,53]. For example, Sillani et al. calculated that the molecular weight of recycled PA-12 powder (50:50 refresh ratio) was 163% greater than virgin powder after only one build cycle [2]. Similarly, Wudy and Drummer used gel-permeation chromatography to show that the M_w of PA-12 powder almost doubles after re-use for five build cycles [28]. In both cases, the increase in M_w was attributed to the polycondensation process. Furthermore, storage of PA-12 powder at elevated temperatures during LS can cause intermolecular cross-linking, leading to the formation of "tie-chains" within the polymer structure (Figure S1). A tie chain represents a portion of a disordered molecule, which is able to traverse the free space between adjacent crystalline lamellae unobstructed [54]. This forms a connection between lamellae structures, restricting the mobility of the inter-lamellae amorphous chains, leading to increases in melt viscosity [9,32,33]. As a result, cross-linking may also contribute to the observed reduction in melt flow and particle coalescence.

The change in coalescence behaviour of PA-12 powder is also evident in Figure 2. In virgin material, at a constant 'sintering' temperature of 200 °C, two separate particles can combine to form one 'fused' particle within 10 s, exhibiting fast coalescence. However, in used powder, at the same temperature of 200 °C, particle coalescence is much slower. With increased powder re-use, a progressively higher thermal exposure time is required to ensure consolidation of the two particles, emphasising a reduction in the flowability and coalescence of the molten material. This experiment was also repeated at temperatures of 195 °C and 205 °C (Figure S4). The coalescence rate of PA-12 powders, as a function of powder re-use and isothermal temperature, was quantified using relative roundness, which describes the time it takes for two separate particles to merge into one "combined" particle (Figure 3). Relative roundness (Equation (4)) is measured using ImageJ and provides a qualitative, dimensionless value, whereby 0.0 relates to two separate particles and 1.0 correlates to a fully consolidated, single, 'round' particle.

$$4 \times \frac{[\text{Area}]}{\pi \times [\text{Major axis}]^2} \tag{4}$$

The data presented in Figure 3 demonstrate that the coalescence behaviour is heavily dependent on temperature. Independent of powder type, with increased hold temperature, the rate of particle coalescence increases. For example, at 205 °C, powder recovered from build 6 reaches a relative roundness of 1.0 after only 120 s, significantly faster than the 250 s required at a temperature of 195 °C for the same powder type. However, relative roundness is also affected by powder re-use. At all temperatures, the coalescence time for virgin powder is less than 15 s, but with increased build number, the time required for full particle coalescence progressively rises.

If the results shown in Figures 1–3 are contextualised to the scale of a laser sintering build chamber, then the impact of reduced melt flowability and restricted coalescence become increasingly relevant. Assuming that build processing parameters remain constant, the higher melting points of used powder (Figure 1) may cause incomplete melting during the sintering process, resulting in the presence of un-molten particle cores [4,33]. In addition, the influence of temperature on the coalescence behaviour of virgin and re-used powder is significant when considering the layer-by-layer nature of the LS process. To ensure sufficient coalescence of particles within each layer, and vertically between layers, a high melt flowability is required. However, as the number of layers for a particular component progressively increases, the bottom layers of that part become further away from the heat-source, causing a decrease in the localised temperature. Similarly, it is possible for large temperature gradients to form within a LS build chamber [33,55]. Figure 3 demonstrated that reductions in temperature, and increased powder re-use, hinder the rate of particle coalescence. As such, there may be limited thermal energy, or insufficient time, for complete coalescence in the lower layers of the build chamber, which prevents full interlayer bonding and increases pore density within vertically orientated parts. This could contribute to the anisotropic behaviour often observed in LS parts, whereby vertically orientated samples have the weakest tensile strength and are the most brittle [53,56]. Overall, un-molten particles and reduced melt flow rate cause reduced material coalescence, resulting in surface defects (orange peel) [34] and increased porosity [32,35,57] within parts fabricated from re-used powder. Increased porosity reduces part strength and induces embrittlement [32,36], which could render LS parts un-fit for purpose, particularly in critical end-use applications.

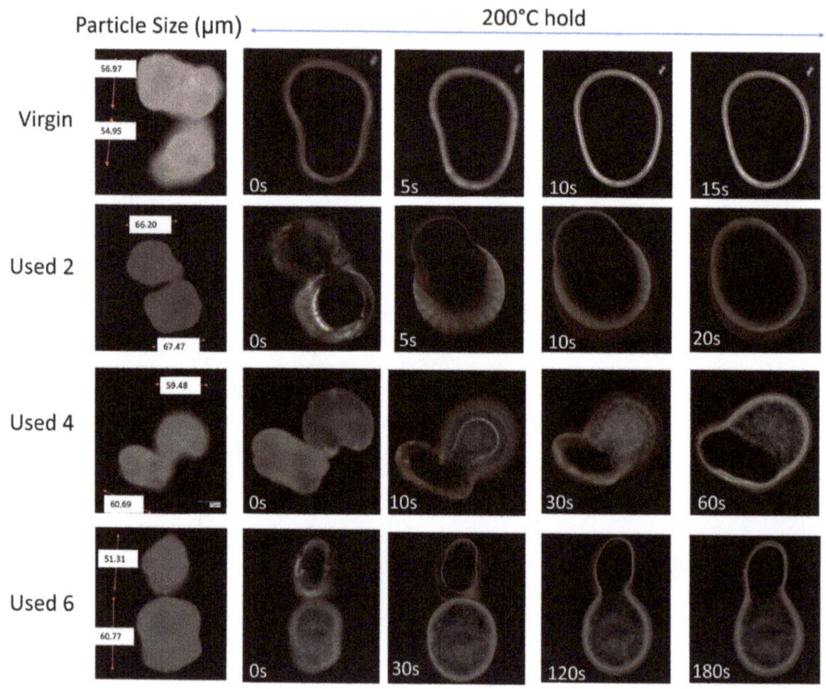

Figure 2. The coalescence behaviour, at 200 °C, of two virgin powder particles, and powder recovered from different LS build cycles.

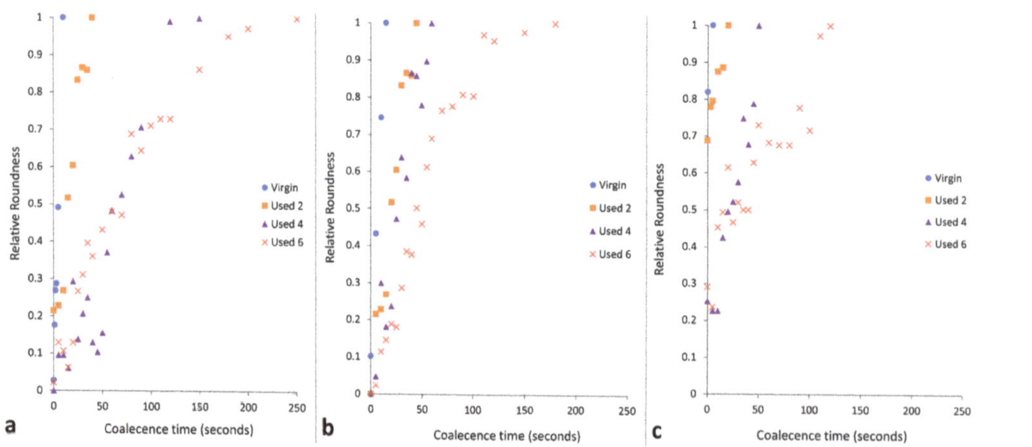

Figure 3. The relative roundness of two powder particles coalescing into one consolidated, melted 'particle' as a function of time when holding at coalescence temperatures of (**a**) 195 °C, (**b**) 200 °C, and (**c**) 205 °C.

3.2. The Effect of Powder Re-Use on Crystallisation Behaviour of PA-12

The mechanical and physical properties of LS parts are also heavily dependent on the crystallisation process and the crystalline morphology that develops. Therefore, isothermal crystallisation experiments were conducted using DSC to understand the effect of powder re-use and subsequent aging processes on the crystallisation behaviour of PA-12. During these experiments, values for absolute crystallinity, measured from the isothermal crystallisation exotherm and peak melting temperature, recorded on the immediate re-heat after the isothermal hold, were obtained (Figure 4). Independent of powder aging state, with increased isothermal T_c, there is an increase in peak melting temperature because diffusion and growth dominate nucleation, so thicker and more perfect lamellae can develop [40,56]. More importantly, at each isothermal crystallisation temperature, used powder samples display a higher T_m than the respective virgin sample, indicating that powder re-use increases melting temperature. Previous literature suggests that an increase in the T_m of polyamides can be caused by polycondensation [11,30,53], secondary crystallisation [9,28,58], and cross-linking [30,59]. However, the increases in T_m observed in this experiment cannot be explained by secondary crystallisation because all traces of crystallinity are removed on the heating segment prior to the isothermal hold. On the other hand, although polycondensation is typically a reversible reaction, an LS build chamber combines high temperatures with an inert atmosphere, which forces the reaction towards continuous water removal. As such, under the conditions present during LS, un-sintered PA-12 powder is effectively exposed to a non-reversible polycondensation reaction. Cross-linking is also irreversible, so both of these processes cause permanent structural changes to PA-12. As such, polycondensation and cross-linking appear to be the dominant cause for the observed increase in T_m with successive powder re-use, and this may have significant consequences for the quality of final LS parts, as explained in Section 3.1.

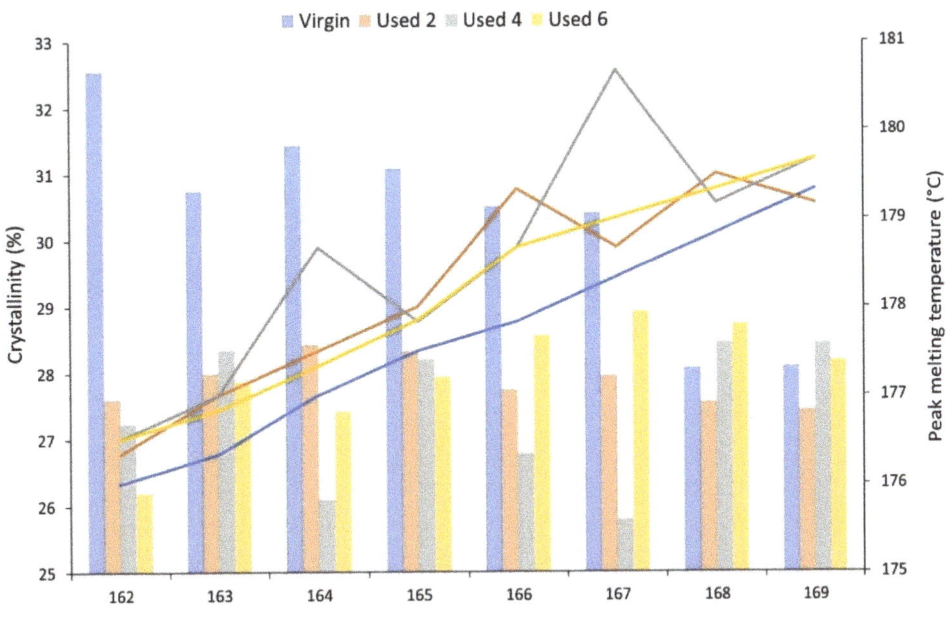

Figure 4. The change in absolute crystallinity (columns), measured from the exothermal crystallisation peak, and peak melting temperature (lines), measured on the subsequent re-heat, as a function of isothermal crystallisation temperature for virgin and used powder samples.

Additionally, for isothermal T_c's between 162–167 °C, crystallinity development is higher within virgin powder samples than re-used material, which provides further evidence for the occurrence of polycondensation and cross-linking during LS (Figure 4). These aging processes result in structural defects, which reduce chain mobility and prevent amorphous fractions from re-ordering. This hinders the crystallisation process, resulting in reduced crystallinity development within used powder samples during isothermal crystallisation. The reported values for crystallinity were calculated using the enthalpy of crystallisation (ΔH_c) from the isothermal crystallisation exotherm. To ensure that no reorganisation processes were occurring during re-heating, enthalpy of fusion (ΔH_f) was also measured from the melting endotherm, and, within the limits of experimental variability, the change in ΔH_f is equal to the change in ΔH_c (Figure S5). This confirms that the observed changes in thermal properties and crystallisation behaviour, as a function of powder re-use, were a result of aging rather than crystal reorganisation during re-heating. Furthermore, the most significant increase in T_m, and the largest reduction in crystallinity, occurs over the first two build cycles (Figure 4). For example, at an isothermal T_c of 162 °C, there is a 3.45% reduction in absolute crystallinity between virgin powder and material recovered from build 2; similarly, T_m increases by 0.9 °C. However, these changes are followed by a plateau, whereby the difference across the next four build cycles is minimal. It has been suggested previously that polycondensation occurs more rapidly in thermally unstressed material, i.e., virgin PA-12 powder; in used, aged powder, further chain growth is hindered by a reduced availability of end groups [6,12,30,57].

It is generally agreed that the glass transition temperature (T_g) of linear polymers is dependent upon the average M_w of the system; polymers with a greater M_w usually have a higher T_g [60–63]. This relationship is related to the free-volume present in the system, whereby a reduction in free volume limits the mobility of amorphous chains, therefore causing an increase in T_g [54,60]. Polycondensation [2,6,8,28,52] and cross-linking [9,30,32] are both thought to increase the M_w of PA-12. These aging processes also cause an increase in chain-end and cross-link density, which decreases the free-volume present in the inter-lamellae amorphous phase. With increased powder re-use, there is an increase in the onset and midpoint of T_g, thus suggesting an increase in M_w (Figure 5). This provides further evidence of polycondensation and cross-linking occurring with successive re-use of PA-12 within LS.

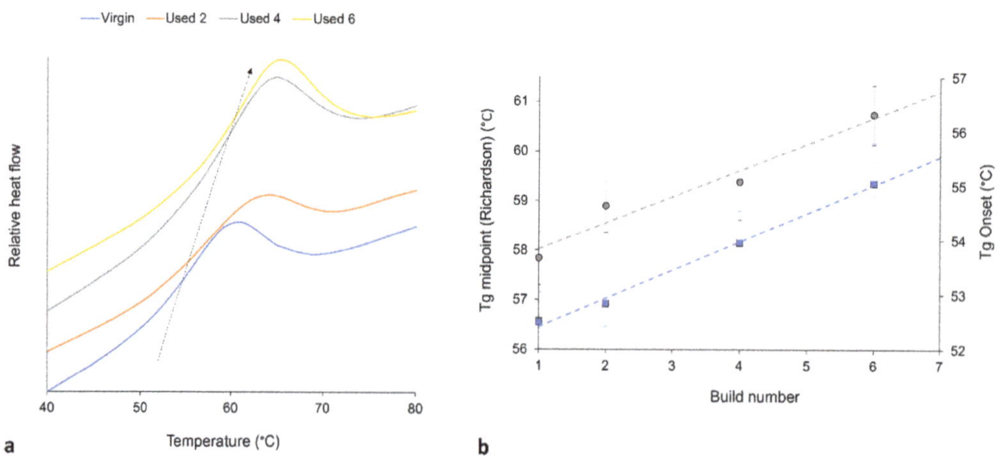

Figure 5. (**a**) With increased powder re-use, there is a shift in T_g to higher temperatures (represented by arrow), as recorded by DSC; (**b**) shows the increase in T_g onset (blue datapoints) and T_g midpoint (gray datapoints), whereby T_g is measured using the Richardson approach, and the plotted values are taken as an average from 10 repeats.

3.3. Avrami Analysis of Isothermal Crystallisation

Sections 3.1 and 3.2 have provided evidence that shows polycondensation and cross-linking are the key aging processes occurring when PA-12 is recycled across multiple LS builds. In addition to hindering material coalescence, reducing melt flowability, and decreasing the extent of crystallinity development, these processes also influence the rate of isothermal crystallisation of PA-12. Avrami is the most commonly used kinetic model for describing the crystallisation behaviour of polymeric materials. However, the effect of powder re-use, and related aging processes, on the applicability of the Avrami model is yet to be determined. Therefore, isothermal crystallisation experiments were conducted, for each powder type, in the temperature range of 162–169 °C, and the raw data are presented in Figure S6. In both virgin and used powder, as the T_c increases, there is a reduction in the rate of crystallisation, as reported previously [14,40,48]. Using the Avrami model, the raw DSC data were converted into double log plots, whereby the experimental data are limited to primary crystallisation alone (Figure S7). Data are restricted to the linear region so that the Avrami crystallisation kinetic parameters, n and k, can be accurately estimated from the gradient and y-intercept, respectively. The estimated kinetic parameters are displayed in Table S1. These data were used to create the plots presented in Figure 6, highlighting that with increased powder re-use, there is a reduction in the crystallisation kinetics of PA-12. Used powder was found to have a greater crystallisation half-life (t $\frac{1}{2}$) than virgin material (Figure 6a), indicating a reduction in the rate of crystallisation due to morphological changes caused by polycondensation and cross-linking. Chain entanglements, knots, and permanent, non-reversible tie-chains create disorder and steric hindrance, reducing the number of stable nuclei, which can form within the polymer system. These structural changes also restrict chain mobility and prevent chains from re-ordering into a crystalline structure. As such, the rate of both nucleation and growth decreases, reducing the crystallisation kinetics and overall extent of crystallinity development. The reduction in crystallisation rate is further shown in Figure 6b, which, particularly for lower isothermal T_c's, displays a reduction in the Avrami rate constant, k_a.

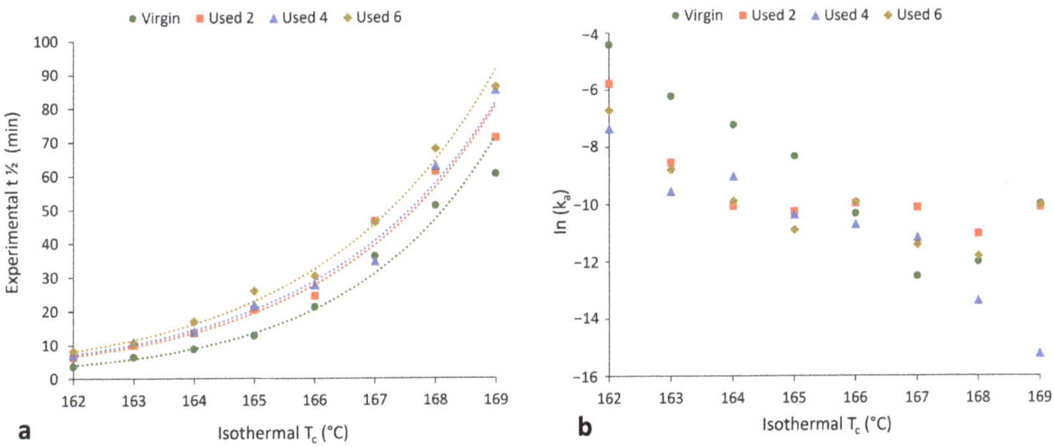

Figure 6. The change in (**a**) crystallisation half-life and (**b**) Avrami rate constant (k_a) as a function of isothermal crystallisation temperature for each powder type.

To better understand the effect of powder re-use on the crystallisation behaviour of PA-12 powder at every isothermal crystallisation temperature, the kinetic parameters (n_a and k_a) presented in Table S1 were applied to create multiple relative cumulative crystallinity curves using the Avrami equation (Figure 7). These plots track the progression of phase transformation as a function of time for each powder type, providing information about the crystallisation process and the fit of the Avrami model. In virgin powder, at every crystallisation temperature, the experimental data form a typical sigmoidal curve. Initially, there is a crystallisation induction time, followed by accelerated primary crystallisation at a constant rate before retardation of the crystallisation process. In the latter stages of crystallisation, the rate of change in relative crystallinity with time is very small, and a pseudo-equilibrium level of crystallinity is obtained, which represents completion of the process [54]. In virgin material, the Avrami model almost parallels the experimental data (Figure 7a), and the constant, linear increase in crystallinity, as a function of time, continues until an almost complete phase transformation. This illustrates that the crystallisation behaviour of virgin PA-12 powder adheres to the Avrami model, and termination proceeds via spherulite impingement. This is further supported by co-efficient of determination (R^2) values of >0.995 for the virgin samples (Table S2).

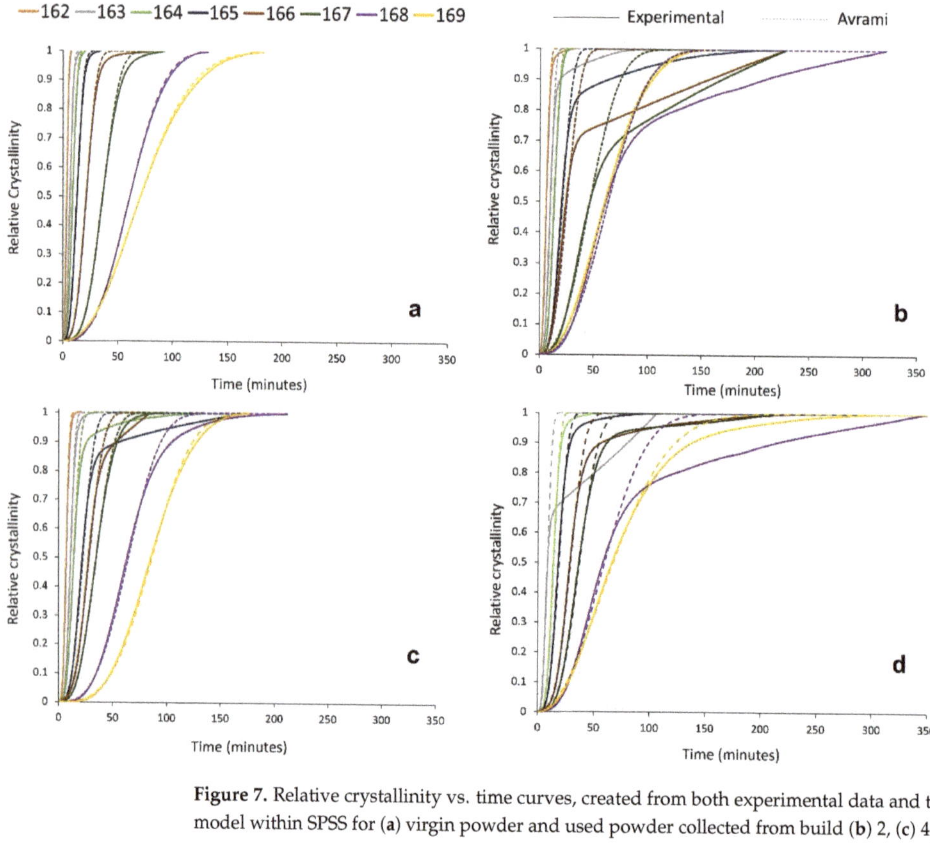

Figure 7. Relative crystallinity vs. time curves, created from both experimental data and the Avrami model within SPSS for (**a**) virgin powder and used powder collected from build (**b**) 2, (**c**) 4, and (**d**) 6.

However, the used powder experimental data deviate away from the Avrami model at significantly lower levels of phase transformation. This is most clearly visualised in Figure 7b, whereby, at multiple isothermal T_c's, used powder recovered from build 2 only conforms to the Avrami theory for up to ~60–70% of transformation. Beyond this point, there are considerable deviations, and the crystallisation rate reduces significantly. A similar trend is observed in used powder recovered from builds 4 and 6; however, the difference between the experimental data and the Avrami model is most significant with used 2 powder. This is also reflected by the R^2 values for used material (Table S2), which are substantially lower than virgin samples, indicating that the Avrami theory can only accurately model a portion of the crystallisation process within aged powder.

Previous work from Mandelkern et al. suggests that deviations from the Avrami expression occur as M_w increases [54,64]. The influence of M_w may explain why, within re-used powder, the Avrami expression can accurately explain the early stages of crystallisation but fails as phase transformation progresses. The observation that Avrami is unable to fit the experimental data at higher levels of phase transformation indicates that the crystallisation process is no longer obliging to the Avrami theory of termination via spherulite impingement [54]. Instead, other factors, such as microstructural defects, must become involved as crystallisation progresses. This paper has provided significant evidence to suggest that polycondensation and cross-linking occur when PA-12 powder is re-used across multiple LS build cycles. Both processes cause an increase in M_w, which alters the morphology and microstructure of the polymer; therefore, they are the most likely cause for the observed deviation from the Avrami model in used powder samples. Polycondensation and the lengthening of non-crystalline amorphous polymer chains causes an increase in the concentration of permanent structural defects, such as chain entanglements and knots, which are unable to participate in the crystallisation process. Similarly, intermolecular cross-linking results in tie-chains, and these structures cannot be reversed or dissipated during melting or crystallisation. Therefore, there are fewer crystallisable units available within the polymer system, and the rate of nucleation and growth cannot continue at the assumed constant rate. Therefore, at this stage, there is no longer a linear increase in relative crystallinity as a function of time, so the progression of primary crystallisation is impeded [54], causing the experimental data to deviate away from the Avrami model at earlier stages of phase transformation. Furthermore, the level of deviation is most significant in powder recovered from build 2, which can be explained by the suggestion that polycondensation occurs more readily within thermally unstressed material. This supports the crystallisation behaviour displayed in Figure 4, whereby the largest change in crystallinity development and T_m occurred over the first two build cycles, before a plateau was observed.

3.4. Modelling the Crystallisation Kinetics of Re-Used PA-12 Powder

Although the most commonly used model for analysing the isothermal crystallisation kinetics of polymers, the Avrami theory can only accurately model a portion of the crystallisation process within aged material. Consequently, Avrami is unsuitable for modelling the overall crystallisation behaviour of re-used PA-12 powder, and a more appropriate model is required for used, aged powder. As explained in Section 2.1.2, the Avrami, simplified Hillier, Tobin, and Hay crystallisation kinetic parameters were obtained from traditional double log plots, whilst the Malkin parameters were calculated from Avrami data (Tables S3–S5). To investigate the suitability of each kinetic model as a function of powder re-use, SPSS software determined the fit of these kinetic parameters to the experimental data for each powder type at a constant crystallisation temperature of 165 °C (Figure 8). In this case, all cumulative data were converted to actual fractional crystallinity using Equation (2). A temperature of 165 °C was selected because it is in the middle of the crystallisation temperature range studied, so there is a compromise in terms of crystallisation rate. This ensures that crystallisation does not occur too rapidly, yet fast enough that the small changes in heat flow, with time, can still be consistently measured accurately. Nonetheless, similar behaviour was observed at every isothermal crystallisation temperature.

In virgin powder, the Avrami model sufficiently describes the full crystallisation process, emphasised by an R^2 value of 0.999, whilst the simplified Hillier and Malkin models also produce high R^2 values. Since virgin powder has not been exposed to aging processes, the experimental data do not deviate away from the constant primary crystallisation rate assumed within the Avrami model. Conversely, the Hay and Tobin models have lower R^2 values and are unable to accurately track the whole crystallisation process. Therefore, the traditional Avrami theory is the most suitable model for describing the crystallisation behaviour of virgin LS-grade PA-12 powder. On the other hand, Figure 8 emphasises that Avrami is unable to precisely model the full phase transformation for re-used PA-12 powder, supporting the results shown in Figure 7. The Avrami equation sufficiently defines primary crystallisation; however, it cannot accurately track the region where the rate of change in fractional crystallinity significantly reduces. As such, relative to virgin PA-12 powder, the Avrami model has substantially lower R^2 values for re-used powder.

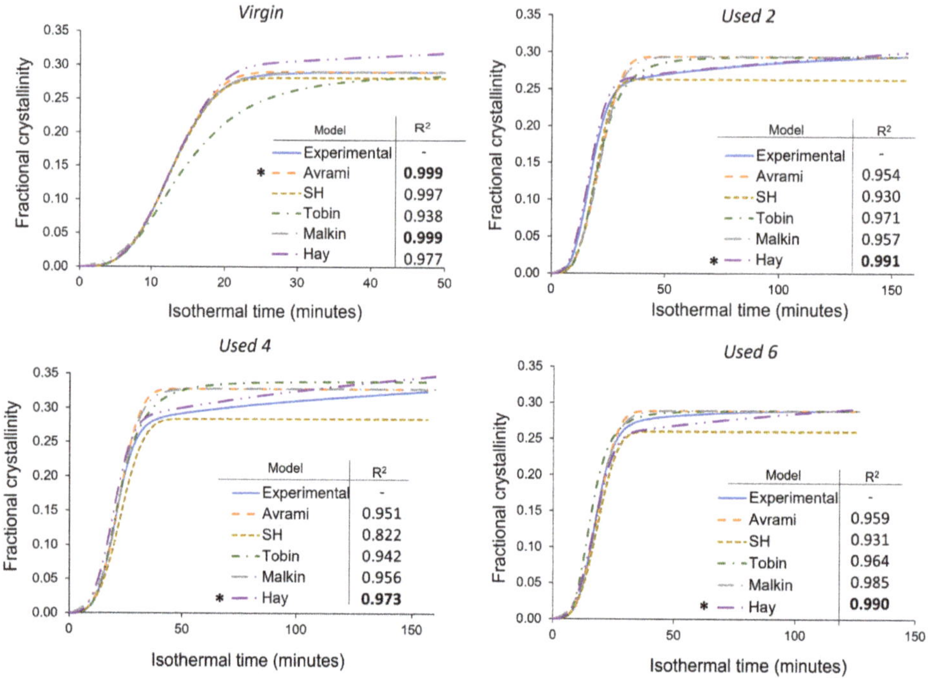

Figure 8. Comparison of the Avrami, simplified Hillier, Tobin, Malkin, and Hay models with experimental data obtained for PA-12 at 165 °C for each powder type. Fractional crystallinity curves were fabricated by inputting the crystallisation kinetic parameters (Tables S3–S5) into the corresponding models. Superimposed on each plot is a key and the respective co-efficient of determination (R^2) value for each model, * represents the model with the highest R^2 in each plot.

As mentioned in Section 3.3, deviations away from the Avrami model, within used powder, occur when primary crystallisation becomes limited due to restricted mobility of long, entangled, and knotted polymer chains. As such, re-used powder contains more "non-crystallisable units" that are unable to contribute to the primary process, so the rate and extent of primary crystallisation reduces. Therefore, for full phase transformation to occur, a greater contribution from secondary crystallisation is required. It is thought that secondary crystallisation is a thickening process that adheres to either a log [23–25] or a root time [18,47] dependence, so it occurs relatively slowly compared to primary crystallisation. For the secondary process to occur, rearrangement of relatively immobile, highly entangled inter-lamellar amorphous regions is required. Additionally, polycondensdation and cross-linking also reduce the rate of secondary crystallisation as lamellar thickening becomes restricted by diffusion of structural defects away from the lamellar growth front [21]; this further increases the time required for full phase transformation within re-used powder samples. Hay suggested that the overlap of primary and secondary crystallisation explains the non-integer n values commonly observed when using the Avrami equation in polymer crystallisation [22,51]. Hay also presented evidence to demonstrate that primary and secondary processes occur simultaneously [22,27,51], so the Hay model is designed to include contributions from both primary and secondary crystallisation throughout the entire curve.

As a result, Hay's model is the most effective for describing the full crystallisation behaviour of used powder because it can accurately estimate the latter stages of crystallisation when the contribution from the secondary process is more significant. This is emphasised by the high R^2 values reported when modelling the isothermal crystallisation of used 2, used 4, and used 6 datasets with the Hay theory (Figure 8). Using SPSS software, the experimental data were also curve-fitted to each respective model via nonlinear, multivariable regression analysis (Figure 9). These curves revealed almost identical results and similar R^2 values for both virgin and re-used powder. This further demonstrates that although the Avrami model is suitable for modelling the behaviour of virgin PA-12 powder, the Hay model is most appropriate for describing the full crystallisation process within re-used, aged material.

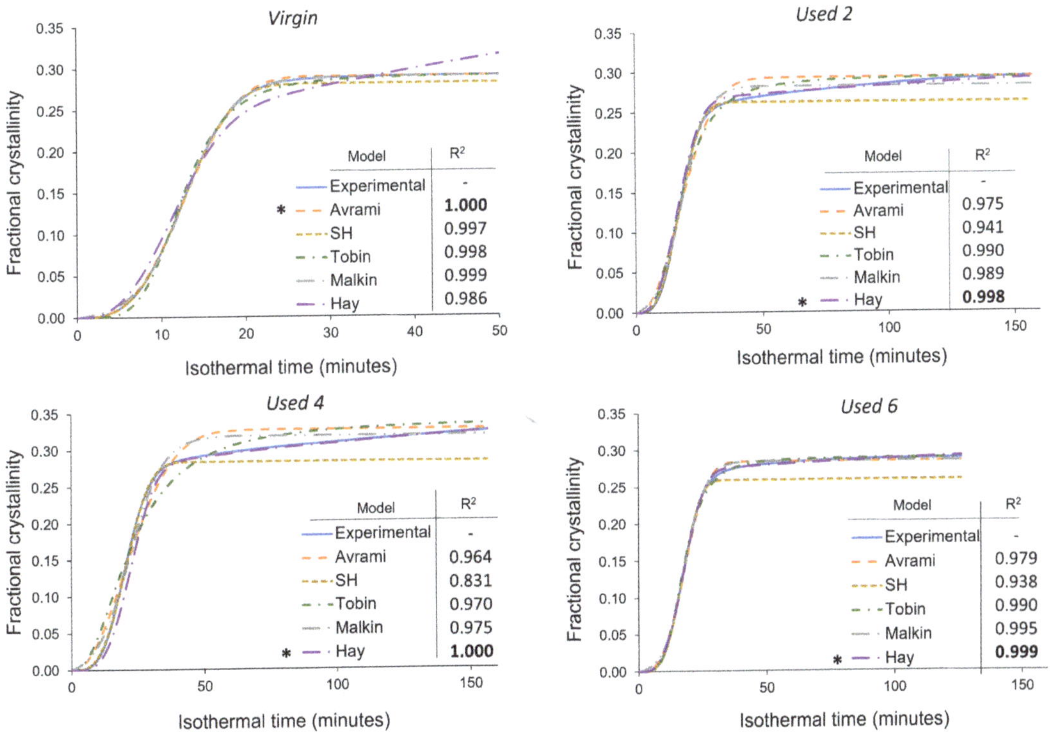

Figure 9. Nonlinear multivariable curve fitting of the Avrami, simplified Hillier, Tobin, Malkin, and Hay kinetic models to the isothermal crystallisation of PA-12 at 165 °C for each powder type. Superimposed on each plot is a key and the respective co-efficient of determination (R^2) value for each model,* represents the model with the highest R^2 in each plot.

A more comprehensive approach for comparing different models is to consider the standard error of regression, s, (Equation (3)) for four individual areas of the cumulative fractional crystallinity curve, whereby the lower the *s* value, the better the fit between the model and experimental data. These four regions resemble the initial stages of crystallisation, the region dominated by primary crystallisation, a transition segment, and the secondary crystallisation region. Other than the initial stage ($0 < X_t < 0.1$), the exact X_t values for each region varies depending on powder type. This method of analysis allows the suitability of each model to be compared at different stages of the crystallisation process. The data presented in Table 1 show that within virgin powder, all models have low s values for the initial and primary regions of crystallisation, suggesting that each model can accurately describe the crystallisation induction period and primary crystallisation process.

However, only the Avrami and Malkin models could successfully model the transition and secondary crystallisation regions. This is because in virgin powder, primary crystallisation dominates, and there is limited involvement of the secondary crystallisation process. The Hay model, for example, expects a rise in fractional crystallinity due to secondary crystallisation; in reality, as the primary process is not limited by hindered chain mobility, almost full phase transformation occurs without much input from the secondary process. As such, the s values shown in Table 1 emphasise that the Avrami (and Malkin) models provide the closest fit to the experimental data for each stage of the crystallisation process in virgin powder.

Table 1. Standard error of regression for each isothermal crystallisation kinetic model over four selected regions of the cumulative fractional crystallinity curve.

Powder Type	Region	X_t Range	Avrami		Simplified Hillier		Tobin		Malkin		Hay	
			DLP	CF	DLP	CF	DLP	CF	DLP	CF	DLP	CF
Virgin	Initial	0.0–0.1	0.0003	0.0011	0.0003	0.0003	0.0004	0.0052	0.0044	0.0034	0.0009	0.0058
	Primary	0.1–0.8	0.0010	0.0020	0.0016	0.0004	0.0286	0.0064	0.0018	0.0017	0.0020	0.0109
	Transition	0.8–0.95	0.0053	0.0020	0.0013	0.0008	0.0527	0.0049	0.0007	0.0012	0.0101	0.0173
	Secondary	0.95–1.0	0.0024	0.0019	0.0077	0.0077	0.0232	0.0032	0.0006	0.0005	0.0212	0.0135
Used 2	Initial	0.0–0.1	0.002	0.010	0.003	0.002	0.003	0.005	0.002	0.008	0.002	0.006
	Primary	0.1–0.7	0.037	0.012	0.029	0.002	0.032	0.010	0.037	0.007	0.011	0.011
	Transition	0.7–0.9	0.026	0.013	0.009	0.002	0.020	0.010	0.023	0.009	0.009	0.007
	Secondary	0.9–1.0	0.013	0.013	0.022	0.022	0.010	0.008	0.013	0.008	0.002	0.003
Used 4	Initial	0.0–0.1	0.0012	0.0149	0.0008	0.0016	0.0004	0.0201	0.0058	0.0123	0.0076	0.0019
	Primary	0.1–0.7	0.0121	0.0186	0.0257	0.0018	0.0041	0.0240	0.0045	0.0113	0.0225	0.0023
	Transition	0.7–0.9	0.0293	0.0226	0.0140	0.0058	0.0254	0.0213	0.0317	0.0173	0.0108	0.0014
	Secondary	0.9–1.0	0.0172	0.0169	0.0309	0.0309	0.0253	0.0136	0.0172	0.0107	0.0163	0.0006
Used 6	Initial	0.0–0.1	0.0008	0.0035	0.0005	0.0004	0.0048	0.0054	0.0055	0.0010	0.0018	0.0015
	Primary	0.1–0.8	0.0061	0.0051	0.0193	0.0012	0.0062	0.0038	0.0324	0.0035	0.0023	0.0012
	Transition	0.8–0.95	0.0125	0.0073	0.0157	0.0081	0.0075	0.0055	0.0083	0.0047	0.0012	0.0004
	Secondary	0.95–1.0	0.0050	0.0036	0.0258	0.0258	0.0046	0.0033	0.0031	0.0021	0.0014	0.0005

On the other hand, in re-used PA-12 powder, the Hay model generally has the lowest s values, especially in the transition and secondary crystallisation regions of the fractional crystallinity curve. Compared to virgin material, all models are less successful at tracking primary crystallisation. In virgin powder, the standard error of regression for the primary region was commonly calculated to be less than 0.002, whilst in re-used powder samples, s values are usually greater than 0.01. The higher s values are particularly noticeable in powder recovered from builds 2 and 4. In used powder, polycondensation and cross-linking reduce the rate of primary crystallisation, resulting in an inconsistent growth rate. Therefore, predicting and describing the change in fractional crystallinity as a function of time becomes more difficult. These aging processes also limit the extent of the primary crystallisation process, causing the experimental data to deviate away from the Avrami model. As such, the s values in the transition and secondary regions are also generally greater within used powder than the respective values in virgin samples (Table 1).

The Avrami model is unable to accurately describe the latter stages of phase transformation in re-used powder because Avrami assumes that the rate of crystal growth is linear and constant. However, structural defects caused by polycondensation and cross-linking reduce the availability of crystallisable polymer units; therefore, the rate of nucleation and growth are no longer constant, with respect to the extent of phase transformation [54]. The simplified Hillier, Tobin, and Malkin models are derived from, and closely related to, Avrami and therefore display similar results. Hay is the only model that accounts for primary and secondary crystallisation simultaneously. Similarly, Hay does not presume a constant nucleation and growth rate, so it is better suited to describe the behaviour of aged PA-12 powder, whereby structural defects hinder the crystallisation rate in the latter stages of phase transformation. As such, within re-used powder samples, Hay is considerably more successful at modelling the transition and secondary crystallisation portion of the curves, and s values are typically 0.005 or less.

These results, calculated for an isothermal crystallisation temperature of 165 °C, are highlighted by the data presented in Tables S6 and S7, which show that comparable results were obtained at every isothermal T_c. In these data, there is a focus on the Avrami and Hay models because they were the two most successful models for describing the crystallisation behaviour of virgin and used powder, respectively. Table S6 illustrates that for virgin powder, the Avrami model has the highest R^2 values, whilst the Hay model has higher R^2 values for every batch of re-used powder. This is further shown by Table S7, which compares the s values of virgin and re-used powder at each isothermal T_c for the Avrami and Hay models. At 168 °C and 169 °C, datapoints are missing for the Hay model. This theory is unsuitable at higher temperatures because the crystallisation rate becomes so slow that extended isothermal times of >8 h are required to fully model the whole crystallisation process. It was observed that the DSC is unable to precisely maintain temperature over this extended time period or accurately monitor the small changes in heat-flow. Nonetheless, 162–167 °C is a sufficient temperature range to demonstrate the advantages of using the Hay model to investigate the isothermal crystallisation of aged PA-12 powder.

To fully understand the suitability of the Hay model for describing the crystallisation behaviour of used powder, it is necessary to explore the relationship between powder type and Hay's kinetic parameters. Crystallisation half-time ($t\,\frac{1}{2}$) can also be estimated using the Hay model, whereby half-time is taken to be the time at which $X_t = \frac{X_{p,inf}}{2}$ [51], whereby $X_{p,inf}$ describes the final fractional crystallinity upon completion of the Avrami primary process. Figure 10 demonstrates that with an increased isothermal T_c, and increased powder re-use, there is generally a reduction in Hay's primary rate constant (k_p) and Hay's $t\,\frac{1}{2}$. These same trends were observed when using the Avrami model to calculate k_a and $t\,\frac{1}{2}$ (Figure 6), and the similarity between the Hay and Avrami estimations is shown in Figure 11. Therefore, independent of the kinetic model employed, it is clear that there is a reduction in the primary crystallisation rate as a function of powder re-use. The Hay model offers additional, unique parameters that can reveal more information about the full crystallisation process following the primary region. Although there does not appear to be a significant change in the secondary rate constant (k_s), as a function of powder re-use (Table S8), a reduction in $X_{p,inf}$ is observed (Figure 10a). This shows that primary crystallisation terminates at lower levels of phase transformation within re-used powder, further supporting the relative crystallinity curves presented in Figure 7. Values of $X_{p,inf}$ are also important when considering the estimations of crystallisation half-time using the Hay method. The extent of the increase in $t\,\frac{1}{2}$, as a function of powder re-use, is modest when calculated using Hay compared to Avrami (Figure 11b). In virgin material, the values for $t\,\frac{1}{2}$ are almost identical; however, in used powder, Hay calculates $t\,\frac{1}{2}$ to be lower. This occurs because Hay's estimation of $t\,\frac{1}{2}$ only includes the primary crystallisation process, and with increased powder re-use, the extent of the primary process is reduced, leading to lower values for $t\,\frac{1}{2}$. Overall, the similarity between the values of the kinetic parameters for Avrami and Hay emphasises that both models can successfully model the primary region of crystallisation for all powder types. However, Hay has the unique advantage of being able to estimate the point at which the primary process terminates and can successfully model the secondary crystallisation region. Further comparisons of the Hay and Avrami kinetic parameters, at every isothermal T_c, are provided in Table S8.

The results outlined in this section have emphasised that although the traditional Avrami model is the most suitable for describing the crystallisation behaviour of virgin LS grade PA-12 powder, it is unsuitable and inaccurate for powder that has been re-used in multiple LS builds. However, the Hay model can sufficiently describe the complete crystallisation process for re-used PA-12 powder. To the best of the author's knowledge, this is the first study to establish the most appropriate kinetic model for describing the crystallisation behaviour of re-used and aged PA-12.

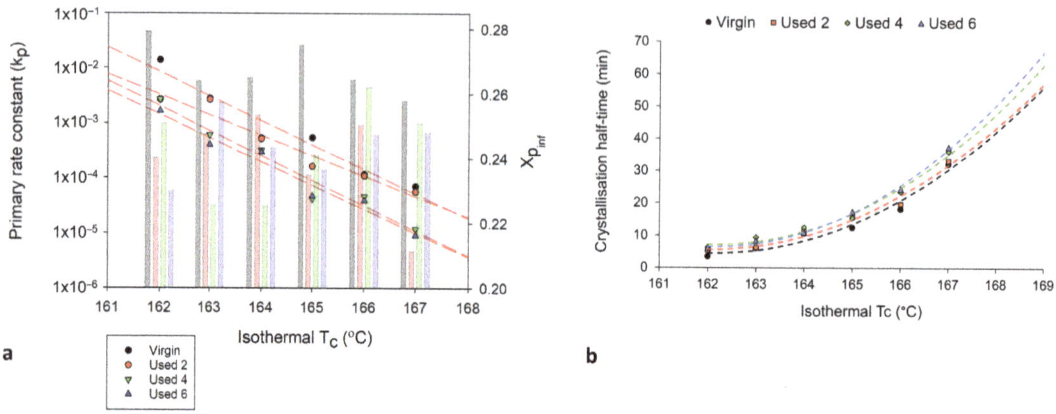

Figure 10. The change in (**a**) the primary crystallisation rate constant, k_p (datapoints) and Xp_{inf} (columns) and (**b**) the crystallisation half-time, whereby the trendline is extrapolated to include isothermal T_c's: 168 °C and 169 °C, calculated using the Hay model, as a function of isothermal T_c and powder type.

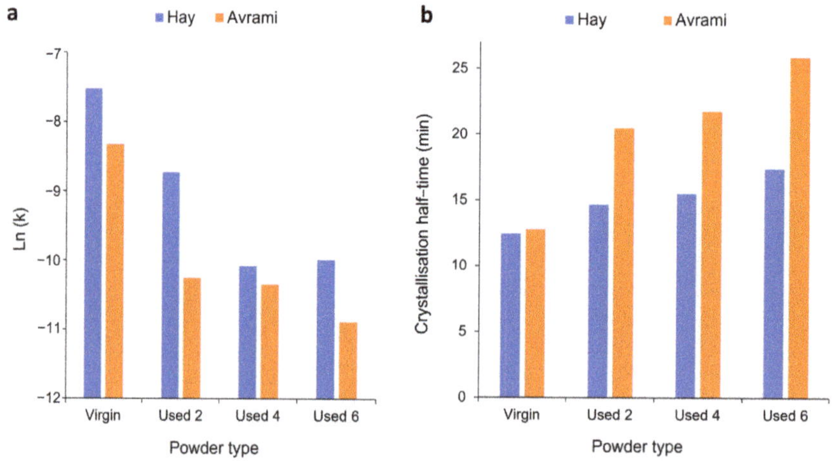

Figure 11. Comparison of the values of (**a**) k_p and (**b**) $t\frac{1}{2}$ using both the Hay and Avrami models, for each powder type, at an isothermal T_c of 165 °C.

4. Conclusions

In this study, PA-12 powder was re-used for a total of seven laser sintering (LS) build cycles, using a 70:30 refresh ratio. Hot-stage microscopy indicated that as a function of powder re-use, there is an increase in melting temperature, a reduction in melt flowability, and restricted particle coalescence. These changes in the thermal behaviour of PA-12, attributed to polycondensation and cross-linking, can cause increased porosity and reduced mechanical properties within LS parts fabricated from re-used powder. The mechanical and structural properties of final parts are also heavily dependent on the crystallisation behaviour of the polymer. In this work, the isothermal crystallisation of PA-12 was studied using a range of models, namely, Avrami, simplified Hillier, Tobin, Malkin, and Hay. For virgin material, the Avrami model was found to be the most successful at modelling the entire crystallisation process. However, in re-used powder, polycondensation and cross-linking cause structural defects that reduce the rate and extent of primary crystallisation, limiting the applicability of the Avrami model. Nonetheless, as a result of accounting for both primary and secondary crystallisation, the Hay model accurately describes the whole crystallisation process and is therefore the optimum method for defining the crystallisation behaviour of aged polymeric material. From an industry perspective, applying the Hay model to better understand the crystallisation behaviour of re-used PA-12 powder could help ensure the fabrication of LS parts with sufficient mechanical properties and a high dimensional precision.

Supplementary Materials: The following supporting information can be downloaded at: https://www.mdpi.com/article/10.3390/polym16050612/s1: Figure S1: Schematic diagram displaying different mechanisms of secondary crystallisation and cross-linking: (a) lamellar thickening, (b) lamellar infilling, and (c) tie-chain formation as a result of cross-linking. Equations (S1)–(S5): Crystallisation kinetics models. Figure S2: The change in (a) peak T_m and (b) T_m range with increased build number, whereby each datapoint is taken as an average of 3 repeats. Figure S3: Schematic diagram representing the polycondensation process, emphasising the entangled, knotted chain structures present in re-used powder. Figure S4: The coalescence behaviour of two virgin powder particles, and powder recovered from different LS build cycles at (a) 195 °C and (b) 205 °C. Figure S5: The change in enthalpy of crystallisation, measured from the exothermic peak during the isothermal segment (column) and enthalpy of fusion, measured from the endothermic melting peak, when re-heating at 10 °C/min (dotted line), as a function of isothermal crystallisation temperature, and powder re-use. Figure S6: Raw data from the DSC displaying the change in the shape and position of the exothermic crystallisation peak, as a function of isothermal crystallisation temperature for (a) virgin, (b) used 2, (c) used 4, and d) used 6 powder types. Figure S7: Avrami double log plots for each powder type, in each case the Xt data range was restricted to the linear region to ensure an $R^2 > 0.99$. Table S1: Avrami crystallisation kinetic parameters and coefficients of determination (R^2) derived from double log plots over iso crystallisation range of 162–169 °C. Xt range restricted to ensure a R^2 value of > 0.99. Table S2: Co-efficient of determination (R^2) values, calculated using SPSS software, for each isothermal crystallisation temperature and powder type. Table S3: Crystallisation kinetic parameters derived from double log plots (n and k) and from non-linear multi-variable regression analysis in SPSS (n* and k*). Table S4: A comparison of the kinetic parameters derived from double log plots (n and k) with the kinetic parameters derived from non-linear multi-variable regression analysis in SPSS (n* and k*) for the Hay model. Table S5: A comparison of the kinetic parameters derived from double log plots (n and k) with the kinetic parameters derived from non-linear multi-variable regression analysis in SPSS (n* and k*) for the Malkin model. Table S6: Co-efficient of determination (R^2) values for the Avrami and Hay models at every isothermal crystallisation temperature. DLP curve is produced from parameters calculated through double log plots, CF curve is produced via curve fitting in SPSS. Table S7: Standard error of regression, s, values for the Avrami and Hay models, at every isothermal crystallisation temperature. DLP curve is produced from parameters calculated through double log plots, CF curve is produced via curve fitting in SPSS. Table S8: The change in kinetic parameters, as a function of powder re-use, calculated using the Hay and Avrami models.

Author Contributions: Conceptualisation, B.S., E.C. and M.J.; methodology, B.S., C.A.K. and M.J.; validation, C.A.K. and M.J.; formal analysis, B.S.; investigation, B.S.; visualisation, B.S.; resources, E.C. and M.J.; data curation, B.S. and C.A.K.; writing—original draft preparation, B.S.; writing—review and editing, E.C., C.A.K. and M.J.; supervision, E.C. and M.J.; project administration, B.S.; funding acquisition, E.C. All authors have read and agreed to the published version of the manuscript.

Funding: The authors would also like to thank ASTM for funding the laser sintering build cycles, which were completed by Prototal UK Ltd.

Institutional Review Board Statement: Not applicable.

Informed Consent Statement: Not applicable.

Data Availability Statement: Data are contained within the article.

Acknowledgments: The authors would like to acknowledge the technical support of the Manufacturing Technology Centre.

Conflicts of Interest: Author Edward Cant is employed by the company The Manufacturing Technology Centre. The remaining authors declare that the research was conducted in the absence of any commercial or financial relationships that could be construed as a potential conflict of interest.

References

1. Goodridge, R.D.; Tuck, C.J.; Hague, R.J.M. Laser sintering of polyamides and other polymers. *Prog. Mater. Sci.* **2012**, *57*, 229–267. [CrossRef]
2. Sillani, F.; Kleijnen, R.G.; Vetterli, M.; Schmid, M.; Wegener, K. Selective laser sintering and multi jet fusion: Process-induced modification of the raw materials and analyses of parts performance. *Addit. Manuf.* **2019**, *27*, 32–41. [CrossRef]
3. Goodridge, R.D.; Ziegelmeier, S. 7—Powder bed fusion of polymers. In *Laser Additive Manufacturing*; Woodhead Publishing: Sawston, UK, 2017; pp. 181–204. [CrossRef]
4. Dadbakhsh, S.; Verbelen, L.; Verkinderen, O.; Strobbe, D.; Van Puyvelde, P.; Kruth, J.-P. Effect of PA12 powder reuse on coalescence behaviour and microstructure of SLS parts. *Eur. Polym. J.* **2017**, *92*, 250–262. [CrossRef]
5. Zarringhalam, H.; Hopkinson, N.; Kamperman, N.F.; de Vlieger, J.J. Effects of processing on microstructure and properties of SLS Nylon 12. *Mater. Sci. Eng. A* **2006**, *435–436*, 172–180. [CrossRef]
6. Josupeit, S.; Schmid, H.-J. Experimental analysis and modeling of local ageing effects during laser sintering of polyamide 12 in regard to individual thermal histories. *J. Appl. Polym. Sci.* **2017**, *134*, 45435. [CrossRef]
7. Paolucci, F.; van Mook, M.J.H.; Govaert, L.E.; Peters, G.W.M. Influence of post-condensation on the crystallization kinetics of PA12: From virgin to reused powder. *Polymer* **2019**, *175*, 161–170. [CrossRef]
8. Wudy, K.; Drummer, D. Aging behavior of polyamide 12: Interrelation between bulk characteristics and part properties. In Proceedings of the 26th Annual International Solid Freeform Fabrication Symposium, Austin, TX, USA, 10–12 August 2016; pp. 17–19.
9. Kuehnlein, F.; Drummer, D.; Rietzel, D.; Seefried, A. Degradation Behavior and Material Properties of PA12-Plastic Powders Processed by Powder Based Additive Manufacturing Technologies. In Proceedings of the 21st International DAAAM Symposium, Vienna, Austria, 20–23 October 2010; Volume 21.
10. Dotchev, K.; Yusoff, W. Recycling of polyamide 12 based powders in the laser sintering process. *Rapid Prototyp. J.* **2009**, *15*, 192–203. [CrossRef]
11. Pham, D.T.; Dotchev, K.D.; Yusoff, W.A.Y. Deterioration of polyamide powder properties in the laser sintering process. *Proc. Inst. Mech. Eng. Part C J. Mech. Eng. Sci.* **2008**, *222*, 2163–2176. [CrossRef]
12. Weinmann, S.; Bonten, C. Recycling of PA12 powder for selective laser sintering. *AIP Conf. Proc.* **2020**, *2289*, 020056. [CrossRef]
13. De Santis, F.; Pantani, R.; Titomanlio, G. Nucleation and crystallization kinetics of poly(lactic acid). *Thermochim. Acta* **2011**, *522*, 128–134. [CrossRef]
14. Neugebauer, F.; Ploshikhin, V.; Ambrosy, J.; Witt, G. Isothermal and non-isothermal crystallization kinetics of polyamide 12 used in laser sintering. *J. Therm. Anal. Calorim.* **2016**, *124*, 925–933. [CrossRef]
15. Kelly, C.A.; Jenkins, M.J. Modeling the crystallization kinetics of polymers displaying high levels of secondary crystallization. *Polym. J.* **2022**, *54*, 249–257. [CrossRef]
16. Wunderlich, B. *Macromolecular Physics: Crystal Nucleation, Growth, Annealing*; Academic Press: New York, NY, USA, 1976; Volume 2.
17. Hoffman, J.D. Role of reptation in the rate of crystallization of polyethylene fractions from the melt. *Polymer* **1982**, *23*, 656–670. [CrossRef]
18. Chen, Z.; Jenkins, M.; Hay, J. Annealing of poly (ethylene terephthalate). *Eur. Polym. J.* **2014**, *50*, 235–242. [CrossRef]
19. Fischer, E.W. Effect of annealing and temperature on the morphological structure of polymers. *Pure Appl. Chem.* **1972**, *31*, 113–132. [CrossRef]
20. Chen, Z.; Hay, J.N.; Jenkins, M. The kinetics of crystallization of poly (ethylene terephthalate) measured by FTIR spectroscopy. *Eur. Polym. J.* **2013**, *49*, 1722–1730. [CrossRef]
21. Chen, Z.; Hay, J.; Jenkins, M. The effect of secondary crystallization on melting. *Eur. Polym. J.* **2013**, *49*, 2697–2703. [CrossRef]
22. Phillipson, K.; Jenkins, M.J.; Hay, J.N. The effect of a secondary process on crystallization kinetics–Poly (ϵ-caprolactone) revisited. *Eur. Polym. J.* **2016**, *84*, 708–714. [CrossRef]
23. Biddlestone, F.; Harris, A.; Hay, J.; Hammond, T. The physical ageing of amorphous poly (hydroxybutyrate). *Polym. Int.* **1996**, *39*, 221–229. [CrossRef]
24. Wang, Z.G.; Hsiao, B.S.; Sauer, B.B.; Kampert, W.G. The nature of secondary crystallization in poly(ethylene terephthalate). *Polymer* **1999**, *40*, 4615–4627. [CrossRef]
25. Verma, R.; Marand, H.; Hsiao, B. Morphological changes during secondary crystallization and subsequent melting in poly (ether ether ketone) as studied by real time small angle X-ray scattering. *Macromolecules* **1996**, *29*, 7767–7775. [CrossRef]
26. Hay, J.N. Secondary crystallization kinetics. *Polym. Cryst.* **2018**, *1*, e10007. [CrossRef]
27. Abo el Maaty, M.I.; Bassett, D.C. Evidence for isothermal lamellar thickening at and behind the growth front as polyethylene crystallizes from the melt. *Polymer* **2005**, *46*, 8682–8688. [CrossRef]
28. Wudy, K.; Drummer, D. Aging effects of polyamide 12 in selective laser sintering: Molecular weight distribution and thermal properties. *Addit. Manuf.* **2019**, *25*, 1–9. [CrossRef]
29. Sanders, B.; Cant, E.; Amel, H.; Jenkins, M. The Effect of Physical Aging and Degradation on the Re-Use of Polyamide 12 in Powder Bed Fusion. *Polymers* **2022**, *14*, 2682. [CrossRef] [PubMed]

30. Riedelbauch, J.; Rietzel, D.; Witt, G. Analysis of material aging and the influence on the mechanical properties of polyamide 12 in the Multi Jet Fusion process. *Addit. Manuf.* **2019**, *27*, 259–266. [CrossRef]
31. Paolucci, F.; Baeten, D.; Roozemond, P.C.; Goderis, B.; Peters, G.W.M. Quantification of isothermal crystallization of polyamide 12: Modelling of crystallization kinetics and phase composition. *Polymer* **2018**, *155*, 187–198. [CrossRef]
32. Wudy, K.; Drummer, D.; Kühnlein, F.; Drexler, M. Influence of degradation behavior of polyamide 12 powders in laser sintering process on produced parts. *AIP Conf. Proc.* **2014**, *1593*, 691–695. [CrossRef]
33. Yang, F.; Jiang, T.; Lalier, G.; Bartolone, J.; Chen, X. A process control and interlayer heating approach to reuse polyamide 12 powders and create parts with improved mechanical properties in selective laser sintering. *J. Manuf. Process.* **2020**, *57*, 828–846. [CrossRef]
34. Cholewa, S.; Jaksch, A.; Drummer, D. Coalescence Behavior of Polyamide 12 as Function of Zero-Shear Viscosity and Influence on Mechanical Performance. In Proceedings of the 2022 International Solid Freeform Fabrication Symposium, Austin, TX, USA, 25–27 July 2022.
35. Alo, O.A.; Otunniyi, I.O.; Mauchline, D. Correlation of reuse extent with degradation degree of PA 12 powder during laser powder bed fusion and mechanical behavior of sintered parts. *Polym. Eng. Sci.* **2023**, *63*, 126–138. [CrossRef]
36. Abbott, C.S.; Sperry, M.; Crane, N.B. Relationships between porosity and mechanical properties of polyamide 12 parts produced using the laser sintering and multi-jet fusion powder bed fusion processes. *J. Manuf. Process.* **2021**, *70*, 55–66. [CrossRef]
37. Li, X.; He, Y.; Dong, X.; Ren, X.; Gao, H.; Hu, W. Effects of hydrogen-bonding density on polyamide crystallization kinetics. *Polymer* **2020**, *189*, 122165. [CrossRef]
38. Liu, S.; Yu, Y.; Cui, Y.; Zhang, H.; Mo, Z. Isothermal and nonisothermal crystallization kinetics of nylon-11. *J. Appl. Polym. Sci.* **1998**, *70*, 2371–2380. [CrossRef]
39. Zhang, F.; Zhou, L.; Xiong, Y.; Liu, G.; Xu, W. Isothermal crystallization kinetics of high-flow nylon 6 by differential scanning calorimetry. *J. Appl. Polym. Sci.* **2009**, *111*, 2930–2937. [CrossRef]
40. Zhao, M.; Wudy, K.; Drummer, D. Crystallization Kinetics of Polyamide 12 during Selective Laser Sintering. *Polymers* **2018**, *10*, 168. [CrossRef] [PubMed]
41. Lorenzo, A.T.; Arnal, M.L.; Albuerne, J.; Müller, A.J. DSC isothermal polymer crystallization kinetics measurements and the use of the Avrami equation to fit the data: Guidelines to avoid common problems. *Polym. Test.* **2007**, *26*, 222–231. [CrossRef]
42. Supaphol, P.; SPRUIELL, J.E. Application of the Avrami, Tobin, Malkin, and simultaneous Avrami macrokinetic models to isothermal crystallization of syndiotactic polypropylenes. *J. Macromol. Sci. Part B* **2000**, *39*, 257–277. [CrossRef]
43. Hillier, I. Modified avrami equation for the bulk crystallization kinetics of spherulitic polymers. *J. Polym. Sci. Part A Gen. Pap.* **1965**, *3*, 3067–3078. [CrossRef]
44. Tobin, M.C. Theory of phase transition kinetics with growth site impingement. I. Homogeneous nucleation. *J. Polym. Sci. Polym. Phys. Ed.* **1974**, *12*, 399–406. [CrossRef]
45. Malkin, A.Y.; Beghishev, V.; Keapin, I.A.; Bolgov, S. General treatment of polymer crystallization kinetics—Part 1. A new macrokinetic equation and its experimental verification. *Polym. Eng. Sci.* **1984**, *24*, 1396–1401. [CrossRef]
46. Hay, J.N. Application of the modified avrami equations to polymer crystallisation kinetics. *Br. Polym. J.* **1971**, *3*, 74–82. [CrossRef]
47. Kelly, C.A.; Hay, J.N.; Turner, R.P.; Jenkins, M.J. The effect of a secondary process on the analysis of isothermal crystallisation kinetics by differential scanning calorimetry. *Polymers* **2019**, *12*, 19. [CrossRef] [PubMed]
48. McFerran, N.L.; Armstrong, C.G.; McNally, T. Nonisothermal and isothermal crystallization kinetics of nylon-12. *J. Appl. Polym. Sci.* **2008**, *110*, 1043–1058. [CrossRef]
49. Avrami, M. Kinetics of phase change. I General theory. *J. Chem. Phys.* **1939**, *7*, 1103–1112. [CrossRef]
50. Ravindranath, K.; Jog, J.P. Polymer crystallization kinetics: Poly(ethylene terephthalate) and poly(phenylene sulfide). *J. Appl. Polym. Sci.* **1993**, *49*, 1395–1403. [CrossRef]
51. Chen, Z.; Hay, J.N.; Jenkins, M.J. The effect of secondary crystallization on crystallization kinetics–Polyethylene terephthalate revisited. *Eur. Polym. J.* **2016**, *81*, 216–223. [CrossRef]
52. Drummer, D.; Wudy, K.; Drexler, M. Influence of Energy Input on Degradation Behavior of Plastic Components Manufactured by Selective Laser Melting. *Phys. Procedia* **2014**, *56*, 176–183. [CrossRef]
53. Cai, C.; Tey, W.S.; Chen, J.; Zhu, W.; Liu, X.; Liu, T.; Zhao, L.; Zhou, K. Comparative study on 3D printing of polyamide 12 by selective laser sintering and multi jet fusion. *J. Mater. Process. Technol.* **2021**, *288*, 116882. [CrossRef]
54. Samulski, E.; Wignall, G.; Koenig, J.; Mark, J.; Ngai, K.; Mandelkern, L.; Graessley, W. (Eds.) The crystalline state. In *Physical Properties of Polymers*, 3rd ed.; Cambridge University Press: Cambridge, UK, 2004; pp. 209–315.
55. Bourell, D.L.; Watt, T.J.; Leigh, D.K.; Fulcher, B. Performance Limitations in Polymer Laser Sintering. *Phys. Procedia* **2014**, *56*, 147–156. [CrossRef]
56. Hoffman, J.D.; Weeks, J.J. Melting process and the equilibrium melting temperature of polychlorotrifluoroethylene. *J. Res. Natl. Bur. Stand. Sect. A Phys. Chem.* **1962**, *66*, 13–28. [CrossRef]
57. Sanders, B.; Cant, E.; Jenkins, M. Re-use of polyamide-12 in powder bed fusion and its effect on process-relevant powder characteristics and final part properties. *Addit. Manuf.* **2024**, *80*, 103961. [CrossRef]
58. Gogolewski, S.; Czerntawska, K.; Gastorek, M. Effect of annealing on thermal properties and crystalline structure of polyamides. Nylon 12 (polylaurolactam). *Colloid Polym. Sci.* **1980**, *258*, 1130–1136. [CrossRef]

59. Gornet, T.J.; Davis, K.R.; Starr, T.L.; Mulloy, K.M. Characterization of Selective Laser Sintering™ Materials to Determine Process Stability. In Proceedings of the 2002 International Solid Freeform Fabrication Symposium, Austin, TX, USA, 5–7 August 2002.
60. Agapov, A.L.; Sokolov, A.P. Does the Molecular Weight Dependence of Tg Correlate to Me? *Macromolecules* **2009**, *42*, 2877–2878. [CrossRef]
61. Roland, C.; Casalini, R. Temperature dependence of local segmental motion in polystyrene and its variation with molecular weight. *J. Chem. Phys.* **2003**, *119*, 1838–1842. [CrossRef]
62. Izuka, A.; Winter, H.H.; Hashimoto, T. Molecular weight dependence of viscoelasticity of polycaprolactone critical gels. *Macromolecules* **1992**, *25*, 2422–2428. [CrossRef]
63. Burfield, D.R.; Doi, Y. Differential scanning calorimetry characterization of polypropylene. Dependence of Tg on polymer tacticity and molecular weight. *Macromolecules* **1983**, *16*, 702–704. [CrossRef]
64. Ergoz, E.; Fatou, J.; Mandelkern, L. Molecular weight dependence of the crystallization kinetics of linear polyethylene. I. Experimental results. *Macromolecules* **1972**, *5*, 147–157. [CrossRef]

polymers MDPI

Article

Investigation of an Optimal Material Addition Rate for Energy Consumption and Dimensional Accuracy in Fused Filament Fabrication of CFR-PEEK

Kyudong Kim [1], Kijung Park [1,*] and Hyun Woo Jeon [2,*]

1 Department of Industrial and Management Engineering, Incheon National University, 119 Academy-ro, Yeonsu-gu, Incheon 22012, Republic of Korea; sooe07@inu.ac.kr
2 Department of Industrial & Management Systems Engineering, Kyung Hee University, Yongin-si 17104, Republic of Korea
* Correspondence: kjpark@inu.ac.kr (K.P.); hwjeon@khu.ac.kr (H.W.J.)

Abstract: The material addition rate (MAR) of fused filament fabrication (FFF) is an indicator of process efficiency varied by process parameter settings, which affects energy consumption and part quality in FFF. This study aims to identify the optimal MAR of FFF using carbon-fiber-reinforced polyether-ether-ketone (CFR-PEEK) by considering a trade-off between energy consumption and the dimensional accuracy of FFF outputs. A design of experiments considering two main process parameters is planned to print three sample types through FFF for CFR-PEEK. Then, the MAR (i.e., deposited material volume per build time) of FFF is obtained to derive individual regression models of energy consumption and the dimensional accuracy measured for each sample type. Furthermore, a trade-off between energy consumption and dimensional accuracy on the MAR is formulated to derive an optimal MAR for each sample type. The results show that FFF for CFR-PEEK has a trade-off between energy consumption and dimensional accuracy; there exists a specific MAR that maximizes the overall performance of energy consumption and dimensional accuracy for each sample type. The optimal MAR is the highest for the small volume sample, whereas it becomes the lowest for the vertical build orientation sample. This study suggests that the optimal MAR should be flexibly adjusted based on a fabricated part. The findings from this study also address the fact that decision-making for optimal FFF operations needs a transition from the identification of specific process parameter settings to the management of a proper process efficiency level in FFF.

Keywords: fused filament fabrication; CFR-PEEK; material addition rate; energy consumption; dimensional accuracy

Citation: Kim, K.; Park, K.; Jeon, H.W. Investigation of an Optimal Material Addition Rate for Energy Consumption and Dimensional Accuracy in Fused Filament Fabrication of CFR-PEEK. *Polymers* **2024**, *16*, 492. https://doi.org/10.3390/polym16040492

Academic Editors: Cristina-Elisabeta Pelin and Anton Ficai

Received: 14 November 2023
Revised: 2 February 2024
Accepted: 7 February 2024
Published: 9 February 2024

1. Introduction

Additive manufacturing (AM) has received increasing attention as an innovative manufacturing technology in that AM provides new manufacturing opportunities with cost-effectiveness and environmental sustainability [1]. Among AM technologies, fused filament fabrication (FFF) is one of the most widely used techniques due to its cost-effective process, fabrication flexibility, and rapid processing for AM [2,3]. An FFF process extrudes a thermoplastic material through a heated nozzle to deposit the material layer by layer [4]. FFF is compatible with a wide range of polymer-based feedstock materials, from low-performance polymers such as polylactic acid (PLA) and acrylonitrile butadiene styrene (ABS) to high-performance polymers such as polyetherimide (PEI) and polyether-ether-ketone (PEEK) [5–7]. FFF has been employed for various applications, including biomedical [8], aerospace [9], and automotive [10] domains.

As the role of AM has been emphasized to realize sustainable manufacturing, recent studies have investigated energy performance in FFF to understand energy behavior and optimal AM operations for energy savings [1]. A concept of the process rate (i.e., added material volume or mass per printing time) of FFF has been employed in recent studies to effectively model energy performance in FFF [11–14]. This paper refers to the process rate of FFF as the material addition rate (MAR). Balogun et al. [15] proposed a general model of energy requirements in FFF, and they showed that the high cycle time and low MAR of FFF cause energy inefficiency. Liu et al. [12] addressed that the specific energy consumption of various AM devices decreases as the MAR increases. Lunetto et al. [13] derived regression models to predict the energy consumption of FFF using ABS and polycarbonate ABS on the MAR, and they claimed that the MAR is a comprehensive measure of complexity in FFF operations to effectively characterize the energy consumption of FFF. Hassan et al. [14] employed a regression model of electric power demand on the MAR of FFF for CFR-PEEK, and they simulated manufacturing scenarios with different FFF machine quantities to analyze a trade-off between energy costs and cycle time. Kim et al. [16] derived the regression models of power demand and energy consumption on the MAR of FFF for CFR-PEEK to simulate the energy performance of an AM system that handles multiple FFF machines and aircraft part designs.

Along with the energy consumption in FFF, the dimensional accuracy of FFF outputs has been considered one of the important aspects of evaluating part quality in AM. For this, existing studies have primarily investigated the effects of FFF process parameters on dimensional accuracy along with other performance measures [17,18]. Wang et al. [19] performed FFF experiments using ABS and identified that build orientation and layer thickness are significant factors for the dimensional accuracy of FFF outputs. Sood et al. [20] considered five FFF process parameters (i.e., layer thickness, part orientation, raster width, air gap, and raster angle) for ABS to identify optimal FFF parameter settings that maximize the overall dimensional accuracy of FFF outputs through the grey Taguchi method. Furthermore, they employed a machine learning approach using an artificial neural network technique to estimate the dimensional accuracy of FFF outputs. Sahu et al. [21] also considered five process parameters for ABS, which were considered in Sood et al. [20], to derive optimal parameter settings for the overall dimensional accuracy of FFF parts through the Taguchi method and the fuzzy logic method. Beniak et al. [22] performed an analysis of variance (ANOVA) to examine the effects of layer thickness and printing temperature on the dimensional accuracy of FFF using PLA and identified that thick layers and high printing temperatures worsen the dimensional accuracy of FFF outputs. Günay et al. [23] derived optimal layer thickness, printing speed, and orientation angle settings through ANOVA and grey relational analysis to maximize the overall dimensional accuracy of FFF outputs. They further identified the reproducibility of FFF through the process capability analysis of the printed samples. Park et al. [24] found that the layer thickness, build orientation, and printing speed of FFF for CFR-PEEK can differently affect the overall dimensional accuracy of FFF outputs depending on the part design.

The MAR of FFF indicates process efficiency in the AM process, which leads to less printing time for a given amount of material deposition at a higher MAR [13]. The FFF process with low process efficiency would negatively impact energy consumption in that the energy consumption of AM rapidly increases as the MAR decreases [25]. On the other hand, a high MAR that is critical for the energy efficiency of FFF may not be suitable for the part quality of FFF outputs. The fact that FFF does not have an adequate dimensional accuracy level for a fabricated part at a high printing speed [23] supports that a high MAR resulting from a high printing speed level can negatively affect the part quality of FFF outputs. Indeed, optimal FFF operations solely for energy consumption with compromised part accuracy are not desirable in practice, and both energy performance and part quality resulting from AM should be jointly considered to improve the overall performance [26].

The consideration of both energy consumption and part quality in FFF becomes more critical for a high-end polymer such as carbon-fiber-reinforced PEEK (CFR-PEEK). CFR-PEEK is one of the most popular high-performance polymers for medical and advanced engineering applications as a metal substitute due to its superior mechanical and chemical properties [27,28]. FFF using CFR-PEEK generally imposes high-level operational requirements (e.g., high processing time and expensive material costs) [24]. Given the operational expense of FFF for CFR-PEEK, a trade-off between energy consumption and part quality of FFF for CFR-PEEK becomes a critical issue; the energy consumption of FFF using CFR-PEEK is improved at a high MAR [16], whereas the dimensional accuracy of FFF outputs using CFR-PEEK deteriorates at a high printing speed [24] that leads to a high MAR. In this regard, understanding a proper process efficiency level for FFF using CFR-PEEK is necessary to have optimal AM operations that satisfy the overall best performance under a trade-off between energy requirements and part quality.

As a response, this study aims to identify an optimal MAR level for the FFF of CFR-PEEK to maximize the overall performance in both energy consumption and dimensional accuracy as a function of the MAR of the FFF. First, a full-factorial experimental design consisting of layer thickness at five levels and printing speed at six levels is planned to fabricate samples in three types, which are used to compare groups for part volume and build orientation. The energy consumption of each experimental sample is collected while the sample is being fabricated, and the overall dimensional error of each fabricated sample from the original dimensions is measured to represent the dimensional accuracy of FFF using CFR-PEEK. The MAR value of each fabricated sample is calculated using the consumed material volume and printing time of each fabrication to build the regression models of energy consumption and dimensional accuracy on the MAR. Then, a performance model that integrates the fitted regression models of energy consumption and dimensional accuracy is derived to identify an optimal MAR level of FFF for CFR-PEEK. The above procedure is performed for three sample types to identify changes in the optimal MAR level and to achieve the overall performance of both energy consumption and dimensional accuracy.

2. Materials and Methods

The following subsections illustrate the main steps to derive an optimal MAR that simultaneously considers both energy consumption and the dimensional accuracy of FFF for CFR-PEEK. Section 2.1 presents an experimental design to collect energy consumption and dimensional accuracy data for three sample types through FFF using CFR-PEEK. Section 2.2 performs regression modeling for energy consumption and dimensional accuracy on the MAR of FFF. Section 2.3 proposes a trade-off analysis procedure to derive an optimal MAR that simultaneously improves both performance measures.

2.1. Design of the Experiment

This study used three hexahedron design cases for experiments (see Figure 1). These sample types were employed as comparison groups for analysis. In Figure 1, Sample A illustrates a cube (i.e., 10 mm × 10 mm × 10 mm) to represent an object with a smaller volume and no build orientation effect (see Figure 1a). Sample B and Sample C represent a larger-volume object group (=2 cm^3) than Sample A. In addition, they represent different build orientations of an object. Sample B and Sample C, respectively, represent horizontal and vertical orientation (i.e., 20 mm × 10 mm × 10 mm cuboid in Figure 1b and 10 mm × 10 mm × 20 mm cuboid in Figure 1c).

All experimental samples were fabricated by Apium P220 (Apium Additive Technologies GmbH, Karlsruhe, Germany) [29], which is an industrial FFF machine compatible with CFR-PEEK. TECAPEEK CF30 (Ensinger, Nufringen, Germany) [27], which has 1.38 g/cm^3 in density, 6000 MPa in tensile modulus, and 112 MPa in tensile strength, was used as a material for the experiments. For each sample type, a full-factorial experimental design was planned to fabricate experimental samples by varying layer thickness and printing speed levels. These two FFF process parameters were chosen due to their significant effects on energy consumption and dimensional accuracy for CFR-PEEK outputs [14,16,24]. All possible combinations of layer thickness at five levels and printing speed at six levels were randomly organized for the experiments of each sample type (see Table 1). It is noted that 0.2 mm layer thickness and 1200 mm/min printing speed are the default process parameter settings for CFR-PEEK fabrication recommended by the manufacturer of the FFF machine.

Experiments for Sample A in Figure 1 were considered a baseline experiment set to analyze a trade-off between energy consumption and dimensional accuracy based on the MAR of FFF for CFR-PEEK; a total of 90 samples (=five levels for layer thickness × six levels for printing speed × three replicates) were fabricated for all the process parameter combinations. In order to observe changes in the trade-off depending on volume variations and build orientation, 30 samples (=five levels for layer thickness × six levels for printing speed) for each type of Sample B and Sample C in Figure 1 were also fabricated, respectively. All process parameters except for the varied process parameters for the experiments were fixed according to the recommended parameter settings of the machine (see Table 1). In addition, a brim was set for each sample fabrication to ensure a complete AM process for each sample without detachment from the bed platform during the process (see Figure 1d). The process parameter settings of each experimental sample were controlled by the Simplify 3D slicing software [30].

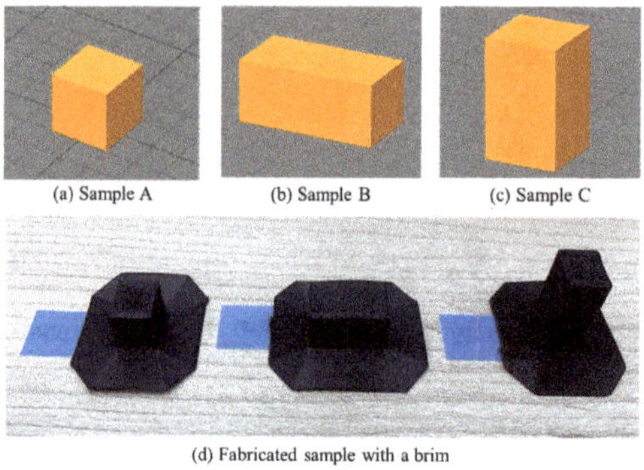

Figure 1. Three sample types for the experiments.

Table 1. Varied and fixed process parameters for experiments.

Type	Process Parameter	Value
Varied	Layer thickness (mm)	0.1 (L1), 0.15 (L2), 0.2 (L3), 0.25 (L4), and 0.3 (L5)
	Printing speed (mm/min)	1000 (P1), 1100 (P2), 1200 (P3), 1300 (P4), 1400 (P5), and 1500 (P6)
Fixed	Bed temperature (°C)	120
	Nozzle temperature (°C)	510
	Nozzle diameter (mm)	0.4
	Perimeter shells (# of layers)	3
	top layers/bottom layers (# of layers)	None
	Extrusion percentage for the first layer (%)	96
	Infill pattern (pattern)	Rectilinear
	Infill density (%)	100

Figure 2 illustrates the experimental environment of this study to collect energy data during each experiment. The energy consumption (W·h: watt-hour) of each experiment was recorded by the Wattman HPM-100A (ADpower, Pyeongtaek, Republic of Korea) [31] during the FFF process. The total energy consumption of each experimental sample during the material extrusion stage of the FFF process was considered for response data since power demands in the heating and cooling stages were not significantly varied across the experiments (see Figure 3).

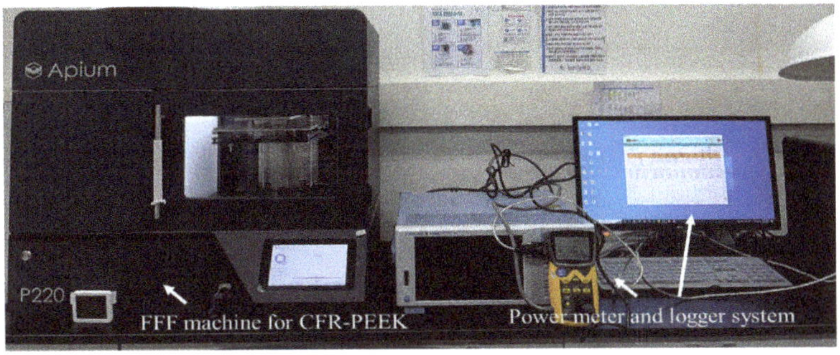

Figure 2. Equipment settings for measuring energy consumption in FFF experiments.

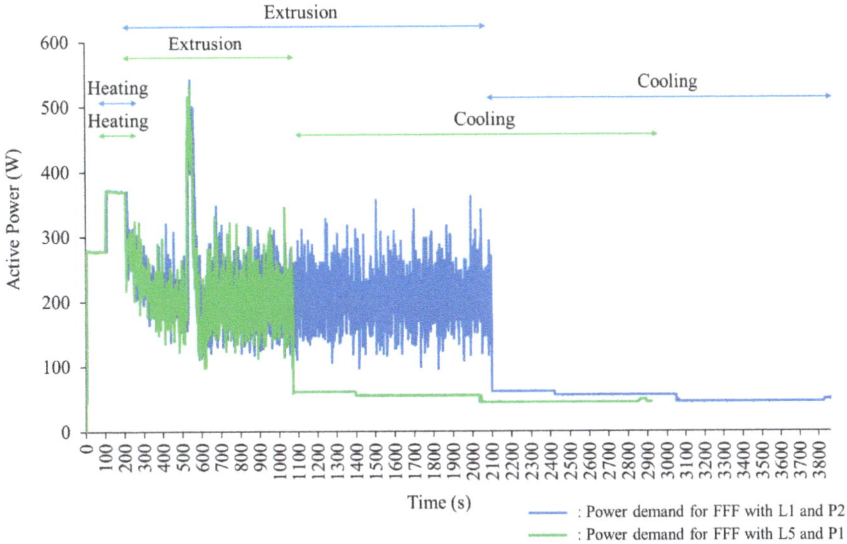

Figure 3. Examples of energy changes during FFF for CFR-PEEK.

The dimensions (i.e., width, length, and height) of the fabricated samples were measured by the Mitutoyo NTD13-P15M digital vernier caliper (Mitutoyo, Kawasaki city, Japan) [32]. The measured three axes of the fabricated samples were used to calculate the overall dimensional accuracy (error) using Equation (1). The root mean square error of each final output was calculated to represent the dimensional accuracy of the experimental sample based on dimensional differences between the original dimensions and the measured dimensions.

$$Dimensional\ Accuracy\ (DA) = \sqrt{\sum_i (M_i - A_i)^2/3}, \quad (1)$$

where M_i is the measured value of dimension i (=x, y, and z-axis) for an experimental sample, and A_i is the original size of dimension i for an experimental sample.

The total build time during the extrusion stage of each experiment and the total volume of the deposited material for each fabricated sample were also collected by referring to time in measured energy profiles and the Apium P220 data system, respectively. Table 2 summarizes the experimental data collected for this study.

Table 2. Collected data from experiments.

Name	Definition	Unit
Build time (t)	The total time taken during the material extrusion stage	s
Deposited material volume (v)	The total volume of deposited material during the material extrusion stage	mm^3
Energy consumption (EC)	The total energy consumed during the extrusion stage	W·h
Dimensional accuracy (DA)	The root mean square error of a final output between original and measured dimensions	mm

2.2. Modeling Energy Consumption and Dimensional Accuracy on MAR

Individual regression models for energy consumption and dimensional accuracy on the MAR of FFF for CFR-PEEK were derived first based on the collected experimental data. The MAR of each experimental sample was calculated by dividing the total deposited volume by the build time ($=v/t$) to represent the amount of deposited material per time. Then, the energy consumption model for FFF in Equation (2) [16] was used to fit the energy consumption data of fabricated samples through linear regression.

$$EC = \beta_{0(EC)} + \beta_{1(EC)} \cdot MAR_{inv}, \tag{2}$$

where $MAR_{inv} = 1/MAR$, $\beta_{0(EC)}$ is an intercept, and $\beta_{1(EC)}$ is a coefficient for the predictor.

The dimensional accuracy of FFF for CFR-PEEK was expressed as a function of MAR through a linear regression model in Equation (3). A linear regression model for dimensional accuracy was assumed by referring to the findings of Park et al. [24], which identified the statistical significance of layer thickness and printing speed on the dimensional accuracy of FFF using CFR-PEEK through ANOVA.

$$DA = \beta_{0(DA)} + \beta_{1(DA)} \cdot MAR, \tag{3}$$

where $\beta_{0(DA)}$ is an intercept, and $\beta_{1(DA)}$ is a coefficient for the predictor.

The above regression models for energy consumption and dimensional accuracy were fitted using MINITAB 20.3 [33]. The statistical significance and prediction performance of the derived regression models were examined by t-test and R-squared, respectively. The modeling procedure in this section was applied to each sample type to derive fitted energy consumption and dimensional accuracy models.

2.3. Optimal MAR Based on the Trade-Off between Energy Consumption and Dimensional Accuracy

A trade-off analysis for each sample type was performed to characterize both energy consumption and dimensional accuracy on the MAR of FFF for CFR-PEEK. Since energy consumption and dimensional accuracy have the same objective direction (i.e., to be minimized) with different value scales, a normalization process in Equation (4) [34] was performed to make them have the same value scale for the experimental dataset of each sample type.

$$y_i\prime = (max(y) - y_i)/(max(y) - min(y)), \tag{4}$$

where $y_i\prime$ is a normalized response value, y_i is an original response value for normalization, $max(y)$ is the maximum value of response variable y, and $min(y)$ is the minimum value of response variable y.

Then, the same regression procedure for the original data in Section 2.1 was performed to derive regression models for normalized energy consumption and dimensional accuracy. Then, regression models constructed for normalized energy consumption and dimensional accuracy were added to characterize the overall AM performance. Equation (5) shows a proposed AM performance model on the MAR of FFF.

$$P = \beta_{0(EC\_Norm)} + \beta_{1(EC\_Norm)} \cdot 1/MAR + \beta_{0(DA\_Norm)} + \beta_{1(DA\_Norm)} \cdot MAR, \tag{5}$$

where P denotes the overall AM performance, $\beta_{0(EC\_Norm)}$ and $\beta_{0(DA\_Norm)}$ are constants for the regression models for normalized energy consumption and dimensional accuracy, and $\beta_{1(EC\_Norm)}$ and $\beta_{1(DA\_Norm)}$ are regression coefficients for normalized energy consumption and dimensional accuracy.

Through the normalization process, both energy consumption and dimensional accuracy are transformed from cost criteria (i.e., the smaller the better) to benefit criteria (i.e., the larger the better). It is assumed that energy consumption and dimensional accuracy are in a trade-off relationship depending on the MAR of FFF; as the MAR increases, the normalized energy consumption performance rapidly increases, whereas the normalized dimensional accuracy performance worsens. Then, the formula for P forms a concave function that reflects the trade-off, where P becomes maximum at a specific MAR. The optimal MAR value that maximizes the overall performance can be obtained by a differential equation of P. That is, the MAR value that satisfies $dP/dMAR = 0$ becomes the optimal MAR considering the trade-off. Changes in the trade-off behavior and the optimal MAR of the sample types were analyzed to find implications regarding the role of the MAR for FFF operations.

3. Results

The regression models for energy consumption and the dimensional accuracy of each sample type are presented in Section 3.1. Section 3.2 shows the results of the trade-off analysis on the MAR.

3.1. Energy Consumption and Dimensional Accuracy Models for FFF Using CFR-PEEK

Table 3 shows the statistical significance of regression models for energy consumption on the MAR of FFF for each sample type. Furthermore, Figure 4 illustrates regression lines of energy consumption on the MAR for each sample type. The regression results of energy consumption for all the sample types reveal that energy consumption is suitably fitted as a linear function of the inverse MAR; the energy consumption of FFF for CFR-PEEK nonlinearly decreases as the MAR increases (see Figure 4b). The inversed MAR is statistically significant in all the derived regression models, regardless of the sample types (see Table 3). Moreover, all the regression models have high explanatory power for a relationship between energy consumption and the inversed MAR.

Table 3. Regression models for energy consumption.

Sample Type	Regression Model	R-Squared	Test for $\beta_{0(EC)}$ (p-Value)	Test for $\beta_{1(EC)}$ (p-Value)
Sample A	$EC = 11.43 + 55.90 \cdot MAR_{inv}$	99.74%	$t = 33.51$ ($p = 0.00$ **)	$t = 183.03$ ($p = 0.00$ **)
Sample B	$EC = 13.49 + 108.36 \cdot MAR_{inv}$	99.82%	$t = 15.60$ ($p = 0.00$ **)	$t = 126.21$ ($p = 0.00$ **)
Sample C	$EC = 15.25 + 109.47 \cdot MAR_{inv}$	99.68%	$t = 12.26$ ($p = 0.00$ **)	$t = 93.24$ ($p = 0.00$ **)

** $p < 0.05$.

In Figure 4a, Sample B and Sample C have almost the same energy consumption pattern, and their energy consumption is always higher than the energy consumption of Sample A for every inverse MAR. This result seems to be due to the part volume required for the FFF of Sample B and Sample C; Sample B and C have twice as large a part volume as Sample A. Accordingly, the regression coefficients for the Sample B and C cases are approximately twice as high as the regression coefficient of the energy consumption model for Sample A (see Table 3). This indicates that the energy consumption required for the FFF of Sample B and Sample C more rapidly decreases than for the FFF of Sample A as the MAR increases (see Figure 4b). The regression results indicate that energy consumption is significantly varied depending on the MAR of FFF and the fabricated part volume.

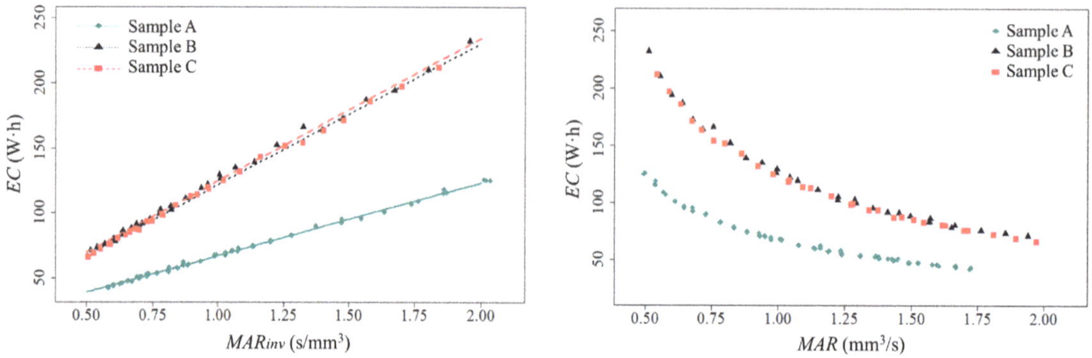

(a) Fitted regression lines for energy consumption (b) Original plots of energy consumption on MAR

Figure 4. Energy consumption on MAR for each sample type.

Table 4 presents the linear regression results of dimensional accuracy on the MAR of FFF for each sample type. Figure 5 shows the linear regression plots of dimensional accuracy on the MAR. The results show that the overall dimensional error of each fabricated sample statistically increases as the MAR of FFF increases. The *t*-test for $\beta_{1(DA)}$ in Table 4 shows that the MAR is a statistically significant linear predictor of the dimensional accuracy measure for each design case. However, the *R*-squared values of all the regression models are at a moderate level, possibly due to dimensional variations added by the manual measuring process using the vernier caliper and uncontrollable experimental factors in this study. Different from the regression results for energy consumption, the fitted regression lines of the Sample A and Sample B cases show a similar pattern. On the other hand, the regression coefficient for Sample C is the highest among the derived regression coefficients for the sample types; the negative effect of the MAR on the overall dimensional error of a fabricated part for Sample C is more critical than in the other cases as the MAR increases (see Figure 5).

Table 4. Regression models for dimensional accuracy.

Sample Type	Regression Model	*R*-Squared	Test for $\beta_{0(DA)}$ (*p*-Value)	Test for $\beta_{1(DA)}$ (*p*-Value)
Sample A	$DA = 0.31 + 0.27 \cdot MAR$	49.94%	$t = 9.55$ ($p = 0.00$ **)	$t = 9.37$ ($p = 0.00$ **)
Sample B	$DA = 0.23 + 0.29 \cdot MAR$	51.00%	$t = 3.36$ ($p = 0.002$ **)	$t = 5.40$ ($p = 0.00$ **)
Sample C	$DA = 0.13 + 0.43 \cdot MAR$	65.02%	$t = 1.73$ ($p = 0.094$)	$t = 7.21$ ($p = 0.00$ **)

** $p < 0.05$.

Sample C, which is a variant of Sample A with an increase in the z-axis dimension from Sample A, has a low dimensional accuracy performance in FFF with process parameters that result in a high MAR. It is noted that the crossing point between the regression lines for Sample A and Sample C in Figure 5 is 1.13 mm^3/s, approximately. The overall dimensional error of Sample C becomes greater than that of Sample A at the MAR, which is greater than 1.13 mm^3/s. On the other hand, the dimensional error of Sample C is expected to be lower than that of Sample A when the MAR is less than 1.13 mm^3/s. This indicates that a decrease in the MAR of FFF using CFR-PEEK can be effective in enhancing the dimensional accuracy of the fabricated part if the part height is increased from its original dimension. Alternatively, a change in the build orientation of Sample C can be an option to improve dimensional accuracy at the same MAR of FFF. Figure 5 shows that Sample B, which has the same part volume as Sample C, is expected to have better dimensional accuracy than Sample C at most MAR levels.

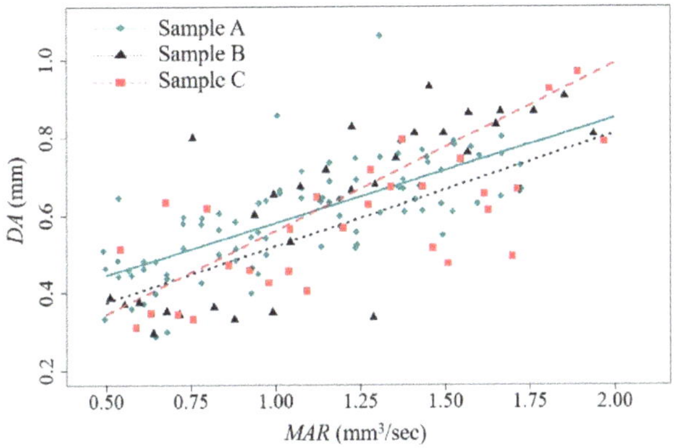

Figure 5. Dimensional accuracy on MAR for each sample type.

The regression models built for the energy consumption and dimensional accuracy of FFF for CFR-PEEK confirm a trade-off between those performance measures on the MAR. Figure 4b supports that energy consumption is in an inverse relationship with the MAR, whereas Figure 5 shows that dimensional error is in proportion to the MAR. The results show that there can be an optimal MAR at which both performance measures are suitably achieved by avoiding extreme losses in either energy consumption or dimensional accuracy performance.

3.2. Optimal MAR Considering Both Energy Consumption and Dimensional Accuracy

Table 5 and Figure 6 show an optimal MAR for each sample type when the trade-off between the energy consumption and dimensional accuracy of FFF for CFR-PEEK is considered to maximize the overall performance. The normalized performance plots shown in Figure 6 clearly address the fact that the FFF of CFR-PEEK for all the sample types has a trade-off between energy and quality performance depending on the MAR.

Table 5. Derived optimal MAR for overall AM performance.

Sample Type	Normalized EC Model	Normalized DA Model	AM Performance Model	Optimal MAR
Sample A	EC_{norm}= 1.37–0.67/*MAR* - Test for β_0(*p*-value): 0.00 - Test for β_1(*p*-value): 0.00 - *R*-squared: 99.74%	DA_{norm}= 0.97–0.34·*MAR* - Test for β_0(*p*-value): 0.00 - Test for β_1(*p*-value): 0.00 - *R*-squared: 49.94%	*P* = 2.34–0.67/*MAR*–0.34·*MAR*	1.40
Sample B	EC_{norm}= 1.36–0.74/*MAR* - Test for β_0(*p*-value): 0.00 - Test for β_1(*p*-value): 0.00 - *R*-squared: 99.82%	DA_{norm}= 1.12–0.44·*MAR* - Test for β_0(*p*-value): 0.00 - Test for β_1(*p*-value): 0.00 - *R*-squared: 51.00%	*P* = 2.48–0.74/*MAR*–0.44·*MAR*	1.30
Sample C	EC_{norm}= 1.35–0.68/*MAR* - Test for β_0(*p*-value): 0.00 - Test for β_1(*p*-value): 0.00 - *R*-squared: 99.68%	DA_{norm}= 1.26–0.67·*MAR* - Test for β_0(*p*-value): 0.00 - Test for β_1(*p*-value): 0.00 - *R*-squared: 65.02%	*P* = 2.61–0.68/*MAR*–0.67·*MAR*	1.01

The optimal MAR for the baseline case (i.e., Sample A) to maximize the overall AM performance of energy consumption and dimensional accuracy is 1.40 mm^3/s, which is the highest process rate of all the sample types. This indicates that Sample A, requiring half the volume of the other sample types, can be optimally fabricated at a relatively higher process rate than the other sample types. In addition, the results in Table 5 and Figure 6 show that Sample B and Sample C require a different optimal MAR, although their designated part volumes are equal. The overall AM performance of Sample B is optimal at the MAR of

1.30 mm^3/s, which is near the optimal MAR for Sample A. On the other hand, the lowest optimal MAR of all the types (=1.01 mm^3/s) is required for Sample C.

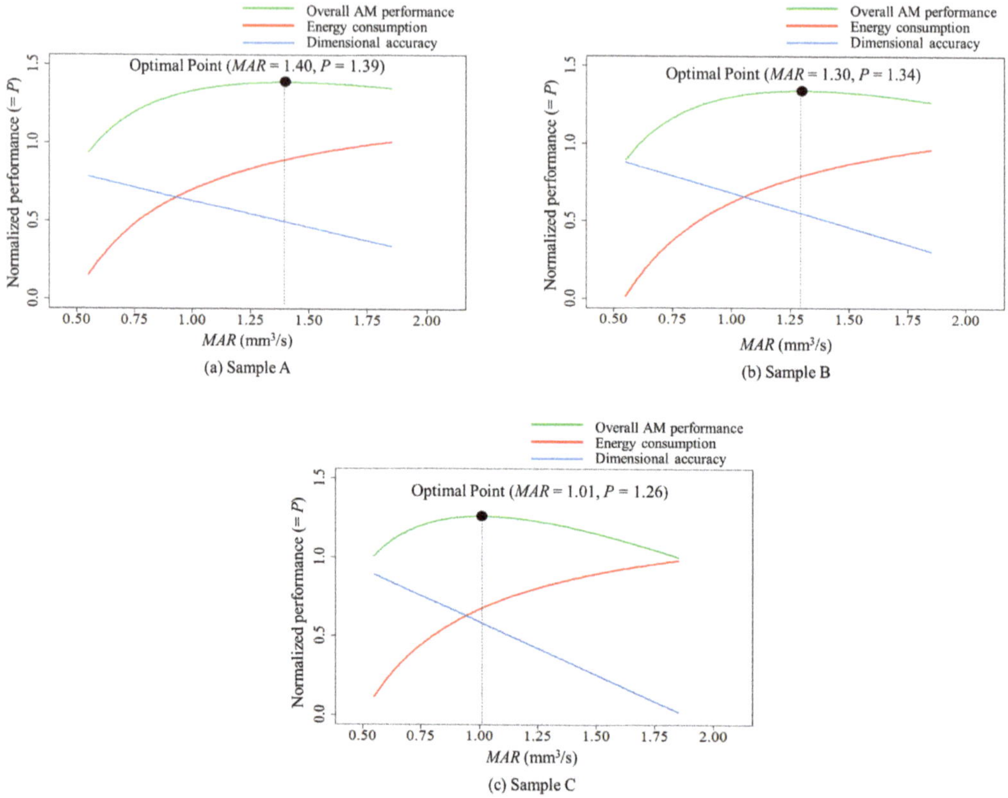

Figure 6. Optimal MAR to maximize overall AM performance.

The optimal MAR for Sample C is lower than that of Sample B, since the dimensional accuracy performance of Sample C decreases faster than that of Sample B (see Figure 6c) without significant variations in energy performance between Sample B and Sample C (see Figure 4b). The findings indicate that the build orientation of a fabricated part can be a critical factor in AM process efficiency. A change to horizontal build orientation to have a wide bottom area can be an effective strategy to pursue higher process efficiency in FFF if mechanical properties (e.g., tensile and compressive strength) affected by the build orientation are not critical for the fabricated part. On the other hand, a part design that has a height size relatively higher than sizes on the x-y plane may require the FFF with a relatively lower MAR to maintain the best energy and quality performance in the FFF if the FFF of the part requires its original build orientation due to mechanical functionality. The above results also suggest that the MAR of FFF for CFR-PEEK should be properly managed to improve the overall AM performance by jointly considering energy consumption and dimensional accuracy, depending on the part designs considered for AM.

Table 6 shows the process parameter combinations of the experiments that lead to the experimental MAR being close to the theoretical optimal value obtained for each sample type. For this, all the MAR values calculated from the experiments were rounded off to the first digit after the decimal point, and then the parameter combinations that result in the MAR value closest to the optimal MAR were identified for each sample type. For Sample A, the process parameter combinations specified in Table 6 are the experimental settings in which all the experimental replicates result in rounded-off MAR values equal to the optimal MAR. It is noted that the energy consumption and dimensional accuracy data specified for Sample A in Table 6 are averaged values for the associated replicates. The results imply that there can be multiple process parameter combinations to satisfy the overall best AM performance of energy consumption and dimensional accuracy. This indicates that the process parameters of FFF for the overall AM performance are not a fixed and single optimal process parameter set but a flexible and selective solution among the alternative parameter combinations leading to the optimal MAR.

Table 6. Suggested process parameter settings.

Suggested Process Parameters	MAR	Measured EC	Measured DA
Sample A	1.40 (optimal)	51.4 (estimated)	0.69 (estimated)
L4 and P4	1.38	52.2	0.66
L4 and P5	1.42	50.9	0.64
L5 and P1	1.37	53.5	0.74
Sample B	1.30 (optimal)	96.8 (estimated)	0.61 (estimated)
L3 and P4	1.27	98.2	0.63
L3 and P5	1.34	93.8	0.67
L4 and P1	1.28	99.2	0.72
Sample C	1.01 (optimal)	123.6 (estimated)	0.56 (estimated)
L2 and P5	0.99	126.3	0.35
L2 and P6	1.04	122.1	0.53
L3 and P1	0.99	129.3	0.65

4. Discussion

The resultant regression models of energy consumption show that the energy consumption of FFF for CFR-PEEK is expressed as a decreasing non-linear function of the MAR, as addressed in the existing studies. Indeed, process parameter combinations that lead to a short build time in this study show a high MAR and a resultant low energy consumption value in that the energy consumption of FFF for CFR-PEEK is significantly affected by printing time [14]. This energy behavior on the MAR is consistently observed in all the sample types, and the results support that an increase in AM process efficiency can greatly decrease the energy consumption of FFF for CFR-PEEK. In particular, Sample B and Sample C, which have the same part volume with different build orientations, show similar energy consumption plots on the MAR; their energy consumption is higher than the energy consumption of Sample A for every MAR. This indicates that the energy consumption of FFF for CFR-PEEK is not significantly varied by the build orientation but is affected by the designated part volume.

On the other hand, the dimensional accuracy of FFF for CFR-PEEK can be modeled as a linear function of the MAR. A high MAR is achieved when layer thickness and printing speed are set at high levels. However, these process parameter settings can negatively impact printing resolution and thereby may cause an increase in the overall dimensional error of a fabricated part. The negative impact of the MAR on dimensional accuracy is stronger for Sample C than for Samples A and B, which have a similar pattern of overall dimensional error on the MAR. A plausible reason for the worsened dimensional accuracy observed in Sample C compared to Sample B, which has the same design and volume as Sample C, is build orientation that increases the number of deposited layers. Sample C is considered a vertically oriented Sample B on the build platform. FFF has greater residual stress and heat shrinkage in the vertical direction (z-axis) than in the horizontal direction (x-y plane), and therefore this can cause more dimensional deviations in a part fabricated in a vertical orientation [35]. The results from the dimensional accuracy models indicate that build orientation is a more critical factor for the dimensional accuracy of FFF for CFR-PEEK than part volume.

The above energy consumption and dimensional accuracy characteristics of FFF for CFR-PEEK clearly show a trade-off on the MAR. The overall AM performance model, through the integration of the normalized energy consumption and dimensional accuracy models, presents that there is an optimal MAR at which FFF for CFR-PEEK is the most effectively operated given the trade-off of energy consumption and dimensional accuracy. Moreover, the fact that the optimal MAR varied by the sample type implies that the FFF for CFR-PEEK should be flexibly operated to achieve an optimal MAR targeted for part volume and orientation; relatively high AM efficiency can be pursued for a small volume part (e.g., Sample A) in the FFF for CFR-PEEK to have the best performance in energy consumption and dimensional accuracy. If the horizontal orientation of a part (e.g., Sample B) is allowed to increase the contract area on the build platform, FFF for CFR-PEEK can be managed to improve process efficiency that is expected to be lower in the original vertical orientation of the part (e.g., Sample C).

5. Conclusions

This study focused on a trade-off between energy consumption and dimensional accuracy of FFF outputs using CFR-PEEK to identify an optimal MAR to maximize the overall AM performance of both energy consumption and dimensional accuracy. For this, an experimental design considering five levels of layer thickness and six levels of printing speed for FFF using CFR-PEEK was planned to manufacture experimental samples of three types. Energy consumption and dimensional accuracy data measured for each experiment were collected as response data, and the MAR of each experiment was calculated to be used as input data. Herein, the MAR was employed to represent the process efficiency of FFF using CFR-PEEK, which is an operational consequence of selected process parameter settings, to effectively model energy consumption and dimensional accuracy for FFF.

Based on the collected data, individual regression models of energy consumption and dimensional accuracy on the MAR for the considered sample types were derived to confirm the impact of the MAR on the performance measures. Then, the original response data were normalized to derive individual regression models for normalized energy consumption and dimensional accuracy again, and they were integrated to represent the overall AM performance model of FFF for CFR-PEEK for the sample types, respectively. Finally, an optimal MAR to maximize the overall AM performance of each sample type given a trade-off between energy consumption and dimensional accuracy performances was identified to address the role of the MAR in successful FFF operation.

The findings of this study suggest that decision-making in FFF operations should be viewed as maintaining optimal process efficiency instead of selecting optimal parameter settings. Previous attempts to suggest optimal FFF operations in the literature focused on identifying specific process parameter settings that maximize a specific performance measure. This was reasonable when the AM decision-maker was only interested in optimizing a single performance measure. However, it would be difficult to determine specific process parameter settings if multiple performance measures with complex trade-offs should be jointly considered to improve the overall AM performance. In addition, optimal process parameter settings for a specific FFF machine cannot ensure the optimal performance of another FFF machine due to different machine specifications. Thus, FFF operations are required to be considered as a generalized problem of process efficiency rather than specific process parameter selection, and any process parameter combination associated with an optimal MAR can be a solution for the best AM operation based on an AM strategy. For example, if an AM decision-maker wants to significantly save energy consumption to fabricate parts by compromising dimensional accuracy to some degree, process parameter combinations generating a higher MAR in any FFF machine would be preferable.

This study contributes to decision-making for AM as a process efficiency-based AM approach to generally model a trade-off between energy consumption and dimensional accuracy. Nevertheless, this study should be further extended to understand the dynamics of process efficiency not only for various design cases but also for more conflicting AM performance measures in practice. In addition, operation strategies to flexibly handle the MAR of FFF machines in a large-scale AM system can provide a new opportunity to maximize the overall performance of the AM system.

Author Contributions: Conceptualization, K.K., K.P. and H.W.J.; methodology, K.K., K.P. and H.W.J.; analysis, K.K. and K.P.; data collection, K.K.; writing—original draft preparation, K.K. and K.P.; writing—review and editing, K.P. and H.W.J. All authors have read and agreed to the published version of the manuscript.

Funding: This work was supported by Incheon National University Research Grant in 2023 (2023-0206).

Institutional Review Board Statement: Not applicable.

Data Availability Statement: The data presented in this study are available on request from the corresponding author. The data is not publicly available due to privacy or ethical restrictions.

Conflicts of Interest: The authors have no conflicts of interest to disclose.

References

1. Peng, T.; Kellens, K.; Tang, R.; Chen, C.; Chen, G. Sustainability of additive manufacturing: An overview on its energy demand and environmental impact. *Addit. Manuf.* **2018**, *21*, 694–704. [CrossRef]
2. Brenken, B.; Barocio, E.; Favaloro, A.; Kunc, V.; Pipes, R.B. Fused filament fabrication of fiber-reinforced polymers: A review. *Addit. Manuf.* **2018**, *21*, 1–16. [CrossRef]
3. Ahn, S.H.; Montero, M.; Odell, D.; Roundy, S.; Wright, P.K. Anisotropic material properties of fused deposition modeling ABS. *Rapid Prototyp. J.* **2002**, *8*, 248–257. [CrossRef]
4. Osswald, T.A.; Puentes, J.; Kattinger, J. Fused filament fabrication melting model. *Addit. Manuf.* **2018**, *22*, 51–59. [CrossRef]
5. Oztan, C.; Ballikaya, S.; Ozgun, U.; Karkkainen, R.; Celik, E. Additive manufacturing of thermoelectric materials via fused filament fabrication. *Appl. Mater. Today* **2019**, *15*, 77–82. [CrossRef]
6. Strano, M.; Rane, K.; Farid, M.A.; Mussi, V.; Zaragoza, V.; Monno, M. Extrusion-based additive manufacturing of forming and molding tools. *Int. J. Adv. Manuf. Tech.* **2021**, *117*, 2059–2071. [CrossRef] [PubMed]
7. Rahmatabadi, D.; Ghasemi, I.; Baniassadi, M.; Abrinia, K.; Baghani, M. 4D printing of PLA-TPU blends: Effect of PLA concentration, loading mode, and programming temperature on the shape memory effect. *JMatS* **2023**, *58*, 7227–7243. [CrossRef]
8. Chohan, J.S.; Singh, R.; Boparai, K.S.; Penna, R.; Fraternali, F. Dimensional accuracy analysis of coupled fused deposition modeling and vapour smoothing operations for biomedical applications. *Compos. B. Eng.* **2017**, *117*, 138–149. [CrossRef]
9. Kumar, L.J.; Krishnadas Nair, C. Current trends of additive manufacturing in the aerospace industry. In *Advances in 3D Printing & Additive Manufacturing Technologies*, 1st ed.; Wimpenny, D.I., Pandey, P.M., Kumar, L.J., Eds.; Springer: Singapore, 2017; pp. 39–54.
10. Liu, Z.; Wang, Y.; Wu, B.; Cui, C.; Guo, Y.; Yan, C. A critical review of fused deposition modeling 3D printing technology in manufacturing polylactic acid parts. *Int. J. Adv. Manuf. Tech.* **2019**, *102*, 2877–2889. [CrossRef]
11. Balogun, V.A.; Oladapo, B.I. Electrical energy demand modeling of 3D printing technology for sustainable manufacture. *Int. J. Eng.* **2016**, *29*, 954–961. [CrossRef]
12. Liu, Z.; Jiang, Q.; Ning, F.; Kim, H.; Cong, W.; Xu, C.; Zhang, H.-C. Investigation of energy requirements and environmental performance for additive manufacturing processes. *Sustainability* **2018**, *10*, 3606–3620. [CrossRef]
13. Lunetto, V.; Priarone, P.C.; Galati, M.; Minetola, P. On the correlation between process parameters and specific energy consumption in fused deposition modelling. *J. Manuf. Process.* **2020**, *56*, 1039–1049. [CrossRef]
14. Hassan, M.R.; Noh, H.; Park, K.; Jeon, H.W. Simulating energy consumption based on material addition rates for material extrusion of CFR-PEEK: A trade-off between energy costs and cycle time. *Int. J. Adv. Manuf. Tech.* **2022**, *120*, 4597–4616. [CrossRef] [PubMed]
15. Balogun, V.A.; Kirkwood, N.D.; Mativenga, P.T. Direct electrical energy demand in fused deposition modelling. *Procedia CIRP* **2014**, *15*, 38–43. [CrossRef]
16. Kim, K.; Noh, H.; Park, K.; Jeon, H.W.; Lim, S. Characterization of power demand and energy consumption for fused filament fabrication using CFR-PEEK. *Rapid Prototyp. J.* **2022**, *28*, 1394–1406. [CrossRef]
17. Mohamed, O.A.; Masood, S.H.; Bhowmik, J.L. Optimization of fused deposition modeling process parameters: A review of current research and future prospects. *Adv. Manuf.* **2015**, *3*, 42–53. [CrossRef]
18. Sheoran, A.J.; Kumar, H. Fused deposition modeling process parameters optimization and effect on mechanical properties and part quality: Review and reflection on present research. *Mater. Today Proc.* **2020**, *21*, 1659–1672. [CrossRef]
19. Wang, C.C.; Lin, T.W.; Hu, S.S. Optimizing the rapid prototyping process by integrating the taguchi method with the gray relational analysis. *Rapid Prototyp. J.* **2007**, *13*, 304–315. [CrossRef]
20. Sood, A.K.; Ohdar, R.K.; Mahapatra, S.S. Improving dimensional accuracy of fused deposition modelling processed part using grey taguchi method. *Mater. Des.* **2009**, *30*, 4243–4252. [CrossRef]
21. Sahu, R.K.; Mahapatra, S.S.; Sood, A.K. A study on dimensional accuracy of fused deposition modeling (FDM) processed parts using fuzzy logic. *J. Manuf. Sci. Prod.* **2013**, *13*, 183–197. [CrossRef]

22. Beniak, J.; Križan, P.; Šooš, Ľ.; Matúš, M. Research on shape and dimensional accuracy of FDM produced parts. In Proceedings of the 9th Thai Society of Mechanical Engineers, International Conference on Mechanical Engineering, Phuket, Thailand, 11–14 December 2018.
23. Günay, E.E.; Velineni, A.; Park, K.; Okudan Kremer, G.E. An investigation on process capability analysis for fused filament fabrication. *Int. J. Precis. Eng. Manuf.* **2019**, *21*, 759–774. [CrossRef]
24. Park, K.; Kim, G.; No, H.; Jeon, H.W.; Kremer, G.E.O. Identification of optimal process parameter settings based on manufacturing performance for fused filament fabrication of CFR-PEEK. *Appl. Sci.* **2020**, *10*, 4630. [CrossRef]
25. Gutowski, T.; Jiang, S.; Cooper, D.; Corman, G.; Hausmann, M.; Manson, J.-A.; Schudeleit, T.; Wegener, K.; Sabelle, M.; Ramos-Grez, J.; et al. Note on the rate and energy efficiency limits for additive manufacturing. *J. Ind. Ecol.* **2017**, *21*, S69–S79. [CrossRef]
26. Alizadeh, M.; Esfahani, M.N.; Tian, W.; Ma, J. Data-driven energy efficiency and part geometric accuracy modeling and optimization of green fused filament fabrication processes. *J. Mech. Des.* **2020**, *142*, 041701. [CrossRef]
27. Ensinger. TECAPEEK CF30 Black. Available online: https://www.ensingerplastics.com/en/shapes/products/peek-tecapeek-cf30-black (accessed on 23 June 2022).
28. Han, X.; Yang, D.; Yang, C.; Spintzyk, S.; Scheideler, L.; Li, P.; Li, D.; Geis-Gerstorfer, J.; Rupp, F. Carbon fiber reinforced PEEK composites based on 3D-printing technology for orthopedic and dental applications. *J. Clin. Med.* **2019**, *8*, 240. [CrossRef] [PubMed]
29. Apium. Apium P220 Datasheet. Available online: https://apiumtec.com/download/apium-p220-datasheet (accessed on 13 July 2022).
30. Simplify3D. Software Feature of Simplify3D. Available online: https://www.simplify3d.com/products/simplify3d-software/features/ (accessed on 14 July 2022).
31. ADPower. HPM-100A. Available online: http://shop2.adpower21com.cafe24.com/product/wattman-hpm-100a/17/category/1/display/2/ (accessed on 14 July 2022).
32. Mitutoyo. Absolute Digimatic Blade Type Caliper. Available online: https://www.mitutoyo.com/products/small-tool-instruments-and-data-management/calipers/digimatic-calipers/absolute-digimatic-blade-type-caliper/ (accessed on 14 July 2022).
33. Minitab. Minitab 20.3.0 Update. Available online: https://www.minitab.com/en-us/support/minitab/minitab-20.3.0-update/ (accessed on 19 July 2022).
34. Masud, A.S.; Ravindran, A.R. Multiple criteria decision making. In *Operations Resesarch and Management Science Handbook*, 1st ed.; Chapter 5; Ravindran, A.R., Ed.; CRC Press: Boca Raton, FL, USA, 2008; pp. 1–35.
35. Caminero, M.A.; Chacon, J.M.; Garcia-Plaza, E.; Nunez, P.J.; Reverte, J.M.; Becar, J.P. Additive manufacturing of PLA-based composites using fused filament fabrication: Effect of graphene nanoplatelet reinforcement on mechanical properties, dimensional accuracy and texture. *Polymers* **2019**, *11*, 799. [CrossRef] [PubMed]

polymers

Article

Durability of Ultem 9085 in Marine Environments: A Consideration in Fused Filament Fabrication of Structural Components

Xiong (Julia) Wang [1], Carly Travis [1], Mark T. Sorna [2] and Dwayne Arola [1,3,*]

[1] Department of Materials Science and Engineering, University of Washington, Box 352120, Seattle, WA 98195-2120, USA; xironw@uw.edu (X.W.)
[2] Naval Undersea Warfare Center, Keyport, WA 98345-7610, USA; mark.t.sorna.civ@us.navy.mil
[3] Department of Mechanical Engineering, University of Washington, Seattle, WA 98195-2120, USA
* Correspondence: darola@uw.edu; Tel.: +1-(206)-685-8158

Abstract: The long-term durability of polymer components produced by additive manufacturing (AM) in marine conditions is poorly understood. Here, fused filament fabrication (FFF) of Ultem 9085 was conducted and accelerated aging was performed. Two printing orientations (−45/45° and 0/90°) and two sample types (ASTM D638 Type 1 and Type 4) were produced and subjected to accelerated aging in either seawater or air. Results from tensile tests showed that the elastic modulus, yield strength and ultimate tensile strength increased after seawater aging, whereas the elongation to failure decreased. Results of thermogravimetric analysis (TGA) and derivative–TGA curves indicated that hydrolysis occurred after seawater exposure to the polycarbonate (PC) component and changes in structure or hydrogen bonds formed in the polyetherimide (PEI) component. Differential scanning calorimetry showed that physical aging occurred after short exposure periods and low temperature. Longer exposures and higher temperatures resulted in increasing plasticization by water and scission of the PC molecules. Results from Raman suggest that hydrolysis of the PC occurred, with a reduction in free volume produced by physical aging or hydrogen bonding with water molecules. These results highlight that Ultem 9085 is susceptible to degradation in marine environments, and there are two primary mechanisms, including physical and chemical aging. Their specific contribution is highly sensitive to the aging temperature and require careful selection in accelerated aging evaluations.

Keywords: accelerated aging; fused filament fabrication; hydrolysis; Ultem 9085

check for updates

Citation: Wang, X.; Travis, C.; Sorna, M.T.; Arola, D. Durability of Ultem 9085 in Marine Environments: A Consideration in Fused Filament Fabrication of Structural Components. *Polymers* **2024**, *16*, 350. https://doi.org/10.3390/polym16030350

Academic Editors: Miguel Ángel López Manchado and Wei Zhang

Received: 5 December 2023
Revised: 21 January 2024
Accepted: 25 January 2024
Published: 28 January 2024

1. Introduction

Additive manufacturing (AM) represents a suite of novel processes for creating three-dimensional physical objects directly from computer-aided design (CAD) models via layer-by-layer construction [1]. Also commonly regarded as 3D printing, AM is considered capable of creating structures with nearly unlimited complexity, with a substantial reduction in time from conception to finished product in comparison with traditional manufacturing processes. Consequently, it has received substantial attention in the last decade. Specific AM processes are being adopted in various industrial sectors, including the aerospace, automotive and medical industries [2].

Fused filament fabrication (FFF) is one of the most common techniques for AM and is used for printing components of thermoplastic polymers, including prototypes and functional end-use parts. In FFF, a polymer filament is extruded from a heated nozzle and deposited onto the printing bed or a previous printed layer in a series of continuous lines with specific orientation [3]. The mechanical properties of components that are produced by FFF depend on the printing parameters, which are optimized to maximize the part quality, the microstructure and overall printing process economy [4,5]. The process physics in FFF involves melting and solidification of the polymer, which can give rise to localized porosity and anisotropic mechanical properties in parts [6]. Nevertheless, the affordability of FFF printing systems and the tremendously large number of feedstock materials available for FFF make it one of the most ubiquitous AM processes overall [4–6].

Underwater and marine structures are an interesting niche application for AM, such as autonomous underwater vehicles, micro underwater vehicles and unmanned surface vehicles [3]. When compared with traditional manufacturing methods, AM could enable these systems to be developed with lower cost and minimal time from conception to final product, substantially increasing the response readiness. However, these applications bring new challenges and concerns, including the durability of 3D-printed products under environmental threats posed by a marine environment [4,7]. Applications that require durability must address the multiple sources for degradation, including exposure to ultraviolet (UV) light, pressure cycles, chemical attack and the potential for hydrolytic damage [4,8–10]. As such, if the components are intended for use in mission-critical applications, these sources of marine environment degradation and their ramifications must be understood.

Polymer feedstock materials for FFF that are intended for applications in marine environments should possess the appropriate resistance to degradation [11,12]. However, the data concerning this aspect of performance are scant. Ultem 9085 is an advanced engineering polymer for FFF that is adopted for applications that require outstanding mechanical and thermal properties and good chemical resistance [6]. It is recognized for its flame retardancy, low toxicity, dimension stability and retention of strength at high temperatures. As such, it is sought for marine, aerospace and automotive applications, especially where low-weight and high-strength materials are required [13]. In addition, its high environmental resistance and high strength-to-weight ratio characteristics make it a promising candidate for the renewable energy and turbomachinery industries [14,15]. Ultem 9085 is a blend of polyetherimide (PEI) with polycarbonate (PC) copolymer, which is added to improve the printability of PEI [5]. Based on the importance of its structural behavior, the mechanical properties of Ultem 9085 have been assessed after FFF through tensile, fatigue, compression and flexural testing [16–23]. Nevertheless, the response to threats related to marine conditions have received limited attention.

PEI is an amorphous engineering thermoplastic with excellent performance. The qualities of PEI include high strength, high stiffness and capacity for high-temperature resistance [24]. There are also limitations to PEI, such as its comparatively high cost and instability in chlorinated solvents. Polycarbonate is an amorphous engineering thermoplastic that is widely used to blend with other materials because of its desirable strength and stiffness, as well as its outstanding compatibility with a series of polymers [25]. However, PC can undergo degradation under exposure to bases and moisture, especially at elevated temperatures [26]. Hence, there are qualities of PEI and PC that suggest that Ultem 9085 could be susceptible to degradation under prolonged exposure to marine conditions.

In recent years, there has been considerable research focused on printing conditions to improve the mechanical performance of 3D-printed Ultem 9085. Shelton et al. developed a chamber to control the printing environment temperature during FFF of Ultem 9085 [27]. It was found that the capacity for inelastic deformation and post-neck yielding increased with build chamber temperature, which had a large effect on the mechanical properties [27]. Zaldivar et al. evaluated the thermal and mechanical behavior of 3D-printed Ultem 9085 with respect to build orientations [6]. Their results demonstrated that the mechanical properties of the printed samples were heavily dependent on the printing orientations [6]. That group also investigated the influence of filament moisture content on the mechanical properties of the printed material, finding that an increase in voids and poor filament orientation arose with increasing moisture content, which were detrimental to the performance [5]. Post-processing of FFF prints can be applied for further improvements. For instance, atmospheric plasma treatment, controlled cooling and post-process annealing have also been explored for enhancing the mechanical performance of Ultem 9085 following FFF [28–31].

Despite the importance of degradation in structure and properties posed by marine applications, only limited assessment of the environmental effects on 3D-printed Ultem 9085 material has been reported. Padovano et al. conducted the thermal aging test on Ultem 9085 by treating samples with high temperature, cyclic climate change and thermal shock [21]. They concluded that the progressive variation in temperature or constant high temperature did not cause significant degradation to the mechanical properties. However, sudden cooling from a high temperature could degrade the mechanical properties of Ultem 9085. A previous study exposed Ultem 9085 to various media and air and water at temperatures from −60 °C to 160 °C (which is highly relevant to industrial applications) [32]. While water immersion for up to 52 weeks had negligible effects on its mechanical performance, exposure to an alkaline cleaner caused degradation after only 1 week of exposure [32]. Gallagher explored the degradation of 3D-printed Ultem 9085 to UV exposure, elevated temperatures and vacuum conditions for up to three months [33]. Results showed that the UV and vacuum environment at elevated temperature caused stiffening and strengthening of Ultem 9085 but a reduction in the strain to fracture [15]. Considerations related to changes in durability with exposure to seawater are just now appearing [15,34].

Environmental degradation over an exposure period can be regarded as aging. In general, there are two categories of "aging" that can happen to polymers, including chemical aging and physical aging, both of which can affect the physical properties [35]. Chemical aging is generally related to the modification of the polymer chain and can result in chain scission or oxidation [35]. It can happen in both processing and applications, where elevated temperatures, air, light, moisture and living organisms and other factors exist [36]. Chemical aging is irreversible and usually causes degradation of the microstructure and mechanical properties of the polymers. Hydrolysis is a common form of chemical aging that involves the breakage of polymer bonds by water molecules, which can occur in marine environments and cause degradation of the performance of polymers.

Contrary to chemical aging, physical aging is a reversible process driven by the non-equilibrium state of polymer chains induced by rapid cooling of the molten polymer in the manufacturing process [35]. During a sudden drop in temperature, the polymer chains do not have ample time to relax and remain in a distorted configuration compared with the chain distribution in an equilibrium state. This phenomenon is especially common in amorphous polymers where the chains pursue densification in an equilibrium state [37]. The reduction in free volume can increase the intermolecular attraction of the molecules, which will increase the strength and stiffness at a macro level but cause lower ductility. Physical aging is enabled when the environmental temperature exceeds that required for mobility of the side chains and localized groups. The propensity and rate of physical aging increases with exposure temperature and its proximity to the glass transition temperature (Tg) [37].

Accelerated aging is a common experimental approach that is adopted to assess the long-term degradation of a material by conducting short-term tests. These are performed using selected conditions that typically involve substantially more aggressive levels of exposure than in natural exposure [38]. The purpose of acceleration is to reduce the required testing time. Polymers, in particular, are susceptible to degradation in a variety of conditions, including exposure to elevated temperatures, moisture, chemicals, ultraviolet light, oxygen, etc. In accelerated aging, the experimental conditions should emulate those of the application to develop data through the responses that are most appropriate for predicting performance. In fact, this is a common approach to support the development of reliability and lifetime prediction models [38]. For example, Davis and Evrard applied accelerating aging to build an Arrhenius model for the degradation of polyurethane in marine environments [39]. Similarly, Tocháček and Vrátníčková utilized accelerated UV aging to build a lifetime prediction model for polypropylene and its copolymers in outdoor exposures [40]. Using results from accelerated thermal and hydrothermal aging, Bergaliyeva et al. simulated the mechanical properties of 3D-printed polylactic acid after 1.5–2.5 years of service [41].

It is important to highlight that accelerated aging evaluations employed to assess the durability of polymers under specific threats (ultraviolet light, moisture, chemicals, etc.) are potentially controversial. The acceleration of aging could illicit mechanisms of degradation or combinations of mechanisms that are quite different from those that are experienced in the natural environment-intended applications [42]. As such, it is important to treat the results and their applications with caution and understand the potential limitations. Nevertheless, accelerated aging can serve as an effective measure of sensitivity to degradation and can be helpful for rating the durability of polymeric materials.

The outstanding properties of Ultem 9085 make it a highly viable material for marine applications. Yet, the durability of this material under prolonged marine environment exposure is relatively unknown. As a result, the objective of this investigation is to evaluate the degradation resistance of 3D-printed Ultem 9085 produced by FFF under exposure to seawater. To achieve this objective, accelerated aging tests were performed in a standardized marine environment. In this investigation, an accelerated aging approach was adopted to evaluate the resistance of 3D-printed Ultem 9085 to degradation under prolonged exposure to seawater. The quality of the printed material was evaluated by optical profilometry, tensile testing and μCT scanning before and after exposure to the seawater environments. Here, results of these accelerated aging evaluations are presented, the contributing mechanisms are assessed and the feasibility of applying accelerating aging tests for Ultem 9085 is discussed.

2. Materials and Methods

2.1. Materials and Printing

The material evaluated in this investigation was Ultem 9085, which was acquired from Stratasys by the Naval Undersea Warfare Center (NUWC, Keyport, WA, USA). Tensile specimens were printed by fused filament fabrication (FFF) using a Fortus 900MC from Stratasys. All printing was performed using the default settings of the machine for Ultem 9085 (which included a layer thickness of 0.254 mm) and an aerospace nozzle tip. The remaining parameters associated with the default conditions of printing Ultem 9085 with this printer are proprietary. To examine the potential influence of printing orientation, two groups of samples were prepared with infill orientations of −45/45° and 0/90°, as shown in Figure 1. These two orientations were included because they are the most common orientations chosen for printing of components. Furthermore, an infill percentage of 100% was used to reduce the relative effects of porosity on the mechanical behavior, and that is the intended form of application.

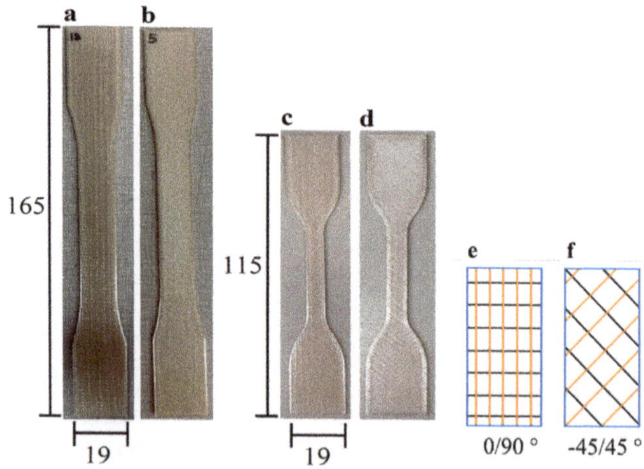

Figure 1. Details regarding the 3D-printed Ultem 9085 tensile bars. (**a**) Type 1, 0/90° orientation; (**b**) Type 1, −45/45° orientation; (**c**) Type 4, −45/45° orientation; (**d**) Type 4, 0/90° orientation; (**e**,**f**) illustration of printing orientations (unit: mm).

The standard ASTM D638 is widely used for evaluating the mechanical properties of polymers and assessing the influence of manufacturing methods on the mechanical behavior [43]. There is presently no additional guideline that defines how the mechanical properties of products manufactured by FFF should be evaluated. Therefore, Type 1 and Type 4 tensile specimens were printed with a 1.59 mm thickness by FFF according to the geometry described in ASTM D638 [44]. A total of 120 specimens were prepared to assess the effects of marine exposure, including 20 specimens with Type 1 geometry for each of the 2 infill orientations and 40 specimens with the Type 4 geometry for each of the orientations ($2 \times 20 + 2 \times 40 = 120$).

2.2. Accelerated Aging

The accelerated aging testing in seawater was performed with multiple digital thermostatic water baths (DK-2000-IIIL-4, Faithful, Huanghua, Cangzhou, China) and with a bath volume of approximately 12 L. The marine environment for aging was designed after prior-published work [39], utilizing standardized artificial seawater that was prepared in accordance with ASTM D114 [45]. A standard seawater sample was chosen as the aging medium (rather than an arbitrary natural seawater sample) to adhere to available guidelines for assessing the importance of seawater exposure to polymer durability and to ensure that the methods could be easily repeated by other investigators. The aging conditions used for evaluating the Ultem 9085 samples are described in Table 1. Ten tensile coupons were immersed within the standardized seawater baths at ambient pressure and at temperatures of 50 °C, 70 °C and 90 °C. It is important to note that these elevated temperatures were not indicative of the application temperatures envisioned for this material and that the conditions did not replicate or mimic the natural service conditions. They were chosen specifically to promote more rapid degradation if the polymer was susceptible while also remaining below critical temperatures for this material. Here, the selected temperature range for this evaluation was chosen because it is a common range used in performing accelerated aging of polymers and it is below the glass transition temperature of Ultem 9085 (186 °C) [38,39]. The specimens were removed from the bath after an exposure period of either 2 or 4 weeks and tensile tests were conducted to evaluate the degradation thereafter.

Table 1. Aging plan for Ultem 9085 in substitute sea water and air.

Temperature (°C)	Aging Period (Days)	Samples in Seawater	Samples in Air
50	14	Type 4, −45/45°, Type 4, 0/90°, Type 1, 0/90°	Type 4, −45/45°, Type 4, 0/90°
50	28	Type 4, −45/45°, Type 4, 0/90°, Type 1, 0/90°	Type 4, −45/45°, Type 4, 0/90°
70	14	Type 4, −45/45°, Type 4, 0/90°	Type 4, −45/45°, Type 4, 0/90°
70	28	Type 4, −45/45°, Type 4, 0/90°	Type 4, −45/45°, Type 4, 0/90°
90	14	Type 4, −45/45°, Type 4, 0/90°, Type 1, 0/90°	Type 4, −45/45°, Type 4, 0/90°
90	28	Type 4, −45/45°, Type 4, 0/90°, Type 1, 0/90°	Type 4, −45/45°, Type 4, 0/90°

Due to the lower number of Type 1 0/90° samples, the aging conditions applied to those samples included 50 °C for 14 days and 90 °C for 14 and 28 days. Pilot experiments were performed on extruded polyurethane sheet as a reference material to confirm the validity of the accelerated aging approach. According to the results of the aging evaluations in seawater, it was identified that complimentary air aging experiments should be conducted to characterize the physical aging that occurs during the aging period at elevated temperatures. Therefore, accelerated aging evaluations were conducted with Ultem 9085 in air under the temperatures and durations outlined in Table 1. These experiments were conducted on additional specimens printed with the two different infill orientations.

2.3. Water Uptake

The uptake of water is an important concern for undersea and marine structures. Therefore, the change in weight of the samples was measured before and after the exposure periods with an electric balance (Model JM-303, RESHY, Huzhou, Zhexing, China) with ±0.001 g precision. After the aging experiments and prior to the weight measurements, excess moisture on the surface of the samples was blotted dry by tissue paper. The extent of water uptake percentage ($w_{wu}\%$) was estimated by taking weight measurements of 5 samples of each sample configuration before (w_o) and after (w_e) of the exposure period according to:

$$w_{wu}\% = \frac{w_e - w_0}{w_0} \times 100\% \quad (1)$$

2.4. *Profilometry*

Profilometry was performed to characterize the surface topography of the printed samples using an optical profilometer (VR-3100 3D, KEYENCE, Osaka, Japan). One sample of each geometry was randomly selected to characterize the surface topography prior to aging. To inspect the surfaces for characteristics that could contribute to the mechanical behavior, topographical images were taken on the printed side of the gauge section, grip section and transition areas of the specimens at 12× magnification. The average roughness (Ra) and ten-point roughness (Rz) of the specimens were estimated using images obtained at higher magnification (×40) on the printed side and the edge of the gauge section using a traverse length of 5.6 mm. The Ra represents the mathematical average deviation of the surface from the mean line over the traverse length, and the Rz is the average deviation of the five highest peaks and valleys from the mean line. The roughness parameters were calculated using the multiline roughness software that is provided with the instrument. An upper cut-off (lc) and lower cut-off (ls) of 0.08 μm and 25 μm were used in treating the height distribution, respectively.

2.5. *Microcomputed Tomography*

To understand water diffusion within the microstructure and its accumulation, microcomputed tomography (μCT) scans were performed on virgin printed samples using a commercial system (X5000, North Star Imaging, Rogers, MN, USA). The potential volume for absorbed liquid in the voids was computed through the areal porosity and the length of the voids. A 3D model for the Type 4 specimen was built by SOLIDWORKS (2020, Dassault Systèmes SOLIDWORKS Corp, Waltham, MA, USA) based on the ASTM D638 standard to estimate the volume of the specimen (2.557 cm^3). The weight of the potential water uptake was calculated by multiplying the volume of the specimens (V), average areal porosity of the specimen (A%) and density of water (ρ) (assumed to be 1 g/cm^3). The approximate weight of specimens prior to aging (w_s) was 2.6 g. Based on those values, the theoretical maximum water uptake range based on filling of the pores ($w_t\%$) was obtained according to the formula

$$w_t\% = \frac{V \times A \times \rho}{W_s} \times 100\% \quad (2)$$

2.6. *Tensile Testing*

Tensile testing of the printed specimens was performed in ambient conditions using a universal testing system (E1000 Dynamic Test instrument, Instron, Norwood, MA, USA) with a 2 kN load cell. The loading was conducted in displacement control at a rate of 5 mm/min according to the ASTM D638 standard guidance for rigid polymers [44]. Four samples each of Type 1 and Type 4 were tested in the unaged condition for each of the two orientations to establish the control responses. Five samples from each aged condition were tested and the stress–strain responses were used in estimating the elastic modulus (E), yield strength (YS), ultimate tensile strength (UTS) and strain at failure. According to guidelines outlined by ASTM D638, the test was only considered valid if failure occurred within the gauge section of the sample; results for samples that failed outside of the gauge section were disregarded. For each property assessed, the average, standard deviation and coefficient of variation were determined. The yield strength was determined from the 0.2% offset method. Two sample *t*-tests with unequal variance were applied to identify significant differences in the mechanical properties between the aged and unaged specimens. Significant differences were identified by $p \leq 0.05$.

2.7. Thermogravimetric Analysis and Differential Scanning Calorimetry

Thermogravimetric analysis (TGA) was performed on the Ultem samples before and after aging using a commercial instrument (TGA Q50, TA Instruments, New Castle, DE, USA). Approximately 5 mg was obtained from the sample of interest and heated within a nitrogen atmosphere from ambient environment temperature to 800 °C at a 10 °C/min heating rate. Conventional TGA as well as D-TGA (derivative TGA) curves were prepared to analyze the changes in molecular conformation and potential degradation that resulted from aging.

Differential scanning calorimetry (DSC) was also performed to estimate the glass transition temperature (Tg), relaxation enthalpy (δ_H) and peak temperature (Tp) of the Ultem samples using a commercial instrument (Q20, TA instruments, New Castle, DE, USA). Samples of approximately 3 mg were sealed in a Tzero aluminum pan with a Tzero aluminum lid. The sealed crucible was heated at a rate of 10 °C/min in a nitrogen environment. The maximum heating temperature was analyzed from the TGA curves up to 300 °C for the samples.

2.8. Raman Spectroscopy

To evaluate the molecular structure for the untreated and treated samples, Raman spectroscopy was performed using a commercial instrument (inVia Raman microscope, Renishaw, Wotton-under-Edge, UK). Raman spectra were generated using a 784 nm laser source with 10 s exposure on the surface of the sample with a 50× objective. The spectra were obtained over wavelengths from 100 to 3200 cm^{-1}. Samples of material for Raman spectroscopy were sectioned from the grip sections of the fractured samples. Baseline correction, normalization and smoothing of the Raman spectra were achieved by data analysis software (Origin, OriginLab, Northampton, MA, USA). In addition, WiRE software (Version 5.2) was used to acquire and compare the Raman spectra for Ultem 1000 to the spectra for PC.

3. Results

3.1. Mechanical Properties

Representative stress–strain curves for the unaged control samples and after accelerated aging in seawater are shown in Figures 2a and 2b, respectively. The aging condition in Figure 2b consists of aging in seawater at 90 °C for 28 days. Results in this figure are presented for both the Type 1 and Type 4 specimens and the two infill orientations. The mechanical properties were determined from these curves and are summarized in Figure 3. The complete data set is listed in the Supplemental Information (SI, Tables S1–S4). It is important to note that, due to printing anomalies and stress concentrations in the transition areas associated with the infill, some of the specimens failed prematurely and outside of the gauge section (Figure 2). As such, the tensile data for those specimens were excluded when estimating the mechanical properties.

The elastic modulus of the samples in Figure 3a revealed that there was an increase in the stiffness of the samples after aging in the artificial seawater regardless of temperature. The yield strength distribution for the control and aged conditions is presented in Figure 3b and shows that the yield strength increased with aging and with respect to the control. Similarly, according to the tensile responses in Figure 3c, the UTS also increased with aging, although not as markedly as the increase in yield strength. The UTS of the Type 1 0/90° samples was lower than that of the UTS of Type 4 0/90° samples in the control condition, which could be attributed to differences in the porosity between the two sample types. Noteworthy, the UTS of the Type 4 samples with −45/45° increased after aging in seawater at all the elevated temperatures with respect to the control. However, the UTS of the Type 4 −45/45° and 0/90° infill samples did not exhibit a distinct trend with increasing temperature or exposure time. For the −45/45° printing orientation, the UTS decreased with increasing exposure time at both 70 and 90 °C. And, for the samples with 0/90° infill at 50 °C and 70 °C, there was a decrease in UTS with extended aging time. Unique from strength, Figure 3d shows that accelerated aging caused a decrease in the strain to failure of the samples in the seawater environment.

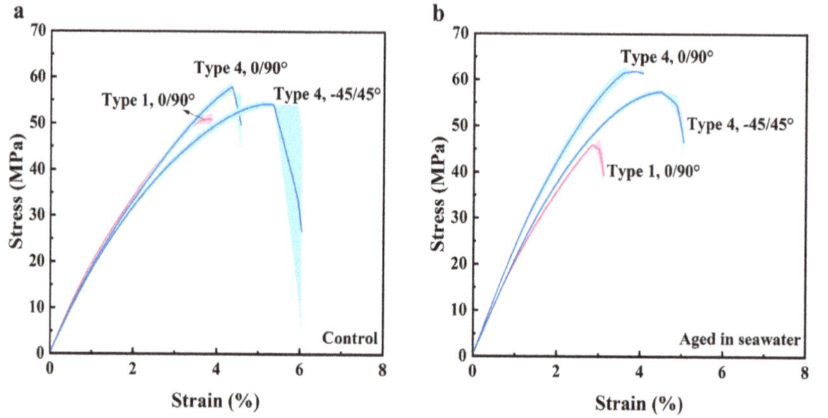

Figure 2. Stress–strain response of Ultem 9085 at different conditions. (**a**) Unaged samples; (**b**) samples submerged in seawater at 90 °C for 28 days.

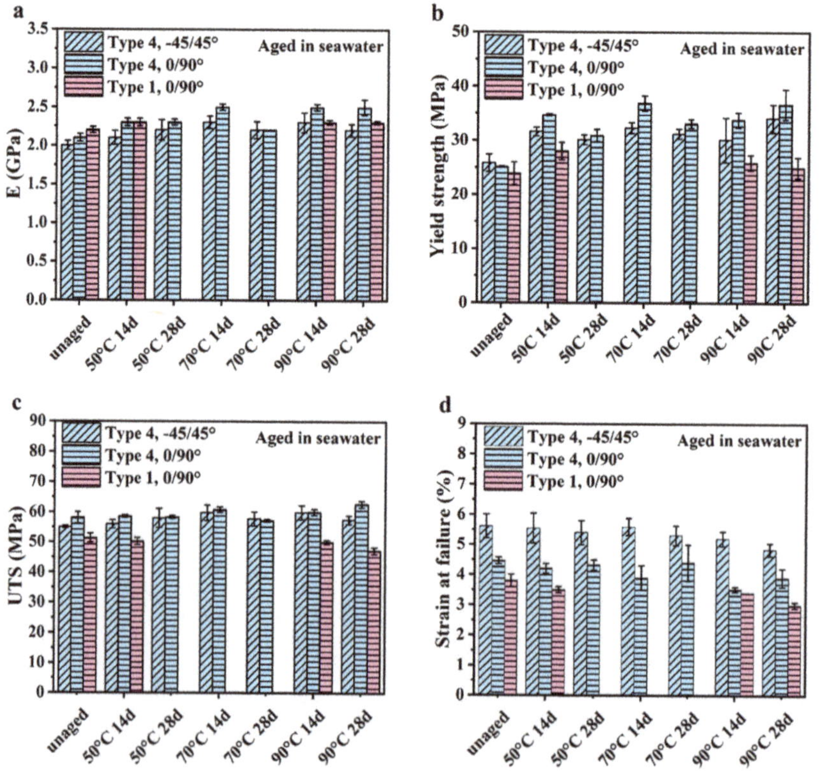

Figure 3. Mechanical properties of the unaged and aged Ultem 9085 samples in seawater determined by their strain–stress curves. (**a**) Elastic modulus (E); (**b**) yield strength; (**c**) ultimate tensile strength (UTS); (**d**) strain at failure. Note: results for the Type 1 −45/45° group were not valid due to failure outside of the gauge section and are not included in the reported properties.

Detailed results for the Student's t-test of the mechanical properties are listed in Tables S1–S4. It is noteworthy that all samples underwent a significant ($p \leq 0.05$) increase in stiffness (reflected by E), except for the Type 1 0/90° specimens aged at 50 °C for 14 days. Only the yield strength of the Type 1 0/90° specimens and Type 4 −45/45° specimens exposed at 90 °C for 14 days did not show a change compared with the control specimens. Regarding UTS, aging at 70 °C for 14 days and at 90 °C had a significant effect on the Type 4 −45/45° samples. However, for the Type 1 0/90° and Type 4 0/90° samples, only aging at 90 °C for 28 days caused a significant change in the UTS of the 3D-printed material. The strain at failure decreased significantly for Type 4 0/90° after the aging at 90 °C for 14 days. At 90 °C, the elongation to failure of both the Type 4 −45/45° and Type 1 0/90° specimens underwent significant decreases after 28 days of accelerated aging.

The coefficient of variation (COV) for the Ultem 9085 samples is summarized in Table S5. Regarding the mechanical properties, the Type 4 −45/45° specimens had the largest COV overall in E, which increased after exposure to the marine environment at elevated temperatures. Likewise, the Type 4 −45/45° specimens had the highest COV in yield strength and UTS. Aging at 90 °C induced a large COV in yield strength for both Type 4 −45/45° and Type 4 0/90° samples, whereas aging at 50 °C with a longer exposure time caused higher COV in the UTS of Type 4 −45/45° samples. The COV for strain at failure increased the most for Type 4 0/90° samples submerged in 70 °C seawater and for Type 4 −45/45° samples submerged in 50 °C seawater. Overall, the Type 4 0/90° samples displayed the greatest COV in strain at failure.

In summary, the E and the yield strength for the 3D-printed Ultem 9085 materials increased significantly under the accelerating aging conditions. The UTS for this material did not increase significantly with aging, except for the longer exposure time (28 days) in seawater at 90 °C. The strain at failure for the Ultem 9085 material underwent a reduction with accelerated aging, but only exposure at 90 °C caused a significant decrease.

3.2. Failure Characteristics and Contributions

Figure 2 presents images from representative specimens after tensile testing to failure. Specifically, selected Type I samples with 0/90° and −45/45° orientations are presented in Figures 4a and 4b, respectively, and corresponding Type 4 samples are shown in Figures 4c and 4d, respectively. All the Type 1 samples with −45/45° infill orientation failed within the transition region between the gauge and grip section, which invalidated the mechanical testing results, whereas the 0/90° samples failed inside the gauge section and at the transition region. For the Type 4 geometry, some of the 0/90° samples broke at the transition from the grip and gauge sections, but notably fewer than occurred for the Type I geometry; none of the 45/45° infill orientation failed outside the gauge section. The documented locations of failure were used to categorize the results and censor the tensile testing response used when estimating the mechanical properties. All results related to failures that did not occur within the gauge section were considered invalid and were not used when evaluating the mechanical properties presented in Figure 3.

Potential contributions to failure in the transition region of the tensile specimens were explored through optical evaluations and profilometry. All the sample configurations exhibited voids in the transition areas, as highlighted in Figure 5, except for the Type 4 samples with −45/45° infill, as evident in Figure 5d. For both the Type 1 and Type 4 samples, voids were identified in the transition areas between the contour and infill, as evident in Figure 5a–c. Indeed, previous studies have commented on how the infill path used for the printing of the ASTM D638 specimens can lead to stress concentrations in the transition region and contribute to the failure process [46–48].

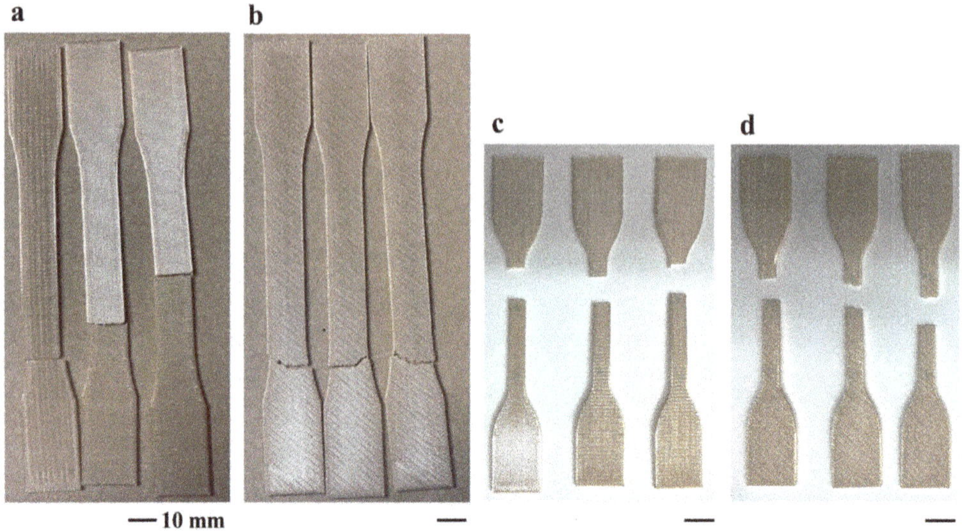

Figure 4. Tested unaged 3D-printed Ultem 9085 samples. (**a**) Type 1, 0/90° orientation; (**b**) Type 1, −45/45° orientation; (**c**) Type 4, 0/90° orientation; (**d**) Type 4, −45/45° orientation.

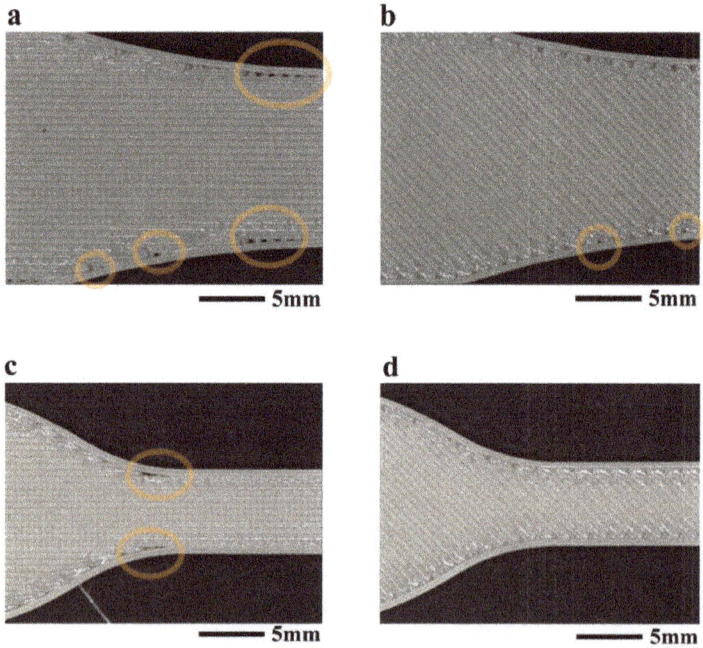

Figure 5. Microscopy images for the Ultem 9085 samples obtained from the optical profilometer. (**a**) Type 1, 0/90° orientation; (**b**) Type 1, −45/45° orientation; (**c**) Type 4, 0/90° orientation; (**d**) Type 4, −45/45° orientation. The yellow circles outline regions of voids at the boundary of contours.

To assess the infill distribution and other aspects of the surface topography that could contribute to failures, the surface texture of the printed samples was evaluated. Profiles were obtained from the gauge section surface (Figure 6) and from along the edge.

Two different orientations of analysis were applied to the specimen faces, as shown in Figure 6a,b. Representative surface roughness profiles taken parallel to the longitudinal axis and (direction of uniaxial tension (Direction 1)) are presented in Figure 6c,e for the 0/90° and −45/45° samples, respectively. Results of the roughness measurements are summarized in Table 2. As evident from these measurements, the samples with −45/45° infill orientation exhibited higher surface roughness in Direction 1, which resulted from gaps between the discrete infill tracks and lower infill density, as evident in Figure 6e. As evident from the surface roughness measurements in Table 2, both the Ra and Rz values of the edges were substantially lower than for the faces, which resulted from the smaller peak-to-valley height variations along the contours.

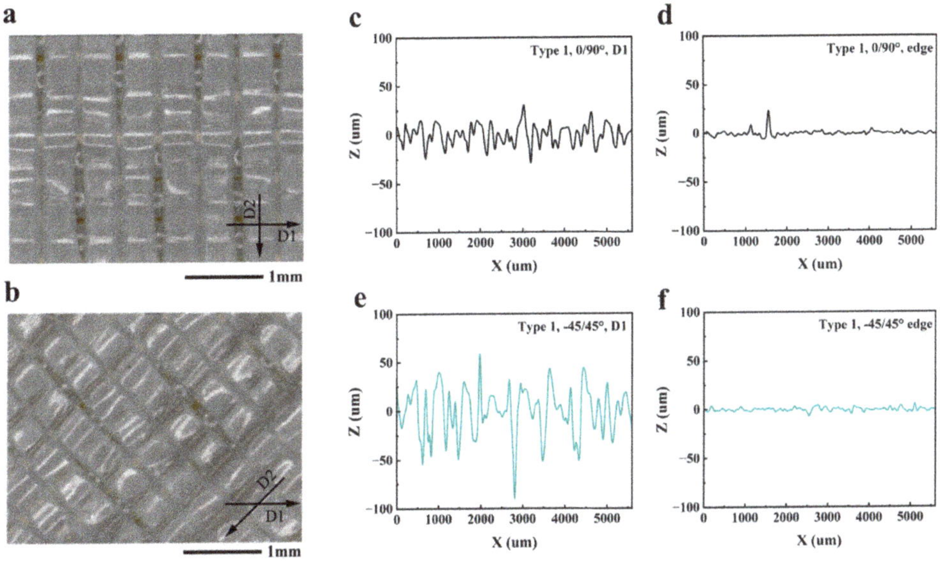

Figure 6. Optical images of the gauge section of the tensile samples with higher magnification and the roughness profile of the unaged Type 1 samples. (**a**) 0/90° printing orientation samples. D1: parallel to the contour, D2: perpendicular to the infill filament. (**b**) −45/45° printing orientation samples. D1: parallel to the contour, D2: perpendicular to the infill filament. (**c**) 0/90° orientation, D1; (**d**) 0/90° orientation, edge; (**e**) −45/45° orientation, D1; (**f**) −45/45° orientation, edge.

Table 2. Roughness of the Ultem 9085 samples produced by FFF.

Samples	Direction 1		Direction 2		Edge	
	Ra (μm)	Rz (μm)	Ra (μm)	Rz (μm)	Ra (μm)	Rz (μm)
Type 1, 0/90°	11	48	10	48	2	11
Type 1, −45/45°	19	102	23	119	2	9
Type 4, 0/90°	12	62	11	66	2	11
Type 4, −45/45°	32	175	33	173	1	7

The porosity in the printed specimens was evaluated from the μCT scans obtained in the transition areas, as shown in Figure 7a. Utilizing the areal porosity estimates in the XZ plane, the maximum water uptake was estimated according to Equation (1) and results are listed in Table 3 as a function of specimen type and infill pattern. Overall, the samples with 0/90° orientation had greater porosity and water uptake than the −45/45° orientation samples. Figure 7b compares the measured water absorption of the samples at different temperatures after 14 and 28 days. For the Type 4 samples, the weight change caused by water inside the pores was estimated from the areal porosity using Equation (1) and is highlighted in the diagram. It is important to note that there appeared to be an increase in the extent of water update and weight increase with aging time and temperature. However, the rate of increase in weight gain decreased over the aging time, as expected, suggesting that the infusion of free water within the pores was reaching saturation.

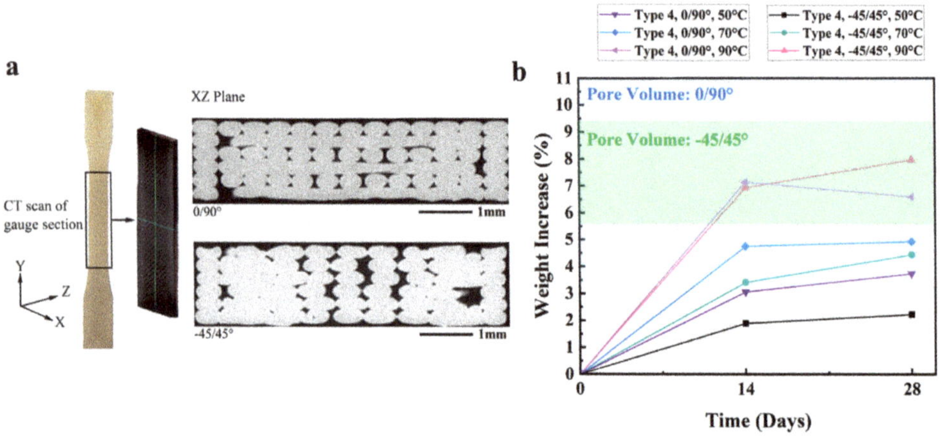

Figure 7. Details of the porosity. (**a**) Type 4 Ultem 9085 dog-bone CT scan result and the cross-section pictures. (**b**) Water uptake for Type 4 samples with respect to printing orientation and seawater temperature.

Table 3. Areal porosity in Type 4 Ultem 9085 samples and potential water uptake range.

Orientation	Areal Porosity (%)	Potential Water Uptake Range (%)
−45/45°	7.34 ± 1.69	7.21 ± 1.66
−45/45°	8.11 ± 1.50	7.98 ± 1.48
0/90°	8.43 ± 5.05	8.29 ± 4.97
0/90°	10.50 ± 5.19	10.33 ± 5.10

3.3. Thermal Properties (TGA and DSC)

The thermal degradation behavior of the 3D-printed Ultem 9085 is shown in Figure 8a for the unaged and post-aging conditions. The D-TGA curve for the unaged samples shows two peaks, which are evident at 506 °C and 594 °C. According to Padovano et al., thermal degradation of Ultem 9085 in a nitrogen environment involved two components, which included the degradation of PC at a lower temperature and PEI at a higher temperature [21]; indeed, that two stage response is noted in the D-TGA curves here.

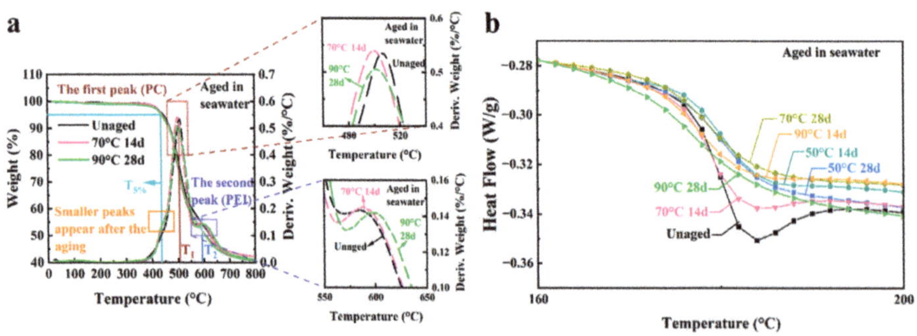

Figure 8. Thermal properties for unaged and aged 3D-printed Ultem 9085 samples in seawater. (**a**) TGA and D-TGA curves; (**b**) DSC curves.

To quantify the degradation of the 3D-printed Ultem 9085 materials due to seawater exposure, the initial decomposition temperature and the temperature for the two maximum degradation peaks were determined from the TGA responses. The initial decomposition temperature here is defined as the temperature with 5% weight loss (T5%); the temperature for the first highest degradation peak (PC) and the second highest degradation peak (PEI) are defined as T1 and T2, respectively. These three values (T5%, T1 and T2) are reported for the Ultem 9085 samples in Table 4 for the experimental aging conditions. The T5% and T1 values for the treated specimens were all lower than the value for the untreated specimens, whereas the T2 values increased after being treated in the marine environment. Also evident in Figure 8a, another small peak appears near 440 °C after the samples underwent accelerated aging.

Table 4. Thermal properties of unaged and aged 3D-printed Ultem 9085 samples before and after seawater exposure. Note that "d" indicates days.

Samples	T5% (°C)	T1 (°C)	T2 (°C)	Tg (°C)	δ_H (J/g)	Tp(°C)
Unaged	444.7	506.4	584.7	180.3	0.148	184
50 °C 14 d	443.2	501.9	589.1	180.7	0.01	188
50 °C 28 d	444.2	502.3	595.9	179.8	0.003	190
70 °C 14 d	441.1	499.9	588	179.1	0.042	184
70 °C 28 d	440.9	502.5	593.2	179.1	0.029	190
90 °C 14 d	435.3	498.8	590.1	178.1	0.028	186
90 °C 28 d	438.6	501.3	597.5	176.8	0	

Relaxation enthalpy (δ_H) was obtained by calculating the enclosed areas between the DSC trace and the baseline fitted through the DSC trace after the glass transition temperatures [49]. The DSC curves for the virgin and aged Ultem 9085 samples are shown in Figures 8b and S2a. Tg, relaxation enthalpy and peak temperature (Tp) before and after seawater exposure are listed in Table 4. It has been widely stated that the Tg for Ultem with over 10% PC blended with PEI will lead to phase separation that is evident in the material [50–53]. As a result, there are often two distinct Tgs observed in the PEI/PC blend product, which are produced by the PC-rich and PEI-rich regions. However, only one Tg was observed in this study for Ultem 9085 at 180.3 °C. That agrees with the Tg reported in a previous work for an Ultem 9085 filament, where the DSC curve exhibited a single peak and Tg of approximately 180 °C [43]. Apart from the samples exposed at 50 °C for 14 days, Tg decreased after aging in seawater, particularly with increasing exposure temperatures. Also evident in Figure 8b, there was an endothermic enthalpy relaxation peak in the signature for the untreated sample and, after accelerated aging, the treated samples had a lower relaxation enthalpy for all the conditions.

3.4. Analysis of Molecular Composition (Raman)

Selected experimental Raman spectra for t Ultem 9085 in the unaged condition and aged at 90 °C for 28 days are shown in Figure 9, ranging from 500 to 3200 cm^{-1}. The spectral properties of Ultem 9085 relies on the two major polymer components, i.e., PEI and PC [54]. According to a comparison of the experimental spectra with the WiRE software (Version 5.2) reference library and the Raman spectra of PC and Ultem 1000 (Figure S3), there was no new peak caused by exposure of Ultem 9085 to accelerated aging in seawater. The Raman spectra for Ultem 9085 are shown in Figure S4a of the SI for samples aged at the different conditions. No new peaks in the spectra were produced after accelerating aging in the artificial seawater, regardless of the aging condition. However, there was some variation in the intensity of different peaks across the aging conditions, which suggests potential differences in the degree of hydrolysis or physical aging reactions.

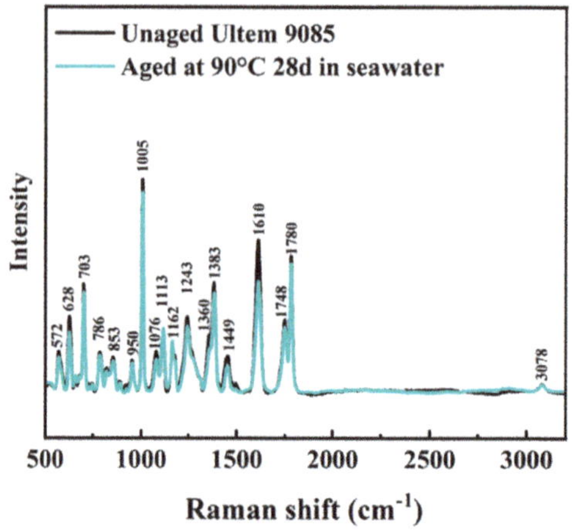

Figure 9. Raman spectra for unaged and aged at 90 °C for 28 days Ultem 9085 material in seawater.

4. Discussion

Results of microscopy and µCT imaging showed that the microstructure of the printed samples exhibited a substantial degree of porosity regardless of the infill pattern (Figure 7a). Considering marine applications, pores within the microstructure are detrimental for at least two reasons, including the potential for premature failures due to the stress concentrations and water ingestion within the pores that facilitates changes in weight and a greater extent of water-related degradation. Previous studies conducted on injection-molded PEI and injection-molded PC reported that the water absorption reached an equilibrium of approximately 0.5 wt% and 0.6 wt% after approximately 4 days [55,56]. Other studies conducted with Ultem 9085 specimens produced by FFF and immersed in a high-pressure saltwater environment and real marine environment at around 15 °C reported that the increase in weight saturated at 9 wt% after 8 days and at 4.5% after 6 months [15,34]. In the present study, the maximum increase in wt% for the Ultem 9085 samples exceeded 7%, which falls within the rather wide reported range. Although the rate of water uptake decreased after 14 days exposure, it was not clear that an equilibrium was reached after 28 days (Figure 7b). Clearly, the water uptake of the Ultem 9085 material exceeded that of both the injection-molded PEI and PC, as expected, and the primary factor was the printed microstructure. Based on the visible porosity, the degree of weight increase was expected to depend on the printing parameters and part geometry, as both could contribute to the extent of porosity in the 3D-printed components. That could also cause differences in the degree of water uptake between benchtop studies with coupons and more completed printed components, an issue of practical importance.

4.1. Property Changes Following Accelerated Aging

The thermal stability of polymers depends on the molecular weight, degree of crosslinking and crystallinity [57]. As such, not all polymers undergo the same degree of aging-related degradation. Hydrothermal aging reportedly does have a significant effect on the properties of PC and is associated with chain scission [58–60], and hydrolytic attack of this polymer reportedly happens more easily in a high pH and high-temperature environment. Indeed, the decrease in T5% and T1 peak values and appearance of the new peak in Figure 8a can be attributed to scission of the PC chains caused by hydrolysis reaction in the artificial seawater. In contrast, the thermal stability of PEI could be improved after prolonged exposure to seawater at elevated temperatures. Crosslinking of PEI occurs above 320 °C, which is much higher than the aging temperature performed here [61]. However, hydrogen bonds can develop that enhance the thermal stability of polymers in general [57]. In PEI, the increased thermal stability can result from the formation of hydrogen bonds, as well as due to changes in the crystallinity or structure of PEI in marine environments at an elevated temperature [62].

The Tg of polymers depends on the polymer chain mobility. The relaxation enthalpy is frequently used to evaluate physical aging of polymers. In the Ultem 9085 samples subjected to accelerated aging, there were four contributions to the polymer chain mobility, which included chain scission by hydrolysis to PC, the formation of interchain hydrogen bonding in PEI, the plasticization posed by water molecules and the physical aging in both PC and PEI. The presence of a relaxation enthalpy peak in the unaged samples revealed that there was already some thermal history on the sample. The process of physical aging involves a competition between aging and rejuvenation [49]. It has been reported that the earlier more dominant rejuvenation process can produce an increase in Tg and Tp and a reduction in δ_H [49]. The increase in Tp and the decrease in δ_H occurring in the aged samples suggested a more significant rejuvenation. For the samples aged at 50 °C for 14 days, the incremental increase in Tg appeared to result from the more severe physical aging driven by the elevation in temperature in comparison with the contribution of water plasticization due to the limited time that evolved for diffusion. With longer exposure periods and higher temperatures, which drive a corresponding increase in water diffusion, the plasticization effect of water and the chain scission of PC can lead to the decrease in Tg, as evident for Ultem 9085 aged at 90 °C for 28 days.

Regarding the Raman spectra in Figure 9, no new peaks were evident after exposure of Ultem 9085 to seawater degradation. Physical aging at an elevated temperature can decrease the free volume, which can lead to changes in infrared spectra intensity [63]. The C-H bonds in both polymers generate peaks around 3000 cm^{-1} [54]. The intensity changes in C=O stretching at 1748 cm^{-1} and 1780 cm^{-1} are reflective of the hydrolysis occurring to PC and the influence of hydrogen bonding in PEI and/or the changes in the molecular structure of PEI [64,65]. To further confirm the development of hydrogen bonds in the system, a wavelength over 3200 cm^{-1} is needed [62], which is of interest but left for future work.

While the changes in thermal properties of Ultem 9085 showed some degree of degradation after accelerating aging in seawater, there was an unexpected improvement in the mechanical properties and the potential mechanisms were not evident in the Raman spectra. As a result, accelerated aging of another group of Ultem 9085 samples was performed in an air environment (sans seawater). This effort was performed using the Type 4 configuration with −45/45° and 0/90° Ultem 9085 sample groups. These two groups were chosen and did not include the Type 1 specimens due to the larger number of Type 1 specimen failures in the out-of-gauge section, as shown in Figure 4. The purpose was to investigate the influence of elevated temperatures on the material independent of the hydrolysis driven by seawater exposure. Akin to the evaluation after seawater aging, tensile tests, TGA, DSC and Raman spectroscopy were performed, and the results were compared to those from seawater aging.

4.2. Comparisons of Air vs. Marine Aging

Stress–strain curves for the Type 4 Ultem 9085 samples after accelerated aging in air at 90 °C for 28 days are shown in Figure 10. The mechanical properties were determined after air aging and are summarized in Figure 11; the complete data set is presented in the Supplemental Information (SI, Tables S6–S9). Akin to results from the seawater aging, some of the specimens failed prematurely in the out-of-gauge section and the tensile data for those failed specimens were excluded when estimating the mechanical properties. All results from the statistical comparisons are detailed in SI Tables S6–S9.

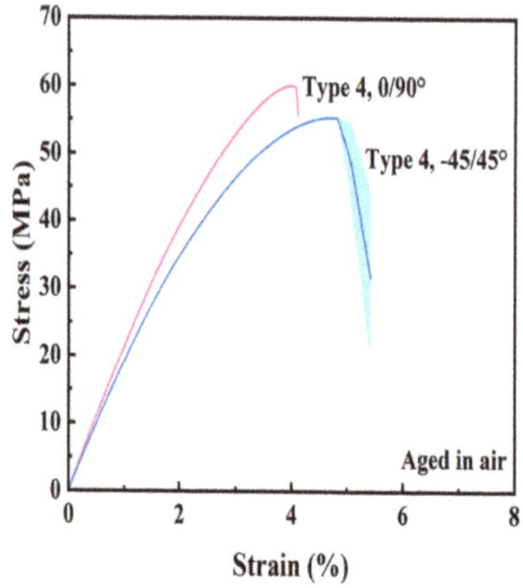

Figure 10. Stress–strain response of Ultem 9085 aged in air at 90 °C for 28 days.

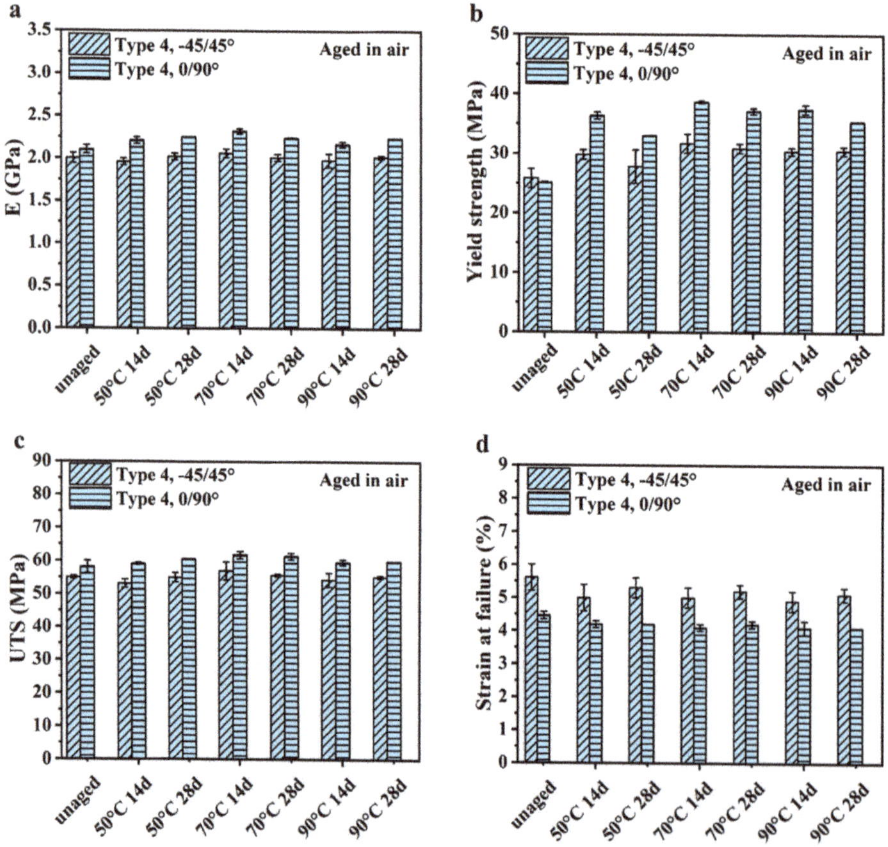

Figure 11. Mechanical properties of different types of unaged Ultem 9085 and aged Ultem 9085 calculated from their strain–stress curves in air. (**a**) Elastic modulus (E); (**b**) yield strength; (**c**) ultimate tensile strength (UTS); (**d**) strain at failure.

Figure 11a presents the elastic modulus of the specimens. As evident from these responses, the E of the Type 4 0/90° specimens increased after aging in air, which was consistent with results of the Type 4 0/90° specimens after aging in seawater. Although there was a very small increase in E of the Type 4 −45/45° specimens after aging at 70 °C for 14 days, there was no significant increase in E for the remainder of the aging conditions. That contrasted results after aging in seawater, where the increase in E was significant for all conditions and temperatures. Results for the yield strength of Ultem 9085 exposed to accelerated aging in air are presented in Figure 11b. Accelerated aging in air caused an increase in the yield strength, which was consistent with seawater aging. The yield strength of both sample groups that underwent air aging exhibited a significant increase, except for the Type 4 0/90° samples aged at 50 °C for 28 days and 90 °C for 28 days and Type 4 −45/45° aged at 50 °C for 28 days. When comparing the increase in yield strength for air and seawater (Figure 7b), air aging caused a higher increase in yield strength of the Type 4 0/90° samples in all conditions, excluding at 90 °C for 28 days. In contrast, the seawater aging led to a higher yield strength of the Type 4 −45/45° samples in all the conditions, excluding at 90 °C for 14 days.

Results for UTS in Figure 11c show that accelerated aging of the Type 4 0/90° specimens caused some moderate increases in UTS, whereas there were no marked changes in the UTS of −45/45° specimens after accelerated air aging. Recall that after exposure to seawater, the UTS of both Type 4 sample configurations increased, especially at the highest temperature. In contrast to aging in seawater, in air, the highest temperature (90 °C) did not result in a significant difference in UTS between the control samples and the aged samples. Rather the UTS of the Type 4 −45/45° samples exposed to air was significantly different from the UTS of the virgin samples at 50 °C for 14 days and at 70 °C for 28 days.

Regarding the strain to failure in Figure 11d, accelerating aging in air caused a reduction in strain to failure regardless of the printing infill orientation, which was consistent with the responses in seawater. In air, a significant decrease in strain at failures was observed for Type 4 0/90° samples aged at 70 °C for 14 days and Type 4 −45/45° samples aged at all temperatures for 14 days. In contrast, in the seawater environment test, the significant decrease in elongation failure of the Type 4 0/90° samples and Type 4 −45/45° samples only occurred at the highest temperature, i.e., at 90 °C for 14 days and for 28 days.

Measures of the COV for the samples aged in air are listed in Table S10. When comparing the responses from the two different aging environments, the specimens aged in seawater at elevated temperatures exhibited a higher variation in the E. Regarding yield strength, the Type 4 0/90° samples exposed to seawater had a larger COV at a higher temperature than the samples exposed to air. Regarding yield strength, the largest COV for the Type 4 −45/45° samples appeared from air aging at a lower temperature, whereas the largest COV for seawater aging happened at the highest temperature. Interestingly, for the Type 4 0/90° samples, the largest COV in UTS occurred for the control condition rather than after aging in either environment, undoubtedly a function of the microstructure. Regarding UTS, the largest COV for the −45/45° samples resulted from aging in seawater compared with air aging. Regarding the Type 4 −45/45° samples, they underwent large variations in elongation failure in both aging environments. In contrast, the largest aging-related variation in the strain at failure of the Type 4 0/90° samples occurred in the marine condition.

The TGA and D-TGA curves of Ultem 9085 after air aging is shown in Figure 12a. Results from the spectra for the parameters used to quantify the thermal degradation (T5%, T1 and T2) are summarized in Table 5. Distinct from the material treated in seawater, no smaller peaks at around 440 °C were produced by aging in air, which suggests that hydrolytic-related degradation in the PC did not occur. In addition, the T5%, T1 and T2 all increased after exposure to air at elevated temperatures.

The DSC curves for the 3D-printed samples are presented in Figures 12b and S2b for the unaged and air-aged conditions. The samples treated in air at 70 °C for the longer duration exhibited an endothermic melting peak at around 220 °C, which suggests that crystallization occurs during air aging. Tg values of the samples aged in air are listed in Table 5 and reflect that there were some small changes, which were related to the rejuvenation and physical aging process. Yet, the variation in Tg from air aging is not as much as that for seawater. At 50 °C, it was obvious that the δ_H and Tp values decreased faster in seawater than those properties in air aging. However, at the higher temperatures (70 °C and 90 °C) in air, the decrease and increase in δ_H and Tp, respectively, and shorter time to reach a zero δ_H, revealed an increase in the rate of physical aging.

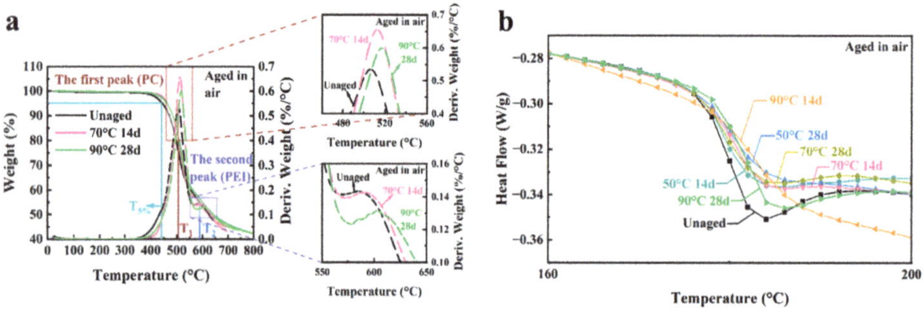

Figure 12. Thermal properties for unaged and aged 3D-printed Ultem 9085 samples in air. (**a**) TGA and D-TGA curves; (**b**) DSC curves.

Table 5. Thermal properties of unaged and aged Type 4 3D-printed Ultem 9085 samples before and after air exposure. Note that "d" indicates days of exposure.

Samples	T5% (°C)	T1 (°C)	T2 (°C)	Tg (°C)	δ_H (J/g)	Tp(°C)
Unaged	444.7	506.4	584.7	180.3	0.148	184
50 °C 14 d	462.2	512	602	179	0.121	184
50 °C 28 d	459	516	598	179.8	0.016	186
70 °C 14 d	464.4	512	592	181.7	0.012	186
70 °C 28 d	464.7	510	598	179.8	0.052	184
90 °C 14 d	464.5	516	610	182.8	0	
90 °C 28 d	462.7	516	602	179.3	0.102	186

A comparison of the Raman spectra for the unaged printed sample and for a representative sample aged in air (90 °C for 28 days) is shown in Figure 13. The Raman spectra for the Ultem 9085 samples at all air-aging conditions are shown in Figure S4b. Identical to Ultem 9085 aged in seawater, no new peaks were identified in the spectra after accelerating aging in air. While there were some changes in the intensity at different peaks that were apparent after air aging, they were not as distinct as those in the marine aging. Indeed, this was expected and confirmatory of the additional contributing mechanisms active in the seawater aging condition, including water plasticization, hydrolytic reactions and physical aging. Only some degree of crystallization and physical aging occurred in air. By comparing the thermal property changes of the Ultem 9085 material after the two aging conditions, it became apparent that hydrolysis and plasticization in the seawater environment had a more pronounced effect at 90 °C, which could also be observed in the Raman spectra.

Accelerated aging in seawater had the most effect on the E and yield strength of Ultem 9085 and was most evident from the responses for the Type 4 samples in Figure 3a,b. Both properties increased with aging time and temperature. Yet, for aging in air, only the yield strength of the Type 4 samples exhibited substantial changes, as noted in Figure 11b; the increase in strength was significant and reached nearly 40%. Through a comparison of the changes in mechanical properties for air and seawater, it can be noticed that E was more affected by accelerating aging in seawater. In addition, the mechanical properties of both Type 4 sample configurations varied differently with the increase in aging period. Nevertheless, overall, the yield strength exhibited the most change after exposure to aging in both conditions. Mechanistically, the increase in strength was associated with the microstructural effects on the initiation of chain sliding. It was hypothesized that chain sliding was resisted by the increase in crystallinity and physical aging. The influence of hydrolytic mechanisms in seawater resulted in a more complex concert of mechanisms.

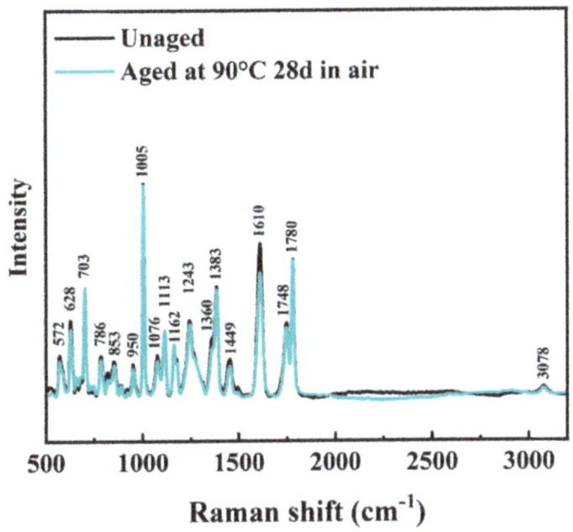

Figure 13. Raman spectra for unaged and aged at 90 °C for 28 days Ultem 9085 material in air.

4.3. Relevant Prior Work

Reported studies on the durability of Ultem 9085 after exposure to marine environments are limited. Padovano et al. [32] evaluated the aging behavior of 3D-printed Ultem 9085 after dry storage at 100 °C, cyclic temperature change from −40 to 90 °C in a humid condition and thermal shock from 70 to −20 °C. They used three-point flexural tests and quantified the E and flexural strength of the aged samples. Interestingly, the samples that underwent gradual thermal cycling in a humid environment underwent a significant increase in E and flexural strength [32]. However, thermal shock caused a decrease in E and strength [32]. Alternatively, Bagsik et al. [21] prepared Ultem 9085 coupons cut from panels printed by FFF and submerged them into real seawater at a low temperature (approximately 15 °C) for 6 months to investigate the changes in mechanical behavior. Results showed that aging in seawater caused an increase in E, yield strength and UTS, as well as a reduction in the strain at failure [21]. These results are largely consistent with those of the present study. However, the contributing mechanisms responsible for the mechanical property variations were not evaluated in these investigations [21,32]. Understanding the mechanisms involved appears critical for distinguishing the applicability of the results to a design environment.

There has been some relevant work performed on changes to the durability of other polymers printed by FFF and then subjected to moist and/or submerged environments. In general, these studies have reported that there is some degree of degradation in the mechanical properties. For instance, Chaudhary et al. immersed a variety of polymers (ABS, PLA, PETG, etc.) printed by FFF in unspecified seawater at temperatures ranging from 30 to 45 °C [66]. After roughly one month of exposure, the changes in E and water diffusivity were evaluated [66]. They found that the 3D-printed polymers had a higher water diffusivity than their traditionally manufactured polymer counterpart [66]. Interestingly, the E of the 3D-printed material exhibited some degradation after the 4-week period, which was assumed to be caused by contributions from the water to the secondary bonds in the respective polymers [66]. El Magri et al. studied the influence of physical aging on the mechanical properties of a 3D-printed blend of PEEK/PEI in dry (120 °C) and 85% relative humidity (70 °C) environments after 42 days of aging [67]. Based on the results of tensile testing, the E of the material decreased considerably in the humid environment, whereas UTS did not appear affected by either the dry or moist aging conditions [67]. They attributed the degradation to plasticization of the polymer chains caused by the water and internal stresses caused by the voids between deposited layers.

Ho et al. performed physical aging experiments on polycarbonate specimens produced by traditional manufacturing methods at temperatures from 60 to 120 °C and for almost 15 days; changes in structure were evaluated with DSC [49]. They found that conformational changes in the polymer chains driven by physical aging led to a more parallel aligned chain distribution, which would foster an increase in crystallinity and improvements in stiffness and yield strength [49,68–70]. In the present accelerating aging study, the significant increase in yield strength could be explained by the contributions of physical aging that was accelerated at elevated temperatures. With respect to the higher E after the seawater aging, water could increase the mobility of the chains, which could contribute to the more significant increase in E within seawater at elevated temperatures [68]. A longer period of storage would likely enable the effects of hydrolytic degradation to become more prominent. As such, accelerated aging studies performed to evaluate the durability of polymers processed by FFF is a useful approach to understand mechanistic contributions to property changes with time. However, the results may not be representative of those that are active in the application environment due to the competing mechanisms and rates of degradation. Although one of the goals of this effort was to develop a model for the aging response, the plurality in active mechanisms, the short aging period and the issues with out-of-gauge failures severely limited the opportunities for modeling. Nevertheless, the investigation was not a failure. Overall, the results demonstrate that degradation of Ultem 9085 can be expected under exposure to marine environments and that it manifests through a reduction in the strain to failure, which results in embrittlement. It must be considered for applications that will involve exposure to marine environments. Furthermore, making adjustments to the design to accommodate this degradation is not sufficient and the degree of degradation is expected to increase over time.

4.4. Limitations

The findings from this investigation showed that there were significant changes in the mechanical properties of Ultem 9085 with accelerated aging and that there were several contributing mechanisms to the degradation. Interestingly, the relative effects of aging were dependent on the sample type (ASTM Type 1 or 4) and infill pattern, which was related to the porosity, differences in water penetration and their importance to the structural behavior. While the findings help to advance our understanding of aging-related degradation in Ultem 9085, there were limitations to this study that should be considered. One obvious concern regarding the aging condition was that a simple laboratory marine environment was used. When applying the material in a real marine environment, there are many additional threats of potential relevance, including UV light, biofouling from marine organisms, different salinity and ions and potential hydrostatic pressure, which could simultaneously and synergistically contribute to deterioration. In addition, while there were some significant changes in mechanical behavior, the extent of overall degradation in strength or strain to failure was rather limited. Admittedly, the accelerated aging protocol involved a limited time, with a maximum of 28 days; more substantial degradation could arise with a longer aging period. Another concern in this study was the specimen type and number of unacceptable failures. It is becoming widely recognized that the ASTM standard specimens for characterizing the tensile properties of polymers are not reliable for FFF [28,34,46–48,71]. Premature failures occurred due to the voids between the infill and contour outlines and the inherent stress concentration produced by the sample design. For instance, results of tension testing for the Type 1 −45/45° samples were not valid. In addition, there was some loss in the data for some samples with Type 1 0/90°, Type 4 0/90° and Type 4 −45/45° configurations due to invalid failure locations. Based on these results, further work should be performed to find the most reliable specimen geometry to be enrolled for future aging evaluations. In addition, longer time periods of aging should be utilized to examine whether the degradation continues to evolve with further time or whether it reaches an equilibrium state.

5. Conclusions

An experimental investigation was performed on 3D-printed Ultem 9085 produced by FFF to characterize the resistance to degradation under accelerated aging involving exposure to seawater. Based on the results and analyses performed, the following conclusions were drawn:

1. Results of accelerated aging in seawater showed there was a significant increase in elastic modulus and both yield and ultimate tensile strength. However, there was a decrease in strain to failure. In air, there was also an increase in yield and ultimate tensile strength. The most substantial change in properties with aging was in the magnitude of yield strength with increasing temperature.
2. Results from TGA revealed that the accelerated aging protocol caused hydrolytic degradation in the PC of Ultem 9085 and appeared most responsible for the decrease in strain to failure. The degradation in strain to failure reached significance when the material was aged in seawater at the highest temperature (90 °C).
3. According to the results of Raman analysis, physical aging occurred in PEI and PC and contributed to the improvement in elastic modulus and strength, with accelerated aging. The process of accelerated aging in air resulted in physical aging without hydrolytic degradation and showed that there was negligible degradation in strain to failure with increasing aging time or temperature.
4. Premature or unacceptable failures occurred in some of the tensile specimens of both types (Type 1 and Type 4) due to the voids located between the infill and contour outlines of the tensile specimens. These internal defects posed inherent stress concentrations and were most detrimental to the Type I tensile specimens. There appears to be a need to modify the specimen geometry of the ASTM standard D638 to support more reliable characterization of the mechanical properties of polymers if they are produced by FFF.
5. When conducting accelerating aging tests of polymers in marine environments, it is important to recognize that physical aging is one of the contributing mechanisms. Consequently, choosing a proper aging protocol that excludes the effects of physical aging may be necessary to isolate the effects of the other aging mechanisms.

Supplementary Materials: The following supporting information can be downloaded at: https://www.mdpi.com/article/10.3390/polym16030350/s1.

Author Contributions: X.W.: formal analysis, investigation, data curation, writing—original draft. C.T.: formal analysis, investigation, data curation, writing—original draft. M.T.S.: methodology, resources, writing—review and editing. D.A.: conceptualization, project administration, funding acquisition, writing—original draft, writing—review and editing. All authors have read and agreed to the published version of the manuscript.

Funding: This work was supported by of the Department of Defense through the Naval Engineering Education Consortium (NEEC) program and Contract No: N00174-20-1-0004.

Institutional Review Board Statement: This work and the manuscript presented here have abided by the ethics guidelines outlined by Polymers.

Data Availability Statement: Data is available upon request.

Acknowledgments: The authors acknowledge the technical support of Nathan Schuyler of the NUWC and Mallory Parker of the UW.

Conflicts of Interest: The authors declare no conflicts of interest.

References

1. Tofail, S.A.M.; Koumoulos, E.P.; Bandyopadhyay, A.; Bose, S.; O'Donoghue, L.; Charitidis, C. Additive Manufacturing: Scientific and Technological Challenges, Market Uptake and Opportunities. *Mater. Today* **2018**, *21*, 22–37. [CrossRef]
2. Shahrubudin, N.; Lee, T.C.; Ramlan, R. An Overview on 3D Printing Technology: Technological, Materials, and Applications. *Procedia Manuf.* **2019**, *35*, 1286–1296. [CrossRef]
3. Mohammed, J.S. Applications of 3D Printing Technologies in Oceanography. *Methods Oceanogr.* **2016**, *17*, 97–117. [CrossRef]
4. Głowacki, M.; Mazurkiewicz, A.; Słomion, M.; Skórczewska, K. Resistance of 3D-Printed Components, Test Specimens and Products to Work under Environmental Conditions—Review. *Materials* **2022**, *15*, 6162. [CrossRef] [PubMed]
5. Zaldivar, R.J.; Mclouth, T.D.; Ferrelli, G.L.; Patel, D.N.; Hopkins, A.R.; Witkin, D. Effect of Initial Filament Moisture Content on the Microstructure and Mechanical Performance of ULTEM® 9085 3D Printed Parts. *Addit. Manuf.* **2018**, *24*, 457–466. [CrossRef]
6. Zaldivar, R.J.; Witkin, D.B.; McLouth, T.; Patel, D.N.; Schmitt, K.; Nokes, J.P. Influence of Processing and Orientation Print Effects on the Mechanical and Thermal Behavior of 3D-Printed ULTEM®9085 Material. *Addit. Manuf.* **2017**, *13*, 71–80. [CrossRef]
7. Montalvão, G.R.; Moshrefi-Torbati, M.; Hamilton, A.; Machado, R.; João, A. Behaviour of 3D Printed PLA and PLA-PHA in Marine Environments. In Proceedings of the IOP Conference Series: Earth and Environmental Science, Singapore, 13–15 August 2020; Volume 424, p. 012013.
8. Kafodya, I.; Xian, G.; Li, H. Durability Study of Pultruded CFRP Plates Immersed in Water and Seawater under Sustained Bending: Water Uptake and Effects on the Mechanical Properties. *Compos. Part B Eng.* **2015**, *70*, 138–148. [CrossRef]
9. Tual, N.; Carrere, N.; Davies, P.; Bonnemains, T.; Lolive, E. Characterization of Sea Water Ageing Effects on Mechanical Properties of Carbon/Epoxy Composites for Tidal Turbine Blades. *Compos. Part A Appl. Sci. Manuf.* **2015**, *78*, 380–389. [CrossRef]
10. Nicholas, J.; Mohamed, M.; Dhaliwal, G.S.; Anandan, S.; Chandrashekhara, K. Effects of Accelerated Environmental Aging on Glass Fiber Reinforced Thermoset Polyurethane Composites. *Compos. Part B Eng.* **2016**, *94*, 370–378. [CrossRef]
11. Abdurohman, K.; Adhitya, M. Effect of Water and Seawater on Mechanical Properties of Fiber Reinforced Polymer Composites: A Review for Amphibious Aircraft Float Development. In Proceedings of the IOP Conference Series: Materials Science and Engineering, Jakarta, Indonesia, 9–10 October 2019; Volume 694, p. 012035.
12. Saroia, J.; Wang, Y.; Wei, Q.; Lei, M.; Li, X.; Guo, Y.; Zhang, K. A Review on 3D Printed Matrix Polymer Composites: Its Potential and Future Challenges. *Int. J. Adv. Manuf. Technol.* **2020**, *106*, 1695–1721. [CrossRef]
13. Gebisa, A.W.; Lemu, H.G. Influence of 3D Printing FDM Process Parameters on Tensile Property of ULTEM 9085. *Procedia Manuf.* **2019**, *30*, 331–338. [CrossRef]
14. Pham, K.D.; O'Brien, W.F.; Case, S.W. Characterizing Static and Dynamic Mechanical Properties for Additive Manufactured ULTEM 9085 Used to Construct Flow Control Devices for Turbomachinery Applications. In Proceedings of the ASME Turbo Expo 2018: Turbomachinery Technical Conference and Exposition, Oslo, Norway, 11–15 June 2018; Volume 6: Ceramics; Controls, Diagnostics, and Instrumentation; Education; Manufacturing Materials and Metallurgy. p. V006T24A008.
15. Murdy, P.; O'Dell, J.; Barnes, D.; McVey, J.R.; Rumple, C. Investigating Marine Environmental Degradation of Additive Manufacturing Materials for Renewable Energy Applications. In Proceedings of the Composites and Advanced Materials Expo, Atlanta, GA, USA, 30 October–2 November 2023.
16. Aguilar, D.; Christensen, S.; Fox, E. *3-D Printed Ultem 9085 Testing and Analysis*; SPH-04-XS-100; National Aeronautics and Space Administration, Ames Research Center: Moffett Field, CA, USA, 2015. Available online: https://ntrs.nasa.gov/citations/20150017060 (accessed on 16 January 2024).
17. Fischer, M.; Schöppner, V. Fatigue Behavior of FDM Parts Manufactured with Ultem 9085. *JOM* **2017**, *69*, 563–568. [CrossRef]
18. Byberg, K.I.; Gebisa, A.W.; Lemu, H.G. Mechanical Properties of ULTEM 9085 Material Processed by Fused Deposition Modeling. *Polym. Test.* **2018**, *72*, 335–347. [CrossRef]
19. Ng, M.B.; Brennan, S.N. Mechanical Performance Analysis of ULTEM 9085 in a Heated, Irradiated Environment. In Proceedings of the ASME International Mechanical Engineering Congress and Exposition, Pittsburgh, PA, USA, 9–15 November 2018; p. V009T12A020.
20. Kaplun, B.W.; Zhou, R.; Jones, K.W.; Dunn, M.L.; Yakacki, C.M. Influence of Orientation on Mechanical Properties for High-Performance Fused Filament Fabricated Ultem 9085 and Electro-Statically Dissipative Polyetherketoneketone. *Addit. Manuf.* **2020**, *36*, 101527. [CrossRef]
21. Padovano, E.; Galfione, M.; Concialdi, P.; Lucco, G.; Badini, C. Mechanical and Thermal Behavior of Ultem® 9085 Fabricated by Fused-Deposition Modeling. *Appl. Sci.* **2020**, *10*, 3170. [CrossRef]
22. Rehman, F.; Diston, J. Material characterization of a high performance additive manufacturing material for efficient component design. *Int. Res. J. Eng. Technol.* **2020**, *7*, 4466–4471.
23. Kobenko, S.; Dejus, D.; Jātnieks, J.; Pazars, D.; Glaskova-Kuzmina, T. Structural Integrity of the Aircraft Interior Spare Parts Produced by Additive Manufacturing. *Polymers* **2022**, *14*, 1538. [CrossRef] [PubMed]
24. Polyetherimide (PEI): A Comprehensive Review. Omnexus. Available online: https://omnexus.specialchem.com/selection-guide/polyetherimide-pei-high-heat-plastic (accessed on 19 October 2023).
25. Polycarbonate (PC). British Plastics Federation. Available online: https://www.bpf.co.uk/plastipedia/polymers/Polycarbonate.aspx (accessed on 19 October 2023).
26. Comprehensive Guide on Polycarbonate (PC). Omnexus. Available online: https://omnexus.specialchem.com/selection-guide/polycarbonate-pc-plastic (accessed on 19 October 2023).

27. Shelton, T.E.; Willburn, Z.A.; Hartsfield, C.R.; Cobb, G.R.; Cerri, J.T.; Kemnitz, R.A. Effects of Thermal Process Parameters on Mechanical Interlayer Strength for Additively Manufactured Ultem 9085. *Polym. Test.* **2020**, *81*, 106255. [CrossRef]
28. McLouth, T.D.; Gustafson, S.M.; Kim, H.I.; Zaldivar, R.J. Enhancement of FDM ULTEM® 9085 Bond Strength via Atmospheric Plasma Treatment. *J. Manuf. Process.* **2021**, *66*, 179–188. [CrossRef]
29. Glaskova-Kuzmina, T.; Dejus, D.; Jātnieks, J.; Kruuv, P.-P.; Zolotarjovs, A.; Einbergs, E.; Vanags, E. Effect of Printing Direction and Post-Printing Conditions on Bending Properties of ULTEM 9085. *J. Compos. Sci.* **2023**, *7*, 316. [CrossRef]
30. Zhang, Y.; Moon, S.K. The Effect of Annealing on Additive Manufactured ULTEM™ 9085 Mechanical Properties. *Materials* **2021**, *14*, 2907. [CrossRef]
31. De Bruijn, A.C.; Gómez-Gras, G.; Pérez, M.A. Thermal Annealing as a Post-Process for Additively Manufactured Ultem 9085 Parts. *Procedia Comput. Sci.* **2022**, *200*, 1308–1317. [CrossRef]
32. Bagsik, A.; Schöppner, V.; Klemp, E. Long-term Ageing Effects on Fused Deposition Modeling Parts Manufactured with Ultem*9085. In Proceedings of the 23rd Annual International Solid Freeform Fabrication Symposium—An Additive Manufacturing Conference, Austin, TX, USA, 6–8 August 2012.
33. Gallagher, W.R. Investigation of Ultem 9085 for Use in Printed Orbital Structures. Master's Thesis, Air Force Institute of Technology, Dayton, OH, USA, 2020.
34. LeBlanc, J.; Shattuck, L.; Warner, E.; Javier, C.; Chenwi, I.; Chu, T.; Shukla, A. Effect of High Pressure Salt Water Absorption on the Mechanical Characteristics of Additively Manufactured Polymers. *Int. J. Lightweight Mater. Manuf.* **2023**, *6*, 379–391. [CrossRef]
35. Cowie, J.M.G.; Arrighi, V. Physical Aging of Polymer Blends. In *Polymer Blends Handbook*, 2nd ed.; Utracki, L.A., Wilkie, C.A., Eds.; Springer: Dordrecht, The Netherlands, 2014; pp. 1357–1394.
36. Elias, H.-G. Chemical Aging. In *Macromolecules: Volume 1—Structure and Properties*; Elias, H.-G., Ed.; Springer: Boston, MA, USA, 1977; pp. 843–859.
37. Sepe, M. The Mystery of Physical Aging, Part 1: Knowing the Difference. Plastic Technology. Available online: https://www.ptonline.com/articles/the-mystery-of-physical-aging-part-1 (accessed on 20 October 2023).
38. Maxwell, A.S.; Broughton, W.R.; Dean, G.; Sims, G. *Review of Accelerated Ageing Methods and Lifetime Prediction Techniques for Polymeric Materials*; DEPC-MPR 016; National Physical Laboratory: London, UK, 2005. Available online: http://eprintspublications.npl.co.uk/id/eprint/3161 (accessed on 18 January 2024).
39. Davies, P.; Evrard, G. Accelerated Ageing of Polyurethanes for Marine Applications. *Polym. Degrad. Stab.* **2007**, *92*, 1455–1464. [CrossRef]
40. Tocháček, J.; Vrátníčková, Z. Polymer Life-Time Prediction: The Role of Temperature in UV Accelerated Ageing of Polypropylene and Its Copolymers. *Polym. Test.* **2014**, *36*, 82–87. [CrossRef]
41. Bergaliyeva, S.; Sales, D.L.; Delgado, F.J.; Bolegenova, S.; Molina, S.I. Effect of Thermal and Hydrothermal Accelerated Aging on 3D Printed Polylactic Acid. *Polymers* **2022**, *14*, 5256. [CrossRef] [PubMed]
42. Frigione, M.; Rodríguez-Prieto, A. Can Accelerated Aging Procedures Predict the Long Term Behavior of Polymers Exposed to Different Environments? *Polymers* **2021**, *13*, 2688. [CrossRef] [PubMed]
43. Tosto, C.; Saitta, L.; Pergolizzi, E.; Blanco, I.; Celano, G.; Cicala, G. Methods for the characterization of polyetherimide based materials processed by fused deposition modelling. *Appl. Sci.* **2020**, *10*, 3195. [CrossRef]
44. *ASTM D638-14*; Standard Test Method for Tensile Properties of Plastics. ASTM International: West Conshohocken, PA, USA, 2014.
45. *ASTM D1141–98*; Standard Practice for Preparation of Substitute Ocean Water. ASTM International: West Conshohocken, PA, USA, 2013.
46. Ahn, S.; Montero, M.; Odell, D.; Roundy, S.; Wright, P.K. Anisotropic Material Properties of Fused Deposition Modeling ABS. *Rapid Prototyp. J.* **2002**, *8*, 248–257. [CrossRef]
47. Milosevic, M.; Stoof, D.; Pickering, K.L. Characterizing the Mechanical Properties of Fused Deposition Modelling Natural Fiber Recycled Polypropylene Composites. *J. Compos. Sci.* **2017**, *1*, 7. [CrossRef]
48. Özen, A.; Auhl, D.; Völlmecke, C.; Kiendl, J.; Abali, B.E. Optimization of Manufacturing Parameters and Tensile Specimen Geometry for Fused Deposition Modeling (FDM) 3D-Printed PETG. *Materials* **2021**, *14*, 2556. [CrossRef]
49. Ho, C.H.; Vu-Khanh, T. Effects of Time and Temperature on Physical Aging of Polycarbonate. *Theor. Appl. Fract. Mech.* **2003**, *39*, 107–116. [CrossRef]
50. Blanco, I.; Cicala, G.; Ognibene, G.; Rapisarda, M.; Recca, A. Thermal Properties of Polyetherimide/Polycarbonate Blends for Advanced Applications. *Polym. Degrad. Stab.* **2018**, *154*, 234–238. [CrossRef]
51. Cicala, G.; Ognibene, G.; Portuesi, S.; Blanco, I.; Rapisarda, M.; Pergolizzi, E.; Recca, G. Comparison of Ultem 9085 Used in Fused Deposition Modelling (FDM) with Polytherimide Blends. *Materials* **2018**, *11*, 285. [CrossRef] [PubMed]
52. Lin, B.; Sundararaj, U. Visualization of Poly(ether Imide) and Polycarbonate Blending in an Internal Mixer. *J. Appl. Polym. Sci.* **2004**, *92*, 1165–1175. [CrossRef]
53. Chun, Y.S.; Lee, H.S.; Oh, T.S.; Kim, W.N. Properties of Blends of Polycarbonate and Polypropylene (I): Crystallization Behavior. *Polymer* **1996**, *20*, 1071–1079. [CrossRef]
54. Luchinsky, D.G.; Hafiychuk, H.; Hafiychuk, V.; Wheeler, K.R. *Molecular Dynamics of ULTEM 9085 for 3D Manufacturing: Spectra, Thermodynamic Properties, and Shear Viscosity*; NASA/TM-2018-220213; National Aeronautics and Space Administration, Ames Research Center: Moffett Field, CA, USA, 2018. Available online: https://ntrs.nasa.gov/citations/20190026629 (accessed on 20 October 2023).

55. Merdas, I.; Thominette, F.; Verdu, J. Humid Aging of Polyetherimide. I. Water Sorption Characteristics. *J. Appl. Polym. Sci.* **2000**, *77*, 1439–1444. [CrossRef]
56. Golovoy, A.; Zinbo, M. Water Sorption and Hydrolytic Stability of Polycarbonates. *Polym. Eng. Sci.* **1989**, *29*, 1733–1737. [CrossRef]
57. Rane, A.V.; Begum, S.A.; Kanny, K. Spectroscopic Analysis of Compatibilized Polymer Blends. In *Compatibilization of Polymer Blends*; Ajitha, A.R., Thomas, S., Eds.; Elsevier: Amsterdam, The Netherlands, 2020; pp. 373–390.
58. Weibin, G.; Shimin, H.; Minjiao, Y.; Long, J.; Yi, D. The Effects of Hydrothermal Aging on Properties and Structure of Bisphenol A Polycarbonate. *Polym. Degrad. Stab.* **2009**, *94*, 13–17. [CrossRef]
59. Fang, M.; Zhang, N.; Huang, M.; Lu, B.; Lamnawar, K.; Liu, C.; Shen, C. Effects of Hydrothermal Aging of Carbon Fiber Reinforced Polycarbonate Composites on Mechanical Performance and Sand Erosion Resistance. *Polymers* **2020**, *12*, 2453. [CrossRef] [PubMed]
60. Benhamada, M.; Bouzid, D.; Saouli, O.; Boyron, O. The Effects of Hydrothermal Aging Characterized by Sec on the Degradations Kinetics of Polycarbonate Calculated through TGA. *Chem. Eng. Trans.* **2015**, *43*, 1183–1188.
61. Amancio-Filho, S.T.; Roeder, J.; Nunes, S.P.; Dos Santos, J.F.; Beckmann, F. Thermal Degradation of Polyetherimide Joined by Friction Riveting (FricRiveting). Part I: Influence of Rotation Speed. *Polym. Degrad. Stab.* **2008**, *93*, 1529–1538. [CrossRef]
62. De Nicola, A.; Correa, A.; Milano, G.; La Manna, P.; Musto, P.; Mensitieri, G.; Scherillo, G. Local Structure and Dynamics of Water Absorbed in Poly(Ether Imide): A Hydrogen Bonding Anatomy. *J. Phys. Chem. B* **2017**, *121*, 3162–3176. [CrossRef] [PubMed]
63. Mzabi, N.; Smaoui, H.; Guermazi, H.; Mlik, Y.; Agnel, S.; Toureille, A. Heating Effects on Structural and Electrical Properties of Polyetherimide. *Am. J. Eng. Appl. Sci.* **2009**, *2*, 120–126.
64. Resta, V.; Quarta, G.; Lomascolo, M.; Maruccio, L.; Calcagnile, L. Raman and Photoluminescence Spectroscopy of Polycarbonate Matrices Irradiated with Different Energy 28Si+ Ions. *Vacuum* **2015**, *116*, 82–89. [CrossRef]
65. Devasahayam, S.; Hill, D.J.T.; Connell, J.W. FT-Raman Studies of a Range of Polyimides Subjected to High-energy Radiations at Room and Elevated Temperatures. *J. Appl. Polym. Sci.* **2006**, *101*, 1575–1582. [CrossRef]
66. Chaudhary, B.; Li, H.; Matos, H. Long-Term Mechanical Performance of 3D Printed Thermoplastics in Seawater Environments. *Results Mater.* **2023**, *17*, 100381. [CrossRef]
67. El Magri, A.; Vaudreuil, S. Effects of Physical and Chemical Ageing on 3D Printed Poly (Ether Ether Ketone)/Poly (Ether Imide) [PEEK/PEI] Blend for Aerospace Applications. *J. Mater Sci.* **2023**, *58*, 1465–1479. [CrossRef]
68. Le Guen-Geffroy, A.; Le Gac, P.-Y.; Habert, B.; Davies, P. Physical Ageing of Epoxy in a Wet Environment: Coupling between Plasticization and Physical Ageing. *Polym. Degrad. Stab.* **2019**, *168*, 108947. [CrossRef]
69. Courvoisier, E.; Bicaba, Y.; Colin, X. Multi-Scale and Multi-Technique Analysis of the Thermal Degradation of Poly(Ether Ether Ketone). *Polym. Degrad. Stab.* **2018**, *151*, 65–79. [CrossRef]
70. Jo, W.H.; Ko, K.J. The Effects of Physical Aging on the Thermal and Mechanical Properties of an Epoxy Polymer. *Polym. Eng. Sci.* **1991**, *31*, 239–244. [CrossRef]
71. Rankouhi, B.; Javadpour, S.; Delfanian, F.; Letcher, T. Failure Analysis and Mechanical Characterization of 3D Printed ABS with Respect to Layer Thickness and Orientation. *J. Fail. Anal. Preven.* **2016**, *16*, 467–481. [CrossRef]

polymers

Article

Experimental Study on Warpage Phenomenon of Wax Parts Manufactured by Fused Filament Fabrication

Muslim Mukhtarkhanov, Essam Shehab and Md. Hazrat Ali *

Department of Mechanical and Aerospace Engineering, SEDS, Nazarbayev University, Astana 010000, Kazakhstan; muslim.mukhtarkhanov@nu.edu.kz (M.M.); essam.shehab@nu.edu.kz (E.S.)
* Correspondence: md.ali@nu.edu.kz

Abstract: Warpage is one of the prominent issues in Fused Filament Fabrication. The cause of this is the rapid cooling of the polymer during extrusion. The residual thermal stresses accumulated within the print part result in a shape distortion and subsequent detachment of the object from the print bed. In this study, both experimental and numerical approaches were used to identify the stresses due to thermal shrinking that occurs in soft polymers such as wax. A temperature sweep test was performed using a rotational rheometer to measure the magnitude of axial forces that are generated due to the thermal shrinking of a thin layer of 3D printable wax. The thermal stresses responsible for warpage were computed analytically and using the FEA. It was found that due to thermal processes, the stress magnitude can reach a value of 1.17 MPa. This value is enough to cause the plastic deformation in the wax part having a thin elongated shape. In addition, Taguchi's robust design has identified two major FFF parameters that impact the warpage in amorphous soft polymers. They are the printing speed and the print bed temperature. To achieve a low level of warpage, it is important to make sure that the layer deposition occurs at medium speeds and the print bed temperature is moderately high according to the findings of this study.

Keywords: warpage; wax; FFF; rheometer

check for updates

Citation: Mukhtarkhanov, M.; Shehab, E.; Ali, M.H. Experimental Study on Warpage Phenomenon of Wax Parts Manufactured by Fused Filament Fabrication. *Polymers* **2024**, *16*, 208. https://doi.org/10.3390/polym16020208

Academic Editor: Babak Eslami

Received: 21 November 2023
Revised: 28 December 2023
Accepted: 9 January 2024
Published: 11 January 2024

1. Introduction

Rapid investment casting (RIC) is the metal manufacturing process where sacrificial patterns of wax or non-wax nature are fabricated through rapid prototyping (RP). This results in reduced production costs due to the elimination of the expenditure spent on designing and manufacturing metal tooling. Over the last three decades, numerous RP methods have been tested for application in RIC [1]. Although it has been shown that most additive manufacturing (AM) technologies are suitable for rapid casting applications, not many techniques can manufacture wax patterns with properties similar to investment casting (IC) wax. As for the non-wax patterns, the quality of printed parts is much superior in comparison to wax patterns. Moreover, the cost of non-wax polymers is substantially lower. However, when it comes to the quality of final cast metal products, the usage of wax patterns yields higher-quality objects. That is because the tough non-wax polymers are difficult to remove from the ceramic mold due to the large thermal expansion of the former. Therefore, the material market for the AM industry has been constantly growing with the introduction of new filament materials that have qualities tailored to the needs of end users.

Among modern RP techniques, three types specialize in manufacturing wax patterns. They are MultiJet Printing (MJP), stereolithography (SLA), and Fused Filament Fabrication (FFF). Even though the technologies are based on layer-by-layer patterns of manufacturing, they are distinguished by the level of accuracy, production speed, and cost. For example, the MJP technology is capable of building objects with layers having a thickness as low as 16 μm in the z-build axis. This implies that high accuracy and detail are possible to achieve [2]. In 2021, Chyuan et al. [3] compared the dimensional accuracies of wax patterns manufactured using MJP and FFF-based rapid tooling. They concluded that the direct manufacturing of wax patterns using MJP technology is costly and, therefore, less expensive manufacturing routes can be selected with the help of rapid tooling. Nevertheless, it should be noted that the implementation of rapid tooling increases the level of complexity since additional processes are included in the sequence of manufacturing.

As for the SLA method, it uses photocurable resins as raw material, which can be rapidly cured with the help of a laser. The technology also produces highly detailed objects with good accuracy though parts with overhanging sections require supplementary structures called support structures. The removal of support structures is carried out manually which often leaves visible marks on the surface of the product. Due to the high cost of commercially available SLA-based castable wax, the production of small-size parts is preferable [4]. With regards to material properties, a photocurable wax formulation requires a burnout of the pattern as opposed to dewaxing which is commonly used in conventional IC manufacturing [5]. Moreover, the selection of the burnout schedule for the SLA castable wax is a nontrivial task since the material decomposes in a non-combustive manner. Thus, if small-size patterns are not difficult to thermally deteriorate [6], working with medium to large-size objects is associated with challenges [7].

The FFF technology employs a rather simple manufacturing process based on the extrusion of molten polymer through a heated small nozzle and depositing it on the print bed. Because the technique is unsophisticated, the quality of the final parts is inferior to MJP and SLA-printed objects. On the other hand, the affordability and availability of FFF machines make the technology the most popular in the AM industry. In addition, recent studies demonstrated that it is possible to manufacture low-cost wax patterns having properties very similar to traditional IC waxes [8,9]. For example, non-tough wax parts having low viscosity in the molten state have been manufactured by the authors of the current study [8]. Nevertheless, 3D printing of wax parts is associated with some challenges. For instance, during the extrusion process, the plastic polymer being deposited undergoes rapid solidification, creating thermal stresses within the body of the part. Those stresses manifest in shrinkage or shape distortion in the final product. In some extreme cases, the accumulated thermal stresses cause the part to detach from the print bed before the 3D printing is over. In this regard, mechanically weak polymers such as wax are more susceptible to warpage due to their inability to withstand large shrinkage stresses. This is visible in Figure 1 where the 3D printed wax part is shown.

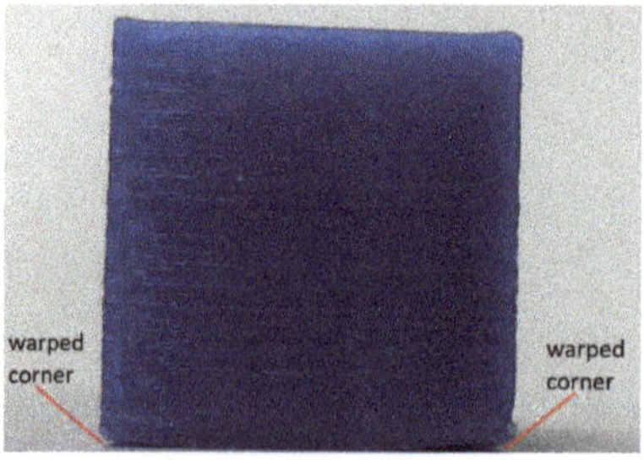

Figure 1. 3D printed cube of 20 × 20 × 20 mm dimension.

Substantial efforts have been put into investigating the issue of warping in AM. Process parameters, material properties, and geometric characteristics are all said to influence the magnitude of the warpage [10]. In addition, numerical and experimental approaches have been proposed to analyze the warping phenomenon. The most common way to measure the warpage magnitude is realized through the evaluation of dimensional deviations in 3D printed benchmarks that represent simple geometric shapes. Table 1 shows the recent works related to the warpage phenomenon relevant to FFF. From Table 1 it is evident that warpage is mostly affected by several process parameters with layer thickness being the most frequently discussed parameter. It is interesting to note that some level of contradiction exists regarding the effect of layer thickness. For instance, while the majority of the publications support the view that a higher level of thickness is preferable to avoid warpage, the works of Antonio et al. [11] and Zhang et al. [12] contradict this view.

Table 1. Warpage issue relevant to FFF.

Material	Part Geometry	Experimental	Analytical/Numerical	Significant Factors Affecting Warpage	Ref.
Acrylonitrile Butadiene Styrene (ABS)	Rectangular plate	+	+	Length of the part, thickness Layer thickness	[11]
	Thin and thickplates	+	+	Print speed, layer thickness	[12]
	Tensile specimen	+	+	Layer thickness	[13]
	Rectangular plate	+		Bed and chamber temperature	[14]
Polylactic acid (PLA)	Thin plate	+		Layer thickness	[15]
	Rectangular block	+		Nozzle temperature	[16]
Polyphenylene sulfide (PPS)	Thin plate	+	+	Coefficient of thermal expansion	[17]
Polyurethane-based shape memory polymer	Thin plate	+	+	Bed temperature, printing speed, layer thickness	[18]

Setting up the higher layer thickness is beneficial for warpage mitigation according to the supporters of this view [13,15]. For instance, it is stated that at a lower thickness of the layer, the less mass of the material is deposited and, consequently, a layer of the extrudate cools at a faster rate [18]. In turn, the rapidly cooling polymer melt has little time to adhere to the print bed surface at the point of deposition. In contrast, the supporters of the opposite view have pointed out that physical mechanisms responsible for this phenomenon are yet to be discussed in the future [11].

The analysis of recent works shows that most of the studies concern the segment of materials representing hard polymers of both amorphous and crystalline kinds. For example, Liu et al. [15] have stressed the substantial work dedicated to amorphous ABS plastics regarding the warpage problem. As for semicrystalline polymers, there is an ongoing search to improve the printability of such plastics since the warpage level is critical in those materials and it is directly proportional to the degree of crystallinity [19]. In turn, the degree of crystallinity is the major factor affecting the mechanical properties.

While it is evident that for hard 3D printing polymers, the subject matter of warpage is largely explored, the issue can be considered unexplored for soft amorphous polymers such as wax. Therefore, the authors of this study find it extremely important to analyze the problem of warpage in soft 3D printable polymers. Especially, for materials intended for RIC application, where the control of dimensional accuracy gains even more significance. In essence, the current study is a continuation of a research project intended to explore the area of 3D printing wax polymers having low melting temperatures for application in RIC. Previously, research has been conducted by the authors to analyze the mechanical, thermal, and rheological properties of 3D printable wax [8]. In turn, the main objective of this paper is to address the prominent issue regarding wax 3D printing, which is warpage. To identify the FFF process parameters that cause the warpage, the design of experiments is performed using Taguchi robust design. In addition, rheological experiments are proposed to assess the shrinkage forces that develop during the rapid solidification of soft 3D printable wax.

2. Methodology

2.1. Materials and Characterization

Commercially available Ø1.75 mm 3D printing wax filament from Filamentarno (Moscow, Russia) in the blue color was purchased for the study. It has a melting temperature of around 100 °C and Young's modulus of 151.5 MPa identified by tensile test [8].

2.2. Assessment of Shrinkage Forces Due to Rapid Cooling Using a Rheometer

Rheological studies have been extensively used to characterize 3D printing polymers [20]. As for this study, we propose a temperature sweep test to understand the shrinkage behavior of the thin wax layer during solidification. A rotational rheometer MCR102 from Anton Paar (Graz, Austria) is used in the experiment with a cone and plate measuring system having Ø50 mm and a truncate angle of 1° as shown in Figure 2. The temperature sweep test allows the measurement of several important characteristics during the temperature change in a polymer sample confined between measuring system plates. Relevant to this study, we are interested in measuring the normal forces that develop in the upper plate of the rheometer as a response to the volumetric shrinkage of wax. It was observed that during the small amplitude oscillatory shear test, a noticeable increase in normal force is detected that acts on the measuring plates as wax transforms from liquid to solid state. This is illustrated in Figure 2 where shrinkage forces are represented as red horizontal arrows pointing towards the center of the cone plate. During the thermal shrinkage of the polymer, friction forces occur between the plate and the material as the latter tries to shrink. Insofar as the gap between plates remains unchanged, the deformation of the wax is suppressed. As a result, the force balance is created between the shrinking wax and the measuring system of the rheometer provided that the plates are under no oscillatory motion. Similarly, during the 3D printing process, a size decrease of the final part evidences the occurrence of thermal shrinkage. Usually, the first layer being deposited adheres to the print bed firmly, provided that the latter is hot enough. As the FFF continues, additional layers are added to the previously deposited layers. Since each individual layer is subject to shrinkage, the thermal stresses accumulate within the body of the solidified part. It is critical to understand that the shrinkage stresses act parallel to the deposition direction. Thus, the elongated objects are more prone to size contraction.

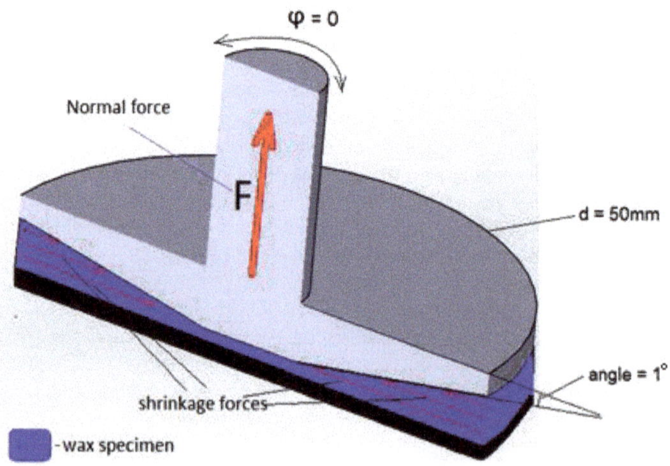

Figure 2. Vertical cross-section view of the cone/plate system with red arrows on the sample representing the forces that occur because of thermal shrinkage.

During testing, small granules of wax filament are placed on the lower plate of the equipment. Then, the plate is heated up to 110 °C to melt the material. After melting the sample, the upper plate is closed to reach the gap of 0.5 mm between the plates. To attain thermal equilibrium, the upper plate is preheated with the help of a heating hood accessory. Once sufficient time is given to reach thermal equilibrium, the temperature is lowered to 90 °C. The lower limit of 90 °C is chosen because the wax has a dropping point at 95 °C, thus, the material turns from liquid to solid state within the given temperature range. Moreover, to prevent an excess in normal force, it was decided to limit the lower value at 90 °C level. It should be noted that the measuring plates were subject to no oscillatory or rotational motion throughout the test. Since the sample has low thickness, the plain strain and plain stress conditions apply.

2.3. Design of Experiments for Identification of FFF Parameters Causing Warpage

To analyze the effect of FFF process parameters on warpage occurrence, different kinds of experimental designs have been proposed previously [21]. Taguchi's design of experiment (DoE) is one of the popular techniques that is broadly used to identify optimal process conditions for different manufacturing methods. An orthogonal array (OA) method was used to identify the effects of three FFF parameters on the level of warpage of a 3D printed part having the shape of a flat bar. Namely, the temperatures of the nozzle and print bed along with printing speed are selected for the experiment. The part has dimensions of 60 × 5 × 20 mm (L × W × H) and the 3D printed sample is shown in Figure 3.

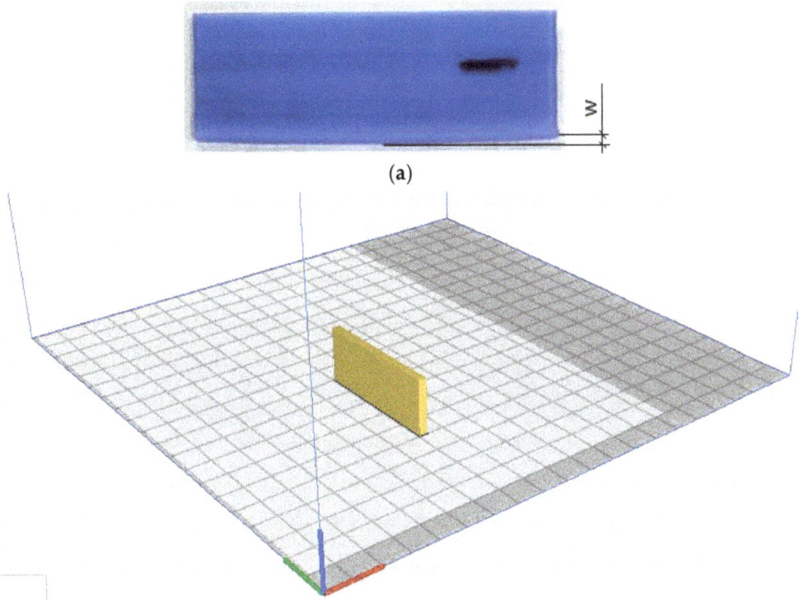

(a)

(b)

Figure 3. A benchmark for the warpage analysis (**a**); the position of the part on the print bed (**b**).

The combinations of three FFF parameters at different levels require nine experiments according to L9 OA Taguchi analysis requirements. To increase the confidence in the results of the tests, each experiment is repeated three times, and an average value is used for calculations. By finding the signal-to-noise (S/N) ratio, it is possible to see the influence level of each parameter. Since warpage is mostly due to thermal processes, two out of three selected parameters relate to the temperature of the extrudate and the platform upon which the material is deposited. As for the printing speed, it influences the rate of solidification and interlayer bonding. Table 2 shows the process parameter values at different levels used in Taguchi analysis.

Table 2. Selected parameters and levels for Taguchi analysis.

Input Parameters	Symbol	Level 1	Level 2	Level 3
Nozzle temperature, °C	I	100	105	110
Nozzle speed, mm/s	II	20	45	70
Bed temperature, °C	III	45	55	65

Minitab 19 software was used to perform Taguchi analysis calculations. To measure the warpage magnitude W indicated in Figure 3a, the coordinate measuring machine (CMM) Duramax from ZEISS (Zeiss Group, Oberkochen, Germany) was used. The CMM machine uses a stylus system to scan the model, therefore, challenges arise when working with soft materials such as wax. To tackle the problem, the warpage level was measured on the ceramic molds shown in Figure 4b instead, which represent a complete replica of the artifact.

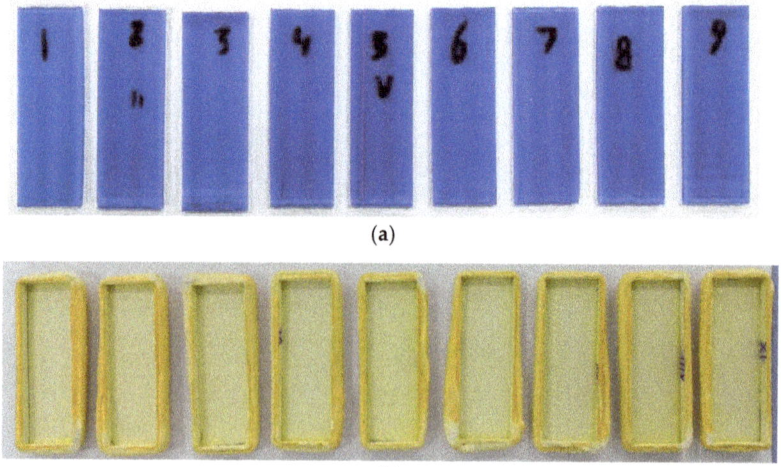

(a)

(b)

Figure 4. 3D printed wax benchmarks for the Taguchi analysis (**a**); ceramic molds fabricated using the wax benchmarks (**b**).

The ceramic shell was created on the surface of wax bars by dipping the latter in the ceramic slurry that is used in investment casting manufacturing. The wax bars and their corresponding ceramic molds are shown in Figure 4. Four layers of ceramic material were enough to build a strong shell with the help of silica-based ceramic slurry Suspendaslurry® FS from Ransom & Randolf (Maumee, OH, USA). After the application of one layer, the parts are left to dry for three hours. To prevent the emergence of drying cracks, each layer was reinforced with a fused silica mesh 50/100. After the ceramic shell construction was completed, the wax models were manually removed to reveal the negative imprint of the benchmark.

The warpage magnitude is measured as the cartesian distance between the lowest and the highest point that lies on a curved line scanned from the warped plane as shown in Figure 5.

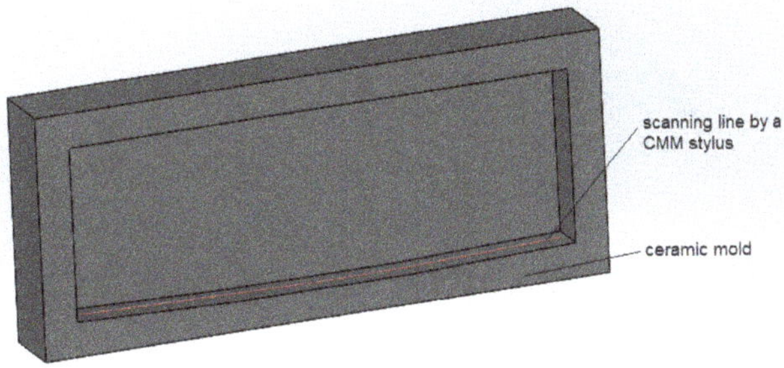

Figure 5. A location where measurements were performed by the CMM measuring tool.

3. Results and Discussion

3.1. Identified Stresses Due to Thermal Shrinkage

The result of the temperature sweep test is shown in Figure 6. The maximum developed normal force within the temperature range of 90–110 °C was measured to be equal to 30.57 N. The cooling rate was identified to be approximately 7.7 °C/min.

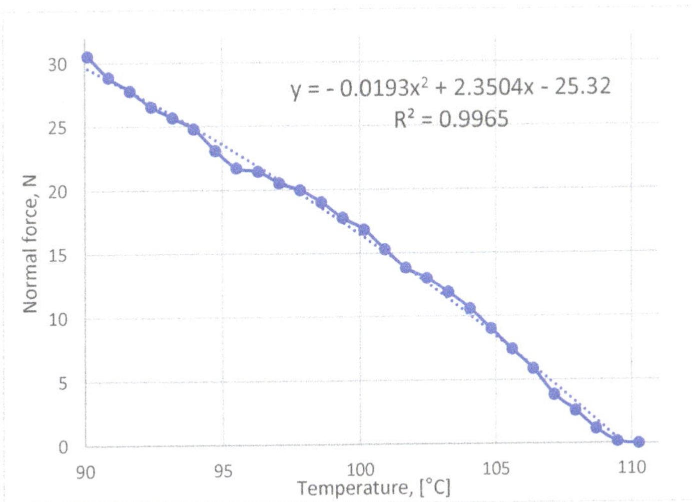

Figure 6. Normal force vs. temperature diagram as per rheometer test.

After finding the magnitude of the normal force caused by the shrinkage of wax, it is possible to evaluate the stress magnitude within the body of the wax sample. To accomplish the task, a Finite Element Analysis (FEA) can be used to implement the static structural module of the SolidWorks 2016 Simulation software. The boundary conditions for the static structural analysis can be established as follows. The wax sample in the form of a thin film having a circular shape with a diameter of 50 mm and thickness of 0.5 mm is subjected to a normal force of 30.57 N. This force is derived from the actual experiment described earlier and its direction coincides with the one identified experimentally. It should be noted that due to the circular shape of the model, it is symmetrical and, thus, for the FEA only a quarter of the model was considered as shown in Figure 7 and the magnitude of applied normal force was reduced accordingly. The wax sample is represented with a blue color and a fixed support indicated with green arrows was applied to the bottom side of the film. The purple arrows shown in Figure 7 represent the normal force. It should be noted that despite the plate having a conical shape, it is assumed flat due to a small truncate angle (1°). It should be noted that additional assumptions were made to simplify the analysis:

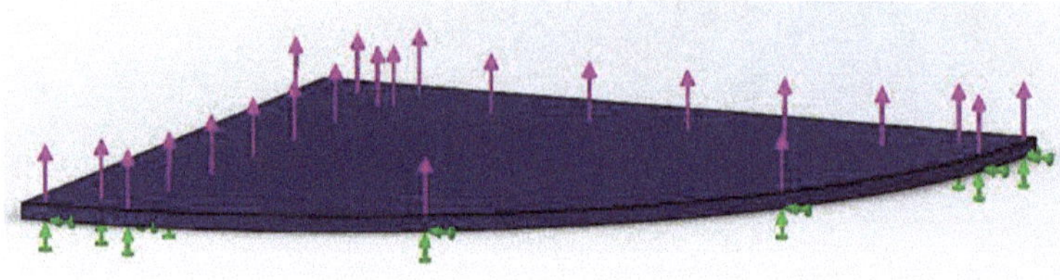

Figure 7. Boundary conditions for the FEA.

(a) The thermal effects including polymer relaxation were neglected. During solidification of the polymer, the mechanical properties change significantly. Although this type of transformation is neglected, it is assumed that all the stresses due to the shrinkage of the material accumulate as residual stresses in the solidified body.
(b) The polymer is assumed to have isotropic properties.
(c) During the FFF, the material is deposited in a layer-by-layer fashion. However, in FEA the entire body of thin film is assumed to be shrinking at once. Since it is observed that the warpage manifests with some delay in time for 3D printed parts, the stress accumulation within the entire body is likely the major factor for warpage occurrence.

A young's modulus of 151.5 MPa and Poisson's ratio equal to 0.4 were assigned as material properties for the simulation. Finally, after assigning the material properties for wax, the model can be meshed. For this analysis, a 0.15 mm mesh size was chosen, and each mesh cell represents a tetrahedral solid body. The results of the FEA in the form of Von Mises's stress are shown in Figure 8. Figure 8 shows the lateral component of Von Mises's stress pointing toward the horizontal Z-axis. As can be seen from Figure 8, a maximum stress of 0.037 MPa is present in the thin film of wax, and it is responsible for thermal contraction since its direction of action is in the horizontal plane.

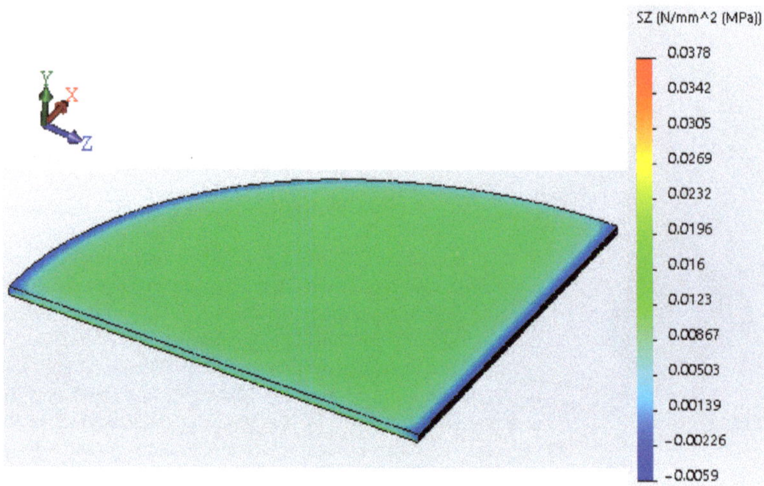

Figure 8. FEA results show a stress component in the lateral direction (load = 30.57 N).

Figure 9 shows the FEA results when the higher force of 37.81 N was applied, which corresponds to the plate temperature of 40 °C. This value of normal force was found by extrapolation using the line equation shown in Figure 5. This is carried out to identify the magnitude of the force at lower temperatures. As was mentioned earlier, due to some restrictions, the experiment was run at a temperature range of 90–110 °C. However, we are interested in finding the force value at the range of 40–110 °C where the material starts to respond to heating as was identified by the DSC analysis [8]. Therefore, the line equation was useful to identify the magnitude of the normal force when a wider temperature range is used. From Figure 9, it is visible that the maximum stress magnitude in the lateral direction increased to 0.046 MPa.

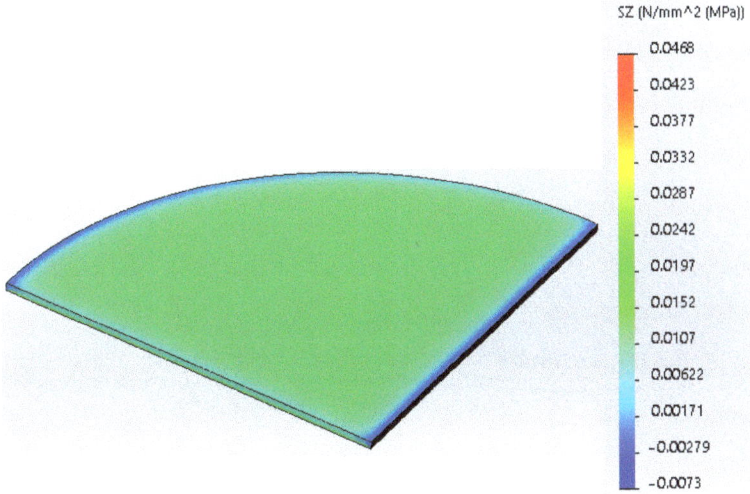

Figure 9. FEA results show a stress component in the lateral direction (load = 37.81 N).

From found values of maximum stresses, it is evident that a 0.5 mm thick layer of deposited wax can produce a stress magnitude that is approximately 40 times less than the Yield strength (σ_y) of the material which is equal to 1.9 MPa [22]. However, since the printed part represents a multi-layer structure, stress accumulation is inevitable. Thus, by dividing the σ_y by the maximum thermal stress of 0.046 MPa, we obtain a value of 41. Since the thickness of the studied layer is equal to 0.5 mm, 0.5×41 rounds up to 20.5 mm. Therefore, the stress accumulation up to the critical value of σ_y is possible when the thickness of the part reaches 20.5 mm. This is possible provided that the bonding force between the material and the print bed has been overcome. It should be noted that for rigid polymers such as ABS, much thinner parts have experienced the warpage issue [12].

The results of the FEA show that the magnitude of the residual stresses is 41 times less than σ_y of the wax material. Therefore, for the residual stresses to reach critical levels, the 3D-printed sample should be of a thicker size. However, it has been demonstrated that warpage occurs even in thin samples [15]. Thus, it is expected that the thermal stresses must be higher than those identified by the FEA method. Moreover, it should be noted that the FEA analysis applied several assumptions. For instance, the plate geometry was considered flat, whereas the real shape of the upper plate is conical as shown in Figure 2. The analytical model can be used to account for the geometry of the measuring system. For example, Figure 10 shows the force balance in the measuring system and the wax sample under thermal stresses. As can be seen from Figure 10, the forces developed in the wax sample due to shrinkage forces F_s, are balanced by the normal forces F_n developed in the measuring system of the rheometer. Since the normal forces are known from the experiment, it is not difficult to compute the thermal stresses. Here, Equation (1) can be used to identify the shrinkage forces Fs:

$$F_n = F_s \, Sin(truncate\ angle^{\circ}) \quad (1)$$

where the truncate angle is 1°. The thermal stresses σ_W present in the wax sample can be computed by dividing the computed F_s values by the area A_c that envelopes the inclined portion of the cone: $\sigma_W = F_s/A_c$.

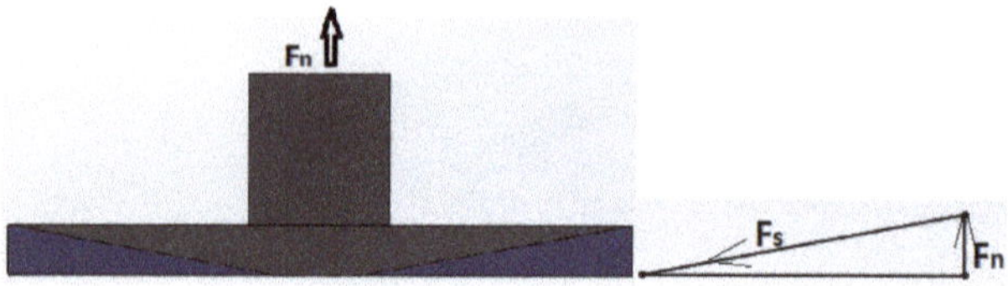

Figure 10. Force balance in cone plate measuring system: schematics of the cp measuring system (**left**); force balance diagram (**right**).

Analytically found values of thermal stresses are shown in Table 3 along with findings from the FEA. From Table 3, it is clear that the calculated values of thermal stresses are much higher compared to the values found by the FEA. Thus, the plastic deformation caused by the warpage might occur in samples considerably thinner than 20.5 mm. The magnified side view of the 3D printed wax benchmark having an overall height of 20 mm is shown in Figure 11. From Figure 11, it is clear that the vertical walls at two ends of the sample have a wavy shape. That is the evidence of the occurrence of plastic deformation caused by the thermal shrinkage of the wax. By looking at the bottom of the part, it is seen that the first layers have not undergone a size contraction. However, as new layers are added, the shrinkage becomes more evident. Once the residual stresses diminish owing to the stress relaxation, newly deposited layers are restricted from shrinking by the bonding forces that hold the adjacent layers together. As the residual stresses build up more, the size contraction takes place again shaping the wavy pattern of the wall. Therefore, it can be concluded that the analytically found stress values are more accurate. Considering the stress accumulation phenomenon, the sample having several layers of thickness, each being equal to 0.5 mm, is enough to initiate the plastic deformation of the wax owing to the weak mechanical properties of the latter (σ_y = 1.9 MPa).

Table 3. Magnitudes of stresses identified analytically.

Temperature Range, [C]	Thermal Stress, σ_W [MPa]	Max Stress Magnitude Found by the FEA, [MPa]
110–90	0.94	0.037
110–40	1.17	0.046

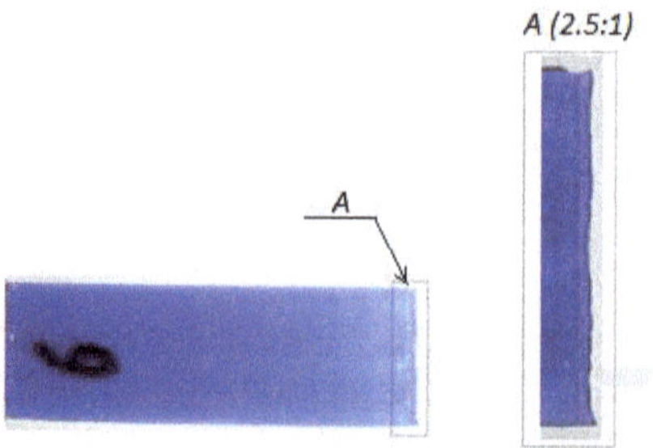

Figure 11. 3D printed sample (**left**) with a magnified view of the wall (**right**).

3.2. Results of Taguchi Analysis

A Signal-To-Noise (S/N) ratio identified by the Taguchi approach is a parameter that was used to evaluate the effect of input parameters on the warpage magnitude. "The lower the better" response was chosen because the preference is to have a low warpage value. An Equation (2) was used to find the S/N ratio as follows:

$$\frac{S}{N} = -10\log_{10}\left[\frac{1}{n}\sum_{i=1}^{n} y_i^2\right] \quad (2)$$

where n is the number of repetitions and y is the characteristic under consideration. The main effect plot for the S/N ratio is shown in Figure 10 and the corresponding response table for S/N ratios is shown in Table 4. The main effect plots evidence that the warpage degree is mostly influenced by the nozzle speed and bed temperature. In contrast, the nozzle temperature was found to be insignificant. As can be seen from Figure 11, the maintenance of higher bed temperature is important to minimize the level of warpage. That is because the low temperature of the bed increases the temperature gradient between the molten polymer extrudate and the environment. Since warpage manifests as the detachment of the part's outer edges from the print bed, the very first layer of the polymer must have a strong bond with the print bed [23]. This can be achieved by increasing the bed temperature.

Table 4. Response table for the S/N ratios.

Level	Nozzle Temperature	Nozzle Speed	Bed Temperature
1	−32.36	−51.78	−48.07
2	−39.69	−27.19	−30.61
3	−37.27	−30.36	−30.65
Delta	7.33	24.59	17.47
Rank	3	1	2

FFF at low and high speeds is not desirable, thus, a middle value around 45 mm/s is a better choice according to the main effect plots. It has been demonstrated by Kwon et al. [24] that the influence of bed temperature on the bonding quality is more prominent than the deposition rate. However, in the case of FFF of soft polymers, it is visible that the nozzle speed has slightly higher importance. Another important aspect is the interlayer bonding. The stronger interlayer bonding is achieved by heat-driven diffusion according to Gao et al. [25]. In weakly connected layers, the level of porosity is higher, therefore, less dense parts have weak mechanical properties [26]. To increase the level of bonding, the layers of polymer should be deposited at a higher rate to ensure a strong connection to the previously placed layer. In this regard, a high printing temperature is required to ensure higher strength of the final part [27]. On the other hand, it can be inferred that the thermal stresses transfer more efficiently between layers if the layers are densely packed [27]. Therefore, 3D printing at high speeds influences the level of warpage and it is supported by the main effect plots shown in Figure 12.

Analysis of variance for S/N ratios is shown in Table 5. It can be concluded that the nozzle speed has the highest impact on the warpage magnitude with a contribution of 59%. After that, the bed temperature appears to be important too. p and F values of the ANOVA test at a 95% confidence range are used to observe the relevance of each process parameter. A 'p' value of less than 0.05 indicates that the parameters under consideration had a substantial impact on the experimental result. It has been determined that the most important input parameter for warpage occurrence is the printing speed with a 'p' value of 0.045. Additionally, the print bed temperature can be regarded as significant too because its p value is only slightly larger than 0.05.

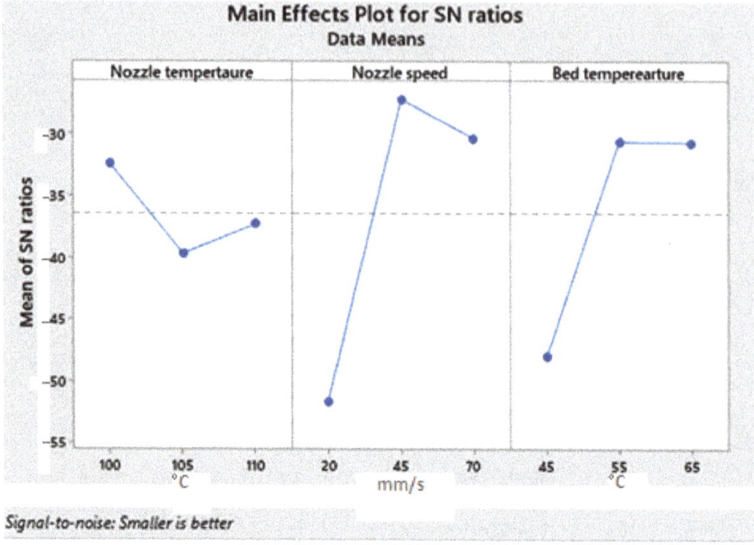

Figure 12. Main effects plot for S/N regarding part's warpage magnitude.

Table 5. ANOVA S/N ratios.

Source	DF	Adj SS	Adj MS	F	p	Significance	Contribution
Nozzle temperature	2	83.76	41.88	1.65	0.378	Nonsignificant	4
Nozzle speed	2	1073.55	536.78	21.10	0.045	Significant	59
Bed temperature	2	608.53	304.27	11.96	0.077	Nonsignificant	32
Residual Error	2	50.89	25.44				
Total	8	1816.73					

4. Conclusions

The results of experimental work performed in this study have provided some insights into the warpage phenomenon that occurs during FFF manufacturing soft wax polymer. The majority of previous research in this area concerns rigid plastics such as ABS or PLA. However, the issue of warpage in soft polymers is not studied in detail. Due to the soft nature of the wax polymer, it was possible to perform a temperature sweep test on a thin film wax sample utilizing rheometer equipment. Temperature sweep test allows for the measurement of normal forces that occur as a response to the thermal shrinkage of the polymer. These forces develop within the measuring system of the rheometer as a thin sample of wax shrinks within the range of temperatures between 110 and 40 °C. The results of the thermo-mechanical experiment were used in the FEA and analytical computation to identify the shrinkage stresses that ultimately cause the warpage in parts manufactured by FFF. It was found that the stress values computed by the analytical method yielded more accurate results. With the level of thermal stress equal to 1.17 MPa, it is estimated that a part having an elongated shape with a small thickness in a horizontal direction will be subject to plastic deformation, which might manifest as a shape distortion.

In addition to the work dedicated to the estimation of shrinkage stresses, Taguchi's design of experiments was performed to identify the significant process parameters of FFF that contribute to the warpage. It was discovered that the impact of 3D printing velocity is significant and to mitigate the warpage issue, the printing speed should be in the vicinity of 40 mm/s. Another significant factor is the print bed temperature. 3D printing using a lower temperature of the print bed is not desirable since it contributes to rapid cooling of the extrudate leading to weak adhesion of the material to the print bed. As a result, the shrinkage stresses easily overcome the bonding force between the polymer and the print bed.

Author Contributions: Conceptualization, M.M. and M.H.A.; methodology, M.M.; investigation, M.M.; resources, M.H.A.; data curation, M.M.; writing—original draft preparation, M.M.; writing—review and editing, M.M., M.H.A. and E.S.; supervision, M.H.A.; funding acquisition, M.H.A. All authors have read and agreed to the published version of the manuscript.

Funding: This research is funded by the Ministry of Education and Science (MES), Republic of Kazakhstan, Grant No AP13068039. The source code of funding is 055.01.01.

Institutional Review Board Statement: Not applicable.

Data Availability Statement: The data presented in this study are available on request from the corresponding author.

Conflicts of Interest: The authors declare no conflicts of interest.

References

1. Chhabra, M.; Singh, R. Rapid casting solutions: A review. *Rapid Prototyp. J.* **2011**, *17*, 328–350. [CrossRef]
2. Raoufi, M.A.; Bazaz, S.R.; Niazmand, H.; Rouhi, O.; Asadnia, M.; Razmjou, A.; Warkiani, M.E. Fabrication of unconventional inertial microfluidic channels using wax 3D printing. *Soft Matter* **2020**, *16*, 2448–2459. [CrossRef] [PubMed]
3. Kuo, C.-C.; Tasi, Y.-R.; Chen, M.-Y.; Yan, Z.-Y. Development of a cost-effective technique for batch production of precision wax patterns using 3D optical inspection and rapid tooling technologies. *Int. J. Adv. Manuf. Technol.* **2021**, *117*, 3211–3227. [CrossRef]
4. Mandolini, M.; Sartini, M.; Favi, C. Techno-Economic Analysis for Comparing Stereolithography and Wax Injection for Pattern Manufacturing in Investment Casting. In Proceedings of the ASME International Mechanical Engineering Congress and Exposition, Columbus, OH, USA, 30 October 2022.
5. Mukhtarkhanov, M.; Perveen, A.; Talamona, D. Application of Stereolithography Based 3D Printing Technology in Investment Casting. *Micromachines* **2020**, *11*, 946. [CrossRef] [PubMed]
6. Badanova, N.; Perveen, A. Study of SLA Printing Parameters Affecting the Dimensional Accuracy of the Pattern and Casting in Rapid Investment Casting. *J. Manuf. Mater. Process.* **2022**, *6*, 109. [CrossRef]
7. Wine, L.N. Applications of Investment Casting as a Manufacturing Process for Injection Mold Tooling. Bachelor's Thesis, Massachusetts Institute of Technology, Boston, MA, USA, 2022.
8. Mukhtarkhanov, M.; Shehab, E.; Ali, M. Process Parameter Optimization for 3D Printed Investment Casting Wax Pattern and Its Post-Processing Technique. *Appl. Sci.* **2022**, *12*, 6847. [CrossRef]
9. Szabó, L.; Deák, G.; Nyul, D.; Kéki, S. Flexible Investment Casting Wax Patterns for 3D-Printing: Their Rheological and Mechanical Characterizations. *Polymers* **2022**, *14*, 4744. [CrossRef] [PubMed]
10. Samy, A.A.; Golbang, A.; Harkin-Jones, E.; Archer, E.; Tormey, D.; McIlhagger, A. Finite element analysis of residual stress and warpage in a 3D printed semi-crystalline polymer: Effect of ambient temperature and nozzle speed. *J. Manuf. Process.* **2021**, *70*, 389–399. [CrossRef]
11. Armillotta, A.; Bellotti, M.; Cavallaro, M. Warpage of FFF parts: Experimental tests and analytic model. *Robot. Comput. Integr. Manuf.* **2018**, *50*, 140–152. [CrossRef]
12. Zhang, Y.; Chou, K. A parametric study of part distortions in fused deposition modelling using three-dimensional finite element analysis. *Proc. Inst. Mech. Eng. B J. Eng. Manuf.* **2008**, *222*, 959–968. [CrossRef]
13. Syrlybayev, D.; Zharylkassyn, B.; Seisekulova, A.; Perveen, A.; Talamona, D. Optimization of the warpage of fused deposition modeling parts using finite element method. *Polymers* **2021**, *13*, 3849. [CrossRef] [PubMed]
14. Kuo, C.-C.; Wu, Y.-R.; Li, M.-H.; Wu, H.-W. Minimizing warpage of ABS prototypes built with low-cost fused deposition modeling machine using developed closed-chamber and optimal process parameters. *Int. J. Adv. Manuf. Technol.* **2019**, *101*, 593–602. [CrossRef]
15. Liu, X.; Li, S.; Liu, Z.; Zheng, X.; Chen, X.; Wang, Z. An investigation on distortion of PLA thin-plate part in the FFF process. *Int. J. Adv. Manuf. Technol.* **2015**, *79*, 1117–1126.
16. Alsoufi, M.S.; Alhazmi, M.W.; Suker, D.K.; Alghamdi, T.A.; Sabbagh, R.A.; Felemban, M.A.; Bazuhair, F.K. Experimental characterization of the influence of nozzle temperature in FFF 3D printed pure PLA and advanced PLA+. *Am. J. Mech. Eng.* **2019**, *7*, 45–60. [CrossRef]
17. Fitzharris, E.R.; Watanabe, N.; Rosen, D.W.; Shofner, M.L. Effects of material properties on warpage in fused deposition modeling parts. *Int. J. Adv. Manuf. Technol.* **2018**, *95*, 2059–2070. [CrossRef]
18. Akbar, I.; El Hadrouz, M.; El Mansori, M.; Lagoudas, D. Continuum and subcontinuum simulation of FFF process for 4D printed shape memory polymers. *J. Manuf. Process.* **2022**, *76*, 335–348. [CrossRef]
19. Verma, N.; Awasthi, P.; Gupta, A.; Banerjee, S.S. Fused Deposition Modeling of Polyolefins: Challenges and Opportunities. *Macromol. Mater. Eng.* **2023**, *308*, 2200421. [CrossRef]

20. Das, A.; Gilmer, E.L.; Biria, S.; Bortner, J. Importance of Polymer Rheology on Material Extrusion Additive Manufacturing: Correlating Process Physics to Print Properties. *ACS Appl. Polym. Mater.* **2021**, *3*, 1218–1249. [CrossRef]
21. Li, Z.; Liu, Y.; Liang, Z.; Liu, Y. The Influence of Fused Deposition Modeling Parameters on the Properties of PA6/PA66 Composite Specimens by the Taguchi Method and Analysis of Variance. *3D Print. Addit. Manuf.* **2023**. *ahead of print*. [CrossRef]
22. Mukhtarkanov, M.; Shehab, E.; Tanveer, M.H.; Araby, S.; Ali, M.H. Investigation of Multi-material Composite Parts Manufactured by Multi-extrusion FFF Printer. In *Flexible Automation and Intelligent Manufacturing: The Human-Data-Technology Nexus, Proceedings of FAIM 2022, Detroit, MI, USA, 19–23 June 2022*; Lecture Notes in Mechanical Engineering; Kim, K.Y., Monplaisir, L., Rickli, J., Eds.; Springer: Cham, Switzerland, 2022.
23. Yin, J.; Lu, C.; Fu, J.; Huang, Y.; Zheng, Y. Interfacial bonding during multi-material fused deposition modeling (FFF) process due to inter-molecular diffusion. *Mater. Des.* **2018**, *150*, 104–112. [CrossRef]
24. Kwon, N.; Deshpande, H.; Hasan, M.K.; Darnal, A.; Kim, J. Multi-Ttach: Techniques to Enhance Multi-Material Attachments in Low-cost FFF 3D Printing. In Proceedings of the SCF 2021: ACM Symposium on Computational Fabrication, Virtual Event, 28 October 2021.
25. Gao, X.; Zhang, D.; Qi, S.; Wen, X.; Su, Y. Mechanical properties of 3D parts fabricated by fused deposition modeling: Effect of various fillers in polylactide. *J. Appl. Polym. Sci.* **2019**, *136*, 47824. [CrossRef]
26. Dawoud, M.; Taha, I.; Ebeid, S.J. Mechanical behavior of ABS: An experimental study using FFF and injection molding techniques. *J. Manuf. Process.* **2016**, *21*, 39–45. [CrossRef]
27. Ansari, A.A.; Kamil, M. Effect of print speed and extrusion temperature on properties of 3D printed PLA using fused deposition modeling process. *Mater. Today Proc.* **2021**, *45*, 5462–5468. [CrossRef]

polymers

Review

Fabrication and Performance of Continuous 316 Stainless Steel Fibre-Reinforced 3D-Printed PLA Composites

Alison J. Clarke [1,*], Andrew Dickson [2] and Denis P. Dowling [1]

[1] I-Form Centre, School of Mechanical & Materials Engineering, University College Dublin, Belfield, D04 C1P1 Dublin, Ireland; denis.dowling@ucd.ie
[2] Infraprint, Nova UCD, Belfield, D04 C1P1 Dublin, Ireland; andrew.dickson@infraprint.com
* Correspondence: alison.clarke1@ucdconnect.ie; Tel.: +353-(0)86-384-7557

Abstract: This study investigates the feasibility of 3D printing continuous stainless steel fibre-reinforced polymer composites. The printing study was carried out using 316L stainless steel fibre (SSF) bundles with an approximate diameter of 0.15 mm. This bundle was composed of 90 fibres with a 14 µm diameter. This fibre bundle was first coated with polylactic acid (PLA) in order to produce a polymer-coated continuous stainless steel filament, with diameters tailored in the range from 0.5 to 0.9 mm. These filaments were then used to print composite parts using the material extrusion (MEX) technique. The SSF's volume fraction (V_f) was controlled in the printed composite structures in the range from 4 to 30 V_f%. This was facilitated by incorporating a novel polymer pressure vent into the printer nozzle, which allowed the removal of excess polymer. This thus enabled the control of the metal fibre content within the printed composites as the print layer height was varied in the range from 0.22 to 0.48 mm. It was demonstrated that a lower layer height yielded a more homogeneous distribution of steel fibres within the PLA polymer matrix. The PLA-SSF composites were assessed to evaluate their mechanical performance, volume fraction, morphology and porosity. Composite porosities in the range of 2–21% were obtained. Mechanical testing demonstrated that the stainless steel composites exhibited a twofold increase in interlaminar shear strength (ILSS) and a fourfold increase in its tensile strength compared with the PLA-only polymer prints. When comparing the 4 and 30 V_f% composites, the latter exhibited a significant increase in both the tensile strength and modulus. The ILSS values obtained for the steel composites were up to 28.5 MPa, which is significantly higher than the approximately 13.8 MPa reported for glass fibre-reinforced PLA composites.

Keywords: 3D printing; thermoplastic polymers; mechanical properties

Citation: Clarke, A.J.; Dickson, A.; Dowling, D.P. Fabrication and Performance of Continuous 316 Stainless Steel Fibre-Reinforced 3D-Printed PLA Composites. *Polymers* **2024**, *16*, 63. https://doi.org/10.3390/polym16010063

Academic Editor: Rong-Ho Lee

Received: 21 November 2023
Revised: 15 December 2023
Accepted: 19 December 2023
Published: 24 December 2023

1. Introduction

Additive manufacturing (i.e., 3D printing) enables components to be fabricated by adding material layer by layer. Several 3D printing materials are available to fabricate components including polymers, metals, ceramics and composites [1–12]. Polymer components which are 3D printed exhibit relatively poor mechanical performance compared with those obtained by injection molding, and hence the interest in adding reinforcing materials [13]. This study investigates the use of a continuous steel fibre bundle as reinforcement for 3D-printed polymer composites.

One of the most widely used 3D printing techniques is material extrusion (MEX) [14–16]. Polymer composites are fabricated through the addition of fibres (short or continuous) or alternatively powder particles, beads and pellets [13,17–20]. The reinforcing fibres available for MEX-printed composite reinforcing fibres are vast and include glass, metal, carbon and basalt [21–23]. The most commonly used thermoplastic feedstocks include polylactic acid (PLA), polycarbonate (PC), polyamide (PA or nylon) and acrylonitrile butadiene styrene (ABS) [13,16–20,24].

The two approaches which are routinely used to fabricate composites through MEX are called ex situ prepreg or in situ fusion [25–27]. The latter method is the one most widely used for the fabrication of continuous fibres and involves direct fibre integration into the print nozzle during 3D printing [12,13,28,29]. Ex situ prepreg production is a two-part process involving fabricating the composite filament and then using the composite filament in the 3D printing process [13]. In situ fusion provides a more rapid route to fabricating composite parts compared with ex situ prepreg [30]. However, due to the short dwell time, along with insufficient pressure from the nozzle and large temperature gradients, there can be issues with poor infusion and adhesion of the polymer matrix around the fibres [27,31]. An example of the ex situ prepreg process was investigated by Hu et al. [27] for the printing of PLA and continuous carbon fibre filaments. Chen et al. [32] also used the ex situ prepreg method to 3D print PLA containing continuous glass fibre (PLA-cGF). It was reported that there was a high level of impregnation of the PLA between and into the fibres.

Heidari-Rarani et al. [33] reported on the development of an in situ fusion 3D printer extruder head designed to reduce nozzle blockage during the fabrication of filaments of PLA reinforced with continuous carbon fibre (PLA-cCF). To improve PLA and cCF bonding, the fibres were pretreated or sized with a PVA solution, resulting in composite parts with a volume fraction of 28.2% and void content of 9.1%. Li et al. [34] also investigated the 3D printing of PLA-cCF, including the use of PLA sizing agents on the fibres. The continuous carbon fibre bundle contained up to a maximum of 1000 individual fibres. The resulting composite exhibited a volume fraction (V_f) of 34% in a unidirectional printing pattern, along with a tensile strength of up to 91 MPa. Rimašauskas et al. [35] investigated using polymer solutions of PLA, PC and ABS for pre-preg sizing to dry cCF at different concentrations by using in situ fusion 3D printing with respective 1.75 mm filaments. A 10% pre-preg concentration resulted in a 25.2 V_f% and tensile strength of 165 MPa. Maqsood et al. [36] used the in situ fusion method along with the PLA pre-impregnated cCF tow to reinforce the PLA (PLA-cCF) and a cCF tow to reinforce the PLA containing short carbon fibres (PLA-SCF-cCF). The resultant composites exhibited 18.5 V_f% and 21.8 V_f% respectively. The composite with the PLA-continuous carbon fibre exhibited a tensile strength of up to 245.4 MPa, but the PLA-SCF-cCF composite was lower at 227.6 MPa.

Caminero et al. [37] evaluated the effects on the interlaminar bonding due to layer thickness and various fibre volume fractions of 3D-printed nylon reinforced with continuous fibres of Kevlar, glass and carbon supplied by Markforged. The interlaminar shear strength was evaluated for nylon-Kevlar, nylon-glass and nylon-CF, with resultant strengths of 14.3, 21.0 and 31.9 MPa, respectively. The fibre content plays an important role in determining the properties of MEX composite filaments, such as with the tensile strength generally increasing with increasing fibre content [13]. One difficulty, however, is that composite filaments, with a high fibre content, can be quite difficult to print, arising from issues with nozzle clogging, in addition to the excessive viscosity of the melted composite filament [24,33,38–42].

There have been very few reports on the incorporation of continuous steel fibres into 3D-printed composites. Quan et al. [43] evaluated the use of a laminating approach, in which an epoxy sandwich containing SSF with carbon fibre was fabricated. The addition of the steel fibre was reported to yield a significant increase in interlaminar and fracture toughness compared with carbon fibre-reinforced plastics. Ibrahim et al. [18,30] and Saleh et al. [19] successfully 3D printed continuous wire polymer composites for sensor applications. These authors used in situ fusion to combine 75 μm nickel-chromium wires and 75 μm copper wire with PLA filament. Only a small increase in the mechanical properties of the PLA was reported through incorporation of the metal wires.

Hamidi et al. [44] investigated 3D printing PLA and bright finish conductive copper wire, as well as the polymer combined with spring-back 302/304 stainless steel wire with diameters of 0.130 mm and 0.178 mm, respectively, as reinforcement for bioinspired joint fabrication. Both samples were printed using in situ fusion with filament 1.75 mm in diameter. It was reported that the copper wire broke when printed in a concentric pattern due to the traction force being too large. Increasing the copper wire's V_f above 0.4% resulted in little improvement in the tensile modulus, and the steel wire showed poor adhesion and bonding with the PLA during tensile testing. Gunes et al. [45] evaluated the tensile strength of nylon reinforced with continuous stainless steel wire (50 μm in diameter) and demonstrated an increase in nylon tensile strength of 7.6 times when 3D printed in a concentric pattern.

The focus of the current study is to investigate the feasibility of printing continuous SSF-reinforced polymer composites. A particular focus of the investigation is to achieve a high fibre volume fraction along with low porosity within the printed composite parts in order to enhance mechanical performance.

2. Materials and Methods

2.1. Materials

The starting points for this study involved the fabrication of stainless steel-reinforced polymer filaments. A continuous 316 L stainless steel fibre (SSF) bundle (Figure 1a) was obtained from NV Bekaert SA (Brussels, Belgium) [31,46]. The SSF was fabricated with a 'bundle wire drawing' manufacturing process, which resulted in a hexagonal fibre cross-section [47,48]. The fibre bundle had a diameter of approximately 0.15 mm and consisted of 90 fibres per bundle, each with a diameter of 14 μm. The bundle had a linear density of 110 decitex (TEX), torsion per cm of 1 and Young's modulus of 200 GPa [46]. The elemental composition of the steel fibre was determined using energy-dispersive X-ray spectroscopy (EDAX), and the results are detailed in Table 1.

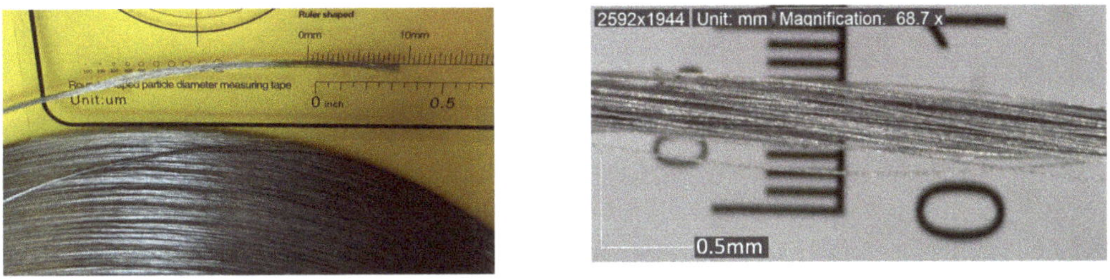

(**a**) (**b**)

Figure 1. Stainless steel fibre bundle: (**a**) spool and (**b**) 90 fibres per bundle.

The PLA polymer used to coat the steel to form the filament was obtained as pellets from Nature Works, with a product reference of Ingeo^{TM} Biopolymer D4043D [1]. The PLA pellets were dried at 55 °C for a minimum of 24 h prior to extrusion.

Table 1. EDAX analysis elemental composition of the stainless steel fibre.

Element	Fe	Cr	Ni	Mo	Si	Al
Weight %	69.6	17.9	5.6	2.1	0.5	0.4

2.1.1. PLA-Reinforced SSF Filament

The fabrication of a PLA-SSF filament was carried out using a 3devo laboratory-scale filament maker (Utrecht, Netherlands) [49]. This was modified to facilitate the introduction of the fibre into the molten polymer during filament extrusion, as shown in Figure 2. The SSF was introduced into the molten polymer using a custom-designed extrusion die similar to those described in the literature [50–52]. The die design minimised the forces on the fibre bundle during co-extrusion by helping to entrain the fibre bundle in the polymer flow. As highlighted by previous authors, in this die design, where the fibre bundle enters the polymer flow, there is a region of maximum velocity, with the lowest inter-material shearing and pressure as shown in Figure 2c [53–57]. As the SSF bundle travels through the die, it undergoes preheating before entering the molten polymer, decreasing the temperature gradient between the materials [58].

Fans cool the polymer-coated SSF bundle after extrusion. We note that rapid cooling can cause polymer melt fracturing or partially solidify the polymer in the extrusion die, resulting in teardrop shapes forming along the SSF [51]. DevoVision-win32-v0.2.0, used in conjunction with the 3devo filament maker, recorded the filament diameter after the cooling zone and before spooling. It was observed that a reduction in the filament diameter below 0.4 mm was indicative of issues, such as poor polymer adherence to the SSF.

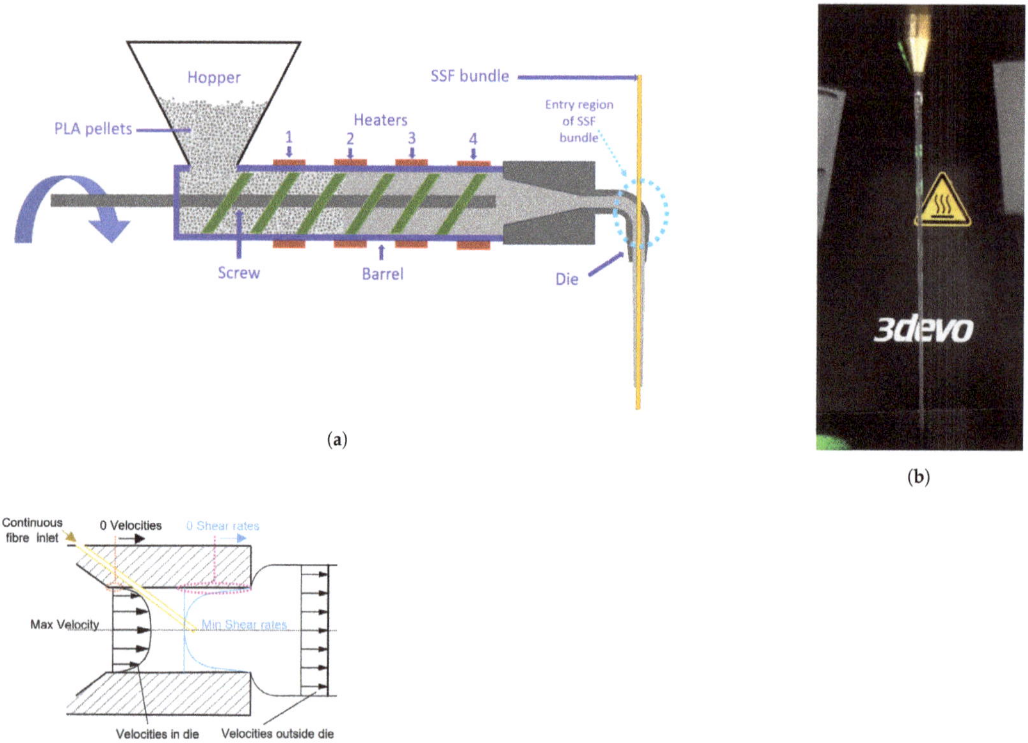

(**a**) (**b**) (c)

Figure 2. Schematic of 3devo filament making. (**a**) Schematic with the fibre introduction region indicated by the blue dashed ovals. (**b**) Photograph of the PLA-SSF filament co-extrusion. (**c**) Velocity and shearing profile of a non-Newtonian polymer-fibre co-extrusion die.

The 3devo filament maker processing parameters used for fabrication of the PLA-SSF filaments are given in Table 2. In order to achieve lower layer heights, it was found to be necessary to decrease the filament diameters from 0.9 to 0.5 mm, as demonstrated in Table 2. Reducing the extrusion speed slowed the extrusion barrel revolution and velocity of the polymer entering and exiting the die.

Table 2. The 3devo filament maker processing parameters. Note that T_1–T_4 relate to the extrusion barrel heater temperatures, as indicated in Figure 2a.

Filament	Filament Diameter (mm)	Fans (%)	Extruder Speed (rpm)	T_1 (°C)	T_2 (°C)	T_3 (°C)	T_4 (°C)
PLA-SSF	0.90	45	2.7	177	187	186	177
PLA-SSF	0.60–0.70	20	2.6	177	187	186	170
PLA-SSF	0.50–0.65	33	2.2	179	186	193	191

2.1.2. Three-Dimensional Printing Continuous PLA-SSF Composite Components

The 3D printing using the PLA-SSF filaments was carried out using a modified Anycubic i3 Mega polymer extrusion printer (Shenzhen, China). The modifications included design change innovations to the polymer-continuous fibre printing head nozzle in order to facilitate the removal of *'excess'* polymer during printing. This was achieved by incorporating a *'polymer pressure vent'* into the print head nozzle, as shown in Figure 3. Excess polymer around the continuous stainless steel wire was allowed to exit through this nozzle 1 mm in diameter during printing. This excess polymer release vent was positioned at the end of the hot zone in the print head as at this location, the polymer was likely to exhibit a lower viscosity. The polymer material was routinely removed during printing to prevent it from falling onto the print surface. As the distance from the print head to the substrate was reduced, there was a change in the level of pressure in the print head, which was relieved by the removal of excess polymer through the polymer pressure vent. Thus, the rate of removal of excess polymer was directly related to the print layer height, with higher levels of removal occurring at lower heights. This thus facilitated control of the metal fibre content within the printed composites. One of the advantages of the pressure release outlet was that it helped to reduce filament failure during printing. By overcoming the backpressure, it was also found to improve the composite surface finish.

An additional printer design innovation is the flattened print head illustrated in Figure 3. This 4.5 mm in diameter 'flat' nozzle head appeared to have an 'ironing' effect on the print, resulting in an increase in the head dwell time during printing, which should facilitate greater polymer diffusion between the fibres in the steel bundle. A further modification was the addition of a soft-wheeled filament feeder, and this was found to help reduce the level of continuous filament breakage during printing.

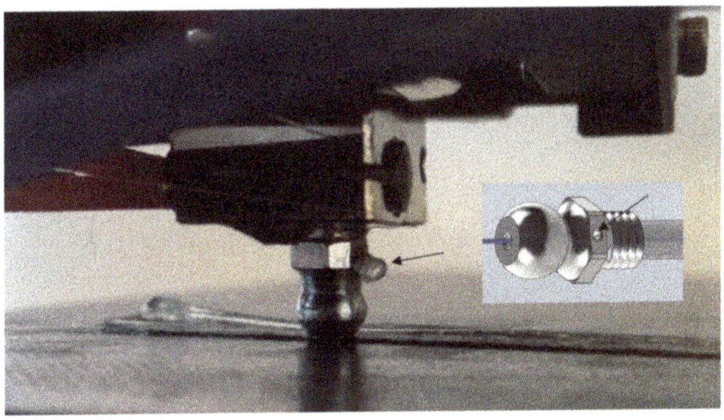

Figure 3. Photograph of the printing head with an insert showing a schematic of the 1 mm in diameter polymer pressure vent (arrow), which facilitated the removal of excess polymer during printing. Note also the 4.5 mm flat area on the print head, which was found to facilitate 'ironing' of the part during printing.

The print geometries were created using a computer-aided design (CAD) file (Solidworks 2021 software, version 23.3.0.0059,) and then exported as a stereolithography (STL) file. The open-source slicing software PrusaSlicer v.2.6.1 was used to slice the STL file generating the G-code for 3D printing. All test samples were 3D printed in a unidirectional continuous pattern (0°). To achieve a continuous printing path with no cutting of the fibres between layers or at the end of each path, only perimeters were selected in the slicing software v.2.6.1.

Printing at layer heights below the PLA-SSF filament diameter (minimum of 0.5 mm) meant there was an excess of polymer in the system, as the maximum quantity of polymer on the print bed was deposited. The excess polymer was either deposited on the surface, resulting in a poor surface finish, or returned up the printing nozzle, generating backpressure and clogging.

A variety of processing parameters and geometry modifications were optimised. This included adjusting the temperatures of the print head heating, print bed and cooling to manage temperature gradients. Additionally, the printing path height and hatch spacing (Figure 4a) were altered, and the deposition speeds of the perimeters and between layers were varied. The print heights were varied between 0.48 and 0.22 mm. In the case of the lowest layer height of 0.22 mm, the print head temperature was increased, as demonstrated in Table 3. The elevated temperature was required to reduce the viscosity of the polymer and to facilitate better flow of the molten polymer as the PLA-SSF exited the printing head. The higher temperature was also found to prevent the polymer from cooling prematurely and solidified within the printing head.

The relatively high thermal conductivity of the stainless steel was likely to assist the composite in retaining a temperature close to the bed temperature during printing, giving the SSF additional dwell time. Thus, the coupling of the composite filament feed rate with the deposition rate and the deposited material's elevated temperature were likely to reduce internal stresses as well as warping.

Experimental 3D prints carried out with various print radii and angles demonstrated that the PLA-SSF filament could achieve filament turns of up to 180° without breakage or degradation of the fibres. An example is shown in Figure 4a. This is in contrast with the behaviour of continuous carbon fibre filaments, which were reported to fail when high filament turn angles are used [37].

Figure 4b shows a close-up at the end of a single tensile sample. Where each line of a continuous SSF bundle's continuous turning path is displayed, a 180° turn is evident along the centre of the print due to the flexibility of the fibres in the bundles. In the case of the 90° corners, however, the 'clear' polymer evident at the right edge of the print indicates that it was not reinforced. The printing path executed the G-code to the corner, depositing the continuous PLA-SSF filament to the 90° corner. However, the polymer solidified at a slower rate than the print heads travelled, which resulted in the SSF drawing through the polymer curving around the corners. Additional cooling and reducing the print spread during cornering helped to minimise this effect in the non-reinforced region. This low turning angle ability of the stainless steel facilitated the direct printing of tensile samples (Figure 4c).

Figure 4. Three-dimensional printing continuous PLA-SSF composite. (**a**) Geometry shape trial prints. (**b**) Close-up of PLA-SSF continuous 3D printing path. (**c**) Tensile sample printed at a layer height of 0.22 mm, (10 mm scale bar).

Based on varying a range of printing parameters, the optimised printing conditions were identified, and they are given in Table 3. It was observed that lower print layer heights generally yielded superior printed part performance, as observed previously by other authors [59]. The optimum obtained print layer height of 0.22 mm was similar to that recommended by manufacturers for commercial filaments of PLA reinforced with short carbon fibres [33]. An advantage of the ex situ prepreg approach is an increase in material cohesion.

Table 3. PLA-SSF 3D print parameters.

Sample Set Layer Height (h) (mm)	Hatch Spacing (w) (mm)	Filament Diameter (mm)	Speed (mm/s)	T_{nozzle} (° C)	T_{bed} (°C)
0.48	0.60	0.50–0.65	3	209	40
0.35	0.40	0.50–0.65	6	209	40
0.22	0.40	0.50–0.65	4	215	55

2.2. *Composite Characterisation*

The print morphology was examined using an inverted metallurgical microscope (Olympus GX51, (Olympus Corporation, Tokyo, Japan) along with a Tabletop Hitachi tm1000 scanning electron microscope (SEM), (Hitachi High-Technologies Corporation, Tokyo, Japan, sourced from Hitachim UK). The specimen dimensions were obtained using Digi Plus Line digital vernier callipers.

2.2.1. Polymer-SSF Filament

Optical and SEM microscopy were used to investigate the morphology, the encapsulation of the SSFs as well as the level of polymer impregnation between the SSFs. This was facilitated by mounting the composite filament in a two-part cold cure Polytek Easyflow resin. The cure temperature was below the glass transmission temperature of the polymer so as to not distort its structure. The samples were ground and polished with abrasive silica paper down to 2000 grit and then polished with a 3 µm diamond suspension.

2.2.2. Three-Dimensional Printed Composites

The porosity and internal structure of the 3D-printed components was evaluated using a GE Nanotom X-ray micro computed tomography (μCT) scanner (GE Sensing & Inspection Technologies GmbH, Wunstorf, Germany), which has up to an 8 µm resolution. Analysis of the μCT scans was carried out using VG Studios software, version 3.5. The μCT scan results were cross-referenced with images obtained using microscopy. The latter images were imported for measurement using image processing software (ImageJ, National Institutes of Health, Bethesda, MD, USA, version Java 1.8.0_345). The volume fraction (V_f) and the number of fibres in the PLA-SSF 3D-printed structure were determined from the μCT scans and ImageJ analysis. The volume fraction (V_f) was calculated using Equation (1) [60]:

$$V_f\ (\%) = \frac{V_{SSF}}{V_T} \quad (1)$$

where: V_{SSF} is the volume of the stainless steel fibres in the bundle and V_T is the overall volume of the composite. The porosity and V_f analysis used different threshold settings to identify the materials. The CT scan images included multiple cross-sections of the PLA-SSF filament obtained from at least three separate 3D-printed parts.

2.3. *Mechanical Performance*

The mechanical performance of the PLA-SSF composites was assessed based on both the interlaminar shear strength according to ASTM D2344 along with the tensile properties following ASTM D5082 [61,62]. Tests were conducted using Zwick Roell Z005

(The ZwickRoell Group, Ulm, Germany. Sourced from ZwickRoell Ltd., Worcester, UK) and Lloyd 6000S (AMETEK, Inc., Berwyn, PA, USA) mechanical testers with 10 and 30 kN load cells, respectively. The data analysis followed the guidelines outlined in ASTM D2344 and D5082 to determine the sample mean, standard deviation and coefficient of variation (expressed as a percentage).

2.3.1. Interlaminar Shear Strength

The interlaminar shear strength (ILSS) was determined through short beam strength testing and was used to investigate the cohesion between different material combinations [61]. The short beam test was performed at a speed of 1 mm/min. The ILSS sample size was defined by the sample thickness (h), the width (b) was two times that thickness, and the length (l) was six times that thickness, with each measuring 3 × 6 × 18 mm. The span of the lower beams S was also defined by the thickness as 4 × (h), measuring 12 mm. The samples were cut to the length using a precision saw (Buehler Isomet™ High-Speed Pro (Bluff, IL, USA)) and shaped to the required dimensions using a water grinding wheel. The ILSS strength τ_{ILSS} (MPa) was calculated using Equation (2) [61]:

$$\tau_{ILSS} = \frac{3}{4} \times \frac{P_m}{b \times h} \quad (2)$$

where P_m is the maximum load observed or failure during the test (N), b is the specimen width (mm) and h is the specimen thickness (mm).

2.3.2. Tensile Testing

The tensile test samples had dimensions of 3× 15 × 175 mm. Sample preparation included grinding all sets to the correct dimensions and bonding steel tabs with dimensions of 1.5 × 15 × 56 mm with an angle of 30° to each end and side of the tensile samples using Loctite 480 Cyanoacrylate (Henkel Adhesive Technologies, Dusseldorf, Germany. Supplied by RS-Ireland. The tensile test was executed at a speed of 5 mm/min. The tensile strength and tensile modulus properties were determined. The PLA-SSF tensile strength (F_{tu}) was calculated using Equation (3) [62]:

$$F_{tu} = \frac{P_{max}}{A} \quad (3)$$

where P_{max} is the maximum force at failure and A is the cross-sectional area. A minimum of five test samples were evaluated under each test condition.

3. Results and Discussion

3.1. Filament Characterisation

The composite polymer-SSF filament morphology, cohesion and dimensions were captured using both optical and SEM microscopy. A cross-sectional SEM image of a PLA-SSF filament 0.7 mm in diameter is displayed in Figure 5a. The fibres were tightly grouped, and the bundle retained an ovular or circular cross-sectional shape, a similar shape to that before SSF's addition to the polymer. The fibre's hexagonal cross-section was a result of the bundle wire-drawing process, as each fibre's flat side aligned with the next fibre, restricting the polymer diffusion to the inner fibres. The results indicated low levels of PLA impregnation within the bundle and porosity around the circumference of the bundle, indicating poor diffusion or adhesion between materials.

In contrast to the filament 0.7 mm in diameter, Figure 5b shows a 0.5 mm PLA-SSF filament cross-section. In this case, the fibres were dispersed across a wider area within the polymer matrix, with a high level of polymer diffusion through the fibres and a low level of porosity. In more commercial applications, the filament diameter can be reduced further.

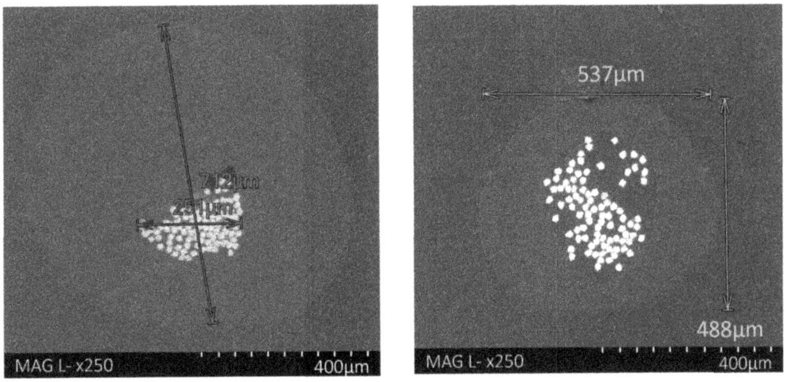

(a) (b)

Figure 5. Cross-section SEM images of PLA-SSF filaments mounted in resin: (**a**) 0.7 mm PLA-SSF filament and (**b**) 0.5 mm PLA-SSF filament. Note the greater dispersion of fibres in the bundle for the filament with the smaller diameter.

3.2. *Printed Composite Evaluation*

The continuous steel fibre-reinforced composites' morphology was first evaluated based on a combination of cross-sectional examination and μCT analysis. This was followed by an evaluation of the composite's mechanical properties based on ILSS and tensile testing.

Examples of μCT scan images of the 3D-printed composite sample cross-sections printed at layer heights of 0.35 and 0.22 mm are given in Figure 6c,d, respectively. The large number of fibres within each SSF bundle is clearly evident within the images, and their high surface areas led to enhanced contact between the fibres and the polymer. Note that the 0.22 mm print had a significantly enhanced stainless steel content along with a more homogeneous distribution of fibres. As detailed in Section 2.1.2, the higher steel fibre content for this composite was facilitated by the removal of the excess polymer through the 'polymer pressure vent ' nozzle on the print head.

As illustrated in Figure 6d, for the 0.22 mm layer height, the SSF in the composite structure indicates a well-ordered, relatively homogenous steel fibre arrangement achieved by increasing the geometrical dimensional widths by 0.4 mm. After structural analysis of the 0.35 mm sample set (Figure 6c), a distortion along the centre seam of the samples was observed. The distortion in the structure was due to an uneven number of printed concentric perimeters caused by the hatch spacing of 0.4 mm being an uneven multiple of the 0.35 mm sample's designed width. The uneven printer head travel path is shown by the orange arrows in Figure 6a along with the blue line in the middle of the geometry. The G-code was compiled for uneven perimeters to print the start of each layer at 0.4 mm from the centre, generating a void (blue line in Figure 6a), and the last line of material deposited was in this space. It was assumed from the polymer 3D printing that there was no material in this region. However, the excess polymer in this polymer-SSF system flowed into the void from the first perimeter deposition and solidified. Obstructing the final line of material deposition caused a structural distortion. The adjustment to the CAD geometry altered the G-code to generate an even number of perimeters (Figure 6b), and this allowed the more homogeneous laying down of fibres in Figure 6d for the height of 0.2 mm. The samples were prepared for mechanical testing as outlined in Section 2.3.

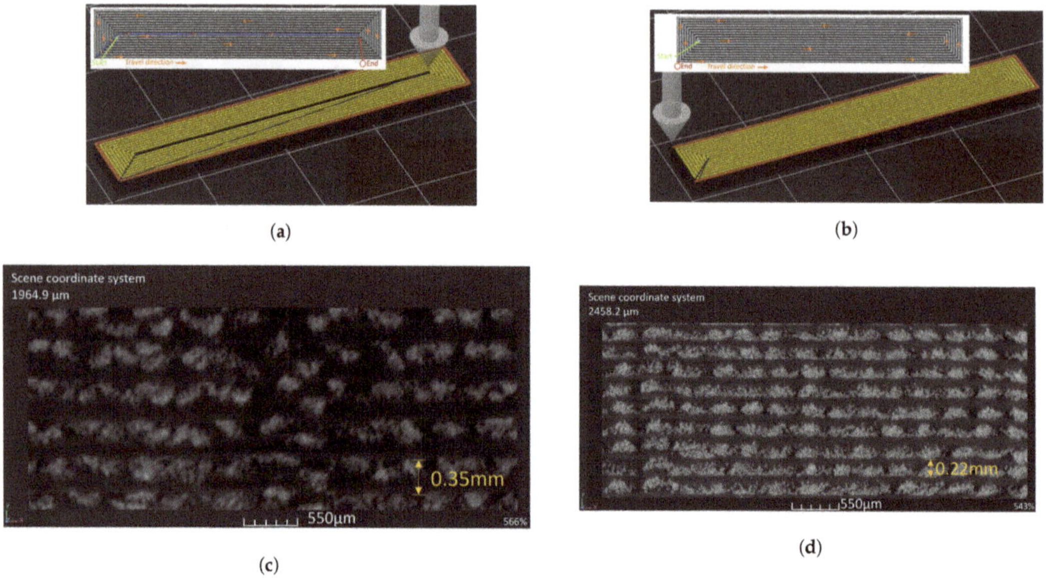

Figure 6. The μCT scans of PLA-SSF 3D-printed part cross-sections (scale bar = 550 µm). (**a**) Printing head travel path, with uneven perimeters of the 0.35 mm layer height samples. (**b**) Printing head travel path, with even perimeters of the 0.22 mm layer height samples. (**c**) Composite (12 V_f%) printed with layer height of 0.35 mm. (**d**) Composite (30 V_f%) printed with layer height of 0.22 mm.

SEM examination of the cross-section of the composite printed at a layer height of 0.22 mm demonstrated the high level of impregnation of the PLA into the steel fibre bundle (Figure 7).

The results for the μCT porosity are given in Table 4, which demonstrates that there was a significant reduction in porosity, as the print layer height was reduced between 0.45 and 0.22 mm. This reduction in porosity was similar to the results reported by other authors, being associated with layer height thickness reduction [20]. In addition to the layer height, other factors which contributed to the porosity reduction included improvements in the travel path, which assisted in creating a more symmetrical, ordered structure as outlined in Section 3.2. Associated with the print layer height reduction was an increase in the SSF content, with the V_f increase being from 6 to 30%.

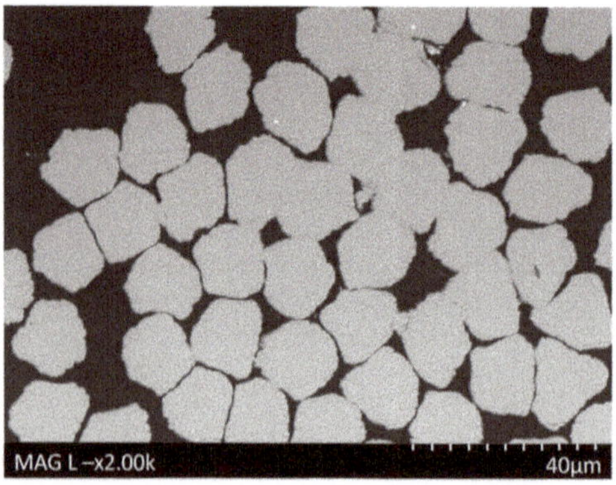

Figure 7. SEM cross-section image of the PLA-SSF composite matrix printed using a 0.22 mm layer height, demonstrating good diffusion of the PLA between the fibres in the SSF bundle.

Table 4. Typical volume fraction and porosity results obtained for the printed PLA-SSF composites obtained using μCT analysis.

Sample Set Layer Height (mm)	Volume Fraction (V_f) (%)	Porosity (%)
0.48	6	21
0.35	12	6
0.22	30	2

3.3. *Mechanical Testing*

3.3.1. Interlaminar Shear Strength

The results of the interlaminar shear strength (τ_{ILSS}) tests for the four layer heights investigated are given in Figure 8. For comparison purposes, this graph includes reports from the literature for 3D-printed continuous glass fibre-reinforced PLA (PLS-cGF) composites reporting a τ_{ILSS} of 13.8 MPa (±1.2) [32]. The measurements for the PLA-only and PLA continuous basalt fibre composites were obtained based on in-house measurements [15].

The PLA-only parts had a τ_{ILSS} of 14.6 MPa (±0.3), and they exhibited a failure mode by tensile fracture with little or no shearing between the layers. The ILSS strength of the PLA-SSF 0.22 mm layer height prints was found to be 28.5 MPa (±2.0), achieving a twofold improvement in ILSS strength. Factors which were likely to influence the performance of the composite are the adhesion between the steel fibres and the polymer, the level of impregnation of the PLA within the SSF bundle and the quantity of SSF in the sample along with the level of porosity.

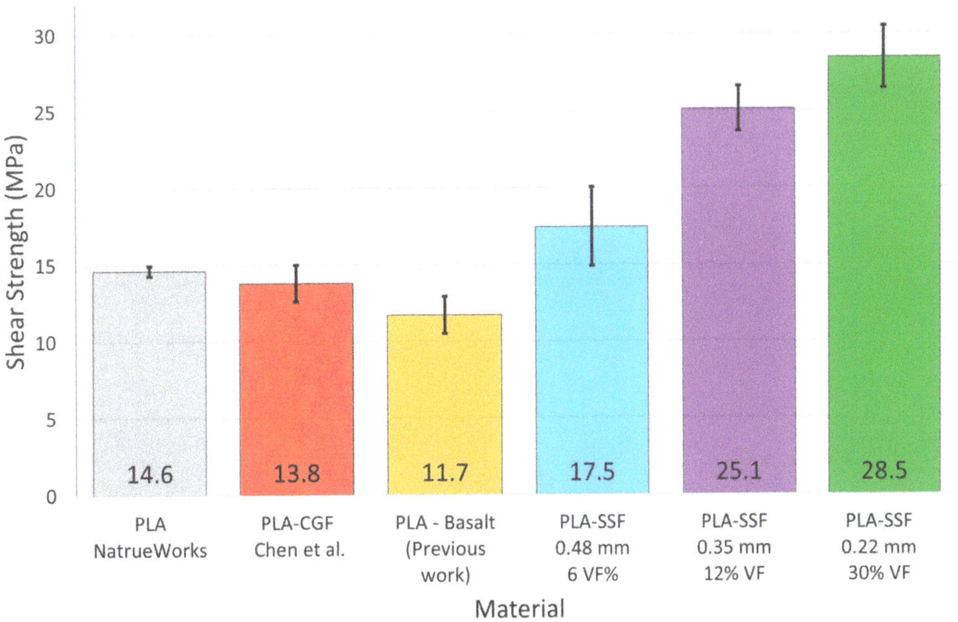

Figure 8. Interlaminar shear strength results for the PLA 0.22, 0.35 and 0.48 mm layer height prints, along with the results reported in the literature for PLA-glass fibres (cGF) and a PLA-Basalt composite investigated in a previous study at UCD (not reported) [32].

With respect to the report from the literature on the ILSS value for PLA-cGF included in Figure 8, no details were provided on the fibre dimensions used to produce the 1 mm diameter filament, which were used for this print [32]. The composite in this case was manufactured from a PLA solution within which the fibre was immersed. In the case of the ILSS test samples, they contained a 4.8wt% fibre content. A more detailed statistical analysis of the PLA-SSF ILSS results from the current study is plotted in Figure 9, which includes the corresponding coefficient of variation (CV) for the sample sets. As expected for the 0.48 mm layer print height samples, a relatively high level of variation in the ILSS results was obtained (CV of 15%). In the case of the 0.35 mm and 0.22 mm layer height prints, the coefficient of variation was in the range of 5–10%.

SEM examination of the fractured samples after ILSS testing was used to help evaluate the mode of failure which, as illustrated in Figure 10a, was found to be due to tensile fracturing with interlaminar shearing [61]. The crack propagation moved through each layer, shearing from the next layer's interface to the next tensile fracture's initiation point, resulting in a jagged failure crack [63]. Also illustrated are the individual SS fibre pull-out, SS fibre necking and bounce back. The first one is where a fibre pulls out of the matrix at the point of failure. The necking of individual fractured fibres is shown in Figure 10b, where a portion of the SSF fibres which fractured close to the PLA's fractured surface and others had been pulled out of the matrix before failure.

Comparing the fractured 0.22 and 0.35 mm print composites after ILSS testing demonstrated that the former presented with less fibre pull-out and crack propagation through the layers. Observing the crack for the 0.22 mm sample indicated that there was a lower level of shearing between the layer interfaces, indicating a stiffer composite matrix.

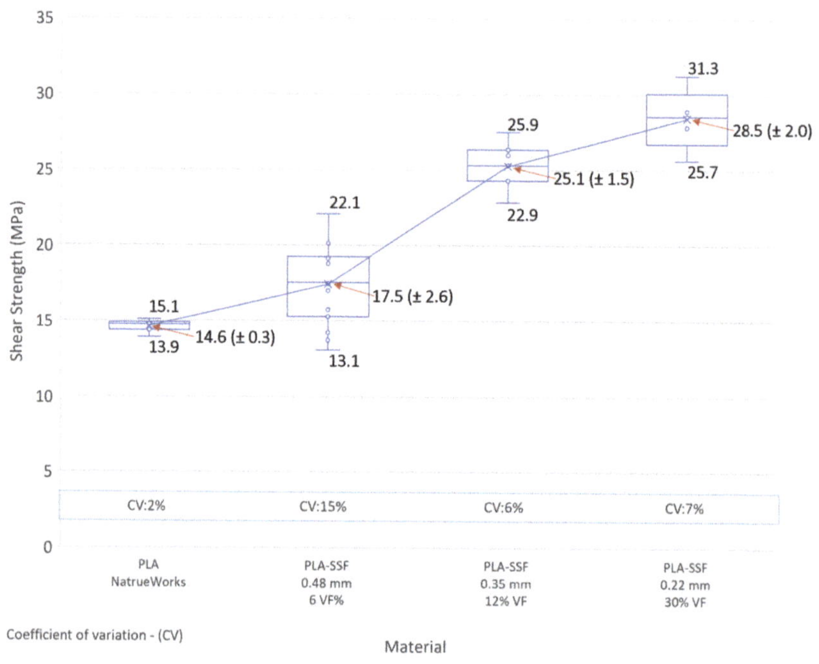

Figure 9. Interlaminar shear strength results for the 0.22, 0.35 and 0.48 mm layer heights investigated with statistical analysis.

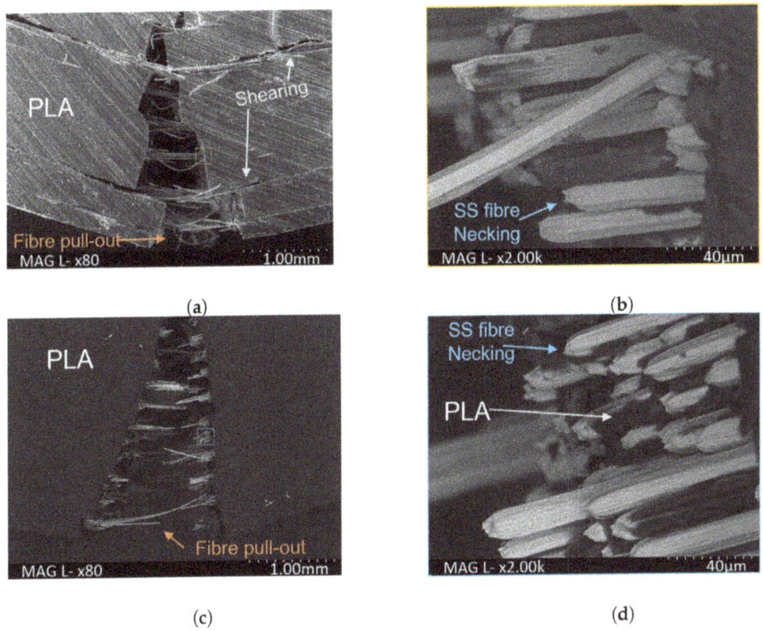

Figure 10. Examination of PLA-SSF composites after ILSS testing. (**a**) Fractured composite printed with a layer height of 0.35 mm, indicating interlaminar shearing and tensile fracture. The yellow dashed square indicates the region at higher magnification in (**b**). (**b**) The fibre pull-out and necking for this sample given in a higher magnification. (**c**,**d**) Corresponding images for a composite printed with a layer height of 0.22 mm, demonstrating similar fracturing along with fibre pull-out and SSF necking. The dashed blue square in (**c**) indicates the higher magnification image in (**d**).

3.3.2. Tensile Properties

The tensile performance results for both the PLA and the PLA-SSF are plotted in Figure 11. Also included in this figure are the results reported for a number of other authors who investigated 3D-printed PLA composites, including studies involving natural fibres, metals wires and carbon fibres (continuous and short) [17–20].

The 3D-printed tensile sample set at a 0.48 mm layer height resulted in a tensile strength of 102.0 MPa (±3.7) and modulus of 5.8 GPa (±0.7). From characterisation and geometry analysis, the structure had a high level of porosity. The filament diameter of 0.50–0.65 mm was close to the printing layer height with a hatch spacing of 0.6 mm, thus likely yielding poor interline bonding.

The PLA-SSF tensile strength printed at a layer height of 0.22 mm was 249.8 MPa (±13.5), and the tensile modulus was 14.3 GPa (±1.2). These values are four and seven times higher, respectively, compared with those obtained for the non-reinforced polymer. The mechanical results displayed a broadly linear pattern where as the fibre volume fraction in the print increased, there was an associated increase in the composites' ILLS, tensile strength, and modulus. A summary of the volume fractions and tensile strengths of the 3D-printed PLA composites are tabulated in Table 5.

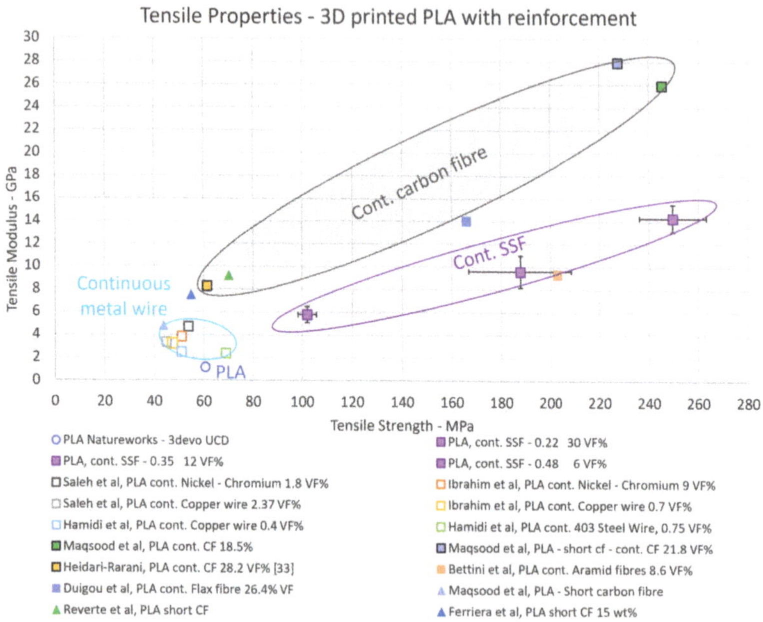

Figure 11. Tensile properties of PLA composite samples printed at the four print head heights in this study (purple squares), with examples of metal- and carbon fibre-reinforced composites reported in the literature [17,19,20,30,33,36,41,44,64].

Table 5. PLA reinforcement materials, volume fractions and tensile properties.

PLA Reinforcement	V_f(%)	Tensile Strength (MPa)	Reference
Continuous Stainless steel fibre bundle 0.22 mm	30	249.8	This work
Continuous Stainless steel fibre bundle 0.35 mm	12	187.8	This work
Continuous Stainless steel fibre bundle 0.48 mm	7	102.0	This work
Continuous Nickel-Chromium wire Wire diameter = 75 µm	9.0	51.2	[30]
Continuous nickel-chromium wire Wire diameter = 75 µm	1.8	53.8	[19]
Continuous copper wire Wire diameter = 75 µm	2.4	44.9	[19]
Continuous copper wire Wire diameter = 75 µm	0.7	47.5	[30]
Continuous copper wire Wire diameter = 0.127 mm	0.4	59.0	[44]
Continuous spring-back 304 stainless steel wire Wire diameter = 0.1778 mm	0.8	61.0	[44]
Continuous carbon fibre	18.5	245.4	[36]
Continuous carbon fibre, (short CF PLA matrix)	21.8	227.6	[36]
Continuous carbon fibre	28.2	61.4	[33]
Continuous carbon fibre	34.0	80.0	[34]
Continuous aramid fibre	8.6	203	[41]
Continuous flax fibre	26.4	166.0	[20]
Short carbon fibre, ColorFabb XT-CF20	20wt%	43.6	[36]
Short carbon fibre, CarbonX™ filament	-	70.3	[17]
Short carbon fibre, Proto-Pasta filament	15wt%	53.4	[64]

SEM images of the cross-section of the 0.35 mm layer height print after tensile testing are given in Figure 12. An SSF bundle is highlighted within the dashed orange oval in Figure 12a. PLA is stiffer than SSF, and due to its brittle failure, it appeared as flat sections on the fractured surface of the composite. As observed for the previous cross-section studies, there was good impregnation of the PLA into the middle of the fibre bundle. As illustrated in Figure 12b, the fibres failed mainly at the fracture face, indicating good interfacial adhesion between the two materials. All the stainless steel fibres exhibited necking prior to ultimate failure [63]. PLA 'stringing', which indicates PLA necking, appeared as a 'cobweb'-like structure over the fractured surface [65]. The gaps in the PLA observed around a number of the SS fibres were possibly associated with elongation and cross-sectional shrinkage of the steel fibre experienced during necking before ultimate failure. The use of metal fibres increases the contact area between the matrix-reinforcing fibre interface, and this bonding can contribute to improving the mechanical properties [66].

As demonstrated in Figure 11, as the stainless steel fibre volume fraction increased, there was a corresponding enhancement in the tensile properties. This observation is similar to the results obtained by other authors [13]. We note that the PLA-cCF was not found to follow this trend with an increased volume fraction. However, for these composites, the different sizing agents used may also have influenced the composite mechanical properties [33,36]. Table 5 demonstrates that the tensile strength and modulus of the PLA-SSF almost doubled when comparing the samples with layer heights of 0.48 and 0.35 mm, with only a 5% increase in the volume fraction.

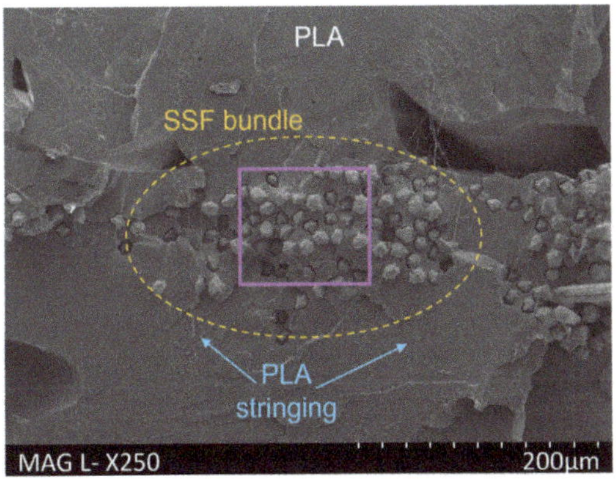

(**a**) (**b**)

Figure 12. SEM images of PLA-SSF composite (35 mm layer height) after tensile testing. The image on the right is a higher magnification of the region in the purple box in the left image.

These results indicate that as the layer height of the print increased, the impregnation or diffusion of polymer within the SSF bundle was relatively poor due to the shape of the SSF, as discussed in Section 3.1, which thus gave rise to an increase in porosity. The PLA-SSF requires a slow printing speed to achieve polymer solidification, retaining the SSF in position as the print head moves and leading to long print times and potential fibre breakage at higher speeds. A potential advantage of this process and technology is its use in bespoke conductive textiles, along with high impact resistance after-market 3D-printable repairs. A further advantage is sustainable sourcing and potential recyclability for compatible remanufacturing.

The PLA-SSF tensile strength and modulus values obtained were the highest reported for metal-reinforced composites to date. The tensile strengths obtained for the composites with printed layer heights of 0.22 mm were approximately four times higher when compared with continuous spring-back 304 stainless steel wire with a 0.127 mm diameter [44].

When comparing the tensile strengths of the PLA-SSF composite with 7 V_f% with that reported in the literature for a PLA-nickel chromium wire with 9 V_f%, the former exhibited strengths which were 50% higher [39]. Amongst the factors which are likely to have contributed to this improvement in mechanical performance in this study are the use of continuous fibres in the prints, the good impregnation of the PLA between the fibres 14 µm in diameter in the 0.15 mm SSF bundle, the relatively low porosity (down to 2%) and the relatively high SSF content (V_f up to 30%).

4. Conclusions

Filaments of continuous stainless steel fibre bundles within a polylactic acid (PLA) polymer were fabricated using a laboratory-scale extrusion system. By systematically controlling the 3D printing conditions, along with the use of a novel 'polymer pressure vent' within the printer nozzle, 3D-printed composites with fibre volume fractions between 4 and 30% were achieved. Good impregnation and adhesion of the PLA matrix into the stainless steel fibre were found based on CT analysis, with the porosity of the resulting composites being in the range of 2–21%. The interlaminar shear strength ($\tau ILSS$) of the PLA-SSF with a volume fraction of 30% was found to be 28.5 MPa (±2.0), which was twice that of the PLA-only parts. Both the interlaminar shear strength and tensile strength properties of the composites were found to increase significantly as the stainless steel volume fraction (V_f) increased from 6 to 30%. The PLA-SSF composites exhibited tensile strengths of up to 249.8 MPa (±13.5), along with tensile modulus values of 14.3 GPa (±1.2). The tensile strengths obtained with the highest stainless steel V_f in this study were approximately four times higher than those reported for other printed metal fibre-reinforced composites in the literature.

Author Contributions: Conceptualisation, A.J.C. and A.D.; methodology, A.J.C. and A.D.; software, A.J.C. and A.D.; validation, A.J.C., A.D. and D.P.D.; formal analysis, A.J.C., A.D. and D.P.D.; investigation, A.J.C.; resources, D.P.D.; data curation, A.J.C.; writing—original draft preparation, A.J.C.; writing—review and editing, D.P.D.; visualisation, A.J.C.; supervision, D.P.D.; project administration, A.J.C.; funding acquisition, D.P.D. All authors have read and agreed to the published version of the manuscript.

Funding: This publication emanated from research supported in part by a research grant from Science Foundation Ireland (SFI) under Grant Number 16/RC/3872. For the purpose of open access, the authors applied a CC by public copyright license to any author accepted manuscript version arising from this submission.

Institutional Review Board Statement: Not applicable.

Informed Consent Statement: Not applicable.

Data Availability Statement: The data presented in this study are available upon request from the corresponding author.

Conflicts of Interest: The authors declare no conflicts of interest.

Abbreviations

The following abbreviations are used in this manuscript:

3D	Three-dimensional
°C	Degrees Centigrade
b	Specimen width
h	Specimen thickness
kN	Kilonewton
μCT	Micro computed tomography
µm	Micrometer
mm	Millimeter

mm/s	Millimeters Per second
wt%	Weight percent
F_{tu}	Tensile strength
P_m or P_{max}	Maximum load or failure
rpm	Revolutions per minute
T_1	1st Extrusion barrel heater temperature
T_2	2st Extrusion barrel heater temperature
T_3	3st Extrusion barrel heater temperature
T_4	4st Extrusion barrel heater temperature
ABS	Acrylonitrile butadiene styrene
CAD	Computer-aided design
cCF	Continuous carbon fibre
EDAX	Energy-dispersive X-ray spectroscopy
ILSS or τ_{ILSS}	Interlaminar shear strength
L	Loop 3D print geometry
GPa	Gigapascal
MEX	Material extrusion
MPa	Megapascal
PC	Polyamide or nylon
PC	Polycarbonate
PLA	Polylactic acid
PLA-cCF	Polylactic acid containing continuous carbon fibre
PLA-cGF	Polylactic acid containing continuous glass fibre
PLA-SCF-cCF	Polylactic acid containing short carbon fibres and continuous carbon fibre
PLA-SSF	Polylactic acid containing continuous stainless steel fibre bundle composite
SEM	Scanning electron 255 microscope
SSF	316 L stainless steel fibre bundle
ATL	Stereolithography
TEX	Decitex
V_f	Volume fraction
V_{SSF}	Volume of continuous stainless steel fibre
V_T	Overall volume

References

1. NatureWorks. *Ingeo™ Biopolymer 3D850 Technical Data Sheet 3D Printing Monofilament—General Purpose Grade*; Technical Report 4; NatureWorks LLC: Fridley, MN, USA, 2018.
2. Luo, M.; Tian, X.; Shang, J.; Yun, J.; Zhu, W.; Li, D.; Qin, Y. Bi-scale interfacial bond behaviors of CCF/PEEK composites by plasma-laser cooperatively assisted 3D printing process. *Compos. Part A Appl. Sci. Manuf.* **2020**, *131*, 105812. [CrossRef]
3. Dutra, T.; Ferreira, R.T.L.; Resende, H. Interlaminar shear strength of continuous carbon fiber reinforced thermoplastic composites manufactured by 3D printing. In Proceedings of the 24th ABCM International Congress of Mechanical Engineering, Lisbon, Portugal, 17–19 September 2018. [CrossRef]
4. Xu, X.; Mi, G.; Luo, Y.; Jiang, P.; Shao, X.; Wang, C. Morphologies, microstructures, and mechanical properties of samples produced using laser metal deposition with 316 L stainless steel wire. *Opt. Lasers Eng.* **2017**, *94*, 1–11. [CrossRef]
5. Bhandari, S.; Lopez-Anido, R.A.; Gardner, D.J. Enhancing the interlayer tensile strength of 3D printed short carbon fiber reinforced PETG and PLA composites via annealing. *Addit. Manuf.* **2019**, *30*, 100922. [CrossRef]
6. Zhang, H.; Liu, D.; Huang, T.; Hu, Q.; Lammer, H. Three-dimensional printing of continuous flax fiber-reinforced thermoplastic composites by five-axis machine. *Materials* **2020**, *13*, 1678. [CrossRef] [PubMed]
7. Li, J.; Durandet, Y.; Huang, X.; Sun, G.; Ruan, D. Additively manufactured fiber-reinforced composites: A review of mechanical behavior and opportunities. *J. Mater. Sci. Technol.* **2022**, *119*, 219–244. [CrossRef]
8. Hao, W.; Liu, Y.; Zhou, H.; Chen, H.; Fang, D. Preparation and characterization of 3D printed continuous carbon fiber reinforced thermosetting composites. *Polym. Test.* **2018**, *65*, 29–34. [CrossRef]
9. Sodeifian, G.; Ghaseminejad, S.; Yousefi, A.A. Preparation of polypropylene/short glass fiber composite as Fused Deposition Modeling (FDM) filament. *Results Phys.* **2019**, *12*, 205–222. [CrossRef]
10. Mirabedini, A.; Foroughi, J. *Carbon Nanotube-Graphene Composites Fibers*; Elsevier Ltd.: Amsterdam, The Netherlands, 2018; pp. 61–86. [CrossRef]
11. Wu, B.; Pan, Z.; Ding, D.; Cuiuri, D.; Li, H.; Xu, J.; Norrish, J. A review of the wire arc additive manufacturing of metals: Properties, defects and quality improvement. *J. Manuf. Process.* **2018**, *35*, 127–139. [CrossRef]
12. Rafiee, M.; Farahani, R.D.; Therriault, D. Multi-Material 3D and 4D Printing: A Survey. *Adv. Sci.* **2020**, *7*, 1902307. [CrossRef]

13. Dickson, A.; Abourayana, H.; Dowling, D. *3D Printing of Fibre Reinforced Thermoplastic Composites Using Fused Filament Fabrication—A Review*; Technical report; University College Dublin: Dublin, Ireland, 2020.
14. Dickson, A.N.; Ross, K.A.; Dowling, D.P. Additive manufacturing of woven carbon fibre polymer composites. *Compos. Struct.* **2018**, *206*, 637–643. [CrossRef]
15. Dowling, D.P.; Abourayana, H.M.; Brantseva, T.; Antonov, A.; Dobbyn, P.J. Enhancing the mechanical performance of 3D-printed basalt fiber-reinforced composites using in-line atmospheric plasma pretreatments. *Plasma Process. Polym.* **2020**, *17*. [CrossRef]
16. Tekinalp, H.L.; Kunc, V.; Velez-Garcia, G.M.; Duty, C.E.; Love, L.J.; Naskar, A.K.; Blue, C.A.; Ozcan, S. Highly oriented carbon fiber-polymer composites via additive manufacturing. *Compos. Sci. Technol.* **2014**, *105*, 144–150. [CrossRef]
17. Reverte, J.M.; Caminero, M.á.; Chacón, J.M.; García-Plaza, E.; Núñez, P.J.; Becar, J.P. Mechanical and geometric performance of PLA-based polymer composites processed by the fused filament fabrication additive manufacturing technique. *Materials* **2020**, *13*, 1924. [CrossRef] [PubMed]
18. Ibrahim, Y. 3D Printing of Continuous Wire Polymer Composite for Mechanical and Thermal Applications. Ph.D. Thesis, YORK UNIVERSITY: York, UK, 2019.
19. Saleh, M.A.; Kempers, R.; Melenka, G.W. 3D printed continuous wire polymer composites strain sensors for structural health monitoring. *Smart Mater. Struct.* **2019**, *28*, 105041. [CrossRef]
20. Le Duigou, A.; Chabaud, G.; Matsuzaki, R.; Castro, M. Tailoring the mechanical properties of 3D-printed continuous flax/PLA biocomposites by controlling the slicing parameters. *Compos. Part B Eng.* **2020**, *203*, 108474. [CrossRef]
21. Hine, P.J.; Rudolf Lusti, H.; Gusev, A.A. Numerical simulation of the effects of volume fraction, aspect ratio and fibre length distribution on the elastic and thermoelastic properties of short fibre composites. *Compos. Sci. Technol.* **2002**, *62*, 1445–1453. [CrossRef]
22. Materials, P.; Materials, E.I.; Matrix, P.; Materials, C.; Specimens, P. *Standard Test Method for Tensile Properties of Plastics 1*; Technical Report January 2004; ASTM International: West Conshohocken, PA, USA, 2006. [CrossRef]
23. Beckman, I.P.; Lozano, C.; Freeman, E.; Riveros, G. Fiber selection for reinforced additive manufacturing. *Polymers* **2021**, *13*, 2231. [CrossRef]
24. Matsuzaki, R.; Ueda, M.; Namiki, M.; Jeong, T.K.; Asahara, H.; Horiguchi, K.; Nakamura, T.; Todoroki, A.; Hirano, Y. Three-dimensional printing of continuous-fiber composites by in-nozzle impregnation. *Sci. Rep.* **2016**, *6*, 23058. [CrossRef]
25. Goh, G.D.; Dikshit, V.; Nagalingam, A.P.; Goh, G.L.; Agarwala, S.; Sing, S.L.; Wei, J.; Yeong, W.Y. Characterization of mechanical properties and fracture mode of additively manufactured carbon fiber and glass fiber reinforced thermoplastics. *Mater. Des.* **2018**, *137*, 79–89. [CrossRef]
26. Botelho, E.C.; Figiel, Ł.; Rezende, M.C.; Lauke, B. Mechanical behavior of carbon fiber reinforced polyamide composites. *Compos. Sci. Technol.* **2003**, *63*, 1843–1855. [CrossRef]
27. Hu, Q.; Duan, Y.; Zhang, H.; Liu, D.; Yan, B.; Peng, F. Manufacturing and 3D printing of continuous carbon fiber prepreg filament. *J. Mater. Sci.* **2018**, *53*, 1887–1898. [CrossRef]
28. Pandelidi, C.; Bateman, S.; Piegert, S.; Hoehner, R.; Kelbassa, I.; Brandt, M. The technology of continuous fibre-reinforced polymers: A review on extrusion additive manufacturing methods. *Int. J. Adv. Manuf. Technol.* **2021**, *113*, 3057–3077. [CrossRef]
29. Zhuo, P.; Li, S.; Ashcroft, I.A.; Jones, A.I. Material extrusion additive manufacturing of continuous fibre reinforced polymer matrix composites: A review and outlook. *Compos. Part Eng.* **2021**, *224*, 109143. [CrossRef]
30. Ibrahim, Y.; Melenka, G.W.; Kempers, R. Fabrication and tensile testing of 3D printed continuous wire polymer composites. *Rapid Prototyp. J.* **2018**, *24*, 1131–1141. [CrossRef]
31. Mahltig, B. *Introduction to Inorganic Fibers*; Elsevier Ltd.: Amsterdam, The Netherlands, 2018; pp. 1–29. [CrossRef]
32. Chen, K.; Yu, L.; Cui, Y.; Jia, M.; Pan, K. Optimization of printing parameters of 3D-printed continuous glass fiber reinforced polylactic acid composites. *Thin-Walled Struct.* **2021**, *164*, 107717. [CrossRef]
33. Heidari-Rarani, M.; Rafiee-Afarani, M.; Zahedi, A.M. Mechanical characterization of FDM 3D printing of continuous carbon fiber reinforced PLA composites. *Compos. Part B Eng.* **2019**, *175*, 107147. [CrossRef]
34. Li, N.; Li, Y.; Liu, S. Rapid prototyping of continuous carbon fiber reinforced polylactic acid composites by 3D printing. *J. Mater. Process. Technol.* **2016**, *238*, 218–225. [CrossRef]
35. Rimašauskas, M.; Kuncius, T.; Rimašauskienė, R. Processing of carbon fiber for 3D printed continuous composite structures. *Mater. Manuf. Process.* **2019**, *34*, 1528–1536. [CrossRef]
36. Maqsood, N.; Rimašauskas, M. Characterization of carbon fiber reinforced PLA composites manufactured by fused deposition modeling. *Compos. Part C Open Access* **2021**, *4*, 100112. [CrossRef]
37. Caminero, M.A.; Chacón, J.M.; García-Moreno, I.; Reverte, J.M. Interlaminar bonding performance of 3D printed continuous fibre reinforced thermoplastic composites using fused deposition modelling. *Polym. Test.* **2018**, *68*, 415–423. [CrossRef]
38. Kabir, S.M.; Mathur, K.; Seyam, A.F.M. A critical review on 3D printed continuous fiber-reinforced composites: History, mechanism, materials and properties. *Compos. Struct.* **2020**, *232*, 111476. [CrossRef]
39. Ibrahim, Y.; Melenka, G.W.; Kempers, R. Additive manufacturing of Continuous Wire Polymer Composites. *Manuf. Lett.* **2018**, *16*, 49–51. [CrossRef]
40. Sugiyama, K.; Matsuzaki, R.; Malakhov, A.V.; Polilov, A.N.; Ueda, M.; Todoroki, A.; Hirano, Y. 3D printing of optimized composites with variable fiber volume fraction and stiffness using continuous fiber. *Compos. Sci. Technol.* **2020**, *186*, 107905. [CrossRef]

41. Bettini, P.; Alitta, G.; Sala, G.; Di Landro, L. Fused Deposition Technique for Continuous Fiber Reinforced Thermoplastic. *J. Mater. Eng. Perform.* **2017**, *26*, 843–848. [CrossRef]
42. Liu, G.; Xiong, Y.; Zhou, L. Additive manufacturing of continuous fiber reinforced polymer composites: Design opportunities and novel applications. *Compos. Commun.* **2021**, *27*, 100907. [CrossRef]
43. Quan, D.; Flynn, S.; Artuso, M.; Murphy, N.; Rouge, C.; Ivanković, A. Interlaminar fracture toughness of CFRPs interleaved with stainless steel fibres. *Compos. Struct.* **2019**, *210*, 49–56. [CrossRef]
44. Hamidi, A.; Tadesse, Y. Single step 3D printing of bioinspired structures via metal reinforced thermoplastic and highly stretchable elastomer. *Compos. Struct.* **2019**, *210*, 250–261. [CrossRef]
45. Gunes, M.; Cayiroglu, I. Mechanical Behaviour of 3D Printed Parts with Continuous Steel Wire Reinforcement. *El-Cezeri J. Sci. Eng.* **2022**, *9*, 276–289. [CrossRef]
46. Bekaert. *Metal Fiber Composite Reinforcement*; Technical report; Bekaert: Zwevegem, Belgium, 2018.
47. Küster, K.; Barburski, M.; Lomov, S.V.; Vanclooster, K. Metal Fibers-Steel. In *Inorganic and Composite Fibers: Production, Properties, and Applications*; Elsevier Ltd.: Amsterdam, The Netherlands, 2018; pp. 219–241. [CrossRef]
48. Verlinden, B.; Driver, J.; Samajdar, I.; Doherty, R. *Thermo-Mechanical Processing of Metallic Materials*; Pergamon Materials Series; Pergamon: Bergama, Turkey, 2007; Volume 11, pp. 7–30. [CrossRef]
49. 3devo. *Composer 350*; Technical report; 3devo: Utrecht, The Netherlands, 2020.
50. Lafleur, P.G.; Vergnes, B. Wire Coating and Cable Insulation. In *Polymer Extrusion*; John Wiley & Sons, Incorporated: Somerset, UK, 2014; Volume 1, pp. 305–333. [CrossRef]
51. Rauwendaal, D.C. *Die Design Handbook*, 5th ed.; Carl Hanser Verlag GmbH & Co. KG: Munich, Germany, 2014; Volume 65, pp. 1–17. [CrossRef]
52. McCullough, K. US6863729B2 Nozzle Insert for Long Fiber Compounding. 2005 Available onlime: https://worldwide.espacenet.com/patent/original-document?channel=espacenet_channel-c9b6335c-1fad-438a-ac13-c4c4532c2a05(accessed on 22 November 2023) .
53. Fang, Q.; Hanna, M.A. Rheological properties of amorphous and semicrystalline polylactic acid polymers. *Ind. Crop. Prod.* **1999**, *10*, 47–53. [CrossRef]
54. Fratte, E.D.; D'hooge, D.R.; Eeckhout, M.; Cardon, L. Principles and Guidelines for In-Line Viscometry in Cereal Extrusion. *Polymers* **2022**, *14*, 2316. [CrossRef]
55. Hussain, A.; Ameer, S.; Javed, F.; Malik, M.Y. Rheological analysis on non-Newtonian wire coating. *J. Braz. Soc. Mech. Sci. Eng.* **2019**, *41*, 115. [CrossRef]
56. Skelland, A.H.P. *Non-Newtonian Flow and Heat Transfer*; Wiley: New York, NY, USA, 1967; pp. 12–14.
57. Shah, R.A.; Islam, S.; Siddiqui, A.M.; Haroon, T. Heat transfer by laminar flow of a third grade fluid in wire coating analysis with temperature dependent and independent viscosity. *Anal. Math. Phys.* **2011**, *1*, 147–166. [CrossRef]
58. Rauwendaal, C. Extruder Screw Design. In *Polymer Extrusion*; Carl Hanser Verlag GmbH & Co. KG: Munchen, Germany, 2014; pp. 509–652. [CrossRef]
59. Yao, Y.; Li, M.; Lackner, M.; Herfried, L. A continuous fiber-reinforced additive manufacturing processing based on PET fiber and PLA. *Materials* **2020**, *13*, 3044. [CrossRef] [PubMed]
60. Naranjo-Lozada, J.; Ahuett-Garza, H.; Orta-Castañón, P.; Verbeeten, W.M.; Sáiz-González, D. Tensile properties and failure behavior of chopped and continuous carbon fiber composites produced by additive manufacturing. *Addit. Manuf.* **2019**, *26*, 227–241. [CrossRef]
61. ASTM D2344; ASTM D2344/D2344M: Standard Test Method for Short-Beam Strength of Polymer Matrix Composite Materials and Their Laminates. In *Annual Book ASTM Standard*; ASTM: West Conshohocken, PA, USA, 2003; Volume 3, pp. 136–140.
62. ASTM. *Standard Test Method for Tensile Properties of Reinforced Thermosetting Plastics*; ASTM Standards: West Conshohocken, PA, USA, 2002; pp. 1–9. [CrossRef]
63. Callens, M.G.; Gorbatikh, L.; Verpoest, I. Ductile steel fibre composites with brittle and ductile matrices. *Compos. Part A Appl. Sci. Manuf.* **2014**, *61*, 235–244. [CrossRef]
64. Ferreira, R.T.L.; Amatte, I.C.; Dutra, T.A.; Bürger, D. Experimental characterization and micrography of 3D printed PLA and PLA reinforced with short carbon fibers. *Compos. Part B Eng.* **2017**, *124*, 88–100. [CrossRef]
65. Ahmad, N.D.; Kusmono.; Wildan, M.W.; Herianto. Preparation and properties of cellulose nanocrystals-reinforced Poly (lactic acid) composite filaments for 3D printing applications. *Results Eng.* **2023**, *17*, 100842. [CrossRef]
66. Callens, M.G.; Gorbatikh, L.; Bertels, E.; Goderis, B.; Smet, M.; Verpoest, I. Tensile behaviour of stainless steel fibre/epoxy composites with modified adhesion. *Compos. Part A Appl. Sci. Manuf.* **2015**, *69*, 208–218. [CrossRef]

polymers

Article

Quality Analysis of Micro-Holes Made by Polymer Jetting Additive Manufacturing

Razvan Udroiu

Department of Manufacturing Engineering, Transilvania University of Brasov, 29 Eroilor Boulevard, 500036 Brasov, Romania; udroiu.r@unitbv.ro; Tel.: +40-268-421-318

Abstract: Material jetting technology is gaining popularity, especially in polymer science, because of their high accuracy for additive manufacturing (AM) products. This paper aims to investigate the quality of micro-holes that are oriented in three basic directions, and manufactured by the material jetting AM process. This paper proposes a novel methodology to evaluate the accuracy of micro-holes features by using a transparent artifact. A test artifact with horizontal and vertical micro-holes in it, with industrial applications, was designed. Micro-holes were placed on planar and curve surfaces. Samples were manufactured by PolyJet technology from a translucent photopolymer resin which allows a facile investigation (by microscopy) of the inner structure of the micro-holes. The features of ten micro-holes printed in matte and glossy finish type, with diameters in coarse and medium options, according to ISO/ASTM 52902, were analyzed. Quality analysis of the micro-holes features was performed by microscopy investigations. The effects of main factors on the deviation of the micro-hole diameter were investigated by using the statistical design of experiments, and four control factors were considered. The best results were obtained for sample printed in matte finishing with the micro-holes oriented along the *x*-axis and *z*-axis. The smallest diameter of the micro-holes obtained by PolyJet technology on an EDEN 350 machine was 0.5 mm, but in industrial applications for a facile post-processing, a higher diameter is recommended to be used. A confirmatory experiment on a wing sample, with a number of micro-holes of the same diameter and a large length to diameter ratio of the micro-holes, was performed, and the repeatability of the results was confirmed.

Keywords: micro-hole; quality analysis; dimensional accuracy; additive manufacturing; 3D printing; polymers; material jetting

check for updates

Citation: Udroiu, R. Quality Analysis of Micro-Holes Made by Polymer Jetting Additive Manufacturing. *Polymers* **2024**, *16*, 32. https://doi.org/10.3390/polym16010032

Academic Editor: Loïc Dupont

Received: 29 November 2023
Revised: 15 December 2023
Accepted: 19 December 2023
Published: 21 December 2023

1. Introduction

Additive manufacturing (AM), or 3D printing, is a key technology in the actual global market characterized by a grown competition, and is based on its ability to fabricate complex products directly from computer-aided design (CAD) files, and reduce the production cost and time [1]. The ISO/ASTM 52900-15 [2] standard has defined seven main types of additive manufacturing process, as follows: vat photo-polymerization (VP), binder jetting (BJ), material extrusion (ME), material jetting (MJ), sheet lamination (SL), powder bed fusion (PBF), and directed energy deposition (DED).

Material jetting is considered to be one of the most popular and mature processes used in additive plastics manufacturing [1], on the basis of its high accuracy [3–5], low surface roughness [6–8], and the good mechanical properties of the parts [9,10]. Polymer jetting (PolyJet) and multi-jet printing (MJM) are the most well-known technologies from the MJ category.

It is mentioned in [11] that the macro-sized holes have a diameter larger than 3 mm and the micro-sized holes are characterized by a diameter less than around 1 mm.

Micro-holes are made in different components, and are incorporated into electronic equipment, medical instruments, panel absorbers, micro-robots flight, microfluidic devices [12,13], and structures embedded with fiber-optic sensors. The small diameter

of acrylic monomer, urethane acrylate oligomer, epoxy acrylate, and photoinitiator. The main properties of Objet Fullcure 720 resin are presented in [34,36]. The support material, FullCure 705 resin, consists of acrylic monomer, polyethylene glycol 400, propane-1, 2-diol, glycerol, and photoinitiator [37,38].

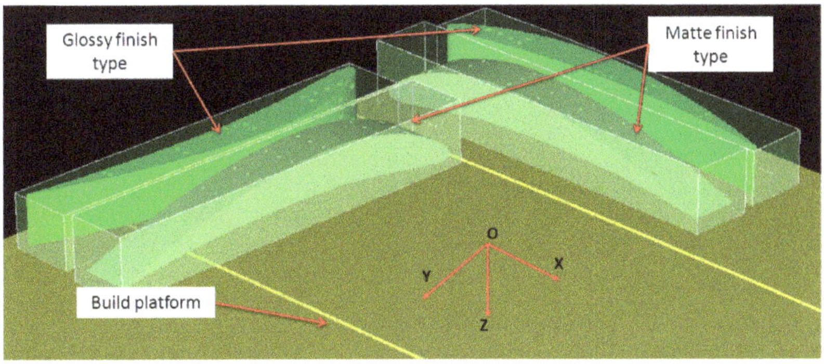

Figure 4. Layout of artifacts orientations on the build platform. The bounding box of samples is displayed.

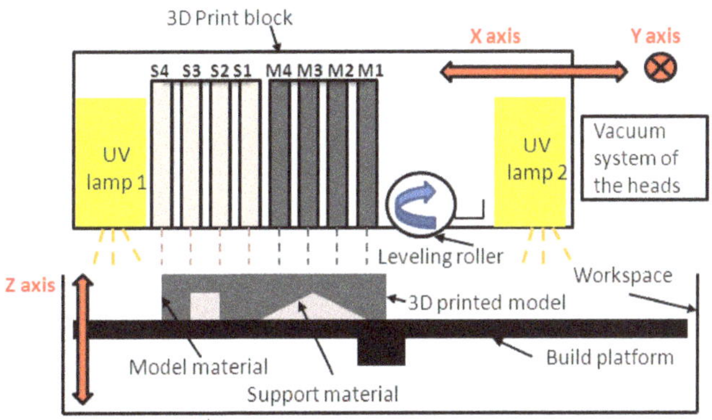

Figure 5. The basic principle of PolyJet technology; the 3D print block mainly contains four support inkjet heads (S1 to S4), four model inkjet heads (M1 to M4) and two UV lamps.

The main parameters of the process were the print head temperature of 72 °C, and a print head vacuum of 6.2 atm. The manufacturing process was performed under a controlled laboratory temperature of 20 °C and a relative humidity of 30%.

All the micro-holes of the 3D samples printed in matte finishing type were filled by support material. The horizontal micro-holes 3D printed in glossy finishing were also filled by support material. The samples needed post-processing in order to remove the support material from micro-holes. The specimens were immersed in a sodium hydroxide solution and were washed with pressurized water afterwards. Also, thin pins with smaller diameters than the micro-hole diameters were used to manually clean the support from the micro-holes. Thus, syringe needles with a 0.1 mm increment in the range diameter of 0.3 mm (codification 30G) to 1.2 mm (18G) were used to remove the support material from the micro-holes.

The investigation of micro-holes features consisted of visual inspection, checking the diameter with "go and no go" pin gauge, and microscopy measurements. Visual tests with no required equipment were performed for all four types of samples. The micro-holes were labeled according to the control factors, as is shown in Table 2.

Table 2. Notation of 3D printed micro-holes.

Notation	Micro-Hole Diameter [mm]	Artifact Orientation and Finish Type	Micro-Hole Orientation	Position of the Micro-Hole End
Dval_GX_Z_Up	Dval	GX (X, glossy)	Z (along z-axis)	Up (on the upper surface)
Dval_GX_Z_Low	Dval	GX (X, glossy)	Z (along z-axis)	Low (on the lower surface)
Dval_GX_Y	Dval	GX (X, glossy)	Y (along y-axis)	-
Dval_GY_Z_Up	Dval	GY (Y, glossy)	Z (along z-axis)	Up (on the upper surface)
Dval_GY_Z_Low	Dval	GY (Y, glossy)	Z (along z-axis)	Low (on the lower surface)
Dval_GY_X	Dval	GY (Y, glossy)	X (along x-axis)	-
Dval_MX_Y	Dval	MX (X, matte)	Y (along y-axis)	-
Dval_MX_Z	Dval	MX (X, matte)	Z (along z-axis)	-
Dval_MY_X	Dval	MY (Y, matte)	X (along x-axis)	-
Dval_MY_Z	Dval	MY (Y matte)	Z (along z-axis)	-

A pin gauge size, which is considerably smaller than the nominal dimension should be, was first tested, subsequently going up in increments of 0.1 mm.

The microscopy measurements were performed by ISM-PRO software, using an ISM-PM200SB digital microscope (Insize, Boituva, Brasil) with an accuracy of 10 microns at a magnification of 150X. In order to achieve high accuracy, calibration using a calibration rule was done before the measurements. The primary measurement for the resolution hole features is the diameter of the micro-hole. The vertical holes oriented along the z-axis are characterized by a start hole diameter, at the lower surface and an end hole diameter at the upper surface. The horizontal holes oriented along the x-axis and y-axis are characterized by left and right end diameters of the hole and right hole diameter. The mean hole diameter was calculated based on the arithmetic mean of the hole ends.

The artifacts printed in glossy finishing are clear, transparent and allow the holes to be investigated in depth. Three measurements were performed on the deep of the clear transparent holes—on the top, middle and bottom. The mean arithmetic diameter was considered. The artifacts printed in matte finishing are not clear or transparent and the micro-holes cannot be investigated in depth.

A series of defects that occurred during layers formation by photo-polymerization of the resin were investigated, namely the photo-polymerization affected zone (PAZ), and array affected zone (AAZ). PAZ has similar parameters to HAS (heat affected zone), and is reported in the subtractive conventional and unconventional methods of processing holes. The PAZ width parameter is defined on the basis of relation (1).

$$\text{PAZ width [mm]} = \frac{(\text{PAZ diameter}) - (\text{Measured diameter of the hole})}{2} \quad (1)$$

PAZ is actually an area adjacent to the micro-hole surface, which is affected by the photo-polymerization process in glossy finishing printing. PAZ width is only measured on the top surface. AAZ is an area around a micro-hole array.

2.4. *Case Study as Confirmatory Experiment*

An experimental wing with an array of horizontal and vertical micro-holes 1 mm in diameter was designed in SolidWorks, as is shown in Figure 6.

The horizontal micro-holes traverse the span of the wing at a depth of 120 mm, and vertical micro-holes with the same diameter are distributed on the top and bottom surface of the aerofoil. This results in a large micro-hole length-to-diameter ratio of L/D = 120.

Each horizontal micro-hole is connected with a vertical micro-hole, having applications in experimental aerodynamics. The connected micro-holes create a channel that allows the static pressure on the upper and lower surface of the wing in different locations to be captured during experimental aerodynamic tests, meaning it is possible to experimentally calculate the lift and drag forces developed on a wing at different angles of incidence.

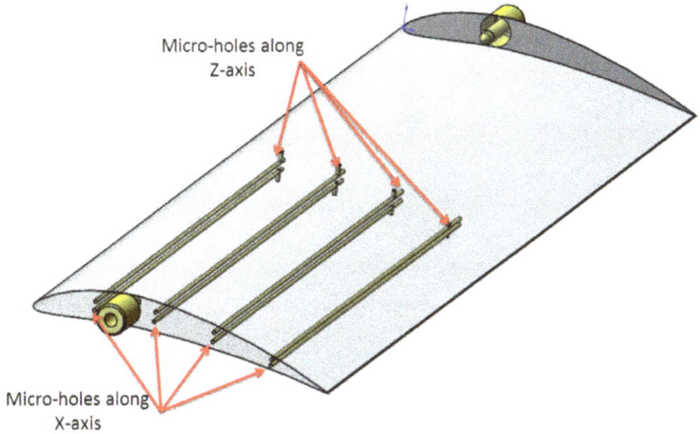

Figure 6. Experimental wing model that contains connected horizontal and vertical micro-holes.

This new design of connected holes is a solution that is practically impossible to be manufactured using any other traditional manufacturing method, such as deep drilling, than additive manufacturing. Five samples of the proposed wing were manufactured. Each sample was positioned in the optimal 3D-printing configuration, XY matte [5]. The micro-holes diameter was measured and analyzed.

3. Results

The layout of the four samples was manufactured in 2 h and 8 min, using 38 g of model material and 29 g of support material.

3.1. Quality Analysis of the Micro-Hole Features

Two kinds of visual inspections were performed, before and after post-processing of the samples. The artifacts placed on the build platform before post-processing are shown in Figure 7. The artifacts printed in glossy finishing present vertical striations of model material on the lateral surface, and support material deposited in the area of the micro-holes on the vertical surface. It was identified that support material of a rectangular shape was deposited on the area of the three micro-holes, as is shown in Figure 7a,b.

The samples printed in matte finishing were covered by support material on the entire surface, as is shown in Figure 7c. Also, all the holes were filled with support material.

After post-processing of the samples, surface errors on the vertical walls of the glossy samples were determined at both ends of horizontal holes, as is shown in Figure 8a. These kinds of errors consist in a rectangular area around the pattern of horizontal holes. These errors were detected both on the glossy sample oriented along the x-axis and on the one oriented along the y-axis. Thus, flat areas placed around the arrays of holes were found. These types of errors are caused by support material deposited inside the horizontal holes. There were no surface defects detected on the samples printed in matte finishing type after post-processing, as is shown in Figure 8b.

A microscopy analysis was performed for quality investigation of the micro-holes for the samples GX, GY, MX, and MY.

The horizontal micro-holes oriented along the x-axis and y-axis were analyzed, and so were the vertical micro-holes along the z-axis.

The upper end of the glossy micro-holes oriented along the z-axis is characterized by a rounded zone (PAZ), as is shown in Figure 9. The average PAZ value detected on the upper surface of the glossy micro-holes is 0.176 mm. It was detected that the glossy micro-holes of 0.25 mm in diameter were absent, as is shown in Figure 9j. A mark was present on the upper surface of these micro-holes, but the hole is absent on the lower surface. Similar morphologies for z-axis micro-holes orientation were found for both orientations of the glossy artifacts.

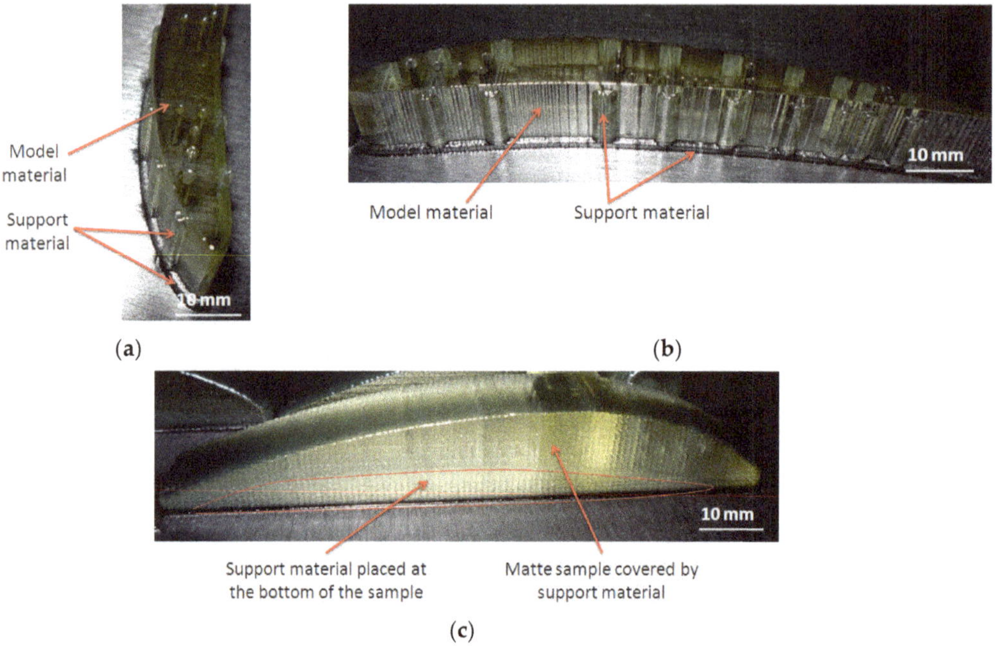

Figure 7. The artifacts on the build platform before post-processing: (**a**) oriented along y-axis, in glossy finishing (GY); (**b**) oriented along x-axis, in glossy finishing (GX); (**c**) oriented along x-axis, in matte finishing (MX)—a similar surface was obtained for a MY artifact.

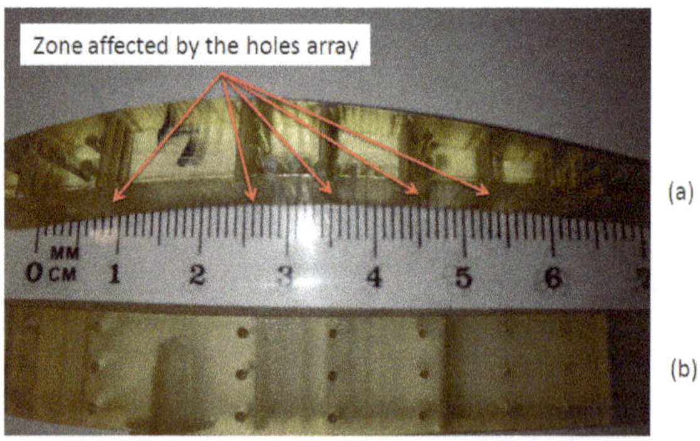

Figure 8. The artifacts after post-processing: (**a**) Zones affected by the three micro-holes' array for glossy artifacts; (**b**) Smooth surface obtained for matte artifacts.

The lower end of the glossy finishing micro-holes oriented along the z-axis were affected by the support material and presented a specific shape, as is shown in Figures 10 and 11. Some small pieces of support material were observed in the micro-holes because the holes were not completely cleaned after the post-processing stage. The micro-holes were not perfectly round, and the measurement therefore gave the maximum inscribed diameter.

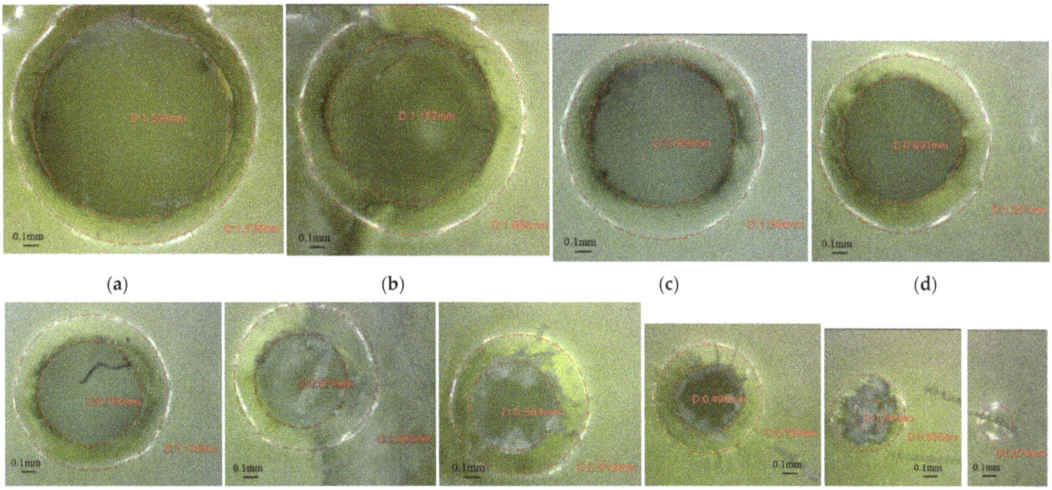

(**a**) (**b**) (**c**) (**d**)

(**e**) (**f**) (**g**) (**h**) (**i**) (**j**)

Figure 9. Microscopic image of the upper end of the micro-holes oriented along the *z*-axis (GX sample), which had different diameters: (**a**) 1.4 mm (D1.4_GX_Z_Up), (**b**) 1.2 mm (D1.2_GX_Z_Up), (**c**) 1 mm (D1_GX_Z_Up), (**d**) 0.9 mm (D0.9_GX_Z_Up), (**e**) 0.8 mm (D0.8_GX_Z_Up), (**f**) 0.7 mm (D0.7_GX_Z_Up), (**g**) 0.6 mm (D0.6_GX_Z_Up), (**h**) 0.5 mm (D0.9_GX_Z_Up), (**i**) 0.3 mm (D0.3_GX_Z_Up) and (**j**) 0.25 mm (D0.25_GX_Z_Up).

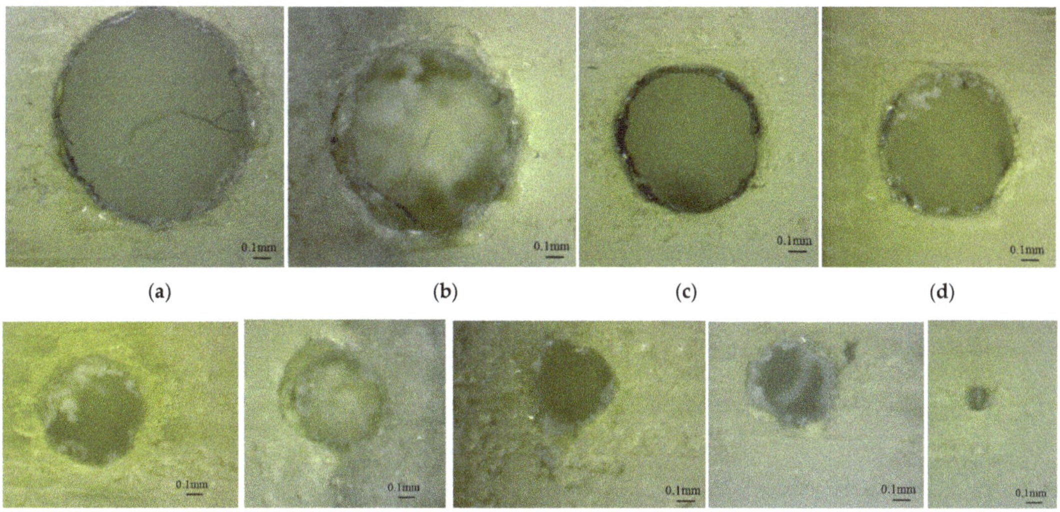

(**a**) (**b**) (**c**) (**d**)

(**e**) (**f**) (**g**) (**h**) (**i**)

Figure 10. Microscopic image of the lower end of the micro-holes oriented along the *z*-axis (GX sample), with different diameters: (**a**) 1.4 mm (D1.4_GX_Z_Low), (**b**) 1.2 mm (D1.2_GX_Z_Low), (**c**) 1 mm (D1_GX_Z_Low), (**d**) 0.9 mm (D0.9_GX_Z_Low), (**e**) 0.8 mm (D0.8_GX_Z_Low), (**f**) 0.7 mm (D0.7_GX_Z_Low), (**g**) 0.6 mm (D0.6_GX_Z_Low), (**h**) 0.5 mm (D0.9_GX_Z_Low), (**i**) 0.3 mm (D0.3_GX_Z_Low).

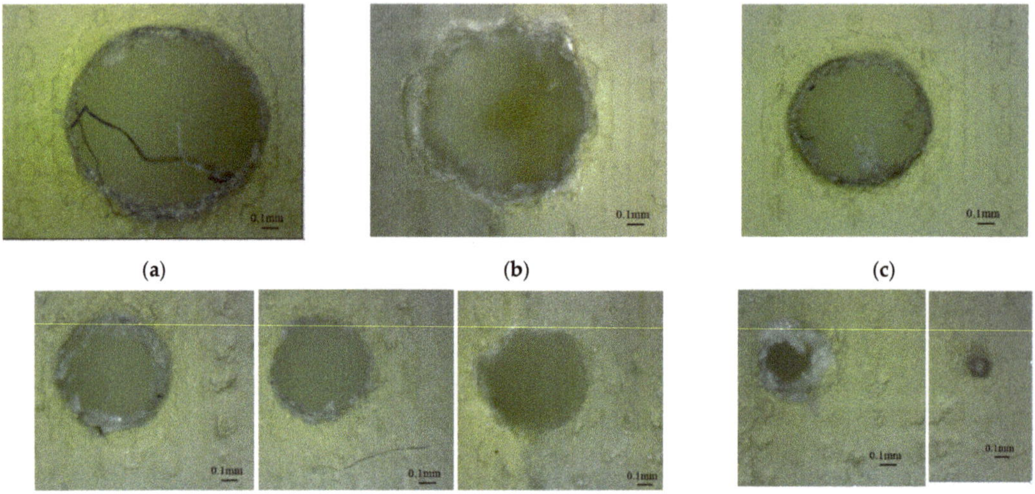

(a) (b) (c)

(d) (e) (f) (g) (h)

Figure 11. Microscopic image of the lower end of the micro-holes oriented along z-axis (GY sample), with different diameters: (**a**) 1.4 mm (D1.4_GY_Z_Low), (**b**) 1.2 mm (D1.2_GY_Z_Low), (**c**) 1 mm (D1_GY_Z_Low), (**d**) 0.9 mm (D0.9_GY_Z_Low), (**e**) 0.8 mm (D0.8_GY_Z_Low), (**f**) 0.7 mm (D0.7_GY_Z_Low), (**g**) 0.6 mm (D0.6_GY_Z_Low), (**h**) 0.5 mm (D0.9_GY_Z_Low).

The morphology of the lower end of the glossy finishing micro-holes oriented along the *z*-axis seems to be similar for both artifact orientations, but the train of resin droplets is deposited in different directions (Figures 10 and 11).

The clear transparent upper surface of the glossy artifact makes it possible to investigate the depth horizontal micro-holes (Figure 12). The diameter of the horizontal micro-holes was measured in three sections and the mean diameter was calculated. Examples of measured diameters of the holes (from the glossy artifact printed in the X direction) are shown in Figure 12. It was determined that the horizontal micro-holes are present in the diameter range extending from 0.5 mm to 1.4 mm. The 0.3 mm diameter micro-holes were partially filled at the hole end. The 0.25 mm diameter holes were closed on the entire depth, and were filled by model material (Figure 12j). These are failed micro-holes that, even though they have a surface geometry, are solid below the surface. The lateral surface of the holes has a specific texture, as is shown in Figure 12. Similar morphologies for horizontal glossy micro-holes oriented along the *x*-axis and *y*-axis were also found.

The results of the microscopy investigations for the matte samples oriented along the *x*-axis and *y*-axis showed similar morphologies, and similar characteristics were found for the left and right ends of the micro-holes. The top and bottom ends of the micro-holes also had the same characteristics. The horizontal micro-holes oriented along the *x*-axis appeared to be properly built into the diameter range of 0.3 mm to 1.4 mm, as is shown in Figure 13. The micro-holes 0.25 mm in diameter appeared to be filled, as is shown in Figure 13j. Similar results were found for the horizontal micro-holes oriented along the *y*-axis.

It was found that the matte micro-holes oriented along the *z*-axis are properly built in the diameter range that extends from 0.5 mm to 1.4 mm. The 0.3 mm and 0.25 mm diameter micro-holes were closed holes. The micro-holes oriented along the *z*-axis were determined to be irregular in shape, as is shown in Figure 14. This could be explained by the lower resolution of the printer in the X and Y directions, compared to the Z direction.

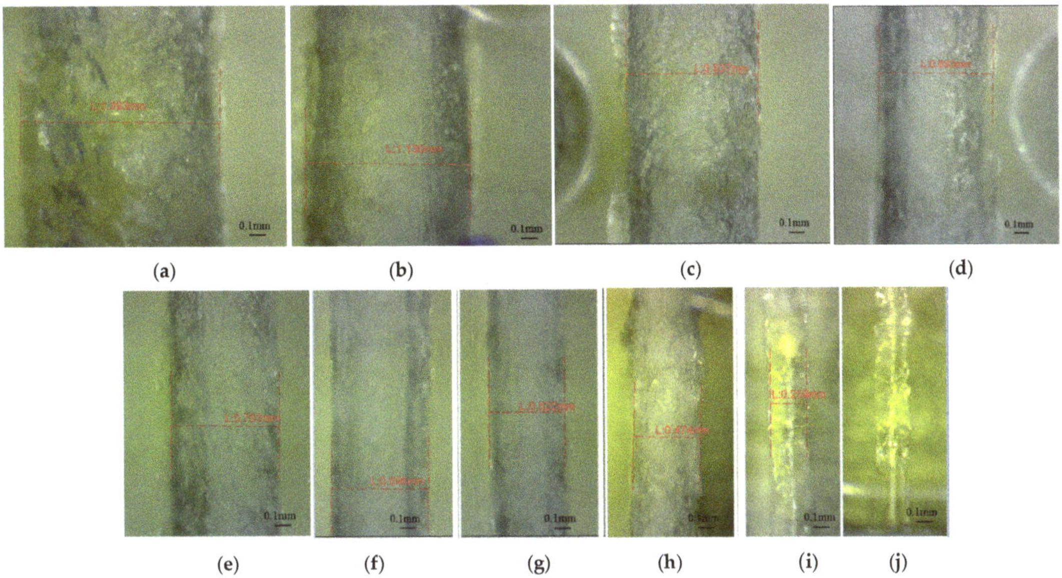

Figure 12. Microscopic image of the horizontal micro-holes oriented along the *y*-axis (GX sample), with different diameters: (**a**) 1.4 mm (D1.4_GX_Y), (**b**) 1.2 mm (D1.2_GX_Y), (**c**) 1 mm (D1_GX_Y), (**d**) 0.9 mm (D0.9_GX_Y), (**e**) 0.8 mm (D0.8_GX_Y), (**f**) 0.7 mm (D0.7_GX_Y), (**g**) 0.6 mm (D0.6_GX_Y), (**h**) 0.5 mm (D0.9_GX_Y), (**i**) 0.3 mm (D0.3_GX_Y), (**j**) 0.25 mm (D0.25_GX_Y).

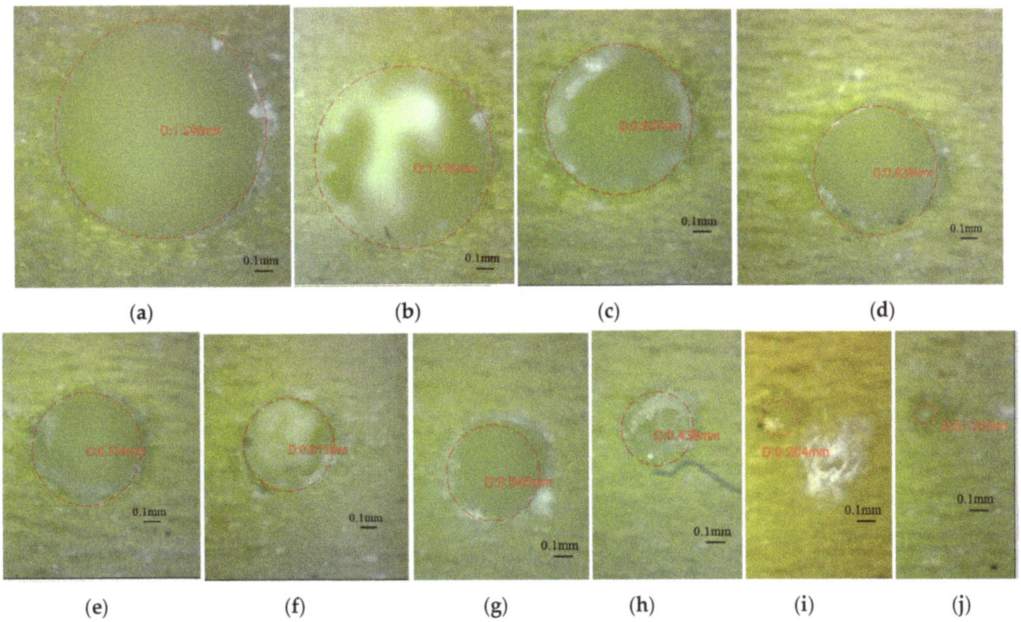

Figure 13. Microscopic image of the micro-holes oriented along the *x*-axis (MY sample), with different diameters: (**a**) 1.4 mm (D1.4_MY_X), (**b**) 1.2 mm (D1.2_MY_X), (**c**) 1 mm (D1_MY_X), (**d**) 0.9 mm (D0.9_MY_X), (**e**) 0.8 mm (D0.8_MY_X), (**f**) 0.7 mm (D0.7_MY_X), (**g**) 0.6 mm (D0.6_MY_X), (**h**) 0.5 mm (D0.9_MY_X), (**i**) 0.3 mm (D0.3_MY_X) and (**j**) 0.25 mm (D0.25_MY_X).

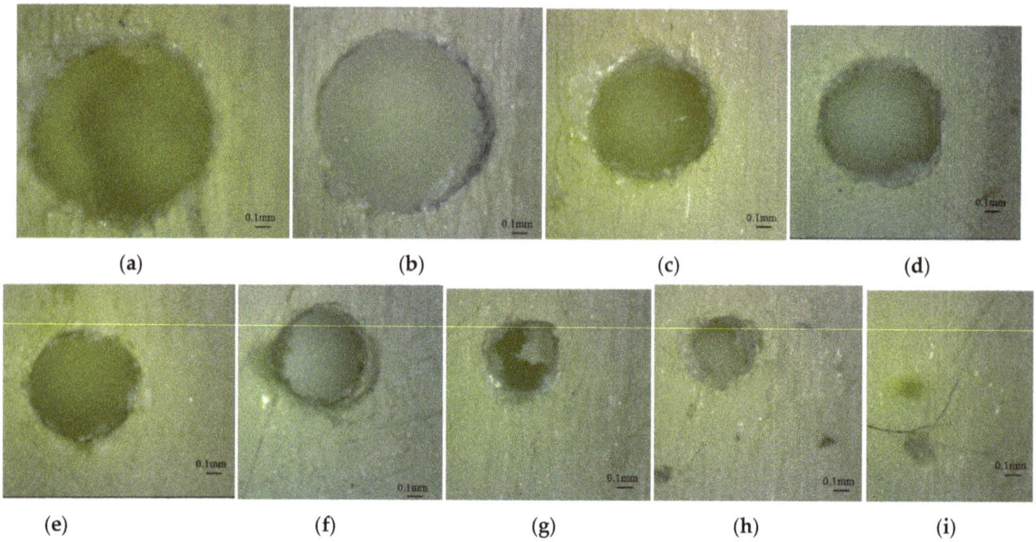

Figure 14. Microscopic image of the micro-holes oriented along the *z*-axis (MY sample), with different diameters: (**a**) 1.4 mm (D1.4_MY_Z), (**b**) 1.2 mm (D1.2_MY_Z), (**c**) 1 mm (D1_MY_Z), (**d**) 0.9 mm (D0.9_MY_Z), (**e**) 0.8 mm (D0.8_MY_Z), (**f**) 0.7 mm (D0.7_MY_Z), (**g**) 0.6 mm (D0.6_MY_Z), (**h**) 0.5 mm (D0.9_MY_Z), and (**i**) 0.3 mm (D0.3_MY_Z).

Deep investigations of the shape of the micro-holes are necessary to understand their characteristics. Thus, the rectangular area around the pattern of horizontal holes (AAZ) detected by visual investigation of the ends of the glossy micro-holes were analyzed, as is shown in Figure 15. The width of this rectangular area (array-affected zone) is in the range of 0.88 mm (D0.3 mm), and 1.786 mm (D1.4 mm) for the artifact printed in glossy mode that is oriented along the *x*-axis (Figure 15a). A relative constant width (AAZ) of around 2.26 mm was found for the artifact oriented along the *y*-axis (Figure 15b). The artifacts printed in matte finishing show a smooth surface around the array of holes oriented on the *z*-axis. No surface defects were detected in this case.

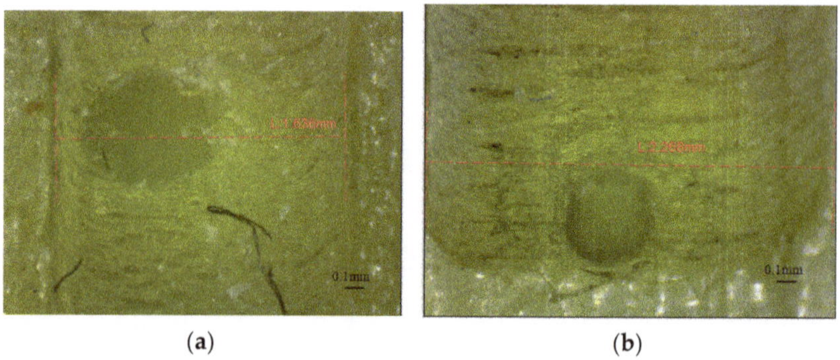

Figure 15. Array affected zone (AAZ) of the horizontal micro-hole, printed in glossy finishing for: (**a**) artifact oriented along the *x*-axis; (**b**) artifact oriented along the *y*-axis.

The ends of micro-holes printed in glossy mode present specific shapes which are perpendicular on the artifact surface, as is shown in Figure 16a–c. The average height of the ends is around 0.34 mm. Filled micro-hole ends are shown in Figure 16d.

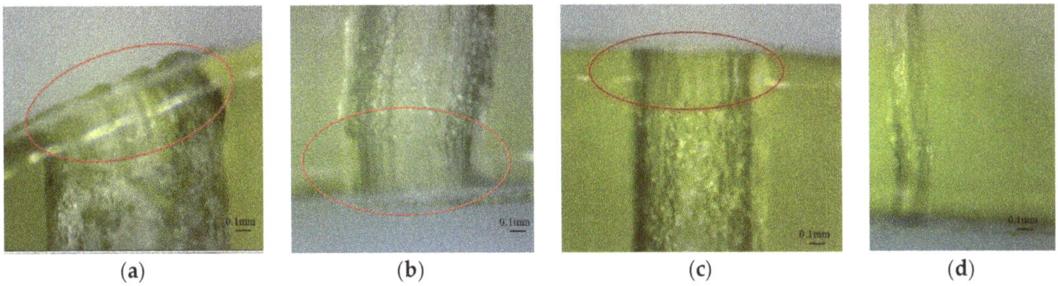

Figure 16. The ends of horizontal micro-holes printed in glossy finishing. The red circle shown the end area of the micro-hole: (a) The micro-hole end on a high sloped surface; (b) The micro-hole end on a low sloped surface; (c) The micro-hole end on a horizontal surface; (d) a closed micro-hole on the entire depth.

The investigation of the diameter of the depth of micro-holes oriented in the z-axis was difficult to perform by microscopy because of lower transparency of the lateral walls of the artifact, as is shown in Figure 17. The internal structure of the matte micro-holes is difficult to investigate based on the lower transparency of the artifact surfaces. The best transparency was obtained for the horizontal surface of the artifact printed in a glossy mode.

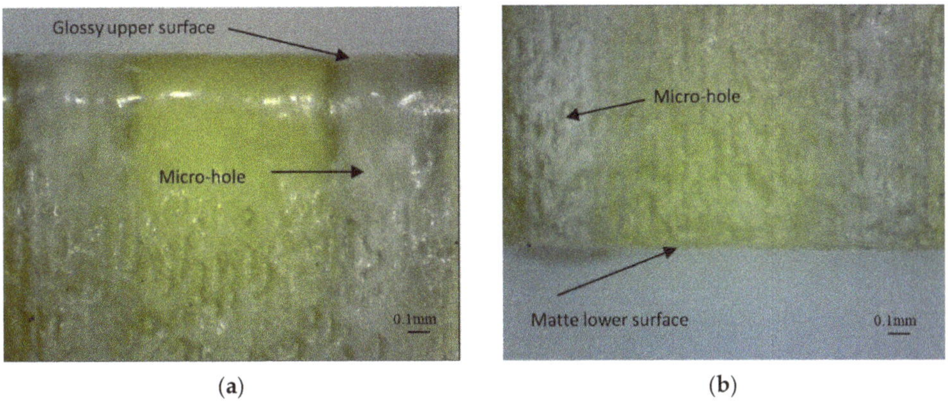

Figure 17. Images of the micro-holes oriented along the z-axis and printed in glossy finishing: (a) Upper surface; (b) Lower surface affected by the support material.

3.2. Results of Statistical Analysis

The results of the ANOVA analysis are shown in Table 3. The hole diameter, finish type and hole orientation had a more significant influence on the diameter deviation, as long as the p-value was lower than the significance level of 0.001. The results were significant at the 0.1% significance level.

The artifact orientation factor had a lower percentage contribution. The most significant influence on the diameter deviation (Dev) was the hole diameter, which explained 27.98% of the total variation. The R-squared value of 95.72% indicates that the model explains all the variability of the response data around its mean.

The interactions of hole orientation with hole diameter and artifact orientation with hole diameter had a significant effect on the diameter deviation, whereby PC% was 15.91% and 10.02%, respectively. The interactions of artifact orientation with finish type, finish type with hole orientation, and finish type with hole diameter had no significant effect on the diameter deviation ($p > 0.05$).

Table 3. The percentage contribution ratio based on the generalized linear model (GLM); The symbol "*" signifies the interaction between factors.

Source	DF	Seq SS	Seq MS	F_{exp}	$F_{0.1\%}$	p	PC (%)
Artifact orientation	1	0.006622	0.006622	19.97	17.14	0.001	6.10%
Finish type	1	0.014793	0.014793	44.61	17.14	<0.001	13.64%
Hole orientation	2	0.013004	0.006502	19.61	9.72	<0.001	11.99%
Hole diameter	7	0.030356	0.004337	13.08	7.07	<0.001	27.98%
Artifact orientation * Finish type	1	0.001131	0.001131	3.41	17.14	0.086	1.04%
Artifact orientation * Hole diameter	7	0.010874	0.001553	4.69	7.07	0.007	10.02%
Finish type * Hole orientation	2	0.00267	0.001335	4.03	9.72	0.042	2.46%
Finish type * Hole diameter	7	0.004573	0.000653	1.97	7.07	0.133	4.22%
Hole orientation * Hole diameter	14	0.017266	0.001233	3.72	5.92	0.01	15.91%
Artifact orientation * Finish type * Hole diameter	7	0.002562	0.000366	1.1	7.07	0.412	2.36%
Error	14	0.004642	0.000332				4.28%
Total	63	0.108491					100.00%

The influence of control factors on diameter deviation was statistically evaluated using graphical methods. Thus, the following graphs were plotted: the main effects plot for deviation, interaction effects plot for deviation, and interval plot of deviation versus the control factors. The results show that artifact orientation at level 1 (0°), finish type at level 1 (matte), hole orientation at level 3 (*z*-axis), and hole diameter at level 1 (1.4 mm) had the main effects plot for deviation (Figure 18). The interaction of hole diameter factor with the other factors had a significant influence on the diameter deviation, as shown in Figure 19.

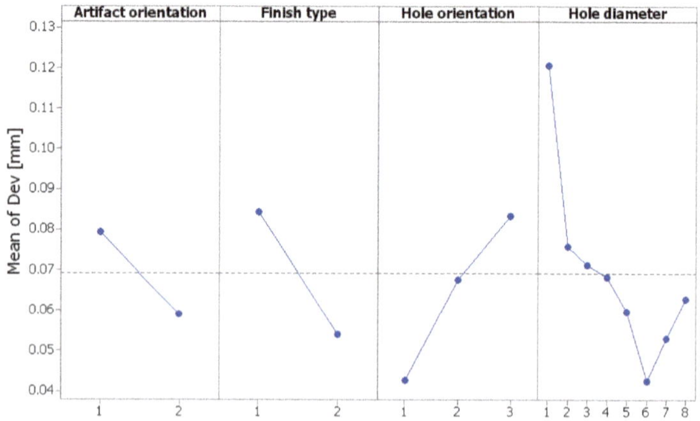

Figure 18. Main effects plot for diameter deviation (Dev).

The interval plots with the standard error bars of each factor versus the diameter deviation (Dev) are shown in Figure 20. The difference for Dev in the hole diameter was probably not significant for the diameters of 1.2 mm, 1 mm, 0.9 mm, and 0.8 mm because all the interval bars easily overlapped (Figure 20a). Also, the hole diameter had an influence on the diameter deviation and it seems that, for hole diameter of 1.4 mm, the mean of Dev was higher, but lower for the hole diameter of 0.7 mm.

The difference between the means for diameter deviation in artifact orientation, finish type, and hole orientation were significant because the interval bars did not overlap (Figure 20a–c). The means of Dev were lower for the artifact orientation along the *y*-axis, the glossy finish type, and the hole orientation along the *x*-axis.

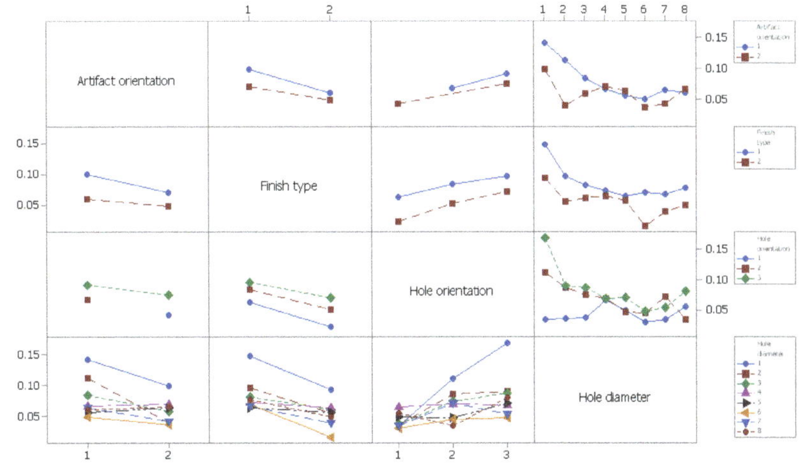

Figure 19. Interaction effects plot for diameter deviation (Dev).

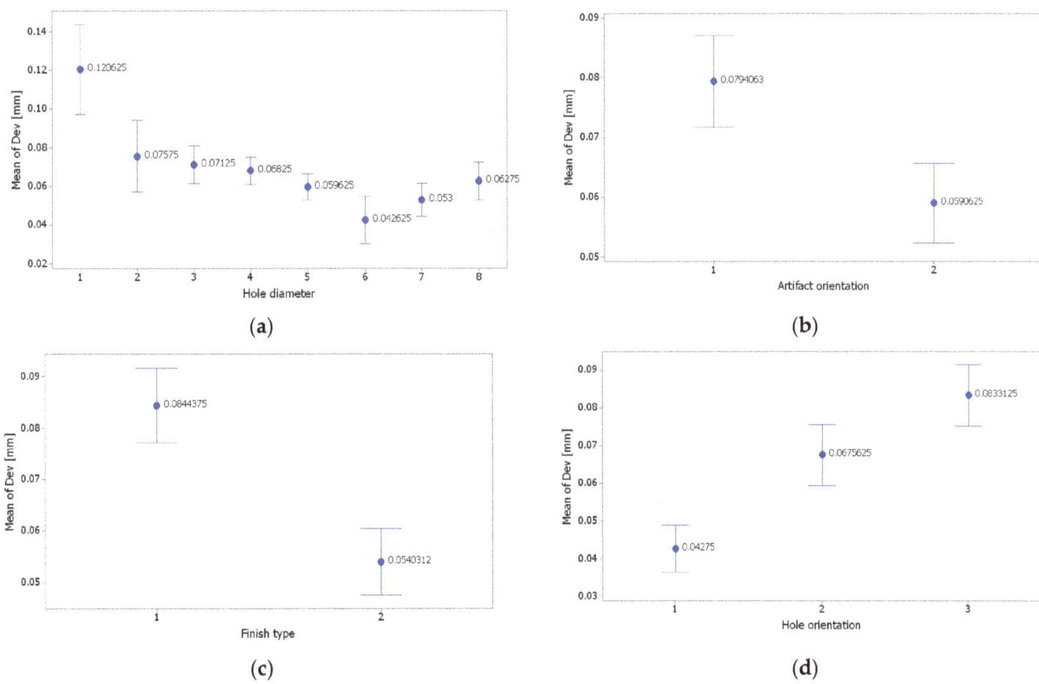

Figure 20. Interval plot of: (**a**) hole diameter; (**b**) artifact orientation; (**c**) finish type; and (**d**) hole orientation. Individual standard deviations were used to calculate the intervals plot. Bars are standard errors of the mean.

The generalized linear model was checked for model adequacy by using the normal probability plots of residuals [27]. The residuals were normally distributed, as is shown in Figure 21.

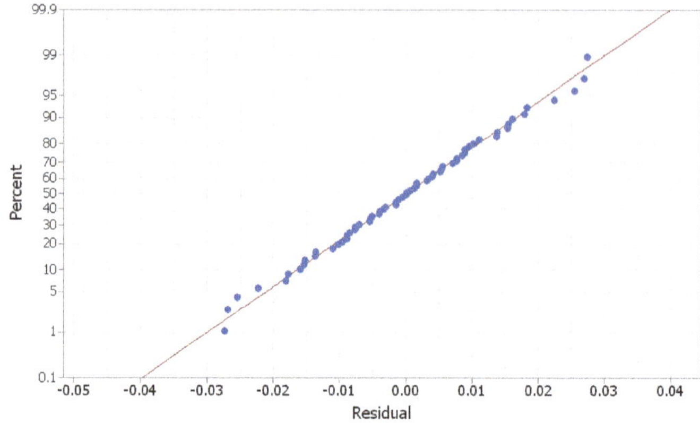

Figure 21. Normal probability plots of residuals for deviation (Dev).

3.3. *Case Study as Confirmatory Experiment*

A theoretical model of the hole shape is proposed by the author, as shown in Figure 22. The train of resin droplets grouped in cylindrical shapes, deposited by each nozzle in the X direction, forms a layer. The intersection between the micro-hole and layers forms the theoretical shape of the 3D-printed hole. The surface of holes oriented along the *y*-axis or *z*-axis results from the intersection of several cylinders from different layers, which are normal for the hole direction, resulting in a rough surface. The hole oriented along the *x*-axis consists of an intersection of the cylindrical train of droplets, which is parallel to the hole axis direction. Thus, the surface of the deep hole oriented along the *x*-axis is more homogeneous and precise than the other two types of holes. The hole surface oriented along the *x*-axis is smooth and continuous (Figure 22).

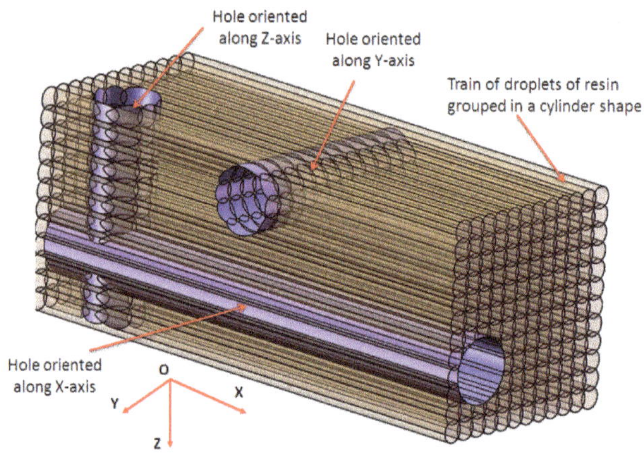

Figure 22. Theoretical model of the micro-holes shape.

Five experimental wings with eight horizontal micro-holes and eight vertical holes that were 1 mm in diameter (D1) were manufactured by PolyJet technology (Figure 23). On the basis of the previous experimental study and the theoretical model, the horizontal deep micro-holes from the wing were oriented along the *x*-axis. The manufacturing time for an experimental wing printed in matte finishing along the *x*-axis was 1 h and 24 min, and the materials consumption was 168 g of model material and 100 g of support material.

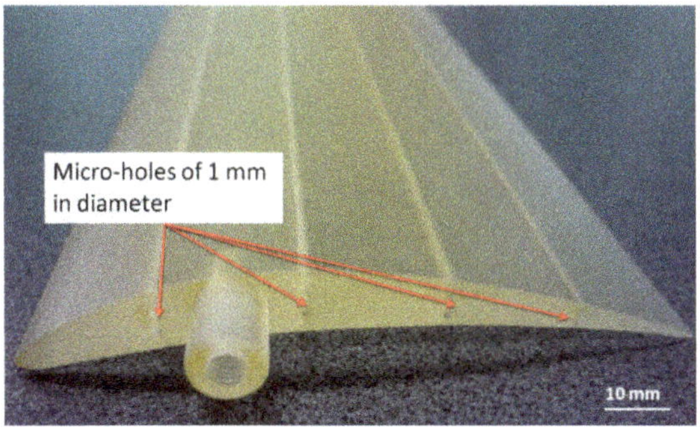

Figure 23. Experimental wing with eight horizontal micro-holes and eight vertical holes that are 1 mm in diameter.

The results of the statistical analysis of the 1 mm diameter micro-holes have shown that the coefficients of variation are lower than 10%, which confirms data heterogeneity and expresses the repeatability of the experiments, as shown in Table 4.

Table 4. Statistics of 1 mm diameter micro-holes.

Micro-Hole Orientation	Mean Value [mm]	Standard Deviation [mm]	Coefficient of Variation [%]
X	0.932	0.01129	1.21
Z	0.899	0.01274	1.42

The interval plots of the D1_X and D1_Z diameter of the micro-holes are shown in Figure 24. Individual standard deviations were used to calculate the interval plot. The highest values for the D1-measured diameter were found for the x-axis orientation of the wing. This was explained by the higher resolution of 0.016 mm in the z-axis. The smallest values of the D1-measured diameter on the z-axis direction (Figure 24) were caused by the lower resolution of 0.042 mm in the x-direction and y-direction. The diameter deviation along the x-axis is therefore smaller than the deviation along the other two axes.

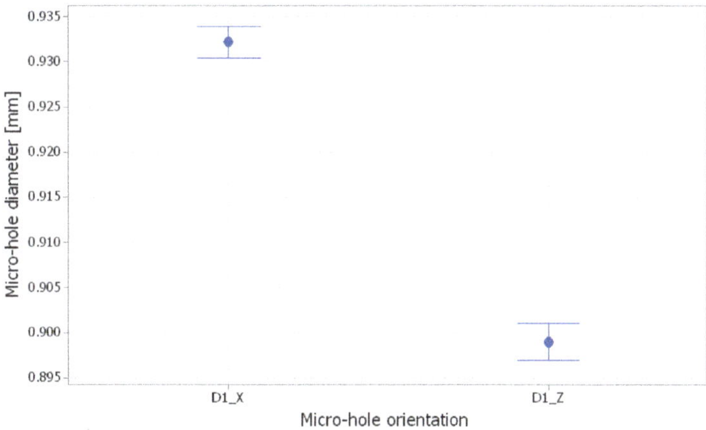

Figure 24. Interval plot of 1 mm diameter micro-holes (D1) for the X and Z orientations; The bars are one standard error from the mean; individual standard deviations were used to calculate the intervals.

4. Conclusions

This paper investigated the quality of micro-holes made by the material jetting AM process and characterized the features of micro-holes built in three basic directions, along the *x*-axis, *y*-axis, and *z*-axis based on the proposed methodology. The following conclusions can be drawn:

- An artifact made by translucent polymer resin allows a facile investigation (by microscopy) of the inner structure of the micro-holes.
- The experimental research demonstrated that hole diameter, finish type, and hole orientation were the most influential factors on diameter deviation of the micro-holes.
- Micro-holes are smaller than nominal. The micro-holes made by the PolyJet technology that are oriented along the *x*-axis are more accurate than the micro-holes along the *y*-axis and *z*-axis, and have an average deviation of 0.042 mm, compared with 0.067 mm and 0,083 mm, respectively.
- The experimental deviations of the diameter of the micro-holes manufactured by PolyJet technology in the glossy finish mode were lower than those obtained in the matte finish mode. After uniform surface and the quality issues of glossy mode were taken into account, the micro-holes printed in matte finishing gave the best results.
- Three situations of micro-holes construction by the AM systems can be noted, names of holes that appear properly built, holes that appear partially filled, and holes that appear absent. The smallest diameter of the properly-built micro-holes obtained by PolyJet technology using the EDEN 350 machine was 0.5 mm. After facile post-processing is taken into account, it is recommended that 1 mm diameter micro-holes should be chosen for industrial application.

The proposed methodology can be used to evaluate the accuracy of the micro-holes made by different materials, and to compare different additive manufacturing processes, in a way that considers the specific characteristics of each AM technology and chooses the proper control factors.

Funding: This research received no external funding.

Institutional Review Board Statement: Not applicable.

Data Availability Statement: Data are contained within the article.

Acknowledgments: The author acknowledges that Transilvania University of Braşov provided the infrastructure used in this paper.

Conflicts of Interest: The author declares no conflicts of interest.

References

1. Tofail, S.A.M.; Koumoulos, E.P.; Bandyopadhyay, A.; Bose, S.; O'Donoghue, L.; Charitidis, C. Additive manufacturing: Scientific and technological challenges, market uptake and opportunities. *Mater. Today* **2018**, *21*, 22–37. [CrossRef]
2. *ISO/ASTM 52900-15*; Standard Terminology for Additive Manufacturing—General Principles—Terminology. ISO: Geneva, Switzerland, 2015.
3. Yap, Y.L.; Wang, C.; Sing, S.L.; Dikshit, V.; Yeong, W.Y.; Wei, J. Material jetting additive manufacturing: An experimental study using designed metrological benchmarks. *Precis. Eng.* **2017**, *50*, 275–285. [CrossRef]
4. Gülcan, O.; Günaydın, K.; Tamer, A. The State of the Art of Material Jetting—A Critical Review. *Polymers* **2021**, *13*, 2829. [CrossRef] [PubMed]
5. Udroiu, R. New Methodology for Evaluating Surface Quality of Experimental Aerodynamic Models Manufactured by Polymer Jetting Additive Manufacturing. *Polymers* **2022**, *14*, 371. [CrossRef] [PubMed]
6. Leach, R. Metrology for additive manufacturing. *Meas. Control* **2016**, *49*, 132–135. [CrossRef]
7. Umaras, E.; Tsuzuki, M.S.G. Additive manufacturing-considerations on geometric accuracy and factors of influence. *IFAC-PapersOnLine* **2017**, *50*, 14940–14945. [CrossRef]
8. Świderski, J.; Makieła, W.; Dobrowolski, T.; Stępień, K.; Zuperl, U. The study of the roundness and cylindricity deviations of parts produced with the use of the additive manufacturing. *Int. J. Adv. Manuf. Technol.* **2022**, *121*, 7427. [CrossRef]
9. Kim, G.D.; Oh, Y.T. A benchmark study on rapid prototyping processes and machines: Quantitative comparisons of mechanical properties, accuracy, roughness, speed, and material cost. *Proc. Inst. Mech. Eng. Part B J. Eng. Manuf.* **2008**, *222*, 201–215. [CrossRef]

10. Mueller, J.; Shea, K.; Daraio, C. Mechanical properties of parts fabricated with inkjet 3D printing through efficient experimental design. *Mater. Des.* **2015**, *86*, 902–912. [CrossRef]
11. Geier, N.; Patra, K.; Shankar Anand, R.; Ashworth, S.; Blasz, B.; Davim, P. A critical review on mechanical micro-drilling of glass and carbon fibre reinforced polymer (GFRP and CFRP) composites. *Compos. Part B* **2023**, *254*, 110589. [CrossRef]
12. Kara, A.; Vassiliadou, A.; Ongoren, B.; Keeble, W.; Hing, R.; Lalatsa, A.; Serrano, D.R. Engineering 3D Printed Microfluidic Chips for the Fabrication of Nanomedicines. *Pharmaceutics* **2021**, *13*, 2134. [CrossRef] [PubMed]
13. Gonzaleza, G.; Roppoloa, I.; Pirria, C.F.; Chiapponed, A. Current and emerging trends in polymeric 3D printed microfluidic devices. *Addit. Manuf.* **2022**, *55*, 102867. [CrossRef]
14. Endo, H.; Marui, E. Small-hole drilling in engineering plastics sheet and its accuracy estimation. *Int. J. Mach. Tools Manuf.* **2006**, *46*, 575–579. [CrossRef]
15. Singh, M.; Singh, S.; Arora, J.K.; Antil, P.; Oza, A.D.; Burduhos-Nergis, D.D.; Burduhos-Nergis, D.P. Multi Response Optimization of ECDM Process for Generating Micro Holes in CFRP Composite Using TOPSIS Methodology. *Polymers* **2022**, *14*, 5291. [CrossRef] [PubMed]
16. Gomez-Gras, G.; Pérez, M.A.; Fabregas-Moreno, J.; Reyes-Pozo, G. Experimental study on the accuracy and surface quality of printed versus machined holes in PEI Ultem 9085 FDM specimens. *Rapid Prototyp. J.* **2021**, *27*, 1–12. [CrossRef]
17. Tiwary, V.K.; Arunkumar, P.; Vinayak, R.M. An overview on joining/welding as post-processing technique to circumvent the build volume limitation of an FDM-3D printer. *Rapid Prototyp. J.* **2021**, *27*, 808–821. [CrossRef]
18. Popescu, D.; Amza, C.G.; Marinescu, R.; Iacob, M.C.; Căruţaşu, N.L. Investigations on Factors Affecting 3D-Printed Holes Dimensional Accuracy and Repeatability. *Appl. Sci.* **2023**, *13*, 41. [CrossRef]
19. Zhang, W.; Lin, X.; Jiang, J. Dimensional accuracy of 3D printing navigation templates of chemical-based sterilization. *Sci. Rep.* **2022**, *12*, 1253. [CrossRef]
20. Olasek, K.; Wiklak, P. Application of 3D printing technology in aerodynamic study. *J. Phys. Conf. Ser.* **2014**, *530*, 012009. [CrossRef]
21. Junka, S.; Schröder, W.; Schrock, S. Design of additively manufactured wind tunnel models for use with UAVs. *Procedia CIRP* **2017**, *60*, 241–246. [CrossRef]
22. Beltrán, N.; Carriles, F.; Álvarez, B.J.; Blanco, D.; Rico, J.C. Characterization of factors influencing dimensional and geometric errors in PolyJet manufacturing of cylindrical features. *Procedia Eng.* **2015**, *132*, 62–69. [CrossRef]
23. Nagendra, K.M.; Rastogi, V.; Singh, P. Comparative Study and Measurement of Form Errors for the Component Printed by FDM and PolyJet Process. *Instrum. Mes. Métrologie.* **2019**, *18*, 353–359. [CrossRef]
24. Pilipović, A.; Baršić, G.; Katić, M.; Rujnić Havstad, M. Repeatability and Reproducibility Assessment of a PolyJet Technology Using X-ray Computed Tomography. *Appl. Sci.* **2020**, *10*, 7040. [CrossRef]
25. Tee, Y.L.; Tran, P.; Leary, M.; Pille, P.; Brandt, M. 3D Printing of polymer composites with material jetting: Mechanical and fractographic analysis. *Addit. Manuf.* **2020**, *36*, 101558. [CrossRef]
26. *ISO/ASTM 52902-15*; Additive Manufacturing—Test Artifacts—Geometric Capability Assessment of Additive Manufacturing Systems. ISO: Geneva, Switzerland, 2019.
27. Udroiu, R.; Braga, I.C.; Nedelcu, A. Evaluating the Quality Surface Performance of Additive Manufacturing Systems: Methodology and a Material Jetting Case Study. *Materials* **2019**, *12*, 995. [CrossRef] [PubMed]
28. Chen, Y.; Lu, J. RP Part Surface quality versus build orientation: When the layers are getting thinner. *Int. J. Adv. Manuf. Technol.* **2013**, *67*, 377–385. [CrossRef]
29. Udroiu, R.; Braga, I.C. System Performance and Process Capability in Additive Manufacturing: Quality Control for Polymer Jetting. *Polymers* **2020**, *12*, 1292. [CrossRef]
30. Moylan, S. Progress toward standardized additive manufacturing test artifacts. In Proceedings of the ASPE 2015 Spring Topical Meeting Achieving Precision Tolerances in Additive Manufacturing, Raleigh, NC, USA, 26–29 April 2015; pp. 100–105.
31. Krolczyk, G.; Raos, P.; Legutko, S. Experimental analysis of surface roughness and surface texture of machined and fused deposition modelled parts. *Teh. Vjesn.* **2014**, *21*, 217–221.
32. Montgomery, D.C. *Design and Analysis of Experiments*; John Wiley & Sons: Hoboken, NJ, USA, 2017; ISBN 9781119113478.
33. Minitab. Getting Started with Minitab 17. Available online: https://www.minitab.com (accessed on 15 March 2022).
34. Objet Geometries. Eden 500V/350V/350 3-D Printer System. In *User Guide*; Objet Geometries Ltd.: Rehovot, Israel, 2007.
35. *ISO/ASTM 52921-13*; Standard Terminology for Additive Manufacturing—Coordinate Systems and Test Methodologies. ISO: Geneva, Switzerland, 2013.
36. Stratasys. PolyJet Materials Data Sheet. Available online: http://www.stratasys.com (accessed on 10 September 2015).
37. Kepler, J. Investigation of Acrilic Based Systems for 3D Polyjet Printing. Master's Thesis, University of Linz, Linz, Austria, 2018.
38. Derby, B. The inkjet printing of functional and structural materials: Fluid property requirements, feature stability, and resolution. *Annu. Rev. Mater. Res.* **2010**, *40*, 395–414. [CrossRef]

polymers

Article

Mechanical and Wear Studies of Boron Nitride-Reinforced Polymer Composites Developed via 3D Printing Technology

Ramaiah Keshavamurthy [1], Vijay Tambrallimath [2,*], Swetha Patil [1], Ali A. Rajhi [3], Alaauldeen A. Duhduh [4] and T. M. Yunus Khan [3]

[1] Department of Mechanical Engineering, Dayananda Sagar College of Engineering, Bangalore 560078, India; keshavamurthy.r@gmail.com (R.K.); patilshweta999@gmail.com (S.P.)
[2] Department of Automobile Engineering, Dayananda Sagar College of Engineering, Bangalore 560078, India
[3] Department of Mechanical Engineering, College of Engineering, King Khalid University, Abha 62529, Saudi Arabia; arajhi@kku.edu.sa (A.A.R.)
[4] Department of Mechanical Engineering Technology, CAIT, Jazan University, Prince Mohammed Street, P.O. Box 114, Jazan 45142, Saudi Arabia; alaaduhduh@jazanu.edu.sa
* Correspondence: vijay-au@dayanandasagar.edu or hod-au@dayanandasagar.edu

Abstract: In the realm of 3D printing, polymers serve as fundamental materials offering versatility to cater to a diverse array of final product properties and tailored to the specific needs of the creator. Polymers, as the building blocks of 3D printing, inherently possess certain mechanical and wear properties that may fall short of ideal. To address this limitation, the practice of reinforcing polymer matrices with suitable materials has become a common approach. One such reinforcement material is boron nitride (BN), lauded for its remarkable mechanical attributes. The integration of BN as a reinforcing element has yielded substantial enhancements in the properties of polylactic acid (PLA). The central objective of this research endeavor is the development of polymer composites based on PLA and fortified with boron nitride. This study undertakes the comprehensive exploration of the compatibility and synergy between BN and PLA with a keen focus on examining their resultant properties. To facilitate this, various percentages of boron nitride were incorporated into the PLA matrix, specifically at 5% and 10% by weight. The compounding process involved the blending of PLA and boron nitride followed by the creation of composite filaments measuring 1.75 mm in diameter and optimized for 3D printing. Subsequently, test specimens were meticulously fabricated in adherence with ASTM standards to evaluate the ultimate tensile strength, dimensional accuracy, wear characteristics, and surface roughness. The findings from these assessments were systematically compared to the wear properties and mechanical behavior of PLA composites reinforced with boron nitride and the unreinforced PLA material. This study serves as a foundational resource that offers insights into the feasibility and methodologies of incorporating boron nitride into PLA matrices, paving the way for enhanced polymer composite development.

Keywords: fused deposition modeling; boron nitride; polylactic acid; tensile strength; dimensional accuracy; wear

Citation: Keshavamurthy, R.; Tambrallimath, V.; Patil, S.; Rajhi, A.A.; Duhduh, A.A.; Khan, T.M.Y. Mechanical and Wear Studies of Boron Nitride-Reinforced Polymer Composites Developed via 3D Printing Technology. *Polymers* **2023**, *15*, 4368. https://doi.org/10.3390/polym15224368

Academic Editors: Anton Ficai and Cristina-Elisabeta Pelin

Received: 12 September 2023
Revised: 27 October 2023
Accepted: 31 October 2023
Published: 9 November 2023

1. Introduction

In recent years, there has been a significant surge in the utilization of materials derived from renewable sources. Biopolymers, as an eco-friendly alternative to petroleum-based polymers, have gained prominence due to their biocompatibility, biodegradability, and renewability [1]. Among the myriad biopolymers available, polylactic acid (PLA) has emerged as a prominent choice. PLA is a thermoplastic aliphatic polyester produced from renewable resources like corn, wheat, cellulose, sugar cane, and starch. It distinguishes itself with its exceptional barrier properties, high strength, biocompatibility, elevated modulus, low toxicity, heat resistance, improved processability, and transparency. PLA finds applications in a wide range of fields, including medicine (porous meshes, scaffolds, drug delivery systems, and implants), automotive, packaging, and even in the creation of urinary bladder scaffolds.

However, the practical utility of these materials has been hindered by their limited thermal stability, rigidity, brittleness, and slow crystallization rate [2]. Nevertheless, the favorable biodegradability and mechanical properties of PLA have spurred extensive research into its production. Numerous endeavors have been undertaken to enhance PLA's properties by introducing various fillers. The literature demonstrates the use of substances such as clay [3], cellulose crystals [4], natural fibers [5], and graphene [6] as well as oxides, carbides, and nitrides to enhance specific attributes, rendering them suitable for a diverse array of applications. The addition of particles to PLA has proven to be an effective method for altering its physical and mechanical properties [7]. Boron nitride (BN), with its impressive thermal conductivity and mechanical properties, is a potential polymer-reinforcement material. The introduction of BN as a reinforcement for PLA will significantly improve their properties, making them more suitable for electronic product packaging. According to the available literature, PLA/BN composites have never been the subject of exhaustive research [7–9]. PLA ranked as the second most consumed natural plastic in the world in 2010, but it is not yet a standard polymer. Multiple physical and processing limitations have hindered the widespread deployment of PLA [10]. Rapid prototyping, also known as additive manufacturing, is a technique for producing prototypes and end-use parts directly from three-dimensional digital data. One of the major benefits of these technologies is the ability to produce personalized products with minimal lead time and expense. In contrast to subtractive and joining techniques used in conventional production systems, additive manufacturing operates by stacking material layer by layer. Feeders or extruders feed filaments or wires into the nozzles and heat them to a liquid or molten state as part of the FDM process. The material is extruded through nozzles when it has been heated slightly above its melting point. The final product consists of layers of molten material deposited in layers and rapidly cooled during deposition [11]. Various studies have investigated various filament types for FFF. Gray et al. [12] made an attempt to develop thermotropic liquid crystalline polymers (TLCPs), which were combined with a polypropylene (PP) matrix to improve the tensile properties of polymer composites so that they could be used as functional components in FFF. A morphological and tensile analysis of filaments extracted via dual extrusion was performed in this study. Zhong et al. [13] conducted a successful study on the use of short reinforcement fiber in an ABS matrix. The composite filament could be used for FFF due to its increased strength. The tensile strength and morphology of the filaments extracted via dual extrusion were analyzed as properties. Shofner et al. [14] manufactured a polymer composite by combining an ABS matrix with single-walled carbon nanotubes. G. Melenka et al. [15] conducted an experiment to develop a polymer composite by adding continuous Kevlar fibers to an ABS matrix. As part of the study, the filler material was altered to determine whether or not the tensile properties varied. The amount of Kevlar fibers added ranged from 4.04 to 8.08 to 10.1%. Along with the strain and Young's modulus, the material's tensile strength increased as it was gradually reinforced. Perez et al. [16] analyzed the effect of various filler materials on the tensile strength of composites. ABS was chosen as the matrix material for the preparation of the samples. Separately incorporated thermoplastic elastomer, jute fiber, and TiO_2 samples were compared to ABS-only samples. ABS-TiO_2 had a greater tensile strength than jute fiber and TPE, whereas ABS-TiO_2 had a greater tensile strength than jute fiber and TPE. When two polymers are combined, there will be an improvement in certain properties. The most commonly employed blends are polycarbonate and acrylonitrile butadiene styrene. By combining these materials, superior impact resistance and thermal stability are achieved. As a result of its enhanced mechanical properties and stability [17], it is an excellent material for the polymer industry. Automobile and home appliance manufacturing are major industries that employ these substances [18]. In a fascinating study, additive manufacturing was used for the first time to produce an electrically insulated, thermally conductive material. Using a 3D printing technique, the mechanical and thermal properties of boron nitride (BN)–

acrylonitrile butadiene styrene (ABS) composites were investigated. In terms of thermal conductivity, a 45–45° infill had slightly better flexural strength than a 0–90° infill, but the thermal conductivity was insensitive to infill orientation. The mechanical strength and anisotropic heat conductivity of 3D-printed composite materials are frequently diminished. Adding BN flakes modestly increased the flexural modulus [19]. When looking into PLA as a matrix material to be used along with BN as reinforcement, studies related to mechanical and wear characterization are very minimal. Among the other available reviews, natural fiber was added to the PLA matrix in place of synthetic fiber to study heat treatment and hygrothermal ageing. Greater absorption of moisture through the natural fiber played a detrimental role in the reduction in mechanical properties [20]. Using the technique of solvent cast processing, Bindhu et. al. [9] developed a PLA composite reinforced with 4 wt% BN to study the thermal and mechanical properties. In various tests, the tensile strength was noted to increase with the addition of BN as a reinforcement. Tensile strength and wear studies using FDM technology for BN-reinforced PLA have not been carried out by many researchers, and hence this provides a way to explore the desired properties developed using FDM technology. Though 3D printing is an advanced and sought-after technology for manufacturing complex parts with sustainable goals, the challenge lies in the development of a composite filament with desired reinforcement properties and its effective utilization in fabricating a part. Apart from these, a different set of anomalies and defects would arise with a new polymer composite, which needs to be overcome by choosing optimized printing parameters. The most common forms of defects found in printing polymer composites were warping, distortions, voids, uneven reinforcement, and reduced bonding, to mention a few [21]. In another research study, an investigation involving the extrusion and subsequent 3D printing of biocomposite filaments was carried out. These filaments were crafted from poly(lactic acid) and infused with sub-micrometric silicon particles, and the process employed was fused deposition modeling (FDM). Filaments were developed with various weight percentages of silicon particles ranging from 0% to 7%, all with the goal of enhancing the performance of poly(lactic acid) (PLA). The comprehensive analysis involved an in-depth examination of the mechanical and tribological properties of the printed biocomposites using well-established techniques. The findings unveiled notable enhancements in key attributes. For instance, the Shore D hardness and tensile strength reached 55 MPa and 60 MPa, respectively, while the impact strength stood at 5.9 kJ/m^2 and the compression strength at 11.8 MPa. These values far surpassed those of the unaltered PLA, which exhibited the lowest results. [22]. In yet another study, the primary objective was to enhance the fused deposition modeling (FDM) process parameters for the improved wear performance of polylactic acid (PLA). The research delved into the investigation of key process variables such as layer thickness, orientation, and extruder temperature. Specifically, these parameters were examined to create wear specimens adhering to ASTM G99 standards [22] through the FDM printing process. The primary focus of the study was to analyze the wear characteristics of a polymer pin under the conditions of a low sliding speed. The results of this investigation highlighted the substantial impact of build orientation on the wear performance of the polymer pin [23].

Considering the insights from prior research, our current study is centered on the development of boron nitride (BN)-reinforced poly(lactic acid) (PLA) through the application of fused deposition modeling (FDM). The primary objective of this research is to innovate in the realm of polymer composites with a specific focus on their production via FDM technology. The aim is to address existing limitations, such as issues related to print quality, compounding processes, pore formation, and dispersion concerns.

This article is primarily dedicated to exploring the potential for fabricating specimens that exhibit reduced porosity, optimum extrusion temperatures, and an ideal dispersion of BN within the PLA matrix. To achieve this, we developed filaments suitable for FDM applications by incorporating 5% and 10% by weight of BN into a PLA matrix. Our experimental investigations predominantly concentrated on assessing the tensile and wear characteristics of FDM-produced polymer composites while adhering to ASTM standards.

2. Materials and Methods

2.1. Polylactic Acid (PLA)

PLA is a common thermoplastic polymer used as a matrix material. PLA plastics were procured in the form of pellets from Adnano Pvt. Ltd., Shivamogga, India. The PLA obtained was of research grade with 99% purity and a molecular weight of 76.12 g/mol. PLA finds large applications in the industrial sector; it is biodegradable in nature and easily processable.

2.2. Boron Nitride

Fine boron nitride (BN) powder with particle sizes ranging from 40 to 45 µm and 99% purity was purchased from M/s Adnano Pvt. Ltd., India, and was considered for the study. The SEM of the BN is shown in Figure 1, and the EDAX results are shown in Figure 2. A scanning electron microscope (JSM 840a Jeol, Tokyo, Japan) was used for the SEM and EDAX studies at BMS College of Engineering, Bangalore. A sample of the filament was cut using a razor blade to have a flat, planar geometry. A sample was placed on the stub using double-sided tape. The coating of the sample was carried out by spraying Au to make it conspicuous. The images were captured as the final step of the required magnification.

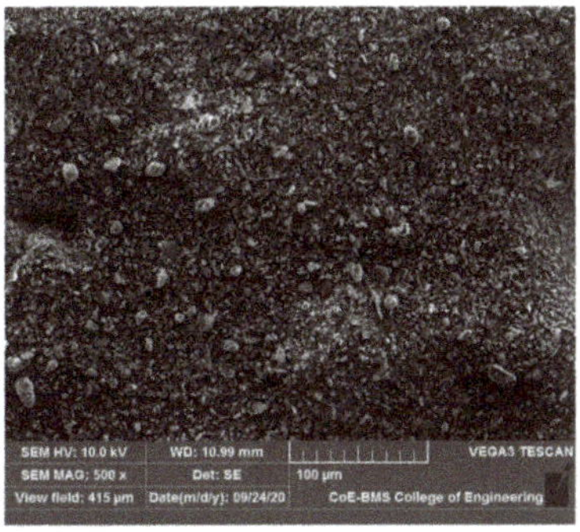

Figure 1. Scanning electron microscopy images of boron nitride.

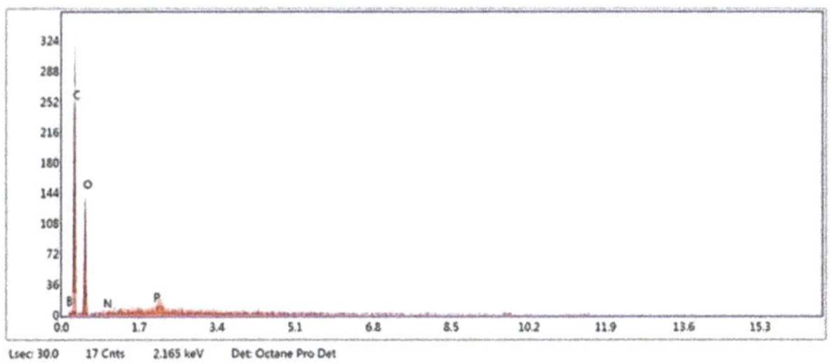

Figure 2. EDAX image of BN.

2.3. Blending and Extrusion

Boron nitride (BN) powder and PLA pellets were dried in a vacuum oven at a temperature of 80 °C for six hours before the processing began. This was done to remove any absorbed water vapor. In order to combine the various raw materials, a microcompounder with twin screws was utilized. Approximately five minutes were spent compounding the substance at a temperature of 220 °C and 100 rpm. For the purpose of 3D printing, the BN was compounded at weight percentages of 5% and 10% in PLA. In order to create samples that could be 3D-printed, compounds were first extruded from the compounder and then pelletized in a consistent manner. The basic mechanism of the extrusion process comprised a screw that transported the raw plastic pellets from a hopper through a heating zone in a metal pipe where the plastic was melted. The raw plastic pellets were gravity-fed from the hopper into the screw. Inside the pipe, the molten plastic was forced through a die at the end of the pipe to form a filament. The scroll length of the twin screw extruder was maintained at 20 mm. The extruder machine (VFX 500) had a maximum temperature of 350 °C with a mica band heater and 4 zones with independent PID controllers. A variety of polymers could be used in this machine. The temperature measurement system used PT100 sensors with a high sensitivity and reliability. The filament variable diameter was from 0.5 mm to 3 mm, and the screw was composed of a chromium–molybdenum alloy. The pelletized material was extruded into a 3D printing filament by GLS Polymers (Bangalore, India) using a screw extruder. An extruder operating at a temperature of 180 °C and a die with a diameter of 1.75 mm were utilized in this process. The filament had a diameter of 1.75 mm ± 0.05 mm, and it was manually drawn and spooled. A comparable analysis was carried out by Vijay et al. [24].

Before the start of the extraction of the filament, the pellets were dried at a temperature of 120 °C for 2 h. Boron nitride and the dried PLA pellets were mixed in the pre-mixer at a rotor speed of 60 rpm at temperature of 225 °C. Compounding and twin-screw extrusion machines were adopted to extract a filament with a 1.75 mm diameter. Filaments with the addition of 5 and 10 wt% BN were developed. A pure PLA filament was also developed by maintaining an extrusion temperature of 220 °C. The twin extrusion setup was utilized by varying the temperature of various zones for smooth extraction. The developed filaments with a 1.75 mm diameter were used in the fused deposition modeling (FDM) machine.

2.4. Fused Deposition Modeling (FDM)

Fused deposition modeling, also known as FDM, is currently one of the most widely used approaches to three-dimensional printing. The development of the 3D model makes use of a CAD file that is saved in STL format. This method can be applied to the production of any complicated component [25]. A Pramaan printer from Global 3D Labs, Karnataka, India, was utilized to fabricate the test specimens for tensile and wear testing. The printer featured an enclosed chamber that had a build volume of 4000 mm^3. The following criteria served as the basis for the development of the models: an infill density of one hundred percent, a layer thickness of 0.1 mm, a shell thickness of 0.4 mm, bottom and top layer thicknesses of 1.2 mm each, a speed of five millimeters per second, a nozzle orientation of forty-five degrees for PLA, a temperature of 240 °C for fabrication, varying temperatures for other proportions, and a bed temperature of 80 °C. After passing through the nozzle and onto the print bed, the filament moved in the same X, Y, and Z directions as the nozzle did. The X-direction was the direction in which the fabrication of the parts took place. It is generally agreed upon that additive manufacturing is a more environmentally friendly alternative to conventional manufacturing, and it is also acknowledged that this technique helps reduce carbon footprints. In the context of the world as a whole, both the number of case studies and the widespread relevance of their findings have been expanding at an exponential rate [26]. The areas of transportation, construction, aerospace, health care, and electronics are among those in which additive manufacturing (AM) has the potential to have a significant impact on society [27].

2.5. Testing and Characterization

The specimen of a dog-bone-shaped bar was produced in accordance with ASTM-D638 [27] for the evaluation of its ultimate tensile strength. Tensile testing was performed on rectangular bar specimens (dog-bone-shaped) according to ASTM standards using the universal testing method (UTM). The testing was carried out at NALRC, Bangalore, India, on an FIE Universal Testing Machine (model number: MV1-PC) with a capacity of 0–6 tonnes.

These samples were subjected to surface roughness and dimensional accuracy tests. Three tests were carried out for each set of specimens, and the resultant values were averaged as an arithmetic mean for the surface roughness and dimensional accuracy. Vernier calipers were used to measure the dimensions. A surface tester was used to measure the surface roughness. The developed parts were also subjected to a wear property assessment. Figure 3 depicts photographs of the PLA and PLA + 5 wt% BN wear specimens created with FDM. The instrument used for the wear analysis was a model TR20LPHM-400 manufactured by DUCOM Instruments, Peenya, Bangalore, India. The ASTM G99 test method was adopted to carry out this experiment. The dimensions of the developed parts were a diameter of 8 mm and a length of 20 mm [28].

PLA PLA+5 wt% BN PLA+10 wt% BN

Figure 3. Photographs of wear specimens.

3. Results and Discussion

3.1. Microstructure Studies

Figure 4 shows SEM images of PLA, PLA with 5% BN, and PLA with 10% BN. The images show that the filaments reinforced with boron nitride powder had a homogeneous dispersion, demonstrating improved BN wettability in the PLA. Furthermore, there are no visible boron nitride debonding or agglomeration defects in the PLA. The excellent surface-to-area volume ratio of the BN, as well as its general properties, played an important role in demonstrating no aggregation. Isometric dispersion can be seen with equal reinforcement distribution and an irregular shape of the filler material. Higher mechanical properties of the composites were achieved with homogenous dispersion of reinforcement in the matrix. Moreover, porosity and cracks are not visible in the SEM images due to the optimal fabricating parameters. The other major factors that influenced the dispersion characteristics were the wettability, size of the reinforcement particles, processing-time density, heat dissipation, nozzle temperature, layer thickness, and build direction. Joy et al. [28] reported in a study of BN nanocomposites that as the content of BN platelets increased, they interacted with one another and formed a network for transporting heat, greatly improving the material's ability to conduct heat. The inclusion of BN platelets improved the thermal stability slightly. Because of the good BN platelet dispersion and improved compatibility between the platelets and matrix, the mechanical properties revealed a noticeably higher impact strength at a high BN loading. The electrical insulation properties of the PLA composites were also unaffected by the BN content. Figure 5 shows the SEM images of parts fabricated via FDM.

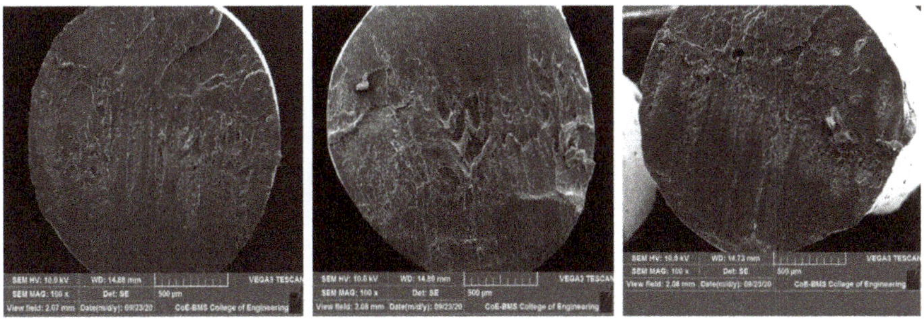

(a) (b) (c)

Figure 4. SEM images of (**a**) PLA, (**b**) PLA + 5 wt% BN, and (**c**) PLA + 10 wt% BN.

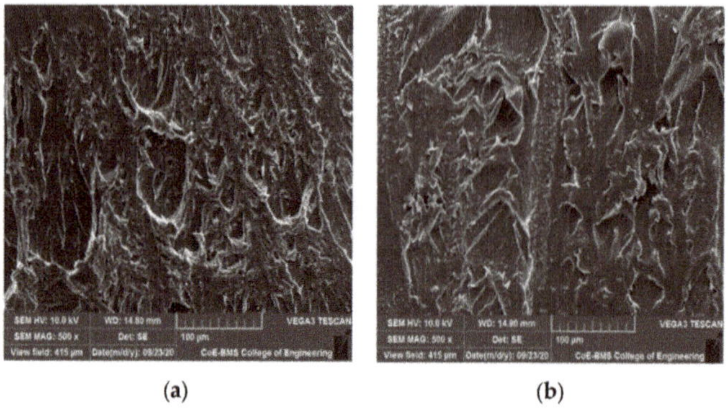

Figure 5. SEM images of BN dispersion: (a) PLA + 5 wt% BN; (b) PLA + 10 wt% BN.

3.2. Surface Roughness

Figure 6 depicts the surface roughness variations in the FDM components of the PLA polymer and the PLA + 5 wt% BN and PLA + 10 wt% BN composite polymers. The figure shows that as the amount of boron nitride powder was increased, the roughness decreased. The study utilized a model SJ-210 surface tester manufactured by Mitutoyo to measure the surface roughness.

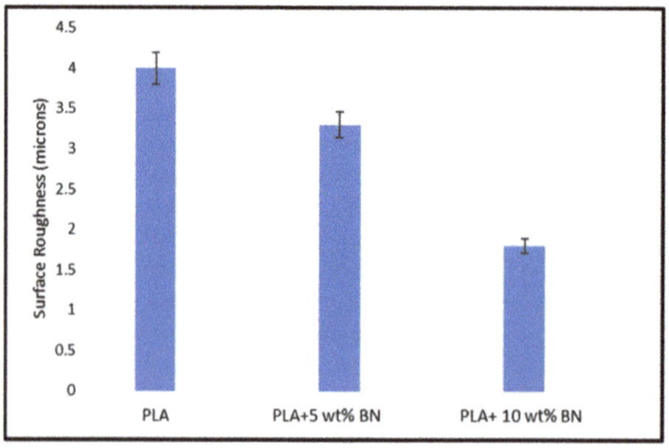

Figure 6. Surface roughness of PLA, PLA + 5 wt% BN composite, and PLA + 10 wt% BN composite.

The presence of boron nitride in the composites, in addition to influencing the printing parameters, contributed significantly to the reduction in the surface roughness. This was attributed to several key factors. Firstly, boron nitride exhibited a high thermal conductivity, which played a crucial role in the fused deposition modeling (FDM) process. It enabled more efficient heat dissipation, ensuring a consistent temperature distribution across the printed layers. This effectively minimized the likelihood of thermal inconsistencies that could lead to surface roughness. Furthermore, the incorporation of boron nitride enhanced the flow characteristics of the composite material. This improvement resulted in a more uniform deposition of the material, ultimately leading to smoother surfaces. Additionally, boron nitride particles served as effective fillers within the composite. As their concentration was increased, they adeptly filled gaps and irregularities in the material, contributing to an overall smoother surface texture.

Moreover, the low surface energy and favorable release properties of boron nitride played a significant role in reducing adhesion between the printed layers. This characteristic promoted a smoother surface finish by minimizing the potential for irregularities caused by layer adhesion.

Additionally, boron nitride's presence could potentially mitigate the material's tendency to shrink or warp during the cooling phase of the FDM process. This in turn led to a final surface that exhibited significantly reduced irregularities.

The FDM components with boron nitride powder had improved surface finishes, which improved the dimensional stability and uniform mixing. The homogeneity of the boron nitride powder distribution improved the structural stability and bulk thermal conductivity of the FDM components. When compared to unreinforced PLA plastic, the surface roughness of the 5 wt% BN and 10 wt% BN decreased by 13.01% and 24.61%, respectively. Among the various available studies, we observed that the improvement in surface quality was influenced majorly by the build orientation and layer thickness [29,30]. The layered nature of part development can be optimized with lower peaks and valleys by including a minimal layer thickness. An increased nozzle temperature leads to increased fluidity of the filament, causing a roundoff at the raster and hence providing a better surface finish [31].

Regarding post-process treatments such as polishing after fused deposition modeling (FDM), no such treatments were carried out. This is because the objective was to assess the wear resistance of the FDM polymer composite in its as-printed condition without any additional modifications. This approach was valuable in understanding the material's performance in its raw state, as it would be directly used in practical applications without further processing.

The decision not to polish before wear resistance testing is likely because polishing would alter the surface characteristics, potentially leading to results that do not accurately represent the material's behavior in its original printed form. However, for all as printed conditions, the surface roughness was recorded without post-printing treatment.

EN 24 steel was used as counterpart material, as it is a widely used engineering steel with good mechanical properties, making it a suitable choice for wear resistance testing. We have not presented details of the counterpart surfaces after testing due to fact that the very minimal damage or identifiable features observed in the steel counterpart disc suggested that the testing conditions may have been more conducive to materials with a lower hardness. This observation further highlighted the relative performance of the FDM polymer composite in comparison to the EN 24 steel under the given testing conditions.

3.3. Dimensional Accuracy

A graph (Figure 7) was plotted for the averaged values of the dimensional accuracy measurements to explain and compare the volumetric shift for the different compositions.

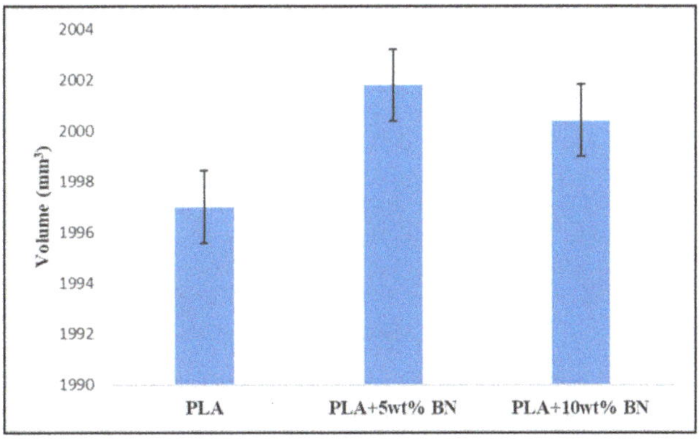

Figure 7. Volume of FDM-processed PLA polymer, PLA + 5 wt% BN composite, and PLA + 10 wt% BN composite with respect to the percentage of reinforcement.

The amount of the dimensional error in PLA plastic can be reduced by increasing the percentage of boron nitride it contains. Components that are constructed using the FDM process benefit from an increased dimensional stability when boron nitride is included in the mix. When compared to the PLA, the dimensional error was found to decrease by 1.041% when BN was added to the PLA at a weight percentage of 5%, and it decreased by 0.494 percent when BN was added at a weight percentage of 10% to the PLA when compared to the PLA + 5 wt% BN. Boron nitride's lower thermal expansion coefficient and higher thermal conductivity both contributed to the fact that its addition resulted in a reduction in the dimensional error. A reduction in the dimensional error could also be attributed to the optimal process parameters, especially an increase in the bond strength between the layers. The layer thickness, nozzle temperature, enclosed printing chamber, and heat transfer capability also majorly determined the dimensional stability of the printed parts [32,33].

3.4. *Ultimate Tensile Strength*

The increased tensile strength of PLA that occurs as a result of the addition of boron nitride is depicted in Figure 8. According to the findings, the ultimate tensile strength of the PLA was significantly improved when the boron nitride powder was included in the manufacturing process. A further observation was that as the percentage of boron nitride in the material increased, the amount of load required for fracture as well as the values for the ultimate tensile strength increased as well. PLA combined with 10 percent BN produced 82 MPa of UTS, while PLA combined with 5 percent BN produced 74 MPa of UTS. In the PLA plastic, increases of 35.504 percent and 46.334 percent were observed after the addition of 5 weight percent and 10 weight percent of boron nitride, respectively. Although pores existed in the parts of the FDM, the specimens showed incremental values of tensile strength with the addition of BN. This was attributed to the molecular organization of the polymer chains through the process of FDM [30]. It has been demonstrated that the fillers, which have mechanical properties that are inherently beneficial, improve the tensile stress performance of composites by acting as a skeletal support structure for the composite [31]. This is because the fillers have mechanical properties that are intrinsically advantageous. By reducing the amount of stress that is concentrated at the interface between the filler and the matrix, a limited or nonexistent agglomeration in the composites helps to prevent stress transfer failure in continuous polymer phases [32]. In addition, the increasing surface contact region between the matrix and the filler makes it possible for the force to be transferred to the network of BN in an efficient manner, which contributes to the improvement in the mechanical strength of such composites [32]. The researchers Zhao et al. [33] investigated what happened when BN was mixed in with aramid fibers. Both the in-plane thermal conductivity and the tensile strength of the multilayer gradient BN/ANF films were found to be significantly higher than average. Voids and interfaces were bound to occur in the FDM-printed parts. The SEM images shown previously indicate indivisible voids in the pure PLA. The addition of reinforcement reduced interfaces and voids, leading to enhanced mechanical properties. This can be clearly seen in Figure 9. The processes of compounding and FDM printing helped in the void reduction and the nonformation of agglomerates, leading to an enhanced ultimate tensile strength [34]. The enhanced mechanical strength of the BN-reinforced composites was also due to the alleviated contact surface area between the matrix and filler, which permitted the load to be effectively transferred to the BN network [35]. The process of building a part through an infill pattern also played an important role in determining the strength of the parts. An infill angle of 90° provided better results during orientation in the load direction than 45° did. Similar results were noted by Li et al. [36].

3.5. *Wear Test*

3.5.1. Constant Load

The relationship between the sliding velocity and wear rate of the FDM parts is illustrated in Figure 10. We observed that the rate of wear on the FDM parts accelerated with an increasing sliding velocity, and this trend held true across all of the investigated compositions. When compared to FDM parts that were not reinforced with a boron nitride composite, however, the FDM parts reinforced with a boron nitride composite showed a significant reduction in wear when subjected to the same testing conditions and at the same sliding velocities. At a rotational speed of 250 rpm, the PLA exhibited the highest wear rate with a value of 0.14 m. At the same rpm, the wear rate for PLA mixed with 5 weight percent BN was measured at 0.12 microns, while the wear rate for PLA mixed with 10 weight percent BN was measured at 0.10 microns. The higher temperature at the interface may have been the result of increased wear brought on by an increase in the sliding velocity. The increased temperature caused the surface of the test samples to become more pliable, which led to an increase in wear.

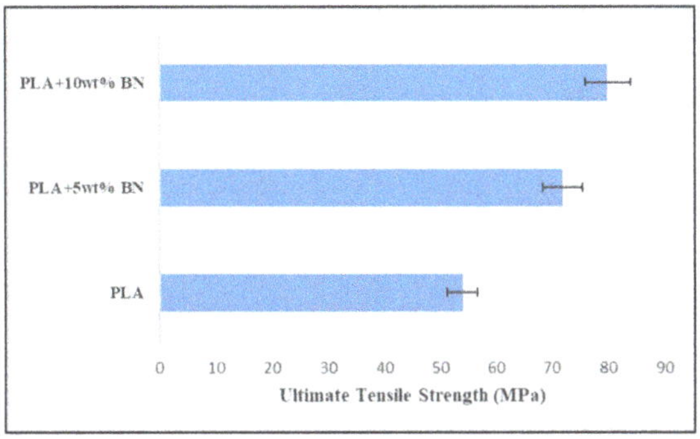

Figure 8. Graph of ultimate tensile strengths.

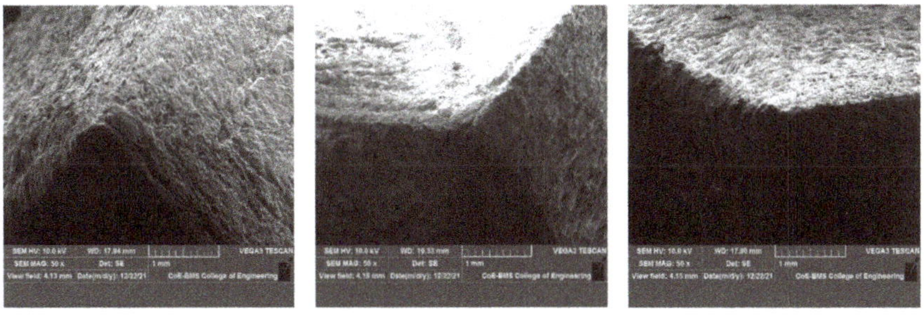

Figure 9. SEM images of specimens fabricated via FDM: (**a**) PLA; (**b**) PLA + 5 wt% BN; (**c**) PLA + 10 wt% BN.

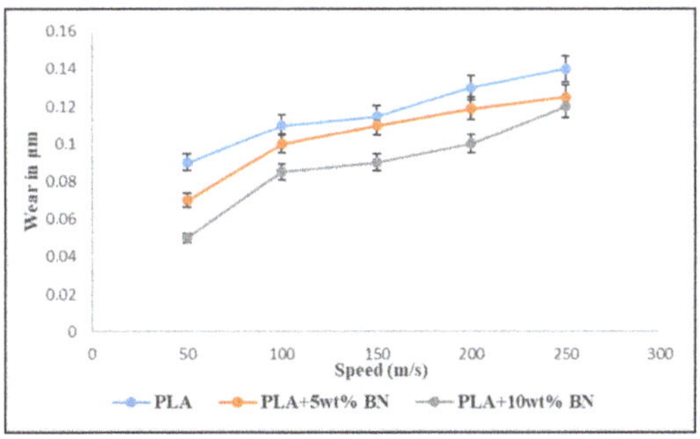

Figure 10. Wear graph of variations in speed while keeping the load constant (10N).

3.5.2. Constant Speed

Figure 11 illustrates the impact that changing the load had on the amount of wear that occurred on different FDM component compositions. When the amount of load that was being applied was steadily raised, we observed that the rate of wear increased regardless of the composition being investigated. The PLA + 10 wt% BN showed a greater capacity for bearing loads by exhibiting a wear rate of 0.06 μm per 50 Newtons of pressure. When subjected to the same loads, the PLA + 5 wt% BN showed a wear rate of 0.10 μm, while the PLA showed 0.12 μm. Plastic deformation was greater at higher loads, which can be attributed to the increase in wear that occurred in conjunction with the increase in load. A greater amount of plastic deformation led to the surface of the test sample cracking, which in turn led to a greater amount of material being lost. Additionally, when greater loads were applied, the transfer film or interfacial film that already existed at the interface between the steel disc and the test sample may have become disturbed and unsteady. This opened the door for the actual surfaces to come into contact with one another, which ultimately resulted in an increase in wear. According to the results of all of the tests, the FDM parts reinforced with boron nitride powder showed significantly less wear than the unreinforced ones.

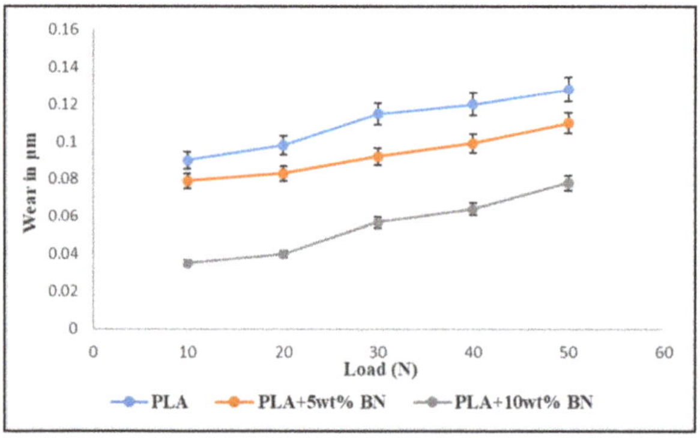

Figure 11. Wear graph of variations in load while keeping the speed constant (50 rpm).

With an incremental load, all three specimens exhibited a similar trend of enhanced wear. Polymers are bound to undergo plastic deformation, and hence this phenomenon was attributed to this process. Deformation led to cracking of the surface, resulting in greater material removal. The interfacial film diluted due to the increased load between the part and disc, resulting in actual physical contact and greater wear [37,38].

3.5.3. SEM of Worn-Out Surface

Amonton's law [39] can be used to calculate friction, which is the force preventing two surfaces from sliding against one another:

$$F = \mu W \quad (1)$$

where the coefficient of friction is denoted as μ, the normal load (N) as 'W', and the friction force as 'N'. As far as tribology is concerned, μ is a significant factor in determining a component's wear rate. COF is mainly determined by the characteristics of the surface and lubrication conditions, aside from the nature of the materials [39,40]. It is therefore possible to optimize component performance by tailoring the surface characteristics and lubrication condition [41,42].

The wear tracks of pure PLA and boron nitride-filled PLA parts tested with constant loads and a constant sliding velocity are shown in Figures 12 and 13, respectively. We observed that when increasing the sliding velocity and load of the pure PLA and the boron nitride-filled PLA, the boron nitride-filled PLA demonstrated damage on the surfaces, and the extent of the damage increased as the sliding velocity and load were increased. The boron nitride-filled FDM parts, in contrast to the pure PLA parts, exhibited astonishingly low levels of surface damage under all loads and sliding velocities that were investigated. When dealing with the pure PLA, a severe plastic deformation could be observed. The PLA components that were filled with boron nitride did not show any signs of visible surface damage or cracks. This is because boron nitride has improved physical and mechanical properties. Both the increased surface hardness and the improved lubrication characteristics were responsible for the significant reduction in wear loss that was observed. The tribological pattern that was observed through experimentation was demonstrably supported by the surface morphology of the noted surfaces.

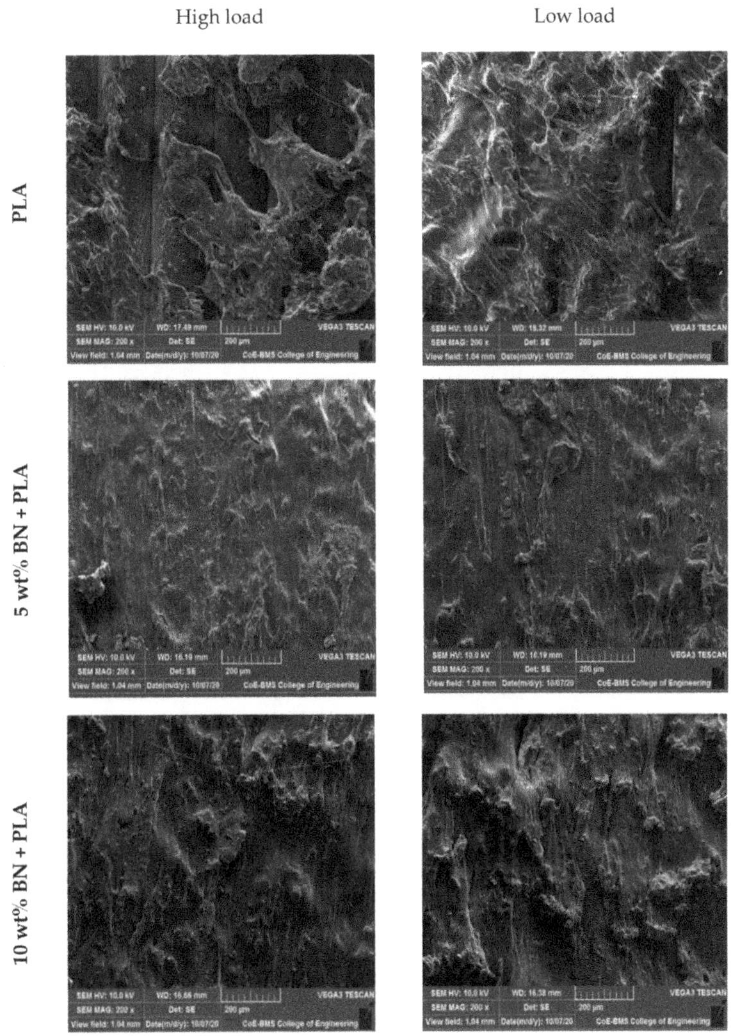

Figure 12. Wear tracks of specimens while keeping the load constant.

The X build orientation of the samples that were developed for the study played an important role in determining the wear behavior. The worn surface of the sample printed in the X orientation had a rough wear pattern and longitudinal lines. The lines, also known as longitudinal lines, were created when the printing layers were applied. The presence of such a structure had the potential to facilitate the distribution of loads across a larger sliding contact area, thereby lowering the wear rate. The tribological characteristics of components that have been 3D printed are significantly influenced by the orientation of the printing process as well as the printing direction. It is possible that the matrix's increased wear resistance and decreased coefficient of friction were the result of its uniform grain distribution, which could be found throughout the matrix. Because of this uniform dispersion between the PLA printed layers, the final product was a structure that was not only harder and more long-lasting but also more resistant to wear. The friction force could be controlled in the end by the internal lubricator that developed as a consequence of the low concentration of BN loading.

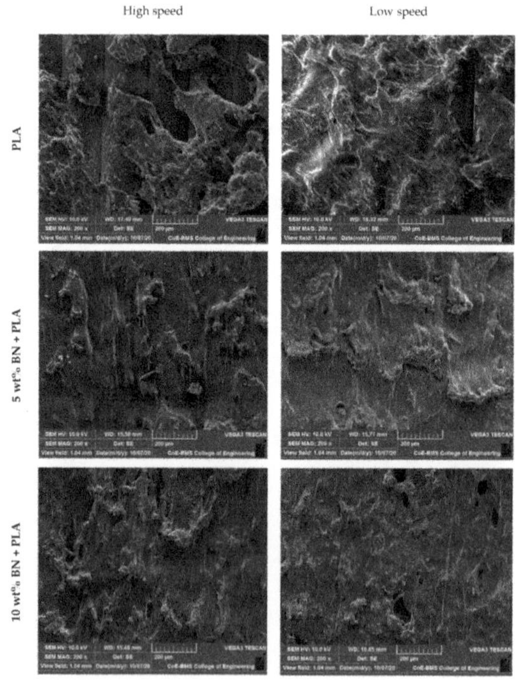

Figure 13. Wear track of specimens while keeping the speed constant.

The most common type of wear is called adhesive wear, and it causes the formation of a significant number of lamellar fragments during the wear process of pure PLA. Additionally, the surface of a PLA matrix is frequently damaged as a result of the lamellar fragments. Because of this, pure PLA has poor properties when it comes to abrasion resistance. When BN was added to the polymer matrix, the surface fragments became smaller, and the polymer surface sustained less damage; as a result, the wear rate of the composite was also reduced. The typical abrasive wear seen on composite surfaces could be identified by the light scuffing and ploughing that occurred across the surface that had been worn. This reinforcing effect was primarily caused by a number of key factors, the most important of which were a high aspect ratio and a high strength of the BN. In PLA matrices that had well-dispersed BN, the polymer molecules were able to interact with the BN surface area over a large area, which helped to facilitate an efficient load transfer to the BN network and suppressed the formation of more wear debris in addition to providing protection against frictional forces (thin transfer films formed as uniform). All of these factors contributed to the increased load-bearing capacities and higher wear resistance of the PLA composites.

In light of this finding, it should come as no surprise that the significant anti-wear effect exhibited by the filler in graphene-modified polymer composites [43,44] is in agreement with the aforementioned conclusion. There is a possibility that the loadings of BN in a PLA matrix reach saturation for nano-composites with a higher BN content (greater than 10 wt%) and that there are some aggregates of BN in these materials. According to the findings, BN aggregates have the potential to act as stress concentration points and reduce the load transmission between the polymer matrix and the BN. As a direct consequence of this, the level of surface damage on the nanocomposite increases significantly, the wear resistance begins a precipitous decline, and a great number of large damage pits begin to appear.

4. Conclusions

Components of PLA and PLA reinforced with boron nitride developed using FDM were subjected to a comparative analysis that proved the provision of enhanced mechanical and wear properties. Using the twin-screw extrusion method, boron nitride-filled PLA filaments were synthesized successfully up to a 10 wt% BN addition. As the percentage of boron nitride increased, the surface roughness and dimensional accuracy of the FDM components improved, providing a positive outcome of desired results. With the addition of 5 and 10% by weight of boron nitride, the ultimate tensile strength of the FDM components was significantly enhanced, showing appropriate dispersion and enhanced bonding. Compared to the unreinforced PLA plastic, the FDM components reinforced with boron nitride possessed superior wear resistance. There was a tendency for the wear rate to increase as the applied load and sliding velocity increased for each of the tested compositions. Boron nitride-filled FDM components have superior mechanical, tribological, and physical properties to FDM components without reinforcement. The scope for utilization of functional BN, a higher percentage of BN, and post-processing of FDM parts remain the future objectives of our research.

Author Contributions: Conceptualization, V.T., R.K. and T.M.Y.K.; methodology, S.P., V.T. and R.K.; validation, R.K., V.T. and A.A.R.; formal analysis, V.T., R.K., S.P. and A.A.D.; investigation, V.T. and T.M.Y.K.; resources, T.M.Y.K. and R.K.; writing—original draft preparation, V.T., R.K. and T.M.Y.K.; writing—review and editing, R.K. and V.T.; project administration, A.A.D. and T.M.Y.K.; funding acquisition, T.M.Y.K.; research support, R.K., V.T., S.P., A.A.R. and T.M.Y.K.; writing—review and funding acquisition, A.A.R., A.A.D. and T.M.Y.K. All authors have read and agreed to the published version of the manuscript.

Funding: This work was funded by King Khalid University under grant number R.G.P 2/282/44.

Institutional Review Board Statement: Not applicable.

Informed Consent Statement: Not applicable.

Data Availability Statement: The data are not available.

Acknowledgments: The author extends his appreciation to the Deanship of Scientific Research at King Khalid University for funding this work through the research groups program under grant number R.G.P. 2/282/44.

Conflicts of Interest: The authors declare no conflict of interest.

References

1. Tambrallimath, V.; Keshavamurthy, R. Glimpses of 3D printing in 21st Century. In *Development, Properties, and Industrial Applications of 3D Printed Polymer Composites*; IGI Global: Hershey, PA, USA, 2023. [CrossRef]
2. Blanco, I.; Cicala, G.; Recca, G.; Tosto, C. Specific Heat Capacity and Thermal Conductivity Measurements of PLA-Based 3D-Printed Parts with Milled Carbon Fiber Reinforcement. *Entropy* **2022**, *24*, 654. [CrossRef] [PubMed]
3. Singh, S.; Ghosh, A.K.; Maiti, S.N.; Raha, S.; Gupta, R.K.; Bhattacharya, S. Morphology and rheological behavior of polylactic acid/clay nanocomposites. *Polym. Eng. Sci.* **2011**, *52*, 225–232. [CrossRef]
4. Pracella, M.; Haque, M.-U.; Puglia, D. Morphology and properties tuning of PLA/cellulose nanocrystals bio-nanocomposites by means of reactive functionalization and blending with PVAc. *Polymer* **2014**, *55*, 3720–3728. [CrossRef]

5. Serizawa, S.; Inoue, K.; Iji, M. Kenaf-fiber-reinforced poly(lactic acid) used for electronic products. *J. Appl. Polym. Sci.* **2006**, *100*, 618–624. [CrossRef]
6. Cao, Y.; Feng, J.; Wu, P. Preparation of organically dispersible graphene nanosheet powders through a lyophilization method and their poly(lactic acid) composites. *Carbon* **2010**, *48*, 3834–3839. [CrossRef]
7. Bindhu, B.; Renisha, R.; Roberts, L.; Varghese, T. Boron Nitride reinforced polylactic acid composites film for packaging: Preparation and properties. *Polym. Test.* **2018**, *66*, 172–177. [CrossRef]
8. Guo, H.; Niu, H.; Zhao, H.; Kang, L.; Ren, Y.; Lv, R.; Ren, L.; Maqbool, M.; Bashir, A.; Bai, S. Highly Anisotropic Thermal Conductivity of Three-Dimensional Printed Boron Nitride-Filled Thermoplastic Polyurethane Composites: Effects of Size, Orientation, Viscosity, and Voids. *ACS Appl. Mater. Interfaces* **2022**, *14*, 14568–14578. [CrossRef] [PubMed]
9. Nagarajan, V.; Mohanty, A.K.; Misra, M. Perspective on Polylactic Acid (PLA) based Sustainable Materials for Durable Applications: Focus on Toughness and Heat Resistance. *ACS Sustain. Chem. Eng.* **2016**, *4*, 2899–2916. [CrossRef]
10. Tambrallimath, V.; Keshavamurthy, R.; Patil, A.; Adarsha, H. Mechanical and Tribological Characteristics of Polymer Composites Developed by Fused Deposition Modelling. In *Fused Deposition Modeling Based 3D Printing, Materials Forming, Machining and Tribology*; Dave, H.K., Davim, J.P., Eds.; Springer: Berlin, Germany, 2021. [CrossRef]
11. GrayIv, R.W.; Baird, D.G.; Bøhn, J.H. Effects of processing conditions on short TLCP fiber reinforced FDM parts. *Rapid Prototype J.* **1998**, *4*, 14–25.
12. Zhong, W.; Li, F.; Zhang, Z.; Song, L.; Li, Z. Short fiber reinforced composites for fused deposition modeling. *Mater. Sci. Eng. A* **2001**, *301*, 125–130. [CrossRef]
13. Shofner, M.L.; Lozano, K.; Rodríguez-Macías, F.J.; Barrera, E.V. Nanofiber-reinforced polymers prepared by fused deposition modeling. *J. Appl. Polym. Sci.* **2003**, *89*, 3081–3090. [CrossRef]
14. Melenka, G.W.; Cheung, B.K.O.; Schofield, J.S.; Dawson, M.R.; Carey, J.P. Evaluation and prediction of the tensile properties of continuous fiber-reinforced 3D printed structures. *Compos. Struct.* **2016**, *153*, 866–875. [CrossRef]
15. Perez, A.R.T.; Roberson, D.A.; Wicker, R.B. Fracture Surface Analysis of 3D-Printed Tensile Specimens of Novel ABS-Based Materials. *J. Fail. Anal. Prev.* **2014**, *14*, 343–353. [CrossRef]
16. Francis, V.; Jain, P.K. A filament modification approach for in situ ABS/OMMT nanocomposite development in extrusion-based 3D printing. *J. Braz. Soc. Mech. Sci. Eng.* **2018**, *40*, 361. [CrossRef]
17. Domingo-Espin, M.; Puigoriol-Forcada, J.M.; Garcia-Granada, A.-A.; Llumà, J.; Borros, S.; Reyes, G. Mechanical property characterization and simulation of fused deposition modeling Polycarbonate parts. *Mater. Des.* **2015**, *83*, 670–677. [CrossRef]
18. Quill, T.J.; Smith, M.K.; Zhou, T.; Baioumy, M.G.S.; Berenguer, J.P.; Cola, B.A.; Kalaitzidou, K.; Bougher, T.L. Thermal and mechanical properties of 3D printed boron nitride—ABS composites. *Appl. Compos. Mater.* **2017**, *25*, 1205–1217. [CrossRef]
19. Jiang, N.; Yu, T.; Li, Y.; Pirzada, T.J.; Marrow, T.J. Hygrothermal aging and structural damage of a jute/poly (lactic acid) (PLA) composite observed by X-ray tomography. *Compos. Sci. Technol.* **2019**, *173*, 15–23. [CrossRef]
20. Tambrallimath, V.; Keshavamurthy, R.; Saravanbavan, D.; Pradeepkumar, G.S.; Harish Kumar, M. Synthesis and Characterization of Graphene Filled PC-ABS Filament for FDM Applications. In Proceedings of the Advances in Polymer Composites: Mechanics, Characterization and Applications, AIP Second International Conference on Polymer Composites (ICPC 2018), Surathkal, India, 15–16 December 2018.
21. Vishal, K.; Rajkumar, K.; Sabarinathan, P.; Dhinakaran, V. Mechanical and Wear Characteristics Investigation on 3D Printed Silicon Filled Poly (Lactic Acid) Biopolymer Composite Fabricated by Fused Deposition Modeling. *Silicon* **2022**, *14*, 9379–9391. [CrossRef]
22. Pant, M.; Singari, R.M.; Arora, P.K.; Moona, G.; Kumar, H. Wear assessment of 3–D printed parts of PLA (polylactic acid) using Taguchi design and Artificial Neural Network (ANN) technique. *Mater. Res. Express* **2020**, *7*, 115307. [CrossRef]
23. Keshavamurthy, R.; Tambrallimath, V.; Rajhi, A.A.; Ahmed R.M., S.; Patil, A.Y.; Khan, T.M.Y.; Makannavar, R. Influence of Solid Lubricant Addition on Friction and Wear Response of 3D Printed Polymer Composites. *Polymers* **2021**, *13*, 2905. [CrossRef]
24. Keshavamurthy, R.; Tambrallimath, V.; Badari, A.; Krishna, R.A.; Kumar, G.P.; Jeevan, M. Friction and wear behaviour of copper reinforced acrylonitrile butadiene styrene based polymer composite developed by fused deposition modelling process. *FME Trans.* **2020**, *48*, 543–550. [CrossRef]
25. Pan, D.; Li, Q.; Zhang, W.; Dong, J.; Su, F.; Murugadoss, V.; Liu, Y.; Liu, C.; Naik, N.; Guo, Z. Highly thermal conductive epoxy nanocomposites filled with 3D BN/C spatial network prepared by salt template assisted method. *Compos. Part B Eng.* **2021**, *209*, 108609. [CrossRef]
26. Joy, J.; George, E.; Haritha, P.; Thomas, S.; Anas, S. An overview of boron nitride based polymer nanocomposites. *J. Polym. Sci.* **2020**, *58*, 3115–3141. [CrossRef]
27. Su, K.-H.; Su, C.-Y.; Shih, W.-L.; Lee, F.-T. Improvement of the Thermal Conductivity and Mechanical Properties of 3D-Printed Polyurethane Composites by Incorporating Hydroxylated Boron Nitride Functional Fillers. *Materials* **2022**, *16*, 356. [CrossRef] [PubMed]
28. Bashir, A.; Maqbool, M.; Lv, R.; Usman, A.; Guo, H.; Aftab, W.; Niu, H.; Liu, M.; Bai, S.-L. Surface modified boron nitride towards enhanced thermal and mechanical performance of thermoplastic polyurethane composite. *Compos. Part B Eng.* **2021**, *218*, 108871. [CrossRef]
29. Otsuki, M.; Matsukawa, H. Systematic Breakdown of Amontons′ Law of Friction for an Elastic Object Locally. *Sci. Rep.* **2013**, *3*, 1586. [CrossRef] [PubMed]

30. Alem, S.; Latifi, R.; Angizi, S.; Mohamadbeigi, N.; Rajabi, M.; Ghasali, E.; Orooji, Y. Development of Metal Matrix Composites and Nanocomposites Via Double-Pressing Double-Sintering (DPDS) Method. *Mater. Today Commun.* **2020**, *25*, 101245. [CrossRef]
31. Anitha, R.; Arunachalam, S.; Radhakrishnan, P. Critical parameters influencing the quality of prototypes in fused deposition modelling. *J. Am. Acad. Dermatol.* **2001**, *118*, 385–388. [CrossRef]
32. Chohan, J.S.; Singh, R. Pre and post processing techniques to improve surface characteristics of FDM parts: A state of art review and future applications. *Rapid Prototyp. J.* **2017**, *23*, 495–513. [CrossRef]
33. Zhao, L.-H.; Wang, L.; Jin, Y.-F.; Ren, J.-W.; Wang, Z.; Jia, L.-C. Simultaneously improved thermal conductivity and mechanical properties of boron nitride nanosheets/aramid nanofiber films by constructing multilayer gradient structure. *Compos. Part B Eng.* **2021**, *229*, 109454. [CrossRef]
34. Kumar, H.R.M.; Benal, M.G.M.; Kumar, G.S.P.; Tambrallimath, V.H.R.G.; Khan, T.M.Y.; Rajhi, A.A.; Baig, M.A.A. Influence of Short Glass Fibre Reinforcement on Mechanical Properties of 3D Printed ABS-Based Polymer Composites. *Polymers* **2022**, *14*, 1182. [CrossRef]
35. Li, S.; Wang, K.; Zhu, W.; Peng, Y.; Ahzi, S.; Chinesta, F. Investigation on the mechanical properties of 3D printed hybrid continuous fiber-filled composite considering influence of interfaces. *Int. J. Adv. Manuf. Technol.* **2022**, *123*, 3147–3158. [CrossRef]
36. Tronvoll, S.A.; Welo, T.; Elverum, C.W. The effects of voids on structural properties of fused deposition modelled parts: A probabilistic approach. *Int. J. Adv. Manuf. Technol.* **2018**, *97*, 3607–3618. [CrossRef]
37. Shaikh, M.; Singh, P.; Kate, K.; Freese, M.; Atre, S. Finite element-based simulation of metal fused filament fab-rication process: Distortion prediction and experimental verification. *J. Mater. Eng. Perform.* **2021**, *30*, 5135–5149. [CrossRef]
38. Kumar, M.S.; Mangalaraja, R.V.; Kumar, R.S.; Natrayan, L. Processing and Characterization of AA2024/Al2O3/SiC Reinforces Hybrid Composites Using Squeeze Casting Technique. *Iran. J. Mater. Sci. Eng.* **2019**, *16*, 55–67.
39. Sarmah, A.; Kar, S.; Patowari, P.K.; Sarmah, A. Surface modification of aluminum with green compact powder met-allurgy Inconel-aluminum tool in EDM. *Mater. Manuf. Process.* **2020**, *35*, 1104–1112. [CrossRef]
40. Tyagi, R.; Pandey, K.; Das, A.K.; Mandal, A. Deposition of hBN + Cu layer through electrical discharge process using green compact electrode. *Mater. Manuf. Process.* **2019**, *34*, 1035–1048. [CrossRef]
41. Huang, T.; Xin, Y.; Li, T.; Nutt, S.; Su, C.; Chen, H.; Liu, P.; Lai, Z. Modified graphene/polyimide nanocomposites: Reinforcing and tribological effects. *ACS Appl. Mater. Interfaces* **2013**, *5*, 4878–4891. [CrossRef]
42. Friedrich, K.; Flöck, J.; Váradi, K.; Néder, Z. Experimental and numerical evaluation of the mechanical properties of compacted wear debris layers formed between composite and steel surfaces in sliding contact. *Wear* **2001**, *251*, 1202–1212. [CrossRef]
43. *ASTM G99*; Standard Test Method for Wear Testing with a Pin-on-Disk Apparatus. ASTM International: West Conshohocken, PA, USA, 2017.
44. *ASTM-D638*; Standard Test Method for Tensile Properties of Plastics. ASTM International: West Conshohocken, PA, USA, 2012.

polymers

MDPI

Review

Interactions of Cells and Biomaterials for Nerve Tissue Engineering: Polymers and Fabrication

Meaghan E. Harley-Troxell [1], Richard Steiner [1,†], Rigoberto C. Advincula [2,3], David E. Anderson [1] and Madhu Dhar [1,*]

[1] Tissue Engineering and Regenerative Medicine, Large Animal Clinical Sciences, College of Veterinary Medicine, University of Tennessee, Knoxville, TN 37996, USA; mharley4@vols.utk.edu (M.E.H.-T.); rcsteiner925@gmail.com (R.S.); dander48@utk.edu (D.E.A.)
[2] Department of Chemical and Biomolecular Engineering, University of Tennessee, Knoxville, TN 37996, USA; radvincu@utk.edu
[3] Oak Ridge National Laboratory, Center for Nanophase Materials Sciences, Oak Ridge, TN 37831, USA
* Correspondence: mdhar@utk.edu
† Current address: Department of Physical Medicine and Rehabilitation, Uniformed Services University of the Health Sciences, Bethesda, MD 20814, USA.

Abstract: Neural injuries affect millions globally, significantly impacting their quality of life. The inability of these injuries to heal, limited ability to regenerate, and the lack of available treatments make regenerative medicine and tissue engineering a promising field of research for developing methods for nerve repair. This review evaluates the use of natural and synthetic polymers, and the fabrication methods applied that influence a cell's behavior. Methods include cross-linking hydrogels, incorporation of nanoparticles, and 3D printing with and without live cells. The endogenous cells within the injured area and any exogenous cells seeded on the polymer construct play a vital role in regulating healthy neural activity. This review evaluates the body's local and systemic reactions to the implanted materials. Although numerous variables are involved, many of these materials and methods have exhibited the potential to provide a biomaterial environment that promotes biocompatibility and the regeneration of a physical and functional nerve. Future studies may evaluate advanced methods for modifying material properties and characterizing the tissue–biomaterial interface for clinical applications.

Keywords: nerve tissue engineering; biocompatibility; polymers; biomaterials; nanomaterials; hydrogels; 3D printing

Citation: Harley-Troxell, M.E.; Steiner, R.; Advincula, R.C.; Anderson, D.E.; Dhar, M. Interactions of Cells and Biomaterials for Nerve Tissue Engineering: Polymers and Fabrication. *Polymers* **2023**, *15*, 3685. https://doi.org/10.3390/polym15183685

Academic Editors: Cristina Elisabeta Pelin and Anton Ficai

Received: 28 July 2023
Revised: 31 August 2023
Accepted: 1 September 2023
Published: 7 September 2023

1. Introduction

The nervous system is divided into two distinct components: the central nervous system (CNS), composed of the brain and the spinal cord, and the peripheral nervous system (PNS), composed of the cranial and spinal nerves that extend from the CNS throughout the body. The normal, healthy nervous system is responsible for sensory, motor, and cognitive functions [1]. Neural injuries and diseases disrupt the signaling pathways responsible for these everyday acts, ultimately declining an individual's quality of life. Affecting millions of people in the United States (US) alone, neural injuries include both CNS injuries, such as traumatic brain injuries (TBIs), spinal cord injuries (SCIs), and neurodegenerative diseases (NDs), as well as peripheral nerve injuries and degeneration (PNI) [2–6]. Each injury has a classification system identifying the nerve damage level [7–9]. In the PNS, a lower degree of injury will typically regenerate independently. However, more severe injuries in the PNS and most injuries in the CNS cannot heal without assistance.

A severe PNI is one in which, in addition to the axon and myelin sheath, the tissues surrounding the various bundles of nerve cells (endoneurium, perineurium, and epineurium) are also damaged, and the size of the PNI gap is more significant than

10 mm. Once the endoneurium (the innermost tissue layer) is affected, the injury worsens when the elasticity causes the proximal and distal nerve ends to retract. Widening the gap makes it more difficult for the developing axon to align accurately and reinnervate the target tissue to restore nerve function. The degenerative process that fragments the cytoskeleton of the axon causes cellular debris to accumulate in the affected area alongside the damaged cells from the initial injury. Secreted pro-inflammatory cytokines, immune cells, and harmful reactive oxygen intermediates (ROIs) in the environment further impair the regenerative ability. The process ends when the nerve stumps develop dense fibrous scar tissue, rendering self-renewal no longer possible [10–15].

While macrophages and Schwann cells help clear debris and direct cell growth in the PNS, their lack of involvement in the brain and spinal cord damages their capacity to regenerate. Like severe PNIs, most CNS injuries result in a sustained inflammatory state filled with cellular debris and impaired by scar tissue. Astrocytes, a supportive glial cell, propagate neuroinflammation and aggregate to form a glial scar preventing self-renewal. The CNS also involves the blood–brain barrier (BBB), whose role is to prevent cells, proteins, and large molecules circulating in the bloodstream from entering the CNS to protect these tissues from harm and shield the sensitive ecosystem. While some CNS injuries damage the BBB with initial insult, others cause secondary injury from the prolonged presence of inflammatory factors that exacerbate cell death, which may occur minutes to months after the original trauma. This disruption allows the peripheral immune cells and exogenous proteins to intrude into the already hostile environment where they are generally absent, further destabilizing the area [11,13–15].

Ultimately, both of these processes have the same result, a microenvironment unsuitable for regeneration, perpetuating disruption of regular nervous system functions [16]. Available treatments include various surgical and non-surgical options, though all are limited in restoring full neural function. The gold standard for a PNI is the use of autografts. Clinical use of autografts fails to restore full function in up to 33% of patients and includes the risk associated with the harvest surgery, loss of innervation, scarring and neuroma formation at the donor site, and limited tissue availability [7,11,17]. CNS treatment options are even more limited, focusing on symptom management to limit the effects of the secondary injury [18,19].

As a result of the limited ability of peripheral nerves to regenerate without assistance and the lack of availability of efficient therapies for nerve repair, tissue engineering and regenerative medicine have become promising areas of study to develop novel, effective treatments. The goal of tissue engineering and regenerative medicine is to use exogenous cells, biomaterials, and/or additional biological factors to promote a more favorable microenvironment after an injury or disease so that the tissue can be repaired, and function restored. Although reaching this goal is complicated by the intricacies of the nervous system, optimal treatment for neural applications would be: (1) tissues obtained from a source that is easy and ethical to access; (2) therapeutic materials that can promote differentiation into specific neural and glial cell types; (3) a method of administration that is safe, minimally invasive, and causes no short-term or long-term adverse effects; and (4) an effective treatment that fully and consistently restores function [15,20,21] (Figure 1). To achieve this goal and optimize the scaffold design, consideration of the biomaterial–cell interactions and the constructs' influence on cell behavior is required. This review will evaluate the natural and synthetic polymers that form the construct, as well as the fabrication methods that modify the material properties that influence cell behavior. This review will also evaluate the methods to measure cell response in vitro and *in vivo*, and the endogenous and exogenous cell types that will interact with the developed construct.

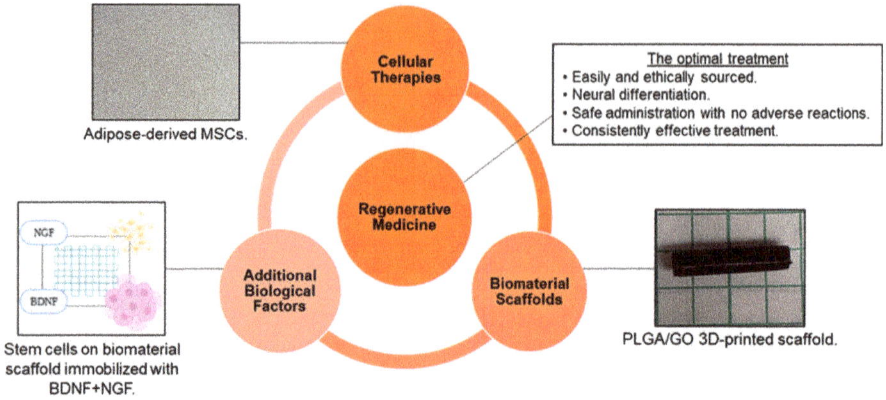

Figure 1. Tissue engineering and regenerative medicine can provide conditions that support physical and functional nerve regeneration. The influence of material and design on cell interactions must be considered to generate the optimal treatment [Images from Harley-Troxell, Dhar, unpublished data; figure made with BioRender and Microsoft PowerPoint].

2. Biocompatibility Assessment

The term biocompatibility is defined by D.F. Williams as "the ability of a material to perform with an appropriate host response in a specific application" [22]. The science community has discussed throughout the conception of biomaterials as to how to characterize biocompatibility of a material and to what degree. Through decades of research in both in vitro and in vivo models, the term can be subdivided into four individual categories which must be reviewed for a material to be labeled as biocompatible: (1) toxicology, (2) reaction to extrinsic microorganisms, (3) mechanical effect, and (4) cell–biomaterial interactions (Figure 2) [23]. This has led to the development of both in vitro and in vivo methods for the assessment of each subdivision, and the expansion of standardized procedures. These methods must answer with high certainty the public concern that a device or material will improve and ensure public safety, taking into consideration the type of biomaterial and biological components being assessed for cytotoxicity. This will determine any potential toxic mechanisms or material property effects that may occur due to reactions with interfering assay reagents, in order to prevent null or false readings. Most of these procedures are standardized by international agencies such as the International Organization for Standardization (ISO) and the American Society for Testing and Materials (ASTM), while other agencies will regulate or enforce these standards, such as the United States Food and Drug Administration (FDA) [24–26].

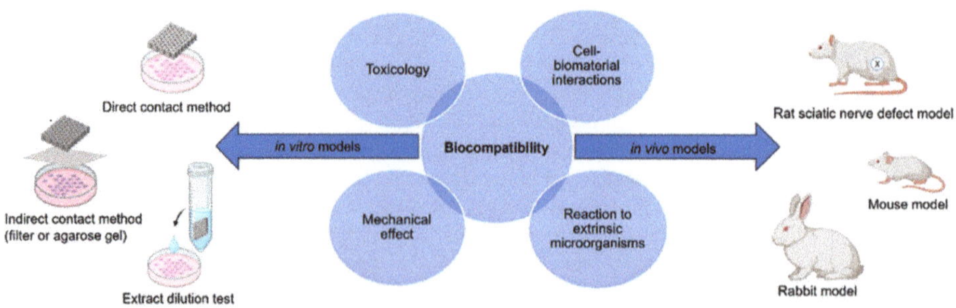

Figure 2. Biocompatibility can be evaluated using the appropriate in vitro and in vivo models to ensure there is no toxicology, mechanical effects, reaction to extrinsic microorganisms, or cell–biomaterial interactions [figure made with BioRender and Microsoft PowerPoint].

2.1. In Vitro Models

The evaluation of cytotoxicity in biomaterials can be simplified by observing how target cells or tissue will react in direct or indirect contact with biomaterials. If, under controlled conditions, cells react to the material with normal cell behavior and function (cell adhesion, proliferation, migration, and differentiation), we can conclude that the cell's physiological system is normal or homeostatic. However, if under these same conditions, cells react with behaviors indicative of cell stress or apoptosis, we can conclude that the material induces some toxic or stressful effect on the cells. In vitro models have a distinct advantage of control over both the environment and the methods employed. Under stable laboratory conditions, specific variables related to cell–biomaterial cytotoxicity can be analyzed under conditions that provide more precise, accurate, and reproducible data to strengthen statistical models. Additionally, in vitro methods can result in both qualitative and quantitative data that assess the state of cellular health. Qualitative methods accomplish this by both evaluating the morphology of cell lines, after exposure to a material, and grading the health of cells based on the amount of cell death, cytoplasmic granule formations, and percentage of cells not adhered to the material. Quantitative methods determine the number of viable cells based on the presence of nucleic acids, metabolic by-products, or the expression of surface adhesion markers. These assessments follow the standards set by ISO, Section 10993, part 5, which stipulates the different methods for collecting cytotoxicity data *in vitro*, and can be further subdivided into more specific analyses of cell–biomaterial interactions [23–27].

The indirect contact method involves either Agar or Filter diffusion tests. In both cases, a barrier diffusion model is created when a monolayer cell culture is grown, and the physical biomaterial is placed in the same culture environment but separated by an Agarose gel or cell filtration film. Over time, the material may diffuse compounds through the membrane and into the culture where it will make direct contact with the cell line. The purpose of this experiment is to observe whether the cell lines can be induced towards a cytotoxic state by providing indirect exposure to potentially toxic materials, protein-loading components, nanomaterial agglomerates, and degradation agents. The advantage of this system is that it is controlled to examine the toxicity of cells specific to the leeched components from the materials. Meanwhile, qualitative and quantitative assays provide additional sensitivity and accuracy parameters for identifying the specific toxic components responsible. The disadvantage of this system is the omission of toxic effects due to mechanical trauma, mechanobiological pathways, and cell–biomaterial interactions. This must, instead, be further analyzed with a direct contact model [23,24,26,27].

The direct contact method involves the direct contact of cells and substrate, with no membrane separation. The cell line is cultured with the material, which adds the cytotoxicity variable of cell–biomaterial reactivity, as well as the physicochemical variables specific to indirect methods of analysis. There is additional variability in the nature of the materials (three dimensional (3D) structure, bioresorption, nanotopography, etc.) being evaluated and the different cell lines used, all of which cause concern in establishing standard testing methodologies for direct contact methods. The higher degree of variability creates a more complex model of analysis to determine what can elicit a toxic cellular response. Direct contact is a rapid, inexpensive, and sensitive method to assess these toxicity factors, but the variability makes it more difficult to determine with certainty what the exact causes of cytotoxicity are. Therefore, it is necessary to confirm the cause of cytotoxicity with additional testing. If the material under a direct contact model does not elicit a cytotoxic effect to the cell lines, then an initial conclusion would be that both the physicochemical and mechanical properties of the material are not cytotoxic to the cell lines. Although further confirmation with repeated direct contact tests utilizing different cytotoxicity assays is needed, the potential statistical power of a single direct contact test performed correctly could conclude with high certainty whether a material is cytotoxic or not [23,24,26,27].

The extract dilution (elution) test involves suspending a sample of the material in a physiological solution to extract elution solvents present in the material during the production phase. During this process, the solution will simultaneously minimize physical and chemical damage to the material and remain compatible with the cell lines. Elution samples are then added to monolayered cell cultures for 24–48 h before toxicity assays are introduced. Similar to the indirect contact method, the elution test often results in a consistent toxicity pattern in cell culture, due to minimal variability of elution solvents that are present in the material during production. Although this analysis is a good first approach to determine the cytocompatibility of the material, it suffers the same disadvantages as the indirect contact method [23,24,26,27].

Qualitative assessment of cytotoxicity in in vitro studies involves the use of categorical variables to determine if cytotoxicity is present in cell lines. This form of assessment will often use stains for viable and non-viable cells under fluorescent microscopy (FM) as a morphological indicator of cell health. Some of the more common viable stains involve high affinity towards specific cell structures, such as the nucleotides within the nucleus, adeno-triphosphate (ATP) and co-enzymes in the mitochondria, the permeability of the cell membrane, or cytoplasmic proteins/cytoskeletons. Fluorescent stains must be sensitive enough to differentiate between cell structures, or can be used in combination, such as the use of nuclear stains coupled with cytoplasmic counter stains. The intensity of the stain is correlated to the number of structures present in the cell, and the combination of stained cells gives an indication of viability for the entire culture. In addition to FM, viability data can be acquired using light microscopy (LM) or scanning electron microscopy (SEM) to determine cell morphology and death based on physical characteristics such as cell shape, size, spreading, or density. Cells suspended in media and incubated for 24–48 h, which exhibit spherical shapes and minimal surface spreading, are often key indicators of apoptotic behavior. While viability stains can confirm that cell integrity, deoxyribonucleic acid (DNA), and cytoplasmic environments are stable, morphological analysis is often categorized as a graded scale assessment of cytotoxicity. Other qualitative methods include real-time tracking of cells, which involves implanting fluorescent protein or radioactive isotope components, or labeling protein/gene sequence markers using fluorochrome-labeled antibodies. These specific markers can then be detected and quantified using spectrofluorometric analysis, scintillation to detect radio isotopes, or imaging software. Caution must be reserved when interpreting qualitative data, as they will often introduce debate among experts as to what the cells' morphology indicates about the health of the cells. It is important to design in vitro models to account for statistical errors based on the sensitivity and accuracy of the assays used. This can be accomplished by including enough replicates to overcome these limitations and using experts to minimize the risk of bias observations. With a categorical data set, it is often difficult to determine whether a correlation exists between the material and variables for cytotoxicity, as compared with continuous data acquired during quantitative tests. This is why it is suggested that qualitative data be complemented by quantitative data to detect specific cytotoxic pathways and determine a correlation between the material variables and any cytotoxic mechanisms [26,28–30].

Quantitative assessment of cytotoxicity in in vitro studies involves methods that can assess the specific factors related to cell death. One of the most common quantitative biomaterial assessments is to count the number of viable cells across a material surface under normal incubation conditions. While the previously described colorimetric assays are common qualitative methods for assessing the proliferation pattern of cells, this method involves molecular compounds designed to fluoresce when exposed to cells with activated cytotoxic pathways. The fluorescent labeled cells can be quantified by counting the number of viable or non-viable cells through a hemocytometer or flow cytometer. The light intensity of the fluorochrome and density of the culture area can then be determined. The fluorescence is detected at a specific wavelength using calibrated light spectroscopy and imaging software (e.g., ImageJ) to determine the optical properties of images taken by FM. This is possible only if the fluorochromes remain bound to the internal cell structures and minimal leaching out of the cell is observed. One of the most reliable and sensitive calorimetry evaluations are tetrazolium-salt-based assays. The oxidation of nicotinamide adenine dinucleotide phosphate (NADPH) is measured to reduce tetrazolium salt into a fluorescent protein known as Formazan. This allows for the determination of mitochondrial metabolic activity with flavoproteins. Soluble Formazan, present in the surrounding media, can then be detected using a spectrometer apparatus where the fluorescence intensity can be correlated to give an estimated number of viable cells that is within an acceptable range. Each of these assays are labeled as 3-(4,5-dimethylthiazol-2-yl)-2,5-diphenyltetrazolium bromide (MTT), with other iterations often removing the extra step of solubilizing the formazan proteins into the media solution (e.g., 3-(4,5-dimethylthiazol-2-yl)-5-(3-carboxymethoxyphenyl)-2-(4-sulfophenyl)-2H-tetrazolium (MTS), water-soluble tetrazolium salt (WST), or 2,3-bis-(2-methoxy-4-nitro-5-sulfophenyl)-2H-tetrazolium-5-carboxanilide (XTT)). Calorimetry assays often consist of simple, efficient protocols, while also providing extremely sensitive and accurate data that correlate to the physical cell numbers. This presents possible limitations to their effectiveness around different cell lineages and any indirect methods of quantifying cell numbers. Some assays could be toxic to certain cell lines, such as those using radioactive isotopes (e.g., 3H-thymidine). Others may not take into consideration outside factors that trigger apoptotic pathways, leading to incorrect interpretation of the data. Biomaterials can also impact the function of these assays by either activating the fluorescence as a false positive or preventing the reagent from binding to its specific target as a false negative. It is therefore essential that the limitations of one assay be supported by the advantages of another assay. This will ensure that the data are interpreted correctly and provide a standard guide of compatible assays for different cell lines and biomaterials that can be utilized in vitro. Lastly, there are methods that involve detecting and quantifying more precise protein/genetic markers indicative of activated apoptotic pathways, such as pro-apoptotic genes (b-cell lymphoma 2 (bcl-2) associated x-protein (bax), bcl-2 antagonist/killer 1 (bak), and cytochrome c), and anti-apoptotic genes (bcl-2, bcl-extra-large (bcl-xl), and myeloid cell leukemia-1 (Mcl-1)). Western blot (WB), quantitative polymerase chain reaction (qPCR), and immunofluorescence (IF) are all methods capable of identifying apoptotic pathways initiated by a specific toxicity profile. The ratio of labeled surface markers, gene sequences, or proteins to the total image area or volume of cell culture can be quantified to indicate whether apoptotic behavior is present along one or more specific pathways [26,27,29,31–33].

2.2. *In Vivo Models*

While in vitro models are more simply designed to control experimental variable interactions and determine the correlations between them, they are incapable of recreating the same biological/physiological environment that the implant will be exposed to in a living organism. Therefore, the full spectrum of measurements of biomaterial performance cannot be assessed. In vivo models are a more complex system in which two or more experimental variables, including factors not used or relevant in the in vitro model, are now exposed. Inflammatory responses, degradation, biochemical kinetics, angiogenesis, and 3D tissue remodeling are complex processes that can only be fully studied *in vivo*. This can make opportunities to observe and identify associations between variables difficult. The goal of an in vivo model is to not only mimic the physiological conditions of a clinical case, but to obtain meaningful data that can be measured and analyzed for clinically relevant differences in material performance. It is prudent to observe overall physiological reactions and compare them to the current gold standard (positive control variable) and clinical care needed to determine the efficacy of the implant towards its intended targeted therapy. This essential step requires careful selection of the model system, variable control, and access, depending on the application of the implant and the current standardized ISO and ASTM guidelines for in vivo testing [34,35].

The use of mammalian physiological systems can accurately represent most of the conditions that biomaterials will be exposed to under practical clinical applications. Therefore, in vivo methods are the most reliable way of assessing how biomaterials will perform in a physiological environment that is similar to that of a human system. In an attempt to standardize in vivo models, we can tightly control variables such as anatomical location of implant site, size of the implant incision, species, gender, age, sample size, and methodologies for assessing material performance. The design of the model should narrow and strengthen cause/effect relationships of physiological events to the interaction of biomaterials. The controlled variables should also be focused towards answering specific questions related to the assessment of biomaterials in one or all domains of in vivo biomaterial assessments (biocompatibility testing, bioactivity testing, or preclinical testing). Meanwhile, these variables should retain the ability to translate these control parameters to alternative or more complex animal models [34,35].

In vivo assessment of biomaterials for neural regenerative applications must therefore fall under the same system of standardization for in vivo models. In an attempt to standardize, there have been numerous in vivo studies that evaluated peripheral nerve repair using scaffolds, creating a varied list of the most common animal model designs and methodologies for assessing nerve recovery. Literature reviews and studies of in vivo neural regeneration models found that the majority of these studies utilized rat models to assess scaffold performance, followed by mice and rabbits, respectively. Of these studies, the majority create defects in the sciatic nerve, with most defect gaps being between 5–10 mm. Additionally, the studies using rat models most commonly used quantitative histology analysis to assess nerve healing, followed by nerve morphometric analysis, electrophysiology, and immunohistology, respectively. The number of animal parameters to consider and the degree of control between model variables has created problems among investigators in establishing a consensus on the best animal model design and methods for assessing nerve recovery [34–36].

3. Polymers for Neural Applications

Numerous materials are available for biological applications in nervous tissues. Arguably, the most important characteristics are those that influence safety and efficacy. For a material to be used in the body, it must be deemed safe, biocompatible, and approved for use by the FDA. The FDA has an established set of guidelines that states that the biocompatibility of each material component should be understood at each step in its formation. This includes the original material before and after polymerization, as well as after sterilization, coatings, and the degradation products. Ultimately, the material should not result in an adverse biological response from the body [37]. A variety of materials have been established as biocompatible, such as the synthetic polymers polyglycolic acid (PGA) and polylactic acid (PLA), which are widely used in biodegradable sutures [38]. Others are more problematic, such as the synthetic polytetrafluorethylene (ePTFE), whose lack of degradability has resulted in long-term foreign body reactions [39]. All polymers have advantages and disadvantages which may differ among natural and synthetic polymers. Those most suitable for nerve tissue engineering will be discussed herein (Table 1).

3.1. Natural Polymers

Characteristics of natural polymers should include biocompatibility and biodegradability, minimal immunogenicity, and, ideally, provide properties that mimic the biological extracellular matrix (ECM) to provide structural and functional support. Replicating these properties allows for high-quality cellular interactions, including cell adhesion, proliferation, and differentiation for tissue regeneration. However, individual natural polymers tend to be less stable, have poor mechanical properties, and the purification method is less standardized. The variation between polymers and among different products of the same polymer is a potential limitation to translating natural polymers to commercial use because reproducibility is a critical factor in developing an optimal treatment. [15,39–43].

Collagen is an abundant natural polymer found in connective tissue and the ECM and is biocompatible, biodegradable, and bioactive. Currently, five FDA-approved collagen nerve conduits are commercially available. Many in vitro studies, in vivo studies, and clinical trials have identified the effective use of collagen support for wound healing, cell adhesion, migration, proliferation, and functional nerve regeneration. Additionally, collagen has been shown to bridge neural injury gaps up to 20 mm in length [13,39,44]. Collagen has tunable physical characteristics that can aid in directing axonal growth. However, collagen has limited mechanical stress resistance, which can be reinforced with additional biomaterials and/or fabricated into a hydrogel. Fabrication can be difficult due to collagen's poor manipulability [41,42]. In some cases, collagen has decreased neuroma formation and scarring, but has not wholly eliminated these complications [19,42]. A recent study assessed mesenchymal stem cell (MSC) spheroid-loaded collagen hydrogels to promote neurogenesis and anti-inflammatory properties. Injuries treated with collagen hydrogels, with and without the MSCs, significantly improved cell viability, neural differentiation, and mediation of inflammatory components compared to negative controls [45].

Hyaluronic acid (HA) is a natural polymer found in the ECM of neuroepithelial-derived tissues. In addition to its biocompatibility, HA is known for reproducible outcomes and efficacy in decreasing scar tissue formation [19,42,46]. HA hydrogels are known to be cytocompatible with no significant immune response when co-cultured with human peripheral blood mononuclear cells in vitro [47]. However, its fast degradation and poor mechanical properties limit its solo use. Fortunately, HA can be combined with additional material to optimize its mechanical properties for neural applications [19,42,46].

Silk fibroin is a polypeptide isolated from the natural polymer silk. It is a strong, stable, biodegradable, and biocompatible material. It has also been shown to support cell attachment and neural and Schwann cell proliferation to promote myelinated axon regeneration. Silk fibroin has desirable elasticity and flexibility to avoid collapse, and effective permeability allowing for nutrient exchange. However, silk fibroin's limitations result from its mechanically weak and fragile properties [42]. Silk fibroin can be processed using several different methods, including 3D bioprinting [39–41,48]. However, silk fibroin can be difficult to purify, as often is a natural polymer. A 2021 study by Lin et al. showed strong immune responses from their silk fibroin. After further experimentation, they discovered endotoxin contamination that adversely affected biocompatibility [47].

Fibrin is a natural polymer whose role in the body involves blood clotting. Fibrin has been shown to improve axonal regeneration and support blood vessel reparation [19]. Fibrin promotes cell adhesion for improved viability, proliferation, and differentiation of neural cells. Fibrin has been shown to prevent fibrous tissue formation and scarring while retaining biocompatibility [42]. Cell viability and neural differentiation have recently been reported after 3D printing (3DP) fibrin-based bioinks containing stem cells. Though rapid degradation limits its effectiveness, with adequate porosity for nutrient and waste exchange, and appropriate mechanical strength, fibrin-based bioink scaffolds achieved supportive results for neural applications [42,49,50].

Chitin is the second most abundant natural polymer on Earth [44]. However, due to its poor solubility, it is often modified to chitosan, maintaining the same chemical structure while adding necessary free amine groups [51]. Chitosan is low-cost, easy to produce, and biocompatible [52]. Although chitosan does not have desirable degradation and mechanical properties, chitosan has ECM-like features that support numerous studies showing significant improvements in functional axon regeneration, cell adhesion, and neural differentiation while decreasing scar formation in vivo [19,39–42,44,48,51]. A recent review on chitosan scaffolds for the treatment of myocardial infarctions has been reported, detailing polymer characteristics, such as elasticity and improved electrical conductivity, that would be translatable and beneficial for neural applications [53]. Meanwhile, a micropatterned, porous chitosan conduit was evaluated in a 10 mm sciatic nerve defect rat model. After three months, the chitosan conduit showed improved functional recovery comparable to or better than an autograft, though there was lesser neural, ECM, and blood vessel expression. The authors hypothesized that since chitosan can be easily modified, immobilizing growth factors through surface adsorption could improve this material facet [54]. One interesting disadvantage of this material is due to its sourcing. Since chitin is a core component of the exoskeleton of crustaceans, it may increase the risk of seafood allergy and inflammatory immune responses in certain patients [44,55].

There are additional natural polymers that, when combined with other materials, have been shown to promote axonal regeneration. Still, these polymers are less efficient when used alone due to weak mechanical resistance, fast degradation rate, and insufficient solubility. These materials include alginate, gelatin, and keratin, all of which, despite these challenges, have been shown to exhibit excellent biocompatibility [19,41,42,48].

3.2. Synthetic Polymers

Many synthetic polymers exhibit biocompatibility without cytotoxicity and can either be biodegradable (also known as bioresorbable), often preferred, or non-degradable. Poor degradation of synthetic polymers limits their use due to the higher potential for chronic inflammation [43]. These polymers also have uniquely beneficial characteristics, including tunability in mechanical characteristics and degradation rate. These polymers are easier to manufacture into complex architectures to better achieve biomimicry of the desired microenvironment, and are readily reproducible. One disadvantage is that these materials can be costlier and more time-consuming to produce. The most common synthetic polymers that support axon regeneration, guidance, and cell migration for improved nerve repair will be discussed [39–42,48,56].

Silicone and ePTFE are synthetic, non-degradable polymers that successfully treat PNIs. A second surgery to remove the implant may be required if complications, such as long-term foreign body reactions, occur. Despite this shortcoming, these polymers remain biocompatible for PNI healing and successful regeneration of myelinated axons [13,39,44].

Probably the most studied synthetic polymers for neural applications are polyesters. Polyesters include PGA, PLA, poly (lactic-co-glycolic acid) (PLGA), polycaprolactone (PCL), and poly(L-lactide-co-ε-caprolactone) (PLCL). These materials are favorable for use because of biocompatibility, tunable degradability, and limited cytotoxicity of degradation by-products. However, if the degradation occurs too rapidly, an acidic, inflammatory environment has been documented, resulting in tissue necrosis. Interestingly, PLGA's degradation can be adjusted by changing the molar ratio of PGA to PLA, while PLCL's copolymerization can neutralize the acidic environment [34,39,44,56,57]. Polyesters have desirable mechanical properties, flexibility, porosity, stability, solubility, hydrophilicity, crystallinity, and processability, making them valuable candidates for bioinks and 3DP methods. Yet, evidence of material deformation has been observed with long-term strain [20]. Many studies have shown their support for axon extension, regeneration, myelination, neural cell proliferation, differentiation, and neurotrophic factor expression [19,39,40,44,46,56–59]. Although the regenerative effects of these polyesters tend to be similar, PLGA has been reported to exhibit accelerated effects for more remarkable functional improvement over PCL and PLA [19].

Polyethylene glycol (PEG) is a biocompatible, synthetic polymer often used to adjust hydrogel swelling and mechanical properties [43,48,56]. PEG is often modified before use due to its easy processability, allowing for chemical modification and cross-linking, as well as its use in manufacturing methods such as 3DP [40]. PEG biodegrades poorly, limiting its use in vivo [48,60]. However, when modified, PEG can improve the microenvironment and support neural cell adhesion, proliferation, and differentiation [19].

Finally, polyurethane (PU) is a versatile synthetic polymer with adjustable mechanical and degradation properties. The Advincula group has reported PU/graphene nanocomposites to have potential for biomedical applications based on relatively stable and low toxicity when co-cultured with mammalian NIH-3T3 cells [61]. However, degradation by-products exhibit potential cytotoxicity [62]. Favorable characteristics of PU include fatigue resistance, flexibility, adhesion, durability, crystallinity, permeability, and anti-thrombotic properties.

Together, these characteristics have been shown to promote myelination, anti-inflammatory responses, and functional axonal regeneration for nerve repair [19,40,43,44,56].

Table 1. Characteristics of natural and synthetic polymers for nerve tissue engineering.

Polymer	Advantages	Disadvantages	Ref.
Natural Polymers			
Collagen	Favorable biocompatibility, biodegradability, bioactivity, and processability.	Low mechanical stress resistance, often requiring additional materials or cross-linking. Poor manipulability can make this difficult. Evidence of neuroma scar formation.	[13,19,39,41,42,44,45]
Hyaluronic Acid (HA)	Favorable biocompatibility. High producibility. Preventative of scar formation.	The fast degradation rate and poor mechanical properties can be adjusted with additional materials.	[19,42,46,47]
Silk Fibroin	Strong, stable, biocompatible, and biodegradable. Favorable elasticity, flexibility, permeability, and processability.	It can be difficult to purify. Often mechanically weak and fragile.	[39–42,47,48]
Fibrin	Desirable biocompatibility and cell adhesion. Preventative of scar formation. Supports blood vessel reparation.	Rapid degradation rate.	[19,42,49,50]
Chitin, Chitosan	Desirable biocompatibility. Abundant, low cost, easy to produce. ECM-like behaviors.	Chitin has poor solubility, improved with Chitosan modification. Weak degradation and mechanical properties. May increase the risk of seafood allergy and inflammatory immune responses in certain patients.	[19,39–42,44,48,51–54]
Synthetic Polymers			
Non-degradable polymers (Silicone, ePTFE)	Biocompatible for PNI healing.	Lack of degradation requires a second removal surgery before long-term foreign body reaction.	[13,39,44]
Polyesters (PGA, PLA, PLGA, PCL, PLCL)	Favorable biocompatibility. Tunable degradability. Easily excreted by-products. Desirable mechanical properties, flexibility, porosity, stability, solubility, hydrophilicity, and crystallinity. Favorable processability.	If it degrades too rapidly, PGA and PLA cause an acidic, inflammatory environment, yet PLCL copolymerization can neutralize an acidic environment. Deformation with long-term strain.	[19,20,39,40,44,46,56–59]
PEG	Favorable biocompatibility. Easy processibility allows for chemical modification and cross-linking to tune hydrogels' swelling and mechanical properties.	Poor degradative properties.	[19,40,43,48,56,60]
PU	Favorable biocompatibility. Tunable mechanical and degradation properties. Favorable fatigue resistance, flexibility, adhesion, durability, crystallinity, permeability, and anti-thrombotic properties.	Degradation by-products exhibit potential cytotoxicity.	[19,40,43,44,56,61,62]

4. Cells That Interact with Biomaterials

Regardless of the material, it is essential to consider the types of cells interacting with the surface when fabricating scaffolds. These cell types include any exogenous cells seeded on the construct and the endogenous cells within the injured area. Regenerative medicine for neural applications will often add primary culture Schwann cells, neural stem cells (NSCs), MSCs, embryonic stem cells (ESCs), or induced pluripotent stem cells (iPSCs) to scaffolds [19,43]. The source of these cells can be autologous, allogeneic, or xenogeneic [20]. Overall, the type of cell and how it will interact with the biomaterial surface influences its ability to repair neural injury. Meanwhile, the body will have systemic and localized reactions to the implanted material. Following the standard pattern of wound healing after a nerve is damaged, the foreign object implanted in the injured area will interact with peripheral immune cells, fibroblasts, progenitor cells, and the neural and glial cells that form the body's nervous system [15,63,64]. The interactions between these cells and different biomaterials will determine if the entire construct is both biocompatible and capable of promoting functional nerve recovery.

Schwann cells have been studied in conduits developed to treat PNIs. They are effective at directing axon elongation, neurite outgrowth, and forming a new myelin sheath, but difficulties in sourcing and purification have limited their use [19,20]. Stem cells are more commonly studied, having exhibited promotion of repair in neural injuries when added to a biomaterial scaffold. iPSCs seeded onto a PLA/PCL co-polymer conduit accelerated peripheral nerve repair [65]. The iPSCs were pre-differentiated into Schwann cell lineage before implanting the conduit into a sciatic nerve defect mouse model. This process removed the issues with sourcing and purification while retaining the benefits for directing axon growth. They found that adding the iPSCs to their material significantly enhanced their mice's functional recovery and promoted neural and myelin tissue histologically [65]. Similar to ESCs, because iPSCs are pluripotent, they have no impact of age or disease on their development. However, this benefit is weighed against the concern that these cells could become tumorigenic and immunogenic [20]. NSCs have been studied due to their potential to differentiate into neural and glial cell types, making them available for broader applications and removing the potential for developing into unwanted cell types that may worsen the injury [19]. Lastly, MSCs have the added characteristic of being easy to access. They can often be obtained from disposable or non-invasive tissues, such as adipose tissue and the umbilical cord, of both autologous and allogenic donors. MSCs also proliferate quickly and provide immunoregulatory factors to promote a microenvironment suitable for nerve repair [20]. Adipose-derived MSCs, 3D-printed into a fibrin-based bioink, expressed immature neural and dopaminergic neural markers *in vitro*, presenting potential therapeutic effects for NDs such as Parkinson's Disease [50].

When a nerve injury occurs, or a scaffold is implanted, many cell types are recruited endogenously to repair the damage. Although inflammation is a natural response to injury, the harm comes when the inflammation is prolonged. Local proteins adsorb to the material surface, and this material–cell interface's properties determine the immune response's extent [66]. If the interactions at the interface are non-immunogenic, the body accepts the material, i.e., it is biocompatible, and the inflammatory response will not last longer than a few days. However, if the material is identified as a foreign body, the secretion of inflammatory cytokines and chemokines will initiate a response from the immune system [66–68]. Phagocytic cells, such as neutrophils, monocytes, local macrophages, and microglia, are attracted to these signals and migrate to the injured area [66–70]. These cells will remove cellular debris and pathogens while releasing reactive oxygen species (ROS) that cause additional damage [68]. However, the recruitment of these cells is necessary to help with regeneration, as they also secrete beneficial growth factors that promote tissue organization and angiogenesis [67,68,70]. The monocytes recruited to the area differentiate into macrophages. These macrophages attempt to engulf the scaffold but cannot phagocytize such a significant material and become stressed. The microenvironment becomes chronically inflamed, and the macrophages begin to morph together and form foreign body giant cells (FBGCs). A fibrous tissue encapsulates the material and attempts to expel the material from the body, rejecting the implant [66,69,70]. Although the innate immune cells are the first line of defense, the adaptive T cells and B cells often complement the system, recognizing specific antigens and aiding in the cascade of inflammatory factors [69–71]. When the inflammation begins to subside around the biocompatible scaffold, neural and glial progenitor cells will move through the area to regenerate and remodel the tissue. Depending on whether the injury occurs in the CNS or PNS, these cells will differentiate into neural and glial cells, including astrocytes, oligodendrocytes, or Schwann cells, to rebuild the nerve. Each cell type has a vital role in the nervous system that helps regulate neural activity and the exchange of nutrients in a healthy microenvironment. Finally, macrophages, fibroblasts, and endothelial cells will reconstruct the surrounding ECM and vasculature [66].

5. Biomaterial Modifications Influence Cell Behavior

Although these polymers have desirable biocompatibility, tunable degradation, and mechanical properties that benefit biological use, most polymers are not used without additional modifications. These polymers are altered to optimize their use in neural applications while retaining the benefits of their material. Nanomaterials, such as graphene or gold, are commonly incorporated due to their conductive capabilities [52,72–77]. Polymers also can be altered for a more specific application. For example, cross-linking polymers to create hydrogels are suitable for use in the CNS, specifically for TBIs, and hollow conduits and wraps are more suitable for use in PNIs and SCIs due to the shape of axons and the spinal cord. While there have been many studies on the effects of polymers, incorporating nanomaterials and cross-linking changes the material's characterization and the behavior of the cells they interact with [56]. In all, the goal of these modifications remains the same, to develop a safe and functional treatment for neural injuries [15].

5.1. Hydrogel Formation

Hydrogels can be used for various applications, including nerve tissue engineering. They are often preferred for CNS injuries, particularly TBIs, where nerve guidance conduits would be too invasive. Defined as 3D cross-linked polymers with high water content and tunable properties, hydrogels can comprise individual polymers or composites of two or more materials. When forming a composite, the strengths and weaknesses of the unique polymers are combined [13]. Overall, hydrogels tend to induce minor inflammation, have favorable biocompatibility and biodegradability, but have weak mechanical properties [43,56,78,79]. The molecular weights and the concentrations of the distinct polymers, alongside the cross-linking method, can be used to alter the characteristics of the hydrogel. Standard cross-linking methods include ultraviolet (UV) light, pH, and temperature. Additionally, each of these factors can alter the physicochemical properties of the materials to produce a treatment optimized for the nervous system [19,42,47,79]. These properties include gelation time, swelling ratio, physical size, shape, structure, mechanical strength, stress, porosity, and surface charge and chemistry.

The gelation time is essentialbecause the hydrogel needs to have the appropriate thixotropic properties (injectability, shear-thinning, etc.) to be appropriately processed (3D printed, injected, etc.). A benefit of hydrogels for their use in CNS injuries is that they can fill irregularly shaped injury sites while providing the appropriate structural support required for nerve development [79]. However, without the appropriate gelation, a hydrogel can take hours or days to obtain the proper mechanical stability characteristic of this modification [80]. In a study using a photocurable, methacrylate, silk fibroin bioink, the raw silk fibroin bioink took several hours to gel, making it a less-than-ideal candidate for clinical use [78]. By adjusting factors such as treating with enzymes or salts, altering conditions including pH and temperature, and the cross-linking method, they decreased the hydrogels' gelation time to minutes. The material will degrade once a hydrogel has solidified and reached its maximum swelling ratio due to water uptake [78,81].

In addition to the gelation kinetics, the physicochemical attributes of the hydrogel significantly influence cellular behavior at the tissue–material interface. Characteristics such as porosity, micro patterns, single vs. multi-channel, filament diameter, and surface roughness can all change cell adhesion, proliferation, axon guidance, and neural differentiation [42,82]. Hydrogels mimic the natural ECM, providing a microenvironment that allows for cell signaling and cell adhesion to promote nerve regrowth [83]. A porous hydrogel creates the space for increased cell–cell interactions and access to nutrients and supportive molecules [78]. Meanwhile, the surface microtopography can increase cell adhesion, by increasing the material surface area while providing guidance cues for axon and neurite growth to their target location [79,82]. One study compared micropatterned chitosan conduits to non-patterned, random conduits of the same material. Their results showed that the micropatterning significantly improved functional recovery as compared with the non-patterned conduit [54]. In another study, a low-stiffness hydrogel greatly improved nerve growth in a sciatic nerve injury rat model compared to more rigid hydrogels of the same material [84]. In general, a Young's Modulus value of 0.1 to 1.0 kPa provides the appropriate stress/strain to promote neural differentiation favorably [79].

These characteristics of hydrogels, and the physical and chemical cues they provide, are proper modes for delivering cells, drugs, and other biological factors to aid nerve repair. The material encapsulates the factors of choice to support further the environment in which it is implanted [82,83]. One study immobilized umbilical cord-derived MSCs and neurotrophic factor fibroblast growth factor (FGF) onto a heparin–poloxamer hydrogel to repair SCI. The study found decreased cell apoptosis and increased mitochondrial function in vivo compared to hydrogels without the additional cells and factors [85]. Other factors can be added to reduce the aversiveness of the injured microenvironment and better support endogenous nerve repair. For example, a 2022 study by Jin et al. observed that urolithin-A-loaded hydrogels in PCL conduits exhibited anti-inflammatory, antioxidant, and immunomodulatory effects [86].

A large field of study is involved in incorporating conductive polymers into hydrogels to improve the electrophysiological function of the regenerating nerve. Nerves have natural electrical activity in their tissue environment, and introducing electrical components before functional nerve regeneration has been shown to improve cell adhesion, self-renewal, proliferation, differentiation, and signaling. The surface charge of the hydrogel, when made highly positive, can improve the electrostatic interactions with the negatively charged cell membrane, to further enhance cell growth and neural differentiation [87]. Electrical stimulation has also been shown to accelerate axon regeneration for faster healing. This occurs as neurons can continue transmitting information, forming new synapses to integrate themselves into functional neural circuits [42,75,88,89]. Poly(3,4-ethylenedioxythiophene) (PEDOT) is a common conductive polymer that is being incorporated for neuroregeneration. Other conductive biomaterials incorporated include polypyrrole, polythiophene, and polyaniline. These polymers are often hard to process and contain due to their brittleness, rigidity, and insolubility, though PEDOT has had the best success thus far [42,74,82,87,90]. A biocompatibility study of electrically stimulated PEDOT in the brain found that significant damage was created at higher current densities and could not be reversed. However, low pulse currents did not significantly diminish cell viability [91].

5.2. Incorporation of Nanomaterials

Nanomaterials are materials composed of particles that have nanoscale dimensions (<100 μm). These can include various materials, among which carbon-based nanomaterials, such as graphene and its derivatives and gold (Au) nanomaterials, are commonly used. These materials improve polymers' strength, stiffness, electrical conductivity, topographical features, and surface area while retaining biocompatibility. However, they are not without their challenges. Nanomaterials can improve the loading efficiency, binding efficiency, and sustained release of biomolecules, such as growth factors for enhanced regeneration, or drugs for therapeutic drug delivery. These particles are often conductive and hydrophobic, so they can be challenging to handle as they aggregate together and poorly disperse [41,75,88–90].

Graphene has two common functionalizations that often are used for neuroregeneration: graphene oxide (GO) and reduced graphene oxide (rGO). GO has been shown to increase secretions of neurotrophic factors, neural differentiation, cell attachment and proliferation, and neurite outgrowth. rGO offers these same regenerative benefits alongside increased electrical conductivity. Graphene can be incorporated in several ways, including blending with a polymer, or coating a surface [19,48,57,74,90,92]. Previous studies evaluated the effect of GO-coated PLGA electrospun nanofibers immobilized with two neural growth factors. They found that adding GO decreased fiber diameter, altered the surface chemistry, enhanced the hydrophilicity, and improved the binding efficiency and sustained release of the two growth factors. These studies also found maintained or increased cell viability, proliferation, and neural differentiation in vitro and in vivo [93]. Graphene may have bacterial inhibiting properties, which not only aid in the influence of neural and glial cell behavior but also impact the injured area by preventing infection and long-term inflammation of the microenvironment [87].

Gold nanoparticles offer the same biocompatibility and enhanced neuroregenerative effects as graphene and its derivatives, while retaining the same issues with aggregation and dispersion [52,94]. The results of a chitosan conduit modified with Au nanoparticles were studied. The material maintained desirable biocompatibility while improving cell viability, adhesion, proliferation, and differentiation by increasing surface roughness, reducing inflammatory response, and influencing signaling factors and electrical conductivity [52].

6. Biomaterial Fabrication Methods

Though there are many fabrication methods when it comes to materials for biomedical applications, 3DP has resulted in significant advancements in tissue engineering. One of the benefits of 3DP is its specificity. Computer-aided design (CAD), automated slicing software (creating a stereolithography (STL) file), and typed geometric code (G-code) allow for the creation of complex, biomimetic designs that match what is seen in tissues in the human body (Figure 3). 3DP creates these designs on demand, allowing for a future in personalized medicine [19,95]. Due to the complexities of the nervous system, 3DP is an excellent method to fabricate nerve guidance conduits and hydrogels for clinical use.

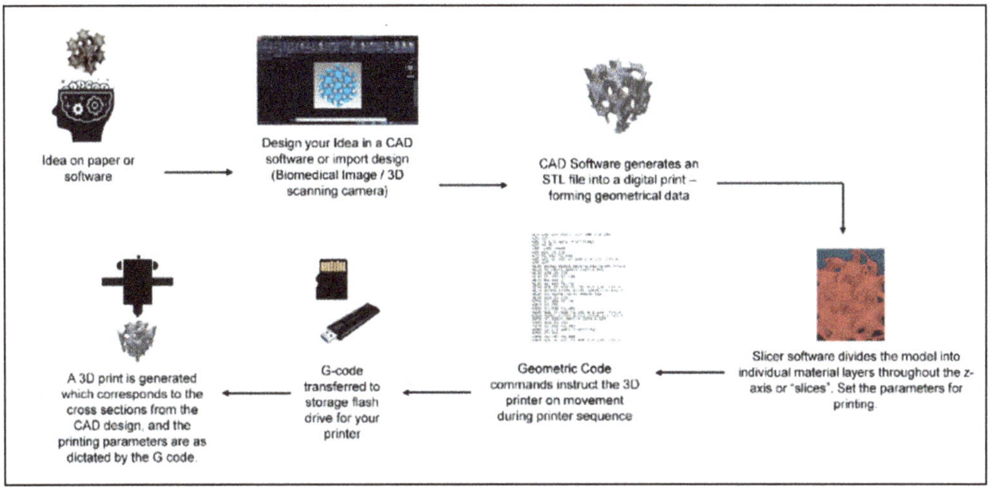

Figure 3. General schematic of the computer-aided 3DP process. Figure adapted from [95]. The use of this figure is governed by the Creative Commons Attribution (CC BY license https://creativecommons.org/licenses/by/4.0/) to the licensee MDPI, Basel, Switzerland.

As discussed in previous sections, many properties contribute to biomaterials' processability and printability. Biomaterials must be adapted for safety and efficacy in the body and withstand the fabrication, sterilization, and implantation processes. There is a similar process in the choice of the 3DP method. Although there are many types, specific techniques are more beneficial for bioprinting, and certain methods show favorable properties for printing live cells within bioinks. For example, the Advincula group has reported a number of studies using biomedically relevant polymer materials and composites for 3DP, mainly via the extrusion-based method of direct ink writing (DIW) [96–100]. The 3DP methods found to be best for nerve tissue engineering are photopolymerization techniques such as SLA and digital light projection (DLP), and extrusion-based techniques such as fused deposition modeling (FDM) and micro-extrusion [19,40,101]. Meanwhile, co-axial printing, an advanced extrusion-based technique, allows for a more complex, layered, multi-print head method that is favorable for live cell bioprinting [102].

SLA uses a UV laser to photosynthesize hydrogels or resins by immersing the stage in liquid and moving to change the layer height and resolution. Laser power, laser size, exposure time, and light wavelength can impact SLA printing. This method can be slow, costly, and dependent on specific material characteristics [19,43,101]. DLP is similar to SLA in that it uses photopolymerization. However, this method uses numerous mirrors to project the printed image, improving printing speed [43]. These photopolymerization methods are ideal for producing hydrogels for neural applications. Recently, a DLP 3D printer made photocurable silk fibroin hydrogels. Based on the data, any material with less than 10% wt./vol. hydrogel composition was difficult to print due to a loss in structural integrity. However, the hydrogel could be adjusted in a higher-range concentration (10–30%) to create physical properties, such as gelation time, swelling ratio, Young's Modulus, porosity, and degradation time, that were optimal for nerve treatments. The authors also discussed maintaining approximately 95% cell viability when encapsulated inside the hydrogel. However, the fabrication process for making the materials, printing the structure, and implanting the material required 2 weeks, limiting application in clinical settings [78]. Another study, using a DLP 3D printer, printed a hydrogel loaded with a small flavonoid molecule, 7,8-dihydroxyflavone, to enhance the effects for a PNI treatment. The drug-loaded bioink was prepared within a day, and exhibited favorable mechanical properties and strong effects of cell adhesion and migration when co-cultured with Schwann cells [103].

FDM creates filaments using temperature changes to extrude material in a semi-liquid state, solidifying on a platform in a layer-by-layer deposition. This method is not ideal for direct cell printing if higher temperatures are required for the polymer(s). Although it is a simple, low-cost process, the prints may result in interlayer weakness depending on layer thickness, filament diameter, direction, and printing speed [19,43,49,101]. This is a standard method for printing nerve guidance conduits for PNI and SCI applications. In a study that used FDM to print PLA scaffolds coated with Au-nanoparticles, specific printing variables were kept standard, including layer height, fill density, fill pattern, extrusion temperature, bed temperature, and print speed. The study assessed variability in prints, including filament diameter, pore size, Young's Modulus, and tensile strength. Slight variations were found, even though overall consistency between prints was considered favorable, while demonstrating biocompatibility and promotion towards neural lineage [94].

Micro-extrusion uses a continuous bioink stream dispensed through nozzles to fabricate structures using pistons, screws, or pneumatic pressures rapidly. Due to speed and force, this method can often exhibit too much shear stress for live cell printing, though adjustments made in these variables can exhibit favorable outcomes [19,49,101]. One study used a micro-extrusion bioprinter to print fibrin hydrogels containing MSCs. The MSCs showed high cell viability immediately after printing and 4, 9, and 12 days *in vitro*. The scaffold even promoted the successful differentiation of MSCs into neural and glial cell types and exhibited electrophysical/functional maturation in vitro [50].

These extrusion-based techniques have advanced to include co-axial printing, composed of multiple printheads with different parameters for different materials, that come together in alternating layers to create a single scaffold. One study cultured 3D ESCs differentiated towards spinal cord motor neurons, and suspended them within a GelMA/alginate hydrogel. First, PCL-based inner and outer hollow tubes were printed. The hydrogel was then printed between the two PCL walls, in alternating layers of ESC-loaded hydrogel, and alginate alone. Alternating materials and varying printing conditions between each material allowed for a construct that promoted cell viability and structural integrity, with properties promoting treatment of SCIs [104]. The alternating tissue layers were found to be more comparable to native tissues, and a beneficial method for bioprinting live cells [102]. Another study used two print heads, one with a fibrin-based bioink, and one with a cross-linker, so that the material was polymerized prior to extruding. This limits the shear stress of the live cells as they pass through the printing needle [50].

7. Future Perspectives

Future studies may look into additional methods to alter a material that could improve functionality for directing desired cell properties. Using existing materials, applications of more surface modification methods could be studied to aid in the control of protein adsorption, chemotaxis, and cell-tissue morphologies. These surface modifications include polymer grafting or self-assembly using chemical and physical adsorption methods [105]. Grafting polymers and polymer brushes could enable a more desirable surface functionality and chemistry, independent of the bulk thermo-mechanical properties of the individual biomaterials, to aid in cell proliferation, guidance of neural cell growth, and axon-dendrite distribution [106–108]. Further investigations of biomaterial surface morphology, thermomechanical properties (e.g., change in elasticity or flexural properties), directional properties, anisotropic properties, and processing methods that influence semi-crystallinity (e.g., polymer domain size) will bring us closer to improved processing and manufacturing methods for clinical applications.

The above methods show how a material can be modified to improve cell behavior to a specific biomaterial. Cell biologists and material scientists can work together to fabricate a surface according to the needs of the cellular applications. For instance, the importance of cell adherence is provided by a material's surface topography and chemical composition. A surface functionalized with -COOH or -NH_3 groups can enhance adsorption of proteins. Understanding and altering a material to influence neural lineage can drastically improve cell behavior for the appropriate applications. Conversely biological tools exist which can guide cells to behave as per their need [109–112].

The use of hyphenated and more state-of-the-art surface-sensitive spectroscopic and microscopic analytical characterization methods, such as advanced IR-AFM (infrared spectroscopy, atomic force microscopy) methods, higher resolution or super-resolution fluorescence microscopy, electrochemical-AFM imaging methods, and better TEM (transmission electron microscopy) and cryo-TEM methods, would advance the process of biomaterial characterization closer to physiological conditions for more accurate assessments [113–115]. At the same time, using probe methods and label or non-labeling methods (e.g., quantum dots) could enhance imaging to provide better and higher-throughput in interpretations [116]. The use of high-contrast imaging agents (dyes, nanoparticles, ligands, etc.) for multimodal and multiplexing is also important [117]. Finally, AI/ML (artificial intelligence, machine learning) methods can be applied to imaging modalities and polymer design methods. AI/ML could improve the testing methods and failure analysis of implanted materials [118].

As summarized in Figure 4, recent progress in team science, technological advances in material science and cell biology, along with the novel opportunities at hand will provide potential tissue engineering products for neural repair.

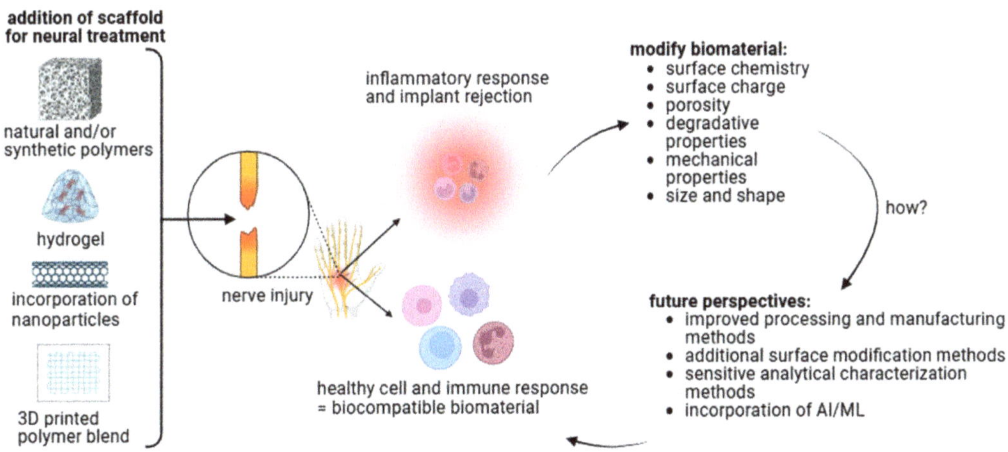

Figure 4. Biomaterial–cell interactions influence biocompatibility. Polymer choice, fabrication method, and additional modifications can alter the biomaterials' physical and chemical properties. These properties affect cell behavior to determine local and systemic immune responses. Future perspectives should be considered to develop a therapeutic treatment for neural injuries [figure made with BioRender].

8. Conclusions

Regenerative medicine and nerve tissue engineering are essential research fields for developing an effective treatment for neural injuries. Numerous natural and synthetic polymers exhibit excellent biocompatible, biodegradable, and bioactive properties, in addition to excellent processability and printability. The type of polymer, method of modification, and method of fabrication can significantly impact a polymer's physicochemical properties. These properties, such as surface chemistry, mechanical strength, and topography, greatly influence cellular adhesion, migration, proliferation, and differentiation, impacting how the neural environment affects biocompatibility and regeneration. Although numerous factors are advantageous or disadvantageous, many of these polymers, modification techniques, and fabrication methods support the safe and effective regeneration of a physical and functional nerve. They are, ultimately, moving closer toward a treatment to improve the quality of life of the millions impacted by neural injuries.

Author Contributions: Each author contributed various sections of this review article. All authors approved the final article. M.E.H.-T. and M.D. were responsible for conceptualizing the idea for the review article. R.S. was responsible for the biocompatibility section. R.C.A. was responsible for the future perspectives section and the review and editing of the biomaterial information. D.E.A. was responsible for the review and editing of the cellular behavior information. All authors have read and agreed to the published version of the manuscript.

Funding: This article was funded by the University of Tennessee: Office of Research, Innovation, and Economic Development Seed Award, 2022 Student/Faculty Research Award, and Comparative and Experimental Medicine Program.

Institutional Review Board Statement: Not applicable.

Informed Consent Statement: Not applicable.

Acknowledgments: The authors acknowledge the support of the University of Tennessee: Office of Research, Innovation, and Economic Development Seed Award, 2022 Student/Faculty Research Award, and Comparative and Experimental Medicine Program.

Conflicts of Interest: The authors declare no conflict of interest.

References

1. Purves, D.; Augustine, G.; Fitzpatrick, D.; Hall, W.; LaMantia, A.; White, L.; Mooney, R.; Platt, M. *Neuroscience*, 6th ed.; Oxford University Press: New York, NY, USA, 2018. Available online: https://www.academia.edu/43014289/Neuroscience_by_Dale_Purves_et_al_eds_z_lib_org_ (accessed on 13 August 2023).
2. Lo, J.; Chan, L.; Flynn, S. A Systematic Review of the Incidence, Prevalence, Costs, and Activity and Work Limitations of Amputation, Osteoarthritis, Rheumatoid Arthritis, Back Pain, Multiple Sclerosis, Spinal Cord Injury, Stroke, and Traumatic Brain Injury in the United States: A 2019 Update. *Arch. Phys. Med. Rehabil.* **2021**, *102*, 115–131. [CrossRef]
3. Houshyar, S.; Bhattacharyya, A.; Shanks, R. Peripheral Nerve Conduit: Materials and Structures. *ACS Chem. Neurosci.* **2019**, *10*, 3349–3365. [CrossRef]
4. Soman, S.S.; Vijayavenkataraman, S. Perspectives on 3D Bioprinting of Peripheral Nerve Conduits. *Int. J. Mol. Sci.* **2020**, *21*, 5792. [CrossRef]
5. Vella, M.A.; Crandall, M.L.; Patel, M.B. Acute Management of Traumatic Brain Injury. *Surg. Clin. N. Am.* **2017**, *97*, 1015–1030. [CrossRef]
6. Modrak, M.; Talukder, M.A.H.; Gurgenashvili, K.; Noble, M.; Elfar, J.C. Peripheral nerve injury and myelination: Potential therapeutic strategies. *J. Neurosci. Res.* **2020**, *98*, 780–795. [CrossRef]
7. Wang, M.L.; Rivlin, M.; Graham, J.G.; Beredjiklian, P.K. Peripheral nerve injury, scarring, and recovery. *Connect. Tissue Res.* **2019**, *60*, 3–9. [CrossRef]
8. Kirshblum, S.; Snider, B.; Eren, F.; Guest, J. Characterizing Natural Recovery after Traumatic Spinal Cord Injury. *J. Neurotrauma* **2021**, *38*, 1267–1284. [CrossRef] [PubMed]
9. Schepici, G.; Silvestro, S.; Bramanti, P.; Mazzon, E. Traumatic Brain Injury and Stem Cells: An Overview of Clinical Trials, the Current Treatments and Future Therapeutic Approaches. *Medicina* **2020**, *56*, 137. [CrossRef] [PubMed]
10. Gordon, T. Peripheral Nerve Regeneration and Muscle Reinnervation. *Int. J. Mol. Sci.* **2020**, *21*, 8652. [CrossRef] [PubMed]
11. Hussain, G.; Wang, J.; Rasul, A.; Anwar, H.; Qasim, M.; Zafar, S.; Aziz, N.; Razzaq, A.; Hussain, R.; de Aguilar, J.G.; et al. Current Status of Therapeutic Approaches against Peripheral Nerve Injuries: A Detailed Story from Injury to Recovery. *Int. J. Biol. Sci.* **2020**, *16*, 116–134. [CrossRef] [PubMed]

12. Cofano, F.; Boido, M.; Monticelli, M.; Zenga, F.; Ducati, A.; Vercelli, A.; Garbossa, D. Mesenchymal Stem Cells for Spinal Cord Injury: Current Options, Limitations, and Future of Cell Therapy. *Int. J. Mol. Sci.* **2019**, *20*, 2698. [CrossRef] [PubMed]
13. Li, Y.; Ma, Z.; Ren, Y.; Lu, D.; Li, T.; Li, W.; Wang, J.; Ma, H.; Zhao, J. Tissue Engineering Strategies for Peripheral Nerve Regeneration. *Front. Neurol.* **2021**, *12*, 768267. [CrossRef] [PubMed]
14. Puleo, D.A.; Bizios, R. (Eds.) *Biological Interactions on Materials Surfaces*; Springer: New York, NY, USA, 2009.
15. Doblado, L.R.; Martínez-Ramos, C.; Pradas, M.M. Biomaterials for Neural Tissue Engineering. *Front. Nanotechnol.* **2021**, *3*, 643507. [CrossRef]
16. Jessen, K.R.; Mirsky, R. The Success and Failure of the Schwann Cell Response to Nerve Injury. *Front. Cell Neurosci.* **2019**, *13*, 33. [CrossRef] [PubMed]
17. Lopes, B.; Sousa, P.; Alvites, R.; Branquinho, M.; Sousa, A.C.; Mendonca, C.; Atayde, L.M.; Luis, A.L.; Varejao, A.S.P.; Mauricio, A.C. Peripheral Nerve Injury Treatments and Advances: One Health Perspective. *Int. J. Mol. Sci.* **2022**, *23*, 918. [CrossRef]
18. Graham, N.S.; Sharp, D.J. Understanding neurodegeneration after traumatic brain injury: From mechanisms to clinical trials in dementia. *J. Neurol. Neurosurg. Psychiatry* **2019**, *90*, 1221–1233. [CrossRef]
19. Bedir, T.; Ulag, S.; Ustundag, C.B.; Gunduz, O. 3D bioprinting applications in neural tissue engineering for spinal cord injury repair. *Mater. Sci. Eng. C Mater. Biol. Appl.* **2020**, *110*, 110741. [CrossRef]
20. Sensharma, P.; Madhumathi, G.; Jayant, R.D.; Jaiswal, A.K. Biomaterials and cells for neural tissue engineering: Current choices. *Mater. Sci. Eng. C Mater. Biol. Appl.* **2017**, *77*, 1302–1315. [CrossRef]
21. Cangellaris, O.V.; Gillette, M.U. Biomaterials for Enhancing Neuronal Repair. *Front. Mater.* **2018**, *5*, 21. [CrossRef]
22. Williams, D.F. Definitions in biomaterials. In Proceedings of the Consensus Conference of the European Society for Biomaterials, Chester, UK, 3–5 March 1986.
23. Buddy, D.; Ratner, A.S.H.; Frederick, J.; Schoen, J.; Lemons, E. (Eds.) *Biomaterials Science: An Introduction to Materials in Medicine*; Elsevier Inc.: Waltham, MA, USA, 2013.
24. Jinku Kim, A.S.; Hollinger, J.O. *An Introduction To Biomaterials*; Hollinger, J.O., Ed.; Taylor & Francis Group: Boca Raton, FL, USA, 2012.
25. Crawford, L.; Wyatt, M.; Bryers, J.; Ratner, B. Biocompatibility Evolves: Phenomenology to Toxicology to Regeneration. *Adv. Healthc. Mater.* **2021**, *10*, e2002153. [CrossRef]
26. Bernard, M.; Jubeli, E.; Pungente, M.D.; Yagoubi, N. Biocompatibility of polymer-based biomaterials and medical devices–regulations, in vitro screening and risk-management. *Biomater. Sci.* **2018**, *6*, 2025–2053. [CrossRef]
27. Savage, D.T.; Hilt, J.Z.; Dziubla, T.D. In Vitro Methods for Assessing Nanoparticle Toxicity. *Methods Mol. Biol.* **2019**, *1894*, 1–29. [CrossRef] [PubMed]
28. Vidal, M.; Granjeiro, J. Cytotoxicity Tests for Evaluating Medical Devices: An Alert for the Development of Biotechnology Health Products. *J. Biomed. Sci. Eng.* **2017**, *10*, 431–443. [CrossRef]
29. De Stefano, P.; Federici, A.S.; Draghi, L. In Vitro Models for the Development of Peripheral Nerve Conduits, Part I: Design of a Fibrin Gel-Based Non-Contact Test. *Polymers* **2021**, *13*, 3573. [CrossRef]
30. Bikmulina, P.; Kosheleva, N.; Efremov, Y.; Antoshin, A.; Heydari, Z.; Kapustina, V.; Royuk, V.; Mikhaylov, V.; Fomin, V.; Vosough, M.; et al. 3D or not 3D: A guide to assess cell viability in 3D cell systems. *Soft Matter* **2022**, *18*, 2222–2233. [CrossRef] [PubMed]
31. Ng, K.W.; Leong, D.T.W.; Hutmacher, D.W. The challenge is to measure cell proliferation in two and three dimensions. *Tissue Eng.* **2005**, *11*, 182–191. [CrossRef]
32. Jiao, G.; He, X.; Li, X.; Qiu, J.; Xu, H.; Zhang, N.; Liu, S. Limitations of MTT and CCK-8 assay for evaluation of graphene cytotoxicity. *RSC Adv.* **2015**, *5*, 53240–53244. [CrossRef]
33. Wang, P.; Henning, S.M.; Heber, D. Limitations of MTT and MTS-based assays for measurement of antiproliferative activity of green tea polyphenols. *PLoS ONE* **2010**, *5*, e10202. [CrossRef]
34. Jahromi, M.; Razavi, S.; Bakhtiari, A. The advances in nerve tissue engineering: From fabrication of nerve conduit to in vivo nerve regeneration assays. *J. Tissue Eng. Regen. Med.* **2019**, *13*, 2077–2100. [CrossRef]
35. Li, A.; Pereira, C.; Hill, E.E.; Vukcevich, O.; Wang, A. In Vitro, In Vivo and Ex Vivo Models for Peripheral Nerve Injury and Regeneration. *Curr. Neuropharmacol.* **2022**, *20*, 344–361. [CrossRef]
36. Vela, F.J.; Martinez-Chacon, G.; Ballestin, A.; Campos, J.L.; Sanchez-Margallo, F.M.; Abellan, E. Animal models used to study direct peripheral nerve repair: A systematic review. *Neural Regen. Res.* **2020**, *15*, 491–502. [CrossRef] [PubMed]
37. Basics of Biocompatibility: Information Needed for Assessment by the FDA. Available online: https://www.fda.gov/medical-devices/biocompatibility-assessment-resource-center/basics-biocompatibility-information-needed-assessment-fda (accessed on 16 August 2023).
38. Göktürk, E.; Erdal, H. Biomedical Applications of Polyglycolic Acid (PGA). *Sak. Univ. J. Sci.* **2017**, *21*, 1237–1244. [CrossRef]
39. Wang, Y.; Zhang, Y.; Li, X.; Zhang, Q. The progress of biomaterials in peripheral nerve repair and regeneration. *J. Neurorestoratology* **2020**, *8*, 252–269. [CrossRef]
40. Sta Agueda, J.R.H.; Chen, Q.; Maalihan, R.D.; Ren, J.; da Silva, I.G.M.; Dugos, N.P.; Caldona, E.B.; Advincula, R.C. 3D printing of biomedically relevant polymer materials and biocompatibility. *MRS Commun.* **2021**, *11*, 197–212. [CrossRef]
41. Kumar, S.S.D.; Abrahamse, H. Advancement of Nanobiomaterials to Deliver Natural Compounds for Tissue Engineering Applications. *Int. J. Mol. Sci.* **2020**, *21*, 6752. [CrossRef]

42. Khan, H.M.; Liao, X.; Sheikh, B.A.; Wang, Y.; Su, Z.; Guo, C.; Li, Z.; Zhou, C.; Cen, Y.; Kong, Q. Smart biomaterials and their potential applications in tissue engineering. *J. Mater. Chem. B* **2022**, *10*, 6859–6895. [CrossRef]
43. Yu, X.; Zhang, T.; Li, Y. 3D Printing and Bioprinting Nerve Conduits for Neural Tissue Engineering. *Polymers* **2020**, *12*, 1637. [CrossRef]
44. Dai, W.; Yang, Y.; Yang, Y.; Liu, W. Material advancement in tissue-engineered nerve conduit. *Nanotechnol. Rev.* **2021**, *10*, 488–503. [CrossRef]
45. He, J.; Zhang, N.; Zhu, Y.; Jin, R.; Wu, F. MSC spheroids-loaded collagen hydrogels simultaneously promote neuronal differentiation and suppress inflammatory reaction through PI3K-Akt signaling pathway. *Biomaterials* **2021**, *265*, 120448. [CrossRef]
46. Li, M.; Wang, Y.; Zhang, J.; Cao, Z.; Wang, S.; Zheng, W.; Li, Q.; Zheng, T.; Wang, X.; Xu, Q.; et al. Culture of pyramidal neural precursors, neural stem cells, and fibroblasts on various biomaterials. *J. Biomater. Sci. Polym. Ed.* **2018**, *29*, 2168–2186. [CrossRef]
47. Lin, C.; Ekblad-Nordberg, A.; Michaelsson, J.; Gotherstrom, C.; Hsu, C.C.; Ye, H.; Johansson, J.; Rising, A.; Sundstrom, E.; Akesson, E. In Vitro Study of Human Immune Responses to Hyaluronic Acid Hydrogels, Recombinant Spidroins and Human Neural Progenitor Cells of Relevance to Spinal Cord Injury Repair. *Cells* **2021**, *10*, 1713. [CrossRef] [PubMed]
48. Yi, S.; Xu, L.; Gu, X. Scaffolds for peripheral nerve repair and reconstruction. *Exp. Neurol.* **2019**, *319*, 112761. [CrossRef] [PubMed]
49. Abelseth, E.; Abelseth, L.; De la Vega, L.; Beyer, S.T.; Wadsworth, S.J.; Willerth, S.M. 3D Printing of Neural Tissues Derived from Human Induced Pluripotent Stem Cells Using a Fibrin-Based Bioink. *ACS Biomater. Sci. Eng.* **2019**, *5*, 234–243. [CrossRef] [PubMed]
50. Restan Perez, M.; Sharma, R.; Masri, N.Z.; Willerth, S.M. 3D Bioprinting Mesenchymal Stem Cell-Derived Neural Tissues Using a Fibrin-Based Bioink. *Biomolecules* **2021**, *11*, 1250. [CrossRef]
51. Xiang, W.; Cao, H.; Tao, H.; Jin, L.; Luo, Y.; Tao, F.; Jiang, T. Applications of chitosan-based biomaterials: From preparation to spinal cord injury neuroprosthetic treatment. *Int. J. Biol. Macromol.* **2023**, *230*, 123447. [CrossRef]
52. Hung, H.S.; Yang, Y.C.; Chang, C.H.; Chang, K.B.; Shen, C.C.; Tang, C.L.; Liu, S.Y.; Lee, C.H.; Yen, C.M.; Yang, M.Y. Neural Differentiation Potential of Mesenchymal Stem Cells Enhanced by Biocompatible Chitosan-Gold Nanocomposites. *Cells* **2022**, *11*, 1861. [CrossRef]
53. Beleno Acosta, B.; Advincula, R.C.; Grande-Tovar, C.D. Chitosan-Based Scaffolds for the Treatment of Myocardial Infarction: A Systematic Review. *Molecules* **2023**, *28*, 1920. [CrossRef]
54. Li, G.; Xue, C.; Wang, H.; Yang, X.; Zhao, Y.; Zhang, L.; Yang, Y. Spatially featured porous chitosan conduits with micropatterned inner wall and seamless sidewall for bridging peripheral nerve regeneration. *Carbohydr. Polym.* **2018**, *194*, 225–235. [CrossRef]
55. Elieh Ali Komi, D.; Sharma, L.; Dela Cruz, C.S. Chitin and Its Effects on Inflammatory and Immune Responses. *Clin. Rev. Allergy Immunol.* **2018**, *54*, 213–223. [CrossRef]
56. Amani, H.; Kazerooni, H.; Hassanpoor, H.; Akbarzadeh, A.; Pazoki-Toroudi, H. Tailoring synthetic polymeric biomaterials towards nerve tissue engineering: A review. *Artif. Cells Nanomed. Biotechnol.* **2019**, *47*, 3524–3539. [CrossRef]
57. The Macrogalleria. Available online: https://pslc.ws/macrog/index.htm (accessed on 16 August 2023).
58. Miri, V.; Asadi, A.; Sagha, M.; Najafzadeh, N.; Golmohammadi, M.G. Poly (L-lactic acid) nanofibrous scaffolds support the proliferation and neural differentiation of mouse neural stem and progenitor cells. *Int. J. Dev. Neurosci.* **2021**, *81*, 438–447. [CrossRef] [PubMed]
59. Min, K.; Kong, J.S.; Kim, J.; Kim, J.; Gao, G.; Cho, D.W.; Han, H.H. Three-Dimensional Microfilament Printing of a Decellularized Extracellular Matrix (dECM) Bioink Using a Microgel Printing Bath for Nerve Graft Fabrication and the Effectiveness of dECM Graft Combined with a Polycaprolactone Conduit. *ACS Appl. Bio Mater.* **2022**, *5*, 1591–1603. [CrossRef] [PubMed]
60. Badekila, A.K.; Kini, S.; Jaiswal, A.K. Fabrication techniques of biomimetic scaffolds in three-dimensional cell culture: A review. *J. Cell Physiol.* **2021**, *236*, 741–762. [CrossRef] [PubMed]
61. Chen, Q.; Mangadlao, J.D.; Wallat, J.; De Leon, A.; Pokorski, J.K.; Advincula, R.C. 3D Printing Biocompatible Polyurethane/Poly(lactic acid)/Graphene Oxide Nanocomposites: Anisotropic Properties. *ACS Appl. Mater. Interfaces* **2017**, *9*, 4015–4023. [CrossRef]
62. Kemona, A.; Piotrowska, M. Polyurethane Recycling and Disposal: Methods and Prospects. *Polymers* **2020**, *12*, 1752. [CrossRef] [PubMed]
63. Anderson, J.M.; Rodriguez, A.; Chang, D.T. Foreign body reaction to biomaterials. *Semin. Immunol.* **2008**, *20*, 86–100. [CrossRef]
64. Whitaker, R.; Hernaez-Estrada, B.; Hernandez, R.M.; Santos-Vizcaino, E.; Spiller, K.L. Immunomodulatory Biomaterials for Tissue Repair. *Chem. Rev.* **2021**, *121*, 11305–11335. [CrossRef]
65. Onode, E.; Uemura, T.; Takamatsu, K.; Yokoi, T.; Shintani, K.; Hama, S.; Miyashima, Y.; Okada, M.; Nakamura, H. Bioabsorbable nerve conduits three-dimensionally coated with human induced pluripotent stem cell-derived neural stem/progenitor cells promote peripheral nerve regeneration in rats. *Sci. Rep.* **2021**, *11*, 4204. [CrossRef]
66. Salthouse, D.; Novakovic, K.; Hilkens, C.M.U.; Ferreira, A.M. Interplay between biomaterials and the immune system: Challenges and opportunities in regenerative medicine. *Acta Biomater.* **2023**, *155*, 1–18. [CrossRef]
67. Lock, A.; Cornish, J.; Musson, D.S. The Role of In Vitro Immune Response Assessment for Biomaterials. *J. Funct. Biomater.* **2019**, *10*, 31. [CrossRef]
68. Boehler, R.M.; Graham, J.G.; Shea, L.D. Tissue engineering tools for modulation of the immune response. *Biotechniques* **2011**, *51*, 239–254. [CrossRef] [PubMed]
69. Mariani, E.; Lisignoli, G.; Borzi, R.M.; Pulsatelli, L. Biomaterials: Foreign Bodies or Tuners for the Immune Response? *Int. J. Mol. Sci.* **2019**, *20*, 636. [CrossRef] [PubMed]

70. Shen, P.; Chen, Y.; Luo, S.; Fan, Z.; Wang, J.; Chang, J.; Deng, J. Applications of biomaterials for immunosuppression in tissue repair and regeneration. *Acta Biomater.* **2021**, *126*, 31–44. [CrossRef] [PubMed]
71. Eppler, H.B.; Jewell, C.M. Biomaterials as Tools to Decode Immunity. *Adv. Mater.* **2019**, *32*, 1903367. [CrossRef] [PubMed]
72. Vijayavenkataraman, S.; Thaharah, S.; Zhang, S.; Lu, W.F.; Fuh, J.Y.H. 3D-Printed PCL/rGO Conductive Scaffolds for Peripheral Nerve Injury Repair. *Artif. Organs* **2019**, *43*, 515–523. [CrossRef] [PubMed]
73. Lawkowska, K.; Pokrywczynska, M.; Koper, K.; Kluth, L.A.; Drewa, T.; Adamowicz, J. Application of Graphene in Tissue Engineering of the Nervous System. *Int. J. Mol. Sci.* **2021**, *23*, 33. [CrossRef]
74. Zhang, F.; Zhang, M.; Liu, S.; Li, C.; Ding, Z.; Wan, T.; Zhang, P. Application of Hybrid Electrically Conductive Hydrogels Promotes Peripheral Nerve Regeneration. *Gels* **2022**, *8*, 41. [CrossRef]
75. Liu, Z.; Wan, X.; Wang, Z.L.; Li, L. Electroactive Biomaterials and Systems for Cell Fate Determination and Tissue Regeneration: Design and Applications. *Adv. Mater.* **2021**, *33*, e2007429. [CrossRef]
76. Tang, M.; Song, Q.; Li, N.; Jiang, Z.; Huang, R.; Cheng, G. Enhancement of electrical signaling in neural networks on graphene films. *Biomaterials* **2013**, *34*, 6402–6411. [CrossRef]
77. Tupone, M.G.; Panella, G.; d'Angelo, M.; Castelli, V.; Caioni, G.; Catanesi, M.; Benedetti, E.; Cimini, A. An Update on Graphene-Based Nanomaterials for Neural Growth and Central Nervous System Regeneration. *Int. J. Mol. Sci.* **2021**, *22*, 13047. [CrossRef]
78. Kim, S.H.; Hong, H.; Ajiteru, O.; Sultan, M.T.; Lee, Y.J.; Lee, J.S.; Lee, O.J.; Lee, H.; Park, H.S.; Choi, K.Y.; et al. 3D bioprinted silk fibroin hydrogels for tissue engineering. *Nat. Protoc.* **2021**, *16*, 5484–5532. [CrossRef] [PubMed]
79. Grimaudo, M.A.; Krishnakumar, G.S.; Giusto, E.; Furlani, F.; Bassi, G.; Rossi, A.; Molinari, F.; Lista, F.; Montesi, M.; Panseri, S. Bioactive injectable hydrogels for on demand molecule/cell delivery and for tissue regeneration in the central nervous system. *Acta Biomater.* **2022**, *140*, 88–101. [CrossRef] [PubMed]
80. Nelson, D.W.; Gilbert, R.J. Extracellular Matrix-Mimetic Hydrogels for Treating Neural Tissue Injury: A Focus on Fibrin, Hyaluronic Acid, and Elastin-Like Polypeptide Hydrogels. *Adv. Healthc. Mater.* **2021**, *10*, e2101329. [CrossRef] [PubMed]
81. Zhang, M.; Li, L.; An, H.; Zhang, P.; Liu, P. Repair of Peripheral Nerve Injury Using Hydrogels Based on Self-Assembled Peptides. *Gels* **2021**, *7*, 152. [CrossRef]
82. Madhusudanan, P.; Raju, G.; Shankarappa, S. Hydrogel systems and their role in neural tissue engineering. *J. R. Soc. Interface* **2020**, *17*, 20190505. [CrossRef]
83. Sharma, P.; Pal, V.K.; Roy, S. An overview of latest advances in exploring bioactive peptide hydrogels for neural tissue engineering. *Biomater. Sci.* **2021**, *9*, 3911–3938. [CrossRef]
84. Liu, Z.; Tong, H.; Li, J.; Wang, L.; Fan, X.; Song, H.; Yang, M.; Wang, H.; Jiang, X.; Zhou, X.; et al. Low-Stiffness Hydrogels Promote Peripheral Nerve Regeneration Through the Rapid Release of Exosomes. *Front. Bioeng. Biotechnol.* **2022**, *10*, 922570. [CrossRef]
85. Li, Y.; Yang, L.; Hu, F.; Xu, J.; Ye, J.; Liu, S.; Wang, L.; Zhuo, M.; Ran, B.; Zhang, H.; et al. Novel Thermosensitive Hydrogel Promotes Spinal Cord Repair by Regulating Mitochondrial Function. *ACS Appl. Mater. Interfaces* **2022**, *14*, 25155–25172. [CrossRef]
86. Jin, X.H.; Fang, J.Q.; Wang, J.G.; Xu, B.; Wang, X.; Liu, S.H.; Chen, F.; Liu, J.J. PCL NGCs integrated with urolithin-A-loaded hydrogels for nerve regeneration. *J. Mater. Chem. B* **2022**, *10*, 8771–8784. [CrossRef]
87. Khan, Z.M.; Wilts, E.; Vlaisavljevich, E.; Long, T.E.; Verbridge, S.S. Electroresponsive Hydrogels for Therapeutic Applications in the Brain. *Macromol. Biosci.* **2022**, *22*, e2100355. [CrossRef]
88. Guo, B.; Ma, P.X. Conducting Polymers for Tissue Engineering. *Biomacromolecules* **2018**, *19*, 1764–1782. [CrossRef] [PubMed]
89. Green, R.; Abidian, M.R. Conducting Polymers for Neural Prosthetic and Neural Interface Applications. *Adv. Mater.* **2015**, *27*, 7620–7637. [CrossRef]
90. Magaz, A.; Li, X.; Gough, J.E.; Blaker, J.J. Graphene oxide and electroactive reduced graphene oxide-based composite fibrous scaffolds for engineering excitable nerve tissue. *Mater. Sci. Eng. C Mater. Biol. Appl.* **2021**, *119*, 111632. [CrossRef] [PubMed]
91. Chen, H.L.; Yang, D.; Chen, C.R.; Tian, G.Z.; Kim, D.H. In situ polymerization of conducting polymers around living neural cells: Cellular effect study. *Colloids Surf. B Biointerfaces* **2022**, *213*, 112410. [CrossRef]
92. Polo, Y.; Luzuriaga, J.; Iturri, J.; Irastorza, I.; Toca-Herrera, J.L.; Ibarretxe, G.; Unda, F.; Sarasua, J.R.; Pineda, J.R.; Larranaga, A. Nanostructured scaffolds based on bioresorbable polymers and graphene oxide induce the aligned migration and accelerate the neuronal differentiation of neural stem cells. *Nanomedicine* **2021**, *31*, 102314. [CrossRef]
93. Pan, S.; Qi, Z.; Li, Q.; Ma, Y.; Fu, C.; Zheng, S.; Kong, W.; Liu, Q.; Yang, X. Graphene oxide-PLGA hybrid nanofibres for the local delivery of IGF-1 and BDNF in spinal cord repair. *Artif. Cells Nanomed. Biotechnol.* **2019**, *47*, 651–664. [CrossRef] [PubMed]
94. Kim, G.J.; Lee, K.J.; Choi, J.W.; An, J.H. Modified Industrial Three-Dimensional Polylactic Acid Scaffold Cell Chip Promotes the Proliferation and Differentiation of Human Neural Stem Cells. *Int. J. Mol. Sci.* **2022**, *23*, 2204. [CrossRef]
95. MacDonald, A.F.; Harley-Troxell, M.E.; Newby, S.D.; Dhar, M.S. 3D-Printing Graphene Scaffolds for Bone Tissue Engineering. *Pharmaceutics* **2022**, *14*, 1834. [CrossRef] [PubMed]
96. Espera, A.H., Jr.; Dizon, J.R.; Valino, A.D.; Chen, Q.; Silva, I.G.; Nguyen, D.V.; Rong, L.; Advincula, R.C. On the 3D printability of silicone-based adhesives via viscous paste extrusion. *MRS Commun.* **2023**, *13*, 102–110. [CrossRef]
97. Advincula, R.C.; Dizon, J.R.C.; Caldona, E.B.; Viers, R.A.; Siacor, F.D.C.; Maalihan, R.D.; Espera, A.H., Jr. On the progress of 3D-printed hydrogels for tissue engineering. *MRS Commun.* **2021**, *11*, 539–553. [CrossRef]
98. Niu, W.; Zhang, Z.; Chen, Q.; Cao, P.-F.; Advincula, R.C. Highly Recyclable, Mechanically Isotropic and Healable 3D-Printed Elastomers via Polyurea Vitrimers. *ACS Mater. Lett.* **2021**, *3*, 1095–1103. [CrossRef]

99. Siacor, F.D.C.; Chen, Q.; Zhao, J.Y.; Han, L.; Valino, A.D.; Taboada, E.B.; Caldona, E.B.; Advincula, R.C. On the additive manufacturing (3D printing) of viscoelastic materials and flow behavior: From composites to food manufacturing. *Addit. Manuf.* **2021**, *45*, 102043. [CrossRef]
100. Chen, Q.; Zhao, J.; Ren, J.; Rong, L.; Cao, P.F.; Advincula, R.C. 3D Printed Multifunctional, Hyperelastic Silicone Rubber Foam. *Adv. Funct. Mater.* **2019**, *29*, 1900469. [CrossRef]
101. Yu, K.; Zhang, X.; Sun, Y.; Gao, Q.; Fu, J.; Cai, X.; He, Y. Printability during projection-based 3D bioprinting. *Bioact. Mater.* **2022**, *11*, 254–267. [CrossRef] [PubMed]
102. Mohan, T.S.; Datta, P.; Nesaei, S.; Ozbolat, V.; Ozbolat, I.T. 3D Coaxial Bioprinting: Process Mechanisms, Bioinks and Applications. *Prog. Biomed. Eng.* **2022**, *4*, 022003. [CrossRef]
103. Wu, W.; Dong, Y.; Liu, H.; Jiang, X.; Yang, L.; Luo, J.; Hu, Y.; Gou, M. 3D printed elastic hydrogel conduits with 7,8-dihydroxyflavone release for peripheral nerve repair. *Mater. Today Bio* **2023**, *20*, 100652. [CrossRef]
104. Hamid, O.A.; Eltaher, H.M.; Sottile, V.; Yang, J. 3D bioprinting of a stem cell-laden, multi-material tubular composite: An approach for spinal cord repair. *Mater. Sci. Eng. C Mater. Biol. Appl.* **2021**, *120*, 111707. [CrossRef]
105. Brittain, W.J..; Advincula, R.C..; Caster, K.C.; Rühe, J. (Eds.) *Polymer Brushes: Synthesis, Characterization, and Applications*; John Wiley & Sons, Inc.: Hoboken, NJ, USA, 2004; p. 501.
106. Dong, R.; Molloy, R.P.; Lindau, M.; Ober, C.K. Direct synthesis of quartenized polymer brushes and their application for guiding neuronal growth. *Biomacromolecules* **2010**, *11*, 2027–2032. [CrossRef]
107. Krishnamoorthy, M.; Hakobyan, S.; Ramstedt, M.; Gautrot, J.E. Surface-initiated polymer brushes in the biomedical field: Applications in membrane science, biosensing, cell culture, regenerative medicine and antibacterial coatings. *Chem. Rev.* **2014**, *114*, 10976–11026. [CrossRef]
108. Ayres, N. Polymer brushes: Applications in biomaterials and nanotechnology. *Polym. Chem.* **2010**, *1*, 769–777. [CrossRef]
109. Viswanathan, P.; Ondeck, M.G.; Chirasatitsin, S.; Ngamkham, K.; Reilly, G.C.; Engler, A.J.; Battaglia, G. 3D surface topology guides stem cell adhesion and differentiation. *Biomaterials* **2015**, *52*, 140–147. [CrossRef]
110. Zhao, X.; Jin, L.; Shi, H.; Tong, W.; Gorin, D.; Kotelevtsev, Y.; Mao, Z. Recent advances of designing dynamic surfaces to regulate cell adhesion. *Colloid Interface Sci. Commun.* **2020**, *35*, 100249. [CrossRef]
111. Cai, S.; Wu, C.; Yang, W.; Liang, W.; Yu, H.; Liu, L. Recent advance in surface modification for regulating cell adhesion and behaviors. *Nanotechnol. Rev.* **2020**, *9*, 971–989. [CrossRef]
112. Majhy, B.; Priyadarshini, P.; Sen, A.K. Effect of surface energy and roughness on cell adhesion and growth–facile surface modification for enhanced cell culture. *RSC Adv.* **2021**, *11*, 15467–15476. [CrossRef]
113. Dazzi, A.; Prater, C.B. AFM-IR: Technology and Applications in Nanoscale Infrared Spectroscopy and Chemical Imaging. *Chem. Rev.* **2017**, *117*, 5146–5173. [CrossRef]
114. Schermelleh, L.; Ferrand, A.; Huser, T.; Eggeling, C.; Sauer, M.; Biehlmaier, O.; Drummen, G.P.C. Super-resolution microscopy demystified. *Nat. Cell Biol.* **2019**, *21*, 72–84. [CrossRef]
115. Fernandez-Leiro, R.; Scheres, S.H. Unravelling biological macromolecules with cryo-electron microscopy. *Nature* **2016**, *537*, 339–346. [CrossRef]
116. Jamieson, T.; Bakhshi, R.; Petrova, D.; Pocock, R.; Imani, M.; Seifalian, A.M. Biological applications of quantum dots. *Biomaterials* **2007**, *28*, 4717–4732. [CrossRef]
117. Rieffel, J.; Chitgupi, U.; Lovell, J.F. Recent Advances in Higher-Order, Multimodal, Biomedical Imaging Agents. *Small* **2015**, *11*, 4445–4461. [CrossRef]
118. Basu, B.; Gowtham, N.H.; Xiao, Y.; Kalidindi, S.R.; Leong, K.W. Biomaterialomics: Data science-driven pathways to develop fourth-generation biomaterials. *Acta Biomater.* **2022**, *143*, 1–25. [CrossRef]

polymers MDPI

Article

Finite Element Analysis of Patient-Specific 3D-Printed Cranial Implant Manufactured with PMMA and PEEK: A Mechanical Comparative Study

Freddy P. Moncayo-Matute [1,†], Efrén Vázquez-Silva [1,*,†], Pablo G. Peña-Tapia [2,†], Paúl B. Torres-Jara [1,†], Diana P. Moya-Loaiza [1,†] and Tony J. Viloria-Ávila [3,†]

1 Grupo de Investigación en Nuevos Materiales y Procesos de Transformación (GIMAT), Universidad Politécnica Salesiana, Sede Cuenca EC010102, Ecuador; fmoncayo@ups.edu.ec (F.P.M.-M.); ptorresj@ups.edu.ec (P.B.T.-J.); dmoyal@ups.edu.ec (D.P.M.-L.)
2 Instituto oncológico SOLCA, Sociedad de Lucha Contra el Cáncer, Cuenca EC010109, Ecuador; bioinfo@institutodelcancer.med.ec
3 Grupo de Investigación en Biotecnología y Ambiente (INBIAM), Universidad Politécnica Salesiana, Sede Cuenca EC010102, Ecuador; tviloria@ups.edu.ec
* Correspondence: evazquez@ups.edu.ec; Tel.: +593-07-4135250
† These authors contributed equally to this work.

Abstract: This article reports on a patient who required a cranial protection system. Using additive manufacturing techniques and surgical planning with the help of bio-models, a patient-specific bone implant solution was proposed that allows aesthetic restoration of the affected area and provides an adequate level of protection. In addition, through a comparative analysis with finite elements, the mechanical response to external actions of the medical device, printed with two materials: polymethylmethacrylate (PMMA) and polyether-ether-ketone (PEEK), is simulated. The tested materials have recognized biocompatibility properties, but their costs on the market differ significantly. The results obtained demonstrate the similarities in the responses of both materials. It offers the possibility that low-income people can access these devices, guaranteeing adequate biomechanical safety, considering that PMMA is a much cheaper material than PEEK.

Keywords: polymethylmethacrylate; polyether-ether-ketone; custom medical device; finite element analysis

Citation: Moncayo-Matute, F.P.; Vázquez-Silva, E.; Peña-Tapia, P.G.; Torres-Jara, P.B.; Moya-Loaiza, D.P.; Viloria-Ávila, T.J. Finite Element Analysis of Patient-Specific 3D-Printed Cranial Implant Manufactured with PMMA and PEEK: A Mechanical Comparative Study. *Polymers* **2023**, *15*, 3620. https://doi.org/10.3390/polym15173620

Academic Editors: Cristina-Elisabeta Pelin and Anton Ficai

Received: 14 July 2023
Revised: 8 August 2023
Accepted: 18 August 2023
Published: 1 September 2023

1. Introduction

Currently, advances are palpable in terms of the application of additive manufacturing techniques and the use of different materials and composites for medical problem solving, such as, for example, the restoration of bone structures that have suffered damage due to cancerous pathologies, or accidents of diverse nature. Among the materials that have been applied are metals and alloys, ceramics, and polymers. Precisely, among the latter are Polyether-ether-ketone (PEEK) and Polymethylmethacrylate (PMMA). These polymers have been investigated based on their biological properties (biocompatibility, osseointegration, and others), and their mechanical properties (ability to emulate natural bone behavior, responses to external loads, and so on). Regarding the good qualities of PEEK and its behavior in contact with the human body, it is possible to consult, for example, results published in [1–15]. These articles reported on the absence of clinical complications when PEEK was used; the improvement of the properties of this material when an adequate surface treatment was carried out; the increase in biological activity and the increase in antibacterial activity thanks to the incorporation of other material in its base matrix; how the biocompatibility improves with the modification of the microstructure of the polymer; the predominance of the Fused Deposition Modelling technique as a manufacturing method; and, in general, about the good postoperative results with patients implanted with PEEK.

In summary, PEEK is suitable for artificial bone replacements because it is biocompatible, non-toxic, and noninflammatory and osteoconductive. Its low molecular weight makes it ideal in orthopaedics for fracture fixation and osteotomies, spinal fusions, and ligament reconstructions. Its high resistance provides it with better mechanical properties compared to other traditionally used materials since it supports the loads generated in the human body. It also helps to evaluate the evolution of fractures, thanks to the fact that it is radiolucent in radiographs. Moreover, it is compatible with Computed Tomography (CT) and Magnetic Resonance Imaging (MRI) technologies, that is, PEEK does not interfere with these imaging techniques.

Before the appearance of PEEK, another polymer used in this type of surgical procedure and treatment of bone conditions was PMMA. In the review article [16], a section is dedicated to the comparison between both materials, in terms of advantages and disadvantages in implantology applications. As for PMMA, it polymerizes through an exothermic reaction that can be harmful to overlying soft tissues [17]. Implants based on this material cannot be infiltrated by new bone tissue due to its lack of porosity, it interferes with osteoconduction and vascularization, it does not interact with the surrounding tissue, and it may be susceptible to higher infection rates [18–21]. On the other hand, four studies [22–25] report similarities between PEEK and PMMA in terms of the success rate of treatment and rate of complications.

This paper reports on a planned and materialized cranioplasty with PMMA in a health institution in southern Ecuador. And as part of the investigation, the performance of the customized bone implant, manufactured with PEEK and with PMMA, is compared at the level of simulation of the mechanical response. This considers the advantage of PMMA over PEEK in terms of cost. A random internet search found the value of PMMA (https://spanish.alibaba.com/g/medical-grade-pmma.html, accessed on 1 July 2023) to be between USD 1.99 and USD 3.80 per kg. While the price of medical-grade PEEK resin (https://spanish.alibaba.com/g/medical-grade-peek.html, accessed on 1 July 2023) is between USD 1450 and USD 1470 per kg. A limitation of PEEK technology is the requirement for support structures, which incurs additional costs. Considering the above, and that the average purchasing power of the inhabitants of this area is low, the PMMA-based variant of implants remains a viable option for many people who do not have the necessary resources to afford the best PEEK-based solution. Regardless of the proven biological restrictions of PMMA (possible triggering of infectious processes caused by allergy to the material), the hypothesis that is put forward in this study is that PMMA continues to be a viable option for people whose countries of origin or residence do not provide them with a public health system that covers their needs, and whose low income prevents them from facing the high costs of PEEK as a material for therapeutic use.

2. Materials and Methods

2.1. Data Acquisition Medical Imaging and Segmentation

A computed tomography of the patient was performed, and the images were saved in a high-resolution Digital Imaging and Communications in Medicine (DICOM) file, with voxels of $512 \times 512 \times Z$, where Z varies from 150 to 520 slices. The images were then processed using the open-source software 3D Slicer (https://www.slicer.org, accessed on 1 July 2023) to generate the Standard Tessellation Language or Stereolithography (STL) file of the anatomy of interest.

The tomographic images were segmented with a specific intensity, called Hounsfield Unit (HU), which measures the attenuation coefficient on the grey scale for compact tissues, in this case, the target anatomical region of the patient.

For the present study, values of 211.73–2755.0 HU are used to segment the cranial compact bone model. In Figure 1, the medical images and segmentation are shown.

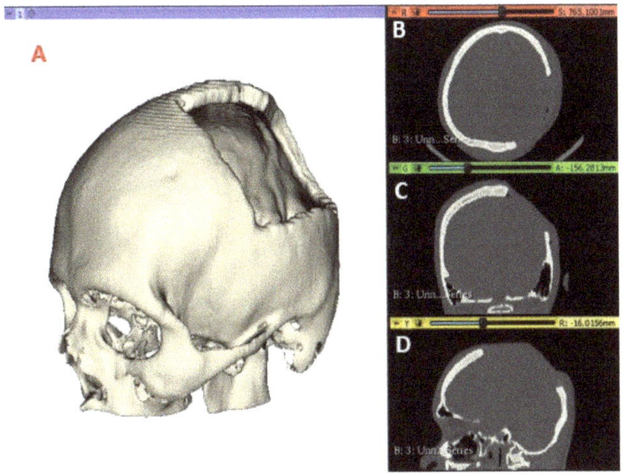

Figure 1. CT Scan Images: segmented model (**A**); axial view (**B**); coronal view (**C**); and sagittal view (**D**).

2.2. *Finite Element Method*

2.2.1. Cranial Model

To obtain the three-dimensional geometry of the patient's cranial model, the STL file was processed by treatment of meshes and cloud of points, with reverse engineering tools provided by the Autodesk Meshmixer (https://www.meshmixer.com, accessed on 1 July 2023), and ANSYS WORKBENCH R 21.1 software (ANSYS Inc., Canonsburg, PA, USA) (Our university has the exploitation license for this software).

This process was applied to the anatomical model of the patient to obtain a precise 3D model of his anatomy, which was then used for finite element analysis (FEM) (Figure 2).

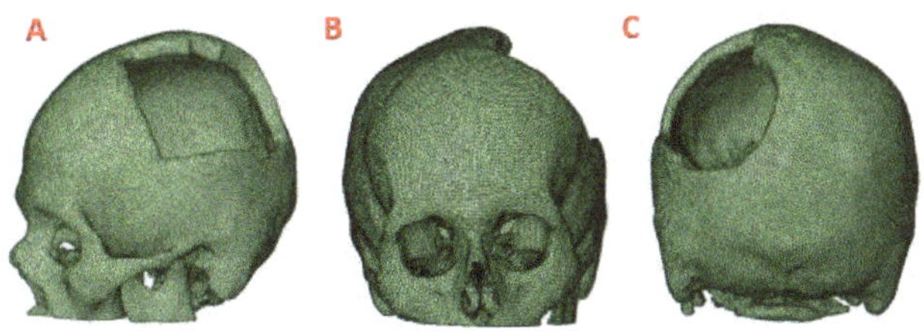

Figure 2. Computational model of the skull: (**A**) Lateral view, showing damage or trauma to the local parietal bone; (**B**) Front view of the cranial model, the defect is observed as well as the lack of symmetry; (**C**) Posterior view of the cranial model.

2.2.2. Customized Implant Model

The post-processed model of the skull with the defect was used for the reconstruction with the cranial implant, using the Autodesk Meshmixer software 3.4. A symmetrical reference plane was created in the three-dimensional model. For the restoration of the affected area (missing cranial bone), the structural symmetry of the human body was assumed. With the editing tools of the software, the healthy side of the structure is inverted, creating a mirror image that can be superimposed on the area to be restored. Both parts of the structure are assembled to fill the cavity. Subsequently, the Boolean subtraction tool is applied to obtain the required custom implant design (Figure 3).

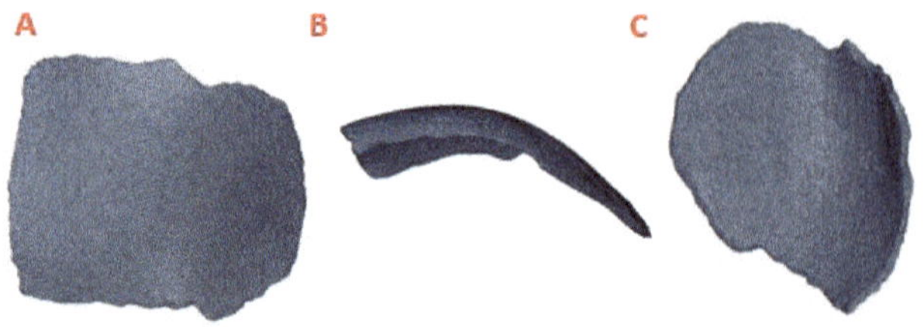

Figure 3. Reconstruction of the cranial 3D model with the help of the customized implant to simulate the clinical scenario: (**A**) Geometric model of the implant to cover the defect in the patient's parietal area; (**B**) Geometric model of the customized implant, coronal contour perimeter zone; (**C**) 3D view of the implant, showing the internal shape and the variation of the angulation of the contact interface along the contour.

2.2.3. Assembly of the Skull Model and Medical Device

Figure 4 shows the reconstructed cranial model of the patient. The design of the implant was carried out in a personalized way, and when the person receiving the medical device requires it (in extreme cases), this model is also useful for simulations. For example, in high-performance sports activity, failures of implanted medical devices have been evidenced, causing structural and physiological damage to the bone. An adequate simulation study can contribute to the determination of the mechanical parameters that personalized medical devices must present to be able to withstand certain static loads, also simulating events that may take place during the performance of these types of activities [26].

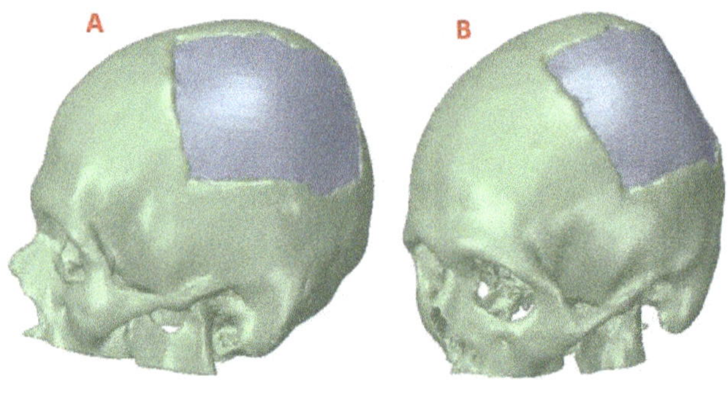

Figure 4. Reconstruction of the skull: (**A**) Lateral view of the skull and customized implant, showing the reconstruction; (**B**) Frontal view, defect covered by the cranial implant.

For the simulation studies in this case, the materials used for the computational analysis were two: one of low cost, widely used in operating rooms for immediate cranioplasties, and in the dental sector, PMMA. And the second of high performance and high cost, although with better biocompatibility properties, industrial-grade PEEK, is used in Latin American countries for medical purposes.

2.2.4. Coupling Details

For the successful fixation of the implant, it is important to detail the interface contours of the cranial profile and the location of the implant. For this, the perimeter area of the damage is studied as well as how the stress transfer occurs under considered external load conditions. In the article [27], the relevance of analysing the contacts of natural bone-implant interfaces is highlighted, in addition, its authors corroborate that the edges of the defect must have a surgical preparation at positive angles, while the edges of the device to be implanted must have opposite angles, in such a way that the "fit" will be the best possible, thus contributing to the coupling safety.

If the implant does not have support on the cranial bone, in the presence of external loads, stresses are transferred to the fixing elements, which endangers the integrity of the system. High stresses would potentially compromise the bone morphology required for anchorage. For the case reported in this paper, the fixation systems were not analysed since full support of the implant was achieved on the cranial damage perimeter (Figure 5).

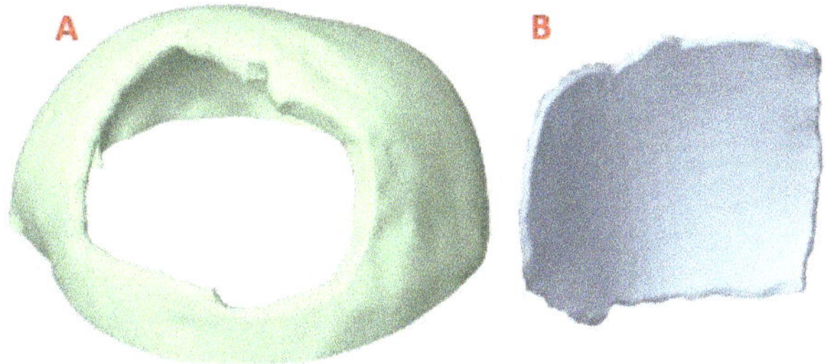

Figure 5. Configuration of the model to simulate the clinical scenario, highlighting the contour of the defect and adjustment of the implant: (**A**) Geometric model of the location of the defect in the left lateral parietal area of the patient; (**B**) Geometric model of the customized implant, showing the contours of the perimeter adaptation to the cranial bone.

2.2.5. Boundary Conditions

The simulation study was developed with the finite element method, applied with the structural static module of the ANSYS software. The loading conditions were assigned on the external surface of the customized implant. An external load of 8000 N was applied to the lower right surface of the implant [28], which simulates the static load caused by a foreign object impact. Likewise, a load of 2000 N was applied to the upper left part of the implant, simulating the rest activity. To reduce the computational load in the simulations, the cranial model was simplified, thereby reducing the meshing and processing time. An embedment condition called "Fixed Support" was also assigned at the base of the axial slice of the skull, with zero displacements and rotations. The external charge states are observed in Figure 6.

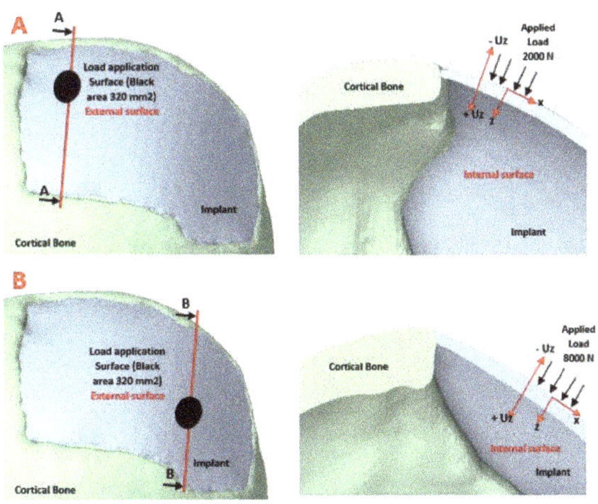

Figure 6. External load application scenarios: (**A**, left) Normal load of 2000 N on the implant surface in the position identified with the black circle with area of 320 mm^2; (**A**, right) Cross section showing the effect of the applied load; (**B**, left) Normal load of 8000 *N* on the surface of the implant in the position identified with the black circle with area of 320 mm^2; (**B**, right) Cross section showing the effect of the applied load.

2.2.6. Mesh Conditions

The simulation was carried out with the help of the ANSYS software. The selected mesh was made up of tetrahedral elements (SOLID185). The mesh refinement tests yielded a 7% convergence guarantee. The model is composed of 329,898 nodes and 178,488 elements of size 5 mm. "Bonded contact" was also considered for border conditions at the implantcortical bone interface, the most realistic according to [29]. Details of the mesh can be seen in Figure 7.

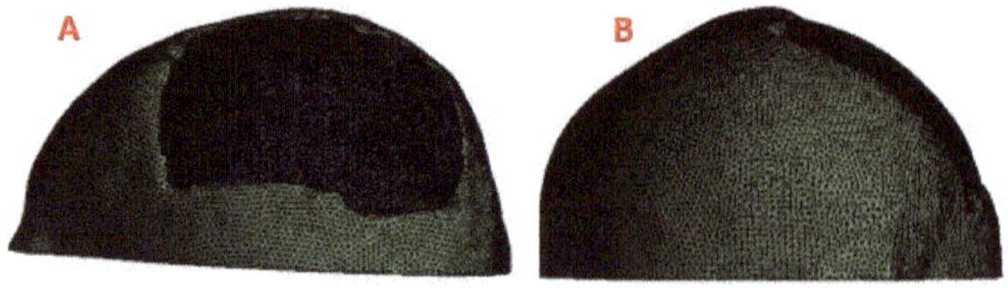

Figure 7. Mesh elements: (**A**) Mesh of the implant with fixation towards the skull; (**B**) The meshing of the cranial model with the implant.

2.3. Material Properties

In the computational model, the materials were assumed to be isotropic, homogeneous and linearly elastic, by [30,31]. The custom medical device was made of PMMA and PEEK. Table 1 presents the mechanical properties of each material used in the simulations.

Table 1. Material properties of components used in the simulation of the assembly model 3D.

Parameter	Cranial Cortical Bone	PMMA	PEEK
Young's modulus	Eb = 15,000 MPa	$EPMMA$ = 3000 MPa	$EPEEK$ = 3600 MPa
Poisson's ratio	μb = 0.3	$\mu PMMA$ = 0.38	$\mu PEEK$ = 0.39
Ultimate tensile	$\sigma u,b$ = 130 MPa	—	$\sigma u,PEEK$ = 172 MPa
Yield strength	—	$\sigma y,PMMA$ = 72 MPa	$\sigma y,PEEK$ = 90 MPa
Reference	[32]	[33]	[33]

Once the mechanical properties of the components were introduced, FEA simulations were carried out to study the behaviour of the cranial model under external static loads, which are useful to understand how the stresses are distributed in the cranial area during the performance of the considered activities.

2.4. Surgical Planning and Manufacturing

In the manufacture of the anatomical test model, for surgical planning, the additive manufacturing technology Fused Deposition Modelling (FDM) was applied, with the help of a Creality CR-X Pro 3D FDM printer (Shenzhen Creality 3D Technology Co., Ltd., Shenzhen, China).

As printing material, Creality's HP PLA 1.75 mm Series filament was used. The STL file was used as the basis for reading, processing and obtaining the route code, while for the 3D printing process, the Creality Slicer 1.2.3 software was used. The support material was removed manually, paying special attention not to affect the surface of the bone structure. Table 2 presents the characteristics of the FDM technology.

The 3D version of the test anatomical model, printed on a 1:1 scale, can be seen in Figure 8. The level of damage in the left parietal area of the patient is considerable, so the results of the finite element analysis offer the necessary support for the medical device to withstand demand loads.

Table 2. FDM additive manufacturing characteristics and parameters.

Characteristics and Manufacturing Parameters	Fused Deposition Modelling
Company and model	Creality CR-X Pro (2019 Updated)
Maximum build envelope	$300 \times 300 \times 400$ mm^3
Nozzle diameter	0.4 mm
Positioning resolution (X/Y/Z)	1.25 µm/1.25 µm/1 µm
Selected layer thickness	0.10 mm
Printed filament line width	0.4 mm

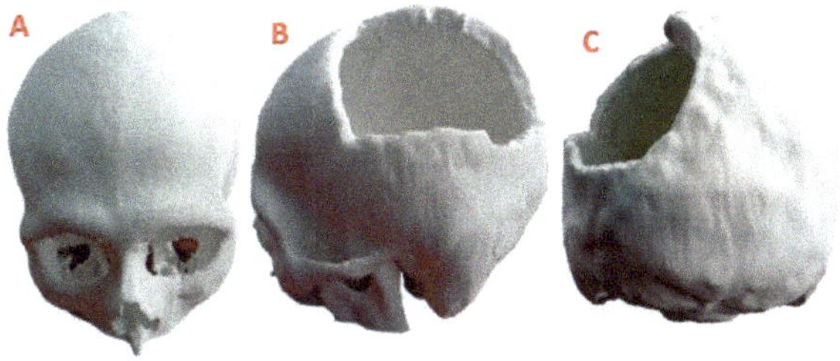

Figure 8. Cranial model for surgical planning: (**A**) Frontal view; (**B**) Lateral view, area of damage; (**C**) Back view.

The trial anatomical model was used by the surgeon in the preoperative evaluation and surgical planning, thus verifying the functionality of the customized device.

The digital model of the custom implant was printed in PLA. With the help of this physical model, the moulds (negative and positive) were created based on Vinyl Polysiloxane (hydrophilic material). The resulting cavities were joined to pour the PMMA solution into the free space between them, which solidifies, acquiring the shape of the final implant. The fixation holes in the final device were drilled with a surgical drill. For its part, the customized implant model, in industrial PEEK, was manufactured with the help of a FUNMAT PRO 410 printer (INTAMSYS. Shanghai, China), applying FDM additive technology. Subsequently, the definitive test of the implant was carried out on the cranial anatomical model, and it was verified that the shape was consistent with the virtual models (Figure 9).

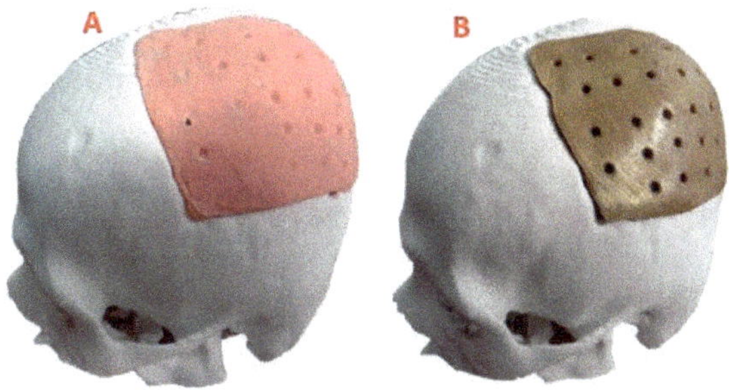

Figure 9. Cranial model with the patient-specific implant: (**A**) Cranial bio-model with the PMMA implant coupled; (**B**) Cranial bio-model with the PEEK implant coupled.

3. Results

Four simulation cum-shots were performed under the "Force" load condition of 2000 N applied to the external surface of the cranial bone-coupled PMMA-based custom implant model, and with the same condition applied also on the external surface of the personalized implant model based on PEEK coupled to the cranial bone. The geometry of the personalized implants, for both materials, presents variable thicknesses. Towards the central point, where the 2000 N load was applied, the thickness is 2.58 mm; while at the point where the 8000 N load is applied, the thickness is 3.02 mm.

Likewise, for the load condition "Force" of 8000 N applied in the central zone of the external surface of the model of the personalized PMMA implant, coupled to the cranial bone. And the same load condition was applied to the central area of the external surface of the custom PEEK implant model, coupled to the cranial bone. This allowed us to obtain the VonMises stress distribution for the implant and the cranial bone, and the total deformations (Table 3).

Table 3. Von Mises stresses and total deformation of the structures *.

Structure	Material	Von Mises Stress [MPa]	Total Deformation [mm]	Load State [N]
Skull	Cortical bone	40 ± 10	0.080 ± 0.001	2000
Implant	PMMA	40 ± 7	0.56 ± 0.02	2000
Skull	Cortical bone	36 ± 8	0.073 ± 0.001	2000
Implant	PEEK	44 ± 5	0.42 ± 0.08	2000
Skull	Cortical bone	83 ± 7	0.210 ± 0.008	8000
Implant	PMMA	103 ± 20	2.71 ± 0.21	8000
Skull	Cortical bone	59 ± 22	0.21 ± 0.01	8000
Implant	PEEK	99 ± 11	2.30 ± 0.03	8000

* Each load state was applied to the external surface of the implant, at the same point. The effort with the deformation generated in the cranial structure is the result of the coupling and transfer of the load on the implant.

In Figure 10, the Von Mises stress distribution for each element can be seen, according to the state of external load. In Figure 11, the total deformations generated by the states of external loads, in the implant and the cranial bone, are detailed.

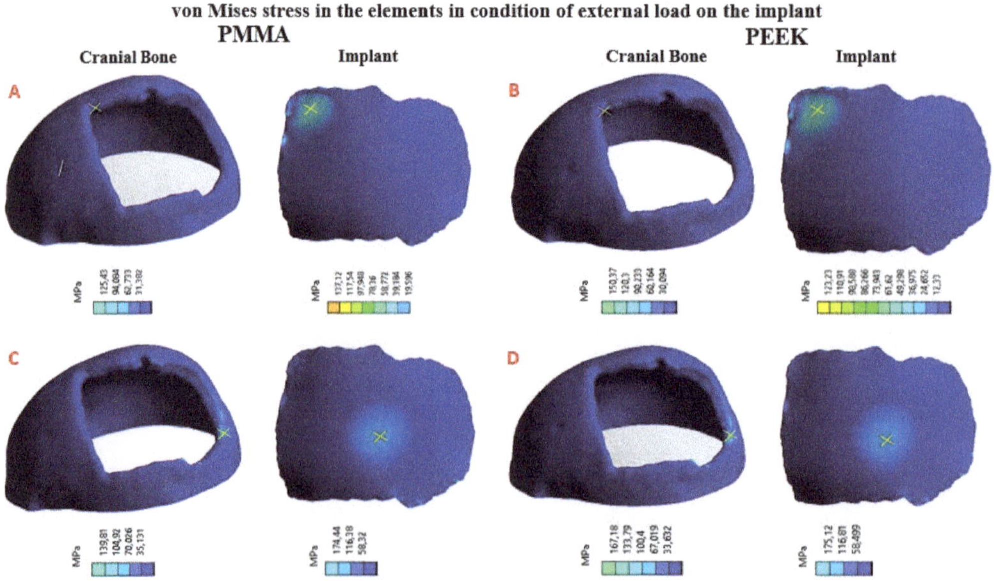

Figure 10. Distribution of Von Mises stresses in the implant and cranial bone (maximum values marked with a yellow X): (**A**) Load condition "Force" 2000 N, implant with PMMA; (**B**) Load condition "Force" 2000 N, implant with PEEK; (**C**) Load condition "Force" 8000 N, implant with PMMA; (**D**) Load condition "Force" 8000 N, implant with PEEK.

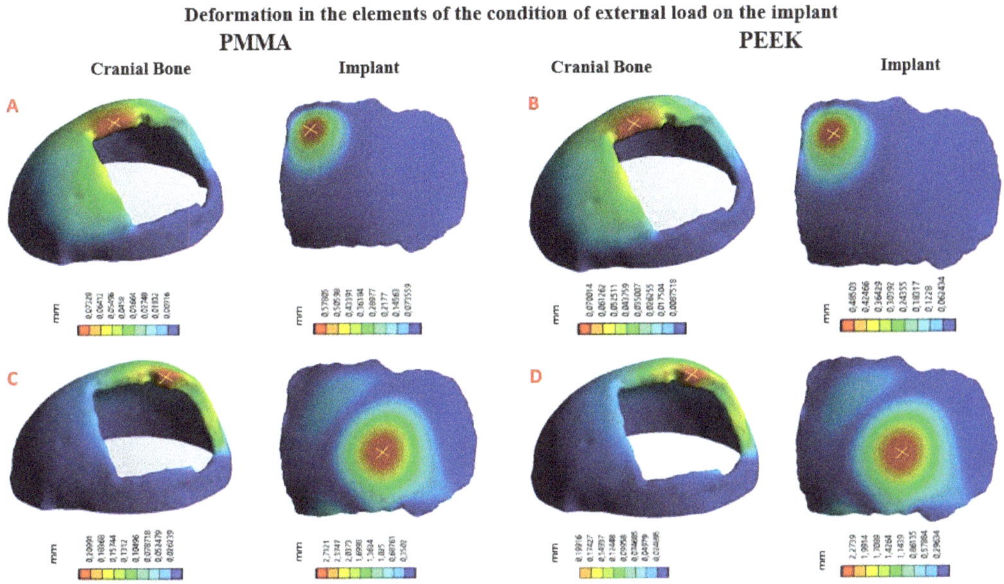

Figure 11. Total deformations in the implant and cranial bone (maximum values marked with a yellow X): (**A**) Load condition "Force" 2000 N, implant with PMMA; (**B**) Load condition "Force" 2000 N, implant with PEEK; (**C**) Load condition "Force" 8000 N, implant with PMMA; (**D**) Load condition "Force" 8000 N, implant with PEEK.

3.1. Statistic Analysis

The analysis of the computational results was carried out by applying a comparison of means, with the help of the SPSS Statistics software 29.0 (SPSS, Inc., Chicago, IL, USA), considering that, for each simulated external load, five runs were performed with each material. The ANSYS software provides the maximum stress and deformation values. In Figure 12A (Figures 12 and 13, respectively, show the box and whiskers diagrams for Von Mises stresses, and for deformations, for each material. In them, the mean values are represented by a red dot in boxes, and with respect to these points are referenced the corresponding standard deviations.), the Von Mises stress levels for the personalized implant with PMMA and PEEK are observed. Figure 12B shows the total deformation of the implant with each material under the load condition "Force" 8000 N.

In this computational study, the effect of using PMMA and PEEK for the protection of the skull against the action of a static load of 8000 N on the surface of the device was examined. The mean and standard deviation of the Von Mises stress analysis for the PMMA implant was 103 $\pm$ 20 MPa; while for the PEEK implant, it was 99 $\pm$ 11 MPa. In this case, no significant differences are observed ($p > 0.05$), with a confidence level of 95%.

The total deformation of the PMMA implant was also examined, where the mean and standard deviation of the analysis were 2.71 $\pm$ 0.22 mm; while for the PEEK implant, 2.30 $\pm$ 0.03 mm was obtained. In this case, significant differences were observed ($p < 0.05$), with a confidence level of 95%.

In Figure 13A, the Von Mises stress levels in the cranial bone are observed, resulting from the couplings of the PMMA and PEEK implants, respectively. Whereas, in Figure 13B, the total deformation of the cranial bone caused by the coupling of the implant is presented. All of the above under the application of a load "Force" 2000 N.

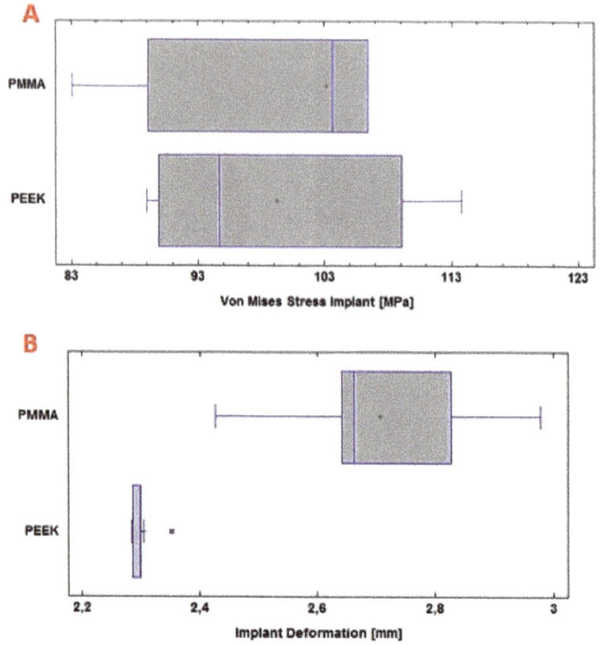

Figure 12. Statistical analysis: (**A**) Von Mises stresses of the PMMA and PEEK implants, respectively, on the abscissa axis, in Megapascals; (**B**) Deformation of the PMMA and PEEK implants, respectively, on the abscissa axis, in millimetres.

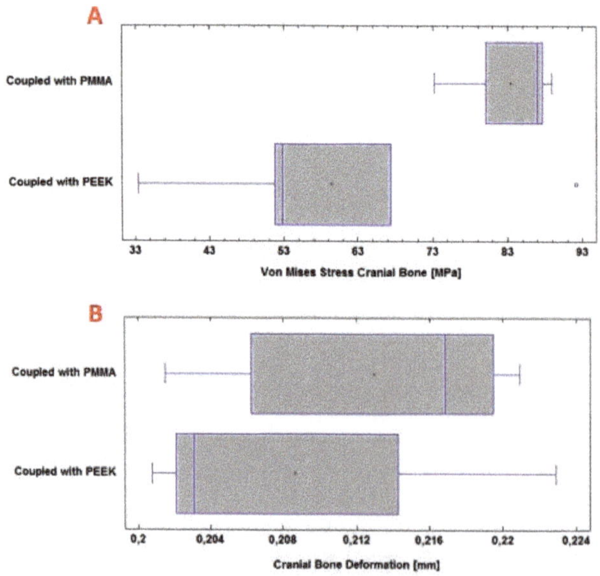

Figure 13. Statistical analysis: (**A**) Von Mises stresses in the cranial bone, on the abscissa axis, in Megapascals; (**B**) Deformation of the cranial bone, on the abscissa axis, in millimetres.

During the computational analysis, the effect of using PMMA and PEEK implants, respectively, coupled to the cranial bone, was examined to achieve the necessary protection of the internal structures and tissues, before a static load of 8000 N on the external surface of the implant. The mean and standard deviation of the Von Mises stress analysis for the cranial bone was 83 ± 7 MPa with PMMA. While with PEEK 59 ± 22 MPa was obtained. In this case, significant differences were observed ($p < 0.05$), with a confidence level of 95%.

The total deformation of the cranial bone when coupled with the PMMA implant was also examined. The values for the mean and standard deviation of the analysis were 0.213 ± 0.008 mm. For PEEK, in the same situation, 0.208 ± 0.009 mm was obtained. No significant differences are observed ($p > 0.05$), with a confidence level of 95%.

The analysis focused only on the case of the greater load because the safety of the personalized implant will depend on this action. The load of 2000 N simulates the incidence during the rest-activity; in such a situation, failures will not occur, since the stresses generated in the implant do not exceed the yield stress of either of the two materials [28,34].

The simulation results show that both devices (manufactured with PEEK and PMMA, respectively) present a similar resistance to the action of considered external loads. But the PMMA-based device suffers from greater deformation. However, for this second material, the fluence limit is not exceeded, that is, there is no risk of damage due to contact of the device with the brain mass. Furthermore, an additional option to increase the safety of the PMMA implant would be to control its thickness in such a way as to reduce deformation.

3.2. Post-Operative

The implant manufactured in PMMA was successfully placed in the patient. The postoperative evolution was satisfactory, there were no complications of any kind, and the person left the hospital facility 48 h after the intervention. In addition, for follow-up, a visit was scheduled 14 days after surgery. In Figure 14, the symmetry achieved with the cranial reconstruction is visible. There were also no difficulties with healing, the aesthetic and functional result was as expected and completely to the patient's satisfaction.

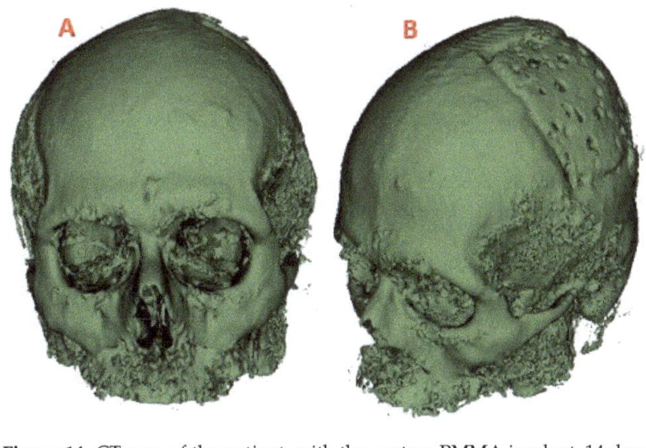

Figure 14. CT scan of the patient, with the custom PMMA implant, 14 days post-surgery: (**A**) Image of the skull with the PMMA implant attached, front view; (**B**) Image of the skull with the attached PMMA implant, 3D view.

Postoperative Follow-Up

Computer-aided design and additive manufacturing have proven to be effective tools in the design of custom implants. The results presented describe the entire custom manufacturing process, surgical planning, finite element analysis, and 3D manufacturing, which facilitated the performance of cranioplasty in a patient affected by skull bone cancer. The general procedure (fabrication of the implant and surgical planning) was carried out according to the proposed methodology and using open-source and commercial software for the segmentation, post-processing and mechanical design stages. The use of free software, whenever possible, was an advantageous factor considering the economic possibilities of the region where the methodology has been applied. In Figure 15, the final results of the entire process can be seen.

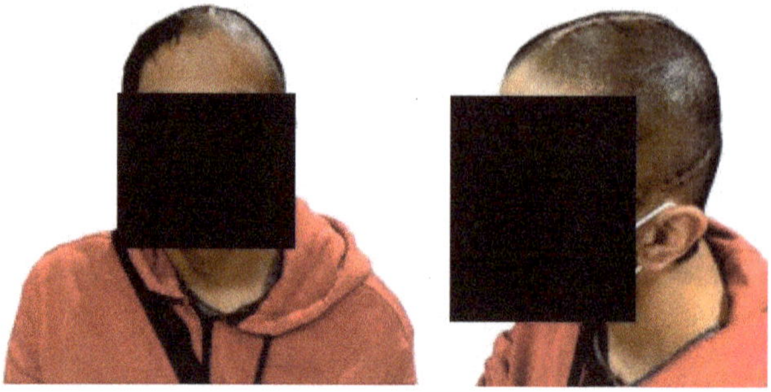

Figure 15. Patient benefited from the personalized PMMA implant.

4. Discussion and Conclusions

Through the presentation of a real case, the usefulness of the proposed method for the analysis of clinical cases in the area of neurosurgery and reconstruction of cranial defects with a complex surgical approach was demonstrated. Through a finite element study, the level of Von Mises stresses and total deformations that take place in the coupling of the bone-implant system under static loads were determined, which is established as one more criterion for the design of medical devices for the cranioplasty treatment with biocompatible materials such as PMMA and PEEK.

Four computational analyses were developed under a static external load of 2000 N, which was applied to the upper left part of the implant, simulating an external action that this area of the head would receive during rest-activity; and a load of 8000 N, applied in the centre of the implant, simulating a static impact action. All of the above to determine the stress levels and total deformations of the implant-cranial bone system. In this investigation it was found that there are no significant differences between the PMMA personalized medical device and the PEEK one, concerning the total deformations caused by the load of 8000 N, considering the state of load as maximum. Significant differences were found in the Von Mises stress distribution.

The decision to know which device to use for cranioplasty treatment will depend on the magnitude of the damage and the patient's financial availability. In Eastern European and Asian countries, PMMA is currently used as an alternative material for cranioplasty. In [35], is reported on the use of locally developed polylactic acid and polymethyl methacrylate moulds to perform cranioplasties for bone defects in technically demanding areas of the skull, while ensuring good aesthetic results and functional recovery. According to the authors, no surgical complications occurred in 14 patients. In addition, the subjective and objective evaluation revealed a significant improvement in the results. There were also no postoperative complications during a 6-month follow-up period, except in one patient who presented a late infection. Studies like this validate the use of acrylic materials as an alternative for cranioplasty treatment.

Late infection rates from the use of PMMA in custom-made bone implants are frequently increased by improper handling of sterilization protocols for manufactured devices, or by bacterial contamination during handling.

It has also happened that, in patients who have undergone successive surgery to remove and replace implants, the capillary tissue has lost healing properties, causing the decubitus effect and with it the proliferation of bacteria that contaminate the implanted device. Even so, the cranioplasty surgeries that have been performed in this area of Ecuador have had an approximate average cost of USD 450. In addition, the preoperative preparation with the anatomical test models has helped to reduce the duration of the interventions and the time the patient remains under the effects of anaesthesia, also reducing the risks of surgical complications. A similar scenario is described in [35,36]. As reported, the investment in a PMMA implant was USD 50.

The manufacturing process with PEEK has an approximate cost of USD 5000, inaccessible to the economy of many people. In the present study, however, the simulation results showed that both materials are equal in terms of security and structural integrity that they provide in the event of static load events: they do not exceed the yield point, so under the same demands, they are capable of responding without failure. Regarding the total deformations suffered, no significant differences were observed. In addition, the possible scope of the finite element method has been verified in terms of optimizing the structural properties of the implant to maximize resistance and durability and minimize weight and volume.

In some public and private hospitals in the Republic of Ecuador, a country located in the northwestern region of South America, the introduction of advanced medical technologies, such as 3D printing, has been fundamentally limited for economic reasons and a lack of investment in research and development. However, in a certain way, these limitations have been overcome by applying viable alternatives. For example, keeping PMMA as a feasible option in terms of biocompatibility and mechanical performance, and advantageous in economic terms. So, considering the above factors in terms of benefit-cost, an immediate alternative for progress could be that, when the solution in terms of manufacturing material for a customized implant be PMMA, prior to device placement, a test of allergy (for example, a patch type test) to try to predict the behaviour of the prosthesis once in contact with the body of the patient who will receive it.

In three health institutions in the region, since July 2020, five surgeries were performed that required customized bone replacement (the last one made in September 2022). Three cranioplasties, one clavicular implant and one for the sternum-clavicle. All medical devices placed in these patients were manufactured with PMMA. Follow-up has been maintained on each of them up to the present, and in none of the cases has there been any complication. These surgical procedures have been performed under the protection of the rules of the National Agency for Health Regulation, Control and Surveillance (ARCSA-acronym in Spanish), RESOLUTION No. ARCSADE0262016YMIH, Article 20, items (a), (b), (c) and (d).

Author Contributions: Conceptualization, F.P.M.-M. and E.V.-S.; methodology, F.P.M.-M.; P.B.T.-J. and E.V.-S.; software, F.P.M.-M.; validation, F.P.M.-M., P.B.T.-J. and P.G.P.-T.; formal analysis, P.G.P.-T.; investigation, F.P.M.-M., P.G.P.-T., P.B.T.-J. and E.V.-S.; resources, D.P.M.-L.; data curation, D.P.M.-L.; writing—original draft preparation, E.V.-S.; writing—review and editing, T.J.V.-Á.; visualization, E.V.-S.; supervision, D.P.M.-L. and T.J.V.-Á.; project administration, E.V.-S.; funding acquisition, D.P.M.-L. All authors have read and agreed to the published version of the manuscript.

Funding: The work was carried out under the auspices of research projects, Development of customized bone implants with reduction of the affected area through the use of surgical guides and additive manufacturing. Phase I: development of the comprehensive methodology. and Development of customized bone implants with reduction of the affected area through the use of surgical guides and additive manufacturing. Phase II: additive manufacturing with PEEK and PEKK, and evaluation of a superficial modification of the implant, financed by the Salesian Polytechnic University. Approving resolutions: RESOLUTION No.002-003-2020-07-15 and RESOLUTION No005-005-2023-05-25, respectively.

Informed Consent Statement: Informed consent was obtained from all subjects involved in the study. Written informed consent has been obtained from the patient(s) to publish this paper.

Data Availability Statement: Data included in article/referenced in the article.

Acknowledgments: The authors of this work appreciate the support provided by the Research Group on New Materials and Transformation Processes (GIMAT) of the Mechanical Engineering Carrier, in terms of the availability of hours to carry out the research processes. The authors also appreciate the support provided by the Neurosurgery Department of "Instituto de lucha contra el Cáncer—SOLCA", in Cuenca city, which provided its facilities for performing the surgery and patient follow-up. The authors also thank Group 1 of the Biomedicine Career at UPS, who during teaching activities corresponding to the Biomaterials subject, in the period 62 (March-August 2023) collaborated with the printing of the test anatomical model shown in Figure 8, in the present work.

Conflicts of Interest: The authors declare no conflict of interest.

References

1. Chepurnyi, Y.; Chernogorskyi, D.; Kopchak, A.; Petrenko, O. Clinical efficacy of peek patient-specific implants in orbital reconstruction. *J. Oral Biol. Craniofacial Res.* **2020**, *10*, 49–53. [CrossRef] [PubMed]
2. Limaye, N.; Veschini, L.; Coward, T. Assessing biocompatibility & mechanical testing of 3D-printed PEEK versus milled PEEK. *Heliyon* **2022**, *8*, e12314. [PubMed]
3. Zheng, Z.; Liu, P.; Zhang, X.; Zou, X.; Mei, X.; Zhang, S.; Zhang, S. Strategies to improve bioactive and antibacterial properties of polyetheretherketone (PEEK) for use as orthopedic implants. *Mater. Today Bio* **2022**, *16*, 100402. [CrossRef] [PubMed]
4. Shi, Y.; Deng, T.; Peng, Y.; Qin, Z.; Ramalingam, M.; Pan, Y.; Chen, C.; Zhao, F.; Cheng, L.; Liu, J. Effect of Surface Modification of PEEK Artificial Phalanx by 3D Printing on its Biological Activity. *Coatings* **2023**, *13*, 400. [CrossRef]
5. Kang, J.; Wang, L.; Yang, C.; Wang, L.; Yi, C.; He, J.; Li, D. Custom design and biomechanical analysis of 3D-printed PEEK rib prostheses. *Biomech. Model. Mechanobiol.* **2018**, *17*, 1083–1092. [CrossRef]
6. Zhao, Y.; Zhao, K.; Li, Y.; Chen, F. Mechanical characterization of biocompatible PEEK by FDM. *J. Manuf. Process.* **2020**, *56*, 28–42. [CrossRef]
7. Oladapo, B.I.; Zahedi, S.A.; Ismail, S.O. Mechanical performances of hip implant design and fabrication with PEEK composite. *Polymer* **2021**, *227*, 123865. [CrossRef]
8. Thieringer, F.M.; Sharma, N.; Mootien, A.; Schumacher, R.; Honigmann, P. *Patient Specific Implants from a 3D Printer—An Innovative Manufacturing Process for Custom PEEK Implants in Cranio-Maxillofacial Surgery*; Springer: Cham, Switzerland, 2018.
9. Honigmann, P.; Sharma, N.; Schumacher, R.; Rueegg, J.; Haefeli, M.; Thieringer, F. In-hospital 3D printed scaphoid prosthesis using medical-grade polyetheretherketone (PEEK) biomaterial. *BioMed Res. Int.* **2021**, *2021*, 1301028. [CrossRef]
10. Rodzeń, K.; Sharma, P.K.; McIlhagger, A.; Mokhtari, M.; Dave, F.; Tormey, D.; Sherlock, R.; Meenan, B.J.; Boyd, A. The direct 3D printing of functional PEEK/hydroxyapatite composites via a fused filament fabrication approach. *Polymers* **2021**, *13*, 545. [CrossRef]
11. Zhang, J.; Su, Y.; Rao, X.; Pang, H.; Zhu, H.; Liu, L.; Chen, L.; Li, D.; He, J.; Peng, J.; et al. Additively manufactured polyetherether ketone (PEEK) skull implant as an alternative to titanium mesh in cranioplasty. *Int. J. Bioprint.* **2023**, *9*, 634. [CrossRef]
12. Rosa-Sainz, A.; García-Romeu, M.; Ferrer, I.; Silva, M.; Centeno, G. On the effective peek application for customized cranio-maxillofacial prostheses: An experimental formability analysis. *J. Manuf. Process.* **2023**, *86*, 66–84. [CrossRef]
13. Bianchi, F.; Signorelli, F.; Di Bonaventura, R.; Trevisi, G.; Pompucci, A. One-stage frame-guided resection and reconstruction with PEEK custom-made prostheses for predominantly intraosseous meningiomas: Technical notes and a case series. *Neurosurg. Rev.* **2019**, *42*, 769–775. [CrossRef] [PubMed]
14. Narciso, R.; Basile, E.; Bottini, D.J.; Cervelli, V. PEEK implants: An innovative solution for facial aesthetic surgery. *Case Rep. Surg.* **2021**, *2021*, 5518433. [CrossRef]
15. Haleem, A.; Javaid, M.; Vaish, A.; Vaishya, R. Three-dimensional-printed polyether ether ketone implants for orthopedics. *Indian J. Orthop.* **2019**, *53*, 377. [CrossRef] [PubMed]
16. Zhang, J.; Tian, W.; Chen, J.; Yu, J.; Zhang, J.; Chen, J. The application of polyetheretherketone (PEEK) implants in cranioplasty. *Brain Res. Bull.* **2019**, *153*, 143–149. [CrossRef] [PubMed]
17. Shah, A.M.; Jung, H.; Skirboll, S. Materials used in cranioplasty: A history and analysis. *Neurosurg. Focus* **2014**, *36*, E19. [CrossRef] [PubMed]
18. Wind, J.J.; Ohaegbulam, C.; Iwamoto, F.M.; Black, P.M.; Park, J.K. Immediate titanium mesh cranioplasty for treatment of postcraniotomy infections. *World Neurosurg.* **2013**, *79*, 207.e11–207.e13. [CrossRef]
19. Goiato, M.C.; Anchieta, R.B.; Pita, M.S.; dos Santos, D.M. Reconstruction of skull defects: Currently available materials. *J. Craniofacial Surg.* **2009**, *20*, 1512–1518. [CrossRef]
20. Kim, B.J.; Hong, K.S.; Park, K.J.; Park, D.H.; Chung, Y.G.; Kang, S.H. Customized cranioplasty implants using three-dimensional printers and polymethyl-methacrylate casting. *J. Korean Neurosurg. Soc.* **2012**, *52*, 541–546. [CrossRef]
21. Turgut, G.; Özkaya, Ö.; Kayal, M.U. Computer-aided design and manufacture and rapid prototyped polymethylmethacrylate reconstruction. *J. Craniofacial Surg.* **2012**, *23*, 770–773. [CrossRef]
22. Nguyen, P.D.; Khechoyan, D.Y.; Phillips, J.H.; Forrest, C.R. Custom CAD/CAM implants for complex craniofacial reconstruction in children: Our experience based on 136 cases. *J. Plast. Reconstr. Aesthetic Surg.* **2018**, *71*, 1609–1617. [CrossRef] [PubMed]
23. Liang, E.S.; Tipper, G.; Hunt, L.; Gan, P.Y.C. Cranioplasty outcomes and associated complications: A single-centre observational study. *Br. J. Neurosurg.* **2016**, *30*, 122–127. [CrossRef] [PubMed]
24. Iaccarino, C.; Viaroli, E.; Fricia, M.; Serchi, E.; Poli, T.; Servadei, F. Preliminary results of a prospective study on methods of cranial reconstruction. *J. Oral Maxillofac. Surg.* **2015**, *73*, 2375–2378. [CrossRef] [PubMed]
25. Paredes, I.; Castaño-León, A.M.; Munarriz, P.M.; Martínez-Perez, R.; Cepeda, S.; Sanz, R.; Alén, J.F.; Lagares, A. Cranioplasty after decompressive craniectomy. A prospective series analyzing complications and clinical improvement. *Neurocirugia* **2015**, *26*, 115–125. [CrossRef]

26. Inchingolo, A.D.; Pezzolla, C.; Patano, A.; Ceci, S.; Ciocia, A.M.; Marinelli, G.; Malcangi, G.; Montenegro, V.; Cardarelli, F.; Piras, F.; et al. Experimental Analysis of the Use of Cranial Electromyography in Athletes and Clinical Implications. *Int. J. Environ. Res. Public Health* **2022**, *19*, 7975. [CrossRef] [PubMed]
27. Ridwan-Pramana, A.; Marcián, P.; Borák, L.; Narra, N.; Forouzanfar, T.; Wolff, J. Structural and mechanical implications of PMMA implant shape and interface geometry in cranioplasty–A finite element study. *J. Cranio-Maxillofac. Surg.* **2016**, *44*, 34–44. [CrossRef]
28. Yoganandan, N.; Pintar, F.A.; Zhang, J.; Baisden, J.L. Physical properties of the human head: Mass, center of gravity and moment of inertia. *J. Biomech.* **2009**, *42*, 1177–1192. [CrossRef]
29. Qi, W.; Yan, Y.B.; Zhang, Y.; Lei, W.; Wang, P.J.; Hou, J. Study of stress distribution in pedicle screws along a continuum of diameters: A three-dimensional finite element analysis. *Orthop. Surg.* **2011**, *3*, 57–63. [CrossRef]
30. Ameen, W.; Al-Ahmari, A.; Mohammed, M.; Abdulhameed, O.; Umer, U.; Moiduddin, K. Design, finite element analysis (FEA), and fabrication of custom titanium alloy cranial implant using electron beam melting additive manufacturing. *Adv. Prod. Eng. Manag.* **2018**, *13*, 267–278. [CrossRef]
31. Şimşek, H.; Zorlu, E.; Kaya, S.; Baydoğan, M.; Atabey, C.; Çolak, A. A new multipartite plate system for anterior cervical spine surgery; finite element analysis. *Br. J. Neurosurg.* **2018**, *32*, 276–282. [CrossRef]
32. Motherway, J.A.; Verschueren, P.; Van der Perre, G.; Vander Sloten, J.; Gilchrist, M.D. The mechanical properties of cranial bone: The effect of loading rate and cranial sampling position. *J. Biomech.* **2009**, *42*, 2129–2135. [CrossRef] [PubMed]
33. Gómez, F.; Elices, M. Fracture of components with V-shaped notches. *Eng. Fract. Mech.* **2003**, *70*, 1913–1927. [CrossRef]
34. Moncayo-Matute, F.P.; Peña-Tapia, P.G.; Vázquez-Silva, E.; Torres-Jara, P.B.; Moya-Loaiza, D.P.; Abad-Farfán, G.; Andrade-Galarza, A.F. Surgical planning and finite element analysis for the neurocraneal protection in cranioplasty with PMMA: A case study. *Heliyon* **2022**, *8*, e10706. [CrossRef] [PubMed]
35. Kaya, İ.; Yakar, H.; Kesen, E. Low-cost 3-d-printer–assisted personalized cranioplasty treatment: A case series of 14 consecutive patients. *World Neurosurg.* **2023**, *175*, e1197–e1209. [CrossRef] [PubMed]
36. Rubio-Pérez, I.; Diaz Lantada, A. Surgical planning of sacral nerve stimulation procedure in presence of sacral anomalies by using personalized polymeric prototypes obtained with additive manufacturing techniques. *Polymers* **2020**, *12*, 581. [CrossRef]

polymers MDPI

Article

Tensile Behavior of Chain Links Made of Polymeric Materials Manufactured by 3D Printing

Bruno Rădulescu [1], Andrei Marius Mihalache [2], Emilian Păduraru [1], Adelina Hriţuc [2,*], Mara Cristina Rădulescu [1], Laurenţiu Slătineanu [2] and Vasile Ermolai [2]

1 Department of Digital Production Systems, "Gheorghe Asachi" Technical University of Iași, 700050 Iași, Romania; bruno.radulescu@academic.tuiasi.ro (B.R.); emilian.paduraru@academic.tuiasi.ro (E.P.); mara.radulescu@academic.tuiasi.ro (M.C.R.)

2 Department of Machine Manufacturing Technology, "Gheorghe Asachi" Technical University of Iași, 700050 Iași, Romania; marius-andrei.mihalache@academic.tuiasi.ro (A.M.M.); slati@tcm.tuiasi.ro (L.S.); vasile.ermolai@student.tuiasi.ro (V.E.)

* Correspondence: adelina.hrituc@student.tuiasi.ro; Tel.: +40-751640117

Abstract: For reduced mechanical stress, some chains with links made of metallic materials could be replaced by chains made of polymeric materials. A lower weight and a higher corrosion resistance would characterize such chains. From this point of view, research on the behavior of chain links made of polymeric materials under the action of tensile stresses can become important. Modeling by the finite element method highlighted some specific aspects of the behavior of a chain link subjected to tensile stresses. Later, we resorted to the manufacture by 3D printing of some chain links from four distinct polymeric materials, with the modification of the size of the chain link and, respectively, of the values of some of the input factors in the 3D printing process. The tensile strength of the chain links was determined using specialized equipment. The experimental results were processed mathematically to determine some empirical mathematical models that highlight the influence of the values of the input factors in the 3D printing process on the tensile strength of the samples in the form of chain links. It thus became possible to compare the results obtained for the four polymeric materials considered and identify the polymeric material that provides the highest tensile strength of the sample in the form of a chain link. The results of the experimental research showed that the highest mechanical resistance was obtained in the case of the links made of polyethylene terephthalate glycol (PETG). According to experimental results, when tested under identical conditions, PETG links can break for a force value of 40.9 N. In comparison, polylactic acid links will break for a force value of 4.70 N. Links printed in the horizontal position were almost 9-fold stronger than those printed in the vertical position. Under the same test conditions, according to the determined empirical mathematical models, PETG links printed in a horizontal position will break for a force of 300.8 N, while links printed in a vertical position will break for force values of 35.8 N.

Keywords: chain link; polymers; tensile testing; influence factors; empirical mathematical models

Citation: Rădulescu, B.; Mihalache, A.M.; Păduraru, E.; Hriţuc, A.; Rădulescu, M.C.; Slătineanu, L.; Ermolai, V. Tensile Behavior of Chain Links Made of Polymeric Materials Manufactured by 3D Printing. *Polymers* **2023**, *15*, 3178. https://doi.org/10.3390/polym15153178

Academic Editor: Chin-San Wu

Received: 30 April 2023
Revised: 12 July 2023
Accepted: 14 July 2023
Published: 26 July 2023

1. Introduction

The chain link is a component of the chain. In turn, the chain is a string of links assembled consecutively between them and which is used to materialize some tying, suspension, tensile, etc., operations. The previous definition of a chain also highlights its main possibilities of use or the stress it is subjected to.

In its simplest form, a chain link is shaped similar to a ring deformed to reduce the width of the chain link and, thus, of the chain. The material and dimensions of a chain link in a cross-section are usually determined considering the mechanical stresses to which the chain composed of links will be subjected.

The classification of chain links can be carried out taking into account the shape of the chain links (circular or oval), the destination of the chain (for limiting the movements of an object or animal, for suspending or towing an object or an assembly/subassembly, for increasing the towing capacity of tires (in the case of anti-skid chains)), etc.

Given the chains' relatively high tensile stress values, their links are made of metallic materials, mainly steel. The need to reduce the chains' weight, especially when the mechanical stresses are somewhat lower, has led to the research and use of chain links made of plastic or composite materials.

Plastic materials are non-metallic materials with an amorphous structure obtained by melting together several constituents, such as resins, plasticizers, dyes, lubricants, fillers, or auxiliary materials. These materials are characterized by the possibility of being easily shaped, thanks to good plasticity, at temperatures usually between 140 and 180 °C. On the other hand, composite materials include a metallic or non-metallic mass and are reinforced with resistance elements from the category of short or very short fibers, long fibers, fabrics, felt, etc.

Plastic or polymer matrix composite chain links are lighter than metal chain links and have higher corrosion resistance. Consistent with the general properties of plastics, such materials are thermally and electrically insulating [1,2].

Suppose the production of chain links from metallic materials is based on bending a workpiece originally as a bar or wire. In that case, the chain links from plastic or composite materials can be obtained through specific processes, such as injection and 3D printing.

The expansion in the last decades of 3D printing processes has facilitated the development of some research on the properties required by the uses of chains with chain links made of plastic materials or composite materials with polymer matrix made even by 3D printing.

Thus, the possibilities of direct manufacturing through 3D printing of interconnected moving parts in the category of which interconnected chain links are also highlighted by researchers [3–5].

Chains with links having more complicated shapes were made by 3D printing (SLM) and studied by Venes et al. [6]. The material of the chain links was an alloy containing titanium, aluminum, and vanadium. It was found that it is possible to ensure a clearance between two moving surfaces of 150 μm.

Wójcicki et al. designed and built a stand for the wear testing of the chain links of the scraper conveyor [7]. A simplified stand prototype was materialized using 3D printing of its components.

The possibility of making chain links from lightweight material through additive manufacturing was highlighted [5,8]. Chains of this type can be used in the mining industry to reduce the weight of some equipment.

A theoretical and experimental analysis of the stress generated in metal chain links under different operating conditions was carried out by Mešić et al. [9]. They considered stress analysis for different chain link positions during mechanical loading.

The possibilities of manufacturing some objects from ceramic materials with polymeric matrix and including some chain links from such materials were investigated by Román-Manso et al. They resorted to 3D printing with microwave-activated polymerization [10].

3D printing of polymer link chains for dog harnesses was addressed by Woodman [11]. He produced a report outlining the possibilities of using 3D printing to create objects that could improve the living conditions of pets or laboratory animals. He thus analyzed the situation of the use of chain links made of plastic materials, appreciating that the plastic material of the links must be hard enough so that the animals do not swallow the particles detached from the links as a result of use and harm their health.

The finite element method was used by Noguchi et al. for a static stress study in the case of a link plate of a roller chain [12]. The study also aimed at identifying some possibilities for reducing the weight of the chain.

Researching the behavior of ship mooring chains on shore has been an important objective for different groups of researchers [13–26]. The static, dynamic, and fatigue stresses occurring in the chain links for different operating conditions were thus highlighted.

The values of some input factors in the additive manufacturing process are expected to influence the physical-mechanical properties of the materials of the parts manufactured by such processes [8,9,27].

The problem of manufacturing chain links from polymeric materials was an objective also addressed in various invention patents [28–30].

From those mentioned above, it can be seen that until now, the research related to the chain links made of metallic and non-metallic materials has focused on the establishment of the dimensions of these chain links [1,7,31–33], the efforts generated in the chain links during operation [8,18,25,26,31], the modeling by the finite element method or by other methods of the stresses and deformations from chain links during their mechanical loading [9,12–15,18,19], investigating the fatigue behavior of chain links [20–22].

It was found that there is relatively little information on the behavior of chain links from polymeric materials manufactured by 3D printing when these chain links are subjected to tensile stress. It is necessary to mention that in the category of chains made of metallic materials that could be replaced by links made of polymers or chains with links made of polymeric materials, some chains for ornaments of monuments, chains for suspending some objects of religious worship, chains of ornaments, chains for leashes used to tie dogs, etc.

The authors of this paper are unaware of results obtained by other researchers regarding the influence exerted by the dimensions of polymer material beads and the factors that characterize the manufacturing conditions of beads by 3D printing on the tensile strength of beads. For low mechanical loads, link chains made of polymeric materials could be characterized by lower mass and superior corrosion resistance. Such chains could be made directly by 3D printing without the need for an assembly operation of separately manufactured chain links.

The finite element method was used to obtain additional information regarding the behavior of the chain links during their tensile testing. In the framework of this paper, the main aspects related to oval-shaped rings manufactured by 3D printing processes from polymeric materials will be briefly analyzed. The conditions for conducting the tensile tests of the 3D printed polymer materials and the obtained experimental results will be presented in the next chapter. Afterward, the emphasis will be on the mathematical processing of the experimental results and the identification, as such, of some empirical mathematical models that characterize the tensile behavior of chain links made of polymeric materials and, respectively, of a composite material with a polymeric matrix. The analysis of the empirical mathematical models will have to allow the formulation of some observations regarding the meaning of variation and the intensity of the influence exerted on the maximum force supported by the chain links when the sizes of the input factors in the investigated process change.

2. Materials and Methods

2.1. The Mechanical Loading of a Chain Link

As mentioned, the chain link is shaped similar to an ovalized ring, with a cross-section through the ring usually revealing a circular shape. During operation, static or dynamic stretching, compression, bending, and torsional stresses can occur in different areas of the chain link.

From the point of view of most uses of the chain link, its tensile strength is of particular interest. However, even during a tensile test, different areas of the chain link may be affected by other categories of stress.

In principle, the tensile testing of the chain link involves using hooks or bolts whose ends are inserted into the gap of the chain link and through which tensile stress is generated in the lateral areas (Figure 1).

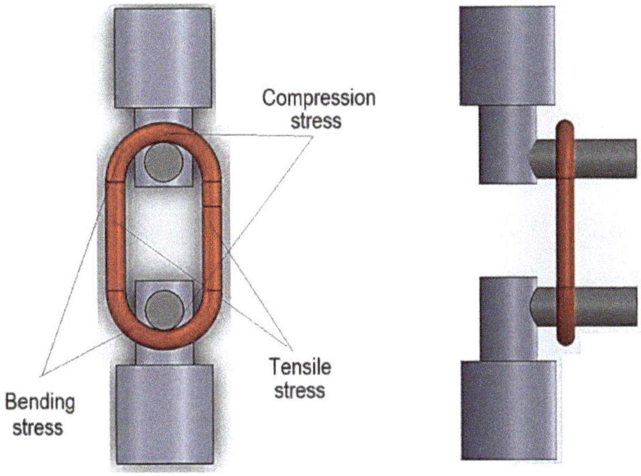

Figure 1. Way to clamp the chain link on the tensile test machine, using two bolts.

The mentioned stress can have a static character. Compressive and crushing stress occurs in the contact areas of the chain link with the two bolts. Bending efforts occur near the contact between the chain link and the bolts through which the chain links are subjected to mechanical stresses. Torsional stresses will still appear in chain links if a twisted chain is considered. Still, there are situations when the loading can also be cyclical, which requires some research on the fatigue resistance of the chain links.

The manufacture of chains from oval-shaped metal chain links may involve operations of cutting straight segments from wire or bar, heating, and descaling, and one or more bending operations, with the eventual joining of chain links, butt welding, shot blasting, marking, thermal treatment, application of surface protection operations, etc. [26]. Chain manufacturing technology also includes some operations to inspect and test the quality of products in different phases of the manufacturing process. Stretch testing and impact testing are such chain quality control operations.

Chain links can be made of polymer or matrix composite materials for low or moderate loads. Thus, chain links can be manufactured by injection molding. As mentioned, the expansion in the last decades of 3D printing processes led, among other results, to ensure the conditions for manufacturing chain links from polymer materials or composite materials with a polymer matrix through additive manufacturing processes. The possibility of changing the values of some parameters specific to the manufacturing conditions of the chain links through 3D printing processes facilitates a deeper investigation of the factors capable of influencing the behavior of the chain link during their operation.

Fused deposition modeling is one of the most widespread groups of techniques that can be used to manufacture parts from polymer materials through 3D printing. In this case, the material of a polymer wire is melted and then deposited to make up the successive layers of material of a particular part. Let us note that fused deposition modeling or 3D printing, in general, can provide conditions for manufacturing the assembled chain links from the beginning so that the chain can be directly obtained.

An improvement in the behavior of chains made of non-metallic materials becomes possible through composite materials with a polymer matrix, in which case the use of certain reinforcement materials could contribute to a significant improvement in the behavior of the chains during their exploitation.

Suppose the manufacture of chain links by fused deposition modeling is considered. In that case, the main groups of factors capable of affecting the behavior of the chain links or chains during their use are the following:

- The sizes and shapes of the chain links;
- The nature and physical-mechanical properties of the polymer material or the composite material with a polymer matrix;
- Some parameters that characterize the conditions for carrying out the 3D printing process;
- The way to load chain links, the nature of these loads, etc.

It is interesting to study the extent to which some of the previously mentioned factors or groups of factors influence the mechanical strength characteristics of chain links made of polymer materials or polymer matrix composite materials.

2.2. Finite Element Modeling of the Behavior of a Plastic or Composite Chain Link

A chain link with the shape and dimensions indicated in Figure 2 was considered. One criterion considered when determining the geometry and dimensions of the 3D printed and tensile-tested chain links was to approximate the dimensions and geometry of existing chain links in practice. Since it was also intended to highlight the influence exerted by the size of the diameter of the za element on the tensile behavior, we resorted to the use of two such diameters, with values close to those existing in the case of some chains encountered in practice.

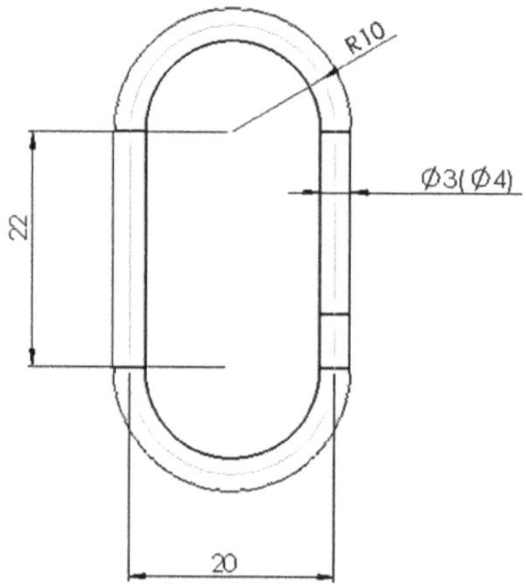

Figure 2. Dimensions of the chain link proposed to be tested.

As a matter of choice, the finite element method (FEM) is to obtain a similar crack pattern as one from the experimental tests. This would give insight into stress distribution just before the crack extends on the whole part length and the fracture occurs. For this purpose, it had to design a similar model with one of the eight test samples used in the experimental tests. It had chosen sample 1 with a 30% infill (see Figure 3a). This option is because sample 1 is printed vertically (see Table 2), making it more difficult to print and analyze than its horizontal peers. The E-moduli obtained using the fused deposition modeling (FDM) printing process is lower than that provided by the manufacturer and depends on the printing direction. Six samples of Ultrafuse ABS-type filament manufactured by BASF (Ludwigshafen, Germany) on Instron 4411 equipment were printed for analyses to be as accurate as possible and tested. It carries a 5 kN load cell with a 5 mm/sec testing speed. The room temperature was 22 °C with a 60% humidity level. It resulted in an E-modulus mean value of 558 MPa. The manufacturer lists in the filament's technical data sheet (TDS) 1608 MPa on ZX direction of printing [34], whereas the Ansys library uses 1628 MPa. The measured value was used inside Engineering Data, Ansys's library of materials. The infill model was replicated in a 3D model and later translated into a Parasolid file for analysis (see Figure 3b).

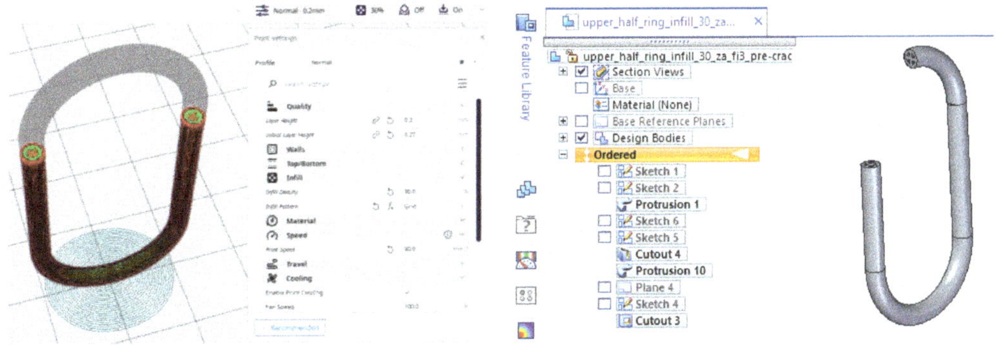

(**a**) (**b**)

Figure 3. Models of the chain link: (**a**)—intermediate step from the preview of the 3D printing process; (**b**)—section view of the 3D model with a 30% infill pattern.

The 3D model also received a crack initiation model because the FEM analyses are based on the prediction of crack propagation using Ansys's SMART Crack Growth, which relies on the energy release rate method. It has opted for the stress intensity factor as the desired fracture criterion inside a static crack growth mechanism. This one predicts a pre-existing crack growth when certain loading conditions are met. The algorithm calculates the maximum stress intensity factor of the crack's front nodes, and if it exceeds the specified criterion, the crack is set to grow. The critical rate was set to 100 MPa $mm^{1/2}$, which will not stop on maximum crack extension. The body was assigned an ABS type of material retrieved from the software's library. It has imposed a virtual topology that uses the edges-only type of behavior. Mesh controls use a patch-conforming method with quadratic element order set for the entire chain link body and a face sizing of 0.15 mm for the two faces of the pre-existing designed crack. Element size was set to 0.25 mm for the entire mesh, with an aggressive mechanical method set for whenever error limits are encountered during analyses. That gave almost 80.000 elements and more than 140,000 nodes (see Figure 4a). The pre-meshed crack was designed to occur before the infill pattern, thus giving us the opportunity to set the front, top and bottom nodes using named selection. As experimental tests revealed in the case of sample 1 the crack initiates at the beginning of the semi-circular top section of the chain link. After several trial and error attempts, it has chosen 6 solution contours and imposed a new Coordinate System set for the crack only for the algorithm to work properly. This system has the X axis set in the direction of the crack extension as the Y sits almost normal on its geometry (see Figure 4b). Conditions use a fixed support for the lower section of the chain link and a force value imposed on the upper section of the chain link in the Z+ direction of the Global Coordinate System (see Figure 4c). Analyses settings are set to 0.00001 sec end step time divided into ten sub-steps.

Just before separation, the value of force required by the software to generate crack growth was 54 N which is close to that registered in the case of sample 1 made of ABS. (Table 2). The difference may come from the fact that the 3D model uses an ideal type of homogeneity, whereas the printed sample suffers from this perspective. It has resulted in 15.025 MPa registered for the equivalent stress evaluated using Von-Mises criteria. The graphical representation shows a distribution along the fracture line with its ends colored in red (see Figure 5a). Equivalent elastic strain peaks at 0.042454 mm/mm, showing that the chain link takes the force solicitation and tries to overcome it by distributing it along its length (see Figure 5b). Finally, it has attached a real-life photo of cracked sample 1 after the experimental test. It can observe a similar direction of propagation as that obtained for the pre-designed crack (Figure 5c).

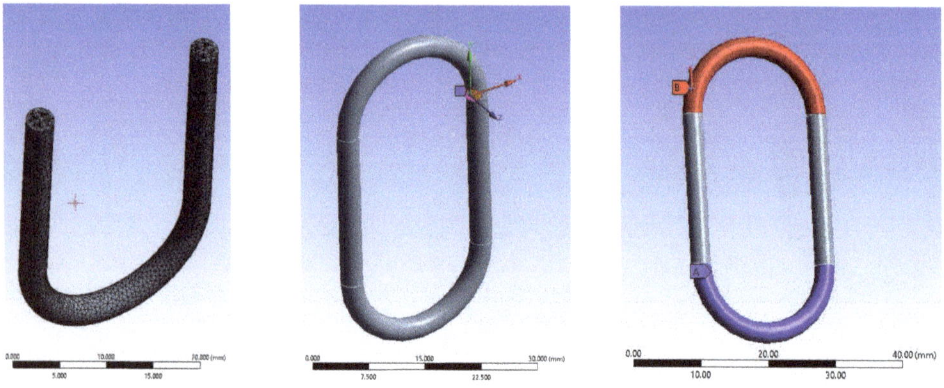

(**a**) (**b**) (**c**)

Figure 4. FEM setup: (**a**)—section view of the meshed model with a 30% infill patterns; (**b**)—view of the pre-meshed crack setup with its coordinate systems; (**c**)—setup conditions inside analyses settings.

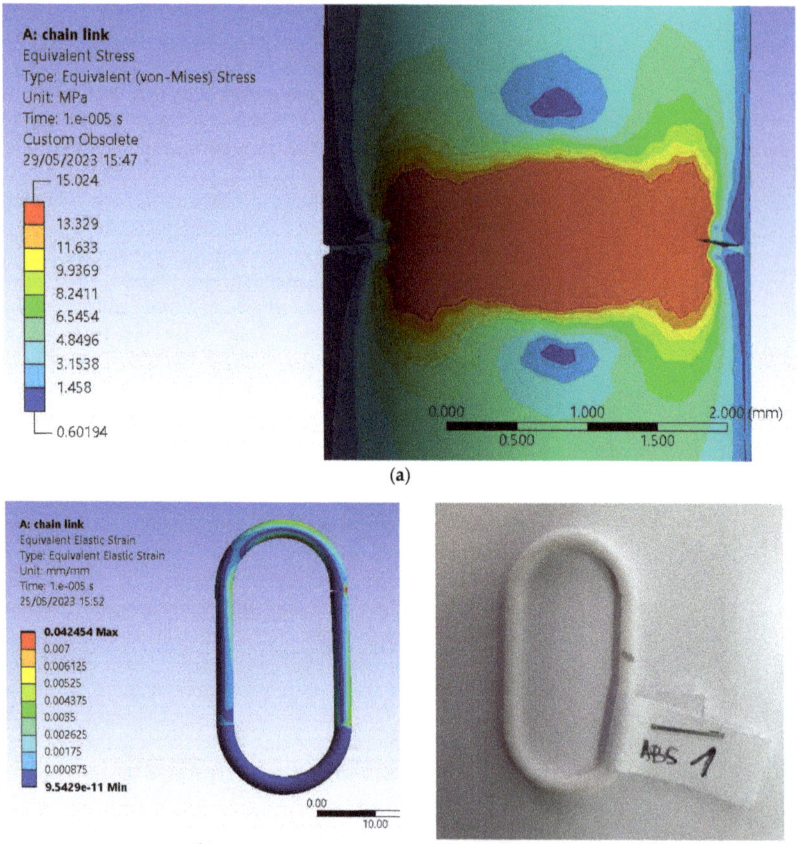

(**b**) (**c**)

Figure 5. Results of FEM and experimental test: (**a**)—side view of the distribution of equivalent (von Mises) stress type of result; (**b**)—front view of the distribution of equivalent elastic strain type of result; (**c**)—front view of the real-life cracked sample 1 made of ABS.

SMART crack growth produces a result similar to that obtained using experimental tests. Before the body gets separated, stress and strain distribution is very useful information for future research, which would consider optimization using 3D printing setup and design. However, results should be used carefully. Further refinement is necessary, considering a finer mesh, more contour plots, or another type of crack growth assessment with an arbitrary or surface-based crack.

2.3. Experimental Conditions

The objective pursued through the experimental research was to highlight the influence exerted by the nature of some polymer materials used for manufacturing chain links, some dimensions of the chain links, and some factors that characterize the conditions of 3D printing of chain links.

Four materials were considered, namely: acrylonitrile butadiene styrene (ABS, Kimya, Nantes, France), acrylonitrile butadiene styrene-Kevlar composite material (ABS-K), polylactic acid (PLA), and polyethylene terephthalate glycol (PETG) (Prusa, Prague, Czech Republic). Information on some of the properties of these materials has been listed in Table 1.

Table 1. Some properties of the materials used in the case of tensile-tested beams.

Properties	ABS	ABS-Kevlar	PLA	PETG
Tensile modulus [GPa]	1.6815 (ISO 527)	1.775 (ISO 527)	2.3 (ASTM D638—Type V)	1.6 ± 0.1 (ISO 527-1—vertical XZ)
Tensile strength [GPa]	0.0436 (ISO 527—at yield)	0.0311 (ISO 527—at yield)	0.0359 (ASTM D638—Type V—at yield)	0.05 (ISO 527-1—vertical XZ)
Elongation	3.5 % (ISO 527—at yield)	2.3 % (ISO 527—tensile strength)	2 % (ASTM D638—Type V—at yield)	5.1 % (ISO 527-1—at yield)
Hardness	97 (Shore A)	65.2 (Shore D)	95 (Shore D)	74 (Shore D)

The chain links were manufactured by 3D printing on a Prusa i3MK3S equipment made in the Czech Republic. The values of some input factors in the 3D printing process have been established, as shown next. The values of the printing speed v, the thickness t of the deposited layer, and those of the infill density i were entered into the 3D printing program. As far as possible, consideration was given to the use of the same input factor values for the experiments performed on each of the four materials from which the chain links test samples were manufactured. Instead, for the printer plate temperature θ_p and the extrusion temperature θ_e, a rewrite of the 3D printing program was required.

The equipment used to determine the mechanical strength of the beads was a tensile testing machine type LRX Plus (Lloyd Instrument Ltd., An AMETEK Company, Hampshire, UK). It uses XLC-5000-Al type of load cell up to 5 kN with a precision of 0.5% according to ASTM E4 and DIN 1221. Data acquisition was made by using equipment's NEXIGEN Data Analysis software (ver. 3.0) solution.

The components involved in the experimental research can be seen in Figure 6.

Table 2. Conditions for performing experimental tests and results of tensile tests.

Part/ Exp no.	Values of the Input Factors							Values of the Output Parameter (Maximum Force F, N)			
	Chain Link Diameter, d, mm	Printing Speed, v, mm/s	Layer Thickness, t, mm	Infill Density, i, %	Temperature Build Plate, θ_p, °C	Extrusion Temperature, θ_e, °C	Printing Position, p	ABS	ABS-K	PLA	PETG
1	3	80	0.2	30	90	245	1	53.809	51.415	4.7005	40.96
2	3	80	0.2	60	110	265	2	350.23	216.41	235.7	414.93

Table 2. *Cont.*

Part/ Exp no.	Values of the Input Factors							Values of the Output Parameter (Maximum Force F, N)			
	Chain Link Diameter, d, mm	Printing Speed, v, mm/s	Layer Thickness, t, mm	Infill Density, i, %	Temperature Build Plate, θ_p, °C	Extrusion Temperature, θ_e, °C	Printing Position, p	ABS	ABS-K	PLA	PETG
3	3	120	0.3	30	90	265	2	459.45	94.09	252.02	414.3
4	3	120	0.3	60	110	245	1	54.519	47.876	302.2	16.923
5	4	80	0.3	30	110	245	2	435.48	208.47	403.73	273.56
6	4	80	0.3	60	90	265	1	120.03	124.44	121.78	80.394
7	4	120	0.2	30	110	265	1	150.47	113.51	141.96	103.43
8	4	120	0.2	60	90	245	2	289.04	233.66	257.23	612.66

Printing position: $p = 1$ for vertical position; $p = 2$ for horizontal position.

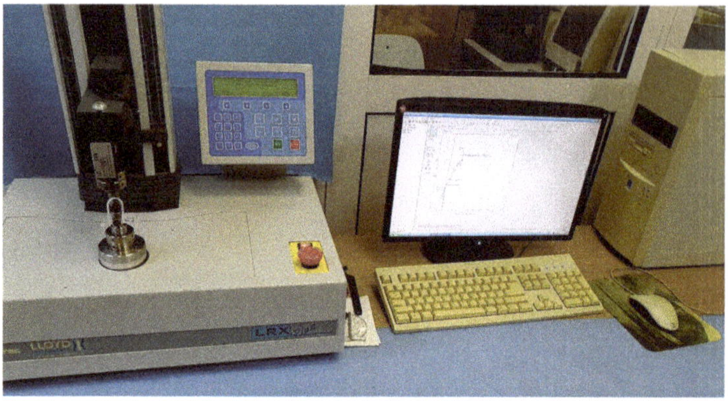

Figure 6. Components of equipment used for tensile testing of chain links.

Clamping of the bars to the tensile testing machine was carried out using 2 bolts (Figure 7).

Figure 7. Clamping the chain link on the tensile testing machine.

The deformation rate of the specimens during the tensile test was 6 mm/min. The LRX Plus Universal Tensile Testing Machine can obtain the force–deformation diagram and all related numerical information directly.

To obtain as complete information as possible regarding the factors that influence the tensile behavior of the chain links manufactured by 3D printing, they resorted to using a Taguchi fractional factorial experiment of type L8 (2^{8-1}), with 7 input factors at two variation levels. Each type of material received an additional sample for testing purposes to proof our L8-1 plan. Such an experimental program also ensures a certain reduction in the number of experimental tests necessary to be performed without the precision of the established empirical mathematical model being significantly affected [35,36].

The 7 input factors considered were the chain link diameter (d_{min} = 3 mm, d_{max} = 4 mm), printing speed (v_{min} = 80 mm/s, v_{max} = 120 mm/s), layer thickness (t_{min} = 0.2 mm, t_{max} = 0.3 mm), infill density (i_{min} = 30%, i_{max} = 60%), build plate temperature (θ_{pmin} = 90°, θ_{pmax} = 110°), extrusion temperature (θ_{emin} = 25 °C, θ_{emax} = 265 °C and, respectively, the way of specimen placement during the 3D printing process (code 1 for vertical printing and code 2 for horizontal printing).

Experimental research has shown that the duration of the printing process of a vertically placed specimen is approximately 50% longer than that corresponding to printing the specimen in a horizontal position. An image made as a screenshot and related to the vertical or horizontal placement of the chain link in the 3D printing process is shown in Figure 8. In this image, the support elements generated by the slicer were highlighted using green color.

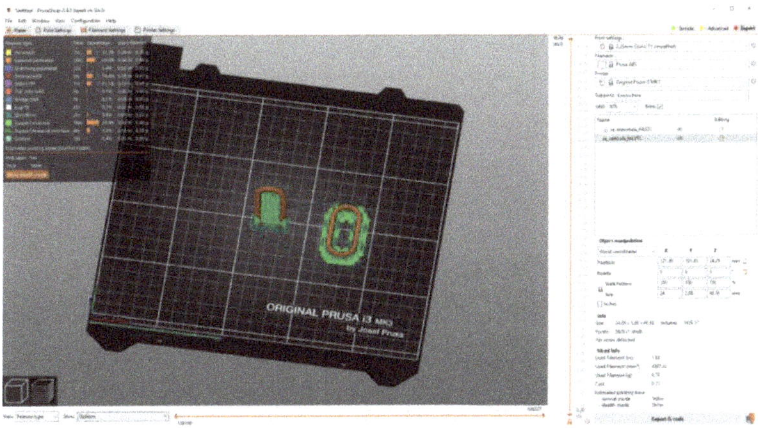

Figure 8. Screenshot showing how to place the printed chain link vertically or horizontally.

The value of the maximum force *F* at which the initiation of the sample damage process is highlighted was used as the output parameter. The values of the *F* forces for each material and experimental test were determined by analyzing the force–deformation diagrams generated by the computer program used by the tensile testing machine.

In Figure 9, the aspect of a chain link can be observed before being subjected to the tensile test and, respectively, after stopping the tensile test.

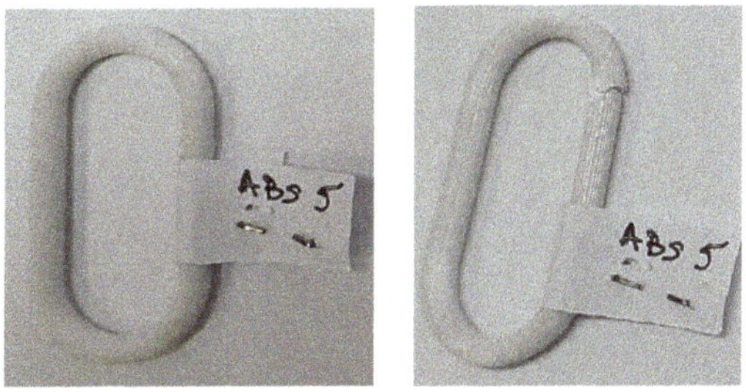

Figure 9. The appearance of a chain link made of ABS polymer material before (**left**) and after (**right**) breaking, respectively.

In Figure 10, the 32 specimens manufactured by 3D printing from different polymer materials can be observed after their rupture by the tensile test.

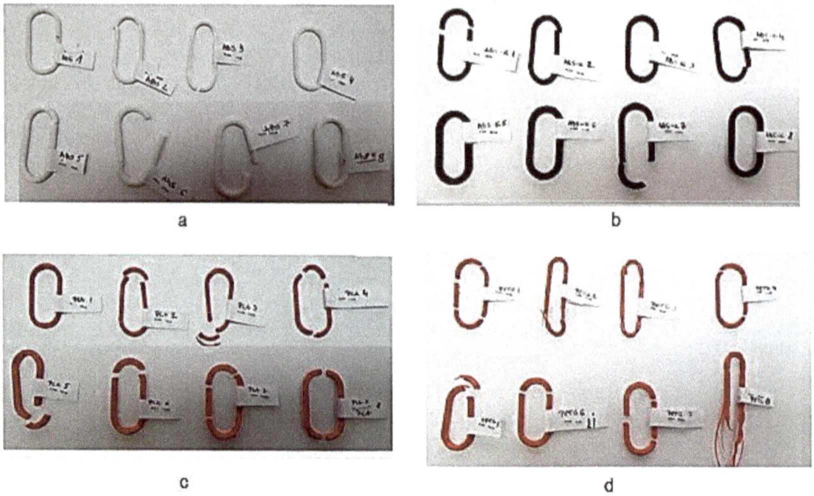

Figure 10. Images of the 32 links manufactured by 3D printing from 4 different polymer materials, after breaking them by the tensile test ((**a**)—ABS links; (**b**)—ABS-K links; (**c**)—PLA links; (**d**)—PETG links).

Some examples of force–deformation curves made under the conditions of testing the specimen made of ABS polymer material can be seen in Figure 11.

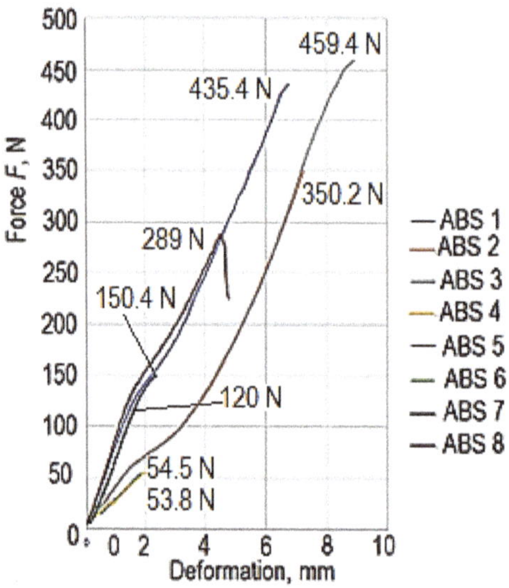

Figure 11. Examples of force–deformation curves in the case of links made of ABS polymer material, corresponding to the 8 experimental tests.

The influence of the diameter d on the magnitude of the maximum force F that determines the damage of a chain link made of different materials can be seen in Figure 12.

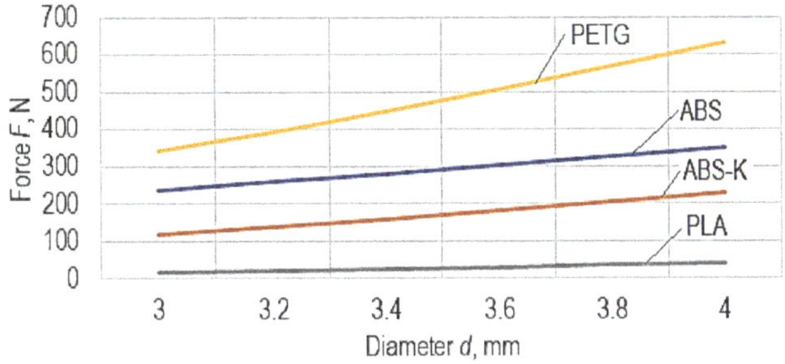

Figure 12. The influence of the diameter d on the magnitude of the maximum force F that determines the damage of a chain link made of different materials (v = 80 mm/min, t = 0.2 mm, I = 30%, $\theta_p = 90°$, $\theta_e = 245°$, p = 2, corresponding to manufacturing in horizontal position).

The conditions for performing the experimental tests were those mentioned in Table 2. Each experimental trial was performed only once, so the experimental trials were not repeated with the same values of the input factors in the 3D printing process. It is mentioned that the design of the experiment method (used to design the experimental trials whose experimental results were included in Table 2) was introduced and used to allow obtaining the maximum information with a minimum of trials [37–40].

The values of the force F determined by analyzing the diagrams corresponding to each chain link made of the 4 materials considered were entered in the last 4 columns of Table 2.

The experimental results were processed using a computer program based on the least squares method [41]. The computer program can select the most appropriate empirical mathematical model from among five models (first-degree polynomial, second-degree polynomial, power-type function, exponential-type function, hyperbolic-type function).

The selection is made by using Gauss's criterion value [41,42]. This value of Gauss's criterion is determined by taking into account, first of all, the sum of the squares of the differences between the values of the ordinates corresponding to the use of the proposed mathematical model and, respectively, the values of the ordinates related to the experimental results, for the same values of the abscissas. The sum of the squares of the mentioned differences is related to the difference between the number of experimental trials and the number of constants in the dependence relationship [41,42]. A lower value of Gauss's criterion characterizes a better approximation of the experimental results to those calculated using a certain empirical mathematical model.

Among the 5 empirical mathematical models that can be determined using the mentioned computer program, the power-type function was preferred, given its use for the characterization of other processes or sizes specific to machine manufacturing (for example, for determining the tool life, cutting force components, and roughness parameters of the machined surfaces). It is necessary to note that power-type functions are preferred when it is considered that there is a monotonous variation in the pursued output parameter to the variation in the input factors in the investigated process. It will be assumed, as such, that in the situations presented in this paper and for the variation intervals considered, the seven input factors do not determine the occurrence of maxima or minima of the value of the force F, and they lead to a monotonous variation in the output parameter. The selection of a factorial experiment of type L8-2^7 to arrive at empirical mathematical models that highlight the influence of different factors on the force F was carried out, taking into account the simplification of the calculations necessary to identify empirical mathematical models [43–46].

A monotonic variation, at least for certain ranges of variation in the values of the input factors in the 3D printing process, was highlighted by the experimental results obtained by other researchers. Thus, Vanaei et al. considered that changes in the crystallization mode could explain the variation in ultimate strength of the liquefier temperature and, therefore, the material's microstructure in the sample [47]. For a temperature variation between 200 and 220 °C, they found an increase in ultimate strength in the case of samples made of polylactic acid. Similar explanations were formulated by Vanaei et al. regarding the influence exerted on ultimate strength by print speed, support temperature, and layer height. The influence of different input factors in the 3D printing process on ultimate strength was investigated by Jackson et al. [48]. They thus considered that increasing the ultimate strength by modifying the deposition angle could be connected with the retraction speed. The 3D printed parts with higher retraction speed presented higher values of the ultimate strength, which means a monotonous variation in the ultimate strength when the retraction speed is increased. Meram and Sözen appreciated that lower layer thickness values lead to higher ultimate strength values, ensuring better adhesion between layers [49]. This means that the two researchers thought there was a monotonous variation in the ultimate strength when changing the value of the thickness of the deposited material layer. A contrary result was obtained by Cho et al., which shows that the higher the thickness of the deposited layer, the higher the mechanical strength of the deposited material [50]. They, therefore, accept a monotonous variation in the mechanical resistance of the material of the specimen made by 3D printing by the thickness of the deposited layer. Müller et al. observed a monotonic variation (an increase) in tensile strength when the infill density increases [51].

An advantage of using empirical mathematical power-type function models is that such models provide direct information on the intensity of influence exerted by an input factor by comparing the value of the exponent attached to that factor in the mathematical power-type function with the values of other exponents. At the same time, a positive value of the exponent in question means that for the range of variation investigated, an increase in the size of the input factor will lead to an increase in the value of the output parameter. In contrast, a negative value of the exponent will cause a decreasing the size of the output parameter.

The mathematical relationships corresponding to the force F and the values of Gauss's coefficients are mentioned next.

Thus, in the case of ABS material, the mathematical relationship was determined: the value of Gauss' criterion being $S_G = 0.115262$.

$$F = 9.436 \cdot 10^{-16} d^{1.366} v^{0.0622} t^{0.288} \mathrm{i}^{-0.322} {\theta_p}^{0.471} {\theta_e}^{6.588} p^{2.144} \quad (1)$$

The empirical mathematical model determined for the ABS-K material has the form:

$$F = 5.166 \cdot 10^{-8} d^{2.276} v^{-0.543} t^{-0.571} i^{0.349} {\theta_p}^{0.692} {\theta_e}^{2.795} p^{1.208} \quad (2)$$

in this case, Gauss's criterion being $S_G = 8.626843 \cdot 10^{-3}$.

For the PLA material, the following form of the empirical mathematical model was arrived at:

$$F = 3.973 \cdot 10^{-32} d^{2.656} v^{2.424} \mathrm{t}^{2.791} i^{1.259} {\theta_p}^{5.856} {\theta_e}^{6.209} p^{1.993} \quad (3)$$

the Gauss criterion having, in this case, the value $S_G = 0.28817645$.

For the PETG material, the empirical mathematical model is of the form:

$$F = 1.158 \cdot 10^{-16} d^{2.137} v^{0.106} t^{-1.118} i^{-0.118} {\theta_p}^{-1.787} {\theta_e}^{8.017} p^{3.071} \quad (4)$$

the Gauss criterion having, in this case, the value $S_G = 0.2780781$.

3. Discussion

The graphical representations in Figures 12 and 13 were developed using empirical mathematical models.

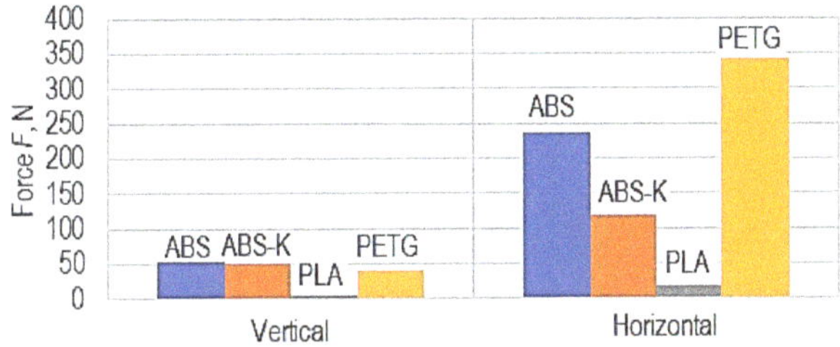

Figure 13. Differences between the maximum force values for the four polymeric materials considered in the case of manufacture in vertical and horizontal positions, respectively (d = 3 mm, v = 80 mm/min, t = 0.2 mm, i = 30%, θ_p = 90 °C, θ_e = 245 °C).

The analysis of the experimental results, the empirical mathematical models, and the graphic representations elaborated on their basis allowed the formulation of the observations mentioned below.

Examining the empirical mathematical models (Equations (1)–(4)) reveals maintenance of the same direction of force magnitude variation only when considering the corresponding diameter d of a cross-section through the chain link and, respectively, the position in which the chain link was manufactured by 3D printing on the printer table.

As expected, an increase in the diameter d of the cross-section through the chain link increases the maximum force F (Figure 11). Still, following the initial hypothesis, manufacturing the chain link in a horizontal position leads to a higher strength of the chain link and a higher value of the maximum force F. This fact can be explained by the better behavior of the applied material layers to the stretching stress right along the main stress direction on the chain link during the tensile test. The continuity of the layers formed by the gradual deposition of the wire of melted polymeric material leads to an increase in the tensile strength of the polymer material.

Less expected results were obtained in the case of chain links made of ABS material with Kevlar when a decrease in the mechanical resistance of the chain links reinforced with one of the most resistant plastic materials (Kevlar) was observed. The analysis of the appearance of the broken test samples as a result of tensile stress reveals a certain separation of the ABS polymer from the Kevlar fibers due to a reduced adhesion between the two materials. Since the Kevlar fibers are not continuous, it is possible that the low adhesion between the materials incorporated in the chain link and the discontinuity of the Kevlar fiber led to the decrease in the mechanical strength of the ABS-K material chain links, compared to the mechanical strength of the chain links made of ABS material only. It was also observed that under the conditions considered for the 3D printing process, the ABS-K type material became brittle, which led to the destruction of the tensile chain link at the appearance of the first cracks. The lower mechanical strength of the chain links made from PLA may be due to the 3D printing conditions. The results of the experimental research showed that the highest mechanical resistance was obtained in the case of the links made of polyethylene terephthalate glycol (PETG). According to the experimental results, when tested under identical experimental conditions (experiment no. 1 in Table 2), PETG links can break for a force value of 40.9 N. In comparison, polylactic acid links will break for a force value of 4.70 N. Links printed in the horizontal position were almost 9-fold stronger than those printed in the vertical position. Under the same test conditions (d = 3 mm, v = 80 mm/s, t = 0.2 mm, I = 30%, θ_p = 90 °C, θ_e = 245 °C, according to the determined empirical mathematical models, PETG links printed in a horizontal position will break for a force of 300.8 N, while links printed in a vertical position will break for force values of 35.8 N.

The chemical compositions of the four polymer materials from which the chain links were fabricated by 3D printing were different. They involved using different input factor values in the 3D printing process. To compare the results obtained in the case of using these different polymer materials, the chain links need to be printed under the conditions of using the same values of the input factors in the 3D printing process. In this way, it was possible that the used values of some of the input factors in the 3D printing process were not the most suitable for making links with high tensile strength. This aspect could be, for example, a cause of the relatively low value of the tensile strength determined in the case of gels made of polylactic acid. These printing conditions were the same for all the polymer materials from which the plates were made, being established in such a way that, as far as possible, they corresponded, to a greater or lesser extent, to the valid recommendations for the manufacture of parts from the respective polymer materials.

According to the determined empirical mathematical models, except for one material (PETG), the values of the maximum force F increase with the increase in the temperature θ_p of the plate on which the chain links were generated. A possible explanation of this situation could be based on a better adhesion of successively deposited material layers when the plate temperature θ_p increases.

It is also found that, except for the ABS-K material, increasing the velocity v leads to an increase in the maximum force F. It is possible that with increasing the velocity v, more favorable heat transfer conditions specific to the cooling of the specimen material are reached and, under these conditions, to achieve better adhesion of successively deposited layers of material, so at a higher value of the magnitude of the maximum force F.

According to the determined empirical mathematical models, a high influence on the magnitude of the maximum force F is exerted by the extrusion temperature θ_e. In all empirical mathematical models, it was found that the exponents attached to the temperature in the extrusion nozzle have large and positive values. This means an increase in the extrusion temperature θ_e will significantly increase the magnitude of the maximum force F. The findings can be explained by the fact that when the extrusion temperature θ_e is increased, there is an increase in the fluidity of the material to be deposited and, as such, a better connection between layers of material added successively.

A validation test of the determined empirical mathematical models was possible by conducting experimental tests for other combinations of the values of some input factors in the 3D printing process. Some such results led to maximum force values close to values determined using empirical mathematical models.

Some of the results determined in the manner previously described were as follows:

- For an ABS chain link, where the variables took the values $d = 4$ mm, $v = 80$ mm/s, $t = 0.2$ mm, $i = 60\%$, $\theta_p = 90$ °C, $\theta_e = 265°$, $p = 2$ (printing in horizontal position), the maximum force F values were 460.84 N for the experimental test and 469 N when the empirical mathematical model was used. The difference between the two values is 1.9%.
- For a chain link made of ABS-K, where the variables took the values $d = 4$ mm, $v = 80$ mm/s, $t = 0.2$ mm, $i = 30\%$, $\theta_p = 90$ °C, $\theta_e = 245°$, $p = 2$ (printing in horizontal position), the values of the maximum force F were 192.66 N for the experimental test and 284.21 N when the empirical mathematical model was used. The difference between the two values is 32.2%.
- For a chain link made of PETG, where the variables took the values $d = 4$ mm, $v = 80$ mm/s, $t = 0.2$ mm, $i = 30\%$, $\theta_p = 90$ °C, $\theta_e = 245°$, $p = 2$ (printing in horizontal position), the maximum force F values were 564.53 N for the experimental test and 556.36 N when the empirical mathematical model was used. The difference between the two values is 1.46%.

Even larger differences were found for other sets of input factor values in the case of additional experimental tests performed. These differences could be explained by a rather large dispersion of the experimental results due, for example, to the variation and other input factors in the 3D printing process and whose values were not followed during the experimental trials. Large differences between the experimental results and those obtained using the empirical mathematical model were also observed in the case of chain links made of PLA. Such differences could also be generated by the fact that the values of some of the input factors in the 3D printing process were not suitable for the PLA material.

The highest value of the maximum force *F* was recorded in the case of the chain link made of PETG material. It should be noted that the value of the maximum force *F* is almost 7-fold higher when the chain link is made in a horizontal position compared to that made in a vertical position. A possible explanation for this was mentioned earlier. PETG polymer material also has the highest tensile strength, as seen in Table 1.

4. Conclusions

The problem of replacing parts made of metallic materials with parts made of plastic materials manufactured by 3D printing can also be formulated in the case of chain links. For reduced mechanical stress, link chains made of polymeric materials have a lower weight and better corrosion resistance than chains made of metallic materials. It should be noted that such chains made of polymer materials could be manufactured directly by 3D printing without assembling separately manufactured chain links. The consultation of specialized literature highlighted research concerns in such a direction. The research presented in this article aimed to highlight the influence exerted by some input factors in the 3D printing process of chain links from four different materials on the tensile behavior of the respective chain links. The four materials from which the chain links were made were acrylonitrile butadiene styrene (ABS), the composite material acrylonitrile butadiene styrene-Kevlar (ABS-K), polylactic acid (PLA) and polyethylene terephthalate glycol (PETG). The experimental tests were conducted according to the requirements of a fractional factorial experiment with seven input factors at two levels of variation. The chain link diameter in a cross-section, printing speed, layer thickness, infill density, the temperature of the printer table, extrusion temperature, and test sample placement during 3D printing were considered input factors. Through the mathematical processing of the experimental results, empirical mathematical power-type function models were determined. These empirical mathematical models provide information on the direction of variation and the intensity of the influence exerted by the input factors considered on the maximum force at which damage to the chain links occurs. The analysis of the experimental results highlighted the fact that the same directions of action of the input factors were recorded only in the case of the diameter of the chain link rod and the way of positioning the sample during printing. As expected, printing the chain link horizontally ensured superior mechanical strength due to the stress development along the deposited layers, thus characterized by a certain continuity. The results of the experimental research showed that chain links printed in a horizontal position are almost 9-fold stronger than chain links printed in a vertical position. Under the same test conditions, according to the determined empirical mathematical models, PETG links printed horizontally will break for a force of 300.8 N. In comparison, links printed in a vertical position will break for force values of 35.8 N. Polyethylene terephthalate glycol (PETG) links have proven the highest mechanical resistance among the materials used. Under the same test conditions, the PETG links broke for a force of 40.6 N, while the polylactic acid rings broke for values of 4.70 N. In the future, it is intended to continue the research by considering other materials for the chain links and identifying more appropriate mathematical and empirical models to illustrate the tensile behavior of some plastic materials incorporated in the chain links. Another research direction could be to directly manufacture complete chains with polymer links assembled by the 3D printing process itself, eliminating the need for an additional operation of assembling separately manufactured links.

Author Contributions: Conceptualization, B.R. and L.S.; methodology, L.S.; software, A.M.M.; validation, E.P. and A.H.; formal analysis, A.H.; investigation, E.P. and V.E.; resources, B.R.; data curation, A.H.; writing—original draft preparation, L.S.; writing—review and editing, A.H.; visualization, A.H., M.C.R.; supervision, L.S.; project administration, B.R.; funding acquisition, B.R. All authors have read and agreed to the published version of the manuscript.

Funding: This research received no external funding.

Institutional Review Board Statement: Not applicable.

Data Availability Statement: The data presented in this study are available on request from the corresponding author. The data are not publicly available due to privacy.

Conflicts of Interest: The authors declare no conflict of interest.

References

1. Fafenrot, S.; Grimmelsmann, N.; Wortmann, M.; Ehrmann, A. Three-dimensional (3D) printing of polymer-metal hybrid materials by fused deposition modeling. *Materials* **2017**, *10*, 1199. [CrossRef]
2. Abdalla, A.; Patel, B.A. 3D Printed Electrochemical Sensors. *Annu. Rev. Anal. Chem.* **2021**, *14*, 47–63. [CrossRef]
3. Kalva, R.S. 3D Printing—The future of manufacturing (the next industrial revolution). *Int. J. Innov. Eng. Technol.* **2015**, *5*, 185–190.
4. Mareș, M. Additive manufacturing technologies. A concise introduction. *Bul. Institutului Politeh. Din Iaşi. Secţia Construcţii De Maşini* **2018**, *64*, 27–57.
5. Wang, L.; Alexander, C.A. Additive manufacturing and big data. *Int. J. Math. Eng. Manag. Sci.* **2016**, *1*, 107–121. [CrossRef]
6. Veness, R.; Andreazza, W.; Gudkov, D.; Miarnau Marin, A.; Samuelsson, S. Metal 3D Additive Machining for in-Vacuum Beam Instrumentation. In Proceedings of the 10th Mechanical Engineering Design of Synchrotron Radiation Equipment and Instrumentation Conference, Paris, France, 25–29 June 2018; pp. 121–124. [CrossRef]
7. Wójcicki, M.; Wieczorek, A.N.; Głuszek, G. Concept of the facility for testing the wear of chain links in the aspect of synergism of environmental factors. *Min. Mach.* **2021**, *39*, 71–81. [CrossRef]
8. Aliakbari, M. Additive Manufacturing: State-of-the-Art, Capabilities, and Sample Applications with Cost Analysis. Master's Thesis, KTH Royal Institute of Technology, Stockholm, Sweden, 2012. Available online: https://www.diva-portal.org/smash/get/diva2:560827/FULLTEXT02.pdf (accessed on 15 May 2023).
9. Mešić, A.; Čolić, M.; Mešić, E.; Pervan, N. Stress analysis of chain links in different operating conditions. *Int. J. Eng. Sci. Invention* **2016**, *5*, 43–49.
10. Román-Manso, B.; Weeks, R.D.; Truby, R.L.; Lewis, J.A. Embedded 3D printing of architected ceramics via microwave-activated polymerization. *Adv. Mater.* **2023**, *35*, 2209270. [CrossRef]
11. Woodman, C. *3D Printing for Animal Welfare. A Report for the Animal Welfare Institute*; Texas A&M University: College Station, TX, USA, 2021; p. 17. Available online: https://awionline.org/sites/default/files/uploads/documents/3d-printing-animal-welfare.pdf (accessed on 15 May 2023).
12. Noguchi, S.; Nagasaki, K.; Nakayama, S.; Kanda, T.; Nishino, T.; Ohtani, T. Static stress analysis of link plate of roller chain using finite element method and some design proposals for weight saving. *J. Adv. Mech. Des. Syst. Manuf.* **2009**, *3*, 159–170. [CrossRef]
13. Das, N. Models to Explain out-of-Plane Bending Mechanism in Mooring Chain Links. Master's Thesis, Delft University of Technology, Delft, The Nederlands, 2016. Available online: http://resolver.tudelft.nl/uuid:f387281a-1365-4121-a2cf-3f0b59448f67 (accessed on 15 May 2023).
14. Berthelsen, K. Out of Plane Bending of Mooring Chains Finite Element. Analysis of a 7-link model. Master's Thesis, Norwegian University of Science and Technology, Trondheim, Norway, 2017. Available online: https://ntnuopen.ntnu.no/ntnu-xmlui/handle/11250/2453087 (accessed on 15 May 2023).
15. Kim, S.; Won, D.-H. Bending behavior of the mooring chain links subjected to high tensile forces. *J. Korean Soc. Steel Const.* **2017**, *29*, 99–110. [CrossRef]
16. Choung, J.; Lee, J.B.; Kim, Y.H. Out-of-plane bending stiffnesses in offshore mooring chain links based on conventional and advanced numerical simulation techniques. *J. Ocean Eng. Technol.* **2018**, *32*, 297–309. [CrossRef]
17. Chung, W.; Kang, H.; Kim, M. Multi-scale approach for chain-mooring OPB-induced failure considering time-varying interlink bending stiffness and fairlead condition. *Appl. Ocean Res.* **2020**, *98*, 102128. [CrossRef]
18. Chung, W.C.; Kim, M.H. Effects of various fairlead-connection parameters on chain-mooring OPB-induced failure. *Mar. Struct.* **2021**, *76*, 102926. [CrossRef]
19. Chung, W.C.; Kang, H.Y.; Kim, M.H. Numerical study on OPB/IPB interlink angle with underwater chain stopper system. In Proceedings of the SNAME 24th Offshore Symposium, Houston, TX, USA, 20 February 2019.
20. Perez, I.M.; Constantinescu, A.; Bastid, P.; Zhang, Y.H.; Venugopal, V. Computational fatigue assessment of mooring chains under tension loading. *Eng. Fail. Anal.* **2019**, *106*, 104043. [CrossRef]
21. Zarandi, E.P. Multiaxial Fatigue Analysis of Offshore Mooring Chains, Considering the Effects of Residual Stresses and Corrosion Pits. Ph.D. Thesis, Norwegian University of Science and Technology, Trondheim, Norway, 2020. Available online: https://www.researchgate.net/publication/348200873_Multiaxial_fatigue_analysis_of_offshore_mooring_chains_considering_the_effects_of_residual_stresses_and_corrosion_pits (accessed on 15 May 2023).
22. Zarandi, E.P.; Skallerud, B.H. Experimental and numerical study of mooring chain residual stresses and implications for fatigue life. *Int. J. Fatigue* **2020**, *135*, 105530. [CrossRef]
23. Zarandi, E.P.; Lee, T.L.; Skallerud, B.H. Data on residual stresses of mooring chains measured by neutron diffraction and hole drilling techniques. *Data Br.* **2020**, *30*, 105587. [CrossRef]

24. Xue, X.; Chen, N.-Z. Fracture mechanics analysis for a mooring system subjected to tension and out-of-plane bending. In Proceedings of the First Conference of Computational Methods in Offshore Technology (COTech2017), Stavanger, Norway, 30 November–1 December 2017; Volume 276, p. 012036. [CrossRef]
25. Xue, X.; Chen, N.-Z.; Pu, Y.; Chen, L.; Wang, L. Fracture mechanics assessment for mooring chain links tensioned over a curved surface. *Appl. Ocean Res.* **2021**, *117*, 102900. [CrossRef]
26. Takeuchi, T.; Utsunomiya, T.; Gotoh, K.; Sato, I. Development of interlink wear estimation method for mooring chain of floating structures: Validation and new approach using three-dimensional contact response. *Mar. Struct.* **2021**, *77*, 102927. [CrossRef]
27. Vanaei, H.R.; Magri, A.E.; Rastak, M.A.; Vanaei, S.; Vaudreuil, S.; Tcharkhtchi, A. Numerical-experimental analysis toward the strain rate sensitivity of 3D-printed nylon reinforced by short carbon fiber. *Materials* **2022**, *15*, 8722. [CrossRef]
28. Egli, E. Link Chain for Transmitting a Mechanical Load. U.K. Patent GB1191225A, 13 May 1970. Available online: https://patents.google.com/patent/GB1191225A/en?oq=GB1191225A (accessed on 15 May 2023).
29. Dietrich, W.; Roelof, M.; Bergsma, O.; Gruber, J.M.; Mehdi, H.; Licup, A.J.; Van Keulen, A.; Van Rijssel, J.; Yalcin, A.O.; Van Eden, G.G.; et al. Polymeric Chain Link. Canada Patent CA2984062A1, 1 December 2016. Available online: https://patents.google.com/patent/CA2984062A1/en?oq=CA2984062A1 (accessed on 15 May 2023).
30. Van Loon, M.A.Y. Chain Link Formed of Polymeric Material. European Patent EP0404224A1 (B1), 27 December 1990. Available online: https://patents.google.com/patent/EP0404224A1/en?oq=EP0404224A1 (accessed on 15 May 2023).
31. Bastid, P.; Smith, S.D. Numerical analysis of contact stresses between mooring chain links and potential consequences for fatigue damage. Paper OMAE2013-11360. In Proceedings of the 32nd International Conference on Ocean, Offshore and Arctic Engineering, Nantes, France, 9–15 June 2013. Available online: https://www.twi-global.com/technical-knowledge/published-papers/numerical-analysis-of-contact-stresses-between-mooring-chain-links-and-potential-consequences-for-fatigue-damage (accessed on 15 May 2023).
32. Hamanaka Group. Chain Manufacturing Process Outline Drawing. Available online: http://www.hamanaka-chain.co.jp/en/products/products3/ (accessed on 16 April 2023).
33. Jha, N.K.; Avinash, G.; Siddharth, V. Design and structural analysis of plastic chain link with polypropylene and polyoxymethylene material. *Mater. Today Proc.* **2021**, *38*, 3066–3076. [CrossRef]
34. Technical Data Sheet. Ultrafuse ABS. Available online: https://www.ultrafusefff.com/wp-content/uploads/2016/05/Ultrafuse_ABS_TDS_EN_v5.2.pdf (accessed on 5 June 2023).
35. Pillet, M. *Introduction Aux Plans D'expériences par la Méthode Taguchi*; Les Éditions d'Organisation: Paris, France, 1992.
36. Chauveau, J.-C.; Chassaing, J.-P. Introduction à la Methode des Plans D'expériences par la Méthode Taguchi. Available online: https://eduscol.education.fr/sti/sites/eduscol.education.fr.sti/files/ressources/techniques/3870/3870-introtaguchi-cned.pdfamanaka-chain.co.jp/en/products/products3/ (accessed on 27 May 2023).
37. Torii, D.; Nagahara, T.; Okihara, T. Suppression of the Secondary Flow in a Suction Channel of a Large Centrifugal Pump. *IOP Conf. Ser. Mater. Sci. Eng.* **2013**, *52*, 032005.
38. Donnelly, T. Efficient simulation using design of experiments (DOE). Available online: https://community.jmp.com (accessed on 27 May 2023).
39. Lakshminarayanan, A.K.; Annamalai, V.E.; Elangovan, K. Identification of optimum friction stir spot welding process parameters controlling the properties of low carbon automotive steel joints. *J. Mater. Res. Technol.* **2015**, *4*, 262–272. [CrossRef]
40. Tinsson, W. *Plans d'expérience: Constructions et analyses statistiques. Mathématiques et Applications*; Springer: Berlin/Heidelberg, Germany, 2010; Volume 67. Available online: https://link.springer.com/book/10.1007/978-3-642-11472-4 (accessed on 27 May 2023). (In French)
41. Crețu, G.; Varvara, G. *Methods of Experimental Research in Machine Manufacturing*; Publishing House Junimea: Iași, România, 1999. (In Romanian)
42. Worthing, A.G.; Geffner, J. *Processing the Experimental Data*; Technical Publishing House: Bucharest, Romania, 1959. (In Romanian)
43. Agarwal, S.; Tyagi, I.; Gupta, V.K.; Jafari, M.; Edrissi, M.; Javadian, H. Taguchi L_8 orthogonal array design method for the optimization of synthesis conditions of manganese phosphate ($Mn_3(PO_4)_2$) nanoparticles using water-in-oil microemulsion method. *J. Mol. Liq.* **2016**, *219*, 1131–1136. [CrossRef]
44. Krishnaiah, K.; Shahabudeen, P. *Applied Design of Experiments and TAGUCHI Methods*; Ghosh, A.K., Ed.; PHI Learning Pvt. Ltd.: New Delhi, India, 2012. Available online: https://eko.staff.uns.ac.id/files/2014/09/Buku-Alternatif.pdf (accessed on 27 May 2023).
45. Mohamed, H.; Lee, M.H.; Sanugi, B.; Sarahintu, M. Taguchi approach for performance evaluation of routing protocols in mobile ad hoc networks. *J. Stat. Modeling Anal.* **2010**, *1*, 10–18.
46. Limon-Romero, J.; Tlapa, D.; Baez-Lopez, Y.; Maldonado-Macias, A.; Rivera-Cadavid, L. Application of the Taguchi method to improve a medical device cutting process. *Int. J. Adv. Manuf. Technol.* **2016**, *87*, 3569–3577. [CrossRef]
47. Vanaei, H.; Shirinbayan, M.; Deligant, M.; Raissi, K.; Fitoussi, J.; Khelladi, S.; Tcharkhtchi, A. Influence of process parameters on thermal andmechanical properties of polylactic acid fabricated by fused filament fabrication. *Polym. Eng. Sci.* **2020**, *60*, 1822–1831.
48. Jackson, B.; Fouladi, K.; Eslami, B. Multi-parameter optimization of 3D printing condition for enhanced quality and strength. *Polymers* **2022**, *14*, 1586. [CrossRef]
49. Meram, A.; Sözen, B. Investigation on the manufacturing variants influential on the strength of 3D printed products. *Res. Eng. Struct. Mater.* **2020**, *6*, 293–313. [CrossRef]

50. Cho, E.E.; Hein, H.H.; Lynn, Z.; Hla, S.; Saw, J.; Tran, T. Investigation on influence of infill pattern and layer thickness on mechanical strength of PLA material in 3D printing technology. *J. Eng. Sci. Res.* **2019**, *3*, 27–37. [CrossRef]
51. Müller, M.; Jirků, P.; Šleger, V.; Mishra, R.K.; Hromasová, M.; Novotný, J. Effect of infill density in FDM 3D printing on low-cycle stress of bamboo-filled PLA-based material. *Polymers* **2022**, *14*, 4930. [CrossRef]

polymers

Article

Effects of Different Polypropylene (PP)-Backbones in Aluminium Feedstock for Fused Filament Fabrication (FFF)

Vahid Momeni [1,*], Zahra Shahroodi [1], Joamin Gonzalez-Gutierrez [1,2], Lukas Hentschel [1], Ivica Duretek [1], Stephan Schuschnigg [1], Christian Kukla [3] and Clemens Holzer [1]

1 Polymer Processing, Montanuniversitaet Leoben, 8700 Leoben, Austria
2 Functional Polymers Research Unit, Materials Research and Technology (MRT) Department, Luxembourg Institute of Science and Technology (LIST), L-4940 Hautcharage, Luxembourg
3 Industrial Liaison Department, Montanuniversitaet Leoben, 8700 Leoben, Austria
* Correspondence: vahid.momeni@unileoben.ac.at

Abstract: The current study presents the effect of the backbone as an important binder component on the mechanical, rheological, and thermal properties of Aluminium (Al) alloy feedstocks. A thermoplastic elastomer (TPE) main binder component was blended with either polypropylene (PP), grafted-maleic anhydride-PP (PPMA), or grafted-maleic anhydride-PPwax (PPMAwax) plus PP, as the backbone. Differential scanning calorimetry (DSC) and thermogravimetric analysis (TGA) tests were performed to investigate the thermal properties of binder systems and feedstocks. Fourier-transform infrared (FTIR) spectroscopy was used to study the chemical interaction between the binder and the Al alloy. After making feedstock filaments, tensile tests, scanning electron microscopy (SEM), and fused filament fabrication (FFF) printing were done. The results showed that although the PP printability was acceptable, the best mechanical properties and printed quality can be achieved by PPMA. TGA test showed that all binder systems in the feedstocks could be removed completely around 500 °C. From FTIR, the possibility of chemical reactions between Al alloy particles and maleic anhydride groups on the grafted PP backbone could explain the better dispersion of the mixture and higher mechanical properties. Tensile strength in PP samples was 3.4 MPa which was improved 1.8 times using PPMA as the backbone.

Keywords: metal material extrusion; fused filament fabrication; aluminium feedstock; printability; binder system; rheological properties

Citation: Momeni, V.; Shahroodi, Z.; Gonzalez-Gutierrez, J.; Hentschel, L.; Duretek, I.; Schuschnigg, S.; Kukla, C.; Holzer, C. Effects of Different Polypropylene (PP)-Backbones in Aluminium Feedstock for Fused Filament Fabrication (FFF). *Polymers* **2023**, *15*, 3007. https://doi.org/10.3390/polym15143007

Academic Editors: Cristina-Elisabeta Pelin and Anton Ficai

Received: 16 June 2023
Revised: 6 July 2023
Accepted: 7 July 2023
Published: 11 July 2023

1. Introduction

Shaping, debinding, and sintering (SDS) encompasses, among others, extrusion-based additive manufacturing, and powder injection molding (PIM) techniques [1]. The SDS process consists of four stages: material preparation (granules or filaments), shaping (injection molding or 3D printing), debinding, and sintering [2]. The feedstock, the raw material for this method, comprises a mixture of metal or ceramic powder and a binder system. Binders are multi-component combinations of several polymers and additives that play a significant role in the SDS processing of components [3]. It has been shown that just one component modification in the binder system may significantly alter feedstock characteristics [4,5]. After the preparation of the feedstock, green parts with the desired shape are produced by injection molding or 3D printing [6]. During the debinding, all components of the binder systems are extracted from the green part gradually using different techniques and without causing any defect [7]. Sintering, the final step, is done in the furnace by heating parts to temperatures lower than the melting point of the metal or ceramic used to achieve a nearly complete density part with high mechanical properties [8,9].

During the injection step in the PIM process, an expensive and unique mold made of durable materials is needed due to high-pressure levels and heating cycles. PIM is a cost-effective manufacturing technique to produce vast quantities of complex-shaped components [10,11]. However, the high cost of the mold makes this approach less viable for small quantities. Therefore, various approaches using additive manufacturing technologies for feedstocks have been developed to manufacture green parts as part of the SDS process [12]. One of these techniques is the extrusion-based AM method called the fused filament fabrication (FFF) process [13].

In contrast to PIM, where the feedstock is often supplied as granules [14], feedstock in the form of filament is required. FFF equipment installation expenses are significantly less than those for PIM equipment [15]. Although the FFF technique produces an almost limitless range of geometries without expensive molds, the quality of final parts should be improved compared to PIM parts by adjusting the process parameters and designing proper binder formulations. However, processing SDS feedstocks is significantly more complex than printing polymers generally used for FFF since they include a high amount of ceramic or metal powder [16]. In addition to appropriate flow properties, FFF filaments must be strong and flexible enough to be easily spooled, de-spooled, and extruded in the printing head, which is quite challenging when there is a high-volume fraction of powder [17]. Consequently, only a few commercial feedstocks are available [18–20]. However, many researchers have used FFF for different metals such as hard metals [21], Fe [22,23], steel [24,25], copper (Cu) [23], Ti-Al-V [26,27], W-Cr [28], Ni-Ti [29], Mg-Al-Zn [30] along with various binder formulations.

The binder systems generally consist of the main component and a backbone component to which various additives like dispersants, plasticizers, and surfactants are added [31]. A thermoplastic polymer with a good combination of stiffness and flexibility is often the main component [32]. Backbones help in shape retention of the green part during the debinding step when the main binder component is removed [33]. Improving the flowability of the feedstock and spreading the powder throughout the polymer matrix to avoid agglomeration could be done by plasticizers and surfactants [34]. It has been demonstrated that a binder system consisting of a thermoplastic elastomer (TPE) as a major component could be utilized successfully for the FFF process because it can provide high flexibility and extrusion reliability for the filaments [35]. In addition, the backbone's type and concentrations determine the binder system's debinding performance. Therefore, choosing a suitable backbone is crucial for developing a proper feedstock for the SDS process.

In addition, Hausnerova [36] reviewed the different binder systems for PIM feedstocks. It is suggested that polypropylene (PP) provides higher mechanical properties than other backbone polymers. In the study of Gökmen and Türker [37], the rheological behavior of the Inconel 625 superalloy feedstock with PP as the backbone was analyzed. They found that using PP as the backbone has the highest impact on the rheology behavior compared to the other binder constituents and improves the injection molding process. This study showed that binder content (especially backbone polymer) strongly influences the rheological properties of feedstocks.

Some studies have been conducted to develop binder systems for the FFF of different metals and ceramics. In the work of Cano et al. [38], a suitable binder formulation was developed for the zirconia powder. The feedstock with the binder system consisting of styrene-ethylene/butylene-styrene (SEBS)/paraffin wax (PW) was printed successfully without any visible defects in printing and after debinding. According to Gorjan et al. [39], ethylene vinyl acetate (EVA) with different contents of stearic acid was successfully applied for the FFF process of alumina. A single-component binder system containing low-density polyethylene (LDPE) was also utilized for the 316L stainless steel powder [40]. The mechanical test findings indicate that the FFF printed and sintered material is equivalent to MIM products. In another study, Nötzel and Hanemann [41] used Polyvinyl butyral (PVB) as the backbone and Polyethyleneglycol (PEG 4000) as the plasticizer, which is soluble in water. The best attainable sinter densities for the complex alumina parts like small gear wheels were 98% of the theoretical value, equivalent to ceramic injection molding results.

Furthermore, in the work of Eickhoff et al. [42], different combinations of LDPE, high-density polyethylene (HDPE), and wax were tried to make Ti6Al4V parts by FFF. It has been revealed that the mechanical and rheological properties may be modified for most of these formulations, which utilize varying grades of polyethylene (PE) as the primary component. Nevertheless, PP is a better candidate than PE as the backbone for metals with low sintering temperatures since PP degrades at lower temperatures than PE.

Aluminium (Al) and its alloys offer several benefits as superior materials in sophisticated and high-tech applications. Increasing demands for high thermal and electrical conductivity [43], low component weight [44], and outstanding specific mechanical properties [45] in the automotive, aerospace, electronics, and military industries have resulted in a surge in the use of Al products [46]. Nevertheless, processing Al powder by SDS is not very common, possibly due to difficulties encountered during the sintering process. One of the challenges is that the sintering temperature of Al is relatively near to the degradation temperature of most polymers. However, certain researchers have attained tremendous achievements. Several studies have been done on the MIM of Al and its alloys [47]. For example, Abdoos et al. [48] investigated the rheological behavior of Al feedstocks for injection molding. The binder system contains PP, paraffin wax (PW), and stearic acid (SA) in various percentages. The results showed that the proper binder system is 60, 35, and 5 vol.% of PW, PP, and SA, respectively.

Furthermore, the binder with extra PW (65 vol.%) had an unstable flow behavior. The main challenge to directly applying MIM Al feedstocks to FFF is that the high wax content makes the filaments unprintable. Therefore, new binder formulations are needed. One way to tailor the formulations is to select an appropriate polyolefin backbone, which was one of the aims of this investigation.

Although polyolefins have been used for the backbones in FFF binder systems for metals [32], there is no investigation to analyze the effects of the type of backbone on the feedstock properties by using a wax and a higher molecular weight PP (i.e., PP polymer vs. PP_{wax}) and grafting. There is one study where a MIM Al feedstock was processed in a screw-based material extrusion additive manufacturing [49], but to our knowledge, there is no study on Al feedstock used in the FFF process in the literature. Additionally, some research has been done on the use of grafted backbones in which a beneficial increase in the polymer-powder interaction was observed [50]. For this reason, grafted polyolefins have been considered in this paper. Therefore, in this study, the rheological behavior, and mechanical and thermal properties of Al feedstocks for FFF were investigated using different PP as backbones to obtain a feedstock that can be reliably used to fabricate specimens in a low-cost FFF printer.

2. Materials and Experimental Methods

2.1. Materials

This study used an Al alloy powder with spherical particles and $d_{90} < 32$ µm to produce FFF feedstocks. The powder was supplied from TLS Technik Spezialpulver KG (Bitterfeld-Wolfen, Germany). It was referred to as AlSi1—three multi-component binder systems with a fixed soluble component and different PP grades as the backbone were utilized. The different grades of modified and unmodified PP consisted of (i) general application PP (Borealis AG, Vienna, Austria) with a density of 0.95 g/cm^3, (ii) grafted-maleic anhydride-PP (PPMA) (BYK-Chemie GmbH, Wesel, Germany) with the density of 0.935 g/cm^3 and, (iii) grafted-maleic anhydride-PP_{wax} ($PPMA_{wax}$) (Clariant AG, Oberhausen, Germany) with the density of 0.93 g/cm^3. A thermoplastic elastomer (TPE) (KRAIBURG TPE GmbH & Co. KG, Waldkraiburg, Germany) with a density of 0.944 g/cm^3 and shore A hardness of 60 was used as the main binder component. According to the melt flow index (MFI) test at 190 °C with a load of 2.16 kg for three repetitions, the MFI of PP and PPMA were 2.27 ± 0.02 g/10 min and 6.37 ± 0.02 g/10 min, respectively. The viscosity of $PPMA_{wax}$ is given by the manufacturer as 0.8–1.4 Pa·s.

The binder and feedstock were mixed in an internal mixer with a chamber volume of 38 cm^3 and counter-rotating roller rotors (Plasti-Corder PL2000, Brabender GmbH & Co. KG, Duisburg, Germany) at the temperature of 200 °C and a rotational speed of 60 rpm.

The mixing was done in an air atmosphere for 45 min to ensure adequate dispersion of powder particles in the binder system, which can be indicated by a constant torque value at the end of blending. For comparison, all the pure materials were processed in the same condition for 45 min. After mixing, the feedstocks were cooled and granulated at room temperature in a cutting mill (SM200, Retsch GmbH, Haan, Germany). Compounded samples were prepared in the ratio where the TPE consists of more than 50 vol.% of the binder composition. High TPE content keeps the filament flexible. After solvent debinding minimizes the amount of polymer that needs to be removed before sintering, which speeds up the thermal debinding process, saving time and energy.

2.2. *Thermal Characterization*

The melting and cooling behavior for the binder systems and feedstocks was characterised using a differential scanning calorimeter (DSC) (Mettler Toledo GmbH, Greifensee, Switzerland) based on ISO 11357. Measurements (3 replicates for each binder and feedstock) were carried out in the temperature range from 25 to 250 °C with a heating rate of 10 K/min and a cooling rate of 20 K/min under nitrogen gas (gas flow rate: 50 mL/min). Melting temperature, crystallization temperature, and degree of crystallinity for binder systems were measured. The degree of crystallinity in each virgin polymer and binder system was calculated according to Equation (1) [51].

$$X_c = \frac{\Delta H_m}{\Delta H_m^0 * w} * 100 \quad (1)$$

where ΔH_m is the enthalpy of melting, ΔH_c is the enthalpy of crystallization, ΔH_m^0 is the enthalpy of 100% crystalized PP, which is 207 J/g [52], and w is the weight fraction of PP in the samples.

For analyzing the thermal stability of binder systems and feedstocks, thermogravimetric analysis (TGA) (Mettler Toledo GmbH, Greifensee, Switzerland) was performed based on ISO 11358. The test was executed to explore the variation of sample weight in wt.% versus temperature in an inert (N_2) atmosphere. One test was done for each sample, and all the samples were heated from room temperature to 800 °C with a heating rate of 10 K/min.

2.3. *Rheological Measurements*

It is also essential to evaluate the rheological characteristics of the feedstocks at high shear rates because the FFF process occurs in this range. Therefore, the high-pressure capillary rheometer Rheograph 2002 (GÖTTFERT Werkstoff-Prüfmaschinen GmbH, Buchen, Germany) was used to investigate the rheological behavior of the feedstocks. The apparent shear viscosity based on ISO 11443 was measured at 200 °C in the apparent shear rate range of 100–1000 s^{-1}; using a die with a diameter and length of 1 and 30 mm, respectively. Three repetitions were done for all samples, and the mean values were reported.

To make specimens for the rotational rheology, disks were manufactured using the hydraulic vacuum press P200PV (COLLIN Lab & Pilot Solutions GmbH, Maitenbeth, Germany) with a pressure of 50 bar at 200 °C. The disks were produced with a diameter of 25 mm and a thickness of about 1 mm. The feedstock sample was preheated between 15 and 20 min before pressing, and the cooling time was fixed at 15 min.

The rheological properties of binder systems and feedstocks at low shear rates were measured using a rotational rheometer MCR 702 MultiDrive (Anton Paar GmbH, Graz, Austria). The experiments were carried out in parallel plate configuration at 200 °C in an inert (N_2) atmosphere with a plate diameter of 25 mm. Strain sweeps were carried out in the strain range (γ) from 0.01 to 10% at a fixed frequency of 1 rad/s to determine the linear viscoelastic region in the samples. Therefore, the frequency sweep tests according to ISO 6721 with the angular frequency range from 0.1 to 500 rad/s for the binder systems and feedstocks were done with γ = 1% and γ = 0.01%, respectively. All the samples had the same recent thermal history. At least three replicates were performed on fresh samples. For better comparison, virgin grades of PP and TPE were also examined.

2.4. Microscopy and Infrared Spectroscopy

Feedstock micromorphology was analyzed using scanning electron microscopy (field emission gun SEM, Zeiss Leo 1525, Carl Zeiss Microscopy, Jena, Germany) using backscattered electrons (15 kV source voltages). The compression molded specimens were mechanically ground and polished with the final step using a suspension of 0.1 μm alumina particles.

The Fourier transform infrared (FTIR) is one of the essential measurements to understand the internal interactions between materials. The FTIR tests used a VERTEX 70v FT-IR spectrometer (Bruker Optics GmbH & Co. KG, Ettlingen, Germany) to analyze the interaction between polymers in the binders and the Al fillers.

2.5. Tensile Testing

Furthermore, filaments were made in the high-pressure capillary rheometer with a die of 1.75 mm diameter and 30 mm length. The filament-making was done at 200 °C and a constant shear rate of 200 s^{-1}. In the next step, the tensile properties of the filaments were measured according to ISO527 using a Zwick Z001 machine with a 1 kN load cell from ZwickRoell GmbH & Co. KG (Ulm, Germany) at room temperature. The initial gauge length and displacement rate were 50 mm and 10 mm/min, respectively. The mechanical properties results were averaged over five samples of each composition.

2.6. FFF Printing Trials

FFF printing was done on a Duplicator i3 v2 (Wanhao, Jinhua, China) machine with a 0.6 mm nozzle diameter. The printed geometry was a rectangular prism with a honeycomb infill at 40% and the dimension of 20 mm × 20 mm × 4 mm. The specimens (3 parts for each feedstock) were printed at a nozzle temperature and a building platform temperature of 240 °C and 100 °C, respectively, to ensure proper extrusion and build platform adhesion. The printing speed was kept at a maximum of 10 mm/min, and the retraction was reduced significantly to avoid excessive shearing of the filament. Furthermore, a low printing speed results in a low volume flow, thus a lower force acting on the filament. Hence, a more stable printing process can be achieved.

3. Results and Discussion

3.1. Thermal Characterization

3.1.1. DSC Analysis

DSC was used to investigate the thermal properties and the degree of crystallinity. The thermographs of pure polymers, binder systems, and feedstocks are depicted in Figure 1, and the thermal properties are reported in Table 1. T_m is the melting temperature, T_c is the crystallization temperature, and X_c is the degree of crystallinity calculated using Equation (1). Processed PP shows the highest T_m, basically around 164 °C. Processed PPMA shows nearly the same T_m as PP. Furthermore, the difference between melting enthalpies of PP and PPMA is not significant. The primary polymer in the binder systems is TPE. Therefore, it would have a big influence on the thermal properties of binder systems. Pure TPE shows a small exothermic peak around 75 °C (Figure 1, inset), which is assumed to be the melting temperature of hard segments of the TPE molecular structure. It is assumed that the TPE is mainly composed of hard and soft segments, each of which provides different properties. The hard segment provides processability, while the soft segment provides elasticity and flexibility. Because the melting enthalpy of TPE is very small, we believe that the melting and crystalline properties of the blends based on TPE and PP would be mainly influenced by the type of PP in the system. After 45 min of processing, TPE at 200 °C the melting peak completely disappeared. By blending TPE, T_m, and the melting enthalpies were reduced.

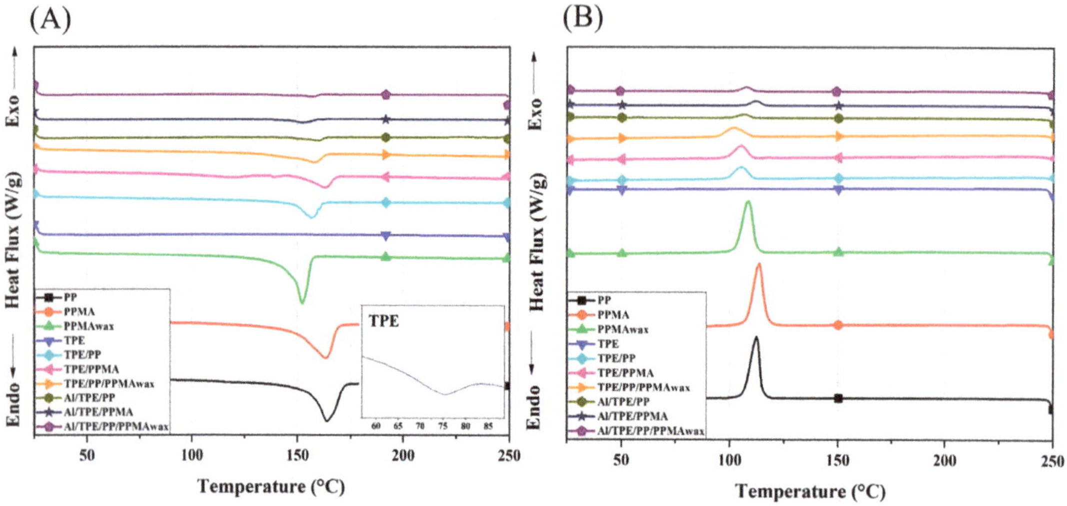

Figure 1. The average DSC curves of polymers, binder systems, and feedstocks, (**A**) second run of melting behavior, (**B**) cooling behavior of samples.

Table 1. The average thermal properties of the binder systems and components.

Sample	Heating		Cooling		
	T_m in °C	ΔH_m in J/g	T_c [°C]	ΔH_c [J/g]	X_c [%]
PP	164.30 ± 1.01	65.52 ± 0.15	112.20 ± 0.78	67.02 ± 0.64	31.65 ± 0.57
PPMA	163.24 ± 0.57	76.58 ± 0.96	113.60 ± 0.43	71.07 ± 0.41	37.00 ± 0.68
TPE	75.10 ± 0.89	0.60 ± 0.12	-	-	-
TPE/PP	161.11 ± 1.51	21.39 ± 0.51	105.15 ± 1.64	22.88 ± 0.65	2.03 ± 0.08
TPE/PPMA	156.29 ± 1.25	21.62 ± 0.67	104.65 ± 1.34	25.45 ± 0.31	5.28 ± 0.01
TPE/PP/$PPMA_{wax}$	156.41 ± 1.78	22.19 ± 0.75	101.50 ± 1.86	23.97 ± 0.80	2.43 ± 0.10
Al/TPE/PP	160.60 ± 1.84	5.72 ± 0.68	106.62 ± 2.06	5.83 ± 0.26	0.76 ± 0.05
Al/TPE/PPMA	153.00 ± 1.68	5.56 ± 0.89	111.77 ± 1.45	6.72 ± 0.13	8.00 ± 0.15
Al/TPE/PP/$PPMA_{wax}$	156.20 ± 1.96	5.02 ± 0.98	107.36 ± 2.64	6.67 ± 0.68	11.38 ± 0.02

As a comparison between TPE/PP and TPE/PP/$PPMA_{wax}$, the reduction in T_m was bigger due to the presence of wax chains that can move and slide easily. Furthermore, the ease in chain movement would enhance the ability of the chain to move and stack on each other and increase the degree of crystallinity [53]. TPE/PPMA sample shows a big reduction in T_m. This would indicate the presence of some chemical interaction among PPMA and TPE chains, which further hinders molecular movements and affects the melting temperature.

Regarding feedstocks, adding Al alloy powder to each binder system decreases T_m, mostly due to the presence of metallic particles that impede the movement of the polymer chain. The presence of Al particles also reduces the melting enthalpy compared to the polymeric binder system.

Regarding the crystallization behavior of samples (Figure 1B), PPMA shows the highest T_c, around 114 °C, resulting from easily packing of PPMA chains and forming a crystalline structure. Adding TPE to PP in binder systems reduces the crystallization temperature, meaning PP chains need more energy reduction to fold on each other and form a crystalline structure. In comparison to TPE/PP binder system, TEP/PP/$PPMA_{wax}$ has a higher T_c, which is mainly by the addition of low molecular weight wax to the system that increases chains movement and makes it easier for chains to fold on each other and make crystalline structures. The same behavior is also observed in TPE/PPMA binder system.

Regarding the degree of crystallinity, by blending TPE with different types of PP, the crystallinity in binder systems reduces significantly. This is mainly due to the higher content of non-crystalline TPE. Furthermore, by adding $PPMA_{wax}$ to the TPE/PP binder system, although the content of PP was lower, a higher degree of crystallinity is observed, which is also due to a higher growth rate in this system because of the shorter wax chains [54].

In feedstocks, by adding Al alloy particles, the T_c for all the samples increased. It can result from the solid metallic particles, which act as a nucleating agent and further would increase the degree of crystallinity. As reported in Table 1, the crystallinity of Al/TPE/PPMA and Al/TPE/PP/$PPMA_{wax}$ increased. However, in Al/TPE/PP system, a reduction in X_c was observed, which should be investigated further since no clear explanation is available. The degree of crystallinity for feedstocks would directly affect mechanical strength, which is addressed in the following sections.

3.1.2. TGA Analysis

Thermal degradation of the polymeric binder system is important in the FFF process of sensitive alloys, particularly Al, because according to [55], the beginning stages of sintering for Al alloys happen at temperatures around 550 °C and most binder polymers degrade up to these temperatures. TGA was used to investigate the degradation further, and the results are presented in Figure 2.

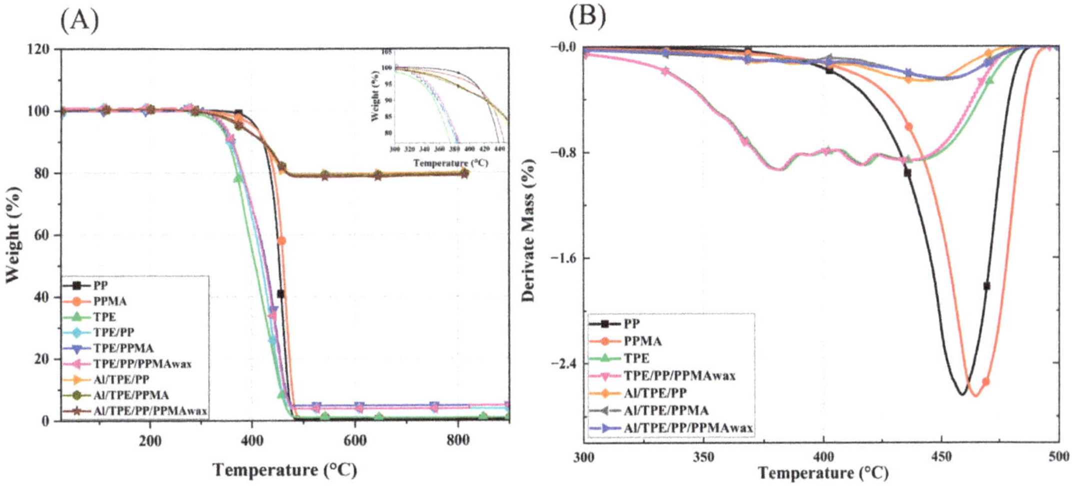

Figure 2. (**A**) TGA representative curve, and (**B**) dTGA curves of processed polymers, binder systems, and feedstocks.

Processed PP has a thermal degradation onset temperature of around 340 °C. PPMA has a lower onset of thermal degradation (around 310 °C), probably due to 1 wt.% maleic anhydride modification in this material. The rate of thermal degradation of the polyolefin backbone can be considered the same. The degradation behavior of binder systems is mainly the same as TPE, with the onset of degradation temperature around 270 °C (the insert of Figure 2). The decrement in degradation temperature is beneficial to decreasing defects and contamination during thermal debinding and sintering.

Furthermore, the degradation rate of processed TPE and binder systems is lower than that of polyolefins. Processed PP and PPMA show an abrupt weight loss in a limited temperature window. By adding TPE to PP and PPMA, the rate of degradation reduces a little bit, and the rate becomes more moderate. Furthermore, the degradation of all binder systems and polymers terminates at around 500 °C. In conclusion, all the binder systems showed good thermal degradation behavior, which is essential for the SDS of Al feedstocks.

To further investigate the effect of the presence of metal particles on the binder system, TGA analyses were also conducted on the feedstocks. The onset of degradation for all feedstocks is around 300 °C. All samples (binder system and feedstock) containing PPMA showed lower thermal degradation temperatures. The degradation rate of feedstocks is smaller than that of binder systems, which is attributed to the presence of Al alloy particles that hinder thermal degradation and slow down the degradation process. Also, in the work of Bek et al. [56], MA chemically reacts with Al particles, which could be why the degradation is slowed down. To further investigate the thermal degradation of samples, dTGA data are depicted in Figure 2B. All the binder systems have the same degradation behavior as TPE. A two-step degradation peak is attributed to the degradation of each soft and hard segment present in TPE. We believe that due to the higher content of TPE in the binder system, the degradation behavior is more affected by the presence of TPE. However, the feedstocks show completely different behavior. The degradation happens in a one-step and at nearly the same temperature as PP and PPMA. Feedstocks degraded in a smaller temperature window which shows the suitability of these binder systems in the SDS process of Al.

3.2. *Rheological Characterization*

3.2.1. High-Pressure Capillary Rheometer

Analyzing the viscosity is very important for FFF feedstocks because it is impossible to print feedstocks with very high viscosity [57]. On the other hand, making printable filaments using low-viscosity feedstocks is impossible since low-viscosity components (mainly waxes) tend to make very fragile filaments, which break during the extrusion process at the printing head.

A high-pressure capillary rheometer (HPCR) followed the flow characteristic of binder systems and feedstocks because, in FFF and PIM processes, the material is forced to flow through a narrow nozzle at high shear rates [58]. The average HPCR results of pure polymer and binder systems are depicted in Figure 3A. All samples show shear-thinning behavior, and the processed TPE shows the highest viscosity among the other polymers. Processed PP and PPMA show nearly the same flow properties and lower viscosity than TPE due to the presence of more free volume, making it easier for polymer chains to slide on each other even at high shear rates. As can be observed in Figure 3, the shear viscosity of TPE/PP is roughly the same as TPE, and TPE/PPMA has an even lower viscosity, which shows that at high shear rates, the binder systems flow more easily than the single binder ingredients. TPE/PP/$PPMA_{wax}$ shows relatively low viscosity among all samples. $PPMA_{wax}$ has a very low viscosity, and its viscosity was not measurable in a capillary rheometer at 200 °C.

The shear viscosity of the feedstocks is depicted in Figure 3B. By adding 55 vol.% Al alloy particles to binder systems, an increment in shear viscosity in comparison to the binder systems is seen, on average 2, 3, and 4.7 times for TPE/PP, TPE/PPMA, and TPE/PP/$PPMA_{wax}$, respectively. The lower viscosity in Al/TPE/PP/$PPMA_{wax}$ is due to the presence of the wax. However, the increment in viscosity in samples containing maleic anhydride-modified PP is higher, which can indicate some chemical interactions between Al alloy particles and maleic anhydride groups on the PP backbone. Therefore, we further analyzed the probable interactions with Fourier transform infrared spectroscopy (FTIR).

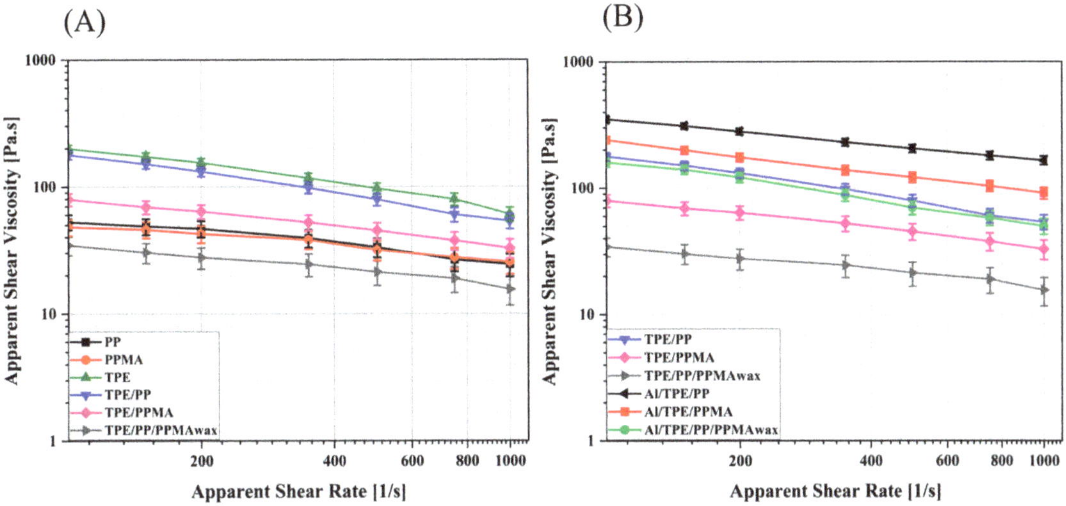

Figure 3. Average HPCR results, (**A**) binder systems and components, (**B**) binder systems and feedstocks.

3.2.2. Rotational Rheology

Investigating the rheological property with rotational rheology is prominent in investigating the structure and interactions at lower frequencies or the flow behavior and processability at higher frequencies. To gain a deeper understanding of the interactions and flow behavior, rotational analysis was done on processed polymers, binder systems, and feedstocks, and the results are depicted in Figures 4 and 5. Processed PP and PPMA show a plateau at lower angular frequencies followed by shear-thinning behavior at higher frequencies. In the range of rheological measurement, PPMA has a higher complex viscosity. PP and PPMA show the typical flow behavior of uncrosslinked thermoplastics (the rheological behavior of pure materials can be seen in the Appendix A—Figure A1). However, processed TPE shows a different behavior. There is no plateau; only shear thinning is observed in the measured angular frequency range.

In the binder systems, similar flow behavior is observed. At all frequencies, all the binder systems have a flow behavior similar to pure TPE but with lower viscosity values compared to TPE and a smaller shear-thinning slope. At all frequencies, TPE/PP shows a higher viscosity. At the medium to high frequencies, TPE/PPMA has a viscosity between TPE/PP and TPE/PP/$PPMA_{wax}$. However, TPE/PP/$PPMA_{wax}$ shows higher viscosity than TPE/PPMA at low frequencies. This is mainly due to the lower viscosity of $PPMA_{wax}$.

In the case of feedstocks, the complex viscosities of samples are higher than binder systems, mainly at low frequencies (or shear rates) due to the presence of a high amount of metallic powder. This behavior is because, at low frequencies, the particles are more prone to make strong networks. In highly-filled composites, the particle-particle interactions have the main effect on the viscoelastic properties. The viscosity upturn in Al/TPE/PPMA and Al/TPE/PP/$PPMA_{wax}$ indicates network formation in the sample, which will be confirmed with scanning electron microscopy analysis (SEM) (see Section 3.3). Furthermore, in the work of Bek et al. [56], it is reported that there is a chemical interaction between polymer chains and Al particles. A higher binding energy was observed in modified polymers, which means stronger adhesion of particles to the polymer backbone, which would hinder the polymer chain movements and lead to an increment in viscosity. Therefore, in the next sections, possible interactions in our system will be analyzed by FTIR.

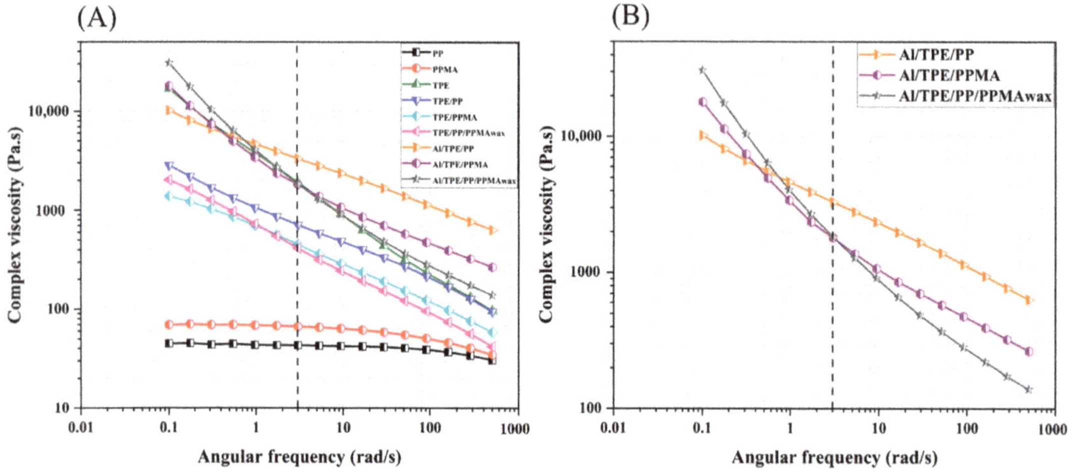

Figure 4. The mean values of complex viscosity, (**A**) processed polymers and binder systems, (**B**) feedstocks.

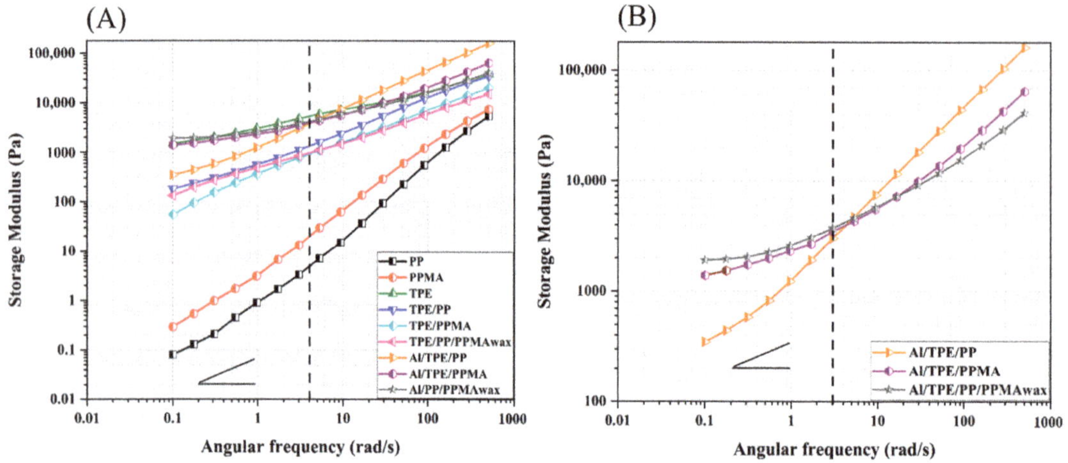

Figure 5. The mean values of storage modulus, (**A**) processed polymers and binder systems, (**B**) feedstocks.

At low frequencies, the difference between the complex viscosity of each feedstock and its binder system is bigger. This difference is constant for TPE/PP sample, but for other systems, the difference decreases with the frequency, which could be related to the presence of agglomerates in Al/TPE/PP. On the contrary, when particles become wet with the appropriate binder system, the particle-polymer interactions increase. This phenomenon is called network formation, which leads to viscosity upturn at lower frequencies. Furthermore, at high frequencies, there is a network breakup, which here is more pronounced in Al/TPE/PP/$PPMA_{wax}$ and resulted in more shear-thinning behavior than Al/TPE/PP and Al/TPE/PPMA.

Roshchupkin et al. [59] reviewed the main ideas for the FFF of metals. They pointed out that the elasticity of melt and flexibility in solid and melt states are two crucial factors for binder systems in the FFF process. Hence, the dependency of storage modulus (G′) on frequency is a useful parameter to evaluate the elastic behavior of polymer melts. Among all processed samples, PP shows the lowest elasticity. The highest elasticity is observed in TPE due to its quasi-elastomeric nature that arises due to its non-covalent interactions between the soft and hard segments (Figure A2 in Appendix A). Furthermore, at low frequencies, the soft segments of TPE would not let G' to decrease very much, and even at lower frequencies ($\omega < 0.1$ rad/s), the flow behavior is like ideal elastic, where G' is independent of frequency.

About the binder systems, the elastic behavior at every frequency is mainly similar to processed TPE. PP and PPMA show terminal behavior, while adding TPE to them, non-terminal behavior is observed, and the curves are shifted to a higher storage modulus. Among all the binder systems, TPE/PP shows the highest G' at almost all frequencies. TPE/PPMA and TPE/PP/PPMA$_{wax}$ have almost the same dependency of elasticity to frequency, except for some deviation at low frequencies in TPE/PP/PPMA$_{wax}$ which further decreases the slope or the dependency of elasticity to frequency. Again, this behavior can be attributed to the presence of PPMA$_{wax}$.

By adding particles, the storage modulus increased, and the behavior of feedstocks is nearly the same as processed TPE, except for Al/TPE/PP sample. More particle-polymer interactions, which result from better particle dispersion and increment in adhesion, have an actual effect on the value of G' (Figure 5). The independency of the storage modulus of frequency, mainly at low frequencies ($\omega < 1$ rad/s), is another indicator of network formation. This is further investigated by SEM results of the feedstocks based on TPE and various PP. The absorption of polymer chains on particles is more pronounced in Al/TPE/PPMA and Al/TPE/PP/PPMA$_{wax}$ feedstocks, and based on FTIR results, it is believed that it may be due to more chemical bonding between particles and grafted polymer chains. Similar results have been reported by Bek et al. [56] and Cano Cano et al. [50]. Based on rheological results, among these feedstocks, TPE/PPMA and TPE/PP/PPMA$_{wax}$ could be better in the FFF process. However, further investigation of the mechanical properties of feedstock filaments and their printability is needed to confirm this hypothesis.

3.3. Morphological Analysis

To further investigate the micromorphology developed after making the feedstocks and the probable presence of agglomerates and networks in the sample, scanning electron analysis was performed on the compressed molded feedstock samples, and some of the results are depicted in Figure 6. The black and white regions are polymer and Al alloy particles, respectively.

A uniform distribution of the polymer phase is essential in the SDS process because if the polymer phase accumulates in one region, cracks will appear during debinding. As discussed in the rheology section, it was hypothesized that in Al/TPE/PP, agglomeration would be observed and, in both Al/TPE/PPMA and Al/TPE/PP/PPMA$_{wax}$, a better dispersion should be observed (due to the upturn in viscosity and the plateau region in storage modulus results). In Figure 6A, an obvious agglomeration can be detected in Al/TPE/PP, where Al particles are directly in contact with each other (inset). Also, the particles are distributed more unevenly in the Al/TPE/PP sample, which can cause some problems like voids and cracking during debinding due to binder-rich zones. Furthermore, even if debinding could be accomplished, the sintering process might cause the development of large pores, leading to poor mechanical properties. In Al/TPE/PPMA sample (Figure 6B) a proper distribution of the Al particles can be observed. The surface detachment is not observed. As shown in Figure 6B (inset), there is a thin polymer layer around Al particles, which confirms the polymer adhesion to the metal particles that increase the viscosity of feedstocks. The Al/TPE/PP/PPMA$_{wax}$ sample has an intermediate dispersive morphology. Even though there are still binder-rich regions, their sizes are smaller than the zones in the Al/TPE/PP sample, and the phase distribution is not uniform. Furthermore, agglomerates can be seen in the sample (Figure 6C).

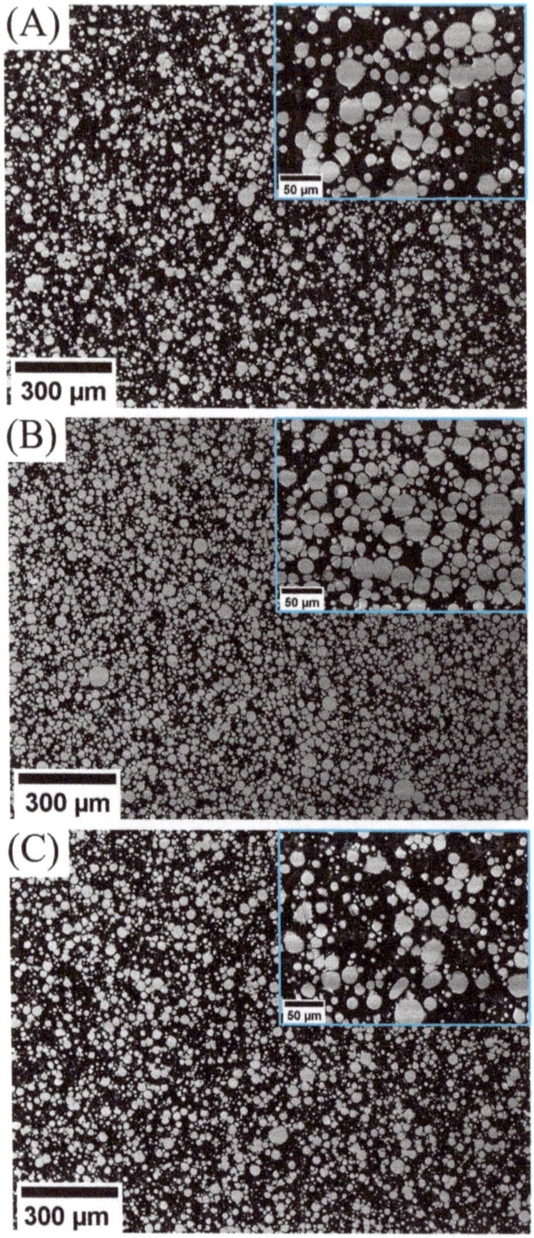

Figure 6. SEM images of (**A**) Al/TPE/PP, (**B**) Al/TPE/PPMA, and (**C**) Al/TPE/PP/PPMA$_{wax}$.

3.4. Fourier-Transform Infrared Spectroscopy (FTIR)

To further investigate the probable chemical interactions in the binder system and feedstocks, Fourier-transform infrared spectrum (FTIR) was used, and some of the results are depicted in Figure 7. The characteristic peaks for TPE are its fingerprints at 690 and 760 cm^{-1}, with peaks around 1730 cm^{-1} attributed to C=O groups. For PPMA also the peak at 1730 cm^{-1} is for C=O and 1210 cm^{-1} for C-O in maleic anhydride. By blending

TPE with PPMA, no new peak appeared, but obviously, the intensities of some peaks were reduced. By the reduction in peaks at 1210 and 1730 cm^{-1}, some chemical interactions can be assumed, which reduced the amount of C-O and C=O groups in the system. For both binder systems, the peak at 1730 cm^{-1} shifts to 1737 cm^{-1}, which would be an actual reason for probable chemical interactions between TPE and PPMA. Different researchers also investigated the probable interaction of the polymer with metallic particles using FTIR. Bek et al. [56] analyzed the AlSi10 Mg surface, and they observed a high amount of Al_2O_3 compared to the elemental Al, which indicates high oxidation of the surface. Oxides on the metal alloy surface can interact with the components of the binder system, particularly with the maleic anhydride (MA). It is reasonable to assume that the MA ring-opened and reacted with the Al filler. A peak shift (associated with Al-O-H vibrations of filler) was observed, indicating the binding of the Al-O surface group located on Al filler material to the binder system and the formation of a covalent bond between Al filler and PPMA.

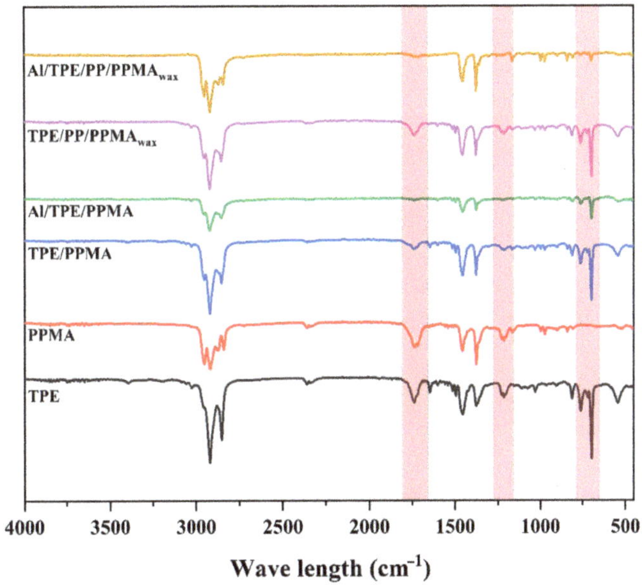

Figure 7. The representative FTIR spectrum of processed polymers, binder systems, and feedstocks.

After preparing the feedstocks, the intensity of fingerprints of TPE lowered, which is due to less TPE in the system, compared to Al particles. However, the characteristic peaks at 1210 and 1730 cm^{-1} almost vanished, indicating chemical interactions between the binder system and Al particles. These results confirm our previous observations in the thermal and rheological sections.

3.5. Mechanical Properties and Printability

To investigate the mechanical properties of the filaments and further produce green parts by 3D printing, the tensile test on filaments and their printability was conducted, and the results are depicted in Figure 8. Some of the printed parts with mean values of the tensile test results can be seen in Figure 8. The mechanical strength of the filaments is crucial because the green part is produced using filaments, and low strength and very brittle material can cause problems during the printing and handling of green parts. Figure 8 shows Al/TPE/PP and Al/TPE/PP/PPMA$_{wax}$ show brittle behavior with low tensile strength around 3.4 and 3.9 MPa, respectively. However, the Al/TPE/PPMA sample shows a 1.6 to 1.8 times improvement compared to the two other feedstocks. This would further justify the suitability of the TPE/PPMA binder system in the FFF process.

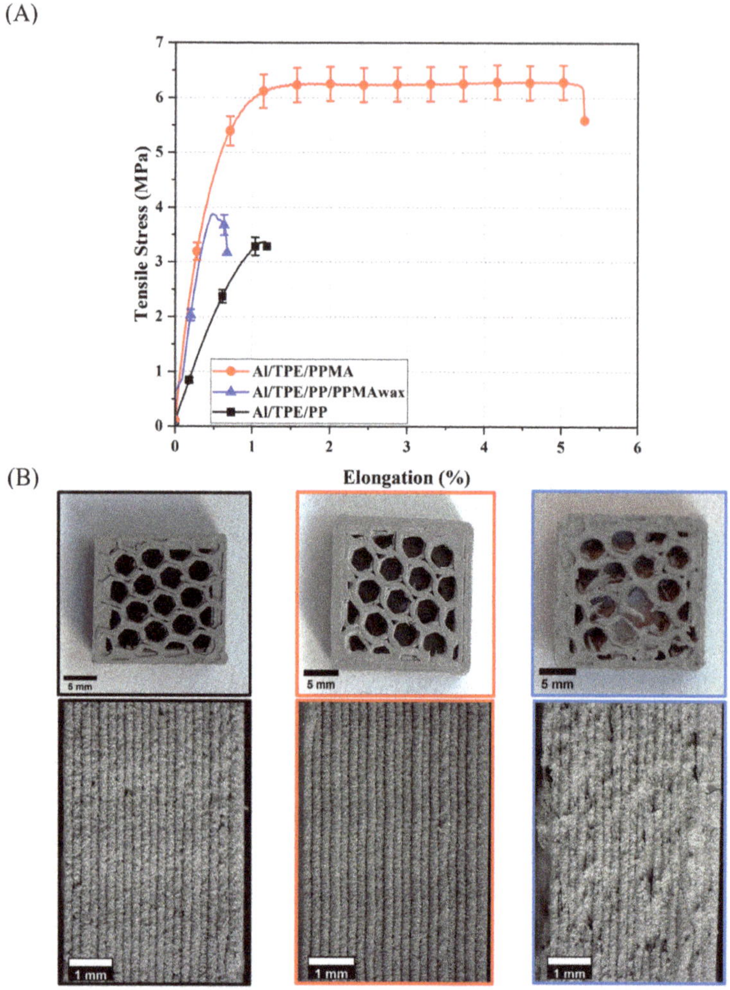

Figure 8. (**A**) The average tensile properties of the filaments, (**B**) pictures of as-printed samples from left to right: Al/TPE/PP, Al/TPE/PPMA, and AL/TPE/PP/PPMA$_{wax}$.

The printed parts can be seen in Figure 8B. For Al/TPE/PP and Al/TPE/PPMA feedstocks, better printability is observed. However, big defects and irregularities are observed in the Al/TPE/PP/PPMAwax printed sample. The lower viscosity at the high printing temperature of 240 °C and the lowest elongation at break in the Al/TPE/PP/PPMAwax feedstock filament makes the extrusion process more unreliable. Hence, the part had the lowest quality (Figure 8B).

4. Conclusions

This study investigated the thermal, mechanical, and rheological properties of the Al feedstock in the FFF process. Different PP-based backbones were used to analyze the impact of backbones in the suitable feedstocks for FFF. According to the DSC results, it can be concluded that the sample with unmodified PP as the backbone has a lower crystallinity value compared to the samples which have maleic anhydride grafted to the PP in the binder systems and feedstocks. In the TGA results, it has been found that samples containing

PPMA show lower degradation temperature, and compared to the PP, PPMA starts to degrade at lower temperatures (around 310 °C) because of the maleic anhydride. From the rheological point of view, all samples show shear-thinning behavior, which is appropriate for the FFF process. In addition, chemical interactions between Al alloy particles and maleic anhydride groups on the PP backbone led to higher changes in viscosity for samples with maleic anhydride-grafted PP. Also, FTIR results showed that there is a possibility of chemical bonding between oxides on the Al particle surface with the maleic anhydride in the binder system. Therefore, TPE/PPMA and TPE/PP/PPMAwax are preferable to the TPE/PP binder system in the FFF process regarding the rheological behavior. However, the filaments of the feedstocks with TPE/PP/PPMAwax as the binder system were brittle, with the lowest elongation at break, as seen in tensile testing results.

Further investigations should be performed to investigate the debinding and sintering behavior of the feedstocks to determine the suitability of the prepared feedstocks to obtain Al alloy specimens. However, this is out of the scope of this investigation.

Author Contributions: Conceptualization, V.M. and Z.S.; methodology, V.M.; software, V.M. and Z.S.; validation, V.M., Z.S. and J.G.-G.; investigation, V.M., Z.S. and J.G.-G.; resources, L.H. and I.D.; writing—original draft preparation, V.M. and Z.S.; writing—review and editing, J.G.-G., C.K., C.H., I.D. and S.S.; supervision, C.H.; project administration, S.S. and C.K.; funding acquisition, S.S. and C.K. All authors have read and agreed to the published version of the manuscript.

Funding: This research was funded by Austrian Research Promotion Agency (FFG), [project name: AlF3, Project No.: 885128].

Institutional Review Board Statement: Not applicable.

Data Availability Statement: The data presented in this study are available on request from the corresponding author. The data are not publicly available due to commercial reasons.

Acknowledgments: The authors especially thank the project partners Fraunhofer Institute for Manufacturing Technology and Advanced Materials (IFAM) and RHP-Technology GmbH.

Conflicts of Interest: The authors declare no conflict of interest.

Appendix A

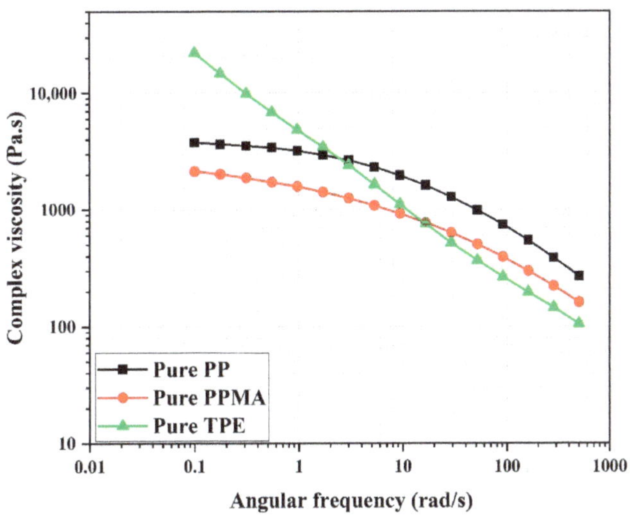

Figure A1. The complex viscosity of the pure binder system components.

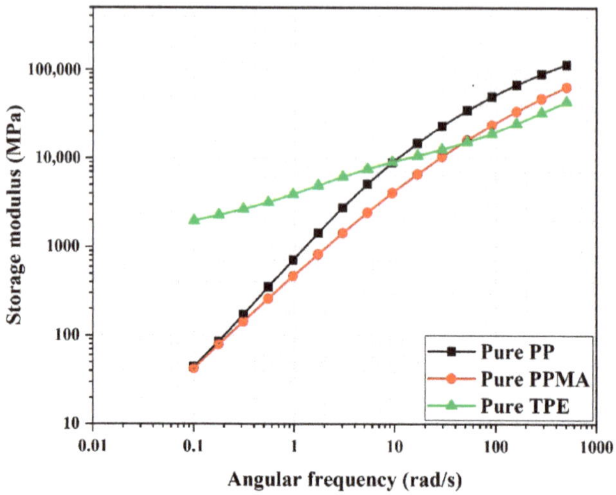

Figure A2. The storage modulus of the pure binder system components.

References

1. Hadian, A.; Fricke, M.; Liersch, A.; Clemens, F. Material extrusion additive manufacturing of zirconia parts using powder injection molding feedstock compositions. *Addit. Manuf.* **2022**, *57*, 102966. [CrossRef]
2. Zhao, Z.; Liu, R.; Chen, J.; Xiong, X. Additive manufacturing of cemented carbide using analogous powder injection molding feedstock. *Int. J. Refract. Met. Hard Mater.* **2023**, *111*, 106095. [CrossRef]
3. Hwang, I.-S.; So, T.-Y.; Lee, D.-H.; Shin, C.-S. Characterization of Mechanical Properties and Grain Size of Stainless Steel 316L via Metal Powder Injection Molding. *Materials* **2023**, *16*, 2144. [CrossRef] [PubMed]
4. Demers, V.; Turenne, S.; Scalzo, O. Impact of binders on viscosity of low-pressure powder injection molded Inconel 718 superalloy. *J. Mater Sci.* **2015**, *50*, 2893–2902. [CrossRef]
5. Momeni, V.; Askari, A.; Allaei, M.H.; Zangi, H. Investigating the Effect of Stearic Acid on the Mechanical, Rheological, and Microstructural Properties of AISI 4605 Feedstock for Metal Injection Molding Process. *Trans. Indian Inst. Met.* **2021**, *74*, 2161–2170. [CrossRef]
6. Fayyaz, A.; Muhamad, N.; Sulong, A.B.; Rajabi, J.; Wong, Y.N. Fabrication of cemented tungsten carbide components by micro-powder injection moulding. *J. Mater. Process. Technol.* **2014**, *214*, 1436–1444. [CrossRef]
7. Naseer, A.; Ahmad, F.; Ali, S.; Haider, W. Powder injection molded nano copper oxide grafted graphene reinforced copper matrix composites. *Powder Technol.* **2022**, *397*, 117101. [CrossRef]
8. Attia, U.M.; Alcock, J.R. Fabrication of hollow, 3D, micro-scale metallic structures by micro-powder injection moulding. *J. Mater. Process. Technol.* **2012**, *212*, 2148–2153. [CrossRef]
9. Momeni, V.; Alaei, M.H.; Askari, A.; Rahimi, A.H.; Nekouee, K. Effect of the Fraction of Steel 4605 Powder in the Load in Injection Molding with the Use of a Polymer-Based Binder. *Met. Sci. Heat Treat* **2020**, *61*, 777–781. [CrossRef]
10. Sidambe, A.T.; Figueroa, I.A.; Hamilton, H.; Todd, I. Metal injection moulding of CP-Ti components for biomedical applications. *J. Mater. Process. Technol.* **2012**, *212*, 1591–1597. [CrossRef]
11. Ghasemi-Mobarakeh, L.; Cano, S.; Momeni, V.; Liu, D.; Duretek, I.; Riess, G.; Kukla, C.; Holzer, C. Effect of Increased Powder-Binder Adhesion by Backbone Grafting on the Properties of Feedstocks for Ceramic Injection Molding. *Polymers* **2022**, *14*, 3653. [CrossRef] [PubMed]
12. Shahrubudin, N.; Lee, T.C.; Ramlan, R. An Overview on 3D Printing Technology: Technological, Materials, and Applications. *Procedia Manuf.* **2019**, *35*, 1286–1296. [CrossRef]
13. Vozárová, M.; Neubauer, E.; Bača, Ľ.; Kitzmantel, M.; Feranc, J.; Trembošová, V.; Peciar, P.; Kritikos, M.; Orlovská, M.; Janek, M.; et al. Preparation of fully dense boron carbide ceramics by Fused Filament Fabrication (FFF). *J. Eur. Ceram. Soc.* **2023**, *43*, 1751–1761. [CrossRef]
14. Momeni, V.; Zangi, H.; Allaei, M.H. Effect of polypropylene as the backbone of MIM feedstock on the micro-structural phase constituents, mechanical and rheological properties of 4605 low alloy steel compacts. *Powder Metall.* **2020**, *63*, 27–34. [CrossRef]
15. Gibson, M.A.; Mykulowycz, N.M.; Shim, J.; Fontana, R.R.; Schmitt, P.; Roberts, A.; Ketkaew, J.; Shao, L.; Chen, W.; Bordeenithikasem, P.; et al. 3D printing metals like thermoplastics: Fused filament fabrication of metallic glasses. *Mater. Today* **2018**, *21*, 697–702. [CrossRef]
16. Momeni, V.; Hufnagl, M.; Shahroodi, Z.; Gonzalez-Gutierrez, J.; Schuschnigg, S.; Kukla, C.; Holzer, C. Research Progress on Low-Pressure Powder Injection Molding. *Materials* **2022**, *16*, 379. [CrossRef] [PubMed]

17. Atatreh, S.; Alyammahi, M.S.; Vasilyan, H.; Alkindi, T.; Susantyoko, R.A. Evaluation of the infill design on the tensile properties of metal parts produced by fused filament fabrication. *Results Eng.* **2023**, *17*, 100954. [CrossRef]
18. Ultrafuse FFF, Metal—Ultrafuse FFF. 2023. Available online: https://www.ultrafusefff.com/product-category/metal/ (accessed on 20 February 2023).
19. Zetamix, Filaments—Zetamix. 2023. Available online: https://zetamix.fr/en/category/filaments-en (accessed on 20 February 2023).
20. Green Polymer Additives, 3D Printing | Green Polymer Additives | Emery Oleochemicals. 2023. Available online: https://greenpolymeradditives.emeryoleo.com/3dprinting/ (accessed on 20 February 2023).
21. Lengauer, W.; Duretek, I.; Fürst, M.; Schwarz, V.; Gonzalez-Gutierrez, J.; Schuschnigg, S.; Kukla, C.; Kitzmantel, M.; Neubauer, E.; Lieberwirth, C.; et al. Fabrication and properties of extrusion-based 3D-printed hardmetal and cermet components. *Int. J. Refract. Met. Hard Mater.* **2019**, *82*, 141–149. [CrossRef]
22. Masood, S.H.; Song, W.Q. Thermal characteristics of a new metal/polymer material for FDM rapid prototyping process. *Assem. Autom.* **2005**, *25*, 309–315. [CrossRef]
23. Zhang, Y.; Poli, L.; Garratt, E.; Foster, S.; Roch, A. Utilizing Fused Filament Fabrication for Printing Iron Cores for Electrical Devices. *3D Print. Addit. Manuf.* **2020**, *7*, 279–287. [CrossRef]
24. Damon, J.; Dietrich, S.; Gorantla, S.; Popp, U.; Okolo, B.; Schulze, V. Process porosity and mechanical performance of fused filament fabricated 316L stainless steel. *RPJ* **2019**, *27*, 1319–1327. [CrossRef]
25. Liu, B.; Wang, Y.; Lin, Z.; Zhang, T. Creating metal parts by Fused Deposition Modeling and Sintering. *Mater. Lett.* **2020**, *263*, 127252. [CrossRef]
26. Zhang, Y.; Bai, S.; Riede, M.; Garratt, E.; Roch, A. A comprehensive study on Fused Filament Fabrication of Ti-6Al-4V structures. *Addit. Manuf.* **2020**, *34*, 101256. [CrossRef]
27. Singh, P.; Balla, V.K.; Tofangchi, A.; Atre, S.V.; Kate, K.H. Printability studies of Ti-6Al-4V by metal fused filament fabrication (MF3). *Int. J. Refract. Met. Hard Mater.* **2020**, *91*, 105249. [CrossRef]
28. Bose, A.; Schuh, C.A.; Tobia, J.C.; Tuncer, N.; Mykulowycz, N.M.; Preston, A.; Barbati, A.C.; Kernan, B.; Gibson, M.A.; Krause, D.; et al. Traditional and additive manufacturing of a new Tungsten heavy alloy alternative. *Int. J. Refract. Met. Hard Mater.* **2018**, *73*, 22–28. [CrossRef]
29. Carreira, P.; Cerejo, F.; Alves, N.; Vieira, M.T. In Search of the Optimal Conditions to Process Shape Memory Alloys (NiTi) Using Fused Filament Fabrication (FFF). *Materials* **2020**, *13*, 4718. [CrossRef]
30. Wolff, M.; Mesterknecht, T.; Bals, A.; Ebel, T.; Willumeit-Römer, R. *FFF of Mg-Alloys for Biomedical Application*; Springer International Publishing: Berlin/Heidelberg, Germany, 2019; pp. 43–49.
31. Schaper, J.G.; Wolff, M.; Wiese, B.; Ebel, T.; Willumeit-Römer, R. Powder metal injection moulding and heat treatment of AZ81 Mg alloy. *J. Mater. Process. Technol.* **2019**, *267*, 241–246. [CrossRef]
32. Wagner, M.A.; Hadian, A.; Sebastian, T.; Clemens, F.; Schweizer, T.; Rodriguez-Arbaizar, M.; Carreño-Morelli, E.; Spolenak, R. Fused filament fabrication of stainless steel structures—From binder development to sintered properties. *Addit. Manuf.* **2022**, *49*, 102472. [CrossRef]
33. Momeni, V.; Zangi, H.; Alaei, M.H. Effect of thermal debinding and sintering parameters on the mechanical properties of 4605 MIM compact using the RSM. *Adv. Mater. Process. Technol.* **2022**, *8*, 3199–3214. [CrossRef]
34. Rolere, S.; Soupremanien, U.; Bohnke, M.; Dalmasso, M.; Delafosse, C.; Laucournet, R. New insights on the porous network created during solvent debinding of powder injection-molded (PIM) parts, and its influence on the thermal debinding efficiency. *J. Mater. Process. Technol.* **2021**, *24*, 117163. [CrossRef]
35. Suwanpreecha, C.; Manonukul, A. A Review on Material Extrusion Additive Manufacturing of Metal and How It Compares with Metal Injection Moulding. *Metals* **2022**, *12*, 429. [CrossRef]
36. Hausnerova, B. Binder systems for powder injection molding: A review. *Adv. Mater. Proc.* **2017**, *2*, 761–768.
37. Gökmen, U.; Türker, M. An Analysis of Rheological Properties of Inconel 625 Superalloy Feedstocks Formulated with Backbone Binder Polypropylene System for Powder Injection Molding. *Arch. Metall. Mater.* **2017**, *62*, 1937–1944. [CrossRef]
38. Cano, S.; Gonzalez-Gutierrez, J.; Sapkota, J.; Spoerk, M.; Arbeiter, F.; Schuschnigg, S.; Holzer, C.; Kukla, C. Additive manufacturing of zirconia parts by fused filament fabrication and solvent debinding: Selection of binder formulation. *Addit. Manuf.* **2019**, *26*, 117–128. [CrossRef]
39. Gorjan, L.; Galusca, C.; Marwah, S.; Sebastian, T.; Clemens, F. Effect of stearic acid on rheological properties and printability of ethylene vinyl acetate based feedstocks for fused filament fabrication of alumina. *Addit. Manuf.* **2020**, *36*, 101391. [CrossRef]
40. Sadaf, M.; Bragaglia, M.; Nanni, F. A simple route for additive manufacturing of 316L stainless steel via Fused Filament Fabrication. *J. Manuf. Process.* **2021**, *67*, 141–150. [CrossRef]
41. Nötzel, D.; Hanemann, T. New Feedstock System for Fused Filament Fabrication of Sintered Alumina Parts. *Materials* **2020**, *13*, 4461. [CrossRef]
42. Eickhoff, R.; Antusch, S.; Baumgärtner, S.; Nötzel, D.; Hanemann, T. Feedstock Development for Material Extrusion-Based Printing of Ti6Al4V Parts. *Materials* **2022**, *15*, 6442. [CrossRef]
43. Tatar, C.; Özdemir, N. Investigation of thermal conductivity and microstructure of the α-Al_2O_3 particulate reinforced aluminum composites (Al/Al_2O_3-MMC) by powder metallurgy method. *Phys. B Condens. Matter* **2010**, *405*, 896–899. [CrossRef]

44. Gökçe, A.; Fındık, F.; Kurt, A.O. Microstructural examination and properties of premixed Al–Cu–Mg powder metallurgy alloy. *Mater. Charact.* **2011**, *62*, 730–735. [CrossRef]
45. Akhlaghi, F.; Zare-Bidaki, A. Influence of graphite content on the dry sliding and oil impregnated sliding wear behavior of Al 2024–graphite composites produced by in situ powder metallurgy method. *Wear* **2009**, *266*, 37–45. [CrossRef]
46. Rahimian, M.; Ehsani, N.; Parvin, N.; Baharvandi, H.R. The effect of particle size, sintering temperature and sintering time on the properties of Al–Al_2O_3 composites, made by powder metallurgy. *J. Mater. Process. Technol.* **2009**, *209*, 5387–5393. [CrossRef]
47. Liu, Z.Y.; Sercombe, T.B.; Schaffer, G.B. Metal injection moulding of aluminium alloy 6061 with tin. *Powder Metall.* **2008**, *51*, 78–83. [CrossRef]
48. Abdoos, H.; Khorsand, H.; Yousefi, A.A. Torque rheometry and rheological analysis of powder–polymer mixture for aluminum powder injection molding. *Iran Polym. J.* **2014**, *23*, 745–755. [CrossRef]
49. Dayam, S.; Tandon, P.; Priyadarshi, S. Development of paste extrusion-based metal additive manufacturing process. *Rapid Prototyp. J.* **2022**, *28*, 1920–1932. [CrossRef]
50. Cano, S.; Gooneie, A.; Kukla, C.; Rieß, G.; Holzer, C.; Gonzalez-Gutierrez, J. Modification of Interfacial Interactions in Ceramic-Polymer Nanocomposites by Grafting: Morphology and Properties for Powder Injection Molding and Additive Manufacturing. *Appl. Sci.* **2020**, *10*, 1471. [CrossRef]
51. Tarani, E.; Arvanitidis, I.; Christofilos, D.; Bikiaris, D.N.; Chrissafis, K.; Vourlias, G. Calculation of the degree of crystallinity of HDPE/GNPs nanocomposites by using various experimental techniques: A comparative study. *J. Mater Sci.* **2023**, *58*, 1621–1639. [CrossRef]
52. Wunderlich, B. *Thermal Analysis*; Elsevier Science: Saint Louis, MI, USA, 1990.
53. Piorkowska, E.; Rutledge, G.C. *Handbook of Polymer Crystallization*; Wiley: Hoboken, NJ, USA, 2013.
54. Thomas, S.; Arif, P.M.; Gowd, E.B.; Kalarikkal, N. (Eds.) *Crystallization in Multiphase Polymer Systems*; Elsevier: Amsterdam, The Netherlands, 2018.
55. Martı, J.; Castro, F. Liquid phase sintering of P/M aluminium alloys: Effect of processing conditions. *J. Mater. Process. Technol.* **2003**, *143*, 814–821. [CrossRef]
56. Bek, M.; Gonzalez-Gutierrez, J.; Kukla, C.; Črešnar, K.P.; Maroh, B.; Perše, L.S. Rheological Behaviour of Highly Filled Materials for Injection Moulding and Additive Manufacturing: Effect of Particle Material and Loading. *Appl. Sci.* **2020**, *10*, 7993. [CrossRef]
57. Hadian, A.; Koch, L.; Koberg, P.; Sarraf, F.; Liersch, A.; Sebastian, T.; Clemens, F. Material extrusion based additive manufacturing of large zirconia structures using filaments with ethylene vinyl acetate based binder composition. *Addit. Manuf.* **2021**, *47*, 102227. [CrossRef]
58. Strano, M.; Rane, K.; Vangosa, F.B.; Di Landro, L. Extrusion of metal powder-polymer mixtures: Melt rheology and process stability. *J. Mater. Process. Technol.* **2019**, *273*, 116250. [CrossRef]
59. Roshchupkin, S.; Kolesov, A.; Tarakhovskiy, A.; Tishchenko, I. A brief review of main ideas of metal fused filament fabrication. *Mater. Today Proc.* **2021**, *38*, 2063–2067. [CrossRef]

polymers

Article

Investigations on the Fatigue Behavior of 3D-Printed and Thermoformed Polylactic Acid Wrist–Hand Orthoses

Diana Popescu [1], Florin Baciu [2], Daniel Vlăsceanu [2,*], Rodica Marinescu [3] and Dan Lăptoiu [4,*]

[1] Department of Robotics and Production Systems, Faculty of Industrial Engineering and Robotics, University Politehnica of Bucharest, 060042 Bucharest, Romania; diana.popescu@upb.ro
[2] Department of Strength of Materials, Faculty of Industrial Engineering and Robotics, University Politehnica of Bucharest, 060042 Bucharest, Romania; florin.baciu@upb.ro
[3] Department of Orthopedics, Carol Davila University of Medicine and Pharmacy, 050474 Bucharest, Romania; rodica.marinescu@umfcd.ro
[4] Department of Orthopedics, Colentina Clinical Hospital, 020125 Bucharest, Romania
* Correspondence: daniel.vlasceanu@upb.ro (D.V.); dan.laptoiu@upb.ro (D.L.)

Abstract: Additively manufactured wrist–hand orthoses (3DP-WHOs) offer several advantages over traditional splints and casts, but their development based on a patient's 3D scans currently requires advanced engineering skills, while also recording long manufacturing times as they are commonly built in a vertical position. A proposed alternative involves 3D printing the orthoses as a flat model base and then thermoforming them to fit the patient's forearm. This manufacturing approach is faster, cost-effective and allows easier integration of flexible sensors as an example. However, it is unknown whether these flat-shaped 3DP-WHOs offer similar mechanical resistance as the 3D-printed hand-shaped orthoses, with a lack of research in this area being revealed by the literature review. To evaluate the mechanical properties of 3DP-WHOs produced using the two approaches, three-point bending tests and flexural fatigue tests were conducted. The results showed that both types of orthoses had similar stiffness up to 50 N, but the vertically built orthoses failed at a maximum load of 120 N, while the thermoformed orthoses could withstand up to 300 N with no damages observed. The integrity of the thermoformed orthoses was maintained after 2000 cycles at 0.5 Hz and ±2.5 mm displacement. It was observed that the minimum force occurring during fatigue tests was approximately −95 N. After 1100–1200 cycles, it reached −110 N and remained constant. The outcomes of this study are expected to enhance the trust that hand therapists, orthopedists, and patients have in using thermoformable 3DP-WHOs.

Keywords: 3D printing; thermoforming; wrist–hand orthosis; mechanical behavior; customization; flexural fatigue; 3D Printing Point-of-Care

Citation: Popescu, D.; Baciu, F.; Vlăsceanu, D.; Marinescu, R.; Lăptoiu, D. Investigations on the Fatigue Behavior of 3D-Printed and Thermoformed Polylactic Acid Wrist–Hand Orthoses. *Polymers* **2023**, *15*, 2737. https://doi.org/10.3390/polym15122737

Academic Editors: Cristina-Elisabeta Pelin and Anton Ficai

Received: 26 May 2023
Revised: 14 June 2023
Accepted: 17 June 2023
Published: 19 June 2023

1. Introduction

The rate of deployment of Additive Manufacturing (AM) applications in healthcare is continuously increasing alongside the research and scientific literature in the field [1,2]. This is explainable as the AM technology favors customization and design freedom, hence addressing patients' anatomical characteristics and individual needs [3]. Moreover, it enables delocalized manufacturing and a simplified supply chain [4], thus allowing production within hospitals and medical centers—the implementations of 3D Printing Points-of-Care (3DP-POCs) being the current focus [5]. All of these advantages recommend 3DP (the colloquial name for the AM technology, usually also designating the process based on material extrusion—MEX) for the orthotics domain, as a modern manufacturing solution replacing the traditional methods of manually producing upper- and lower-limb static orthoses (splints and casts) [6,7].

Tendonitis, sprains and strains, overuse syndrome, contusions, spasticity or muscle weakness, arthritis, wrist instabilities and fractures [8,9] are among the medical conditions that call for immobilization by splints or casts for short periods of time (days, weeks) or permanently in the case of chronic conditions [6,10]. Therefore, besides functionality, it is mandatory to also consider the patient's comfort and adherence to wear. In this regard, the literature covers the advantages of 3DP-WHOs over traditional fiberglass, thermoplastic or plaster orthoses, such as coolness, lighter weight, recyclable material, less skin irritation and itching, better hygiene, less perspiration and odor reduction [7,11,12], easy and rapid placement and a good fit to the patient limb, while satisfying the clinical requirements of maintaining immobilization [13].

However, despite the benefits and the large availability of MEX equipment (3D printers) and feedstock, the use of 3DP-WHOs is not common in clinical practice. Two recent systematic reviews addressing this field [6,14] evidenced a very limited number of level II studies, randomized and prospective clinical trials [8,9,15] aimed at assessing the efficiency of 3DP-WHOs in comparison to traditional orthoses by using functional criteria and patient feedback (VAS—Visual Analog Scale, QUEST rating—Quebec User Evaluation of Satisfaction with Assistive Technology, JTHFT—Jebsen–Taylor Hand Function Test, PRWE—Patient-Rated Wrist Evaluation). The majority of the studies reviewed in [6,14] are case reports, case series or retrospective case series, all showing good functional outcomes, and patients favoring 3DP-WHOs. At the same time, these studies note the dependence of 3DP-WHOs' production on engineering skills, dedicated 3D scanning equipment and software, as well as the long printing times (e.g., 18–20 h [16] or 6 h in [9]). These aspects hamper 3DP-WHOs' clinical use as most hospitals or healthcare centers lack reverse engineering equipment, dedicated 3D modeling software or 24/7 designer support. In this sense, solutions were proposed to semi-automate the design process [17–19]. There is also the flat-shaped orthoses approach [20,21] that offers an alternative to the long manufacturing time and to the patient 3D data acquisition and processing. Moreover, the 3D modeling process of the patient-tailored orthoses can be replaced with a web-based app that uses a parametric design customized according to the patient anatomic dimensions (see Section 1.1).

Equally as important as implementing a fast and efficient manner of producing bespoke 3DP-WHOs is proving their mechanical resistance during tasks specific to daily activities, thus gaining the confidence of orthopedists, hand therapists and patients. In this sense, the literature review conducted in Section 1.2 made clear the scarcity of studies investigating 3DP-WHOs' mechanical behavior, while the flat 3D-printed and then thermoformed WHOs have never been examined from this standpoint. The current paper focuses on this niche by comparing the mechanical performance of the 3DP-WHOs built in the ready-to-use form (as typical for the 3D scanning or CT/MRI based approaches [22], denoted further as 3DP-WHO1, with the mechanical performance of the flat-printed WHOs, denoted further as 3DP-WHO2, thermoformed to fit the patient hand [21]). Flat-shaped WHOs are easier and faster to produce, and, if necessary, allow the embedding of sensors or therapeutic magnets; the research question to answer is whether they provide a similar mechanical resistance as the orthoses directly 3D-printed in the hand-shaped form. In this study, the 3DP-WHOs were subjected to three-point bending tests corresponding to wrist flexion/extension movements, and comparatively analyzed. Moreover, the durability of thermoformed 3DP-WHOs was assessed by conducting fatigue bending tests.

1.1. 3DP-WHOs' Development Approaches

Any AM process requires the existence of the digital 3D model of the object to be manufactured. In AM-based medical applications, 'reverse engineering' the patient to reconstruct his/her anatomy and then designing the orthoses, prostheses or surgical guides are the steps to follow for patient-tailoring the device or product [13,22]. Once the digital model is available, the suitable material and AM process are selected. AM based on MEX and a powder fusion process (SLS—Selective Laser Sintering) are preferred for orthoses manufacturing, with MEX having the advantages of equipment and feedstock affordability. Other AM processes are also reported in the literature, such as Polyjet, for producing multi-material orthoses, as detailed further below when discussing the literature review results. The most common materials for 3DP-WHOs include polylactic acid (PLA), acrylonitrile butadiene styrene (ABS), or polyamide 12 (PA12).

Regarding the patient data acquisition, three approaches can be currently distinguished, with the type of equipment determining the subsequent development steps (Figure 1): 3D scanning-based; medical imaging-based (CT/MRI); caliper-based.

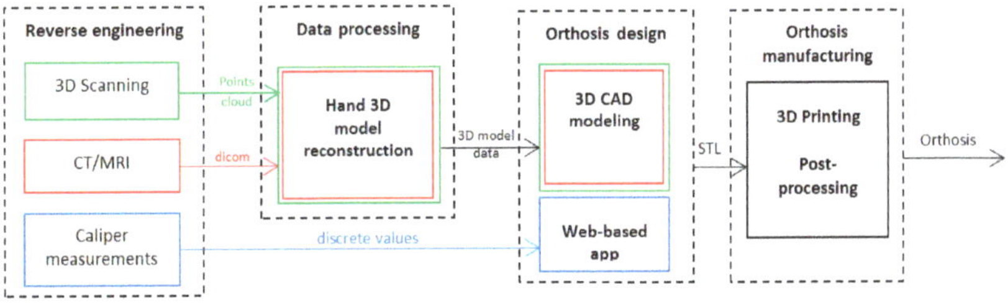

Figure 1. 3DP-WHOs' development flows.

The first two approaches require 3D modeling and medical reconstruction knowledge, and dedicated software and equipment, while the third solution is based on the ability to correctly measure the patient hand and to thermoform the 3DP-WHOs for a good fit on the upper limb, skills which are common for hand specialists. Moreover, in the latter approach, the 3DP-WHO is built in a flat position [21], thus reducing the print time and cost in comparison to the hand-shaped orthoses 3D-printed usually in a vertical position that require support structures for openings or overhang features.

1.2. Literature Review

As mentioned in the introductory section, the recent reviews in the field present relevant information on the state of the art, including measured outcomes related to hand functions, functionality and patient satisfaction while wearing 3DP-WHOs [10], as well as the design approach and software, and production cost and time assessments [6]. However, so far, no data have been gathered on 3DP-WHOs testing for the evaluation of compliance with the mechanical performance criteria (such as bending, fatigue or impact). Therefore, a systematic search in the PubMed, Web of Science and Scopus electronic databases was performed by using the following keywords: ("3D printing" OR "additive manufacturing" OR "material extrusion" or "selective laser sintering") AND (orthosis OR splint OR cast OR brace) AND (hand OR "upper limb" OR wrist) AND (test OR strength OR resistance OR bending OR fatigue OR "finite element" OR "mechanical behavior" OR strength). English language studies up to November 2022 were included. Title and abstract screening were performed by two authors after performing duplicates removal by using Mendeley Desktop 1.19.8 software. The exclusion criteria were the following: non-human studies, dynamic orthoses, spinal cord orthoses, occlusal splints, and studies related to casting manufacturing process. A total of 113 papers were initially selected, 36 papers being kept for a full-text read. Thirteen papers were identified as relevant in providing information on the mechanical behavior of 3DP-WHOs and/or on the use of finite element method (FEM) for investigating mechanical performance.

2. Materials and Methods

Two sets of orthoses were generated and manufactured (3DP-WHO1 and 3DP-WHO2) for this research, each one being typical for a development flow (Figure 1). They were tested by conducting three-point bending tests to investigate their resistance to wrist sagittal movements. Then, the thermoformed 3DP-WHO2s were subjected to flexural fatigue tests, which correspond to wrist flexion and extension. All 3DP-WHOs were made out of polylactic acid—PLA (Devil Design Sp. J., Mikołów, Poland). Acrylo butadiene styrene—ABS (Stratasys Inc., Eden Prairie, MN, USA)—was used for printing the 3DP-

WHOs supports (Mojo 3D Printer, Stratasys Inc., Los Angeles, CA, USA) for the mechanical tests. For the fatigue tests, a bi-material forearm dummy was produced. The 3DP-WHOs and the mold for the dummy were manufactured on a Prusa Replica 3D printer (Prusa Research, Prague, CZ) using the Prusa Slicer 2.5.2 as slicing software.

2.1. Wrist–Hand Orthoses and Forearm Dummy

The 3DP-WHO2 was generated by using the app described in [21] and measuring the key dimensions of the wrist–hand of a male patient (Figure 2a), while the 3DP-WHO1 was based on the 3D hand model of the same person, reconstructed from the CT data (Mimics 10, Materialise NV, Leuven, Belgium). Their thickness was set to 2.3 mm. The 3DP-WHO1 and 3DP-WHO2 models included ventilation pockets with a hexagonal shape, resulting in a 35% weight reduction. A surface of the flat orthosis (the green surface in Figure 2a) was wrapped on the surface extracted from the 3D hand model (as presented in the flow in Figure 2b) by using the CATIA V5 Generative Surface Design workbench (Dassault Systemes, Velizy-Villacoublay, France). This method allowed us to generate similar orthoses in terms of surface area and volume, for a valid comparison.

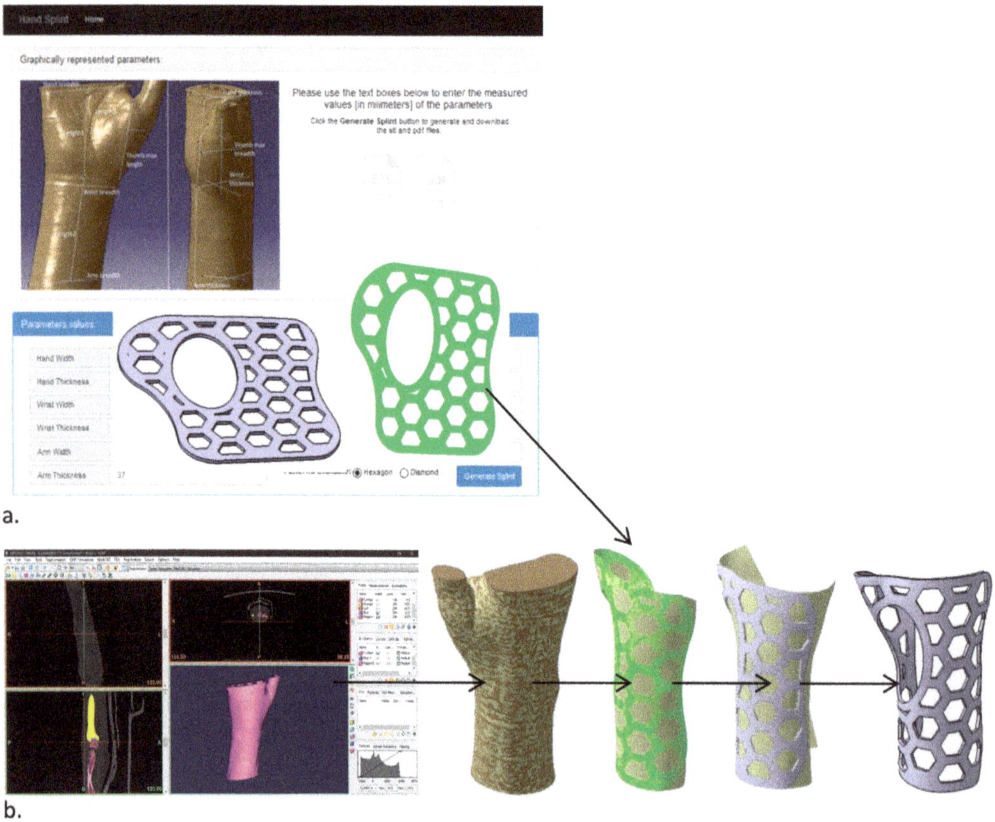

Figure 2. Orthoses design methodology and 3DP-WHOs' development flows for (**a**) flat-printed 3DP-WHO2, and (**b**) hand-shaped 3DP-WHO1.

After segmenting the medical imaging data for tissue, and adding a core composed of the forearm bones, candle gel (with a density of 912 kg/m^3, similar to the density of the human upper limb soft tissue [23]) was cast into a 3D-printed mold. The development flow depicted in Figure 3 was based on the same CT data used for designing the 3DP-WHO1.

To enable the removal of the support structures sustaining the fingers, the two STL files generated using Mimics were 3D-printed separately. The two prints (corresponding to the forearm bones and to the mold) were then assembled together. The bones were 3D-printed with 20% infill density and 5 shells (2 mm, i.e., approximately the cortical bone thickness), while the hand exterior consisted of two shells so that they can be easily cut away after the solidification of the gel, without top/bottom layers, and only brim adhesion support.

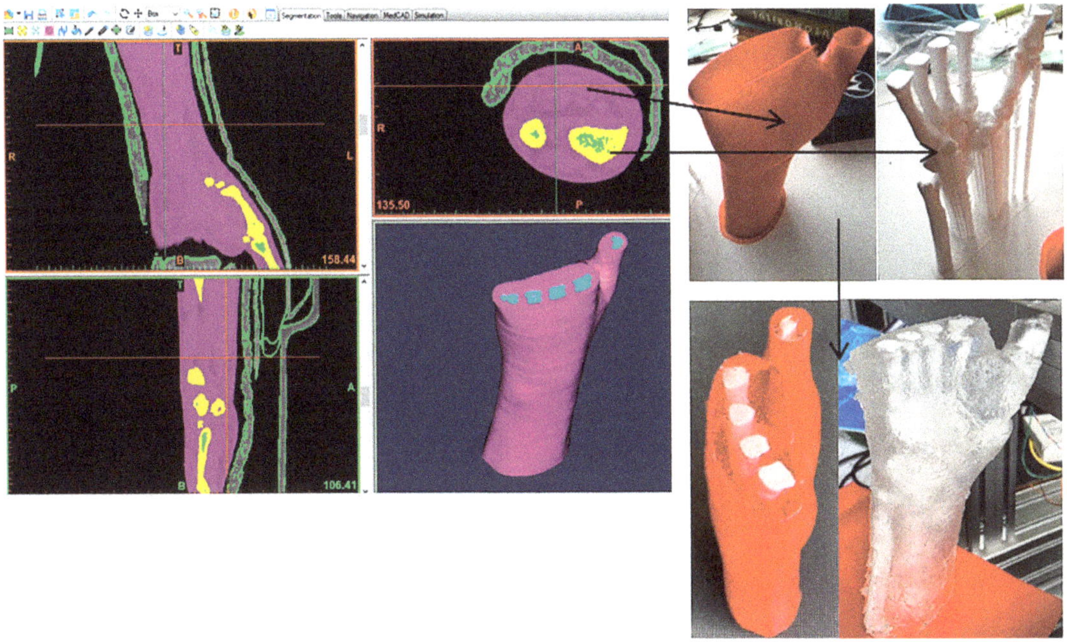

Figure 3. Forearm dummy production process.

2.2. 3D Printing

The main 3DP process parameters are presented in Table 1, which also shows the build orientation and printing times for each WHO. To ensure comparability of the mechanical testing results, it was important to assure not only similar designs, but also similar weights for both types of orthoses. Achieving this required careful consideration of the process parameters' settings. For the vertically printed orthoses, a high infill density, closer to 100%, was necessary due to the small layer thickness (2.3 mm) and the risk of fragility within the interior structure/infill lay between the two perimeters. Furthermore, the orthoses' height (160 mm) increased the likelihood of detachment from the printing platform in case of a smaller density value because of lack of stiffness, as resulted from previous experience. To achieve similar weights (in this case 33 g), we experimented with various combinations of (integer) infill densities for both the vertical and flat orthoses. These experiments were conducted using Ultimaker Cura 4.8.0 (Ultimaker BV, Geldermalsen, NL) simulations. Through this iterative process, infill density values were determined (97% infill density for the vertical orthosis and 40% infill density for the flat orthosis) that resulted in comparable weights for the orthoses. The fact that the infill density (and thus the weight) can be reduced when printing the orthoses as flat, while maintaining the required stiffness as proved by the results of the mechanical tests conducted in this research, is another advantage of our approach.

Table 1. 3DP-WHO2 manufacturing information.

3DP-WHO1	3DP-WHO2	Process Parameters
Hand-tailored form–build orientation 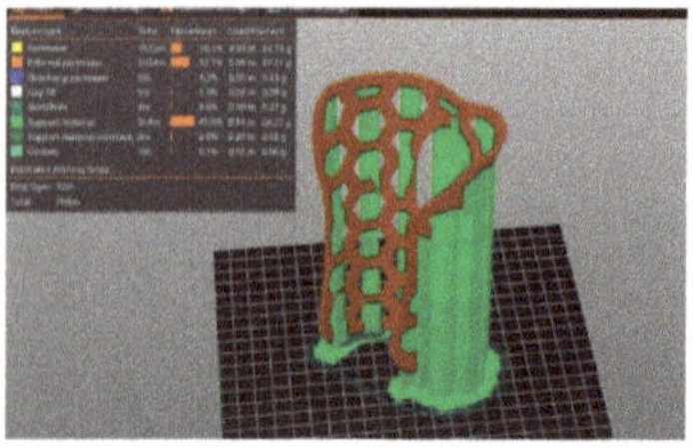 Printing time: 7 h 8 min	Flat form–build orientation 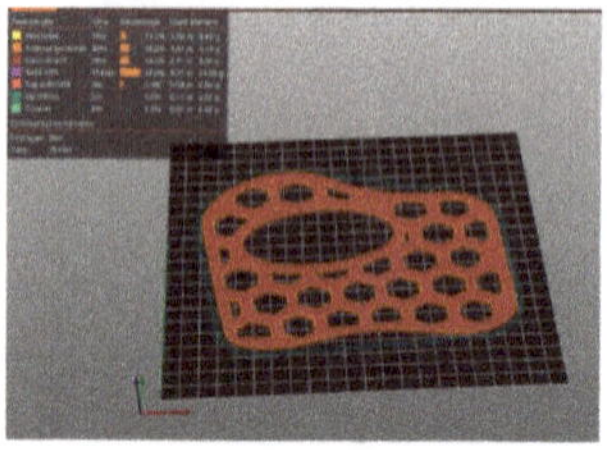 Printing time: 2 h 39 min	Printing temperature: 215 °C Bed temperature: 70 °C Infill pattern: grid 2 shells Top/bottom layers: 2 Infill density WHO1: 40% Infill density WHO2: 97%

Afterwards, the 3DP-WHO2s (Figure 4a) were thermoformed in warm water (at 85 °C) and shaped in the form of the hand by using a 3D-printed mold made out of ABS (Figure 4b).

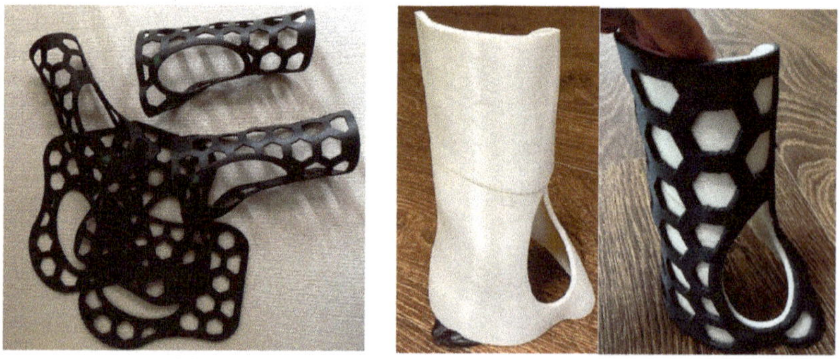

(**a**) (**b**)

Figure 4. (**a**) Flat and hand-shaped 3DP-WHOs with hexagonal pockets for three-point bending tests; (**b**) mold and thermoformed 3DP-WHO2.

2.3. Experimental Tests

To assess the strength of 3DP-WHOs in wrist sagittal movements, a three-point bending test method was chosen. The orthoses were installed on two supports (in the forearm zone and in the palm/fingers zone) with self-locking nylon cable zip ties, leaving the wrist zone unsupported (Figure 5a). Attention was paid to ensure a similar tightening for all WHOs by measuring the free-end length of the zip ties. The load was applied on the orthoses in the wrist region, with the deflection points being separated at 100 mm.

The Instron 8872 Universal Test Machine (Instron Inc., Norwood, MA, USA) was used for the mechanical tests. Three samples of 3DP-WHO1 and three samples of 3DP-WHO2 were subjected to three-point bending tests, and the results were comparatively analyzed. Moreover, 3DP-WHO2s were subjected to a fatigue bending test using tension–compression cyclic loads (Figure 5b) and the dummy forearm (presented in Section 2.1) as the interior support. Each sample underwent 2000 cycles (equivalent to an active wearing of the splint for around three–four weeks [24]) at 0.5 Hz frequency and $\pm$ 2.5 mm displacements (flexion–extension).

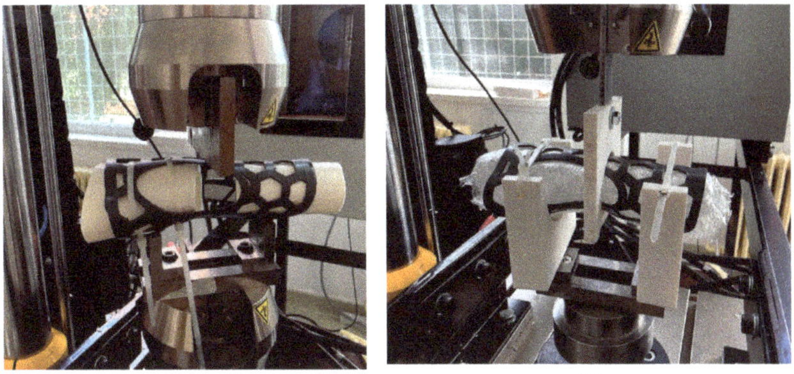

(**a**) (**b**)

Figure 5. (**a**) Experimental set-ups: 3-point bending tests; (**b**) 3-point fatigue bending.

3. Results and Discussion

3.1. Literature Survey Outcomes

Table 4 shows the most relevant data extracted from the full-text reading of the papers. It should be mentioned that only four papers reported the outcomes of experimental investigations on the mechanical properties of 3DP-WHOs. For the mechanical tests, one study [25] used a cadaver forearm, whereas in the other investigations, the 3DP-WHOs were mounted on specially designed supports inside the testing apparatus. A TPU (thermoplastic polyurethane) model of the entire forearm was used in one study as an internal support for the orthoses during the tests [26]. Most of the reviewed studies considered three-point bending tests for simulating the wrist flexion, although Cazon et al. [27] and Hoogervorst et al. [25] focused on all hand movements (radial, ulnar, and flexion).

With the exceptions of Chen Y et al.'s research in which CT imaging was used [28], and Sorimpuk et al. who used a flat model of the orthosis [20], in all the other studies, the 3DP-WHOs are designed starting from 3D scans of the patient hand. Methods to semi-automate the WHO design process were considered in many papers (Table 1) as solutions to decrease the design time and dependence on 3D modeling skills. In the same context of the design and manufacturing time reduction, Kim et al. proposed an interesting solution to make WHOs from two parts from which only the interior frame is patient-customized, while the outer frame is injection-molded and made on sizes corresponding to Korean population forearm data [29]. Long printing times are mentioned as a drawback in the majority of reviewed studies.

Sorimpuk et al.'s study was the only one found on the topic of flat 3DP-WHO mechanical behavior. However, the orthosis model is generated based on the average dimensions of adult Malaysians' hands [20] and it is not patient-customized, while the flat model of WHO was digitally wrapped on the forearm 3D model for performing FEA. No experimental tests are presented for validating the simulation model.

In twelve papers, FEM is used for investigating the mechanical behavior of 3DP-WHOs under different loads occurring in daily use, including simulations of accidental impact [30–32] and thermal analysis [33]. FEM is also used for the topological optimization of 3DP-WHOs' design [18,29]. The reported limitations of the numerical simulations relate mainly to the anisotropic characteristics of 3D prints. The results of the FE-based simulations performed on orthoses with different dimensions, materials and designs proved that 3DP-WHOs are suitable in terms of mechanical resistance and can provide the required immobilization (i.e., the displacement analysis under loads).

In summary, the literature review reveals the following key findings:

- Out of thirteen papers, four conducted experimental assessments on the mechanical behavior of 3D-printed wrist–hand orthoses (3DP-WHOs);

- No research has been conducted to compare the mechanical behavior of 3DP-WHOs manufactured in hand-shaped form with thermoformed 3D-printed orthoses;
- The most commonly used mechanical test for 3DP-WHOs is three-point bending, specifically for wrist flexion movement;
- Six papers focused on discussing WHO production through the MEX process, while the remaining papers discussed the utilization of SLS, SLA and Polyjet processes;
- For MEX orthoses, the following materials were used: PLA, ABS, PA12, HIPS;
- In two studies, multi-materials or hybrid manufacturing were used for producing the orthoses;
- For the purpose of reverse engineering the patient hand and designing the hand-shaped orthoses, eleven out of thirteen papers performed 3D laser scanning and associated data processing.

Table 2. Summary of the studies on 3DP-WHOs' mechanical testing and FE simulations.

Study	Acquisition Method / WHO Design Method/Software	Manufacturing Process/Material	Mechanical Testing	FEA	Observations
Agudelo-Ardila, et al., 2019 [33]	3D scanning, 3D handheld scanner Meshmixer, 2 parts WHO, Voronoi structure, 2.5 mm thickness	SLS process, ProX SLS 500 printer (3D Systems, Inc., Rock Hill, SC, USA, DuraForm ProX PA	No	Stress (load: 60N bending), thermal analysis (40 °C).	Comparison of manufacturing times for 3DP-WHOs vs. conventional WHOs.
Buonamici, et al., 2019 [18]	3D scanning using a dedicated, in-house developed device Customized semi-automatic design software based on Siemens NX 10; 2 parts WHO with zip ties; different designs for the ventilation holes (circular holes, Voronoi, topology optimization)	MEX process, Stratasys F370 (Stratasys Inc., Eden Prairie, MN, USA)—FDM printer, ABS M30	No	FEA used for topology optimization for reducing weight by ventilation area.	Child WHO printing time 7h 21min; male adult 18h 6min; 52 min modeling time.
Cazon, et al., 2017 [27]	3D scanning, ZScanner 800 3D laser scanner Geomagic Studio 2013	Polyjet process, Object Connex printer (Stratasys Inc., Eden Prairie, MN, USA)—multi-material WHO: VeroWhitePlus, TangoBlackPlus	Yes, dedicated support for WHO testing, mechanical tests for radial, ulnar, flexion and extension movements	FEA using Creo 3.0, 4 hand movements: radial, ulnar, flexion and extension, torques (Vanwearingen torques used as reference: 14.8 Nm, 8.4 Nm, 11.4 Nm, and respectively 9.9 Nm) and loads on x and z directions.	Mechanical tests showed displacements of 3.46 mm, 0.97 mm, 3.53 mm, and 2.51 mm for flexors, extensors, radial deviators and ulnar deviators. In ulnar direction, 3DP-WHO had a greater displacement, as resulted in FEA. 3DP-WHO proved suitable for everyday use.
Chen, C.D. et al., 2019 [15]	3D scanning, Creaform Go!Scan50 scanner 2 parts WHO tied with Velcro strips, designed using in-house solutions	MEX process, PLA	No	ANSYS, 3 parameters with 2 design values (WHO thickness, ventilation holes diameter, holes center distance); -flexion 30 N; extension 25 N; radial deviation 30 N; ulnar deviation 30 N -impact with 40 mm diameter steel ball of 0.3768 kg.	Material properties used in FEA were experimentally determined on specimens–ASTM D638, ISO 180.
Chen, Y, et al., 2020 [28]	CT, Mimics 10.01 for forearm reconstruction Solidworks 2015; 2 parts WHO	SLS process, EOS P395 (EOS GmbH, Krailling, Germany), PA2200	No	ANSYS Workbench 18; 6 loading conditions including anterior to posterior (AP), posterior to anterior (PA), medial to lateral (ML), lateral to medial (LM), inward (IR), and outward OR to calculate the displacement and stress; 400 N compression load on the palm along AP, PA, ML, LM; 1 Nm rotation moment toward the IR and OR of the palm, applied to the top end side of cast.	FE model for bone, soft tissue and cast. Immobilization using 3DP-WHO was effective. A total of 60 patients, 20 used 3DP-WHOs, patients satisfaction assessment. Long manufacturing times, not suitable for emergency situations.

Table 3. Summary of the studies on 3DP-WHOs' mechanical testing and FE simulations.

Study	Acquisition Method / WHO Design Method/Software	Manufacturing Process/Material	Mechanical Testing	FEA	Observations
Gorski, et al., 2020 [26]	3D scanning, David SLS-3 optical scanner In-house dedicated design app (AutoMedPrint), 2 parts WHOs with snap fit connection, 4 mm thickness	MEX process, Raise 3D Pro machine (Raise 3D Technologies, Nantong, China); ABS, PLA; PA12, HIPS; different infills, layer thicknesses; vertical and horizontal build orientation	Yes, three-point bending using a TPU phantom of the forearm	No	Different assessment criteria: manufacturing time and cost, strength dependence on process parameters and material, the patient wore the 3DP-WHOs for 15 min for yes/no comfort feedback.
Hoogervorst, et al., 2019 [25]	3D scanning, 3D Structure Sensor infrared scanner Open lattice design	MJF process, HP Multi Jet Fusion Printer (HP Inc., Palo Alto, CA, USA), PA12	Yes, using a cadaver forearm; -flexion and extension of digits (1000 loading cycles (20–100 N tensile force), -pronation and supination of the hand (1000 cycles of torque (−0.5 to 0.5 Nm)), -three-point bending (1000 cycles (50–500 N))	No	Cadaver biomechanical study for assessing the stabilizing properties of 3DP-WHOs in comparison with traditional fiberglass cast. Only three-point bending results were statistically different, but with a very small value of the absolute motion: 0.44 (±0.48)mm.
Kim & Jeong, 2015 [29]	3D scanning 3D CAD software for inner frame, outer frame designed based on population forearm measurements data	Hybrid manufacturing PolyJet, Objet500 Connex (Stratasys Inc., Eden Prairie, MN, USA), ABS, (inner frame, 2 part) and injection molding (outer cover, 2 parts, PC)	No	ANSYS 13, FEA for determining the thickness of WHO outer cover subjected to 200 N impact force	Manufacturing time and cost reduction by customizing just the inner frame and its connection bumps, while the outer frame is available on sizes
Li & Tanaka, 2018 [31]	3D scanning, Sense handheld 3D scanner Semi-automated design using in-house app based on Rhinoceros 5 and Grasshopper 3D; 2–3 parts WHO connected with M3 screws, lattice structure	MEX process, Qidi Tech 1 3D printer (Qidi Tech, Ruian, China), ABS	No	FEA using Fusion 360 software; 30 N loads on the distal edge of the splint and lattice-structure area along three directions for simulating possible hits and stresses.	3DP-WHO tested on 10 healthy subjects, no significant discomfort reported (3 reports of itching)
Lin, et al., 2016 [32]	3D scanning, Artec Eva and Artec Space Spider 3D scanners In-house modeling app for semi-automating the design process, 2 parts WHO, 2 mm thickness	SLA process, RS6000 3D printer (UnionTech, Beijing, China), PP	No	FEA using ANSYS for strength assessment, 3MPa impact pressure applied on the ventilation holes zones.	-

Table 4. Summary of the studies on 3DP-WHOs' mechanical testing and FE simulations.

Study	Acquisition Method / WHO Design Method/Software	Manufacturing Process/Material	Mechanical Testing	FEA	Observations
Lukaszewski et al., 2020 [34]	3D scanning WHO of 4mm thickness	MEX process, FlashForge Creator Pro (Zhejiang Flashforge 3D Technology Co., Ltd., Jinhua, China); ABS; 5 types of samples in different build orientations, 15% infill density, linear pattern, 2 shells	Yes, three-point bending parts for: specimen, part of WHO, part of WHO with ventilation holes, full WHO	FEA using ABAQUS for calculating WHO modulus of elasticity, 100 N loads in the middle of the WHO placed in a horizontal position.	The WHO parts with ventilation holes built in horizontal plane have higher stiffness than those built vertically.
Modi, et al., 2020 [30]	3D scanning, HandySCAN 3D laser scanner Design process using Meshlab, GeoMagic Studio and CATIA V5	SLS process, EOSINT P395 (EOS GmbH, Krailling, Germany), PA 2200	No	FEA using Fusion 360; 100 N on the forearm area near proximal end of splint (as in impact), 30 N loads at the distal end of the splint near fingers as in accidental fingers bending.	The mechanical properties used in FEA were experimentally determined using specimens. Long development process (19 h).
Sorimpuk, et al., 2022 [20]	Average dimensions of the forearms and hand circumference of Malaysian adults applied to a hand model from GrabCAD Flat-designed WHO with different ventilation pockets and specific adaptive pattern for the wrist joint curvature	MEX process, PLA (Ultimaker BV, Geldermalsen, NL)	No	FEA using Inventor 2017; 400 N load in X and Z directions of the cast, and 1 Nm bending moment along the Y direction of the cast.	FEA was performed on the digitally wrapped model of the 3DP-WHO. Results were compared with plaster traditional casts and SLS manufactured cast [28]. Reported printing time 3h 15 min. No process parameters details. No comparison with real 3DP-WHO.

3.2. Results of Mechanical Tests

The results of the three-point bending tests are presented in Figure 6, with the mean force and displacement values listed in Table 5. The force–displacement data indicate that the 3DP-WHOs exhibit similar mechanical stiffness, with almost equally steep curves of up to around 50 N (Figure 6a,c). In cases where greater stiffness (i.e., immobilization strength) is necessary based on medical diagnosis, increasing the thickness of the thermoformable orthoses may be considered.

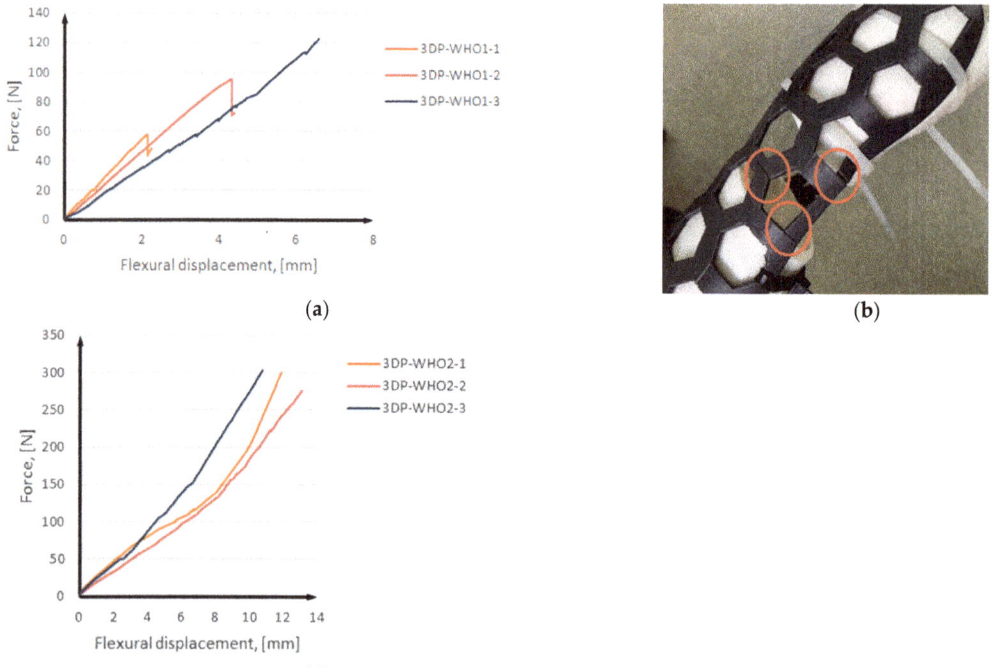

(c)

Figure 6. Results of three-point bending tests: (**a**) 3DP–WHO1s; (**b**) 3DP–WHO1 failure; (**c**) 3DP–WHO2s.

Table 5. Mean values of force and displacement for the tested samples.

Samples	Force at Break, [N]	Displacement at Break, [mm]
3DP-WHO1	81.13	4.41
3DP-WHO2	292.9	11.89

The 3DP-WHO1s samples failed around 60 N, 90 N, and, respectively, 120 N. Inter-layer fractures occurred in all cases (Figure 6b), which was expected considering the vertical build orientation and load direction. The presence of inter-layer defects can explain the differences in the load to failure. It should be noted that printing defects (lack of layer adhesion or inter-layer voids) are more common on tall parts with thin walls [35], as is the case of 3DP-WHO1s. These defects and the fact that the load is applied along the layers leads to the reduction of the structural integrity of these orthoses, thus reducing their life span or use at low loads (<100 N equivalent to 10 Kg). The samples were tested until cracks appeared (Figure 6b).

The experimental tests for 3DP-WHO2s were stopped at an applied load of 300 N (Figure 6c), which was considered as exceeding the typical value for this type of medical device according to the practice and literature data (Table 4). No material changes or breakage was noticed for any 3DP-WHO2. It was observed that the 3DP-WHO2s deformed in the elastic area, with no cracks being visible, and after removing the load, they returned to the original shape.

Cazon et al. comparatively assessed the conventionally manufactured WHOs and multi-material 3D-printed WHOs (obtained by using the Polyjet process), showing that 3DP-WHOs were more rigid in flexion, extension and radial directions [27]. The applied forces corresponded to 8% and 50% of a healthy person load, i.e., 53.7 N flexor, 41.4 N extensor, 56.4 N radial and 48.9 N ulnar for the 50% load, equivalent to the maximum strength of wrist rheumatoid arthritis [27,36]. Chen CD et al. developed an FE model of a PLA 3DP-WHO and applied loads of 30 N for flexion and 25 N for extension [15], but no mechanical tests were performed to validate this model. All the 3DP-WHO samples tested in our study withstand these loads. However, the measured displacements could not be related to the literature data as all the surveyed experimental studies considered full casts and not splints. Moreover, these casts had different pocket designs, masses and larger thicknesses (3 mm in [27], 4 mm in [26]), which would also make the comparison incorrect. Gorski et al. [26] evaluated the mechanical behavior of PLA, nylon, ABS and HIPS (high impact polystyrene) casts, 3D-printed using different process parameters. These WHOs were subjected to 300 N load and the experimental outcomes showed that none of them failed below 750 N [26].

Lukaszewski et al. applied a load of 100 N in the middle of a PLA wrist–hand cast (mass of 49 g) in a three-point bending test [34]. Tests were conducted on different models, and differences in the modulus of elasticity were observed as depending on the samples' build orientation, but also on the presence or lack of ventilation pockets.

Figure 7 shows 3DP-WHO2 samples' cyclic response when subjected to maximum-to-minimal load under different strain values. It can be noted that the hysteresis loops' stabilization took place after 1000 cycles, the loops' shapes being typical for viscoelastic materials. The applied load cycles did not cause the failure of any tested sample. The minimum force occurring during the flexural fatigue test was approximately −95 N, and after approximately 1100–1200 cycles, it reached −110 N and remained constant until the end (Figure 8). A possible explanation for this behavior might be the strain hardening, with more investigations being required in this regard. Additionally, in further research, it will be important to understand the effect of the orthoses' design (thickness, ventilation pockets' shapes and dimensions) on the fatigue limit, as criteria to consider in the design process based on the functional conditions of this medical device. Furthermore, studying the influence of the stress frequency on the fatigue behavior can provide useful data on 3DP-WHOs' durability.

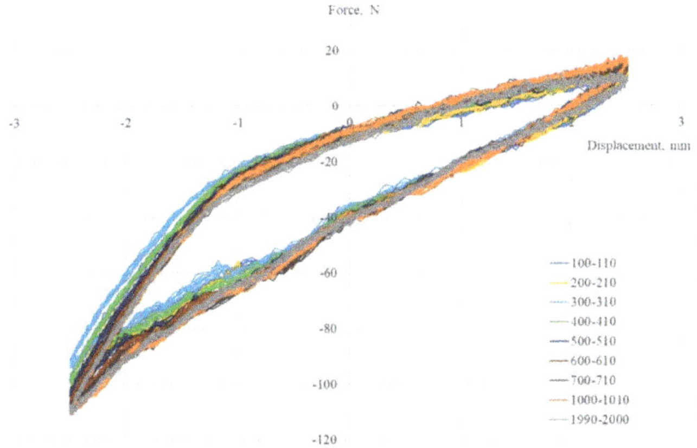

Figure 7. Hysteresis curves in bending fatigue tests for 3DP–WHO2.

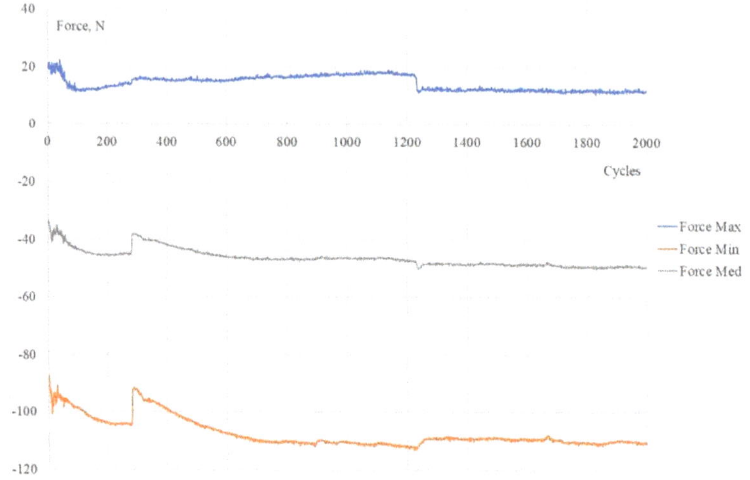

Figure 8. Force-cycle variation for a 3DP–WHO2 sample.

The flexural fatigue behavior of 3D-printed PLA specimens was previously studied for identifying the impact of process parameters on the fatigue strength [37,38]. However, only one reference was found on this type of mechanical test for WHO. A polyamide 3DP-WHO in the form of a cast encasing the forearm was investigated [25] for 1000 cycles between 50–500 N, at a frequency of 0.5 Hz. In their study, a cadaver arm was used, and the displacements were assessed with radiographs.

Both types of tests conducted in this research showed not only that the thermoformable 3DP-WHOs have a better quasi-static flexural behavior compared to 3D-printed orthoses vertically built in their almost ready-to-use shape (support structures' removal being required as well), but they can also sustain long cyclic loads (tension–compression) which might occur during wear.

3.3. Discussion on 3D-Printed PLA Orthoses' Thermoforming

Thermoforming sheets from polymeric material is a common industrial practice to obtain a large range of objects with complex shapes by using mechanical force applied to heated material and using forming tools such as molds and dies. However, this process is not commonly adopted when it comes to 3D-printed thin parts made from polymeric materials, despite interesting potential uses [39,40].

PLA is an aliphatic polyester made from renewable resources, such as sugar, corn, potatoes, and beets, with a glass transition temperature (Tg) ranging between 55 °C and 60 °C, meaning that it becomes soft and easy to deform at lower temperatures in comparison to other thermoplastics. PETG (Polyethylene Terephthalate Glycol-Modified—a thermoplastic polyester derived from petroleum-based sources), for example, exhibits a higher Tg of approximately 85 °C. Thermoforming a 3D-printed flat orthosis should take place at temperatures between Tg and Tcc (cold crystallization temperature of about 100 °C [41]), which allow keeping the material in a semi-solid state with sufficient molecular chain mobility while preventing unwanted crystallization prior to the forming process [42]. Since orthosis thermoforming occurs on the patient's hand, temperatures that are too high can potentially cause burns, making it unsuitable for direct contact with the skin. Conversely, reducing the temperature to a level acceptable for the skin may result in challenges in achieving proper molding on the hand shape, particularly in the palm zone, as depicted in Figure 9 for the PETG orthosis. Within the marked zones in Figure 9, noticeable differences in deformations between the two material types can be observed, with the PETG orthosis failing to conform properly to the shapes of the forearm and palm. Wrapping the patient's hand before thermoforming is advisable. In this research, PLA orthoses were heated at 80 °C, while PETG orthoses at 100 °C.

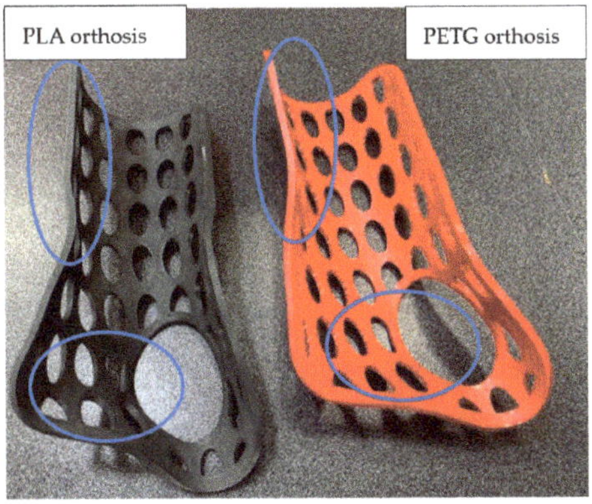

Figure 9. 3D-printed cockup wrist–hand orthoses made from PLA (**left**) and PETG (**right**).

Thermoforming is possible as PLA has rubbery properties above Tg. The phases involved in thermoforming include heating, molding and cooling. Heating determines the molecular movement of the polymer chains facilitated by heat energy, which increases their mobility and decreases intermolecular tensions. Thus, the material becomes soft and ready to be molded, a phase in which the polymer chains align in the direction of the stretching force. Through the cooling process, PLA temperature is lowered below Tg, allowing the polymer chains to regain their rigidity, effectively solidifying the material in the desired shape. In the case of 3D-printed PLA thin objects, such as the orthoses, specific aspects related to infill pattern and infill density can occur, influencing the time required to reach the molted state for forming, as well as the cooling time. Attention should be paid to these characteristics, and they are the subject of further studies.

4. Conclusions and Further Work

In this research, the flexural performance of 3DP-WHOs manufactured by two approaches (i.e., customized upper-limb orthoses manufactured directly in a hand-shaped and flat-shaped form, and then thermoformed orthoses) was assessed in three-point quasi-static and uniaxial fatigue and three-point flexural tests corresponding to the wrist flexion/extension (sagittal) movements. No such investigation had been conducted so far, as the systematic review of the literature showed.

The rationale of this study was related to the observation that the flat-shaped 3DP-WHOs take less time for printing, cost less, and can be generated without necessitating 3D modeling skills. Therefore, they are easier to implement in 3DP-POCs. In this context, the issue of interest was whether their mechanical performance is also suitable for conditions typical for daily uses. The experimental outcomes indicated a positive answer to this research question. Moreover, 3D-printed and thermoformed WHOs proved resistant to cyclic flexural loading up to 2000 cycles, with the orthoses being placed on bi-material phantom aimed at mimicking the human hand during the tests.

Further research will be focused on understanding the dependency between the design, material, process parameters (infill-related, printing temperature, layer thickness), testing conditions, and the flexural strength of these orthoses to gather more data to support hand specialists' decisions in prescribing these 3D-printed medical devices.

Author Contributions: Conceptualization, D.P., D.L. and R.M.; formal analysis, D.P., F.B., D.V. and D.L.; funding acquisition, D.P.; investigation, D.P., F.B., D.V. and D.L.; methodology, D.P., F.B. and R.M.; validation, D.P., F.B. and D.V.; visualization, D.V.; writing—original draft, D.P., F.B., D.L. and R.M.; writing—review & editing, D.P., D.V. and D.L. All authors have read and agreed to the published version of the manuscript.

Funding: This work was supported by a grant of the Ministry of Research, Innovation and Digitization, CNCS–UEFISCDI, project number PN-III-P4-PCE-2021-0070, within PNCDI III.

Institutional Review Board Statement: Not applicable.

Data Availability Statement: Data available on request.

Conflicts of Interest: The authors declare no conflict of interest.

References

1. Li, C.; Pisignano, D.; Zhao, Y.; Xue, J. Advances in Medical Applications of Additive Manufacturing. *Engineering* **2020**, *6*, 1222–1231. [CrossRef]
2. Aimar, A.; Palermo, A.; Innocenti, B. The Role of 3D Printing in Medical Applications: A State of the Art. *J. Healthc. Eng.* **2019**, *2019*, 5340616. [CrossRef]
3. Javaid, M.; Haleem, A. Additive manufacturing applications in medical cases: A literature based review. *Alex. J. Med.* **2018**, *54*, 411–422. [CrossRef]
4. Shree, M.V.; Dhinakaran, V.; Rajkumar, V.; Ram, P.B.; Vijayakumar, M.; Sathish, T. Effect of 3D printing on supply chain management. *Mater. Today Proc.* **2020**, *21*, 958–963. [CrossRef]
5. Daoud, G.E.; Pezzutti, D.L.; Dolatowski, C.J.; Carrau, R.L.; Pancake, M.; Herderick, E.; VanKoevering, K.K. Establishing a point-of-care additive manufacturing workflow for clinical use. *J. Mater. Res.* **2021**, *26*, 3761–3780. [CrossRef]
6. Schwartz, D.A.; Schofield, K.A. Utilization of 3D printed orthoses for musculoskeletal conditions of the upper extremity: A systematic review. *J. Hand Ther.* **2021**, *36*, 166–178. [CrossRef]
7. Skibicki, H.E.; Katt, B.M.; Lutsky, K.; Wang, M.L.; McEntee, R.; Vaccaro, A.R.; Beredjiklian, P.; Rivlin, M. Three Dimensionally Printed Versus Conventional Casts in Pediatric Wrist Fractures. *Cureus* **2021**, *13*, e19090. [CrossRef]
8. Guida, P.; Casaburi, A.; Busiello, T.; Lamberti, D.; Sorrentino, A.; Iuppariello, L.; D'albore, M.; Colella, F.; Clemente, F. An alternative to plaster cast treatment in a pediatric trauma center using the CAD/CAM technology to manufacture customized three-dimensional-printed orthoses in a totally hospital context: A feasibility study. *J. Pediatr. Orthop. B* **2019**, *28*, 248–255. [CrossRef]
9. Kim, S.J.; Kim, S.J.; Cha, Y.H.; Lee, K.H.; Kwon, J.-Y. Effect of personalized wrist orthosis for wrist pain with three-dimensional scanning and printing technique. *Prosthetics Orthot. Int.* **2018**, *42*, 636–643. [CrossRef]
10. Oud, T.A.M.; Lazzari, E.; Gijsbers, H.J.H.; Gobbo, M.; Nollet, F.; Brehm, M.A. Effectiveness of 3D-printed orthoses for traumatic and chronic hand conditions: A scoping review. *PLoS ONE* **2021**, *16*, e0260271. [CrossRef]
11. Graham, J.; Wang, M.; Frizzell, K.; Watkins, C.; Beredjiklian, P.; Rivlin, M. Conventional vs 3-Dimensional Printed Cast Wear Comfort. *Hand* **2020**, *15*, 388–392. [CrossRef] [PubMed]
12. Waldburger, L.; Schaller, R.; Furthmüller, C.; Schrepfer, L.; Schaefer, D.J.; Kaempfen, A. 3D-Printed Hand Splints versus Thermoplastic Splints: A Randomized Controlled Pilot Feasibility Trial. *Int. J. Bioprint.* **2021**, *8*, 474. [CrossRef] [PubMed]
13. Kelly, S.; Paterson, A.M.J.; Bibb, R.J. A Review of Wrist Splint Designs for Additive Manufacture. In Proceedings of the 2015 14th Rapid Design, Prototyping and Manufacture conference (RDPM 14), Loughborough, UK, 15–16 December 2015.
14. Oud, T.; Kerkum, Y.; Groot, P.; Gijsbers, H.; Nollet, F.; Brehm, M. Production time and user satisfaction of 3-dimensional printed orthoses for chronic hand conditions compared with conventional orthoses: A prospective case series. *J. Rehabil. Med. Clin. Commun.* **2021**, *4*, 1000048. [CrossRef] [PubMed]
15. Chen, C.-D.; Chen, C.-H.; Lin, C.-L.; Lin, C.-K.; Wang, C.-T.; Lin, R.-M.; Fang, J.-J. Developments, mechanical property measurements and strength evaluations of the wrist braces for the wrist fracture patients. *J. Mech. Med. Biol.* **2019**, *19*, 1940021. [CrossRef]
16. Keller, M.; Guebeli, A.; Thieringer, F.; Honigmann, P. In-hospital professional production of patient-specific 3D-printed devices for hand and wrist rehabilitation. *Hand Surg. Rehabil.* **2021**, *40*, 126–133. [CrossRef]
17. Formisano, M.; Iuppariello, L.; Casaburi, A.; Guida, P.; Clemente, F. An industrial oriented workflow for 3D printed, patient specific orthopedic cast. *SN Appl. Sci.* **2021**, *3*, 830. [CrossRef]
18. Buonamici, F.; Furferi, R.; Governi, L.; Lazzeri, S.; McGreevy, K.S.; Servi, M.; Talanti, E.; Uccheddu, F.; Volpe, Y. A practical methodology for computer-aided design of custom 3D printable casts for wrist fractures. *Vis. Comput.* **2020**, *36*, 375–390. [CrossRef]
19. Portnoy, S.; Barmin, N.; Elimelech, M.; Assaly, B.; Oren, S.; Shanan, R.; Levanon, Y. Automated 3D-printed finger orthosis versus manual orthosis preparation by occupational therapy students: Preparation time, product weight, and user satisfaction. *J. Hand Ther.* **2020**, *33*, 174–179. [CrossRef]

20. Sorimpuk, N.P.; Choong, W.H.; Chua, B.L. Design of thermoformable three dimensional-printed PLA cast for fractured wrist. *IOP Conf. Ser. Mater. Sci. Eng.* **2022**, *1217*, 012002. [CrossRef]
21. Popescu, D.; Zapciu, A.; Tarba, C.; Laptoiu, D. Fast production of customized three-dimensional-printed hand splints. *Rapid Prototyp. J.* **2020**, *26*, 134–144. [CrossRef]
22. Baronio, G.; Harran, S.; Signoroni, A. A Critical Analysis of a Hand Orthosis Reverse Engineering and 3D Printing Process. *Appl. Bionics Biomech.* **2016**, *2016*, 8347478. [CrossRef] [PubMed]
23. Winter, D.A. (Ed.) Chapter 9: Kinesiological Electromyography. In *Biomechanics and Motor Control of Human Movement*, 3rd ed.; John Wiley & Sons, Inc.: Hoboken, NJ, USA, 2005.
24. Breger-Lee, D.E.; Buford, W.L. Properties of Thermoplastic Splinting Materials. *J. Hand Ther.* **1992**, *5*, 202–211. [CrossRef]
25. Hoogervorst, P.; Knox, R.; Tanaka, K.; Working, Z.M.; El Naga, A.N.; Herfat, S.; Lee, N. A Biomechanical Comparison of Fiberglass Casts and 3-Dimensional–Printed, Open-Latticed, Ventilated Casts. *Hand* **2020**, *5*, 842–849. [CrossRef] [PubMed]
26. Górski, F.; Wichniarek, R.; Kuczko, W.; Żukowska, M.; Lulkiewicz, M.; Zawadzki, P. Experimental Studies on 3D Printing of Automatically Designed Customized Wrist-Hand Orthoses. *Materials* **2020**, *13*, 4091. [CrossRef] [PubMed]
27. Cazón, A.; Kelly, S.; Paterson, A.M.; Bibb, R.J.; Campbell, R.I. Analysis and comparison of wrist splint designs using the finite element method: Multi-material three-dimensional printing compared to typical existing practice with thermoplastics. *Proc. Inst. Mech. Eng. Part H J. Eng. Med.* **2017**, *231*, 881–897. [CrossRef] [PubMed]
28. Chen, Y.; Lin, H.; Yu, Q.; Zhang, X.; Wang, D.; Shi, L.; Huang, W.; Zhong, S. Application of 3D-Printed Orthopedic Cast for the Treatment of Forearm Fractures: Finite Element Analysis and Comparative Clinical Assessment. *BioMed Res. Int.* **2020**, *2020*, 32775455. [CrossRef]
29. Kim, H.; Jeong, S. Case study: Hybrid model for the customized wrist orthosis using 3D printing. *J. Mech. Sci. Technol.* **2015**, *29*, 5151–5156. [CrossRef]
30. Modi, Y.K.; Khare, N. Patient-specific polyamide wrist splint using reverse engineering and selective laser sintering. *Mater. Technol.* **2020**, *37*, 71–78. [CrossRef]
31. Li, J.; Tanaka, H. Rapid customization system for 3D-printed splint using programmable modeling technique—A practical approach. *3D Print. Med.* **2018**, *4*, 1–21. [CrossRef]
32. Lin, H.; Shi, L.; Wang, D. A rapid and intelligent designing technique for patient-specific and 3D-printed orthopedic cast. *3D Print. Med.* **2016**, *2*, 4. [CrossRef]
33. Agudelo-Ardila, C.P.; Prada-Botía, G.C.; Rodrigues, G.P.H. Orthotic prototype for upper limb printed in 3D: A efficient solution. *J. Phys. Conf. Ser.* **2019**, *1388*, 012016. [CrossRef]
34. Łukaszewski, K.; Wichniarek, R.; Górski, F. Determination of the Elasticity Modulus of Additively Manufactured Wrist Hand Orthoses. *Materials* **2020**, *13*, 4379. [CrossRef] [PubMed]
35. Galati, M.; Minetola, P.; Marchiandi, G.; Atzeni, E.; Calignano, F.; Salmi, A.; Iuliano, L. A methodology for evaluating the aesthetic quality of 3D printed parts. *Procedia CIRP* **2019**, *79*, 95–100. [CrossRef]
36. Salaffi, F.; Carotti, M.; Farah, S.; Ceccarelli, L.; Di Carlo, M. Handgrip Strength Features in Rheumatoid Arthritis Patients Assessed Using an Innovative Cylindrical-Shaped Device: Relationships With Demographic, Anthropometric and Clinical Variables. *J. Med. Syst.* **2021**, *45*, 100. [CrossRef]
37. Ezeh, O.; Susmel, L. Fatigue strength of additively manufactured polylactide (PLA): Effect of raster angle and non-zero mean stresses. *Int. J. Fatigue* **2019**, *126*, 319–326. [CrossRef]
38. Jerez-Mesa, R.; Travieso-Rodriguez, J.; Llumà-Fuentes, J.; Gomez-Gras, G.; Puig, D. Fatigue lifespan study of PLA parts obtained by additive manufacturing. *Procedia Manuf.* **2017**, *13*, 872–879. [CrossRef]
39. Mäntyjärvi, K.; Iso-Junno, T.; Mustakangas, A.; Jokelainen, T.; Keskitalo, M.; Järvenpää, A. Exploitation of forming of the 3D printed materials. *AIP Conf. Proc.* **2019**, *2113*, 150007. [CrossRef]
40. Wang, G.; Yang, Y.; Guo, M.; Zhu, K.; Yan, Z.; Cui, Q.; Zhou, Z.; Ji, J.; Li, J.; Luo, D.; et al. ThermoFit: Thermoforming Smart Orthoses via Metamaterial Structures for Body-Fitting and Component-Adjusting. *Proc. ACM Interact. Mob. Wearable Ubiquitous Technol.* **2023**, *7*, 1–27. [CrossRef]
41. Wei, H. Optimisation on Thermoforming of Biodegradable Poly (Lactic Acid) (PLA) by Numerical Modelling. *Polymers* **2021**, *13*, 654. [CrossRef]
42. Milovanovic, S.; Pajnik, J.; Lukic, I. Tailoring of advanced poly(lactic acid)-based materials: A review. *J. Appl. Polym. Sci.* **2021**, *139*, 51839. [CrossRef]

polymers

MDPI

Article

A Regression Approach to Model Refractive Index Measurements of Novel 3D Printable Photocurable Resins for Micro-Optofluidic Applications

Lorena Saitta [1,*,†], Emanuela Cutuli [2,†], Giovanni Celano [1], Claudio Tosto [1], Giovanna Stella [2], Gianluca Cicala [1,3] and Maide Bucolo [2]

1 Department of Civil Engineering and Architecture, University of Catania, Viale Andrea Doria 6, 95125 Catania, Italy; giovanni.celano@unict.it (G.C.); claudio.tosto@unict.it (C.T.); gianluca.cicala@unict.it (G.C.)

2 Department of Electrical Electronic and Computer Science Engineering, University of Catania, Viale Andrea Doria 6, 95125 Catania, Italy; emanuela.cutuli@phd.unict.it (E.C.); giovanna.stella@phd.unict.it (G.S.); maide.bucolo@unict.it (M.B.)

3 INSTM-UDR CT, Viale Andrea Doria 6, 95125 Catania, Italy

* Correspondence: lorena.saitta@phd.unict.it

† These authors contributed equally to this work.

Abstract: In this work, a quadratic polynomial regression model was developed to aid practitioners in the determination of the refractive index value of transparent 3D printable photocurable resins usable for micro-optofluidic applications. The model was experimentally determined by correlating empirical optical transmission measurements (the *dependent variable*) to known refractive index values (the *independent variable*) of photocurable materials used in optics, thus obtaining a related regression equation. In detail, a novel, simple, and cost-effective experimental setup is proposed in this study for the first time for collecting the transmission measurements of smooth 3D printed samples (roughness ranging between 0.04 and 2 μm). The model was further used to determine the unknown refractive index value of novel photocurable resins applicable in vat photopolymerization (VP) 3D printing techniques for manufacturing micro-optofluidic (MoF) devices. In the end, this study proved how knowledge of this parameter allowed us to compare and interpret collected empirical optical data from microfluidic devices made of more traditional materials, i.e., Poly(dimethylsiloxane) (PDMS), up to novel 3D printable photocurable resins suitable for biological and biomedical applications. Thus, the developed model also provides a quick method to evaluate the suitability of novel 3D printable resins for MoF device fabrication within a well-defined range of refractive index values (1.56; 1.70).

Keywords: refractive index measurement; photocurable resin; 3D printing; vat-photopolymerization; micro-optics; microfluidics

check for updates

Citation: Saitta, L.; Cutuli, E.; Celano, G.; Tosto, C.; Stella, G.; Cicala, G.; Bucolo, M. A Regression Approach to Model Refractive Index Measurements of Novel 3D Printable Photocurable Resins for Micro-Optofluidic Applications. *Polymers* **2023**, *15*, 2690. https://doi.org/10.3390/polym15122690

Academic Editors: Cristina-Elisabeta Pelin and Anton Ficai

Received: 10 May 2023
Revised: 5 June 2023
Accepted: 13 June 2023
Published: 15 June 2023

1. Introduction

Since the 1990s, lab-on-a-chip (LOC) has gained great importance [1]. Indeed, by means of this type of device, characterized by the presence of micro-metric channels with disparate geometries, it is possible to analyze and test very small quantities of biological or chemical fluids. As a result, several operations that usually require conventional biochemistry laboratories, such as sample preparation for reactions and biopsies of biological fluids, can be performed [1,2]. Micro-optofluidic (MoF) devices fall within this class of devices, and they are also adopted to optically detect and control two-phase flow flowing within micro-metric channels in several fields of application, such as biomedical and chemical ones [3–7]. The optical approach is the most preferred since it is slightly invasive and allows for many types of measurements [8].

However, it is worthy of attention that when MoF must be used exploiting the optical approach to analyze certain fluids [9–11], they are characterized by a layout where the presence of optical components is properly integrated with microfluidic channels. Thus, care must be taken with the material choice during the design procedure. Indeed, for carrying out this type of analysis performed using MoF devices, it is crucial to confine and transport the beam light coming from a light source towards the small volume of fluid circulating within micro-channels. In the end, properly acquiring the beam of light passing through the fluid makes it possible to collect useful non-specific information, such as its density, viscosity, velocity, and so on [12]. Therefore, to accomplish this purpose, the materials used for manufacturing this class of devices must have excellent optical properties to avoid issues such as high losses. The gold standard material traditionally used for fabricating optical components was glass to fulfill these needs. In the past, processes of etching, deposition, photolithography, and so on were followed to manufacture microfluidic devices in quarts, silicon, and, once again, glass [13–17]. However, new technical strategies were needed because of many experienced drawbacks, such as limited geometries realization, expensive manufacturing processes due to the need to use cleanrooms and high costs for raw materials supply. For this reason, all the aforementioned demands and drawbacks related to more traditional manufacturing processes and raw materials were overcome by a new trend that relies on using polymer materials for 3D printing. Indeed, this strategy allowed the realization of cheaper MoF devices, characterized by complex geometric models that are difficult or impossible to produce by more conventional manufacturing technologies, suitable for mass production. Among the existing 3D printing techniques, which have ensured enhancements both in terms of resolutions and capabilities for microfabrication [18–22], stereolithography (SLA), MultiJet Printing (MJP), and fused deposition modeling (FDM) are the most used ones. Among the latter techniques, the SLA is the most performing one since it permits higher resolution, tighter tolerances, and compatibility with thermoset polymers, which are classified as optical-grade polymers [23] compared to MJP and FDM. For example, *Saitta* et al. [10] previously realized a MoF device for slug-flow detection in one piece by using a one-step manufacturing approach and achieving down-to-micron precision as an alternative to stereolithography. Moreover, the first 3D printed optical components made of plastic, dating back to 2015, were initially realized using thermoplastic materials, such as acrylonitrile butadiene styrene (ABS) and polyethylene terephthalate glycol (PETG). These materials were chosen for their commoditization and easy accessibility. However, their optical performance was found to be very limited due to significant losses in optical quality. This is why novel resins classified as optical-grade polymers (i.e., affected by low optical losses) were introduced [24]. The latter are cyclic olefin polymers (COP), poly(methylmethacrylate) (PMMA), polycarbonate (PC), and poly(styrene) (PS). In the last decade, Poly(dimethylsiloxane) (PDMS) has been selected as the preferred choice [25] due to its suitability for optical detection, which is related to its transparency and quite low refractive index value, i.e., $n = 1.4$. However, this polymer is suitable for casting fabrication techniques, such as soft lithography [26], for realizing micro and nano-structures. But, this type of manufacturing technique does not allow for the realization of devices in one piece, thus requiring the assembly of complex structures, causing issues such as poor bonding, imperfect alignment, etc. In this sense, the 3D printing approach may help overcome the aforementioned drawbacks by providing the opportunity to directly realize 3D objects by exploiting a layer-by-layer approach. For this reason, *Fleck* et al. [27] developed a high-resolution PDMS resin using a methacrylate–PDMS copolymer and the novel combination of a photoabsorber called Sudan I and a photosensitizer known as 2-Isopropylthioxanthone. This formulation exhibited a transmission >75%, which was combined with a digital light processing (DLP) 3D printer to assess the possibility of fabricating truly micron-scale parts with channels as small as 60 μm tall. Other examples of novel 3D printable polymer materials relying on photopolymerization processes have been developed by *Zhu* et al. [28] and *Shi* et al. [29].

Furthermore, there is a need for the development of novel transparent resin formulations that exhibit good optical properties and are compatible with vat photopolymerization (VP) 3D printing techniques. These advancements would enable the manufacturing of micro-parts with high accuracy and precision. The selection of materials with good optical properties, such as high transparency, high transmission percentage, and low refractive index, is essential in LOC to implement optical detection methods for analyzing biological or chemical samples [30–34].

Generally, the VP 3D printing techniques represent optimal manufacturing processes for various use cases across a wide variety of biotechnology industries, such as medical devices [35,36], drug delivery systems [37,38], and microfluidic devices [39,40]. The field of 3D printing offers numerous advantages that contribute to its attractiveness, including high print resolution, relatively high 3D printing speed, fairly low production cost, smooth 3D printed surfaces, and very high accuracy and fabrication precision, allowing for more intricate, exact, and replicable parts [38,41–43]. Among the VP 3D printing methods, digital light processing (DLP) and stereolithography (SLA) have shown an increasing potential for applications in the biotechnology field and, more specifically, in the LOC field due to the aforementioned benefits that distinguish them. DLP is a 3D printing technique whereby the photocurable resin is cured by a projector, which is the light source that projects an image onto the resin by curing the entire projected shape at once. Thus, in this case, the resolution, accuracy, and precision are determined by the projector's resolution (the current standard is 1080P), the optical character of the projection lens, and the tightness of the mechanical XYZ movement [44,45]. In the end, this 3D printing technique generally exploits a bottom-up approach. Hence, the use of support is required to secure the part of the building platform. Conversely, the SLA process exploits a top–down approach where it is not required to use supports in order to secure the bottom surface to the print surface. Furthermore, unlike the DLP 3D printing process, SLA processes use a laser to deliver concentrated light into a vat of resin, thus photopolymerizing the latter to fabricate a solid layer of the final 3D printed object in accordance with the initial design. Hence, XY resolution combines the laser's spot size and the increments by which the laser beam can be controlled [46].

In addition to these, among the VP techniques, an innovative technique has been recently patented (US Patent 9492969), which is named Projection Micro Stereolithography (PμSTL). It has been developed as a hybrid technique combining the advantages of DLP and SLA 3D printing processes to develop a 3D printing technique characterized by a higher level of resolution while simultaneously keeping unchanged the capability to manufacture large parts with a high level of tolerance and ultra-high resolution (down to 10 μm) [47].

Within the LOC field of application, for realizing MoF devices integrating optical elements, it becomes very crucial to determine how the novel 3D printable materials interact with electromagnetic waves. Accessing a knowledge base of optical properties is paramount, even though this information is not generally provided in the technical datasheet (TDS) of novel commercialized resins. In detail, it is important to run quantitative measures of how a material interacts with light. Thus, large efforts must be made to develop simple methodologies to estimate certain types of optical properties of interest, such as the refractive index value. Knowing the accurate value of the latter parameter of a material is crucial to make some predictions (thanks to simulation tools), allowing the prediction of the photon interaction, which may be reflection, absorption, or scattering at the interface [48]. In detail, the refractive index is a measure of the bending of light when moving from one medium into another. It is also defined as the ratio of the velocity of light in a vacuum to the velocity of light in a medium [49].

Many studies have been run over the years to provide refractive index value determination methods to meet this need [50–57]. However, most of them are suitable for analyzing biological or liquid samples. In accordance with the current state-of-the-art, accurate characterization methods such as refractometry, ellipsometry, and prism coupling measurements are commonly employed to assess the optical properties of samples. However, these methods are often expensive and highly demanding regarding the macro- and micro-scale planarity (very smooth surfaces) of the samples [58–61]. To solve this problem, *R. J. Nussbaumer* et al. [62] have proposed an immersion method to determine the refractive index of rough solids. This method relies on simple transmission measurements using an ordinary UV/vis spectrometer, avoiding the complex approach of searching for different liquids matching the solid's refractive index [57,63,64].

Starting from the latter work, the novelty of this experimental study consists of developing a novel polynomial regression model to enable refractive index measurements suitable for transparent 3D printable resins. In addition, to the best of our knowledge, a novel, simple, cost-effective, and efficient experimental setup is presented in this study for the first time. The proposed novel approach relies on ordinary transmission measurements and avoids sample immersion within a liquid to determine the considered optical property. In fact, the transmission measurements were conducted with air as the surrounding medium. This decision was justified by the fact that VA photopolymerization yields a high-quality surface finish characterized by a quite low roughness ranging between 0.04 and 2 μm. Furthermore, even the selected materials to develop the polynomial regression model, optical photocurable adhesives with a known and certified refractive index, presented a very low surface roughness (nanometric scale) [51] once the cross-linking process was completed.

In this work, a *quadratic polynomial regression model* has been developed to correlate the known *refractive index value (R)* of fully cured photocurable adhesives (NOA resins), chosen as an independent variable or predictor, to their empirically measured *transmission value (T)*, selected as a dependent variable or response. As soon as the regression model was empirically developed for a well-defined operative range for the refractive index parameter, i.e., ranging between 1.56 and 1.70, its equation was inverted. This inversion determined the unknown refractive index value of photocurable 3D printable resins, starting from their measured transmission value.

In the end, the optical performances of two different MoF devices were examined as an application in the LOC field. These devices were constructed using novel 3D printable resins and fabricated using an innovative VA 3D printing technique called Projection Micro Stereolithography (PμSTL). The optical performances of the 3D printed MoF devices were evaluated alongside those of a PDMS-based MoF device to provide a basis for comparison. The latter decision is justified by the fact that PDMS is considered the gold standard polymer material for both optical and microfluidic applications.

The obtained results allowed for the identification of a well-defined range of refractive index values for novel resins developed for LOC applications. Additionally, a simple and effective equation was derived to estimate their unknown refractive index value from empirical transmission measurements.

2. Materials and Methods

2.1. Materials

To experimentally determine the regression model that correlates the empirically measured transmission value of full-cured photocurable resins with their refractive index values, four different optical photocurable resins with a known refractive index value were selected:

(i). <u>NOA88</u>—it is a low-viscosity (250 cps) UV-curing adhesive with a refractive index equal to 1.56 at 589 nm, requiring 5 J/cm^2 of energy to fully cure a 25–50 micron bond, and having an absorption range ranging between 315 and 395 nm. According to the provided SDS (safety data sheet), it is a proprietary urethane-related resin-based formulation composed of a mixture of mercapto esters (the content of 50–65%) and triallyl isocyanuarte (the content of 30–55%);

(ii). <u>NOA160</u>—it is a high-viscosity (2200 cps) UV-curing adhesive with a refractive index equal to 1.60 at 589 nm, requiring 6 J/cm^2 of energy to fully cure a 25–50-micron bond. In accordance with the provided SDS (safety data sheet), it is a proprietary urethane-related resin-based formulation composed of a mixture of pentaerythritol tetra 3 (the content of 30–50%) and acrylic monomer (the content of 30–60%);

(iii). <u>NOA1665</u>—it is a UV-curing adhesive characterized by a viscosity of 800–1000 cps, with a refractive index equal to 1.665 at 589 nm, which requires 6 J/cm^2 of energy to fully cure a 25–50-micron bond. In line with the information provided by the SDS (safety data sheet), it is a proprietary urethane-related resin-based formulation composed of a mixture of metal oxide (the content of 30–60%) and acrylated resin (the content of 35–45%);

(iv). NOA170—it is a very high-viscosity (4500–5500 cps) UV-curing adhesive characterized by a high refractive index, i.e., equal to 1.70 at 589 nm, that needs 6 J/cm^2 of energy to fully cure a 25–50-micron bond. Based on the supplied SDS (safety data sheet), it is a proprietary acrylated-related resin-based formulation composed of a mixture of acrylated resin (content of 40–60%) and acrylic monomer (content of 10–30%).

The four selected Norland Optical Adhesives (NOA) for the study were purchased from Edmund Optics LTD (Nether Poppleton, York, UK), and their properties are summarized in Table 1.

Table 1. Norland Optical Adhesives (NOA) main properties.

NOA Commercial Name	Refractive Index Value	UV-Curing Energy
NOA88	1.56 @ 589 nm	5 J/cm^2
NOA160	1.60 @ 589 nm	6 J/cm^2
NOA1665	1.665 @ 589 nm	6 J/cm^2
NOA170	1.70 @ 589 nm	6 J/cm^2

Moreover, by using a casting procedure into a silicon mold, the upper surface (exposed to the air) of the NOA88 fabricated samples was characterized by a surface roughness expressed as average roughness (Ra), root mean square roughness (RMS), and peak-to-peak equal to 12.33 nm, 7.18 nm, and 140.12 nm, respectively. The authors previously determined the latter using atomic force microscopy (AFM) [51]. Similar results for the roughness can be assumed for the other NOA resin specimens since they were manufactured using the same procedure.

Next, three different materials were chosen to manufacture the micro-optofluidic devices for two-phase flow detection. The first device (called *Device 1* henceforth) was fabricated using the SYLGARDTM 184 Silicone Elastomer Kit. The latter presents a chemical composition that is a mixture of both a Poly(dimethylsiloxane) (PDMS) elastomer and a suitable curing agent and is also characterized by a refractive index value of 1.4118 at 589 nm (as declared by the provided TDS). It was purchased from Farnell Italia S.R.L. (Sylgard 184 elastomer kit, Dow Corning, 148 Midland, MI, USA). Moving on, both the second (called *Device 2* from now on) and the third (called *Device 3* henceforward) devices were realized by exploiting the highly innovative micro-precision 3D printing technique named Projection Micro Stereolithography (PμSL) and commercialized by *Boston Micro Fabrication Materials Technology Co., Ltd.* (BMF, Maynard, MA, USA). *Device 2* and *Device 3* were made of photocurable resins named HTL resin and BIO resin, respectively. The first one (HTL) is a high-performance, rigid, and heat-resistant yellow resin proprietary developed by Boston Micro Fabrication (BMF, Maynard, MA, USA).

Moreover, this resin grade is characterized by good temperature stability and mechanical properties, which make it suitable for autoclave sterilization. Next, in agreement with its TDS, this resin shows a heat distortion temperature (HDT) of 140.7 °C (@0.45 MPa) and a glass transition temperature (Tg) of 172 °C. The second photocurable resin (BIO) is a biocompatible resin suitable for non-implantable medical applications and can also undergo sterilization. It has passed several ISO 10993 biocompatibility tests (such as skin irritation and sensitization, toxicity, cytotoxicity, pyrogenicity, and in vitro hemolysis). It has shown a cell culture survival rate in vitro of 75%. The latter-mentioned properties justified the reason it was selected as a raw material for micro-optofluidic devices that find application in health and biomedical engineering. Furthermore, according to its TDS, it is a yellow resin showing a HDT of 85.7 °C (@0.45 MPa), which was developed by Boston Micro Fabrication (BMF, Maynard, Massachusetts) as well.

It must be highlighted that the refractive index of the used PDMS is provided by its TDS. This is because PDMS is a widely used material for optical applications in the state-of-the-art. It is unknown for the two yellow-grade photocurable resins used via PμSL 3D printing.

In the end, the roughness for *Device 1* (PDMS) and *Device 2* (HTL resin) was already determined by the authors using the AFM analysis [10]. The previously collected results are summarized in Table 2. Since *Device 2* (HTL resin) and *Device 3* (BIO resin) were realized using the same 3D printing technique, the latter can be assumed to have the same roughness range values measured for the HTL resin, as declared by the company providing the machine.

Table 2. *Device 1* and *Device 2* roughness measurements [10].

MoF Device	Material	Ra [nm]	RMS [nm]	Peak to Peak [nm]
Device 1	PDMS	1.097	0.763	35.158
Device 2	HTL resin	47.056	37.324	259.121

Specimens Preparation

Specimens having a size equal to (10 × 6 × 4) mm^3 were used for the transmission measurements. Using all the selected NOA materials (see Table 1), the specimens were prepared using the procedure reported in Figure 1. At first, each liquid selected for NOA was micro-injected using a syringe equipped with a 25 gauge needle (step I) within a silicone mold. Secondly, each photopolymer adhesive was irradiated with a universal lamp bulb with UVA emission at 365 nm (step II). In the end, four full-cured bars were obtained by providing the required UV-curing energy (step III).

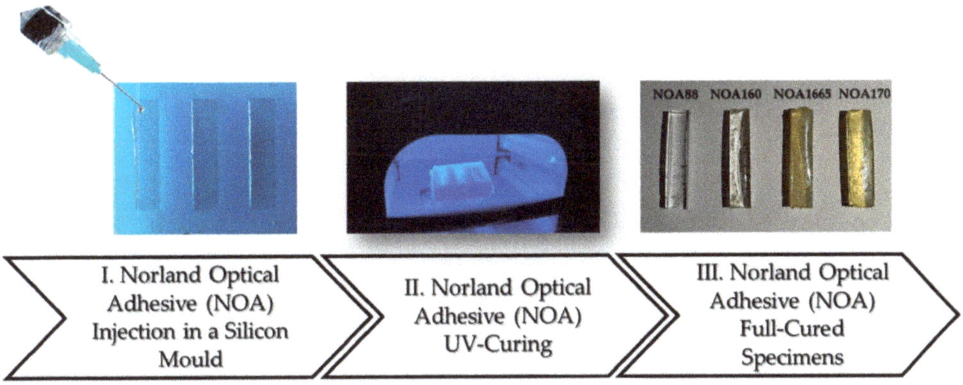

Figure 1. Followed the workflow for the fabrication of the used specimens for transmission measurements.

Next, specimens 4 mm thick made of the two yellow-grade photocurable 3D printable resins (BIO and HTL resins) were 3D printed using a microArch®S140 ultra-high resolution (10 μm) 3D printer (BMF, Maynard, MA, USA) for transmission measurements. The final 3D printed specimens are shown in Figure 2.

Figure 2. 3D printed BIO (**a**) and HTL (**b**) specimens for transmission value measurements.

2.2. Transmission Measurements: Experimental Setup

A block diagram and a schematic representation of the used transmission measurement system are shown in Figures 3a and 3b, respectively. In particular, it is composed of (i) the source light; (ii) the probe; (iii) the material sample; (iv) the spectrophotometer; and (v) a PC for the measurement acquisition through dedicated software.

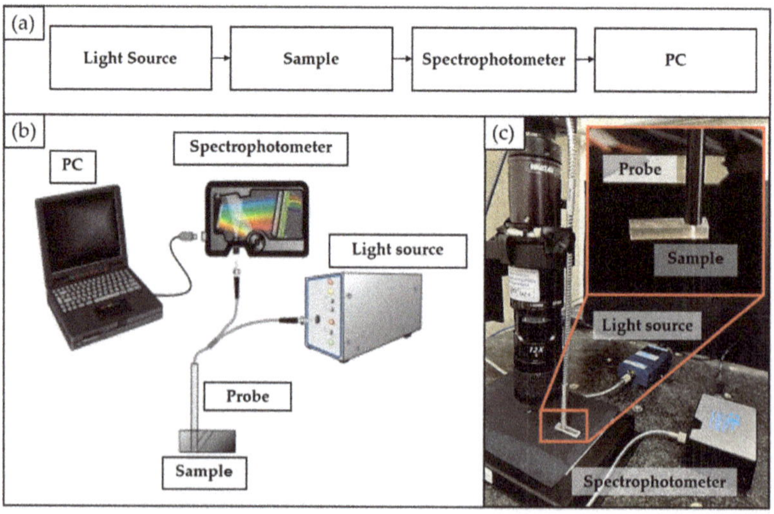

Figure 3. Experimental setup for transmission measurements: (**a**) block scheme; (**b**) schematic representation; (**c**) real picture.

The actual experimental setup is shown in Figure 3c. Specifically, a halogen light source (LS-1 Tungsten Halogen Light Source, Ocean Optics, Dunedin, FL, USA), providing visible light, was connected through a SMA connector to the probe (Avantes, Lafayette, CO, USA). The light beam coming from the light source was sent through an illumination fiber to the sample, and the reflection was measured by using a second fiber in the center of the probe's tip. The latter was positioned in close contact with the sample's surface (as shown in Figure 3c insert) to prevent the existence of different materials between the probe and the sample itself. This setup was implemented to minimize any potential impact on the transmission measurement that could arise from the presence of different materials. The second fiber was coupled through a SMA connector to a spectrophotometer (USB2000, Ocean Optics, Dunedin, FL, USA), configured to the appropriate wavelength range of interest (i.e., 200–1100 nm). The spectrophotometer was connected via a USB cable to the PC to collect the transmission measurements through the Spectra Suite dedicated software 2.0.

Transmission is defined as the percentage of energy passing through a system relative to the amount that passes through a reference sample. In the measurement protocol, to calibrate the instrument, the air was selected as a reference sample with a thickness equal to the investigated material samples (4 mm). The latter was associated with a transmission value of 100%. After the calibration phase, the transmission measurements were performed and acquired statically for each selected material sample (see Table 1), following the replicated general factorial design fully described in Section 2.3.

The collected transmission measurements were exported as a *.ProcSpec* file and then processed in MATLAB (MathWorks®). Specifically, the "*spectrum. wavelength*" and the "*spectrum.processedPixels*" MATLAB (MathWorks®) struct fields, corresponding to the wavelength and the transmission (in percentage), respectively, were considered to track the transmission measurement signal.

The transmission values used for the regression model development were extracted in correspondence with a wavelength equal to 589 nm to directly compare with the known refractive index values of the examined NOA resins. Indeed, according to their TDS, they are in the same condition.

2.3. Regression Model for Refractive Index Estimation

In this experimental study, the functional relationship existing between the known *refractive index value* (*R*) of fully cured photocurable adhesives (NOA resins), chosen as *independent variables* or *predictors*, and their empirically measured *transmission value (T)*, selected as a *dependent variable* or *response*, was estimated. A *quadratic polynomial regression model* has been considered:

$$T_i = \beta_0 + \beta_1 R_i + \beta_{11} R_i^2, \tag{1}$$

where β_0, β_1 and β_{11} are the *regression coefficients*. The matrix notation corresponding to this model is:

$$\underline{\boldsymbol{T}} = \boldsymbol{R}\,\underline{\beta}, \tag{2}$$

where $\underline{\boldsymbol{T}}$ is the (44 × 1) vector of observations, i.e., the measured transmission value; $\boldsymbol{R}$ is the (44 × 3) design matrix; and $\underline{\beta}$ is the (3 × 1) vector of the regression coefficients.

For the test procedure, only one design factor (independent variable) was considered, i.e., the type of material (factor A), which was varied among four different levels ($a = 4$). The $a = 4$ levels correspond to NOA88, NOA160, NOA1665, and NOA170. The measured transmission value for each selected material was considered as the response variable. The number of replications was equal to $n = 11$, for a total of $N = a \cdot n = 44$ runs. The experimental plan is summarized in Table 3.

Table 3. Experimental plan: factors and levels.

Factor	Symbol	Type	Unit	Levels (a = 4)			
Material	A	Categorical	[-]	NOA88	NOA160	NOA1665	NOA170

An analysis of variance (ANOVA) study was run to validate model assumptions in regression.

After developing the empirical regression model for a well-defined operative range for the refractive index parameter, i.e., ranging between 1.56 and 1.70, Equation (1) was properly inverted. This inversion process resulted in the derivation of an empirical equation to predict the unknown *refractive index value* for the photocurable 3D printable resins, starting from their measured *transmission value*:

$$\beta_{11} R^2 + \beta_1 R + \beta_0 - T = 0, \tag{3}$$

The roots of the second-order equation have been determined by exploiting the *roots* function in MATLAB (MathWorks®), which allows calculating the roots of a single-variable polynomial represented by a vector of coefficients. In detail, three acquisitions for the measured transmission value T were run for the two selected photocurable 3D printable resins (i.e., HTL and BIO). Using Equation (3), the estimated refractive index value R was determined for each acquisition. Finally, its estimate was expressed as its mean $\pm$ st. deviation.

2.4. Application: Micro-Optofluidic Device for Slug-Flow Detection

Once the estimated refractive index value was determined for the two novel photocurable 3D printable resins (BIO and HTL resins), this information was used to assess their suitability for micro-optofluidic device fabrication. For this purpose, a micro-optofluidic (MoF) device for two-phase fluid detection, which had been previously designed by the authors in earlier works [9–11], was manufactured through a direct 3D printing process using the two different investigated resins.

Two-phase flow refers to two immiscible fluids, such as gas–liquid, immiscible liquid–liquid, or liquid and microparticles, one dispersed in the other and circulating into the same micro-system [8,65]. Within this context, the optical method is worthy of consideration since it is very strategic for carrying out many less invasive measurements. The MoF device's working principle relies on the absorption phenomenon. This means that when the incident laser beam interacts with the two different phases present in the device, which are characterized by two different refractive index values, the transmission of light exhibits different characteristics. The nature of light transmission strongly depends on the fluid with which it interacts at a precise instant.

The optical properties of the two yellow-grade novel photocurable resins (BIO and HTL), related to their estimated refractive index values, were then compared with the PDMS ones. It is justified by the fact that the latter is the most commonly used polymeric material for optical applications since its refractive index is close to the glass one (1.4118 for PDMS versus 1.52 for glass), which is considered the gold standard material.

In detail, three different devices were fabricated, with variations in the manufacturing technique and the material used for their construction. Their optical performances were compared and tested according to the procedure described in Section 2.4.1. under the same operative conditions. *Device 1*, made of PDMS, was fabricated through an ad hoc master-slave fabrication approach relying on 3D printing, as previously explained elsewhere by the authors [1–3]. Both *Device 2* and *Device 3* were realized by exploiting the highly innovative micro-precision 3D printing technique, PμSL. The three fabricated devices are shown in Figure 4.

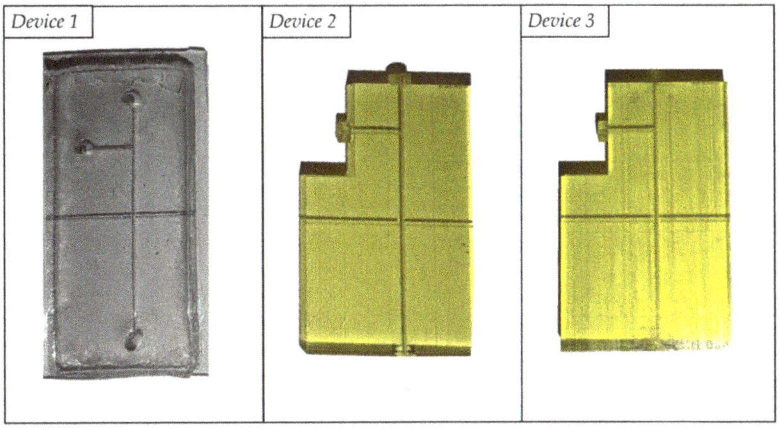

Figure 4. Manufactured micro-optofluidic devices made of PDMS (*Device 1*), HTL (*Device 2*), and BIO (*Device 3*) resins.

The detailed workflow followed for the three devices' manufacturing has already been explained elsewhere by the authors [9,10].

2.4.1. Optical Detection: Experimental Setup and Signals Processing

To assess the optical properties of the investigated photocurable 3D printable resins selected for the realization of the MoF devices, the optical signals were acquired by pumping either deionized water (characterized by a refractive index n_{Water} = 1.3) or air (characterized by a refractive index n_{Air} = 1.0), into the micro-channel. The micro-channels were composed of three different materials: PDMS (*Device 1*), HTL resin (*Device 2*), and BIO resin (*Device 3*). Both air and water were injected within the micro-optofluidic device's micro-channel through a syringe pump system (neMESYS low-pressure module, Cetoni, GmbH, Korbussen, Germany). A laser system (NovaPro 660–125, RGB Lasersystems, Kelhein, Germany) with an emission wavelength of 660 nm was used as a light source to detect the static fluid inside the micro-channel by exploiting three levels of input power, i.e., 1 mW, 3 mW, and 5 mW. The micro-optofluidic device was coupled with two 365 µm diameter optical fibers. The former was connected to the laser system for optical actuation. While, the latter, which was connected to a photodiode (PDA100A, Thorlabs, Newton, NJ, USA) with a gain of 40 dB, was used for the detection,. A PC oscilloscope (Picoscope 2204A, Pico Technology, Cambridgeshire, UK) with a sampling frequency of 1.5 kHz was used to acquire the detected signals. Figure 5 reports the blocking scheme and the real picture of the experimental setup for detecting optical signals.

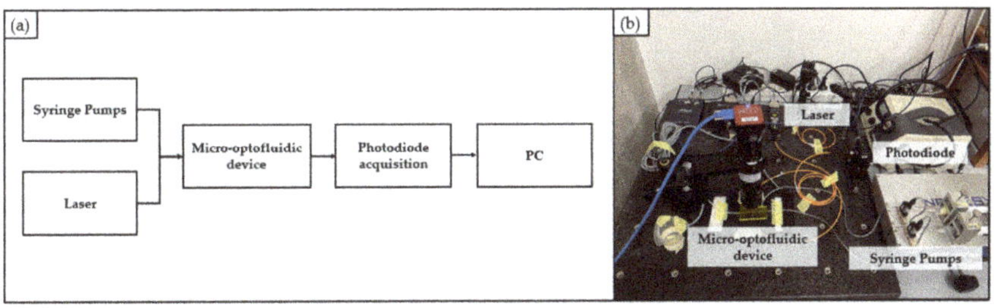

Figure 5. Experimental Setup for Optical Measurements: (**a**) block scheme; (**b**) real picture.

For each investigated scenario, the optical signal was acquired for a period of 30 s. Next, the acquired signals were post-processed in MATLAB (MathWorks®). In particular, a low-pass filter with a 40 Hz cut-off frequency was applied to eliminate high-frequency components. Then, a smoothing procedure was applied to remove the noise from the signal and reveal the main pattern. In this way, a single sample of $V(i)$ observations was obtained for each voltage-acquired signal for $i = 1, \ldots, N_v$, and $N_v = 45{,}000$. After this, the corresponding sample mean values ($\overline{V}_{Water}$ and $\overline{V}_{Air}$) and the sample standard deviations were evaluated for each sample $V(i)$. In the end, the voltage range (ΔV), defined as the difference between the average of the water signal ($\overline{V}_{Water}$) and the average of the air signal ($\overline{V}_{Air}$), was evaluated to prove that the optical part of the device is really able to discriminate between two different fluids (air and water). In fact, due to the difference between their refractive index values (1.3 for water versus 1.0 for air), the signals were detected at two different voltage values using the same laser input power: a higher voltage value for water and a lower voltage value for air.

3. Results and Discussion

3.1. Transmission Measurements Results

The average acquired transmission spectra for the four investigated materials are reported in Figure 6 in correspondence with the whole wavelength range of interest (i.e., 200–1100 nm).

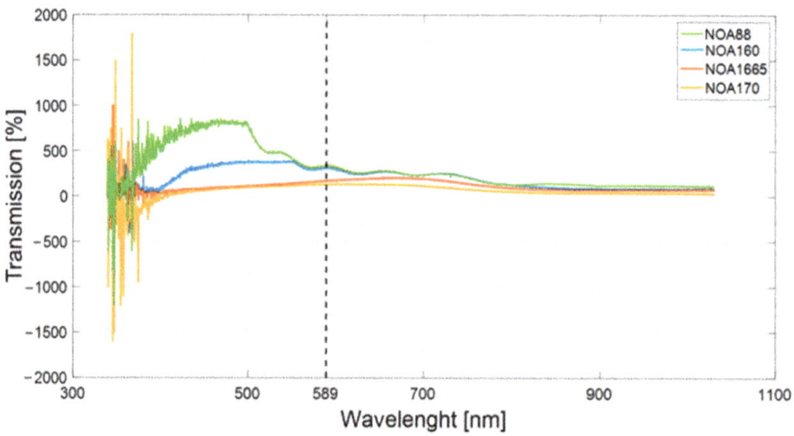

Figure 6. Average acquired transmission spectra for the analyzed Norland Optical Adhesives (NOA) adhesives.

The transmission value at a wavelength of 589 nm was extracted for each acquired spectrum, as reported in Figure 6 with the vertically dashed black line.

The collected measures for each investigated material are reported in the scatter plot in Figure 7a. The calculated average values and standard errors for the photocurable NOA resins are summarized in the bar plot shown in Figure 7b. According to the results, the NOA88 (in green in Figure 7) showed the highest average transmission value, 346.68 ± 7.44. The NOA160 (in blue in Figure 7) presented a decreased value for the transmission (328.57 ± 15.01) of about 5%. Next, for the NOA1665 (in orange in Figure 7), an average transmission value of 183.76 ± 28.93 was measured, reduced by about 44% and 47% compared to the NOA88 and the NOA160, respectively. In the end, the lowest average transmission value was recorded for the NOA170 (in yellow in Figure 7), equal to 122.65 ± 12.75, thus having a lower value of about 65% than the NOA88. The obtained decreasing trend is justified by the photocurable NOAs' refractive index values. The higher the refractive index value of the investigated material, the lower the measured transmission value. Indeed, starting from NOA88, which is fully transparent, the other considered resins became more matt and yellow-like.

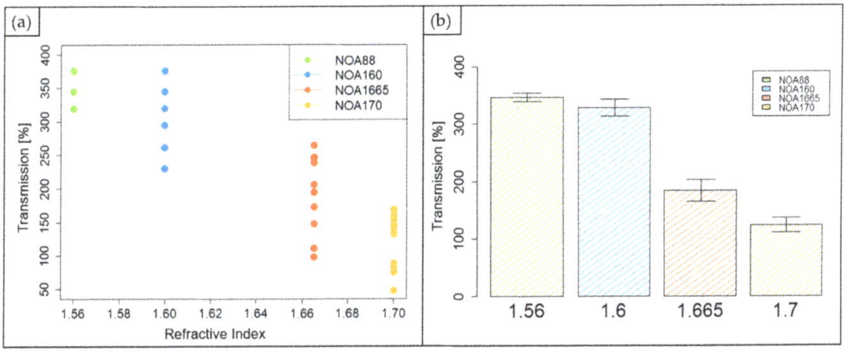

Figure 7. Scatter plot of the collected transmission measurements at 589 nm (**a**); bars plot showing the average transmission value for the four materials. Bar errors represent the standard errors of the collected measurements (**b**).

Furthermore, it must be highlighted that a lower standard error affects the measured transmission value of the NOA88 material sample. This is due to its greater homogeneity and full transparency, as shown in Figure 1.

3.2. Quadratic Polynomial Regression Model for Refractive Index Value Determination

The regression equation unveiling the true functional relationship existing between the known *refractive index value* (R) and the measured *transmission value* (T) of the fully-cured photocurable adhesives (NOA resins) is equal to:

$$T = -15{,}382 + 20{,}903\,R - 6933\,R^2, \quad (4)$$

An ANOVA study was performed to investigate the statistical significance of each term, i.e., both the linear and quadratic ones. The regression model is significant from the ANOVA study shown in Table 4 (p-value < 0.05). Moreover, since the values for R^2 (80.28%) and R^2-adj (79.31%) are quite high, it is possible to assume that the model is adequate since the noise level is very low. As a result, the model fits well with the collected observations. In accordance with the ANOVA results (Table 4), it is possible to notice that the linear term of the model is absolutely significant (p-value < 0.05). On the other hand, the quadratic term is slightly less significant (p-value $= 0.074 > 0.05$). However, it was added to the model since it allows a better fitting of the observations related to the lowest value ($R = 1.56$) of the operative range selected for the refractive index parameter. Figure 8 shows the fitting quadratic regression line. Figure 9 shows the model adequacy check by plotting residuals from the regression model versus fitted values and their normal probability plot. The two plots reveal no anomaly.

Table 4. ANOVA table for the selected quadratic polynomial regression model.

	Source	DF	SS	MS	F	p
Regression	Linear	1	379,065	-	154.76	0.000
	Quadratic	1	7815	-	3.37	0.074
	Error	41	95,059	2319		
	Total	43	481,939			
	S	48.1508				
	R^2	80.28%				
	R^2-adj	79.31%				

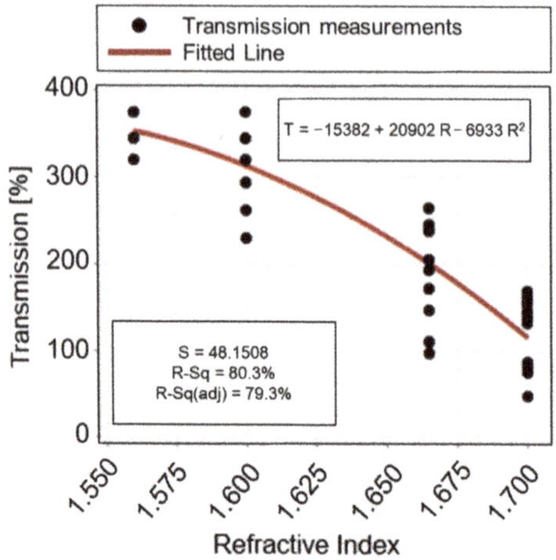

Figure 8. Fitted line plot: selected quadratic polynomial regression model fitting the collected observations (transmission values for NOA materials). R = *Refractive index* value of the *predictor*, and T = *Transmission* value of the investigated *response*.

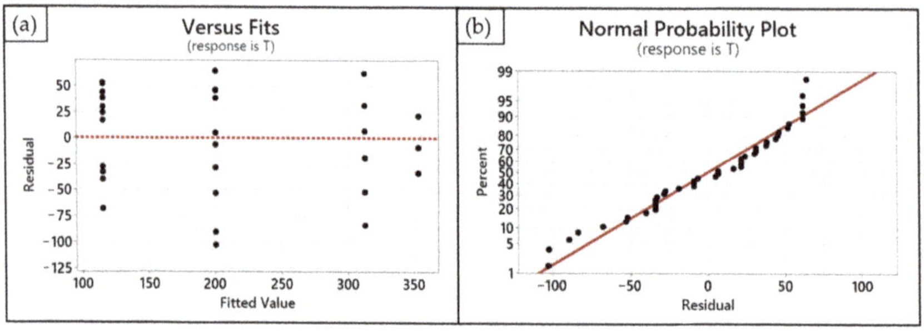

Figure 9. Residual analysis: residual versus fitted value plot (**a**) and normal probability plot for residual values (**b**).

Next, by inverting the regression model Equation (4), an expression is derived to predict the unknown *refractive index value* for the photocurable 3D printable resins, starting from their measured *transmission value*:

$$6933\,R^2 - 20{,}902\,R + (15{,}382 + T) = 0, \tag{5}$$

Three observations of the *transmission value* were collected for the BIO and HTL resins, respectively. The collected observations for the two considered photocurable 3D printable resins are reported in Table 5. Next, the roots of Equation (5) were calculated to determine the estimated refractive index value ($\hat{R}$) for each replication. In detail, two roots were determined for each considered resin; their values are reported in Table 6. The first root ($\hat{R}_1$) was never considered since it is broadly lower than the inferior limit of the operative range for the refractive index value (equal to 1.56). Conversely, the obtained values for $\hat{R}_2$ are strictly close to the upper limit (1.70). Moreover, since the investigated resins BIO and HTL are yellow-like (see Figure 2), as well as the NOA170 resin (see Figure 1), it is reasonable to consider the root $\hat{R}_2$, which is close in value to the refractive index of NOA170 resin.

Table 5. Collected transmission measurements for the photocurable 3D printable BIO and HTL resins for each replication (n = 1, 2, 3).

Resin	Replication (n)	T [%]
BIO	1	151.564
BIO	2	150.701
BIO	3	151.564
HTL	1	104.113
HTL	2	102.558
HTL	3	103.515

Table 6. Estimated refractive index value for BIO and HTL 3D printable resins for each replication (n = 1, 2, 3) of the transmission value measurements (T). Two roots for the second-order Equation (5) were determined for each measured T.

Resin	Replication (n)	$\hat{R}_1$	$\hat{R}_2$
BIO	1	1.3108	1.7040
BIO	2	1.3103	1.7046
BIO	3	1.3106	1.7043
HTL	1	1.3291	1.6858
HTL	2	1.3287	1.6861
HTL	3	1.3291	1.6858

The means of estimated refractive index values for the 3D printable BIO and HTL resins are shown in Figure 10, where they are also compared with the true value for the refractive index of PDMS (provided by its TDS).

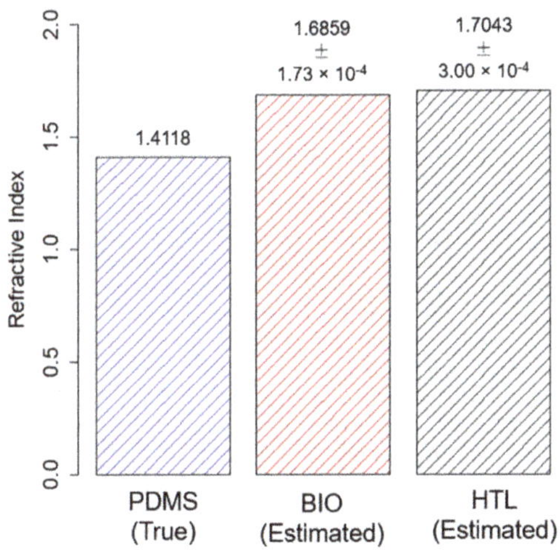

Figure 10. Bars plot for the estimated mean value of the refractive index parameter for photocurable 3D printable resins BIO and HTL, compared to the true value from the technical data sheet (TDS) for the PDMS. Bar errors were not added because they were significantly narrower than the bar height.

3.3. MoF Devices Optical Signals

The acquired optical signals by varying the fluid—air and water—injected within the MoF device's channels and the selected input laser power, 1, 3, and 5 mW, are reported in Figure 11a–f. Based on the obtained results, it was assessed that the estimated refractive index values for the two novel photocurable and 3D printable resins, BIO (n_{BIO} = 1.6859) and HTL (n_{HTL} = 1.7043), are higher than those of PDMS by approximately 19% and 21%, respectively. However, despite this difference, it was concluded that both resins are still suitable for the intended purpose. Using these two polymer resins for manufacturing the MoF devices for two-phase flow detection, it was still possible to acquire valuable optical signals while detecting both air and water in correspondence with the three levels of laser input power, according to the sample mean values reported in Table 7 for each investigated scenario.

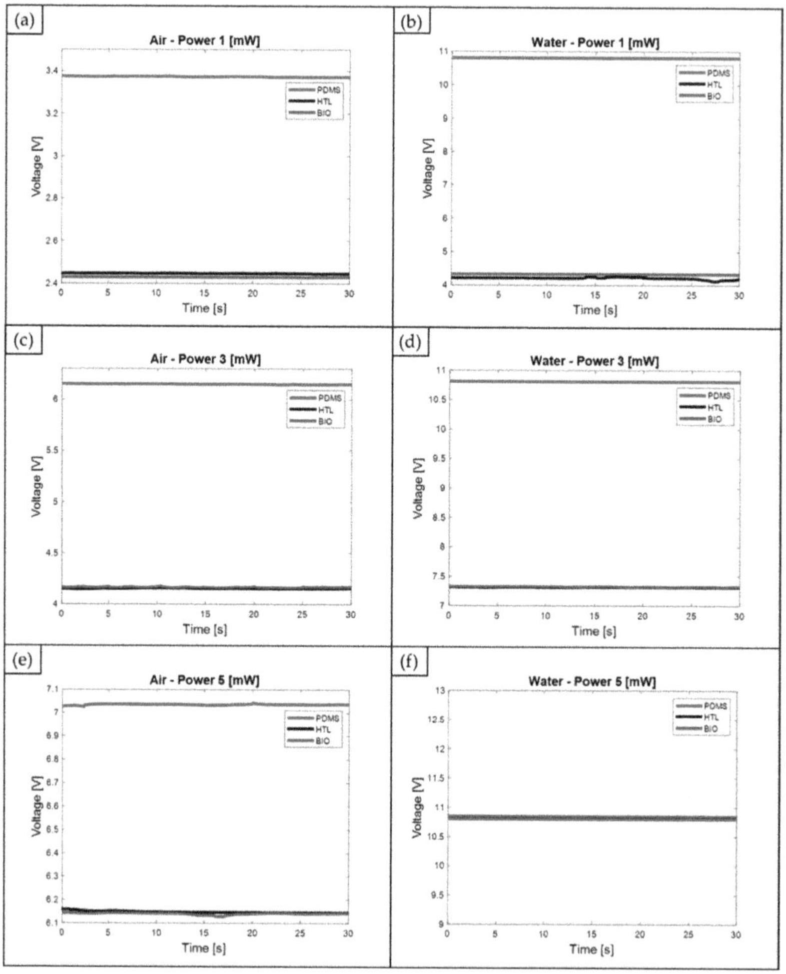

Figure 11. Acquired optical signals by means of the three MoF devices made of PDMS, BIO, and HTL resins: air detection at a laser power of 1 mW (**a**), 3 mW (**c**), and 5 mW (**e**); water detection at a laser power of 1 mW (**b**), 3 mW (**d**), and 5 mW (**f**).

Table 7. Sample mean value ± st. deviation observations [V] and calculated range (ΔV) [V] for all the investigated scenarios. Factors: Fluids = {Air, Water}; Input Laser Power = {1, 5} mW.

	PDMS			BIO			HTL		
	1 [mW]	3 [mW]	5 [mW]	1 [mW]	3 [mW]	5 [mW]	1 [mW]	3 [mW]	5 [mW]
Air	3.3743 ± 1.30×10^{-3}	6.1505 ± 1.10×10^{-3}	7.0920 ± 7.91×10^{-2}	2.4298 ± 1.90×10^{-4}	4.1666 ± 3.30×10^{-3}	6.0996 ± 7.02×10^{-2}	2.4279 ± 5.30×10^{-2}	4.1544 ± 2.10×10^{-3}	6.1471 ± 2.5×10^{-3}
Water	10.8127 ± 2.15×10^{-6}	10.8127 ± 2.15×10^{-6}	10.8127 ± 2.15×10^{-6}	4.3378 ± 2.5×10^{-3}	7.3244 ± 2.8×10^{-3}	10.8494 ± 8.26×10^{-4}	4.1809 ± 4.44×10^{-2}	7.4159 ± 2.4×10^{-3}	10.8487 ± 6.28×10^{-4}
Range	7.4	3.7	3.8	1.9	3.2	4.7	1.8	3.3	4.7

As shown in Figure 11a, the air voltage value at an input laser power of 1 mW is about 28% lower for the MoF devices made of BIO and HTL resins compared to the one detected from the PDMS device under the same conditions. By increasing the laser power to a value of 3 mW (Figure 11c), the difference in the air voltage value between the PDMS device and those made of BIO and HTL resins is reduced by approximately 32% for both of them. Moving to the water voltage values, those detected at a laser power of 1 mW (Figure 11b) for the MoF devices made of BIO and HTL resins are lower than 60% and 61%, respectively, compared to those of the PDMS device. Focusing on the collected voltage values for water detection at 3 mW (Figure 11d), for the BIO and HTL MoF devices, they are lower by 31% and 32%, respectively, than the PDMS one. When the input laser power is set at the highest considered value, i.e., 5 mW (Figure 11e), the difference in the air voltage value between the PDMS device and those made of BIO and HTL resins is reduced by approximately 14% and 13%, respectively.

Moving on to the last operative condition, {Fluid, Input Laser Power} = {water; 5 mW}, the voltage signals are practically equally recognized for the three different selected materials. In fact, the associated curves are almost overlapped, as shown in Figure 11f. This outcome is justified by the 10 V output voltage limit specified in the photodiode TDS. As a result, the saturation value is reached in correspondence with this operative condition (about 11 V), leading to the same water voltage value detected from the three different MoF devices. At this operative condition, it would seem that an optimum operating point has been reached for the two investigated 3D printable resins since it looks like the three devices can discriminate the two investigated fluids similarly. However, the photodiode output voltage limit must be taken into account, which, when the PDMS MoF is used for the detection, takes the detected water voltage signal at the saturation value even when the input laser power is set at 1 mW. Thus, by setting the laser power at 5 mW, it is highly probable that the actual water voltage signal detected from the PDMS device would be higher, providing a higher voltage range value than the resins' ones. This expected result would be consistent with the PDMS better discrimination capability, supported by its optical properties previously discussed.

The discussed signals were obtained through post-processing of the optical signal acquisition, which was conducted for 30 s, following the procedures outlined in Section 2.4.1. Both the calculated average values for a 30 s acquisition time window and the calculated voltage range (ΔV) are reported in Table 7.

The mean values of the acquired optical signals, related to the three devices and calculated in correspondence to an input laser power of 1 mW and 3 mW, are reported in Figure 12. In the operative condition with the laser power at 1 mW, the difference between the water and air signals was higher for the PDMS, being ΔV = 7.4 V, compared to the BIO and HTL devices, which have range values equal to ΔV = 1.9 V and ΔV = 1.8 V, respectively. This result shows that the PDMS device better discriminates the two investigated fluids since the gap between the water and the air values is larger. As illustrated in Figure 4, the full transparency of PDMS compared to BIO and HTL's yellow-like resins justifies this outcome. Raising the laser input power to a value of 3 mW makes the range values comparable for the three devices: ΔV = 3.7 V for PDMS, ΔV = 3.2 V for BIO, and ΔV = 3.3 V for HTL resins. Once again, this result is justified by the fact that, even though the PDMS is projected to have higher voltage values than yellow resins due to its higher transparency (see Figure 4) and lower refractive index value (n = 1.4118), it must be taken into account the photodiode output voltage limit. Indeed, saturation for the water voltage signal is already achieved at the lowest value set for the input laser power (1 mW), which does not increase proportionally with the laser power rise. Contrarily, similar voltage values for the BIO and HTL resins, which result in almost totally overlapped curves, are driven by their nearly identical estimated refractive index values (1.6859 for BIO versus 1.7043 for HTL).

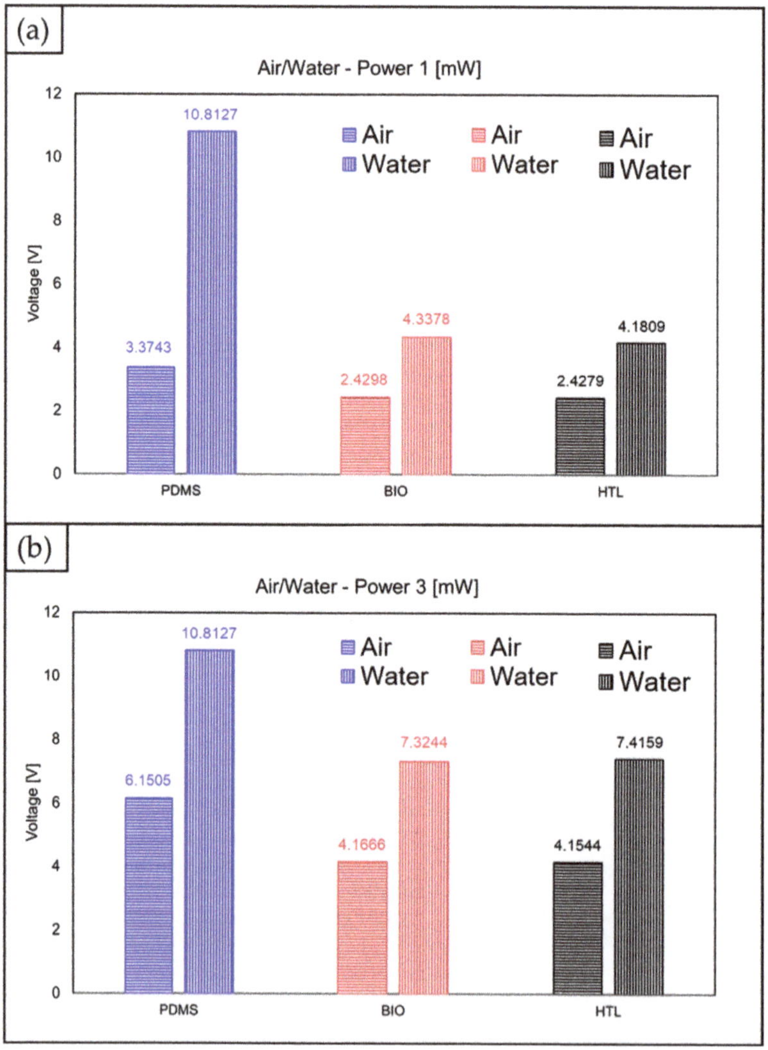

Figure 12. Mean values for each acquired optical signal from the three MoF devices made of PDMS (blue), BIO (red), and HTL (black) resins: detection at a laser power of 1 mW (**a**) and 3 mW (**b**) for either air (horizontal background) or water (vertical background). Bar errors were not added because they were significantly narrower than the bar height.

4. Conclusions

This experimental work proposed a novel method to estimate the refractive index value within an operative range of (n_{min} = 1.56; n_{max} = 1.70) of photocurable 3D printable resins characterized by a smooth surface. It relies on developing a quadratic polynomial regression model showing good performance, i.e., R^2 = 80%. Furthermore, to the best of our knowledge, a novel, simple, cost-effective, and efficient experimental setup is presented in this study for the first time.

The method was implemented to determine the refractive index value for two novel 3D printable resins (commercially named BIO and HTL) compatible with the innovative PμSTL 3D printing technique. The same was further used to manufacture MoF devices, which were first optically characterized to assess their suitability to run two-phase flow detection and control. Next, their optical performances were compared to the PDMS one, which was used as the material to realize the same device. Based on the obtained results, it was demonstrated that the investigated materials were suitable for the intended purpose. They were able to discriminate between two different fluids characterized by different refractive index values (air and water). However, it should be noted that their discrimination capability was lower than that of PDMS polymer, which is considered the gold standard polymer in the LOC application fields with a refractive index of n_{PDMS} = 1.4118. This result is justified because the estimated refractive index values for the BIO and HTL resins (n_{BIO} = 1.6859 and n_{HTL} = 1.7043, respectively) are higher than the PDMS one.

The obtained results paved the way for the opportunity to rapidly assess if novel resins developed for VA 3D printing techniques may be suitable for MoF devices where the optical method for non-invasive detection of biological and chemical fluids is exploited. Indeed, the two tested resins (BIO and HTL) were characterized by estimated refractive index values located near the upper limit of the operative range for the refractive index value, corresponding to the worst condition in terms of an optical property. All the novel 3D printable resins that will be developed in the near future can be considered suitable for MoF device manufacturing (relying on the same working principle) if their estimated refractive index value is lower than 1.70.

In this sense, Equation (5) provided in this paper has twofold utility. First, it represents a useful tool for unknown refractive index estimation of solid, transparent samples showing a smooth surface. Second, it provides a quick method to evaluate the suitability of novel developed 3D printable resins for the fabrication of MoF devices.

In the end, the proposed method for transmission value measurements can be used in future works to design and develop MoF devices with integrated surface plasmon resonance (SPR) sensors [66,67]. The perfect combination of refractive index values between the core and cladding elements is of utmost importance in such applications. Indeed, according to the waveguide working principle, to obtain a performing waveguide, the core must have a higher refractive index value than the cladding [24]. This class of MoF devices would represent a breakthrough in diagnostic technology, allowing for simultaneous *specific* and *non-specific* information collection from biological and chemical samples.

Author Contributions: Conceptualization, G.C. (Gianluca Cicala), L.S. and M.B.; methodology, E.C., G.S. and L.S.; software, E.C. and L.S.; investigation, E.C., G.S. and L.S.; data curation, E.C. and L.S.; writing—original draft preparation, L.S. and E.C.; writing—review and editing, C.T., G.C. (Gianluca Cicala), G.C. (Giovanni Celano), G.S. and M.B.; supervision, G.C. (Gianluca Cicala), G.C. (Giovanni Celano) and M.B.; project administration, G.C. (Gianluca Cicala) and M.B.; funding acquisition, G.C. (Gianluca Cicala) and M.B. All authors have read and agreed to the published version of the manuscript.

Funding: The research was partially funded by the European Union (NextGeneration EU) through the MUR-PNRR project SAMOTHRACE (ECS00000022) and partially funded by the University of Catania under the grant scheme PIACERI with the project MAF-moF "Materiali multifunzionali per dispositivi micro-optofluidici".

Institutional Review Board Statement: Not applicable.

Informed Consent Statement: Not applicable.

Data Availability Statement: Data will be made available on request.

Conflicts of Interest: The authors declare no conflict of interest.

References

1. Scott, S.M.; Ali, Z. Fabrication Methods for Microfluidic Devices: An Overview. *Micromachines* **2021**, *12*, 319. [CrossRef] [PubMed]
2. Zhou, W.; Le, J.; Chen, Y.; Cai, Y.; Hong, Z.; Chai, Y. Recent Advances in Microfluidic Devices for Bacteria and Fungus Research. *TrAC Trends Anal. Chem.* **2019**, *112*, 175–195. [CrossRef]
3. McDonald, J.C.; Duffy, D.C.; Anderson, J.R.; Chiu, D.T.; Wu, H.; Schueller, O.J.; Whitesides, G.M. Fabrication of Microfluidic Systems in Poly(Dimethylsiloxane). *Electrophoresis* **2000**, *21*, 27–40. [CrossRef]
4. Fiorini, G.S.; Chiu, D.T. Disposable Microfluidic Devices: Fabrication, Function, and Application. *Biotechniques* **2005**, *38*, 429–446. [CrossRef]
5. Whitesides, G.M. The Origins and the Future of Microfluidics. *Nature* **2006**, *442*, 368–373. [CrossRef]
6. Convery, N.; Gadegaard, N. 30 Years of Microfluidics. *Micro Nano Eng.* **2019**, *2*, 76–91. [CrossRef]
7. Gale, B.K.; Jafek, A.R.; Lambert, C.J.; Goenner, B.L.; Moghimifam, H.; Nze, U.C.; Kamarapu, S.K. A Review of Current Methods in Microfluidic Device Fabrication and Future Commercialization Prospects. *Inventions* **2018**, *3*, 60. [CrossRef]
8. Minzioni, P.; Osellame, R.; Sada, C.; Zhao, S.; Omenetto, F.G.; Gylfason, K.B.; Haraldsson, T.; Zhang, Y.; Ozcan, A.; Wax, A.; et al. Roadmap for Optofluidics. *J. Opt.* **2017**, *19*, 093003. [CrossRef]
9. Cairone, F.; Davi, S.; Stella, G.; Guarino, F.; Recca, G.; Cicala, G.; Bucolo, M. 3D-Printed Micro-Optofluidic Device for Chemical Fluids and Cells Detection. *Biomed. Microdevices* **2020**, *22*, 37. [CrossRef]
10. Saitta, L.; Celano, G.; Cicala, G.; Fragalà, M.E.; Stella, G.; Barcellona, M.; Tosto, C.; Bucolo, M. Projection Micro-Stereolithography versus Master–Slave Approach to Manufacture a Micro-Optofluidic Device for Slug Flow Detection. *Int. J. Adv. Manuf. Technol.* **2022**, *120*, 4443–4460. [CrossRef]
11. Stella, G.; Barcellona, M.; Saitta, L.; Tosto, C.; Cicala, G.; Gulino, A.; Bucolo, M.; Fragalà, M.E. 3D Printing Manufacturing of Polydimethyl-Siloxane/Zinc Oxide Micro-Optofluidic Device for Two-Phase Flows Control. *Polymers* **2022**, *14*, 2113. [CrossRef] [PubMed]
12. Dawson, H.; Elias, J.; Etienne, P.; Calas-Etienne, S. The Rise of the Om-Loc: Opto-Microfluidic Enabled Lab-on-Chip. *Micromachines* **2021**, *12*, 1467. [CrossRef] [PubMed]
13. Harrison, D.J.; Fluri, K.; Seiler, K.; Fan, Z.; Effenhauser, C.S.; Manz, A. Micromachining a Miniaturized Capillary Electrophoresis-Based Chemical Analysis System on a Chip. *Science* **1993**, *261*, 895–897. [CrossRef] [PubMed]
14. Jacobson, S.C.; Hergenroeder, R.; Koutny, L.B.; Ramsey, J.M. Open Channel Electrochromatography on a Microchip. *Anal. Chem.* **2002**, *66*, 2369–2373. [CrossRef]
15. Jacobson, S.C.; Hergenroeder, R.; Koutny, L.B.; Ramsey, J.M. Fused Quartz Substrates for Microchip Electrophoresis. *Anal. Chem.* **2002**, *67*, 2059–2063. [CrossRef]
16. Matzke, C.M.; Kottenstette, R.J.; Casalnuovo, S.A.; Frye-Mason, G.C.; Hudson, M.L.; Sasaki, D.Y.; Manginell, R.P.; Wong, C.C. Microfabricated Silicon Gas Chromatographic Microchannels: Fabrication and Performance. In Proceedings of the Micromachining and Microfabrication Process Technology IV, Santa Clara, CA, USA, 20–24 September 1998; Volume 3511, pp. 262–268. [CrossRef]
17. Moore, A.W.; Jacobson, S.C.; Ramsey, J.M. Microchip Separations of Neutral Species via Micellar Electrokinetic Capillary Chromatography. *Anal. Chem.* **1995**, *67*, 4184–4189. [CrossRef]
18. Sochol, R.D.; Sweet, E.; Glick, C.C.; Wu, S.Y.; Yang, C.; Restaino, M.; Lin, L. 3D Printed Microfluidics and Microelectronics. *Microelectron. Eng.* **2018**, *189*, 52–68. [CrossRef]
19. Bhattacharjee, N.; Urrios, A.; Kang, S.; Folch, A. The Upcoming 3D-Printing Revolution in Microfluidics. *Lab Chip* **2016**, *16*, 1720–1742. [CrossRef]
20. Macdonald, N.P.; Cabot, J.M.; Smejkal, P.; Guijt, R.M.; Paull, B.; Breadmore, M.C. Comparing Microfluidic Performance of Three-Dimensional (3D) Printing Platforms. *Anal. Chem* **2017**, *89*, 3858–3866. [CrossRef]
21. Waheed, S.; Cabot, J.M.; Macdonald, N.P.; Lewis, T.; Guijt, R.M.; Paull, B.; Breadmore, M.C. 3D Printed Microfluidic Devices: Enablers and Barriers. *Lab Chip* **2016**, *16*, 1993–2013. [CrossRef]
22. Au, A.K.; Lee, W.; Folch, A. Mail-Order Microfluidics: Evaluation of Stereolithography for the Production of Microfluidic Devices. *Lab Chip* **2014**, *14*, 1294–1301. [CrossRef]
23. Stella, G.; Saitta, L.; Ongaro, A.E.; Cicala, G.; Kersaudy-Kerhoas, M.; Bucolo, M. Advanced Technologies in the Fabrication of a Micro-Optical Light Splitter. *Micro* **2023**, *3*, 338–352. [CrossRef]
24. Luo, Y.; Canning, J.; Zhang, J.; Peng, G.D. Toward Optical Fibre Fabrication Using 3D Printing Technology. *Opt. Fiber Technol.* **2020**, *58*, 102299. [CrossRef]
25. Whitesides, G.M. Controlling Flows in Microchannels with Patterned Surface Charge and Topography. *Acc. Chem. Res.* **2003**, *36*, 597–604. [CrossRef]
26. Xia, Y.; Whitesides, G.M. Soft Lithography. *Angew. Chem. Int. Ed.* **1998**, *37*, 550–575. [CrossRef]

27. Fleck, E.; Sunshine, A.; Denatale, E.; Keck, C.; McCann, A.; Potkay, J. Advancing 3d-Printed Microfluidics: Characterization of a Gas-Permeable, High-Resolution Pdms Resin for Stereolithography. *Micromachines* **2021**, *12*, 1266. [CrossRef]
28. Zhu, Y.; Ramadani, E.; Egap, E. Thiol Ligand Capped Quantum Dot as an Efficient and Oxygen Tolerance Photoinitiator for Aqueous Phase Radical Polymerization and 3D Printing under Visible Light. *Polym. Chem.* **2021**, *12*, 5106–5116. [CrossRef]
29. Shi, X.; Zhang, J.; Corrigan, N.; Boyer, C. Controlling Mechanical Properties of 3D Printed Polymer Composites through Photoinduced Reversible Addition-Fragmentation Chain Transfer (RAFT) Polymerization. *Polym. Chem.* **2022**, *13*, 44–57. [CrossRef]
30. Chao, T.-C.; Ros, A. Microfluidic Single-Cell Analysis of Intracellular Compounds. *J. R. Soc. Interface* **2008**, *5* (Suppl. S2), S139–S150. [CrossRef] [PubMed]
31. Eriksson, E.; Sott, K.; Lundqvist, F.; Sveningsson, M.; Scrimgeour, J.; Hanstorp, D.; Goksör, M.; Granéli, A. A Microfluidic Device for Reversible Environmental Changes around Single Cells Using Optical Tweezers for Cell Selection and Positioning. *Lab Chip* **2010**, *10*, 617–625. [CrossRef]
32. Zheng, B.; Roach, L.S.; Ismagilov, R.F. Screening of Protein Crystallization Conditions on a Microfluidic Chip Using Nanoliter-Size Droplets. *J. Am. Chem. Soc.* **2003**, *125*, 11170–11171. [CrossRef] [PubMed]
33. Chiu, D.T.; Jeon, N.L.; Huang, S.; Kane, R.S.; Wargo, C.J.; Choi, I.S.; Ingber, D.E.; Whitesides, G.M. Patterned Deposition of Cells and Proteins onto Surfaces by Using Three-Dimensional Microfluidic Systems. *Proc. Natl. Acad. Sci. USA* **2000**, *97*, 2408–2413. [CrossRef] [PubMed]
34. Huber, D.L.; Manginell, R.P.; Samara, M.A.; Kim, B.-I.; Bunker, B.C. Programmed Adsorption and Release of Proteins in a Microfluidic Device. *Science* **2003**, *301*, 352–354. [CrossRef] [PubMed]
35. Paunović, N.; Bao, Y.; Brian Coulter, F.; Masania, K.; Karoline Geks, A.; Klein, K.; Rafsanjani, A.; Cadalbert, J.; Kronen, P.W.; Kleger, N.; et al. Digital Light 3D Printing of Customized Bioresorbable Airway Stents with Elastomeric Properties. *Sci. Adv.* **2021**, *7*, eabe9499. [CrossRef] [PubMed]
36. Maity, N.; Mansour, N.; Chakraborty, P.; Bychenko, D.; Gazit, E.; Cohn, D. A Personalized Multifunctional 3D Printed Shape Memory-Displaying, Drug Releasing Tracheal Stent. *Adv. Funct. Mater.* **2021**, *31*, 2108436. [CrossRef]
37. Borandeh, S.; van Bochove, B.; Teotia, A.; Seppälä, J. Polymeric Drug Delivery Systems by Additive Manufacturing. *Adv. Drug Deliv. Rev.* **2021**, *173*, 349–373. [CrossRef]
38. Robles-Martinez, P.; Xu, X.; Trenfield, S.J.; Awad, A.; Goyanes, A.; Telford, R.; Basit, A.W.; Gaisford, S. 3D Printing of a Multi-Layered Polypill Containing Six Drugs Using a Novel Stereolithographic Method. *Pharmaceutics* **2019**, *11*, 274. [CrossRef]
39. Bhattacharjee, N.; Parra-Cabrera, C.; Kim, Y.T.; Kuo, A.P.; Folch, A. Desktop-Stereolithography 3D-Printing of a Poly (Dimethylsiloxane)-Based Material with Sylgard-184 Properties. *Adv. Mater.* **2018**, *30*, e1800001. [CrossRef]
40. Au, A.K.; Huynh, W.; Horowitz, L.F.; Folch, A. Mikrofluidik Aus Dem 3D-Drucker. *Angew. Chem.* **2016**, *128*, 3926–3946. [CrossRef]
41. Bao, Y. Recent Trends in Advanced Photoinitiators for Vat Photopolymerization 3D Printing. *Macromol. Rapid Commun.* **2022**, *43*, 2200202. [CrossRef]
42. Yu, C.; Schimelman, J.; Wang, P.; Miller, K.L.; Ma, X.; You, S.; Guan, J.; Sun, B.; Zhu, W.; Chen, S. Photopolymerizable Biomaterials and Light-Based 3D Printing Strategies for Biomedical Applications. *Chem. Rev.* **2020**, *120*, 10695–10743. [CrossRef] [PubMed]
43. Bagheri, A.; Jin, J. Photopolymerization in 3D Printing. *ACS Appl. Polym. Mater.* **2019**, *1*, 593–611. [CrossRef]
44. Chaudhary, R.; Fabbri, P.; Leoni, E.; Mazzanti, F.; Akbari, R.; Antonini, C. Additive Manufacturing by Digital Light Processing: A Review. *Prog. Addit. Manuf.* **2022**, *8*, 331–351. [CrossRef]
45. Zhang, Z.-C.; Li, P.-L.; Chu, F.-T.; Shen, G. Influence of the Three-Dimensional Printing Technique and Printing Layer Thickness on Model Accuracy. *J. Orofac. Orthop.* **2019**, *80*, 194–204. [CrossRef]
46. Martín-Montal, J.; Pernas-Sánchez, J.; Varas, D. Experimental Characterization Framework for SLA Additive Manufacturing Materials. *Polymers* **2021**, *13*, 1147. [CrossRef]
47. Ge, Q.; Li, Z.; Wang, Z.; Kowsari, K.; Zhang, W.; He, X.; Zhou, J.; Fang, N.X. Projection Micro Stereolithography Based 3D Printing and Its Applications. *Int. J. Extrem. Manuf.* **2020**, *2*, 022004. [CrossRef]
48. Zhang, Z.F.; Ye, F.; Ma, X.; Zhao, W.; Wang, H. Dependence of Macro-Bending Loss on Bending Configuration of Multimode Optical Fibers Studied by Ray-Tracing Simulation. *Opt. Rev.* **2020**, *27*, 290–295. [CrossRef]
49. SCAD College of Engineering and Technology; Institute of Electrical and Electronics Engineers. *Comprehensive and Analytical Review on Optical Fiber Refractive Index Sensor*; IEEE Xplore: Piscataway, NJ, USA, 2020; ISBN 9781728155180.
50. Cennamo, N.; Saitta, L.; Tosto, C.; Arcadio, F.; Zeni, L.; Fragalá, M.E.; Cicala, G. Microstructured Surface Plasmon Resonance Sensor Based on Inkjet 3d Printing Using Photocurable Resins with Tailored Refractive Index. *Polymers* **2021**, *13*, 2518. [CrossRef]
51. Saitta, L.; Arcadio, F.; Celano, G.; Cennamo, N.; Zeni, L.; Tosto, C.; Cicala, G. Design and Manufacturing of a Surface Plasmon Resonance Sensor Based on Inkjet 3D Printing for Simultaneous Measurements of Refractive Index and Temperature. *Int. J. Adv. Manuf. Technol.* **2023**, *124*, 2261–2278. [CrossRef]
52. Khan, R.; Gul, B.; Khan, S.; Nisar, H.; Ahmad, I. Refractive Index of Biological Tissues: Review, Measurement Techniques, and Applications. *Photodiagn. Photodyn. Ther.* **2021**, *33*, 102192. [CrossRef]
53. Sequeira, F.; Cennamo, N.; Rudnitskaya, A.; Nogueira, R.; Zeni, L.; Bilro, L. D-Shaped POF Sensors for Refractive Index Sensing—The Importance of Surface Roughness. *Sensors* **2019**, *19*, 2476. [CrossRef]
54. Lee, S.Y.; Jeong, T.Y.; Jung, S.; Yee, K.J. Refractive Index Dispersion of Hexagonal Boron Nitride in the Visible and Near-Infrared. *Phys. Status Solidi B Basic Res.* **2019**, *256*, 1800417. [CrossRef]

55. Liu, M.; Plum, E.; Li, H.; Li, S.; Xu, Q.; Zhang, X.; Zhang, C.; Zou, C.; Jin, B.; Han, J.; et al. Temperature-Controlled Optical Activity and Negative Refractive Index. *Adv. Funct. Mater.* **2021**, *31*, 2010249. [CrossRef]
56. Alsaad, A.M.; Al-Bataineh, Q.M.; Ahmad, A.A.; Albataineh, Z.; Telfah, A. Optical Band Gap and Refractive Index Dispersion Parameters of Boron-Doped ZnO Thin Films: A Novel Derived Mathematical Model from the Experimental Transmission Spectra. *Optik* **2020**, *211*, 164641. [CrossRef]
57. Niskanen, I.; Lauri, J.; Yokota, M.; Heikkila, R.; Hashimoto, T.; Fabritius, T. Determination of the Refractive Index of Particles through the Immersion Solid Matching Method. *IEEE Trans. Instrum. Meas.* **2021**, *70*, 1–5. [CrossRef]
58. Tompkins, H.G.; Jovanovich, H.B.; San, B.; New, D.; London, Y.; Toronto, S.T. *A User's Guide to Ellipsometry*; Courier Corporation: North Chelmsford, MA, USA, 2006.
59. Sharda, T.; Soga, T.; Jimbo, T. Optical Properties of Nanocrystalline Diamond Films by Prism Coupling Technique. *J. Appl. Phys.* **2003**, *93*, 101–105. [CrossRef]
60. Kersten, R. A New Method for Measuring Refractive Index and Thickness of Liquid and Deposited Solid Thin Films. *Opt. Commun.* **1975**, *13*, 327–329. [CrossRef]
61. Sultanova, N.G.; Kasarova, S.N.; Nikolov, I.D. Characterization of Optical Properties of Optical Polymers. *Opt. Quantum Electron.* **2013**, *45*, 221–232. [CrossRef]
62. Nussbaumer, R.J.; Halter, M.; Tervoort, T.; Caseri, W.R.; Smith, P. A Simple Method for the Determination of Refractive Indices of (Rough) Transparent Solids. *J. Mater. Sci.* **2005**, *40*, 575–582. [CrossRef]
63. Ojena, S.M.; De Forest, P.R. Precise Refractive Index Determination by the Immersion Method, Using Phase Contrast Microscopy and the Mettler Hot Stage. *J. Forensic Sci. Soc.* **1972**, *12*, 315–329. [CrossRef]
64. Stoiber, R.E.; Morse, A.S. The Immersion Method. In *Crystal Identification with the Polarizing Microscope*; Springer: Boston, MA, USA, 1994; pp. 49–75.
65. Schembri, F.; Bucolo, M. Periodic Input Flows Tuning Nonlinear Two-Phase Dynamics in a Snake Microchannel. *Microfluid. Nanofluidics* **2011**, *11*, 189–197. [CrossRef]
66. Hassani, A.; Skorobogatiy, M. Design Criteria for Microstructured-Optical-Fiber-Based Surface-Plasmon-Resonance Sensors. *J. Opt. Soc. Am. B* **2007**, *24*, 1423–1429. [CrossRef]
67. Debnath, N.; Live, L.S.; Poudineh, M. A Microfluidic Plasma Separation Device Combined with a Surface Plasmon Resonance Biosensor for Biomarker Detection in Whole Blood. *Lab Chip* **2023**, *23*, 572–579. [CrossRef] [PubMed]

polymers

MDPI

Article

Three-Dimensional Printable Flexible Piezoelectric Composites with Energy Harvesting Features

Mihaela Aradoaei [1], Romeo C. Ciobanu [1,*], Cristina Schreiner [1], Marius Paulet [1], Alina R. Caramitu [2], Jana Pintea [2] and Mihaela Baibarac [3]

[1] Department of Electrical Measurements and Materials, Gheorghe Asachi Technical University, 700050 Iasi, Romania; mihaela.aradoaei@academic.tuiasi.ro (M.A.); cristina-mihaela.schreiner@academic.tuiasi.ro (C.S.); mpaulet@tuiasi.ro (M.P.)
[2] National Institute for Research and Development in Electrical Engineering ICPE-CA, 030138 Bucharest, Romania; alina.caramitu@icpe-ca.ro (A.R.C.); jana.pintea@icpe-ca.ro (J.P.)
[3] National Institute of Materials Physics, Atomistilor Street 405A, P.O. Box MG-7, 077125 Bucharest, Romania; barac@infim.ro
* Correspondence: r.c.ciobanu@tuiasi.ro or rciobanu@yahoo.com

Abstract: The purpose of this work was to obtain an elastic composite material from polymer powders (polyurethane and polypropylene) with the addition of $BaTiO_3$ until 35% with tailored dielectric and piezoelectric features. The filament extruded from the composite material was very elastic but had good features to be used for 3D printing applications. It was technically demonstrated that the 3D thermal deposition of composite filament with 35% $BaTiO_3$ was a convenient process for achieving tailored architectures to be used as devices with functionality as piezoelectric sensors. Finally, the functionality of such 3D printable flexible piezoelectric devices with energy harvesting features was demonstrated, which can be used in various biomedical devices (as wearable electronics or intelligent prosthesis), generating enough energy to make such devices completely autonomous only by exploiting body movements at variable low frequencies.

Keywords: flexible composites with piezoelectric features; 3D printable flexible piezoelectric devices; energy harvesting

Citation: Aradoaei, M.; Ciobanu, R.C.; Schreiner, C.; Paulet, M.; Caramitu, A.R.; Pintea, J.; Baibarac, M. Three-Dimensional Printable Flexible Piezoelectric Composites with Energy Harvesting Features. *Polymers* **2023**, *15*, 2548. https://doi.org/10.3390/polym15112548

Academic Editors: Anton Ficai and Cristina-Elisabeta Pelin

Received: 1 May 2023
Revised: 25 May 2023
Accepted: 29 May 2023
Published: 31 May 2023

1. Introduction

The past 10 years have witnessed enormous interest in the efficient capture of environmental energy through the development of energy-harvesting devices that transform mechanical energy into electricity and greatly reduce our dependence on fossil fuels and CO_2 emissions. The application segments for piezoelectric energy harvesters based on ferroelectrics have dramatically expanded during the past 5 years. For several applications, ferroelectric micro- and nano-crystals with a defined one- (1D), two- (2D), and three-dimensional (3D) shape are of great scientific and technological interest because of their spontaneous polarization as well as their shape- and size-dependent properties. The subject of moldable piezoelectric sensors with energy harvesting features is very actual; research in this area extensively reports mainly thermo-rigid devices based on epoxy, PVDF, or other composite structures, mainly with PZT powders, such as [1–13].

Regarding the analysis of costs vs. benefits of adapted technology and the life cycle of such devices, we note that the basic literature largely supports piezoelectric energy as being applicable in special-purpose niches with high market impact, such as wireless sensor networks (IoT). The harvesting of piezoelectric energy must be interpreted through direct indicators: the power required for the real purpose and the surplus power, which is unused and stored over time. A decision model for analyzing costs vs. benefits would include aspects such as power consumption (both active and stored), physical size/energy density requirements, storage battery technology, and working circuit technology. For example, the CC2630 Zigbee TI controller uses 1.0 μA in passive mode, whereas the CC2538 uses between 1.3 and 600 μA. In "sleep mode", consumption is 0.1 vs. 0.4 μA.

The overall cost is also given by the relevant and precise characterization of the physical conditions that contribute to energy harvesting: the occurrence of dynamic force, vibration frequency, temperature range over time, etc. Depending on the complexity of the equipment, evaluations are made of potential improvements in energy efficiency, price, normalized price, and lifetime. As for the lifetime, the energy harvesting technologies are guaranteed for a minimum of 10 years, being made on the basis of robust materials, a sufficient element to make them very competitive on the renewable energy market, and for the technological design according to the forecasts of microelectronic applications.

From the point of view of technological feasibility, the advantages of piezoelectric generators are mentioned as being simple structures, easy to manufacture on a large scale (mass production), on the one hand, and easy to implement in electronic systems of the 'energy harvesting' type.

The new trend is defined by flexible energy-harvesting devices, which are yet more difficult to manufacture. Some achievements are presented as [14–28], but they are practically never combined with 3D printing technology, which is the interest of our research and the subject of the actual paper. In general, a wide range of composites made of ceramic powders and polymers can be processed by 3D printing technology; however, the receipt at the nanoscale is a key factor in successful part fabrication. The main unmet need is related to the difficulty of adapting any new variant of composite with individual thermo-mechanical characteristics on any 3D printing machine, and there is a must—achieved only by an extended R&D activity—the technological balance between the desired features of filament material, on the one side, and 3D printing capabilities, on the other side, in order to create the best characteristics of the obtained component. The values of material density, viscosity, and surface tension must be correlated. When the ratio is too small, viscous forces predominate, which implies high pressure for ejection; inversely, if this ratio is too large, a continuous column is ejected, which can lead to the formation of satellite drops behind the main drop. The rheology of thermoplastic nano-composite and its behavior at different processing temperatures are problems related to compound melting temperature, stability in the quasi-liquid phase, and final product isotropy. If mixed dissimilar materials are used for nano-composites, e.g., ceramic powder/CNT with a polymer matrix, even if the product is homogenous and stable after extruding—when the filament is generated—it may suffer important physical modifications during the subsequent stage of 3D printing technology. The flowability of powders in quasi-liquid phases is an essential parameter for 3D processing. Sufficient flowability allows the building of high 3D resolution [29–31].

The 3D architectural design of printable piezoelectric components involved a good knowledge of the adsorption of thermoplastic polymers on the surface of ceramic nanoparticles. In the stages prior to the realization of the 3D printable flexible piezoelectric composites with energy harvesting features, the following results were reported by the authors [32–34] as preliminary methods of preparing films containing thermoplastic polymers and their composites with $BaTiO_3$, along with some optical and structural properties.

Our study presents an example of good practice under the circumstances that it is obvious that by an extruding process pellets from thermoplastic composites may be manufactured, even if the inorganic powder content goes towards 80–90%, but it is also well known that the manufacture of rigorous filament in terms of diameter tolerance and reasonable length from thermoplastic composites with more than 10–15% inorganic powders is extremely difficult, mainly due to the fragility of the filament, which in many cases breaks itself under the reeling operation. That is why a tailored compounding of thermoplastic olefin with thermoplastic polyurethane was used. On the other hand, at higher powder contents, the filament homogeneity can suffer important changes under cooling operations due to dilatation, exfoliation, or delamination processes. Another major difficulty noticed by many practitioners refers to the thermal printer adaptation for such composites with a high content of inorganic powders, which makes the actual commercial printers useless for such applications, mainly when higher printing precision (micrometer scale) is compulsory.

The innovation presented by the present study includes a clear demonstration of the conditions imposed on such composites to become 3D printable, i.e., starting from the compounding and filamentation stage, through the 3D thermal deposition of composite filament with no clogging of the nozzle, no problems with multilayer deposition, and no 3D printing defects, to generating structures of rectangular network type with micro-meter precision, along with innovative tests to demonstrate their use as piezoelectric sensors within a dedicated signal processing circuit and the use of virtual instruments to assess their energy harvesting features.

2. Materials and Preparation Methods

2.1. Materials

In order to obtain the composite structures, the following raw materials were used: a thermoplastic olefin TPO (for the presented study, polypropylene TIPPLEN H 318), a thermoplastic polyurethane TPU (Estane 58,887 TPU), and $BaTiO_3$ powder of a maximum 2 micron dimension from Sigma Aldrich (Merck KGaA, Darmstadt, Germany). The present study completes and particularizes the previous research presented in [34] and refers only to TPO:TPU 2:1 type composites with $BaTiO_3$ powder content up to 35% in order to emphasize their piezoelectric features.

2.2. Processing Equipment

The polymers as powders (TPU and TPO—herein polypropylene) and the $BaTiO_3$ powder were homogenized by mixing for one hour in a cylindrical mixer with a 1.3 L capacity TURBULA T2F type with a rubber ring holding device, and the rotation speed was 40 rpm. In this way, we sought to obtain a uniform distribution of the components of the mixtures throughout the structure without using specific additives or adhesives for compatibility. Processing conditions involved a rotation speed (in counter-rotation) of the extruder of 95 rpm and a feed speed from the feed hopper of 450 rpm.

The injection of composite was performed on a Dr. Boy 35A injection machine (Dr. Boy GmbH & Co. KG, Neustadt-Fernthal Germany) with the following characteristics: a screw diameter of 28 mm, an L/D ratio of 18.6 mm, a calculated injection capacity of 58.5 cm^3, a maximum material pressure of 2200 bar, and a real injection capacity minimum of 500 mm. The interface of the injection machine for obtaining the composite materials and the temperature regime on the areas of the cylinders of the injection machine are briefly presented in Figure 1.

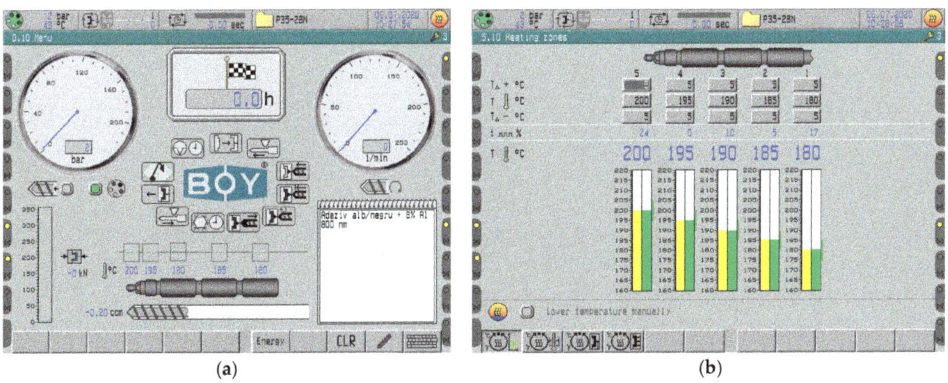

(a) **(b)**

Figure 1. (**a**) The interface of the injection machine, and (**b**) the temperature regime on the areas of the cylinder of the injection machine.

2.3. *Characterization Equipment*

The characterization of the composite materials was carried out with the use of the following equipment:

- The photoluminescence (PL) spectra of composites were recorded with a Fluorolog-3 spectrophotometer, FL3-2.2.1 model, from Horiba Jobin Yvon (Palaiseau, France), with some expertise related in [35].
- SEM optical scanning microscopy was performed with a field emission and focused ion beam scanning electron microscope (SEM) model Tescan Lyra III XMU (Brno—Kohoutovice, Czech Republic).
- The AFM analysis was performed with a WIKO NT 1100 type interferometric microscope in accordance with ISO 10109-7:2001 [36].
- Dielectric features were determined by using the Broadband Dielectric Spectrometer (Novocontrol GMBH), which encompasses an Alpha frequency response analyzer and Quattro temperature controller with tailored measurement cells. The manufactured samples were sandwiched between two copper electrodes of 20 mm diameter and placed inside the temperature-controlled cell [37].
- The piezoelectric features were measured using an Aixact TF Analyzer 2000-Electric Hysteresis Curve Lift System (static and dynamic hysteresis); the voltage that can be applied to the sample is +/− 100 V to +/− 10 kV). The device measures bulk samples (maximum diameter of 20 mm and a maximum thickness of 1.8 mm). The hysteresis curves were raised starting with a frequency of 0.1 Hz at an electric voltage 20% lower than the breakdown voltage of the samples. Simultaneously, the displacement of the sample can be measured with a laser interferometry system.
- TG/DSC thermal analysis, performed on a STA 449 F3 Jupiter TG-DSC simultaneous thermal analyzer, Netzsch (Selb, Germany), working in the temperature range up to 1550 °C, in an inert, oxidizing, reducing, static, or dynamic working atmosphere. The device is provided with a vacuum system with a maximum of 10^{-2} mbar.

3. Results and Discussion at the Composite Level

3.1. *Photoluminescence Spectrum*

The photoluminescence spectrum of thermoplastic composites exhibits a maximum at 462 nm when the excitation wavelength used to record the photoluminescence spectrum is 350 nm. Anisotropy (r) and bond angle (ϕ) of macromolecular compounds, adsorbed on the surface of inorganic particles, can be calculated using anisotropic photoluminescence with the following formulas:

$$r = (I_{vv} - GI_{vH})/(Ivv + 2GI_{vH})$$
$$r = 0.4[(3\cos^2\phi - 1)/2],$$

where I_{vH} corresponds to the photoluminescence intensity when the polarizer is positioned vertically for excitation and horizontally for emission in the spectrophotometer; $G = I_{HV}/I_{HH}$, where I_{HV} corresponds to the light intensity measured when the polarizer is positioned horizontally for excitation and the polarizer is positioned vertically for emission.

The photoluminescence spectra in polarized light of composites with $BaTiO_3$ nanoparticles having concentrations of 12 wt.%, 25 wt.%, and 35 wt.%, respectively, are presented in Figure 2. In Table 1, the values of the intensity of the photoluminescence spectra (PL) are presented, taking into account the way of mounting the polarizers for excitation and emission. According to Table 1, the gradual decrease in the intensity of the PL spectra of TPU:TPO, regardless of the way of mounting the polarizers, as the concentration of inorganic nanoparticles increases, indicates the role of $BaTiO_3$ as quenching agent of polymeric matrix's PL. This result is in agreement with our previous paper [34]. Considering the I_{VV}, I_{VH}, I_{HV}, and I_{VV} values experimentally obtained, which are presented in Table 1, the r and ϕ values in the case of: (i) TPU:TPO 2:1 are equal to 0.302 and 23.70; (ii) TPU:TPO 2:1 + 12% $BaTiO_3$ are equal to 0.0862 and 46.30; (iii) TPU:TPO 2:1 + 25% $BaTiO_3$ equal to

0.078 and 47.10; and (iv) TPU:TPO 2:1 + 35% $BaTiO_3$ equal to 0.029 and 51.80, respectively. In regard to the anisotropy, the composite with 12% $BaTiO_3$ presents the maximum value and is explained by a more homogenous dispersion of particles with different spatial orientations, which is a phenomenon expected for homogenous dispersions with low quantities of particles. In regard to the bond angle, the maximum value is reached by the composite with 35% $BaTiO_3$, which is explained by the composite architecture and will be further analyzed by SEM optical scanning microscopy. The value of r is less than 0.4, a fact that indicates that the excitation and emission transition dipoles are not aligned, which explains the complex dielectric polarization presented further in the paper. This aspect originates in the exchange reaction of the repeating units of the TPU-type polymer with $BaTiO_3$ [33] because it is expected for the dipole of TPU to make the main contribution to both interfacial and dipolar polarizations.

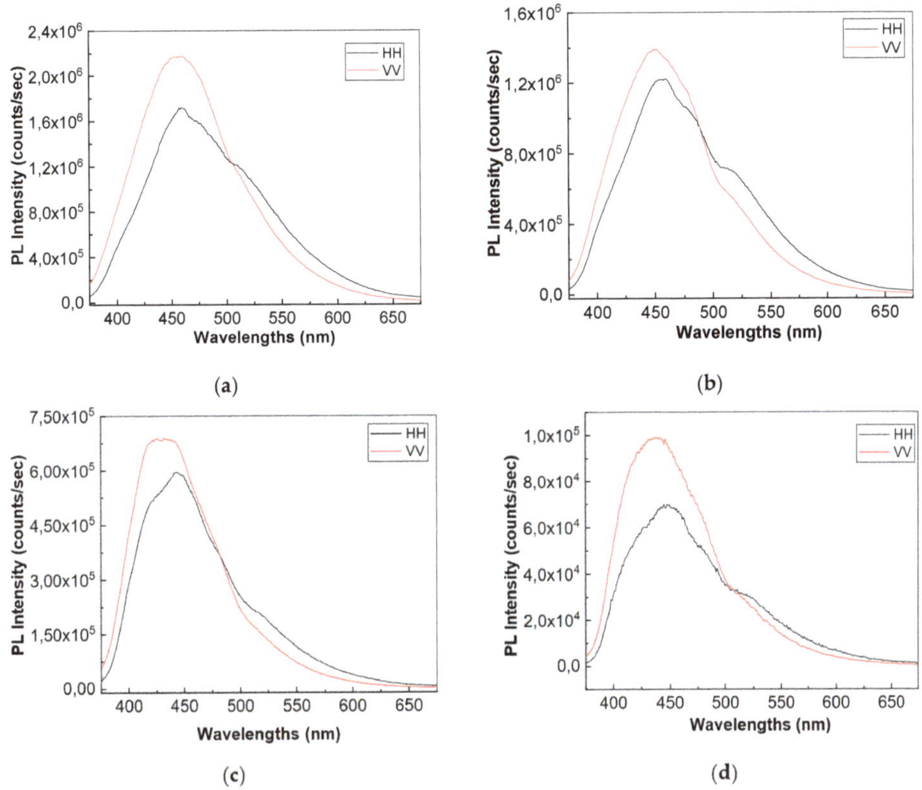

Figure 2. Photoluminescence spectra in polarized light of the following: (**a**) TPU:TPO 2:1; (**b**) TPU:TPO 2:1 + 12% $BaTiO_3$, (**c**) TPU:TPO 2:1 + 25% $BaTiO_3$, and (**d**) TPU:TPO 2:1 + 35% $BaTiO_3$.

Table 1. I_{HH}, I_{HV}, I_{VH}, and I_{VV} values of the PL spectra of the TPO:TPU 2:1 compound and its composites with $BaTiO_3$ nanoparticles.

Compound	I_{HH} (counts/s)	I_{HV} (counts/s)	I_{VH} (counts/s)	I_{VV} (counts/s)
TPO:TPU 2:1	1.72×10^6	1.8×10^6	1.78×10^6	2.17×10^6
TPO:TPU 2:1 + 12% $BaTiO_3$	1.22×10^5	1.17×10^5	1.13×10^5	1.39×10^5
TPO:TPU 2:1 + 25% $BaTiO_3$	5.96×10^5	6.08×10^5	5.38×10^5	6.87×10^5
TPO:TPU 2:1 + 35% $BaTiO_3$	6.99×10^4	8.22×10^4	7.73×10^4	9.91×10^4

3.2. SEM Structural Analyses

SEM structural analyses were performed to highlight the degree of homogeneity of the obtained materials. These analyses were performed on the field emission source and focused ion beam scanning electron microscope. The average dimension of BaTiO3 parti-cles was about 1 µm, (Figure 3 a,b).

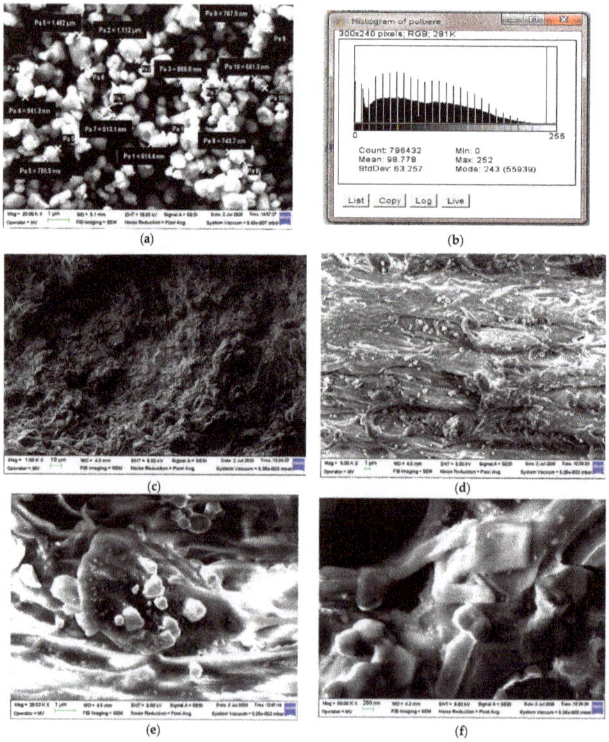

Figure 3. Micrograph of $BaTiO_3$ with 50,000× magnification (**a**), and related $BaTiO_3$ particle histogram (**b**); Micrographs of composite with 35% $BaTiO_3$ with magnifications (**c**) 1000×, (**d**) 5000×, (**e**) 20,000× and (**f**) 50,000×.

It can be generally seen that the filler is well homogenized within the polymer structure with a uniform dispersion, Figure 3c–f. Though slight agglomerations of the filler particles may appear, which may induce a slight inhomogeneity; however, in small, negligible areas, this aspect can be remediated by a longer mixing time and homogenizing the $BaTiO_3$ powder at a higher powder content.

3.3. Atomic Force Microscopy (AFM) Analysis

Basically, the experimental models made of polymer composite materials are very difficult to withstand atomic force microscopy tests because, for the AFM microscopy tests, a surface with nanometric roughness (preferably approx. 50 nm) is required. The roughness was determined as the average value between three measurements performed on the same sample in its central area at a focal distance of 90 µm. Roughness was obtained in the range of 75–450 nm, which gives some minor errors under the circumstances presented as above. In Figure 4, an AFM microscopy image is presented for the composite with 35% $BaTiO_3$, which is considered more relevant due to the higher content of powder, and the resulted roughness was below 100 nm. The roughness of such composites may be relevant in the case of producing filaments for 3D printing purposes, and the obtained value is in line with the requirements for such applications.

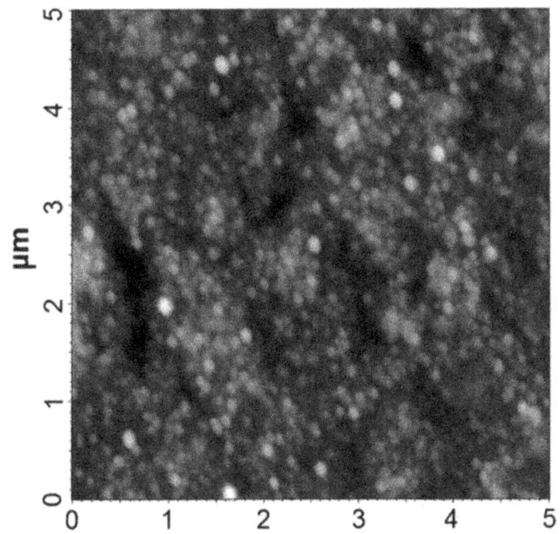

Figure 4. AFM microscopy image of the composite with 30% $BaTiO_3$.

A quasi-uniform distribution of the particles on the surface of the composite material can be observed as a whole, eventually depending on the size of the particle and its concentration. In the case of inorganic powders ($BaTiO_3$), the effect appears more blurred, but there is also a subjective cause given by the optical contract of such powders. These preliminary conclusions from the AFM analysis can be clearly correlated with those from the SEM analysis.

3.4. Dielectric Tests

The dielectric characteristics of the composite materials are presented in Figures 5 and 6, for the blended matrix the thermoplastic composite with 12% $BaTiO_3$, and the thermoplastic composite with 35% $BaTiO_3$.

It is obvious that the addition of $BaTiO_3$ dramatically increases the dielectric permittivity, from about 3.5 of the matrix towards about 13 for an addition of 12% $BaTiO_3$ and finally towards about 28 for an addition of 35% $BaTiO_3$ at 10 Hz frequency. With the increase in frequency, all permittivity values are decreasing, with the largest decrease being for the composite with the higher content of $BaTiO_3$.

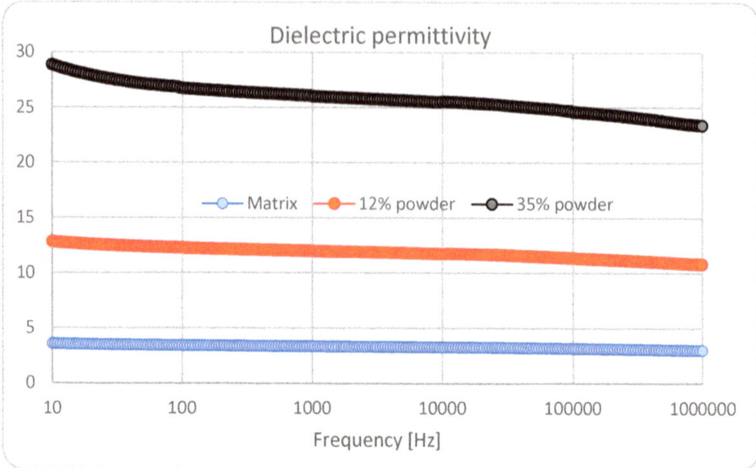

Figure 5. Relative permittivity.

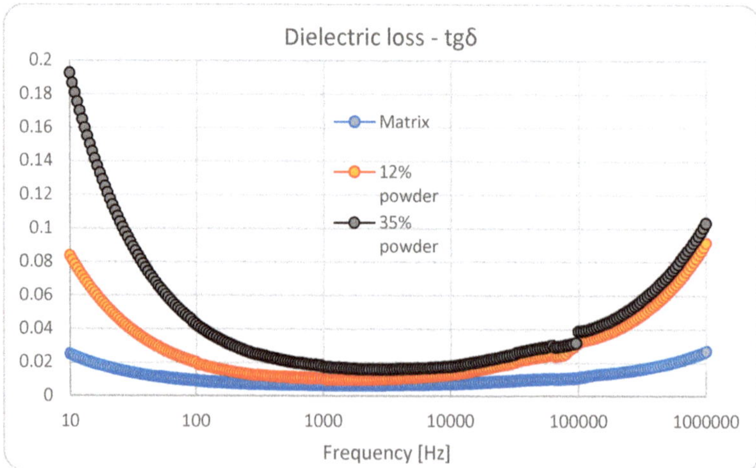

Figure 6. Dielectric loss (Tg Delta).

In regard to the dielectric loss, at lower frequencies, the effect of interfacial polarization is detrimental, with a very significant value for an addition of 35% $BaTiO_3$. Herewith, we speak about three types of complex interfaces: one related to blended matrix and the others related to polymers and $BaTiO_3$. That is why we can notice an increased value at low frequencies for blended matrixes as well. At medium frequencies in the kHz domain, the difference is not so relevant, but at higher frequencies, over 50 kHz, the dielectric polarization effect is obvious, with relevant higher values for the highest content of $BaTiO_3$.

On the other hand, when analyzing the dielectric loss characteristics of the samples with $BaTiO_3$, an interesting variation effect with frequency may be noticed in the vicinity of the 100 kHz frequency domain. Here we can speak about a saltation of the tgδ characteristic, correlated with the architecture of composites, which induces an additional ionic-dipolar conjugated polarization, e.g., a displacement due to the balance between the resonance and anti-resonance frequencies. Such phenomena explain the use of such composites with piezoelectric powders as resonators/filters for tailored applications in the radiofrequency electronic field. In our case, even if such phenomena occur, the low quantity of $BaTiO_3$

powder cannot make the respective composites relevant candidates for such electronic applications. But this phenomenon can be further exploited in relation to 3D printing in order to achieve a quasi-4D effect with frequency. Such an effect may become useful when collecting electrical signals from 3D printable flexible piezoelectric devices to be obtained from such composites for energy harvesting purposes, mainly when they are used for biomedical applications—communications at MHz frequencies.

As long as for reasonable piezoelectric effects the concordance of the highest value of three parameters is needed, i.e., the content of $BaTiO_3$, dielectric permittivity, and dielectric loss at lower frequencies, we may admit that the most relevant sample is the one with 35% $BaTiO_3$, which will be analyzed with priority as follows.

3.5. Piezoelectric Characteristics

The piezoelectric characteristics for the composite with 35% $BaTiO_3$ without preliminary activation under a tailored electric field are presented in Figure 7 at three different frequencies. The respective frequencies were chosen taking into account the potential use of such composites for 3D printable flexible piezoelectric devices with energy harvesting features for biomedical applications, i.e., in conjunction with wearable electronics or intelligent prosthesis.

The values in each table are specific for the hysteresis curve obtained at different frequencies and clearly indicate an alternating current ferroelectricity of the composite.

The values expressed as positive P_{max} and negative P_{max} show the symmetry of the curve at the maximum polarization level. The values expressed as negative Pr and positive Pr show a small difference due to the fact that the hysteresis curves do not fully close, but this aspect is characteristic of $BaTiO_3$. Vc values depend on applied voltage and can be further correlated with the size of the granules, but this is not relevant in our case. Depending on the voltage applied to the samples, different hysteresis curves are obtained, and the maximum polarization is correlated with the applied voltage values. By increasing the frequency value of the applied voltage, a flatter curve is obtained. The angle of inclination of the curve depends on the concentration of the inorganic powder; e.g., an angle of about 45° is observed at 10 Hz for the addition of 35% $BaTiO_3$.

The hysteresis losses, the most relevant parameters for piezoelectric analysis, are correlated with the volume of the hysteresis curve and expressed by W_{loss}. The ideal curve for analyzing the piezoelectric features of the composite is the one at 0.1 Hz, where the W_{loss} value reaches 2556.79 $\mu J/cm^2$.

The results are remarkable for a material without preliminary activation under a tailored electric field, and the relatively high polarization at low frequencies is in line with body reactions, which can be used as a source of energy harvesting via the piezoelectric effect. Further, the features of such composites with activation under a tailored electric field will be presented after the related piezoelectric devices are obtained by 3D printing technology from the filaments achieved with composites with $BaTiO_3$.

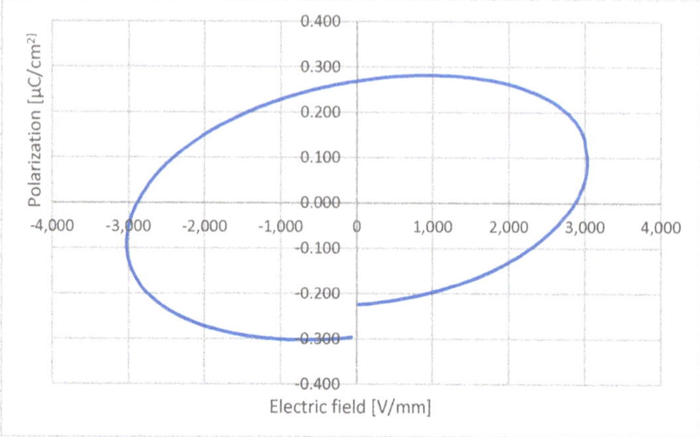

Composite with 35%BaTiO₃
Surface [mm²]: 484
Thickness [mm]: 1
Intensity of current [μA]: 10
Electric field [V/mm]: 3000 V
Frequency [Hz]: 0.1
Vc^+ [V]: 2866.51
Vc^- [V]: −2888.91
Pr^+ [μC/cm²]: 0.26942
Pr^- [μC/cm²]: −0.296397
P_{max+} [μC/cm²]: 0.0886311
P_{max-} [μC/cm²]: −0.0886311
W_{loss} [μJ/cm²]: 2556.79

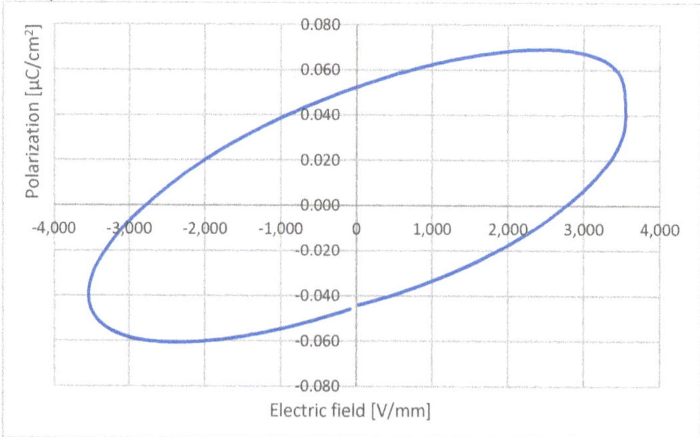

Composite with 35%BaTiO₃
Surface [mm²]: 484
Thickness [mm]: 1
Intensity of current [μA]: 10
Electric field [V/mm]: 3500 V

Figure 7. *Cont.*

Frequency [Hz]: 1
Vc^+ [V]: 2760.88
Vc^- [V]: −2791.09
Pr^+ [μC/cm²]: 0.0524941
Pr^- [μC/cm²]: −0.0458722
P_{max+} [μC/cm²]: 0.0394723
P_{max-} [μC/cm²]: −0.0394723
W_{loss} [μJ/cm²]: 557.933

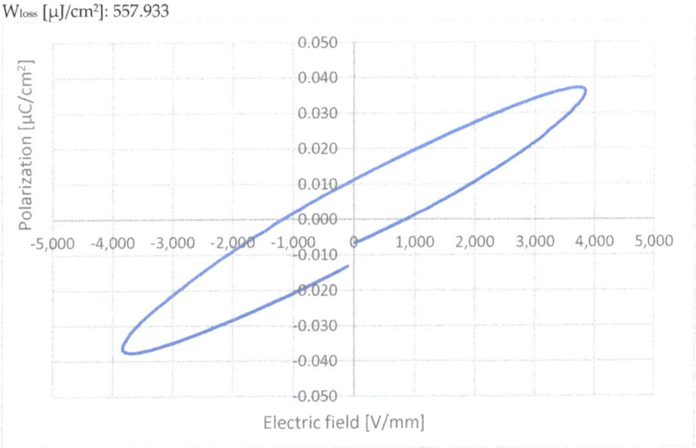

Composite with 35%BaTiO₃
Surface [mm²]: 484
Thickness [mm]: 1
Intensity of current [μA]: 10
Electric field [V/mm]: 3800 V
Frequency [Hz]: 10
Vc^+ [V]: 838.496
Vc^- [V]: −1183.72
Pr^+ [μC/cm²]: 0.011266
Pr^- [μC/cm²]: −0.0131741
P_{max+} [μC/cm²]: 0.0361097
P_{max-} [μC/cm²]: −0.0361097
W_{loss} [μJ/cm²]: 126.353

Figure 7. Polarization characteristics at 0.1, 1, and 10 Hz.

3.6. Thermal Stability

TG-DSC (Differential Scanning Calorimetry) independently measures the heat flow rates between a sample and a reference subjected to the same temperature program (isothermal or dynamic). The difference in heat flow between the sample and the reference, which are heated (or cooled) over a certain temperature range, is then determined, and this difference is plotted as a function of temperature. The direction of the thermal flow towards the DSC sensor is well defined and reproducible [38]. The results for the composite with 35% $BaTiO_3$ are presented in Figure 8.

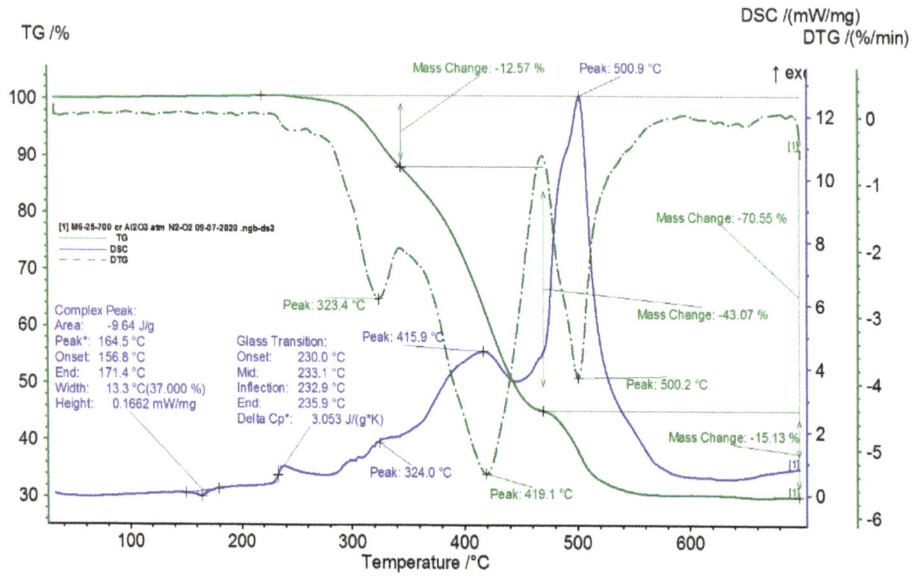

Figure 8. TG and DSC variation curves for the sample with 35% $BaTiO_3$.

For such composite material, in addition to glass transition processes and chemical oxidation processes, there is also a first-order phase transformation process (melting) due to the TPO (PP) melting process. The maximuцm melting temperature is around 170 °C.

4. Development of 3D Printed Flexible Piezoelectric Structures

The development of flexible and elastic piezoelectric energy capture devices based on customized composite elastomers and 3D printing technology involved modeling by fused deposition of composite materials in the form of filaments. The preliminary tests for making specific filaments for 3D printing with a diameter of 1.75 mm were carried out on a small laboratory extruder (Figure 9) from extruded pellets. It is obvious that by an extruding process pellets from thermoplastic composites may be manufactured, even if the inorganic powder content goes towards 80–90%, but it is also well known that the manufacture of rigorous filament in terms of dimeter tolerance and reasonable length from thermoplastic composites with more than 10–15% inorganic powders is extremely difficult, mainly due to the fragility of the filament, which in many cases breaks itself under the reeling operation. On the other hand, at higher powder contents, the filament homogeneity can suffer important changes under cooling operations due to dilatation, exfoliation, or delamination processes. The purpose of the experiment was to develop the optimal extrusion technology, i.e., to account for the optimal temperatures and the speed of pulling the filament to make filaments with rigorously constant dimensions at higher quantities of inorganic powders.

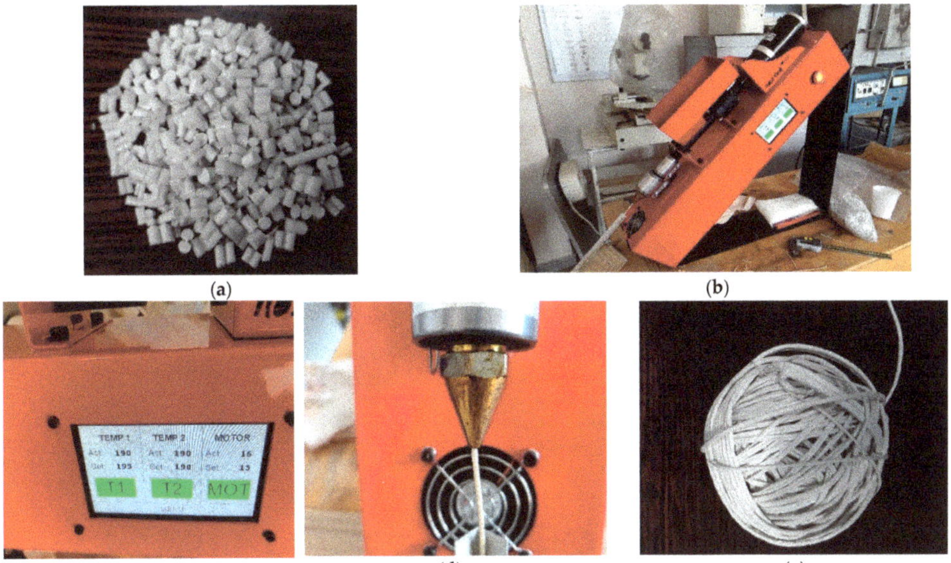

(c) (d) (e)

Figure 9. Specific filaments for 3D printing with a diameter of 1.75 mm from the composite pellets with 35% $BaTiO_3$; (**a**) pellets; (**b**) laboratory extruder; (**c**) temperature control; (**d**) filament generation through the extruder head; (**e**) filament reel.

In this context, adequate thermal control of the extruder was important, with the used temperature being between 180 and 190 °C and more composite recipes being tested [34]. Finally, the filament from the composite material with 35% $BaTiO_3$ resulting from the extrusion operation was elastic, mainly due to the innovative addition of TPO, even close to the elasticity of rubber, but it provided good features to be used for 3D printing applications as well and presented a reasonable variation in a maximum diameter of 10%. The optimal temperatures were around 190 °C with a pulling speed of 15 cm/min. The filament was homogenous, with a smooth surface and no mechanical defects, and had reasonable winding features (Figure 9).

The preliminary 3D printing tests for the realization of structural models were performed using an adapted laboratory 3D thermal printer with a maximum working surface of 120 × 120 mm^2. Other parameters taken into account for printing the models were the viscosity and density of the topical filament. Preliminary testing of the 3D printing parameters is listed in Table 2. A comparison was made between the experimental filaments and the commercial ScotchBlue™ Original Painter's Tape-type PLA filaments and revealed similarities and compatibility by testing the successive depositions of the two filaments; hence, the base deposition support can be made with the commercial one (Figure 10).

Table 2. Setting the parameters of the laboratory 3D thermal printer.

Parameter	Grid
Layer height	0.2 mm
Angle of deposition	90
Deposit density	100%
Print speed	15 mm/s
Nozzle diameter	0.4 mm
Base temperature	60 °C
Extrusion temperature	190 °C
Turns	2
MULTIPLIER	1.2

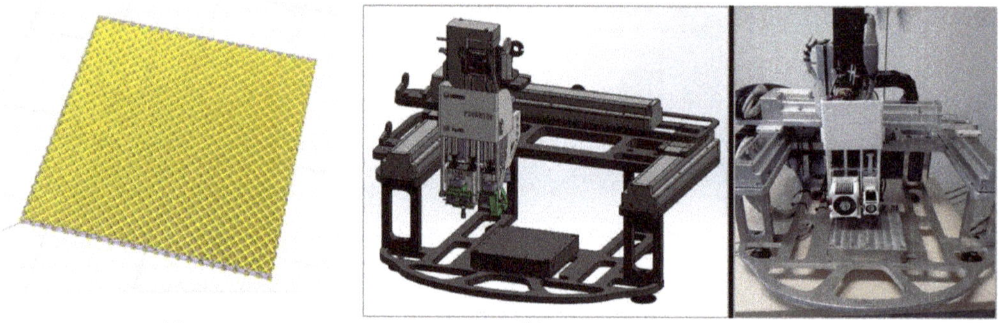

(a) (b) (c)

Figure 10. Three-dimensional printed experimental structural model and the dedicated printer; (**a**) rectangular parallelepiped mesh; (**b**,**c**) images of the tailored printer (face and side view).

Using the Solid Works CAD program or other specialized CAD software, 3D models can be designed, e.g., one of the rectangular parallelepiped network types, which are in fact a 25 × 25 × 1 mm mesh type. After the creation of the CAD model, it is transferred to the laboratory 3D thermal printer software. The experimental structural model is presented in Figure 10.

It was shown that the deposition of composite filament with 35% $BaTiO_3$ was a convenient process with no clogging of the nozzle, no problems with multilayer deposition, and no 3D printing defects. In the end, very good precision structures were obtained experimentally by adjusting the deposition parameters. Even precision mesh structures were made at an angle of 90 compared to the initially planned one of 45 and at deposition densities higher than 50% (Figure 11), to be further tested as basic devices for energy harvesting. As a final comment on this chapter, we can estimate that the mesh structures manufactured by thermal printing provide similar features as homologue structures described in the literature, e.g., in [39], but these are achieved by less productive and more complicated and expensive processes, i.e., stereolithography. Therefore, the advantages of the composite recipe and technology for 3D printing described in this paper are obvious.

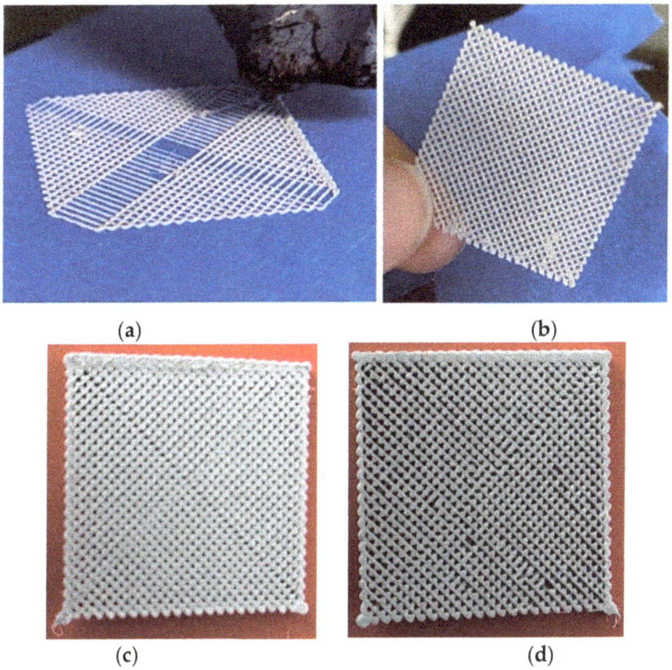

Figure 11. Three-dimensional printed devices for energy harvesting as precision mesh structures; (**a**,**b**) tests with progressive deposition of filament with different angles; (**c**,**d**) final mesh structures with different densities.

5. Testing the Energy Harvesting Features of 3D Printed Devices

For the development and integration of the signal processing circuits and the demonstration of the functionality of the energy harvesting features of 3D printed devices, a special test stand was developed, as in Figure 12, to produce tailored vibrations to activate the devices—here with functionality as piezoelectric sensors. To realize sensorial features, the mesh structures are covered on both sides with self-adhesive copper strips in order to collect the electrical charges obtained by piezoelectric phenomena.

To generate the vibrations, specialized equipment (2) with two windings was used (composed of a fixed part and a mobile part, the mobile part being connected to the fixed one by means of an elastic mechanical suspension), supplied with a constant voltage of 24 V dc from the direct voltage source (5) and also with alternating voltage from the autotransformer (6). The vibration force is proportional to the currents absorbed by the two windings, according to Equation (1):

$$F_{din} = kI_{cc}I_{ca} \tag{1}$$

The vibration frequency was up to 50 Hz, and their amplitude was proportional to F_{din}. When the position of the cursor of the autotransformer is changed, the increase in the alternating voltage applied to the device (2) is achieved, and the mobile part (on which the sensor (1) is placed) will start to vibrate, and thus a signal will be generated at the two terminals of the self-adhesive copper strips of the sensor. Further, this signal is used to charge a capacitor with the role of an energy harvester, and if a resistor R is connected to the output of the sensor, the effective value U of the voltage of the oscillations generated by the sensor can be measured, with the power value to be determined with Equation (2):

$$P = U^2/R \tag{2}$$

Figure 12. The test stand used for the production and acquisition of the signal includes the following: 1—device/sensor; 2—vibration creation equipment; 3—capacitor charging circuit; 4—NI USB acquisition board—6001; 5—direct current source (constant voltage 24 V); 6—autotransformer (alternating voltage supply)—for vibration increase/decrease regime.

In our case, a very common electronic processing module, as presented in Figure 13, was used to collect the energy of the oscillations generated by the sensor and identified in Figure 12 as 3—capacitor charging circuit.

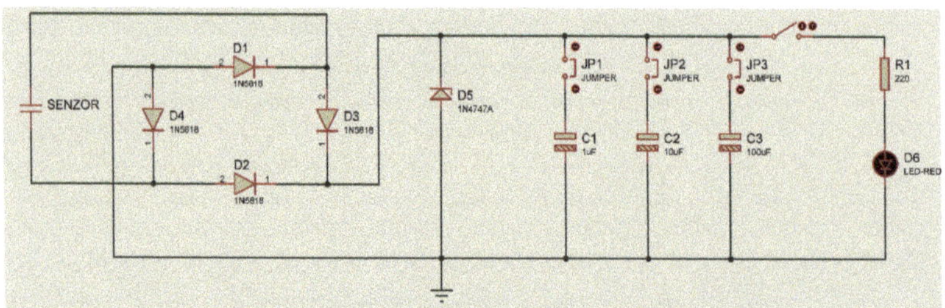

Figure 13. The circuit for acquiring and rectifying the voltage from the sensor.

The output from the sensor is applied to a rectifier bridge made with 4 Shottky diodes of type 1N5818 in order to achieve a continuous charge process only and avoid the reverse bias effect.

The time variation of the voltage $u(t)$ on the capacitor is given by Equation (3), where U is the voltage applied to the capacitor resulting from the rectification and R is the equivalent load resistance of the capacitor.

$$u(t) = U\left(1 - e^{-\frac{t}{RC_H}}\right) \tag{3}$$

The maximum energy E accumulated in the capacitor to which the voltage U is applied is given by Equation (4).

$$E = \frac{C_H U^2}{2} \tag{4}$$

The analog signals when increasing the voltage of the vibration device, i.e., the vibration amplitude level increases, are presented in Figure 14. The signal on the sensor is represented by red oscillations, and the signal on the charging capacitor by white oscillations. The scale is related to signal period, and so the interpretation of the results is difficult without analyzing the real charging phenomenon.

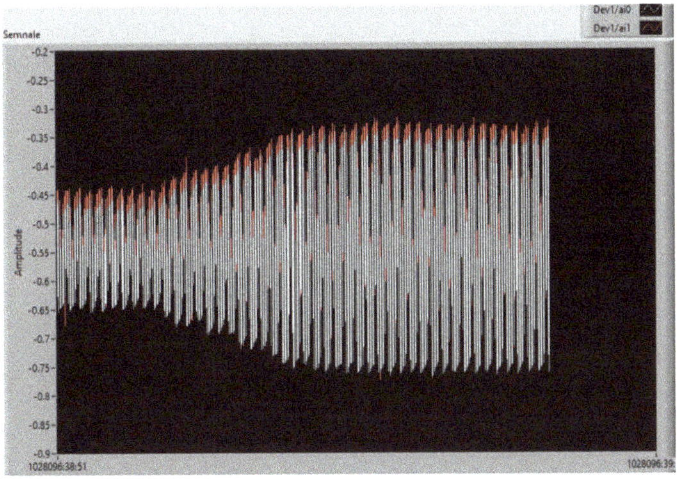

Figure 14. The analog signals by increasing the voltage/vibration level: sensor (red); charging capacitor (white).

That is why intelligent signal processing was embedded by using a virtual instrument (VI) developed under the LabVIEW graphical programming environment [40], briefly presented in Figure 15. After the signal has been acquired with the acquisition board, it is digitized and then transferred to the PC via the USB port. From the USB port, the acquired signal is then processed and displayed using the VI. Block 1 is the data acquisition board setup in subVI. The measured quantity will be set; in our case, it will be "AI Voltage", i.e., the analog voltage. Block 2 is a block used to open a signal acquisition "task". If this subVI is not used, it is possible for the application to start an unsynchronized signal acquisition by itself. If the "DAQmx Start Task VI" subVI and the "DAQmx Stop Task VI" subVI are not used when the "DAQmx Read VI" subVI or the "DAQmx Write VI" subVI are used multiple times (how it would be in a loop) and the acquisition starts and stops repeatedly, starting and stopping a purchase repeatedly reduces app performance. For a continuous acquisition of the signal, the acquisition block (3) is inserted into a "WHILE" loop (9) that will run until an event occurs that will interrupt the running of this loop, either an error or pressing the "STOP" button.

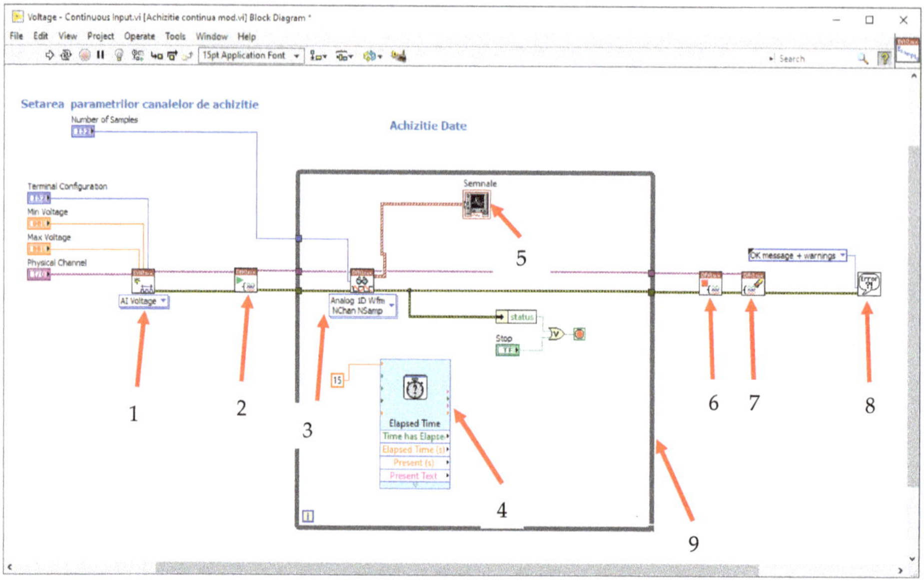

Figure 15. Block diagram of the instrument.

Block 3 is the actual signal acquisition block and has been configured for the acquisition of analog signals. Because we are acquiring signals on several channels, "Multiple Channels" will be selected. Block (4) is a timer used to synchronize the acquisition with the display. After acquiring the signals, these vectors in which the values are stored are displayed value by value on display 5, after which these values are interpolated. These values will be presented on the Y-axis vs. the X-axis, which will show the effective time.

SubVI (6) is used to close the "task" and is opened by block (2). Blocks (2) and (6) will always be used together. The SubVI (7) will clear all values stored in the 1D-type variables to prepare the VI for the next acquisition. The SubVI (8) will display an error message if errors occur. A sample count of 100 has been set, so we will store the values in a 1D vector variable.

The front panel of the signal acquisition instrument was designed to simulate different functional regimes of the energy harvester, i.e., the resistor value, the time, the range (limits) of the input voltage, and the number of signal samples to be processed vs. time.

Two signals are acquired, one directly from the sensor on channel A1 (the waveform is red) and the signal from the capacitors on channel A0 (the waveform is white), the latter being displayed separately for an analysis vs. charging time and charging regime.

The first experiments are performed with the sensor without a preliminary activation with an electric field.

The signals from the sensor output and the charging voltage of the capacitor are shown in Figure 16, where the increase in the amplitude of the signals and the energy of the capacitor according to the value of the applied alternating voltage are presented. The charging of the capacitor starts after the vibrations reach a minimum level, i.e., after a voltage of min. 145 V (case 2) is applied. That is why, in case 1, at 100 V, no charge is noticed. When the voltage exceeds 145 V (case 3), the amplitude of charge on the capacitor increases.

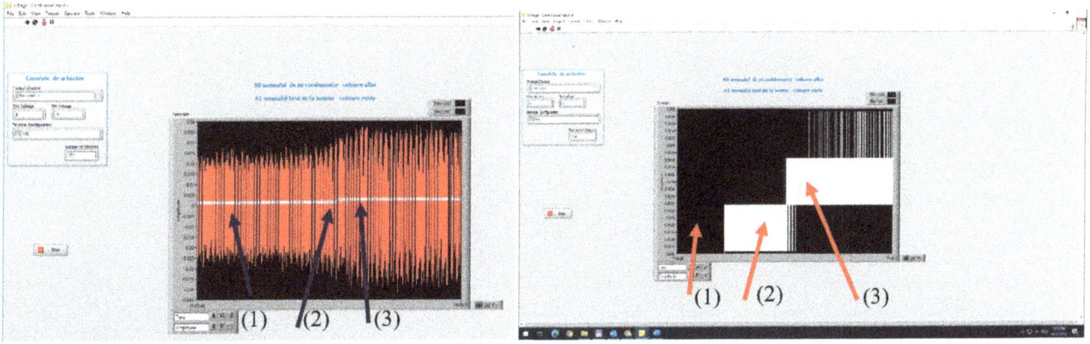

(a) (b)

Figure 16. Increasing vibration level at the test equipment (**a**) and the voltage/the energy level of the capacitor (**b**), for: (1) 100 V; (2) 145 V; (3) 210 V—voltages applied to the test equipment.

In Figure 17, this process is more explicitly explained, where in Figure 17b, 1–represents the individual charging voltage levels, and 2–represents the average charging characteristic of the capacitor. When associating the digital signal on the capacitor with the charging characteristic. Here, 1 represents the instant charging values from the sensor (green), and 2 represents the cumulative charging of the capacitor (red curve).

(a)

Figure 17. *Cont.*

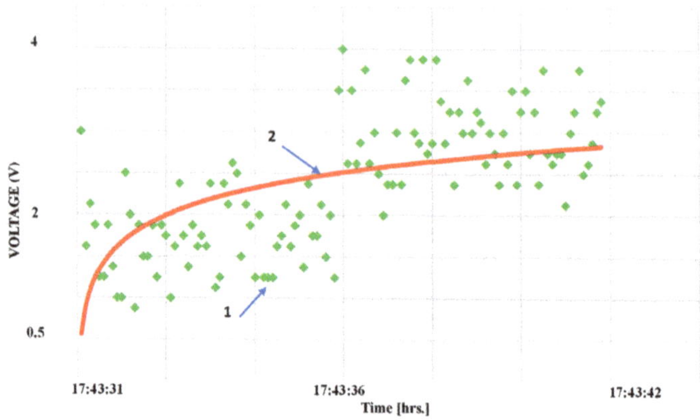

(b)

Figure 17. Digital signal on the capacitor (**a**) and charging characteristic (**b**), for the sensor without activation.

As seen in Figure 17, a minimum charging voltage is needed for the process to start, associated with the amplitude of the vibrations corresponding to 145 V, as explained above. The transient regime depends on the signal applied to the capacitor, and so the capacitor charges exponentially until it is theoretically close to full charge, then it will charge more and more slowly until it is fully charged. The charging characteristic shows the relatively long period of transient regime of capacitor charge, in this case 11 s for the sensor without electrical activation.

The following experiments are made with the sensor after the activation by the electric field. The sensor was activated by applying to its surfaces a continuous voltage of 31.6 V for 72 h to ensure saturation of the hysteresis effect.

Figure 16 shows the evolution of the signal on the capacitor when the vibrations increase.

Here also, the charging of the capacitor starts after the vibrations reach a minimum level, but much earlier compared to a non-activated sensor, i.e., after a voltage of min. 85 V is applied (Figure 18). In this case, even the transient regime is minimal, and the response to the increase in voltage in steps is immediate, as shown by the charging characteristic in Figure 19, where the voltage on the capacitor immediately increases to 5.5 V. In this specified case of excitation in 4 steps, the capacitor reaches a state close to full charge after about 13 s, e.g., 9 V.

At this point in the studies, the feature of 'energy harvesting' mode was demonstrated. By preliminary activating the sensor in the electric field, the value of the electric signal generated by the sensor increases, the efficiency of the sensor is higher, and the charging of the capacitor is faster.

An interesting aspect occurs when comparing the non-activated and activated sensors when the vibrations decrease.

In the case of a non-activated sensor, when the applied voltage goes below 140 V, the sensor becomes passive, and the capacitor is not charged any more. However, in the case of an activated sensor, the charging of the capacitor continues even if the voltage is decreased to 70 V, until saturation (Figure 20) (when the black area becomes compact). In such cases, we can also speak about a mechanical hysteresis in the behavior of electrically activated matrix sensors. The conclusion is also sustained by analyzing the charging characteristic when vibrations decrease (Figure 21), where the charging of the capacitor continues until saturation. Here we practically reached the saturation charge of the capacitor, e.g., 11 V, even under the conditions when the vibrations decreased up to the limit of 1 V on the sensor.

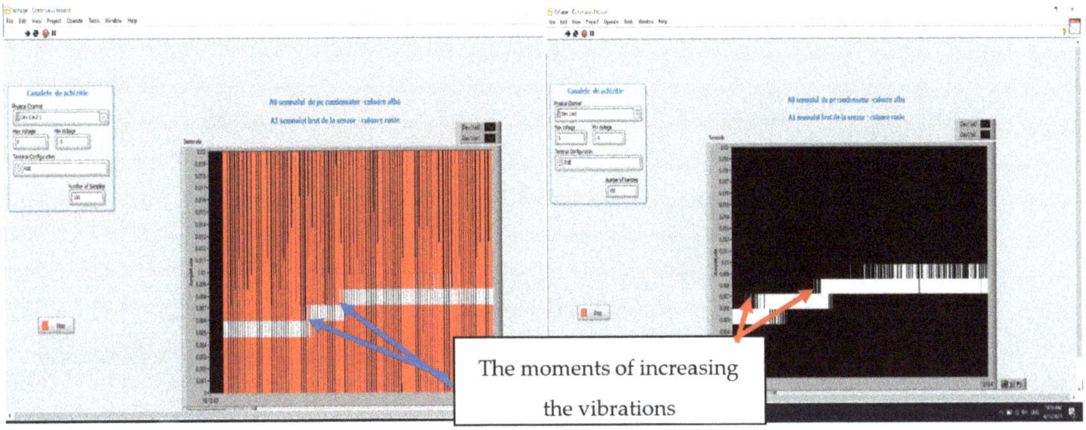

(**a**) (**b**)

Figure 18. Vibration variation of the test equipment (**a**) and its influence upon the voltage on the capacitor (**b**), when the vibration increases on the activated sensor.

(**a**)

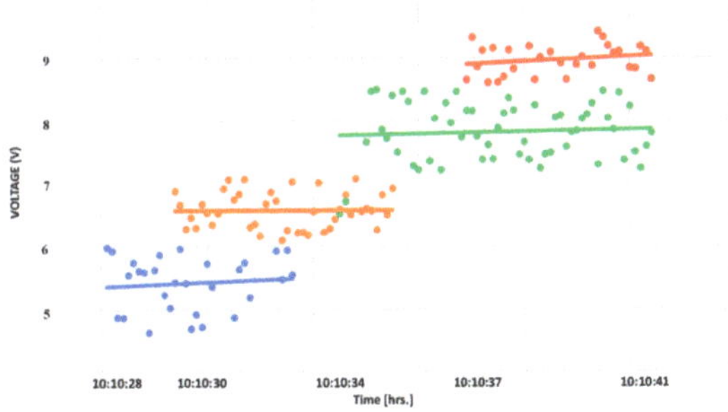

(**b**)

Figure 19. Digital signal on the capacitor (**a**) and charging characteristic (**b**) for the activated sensor.

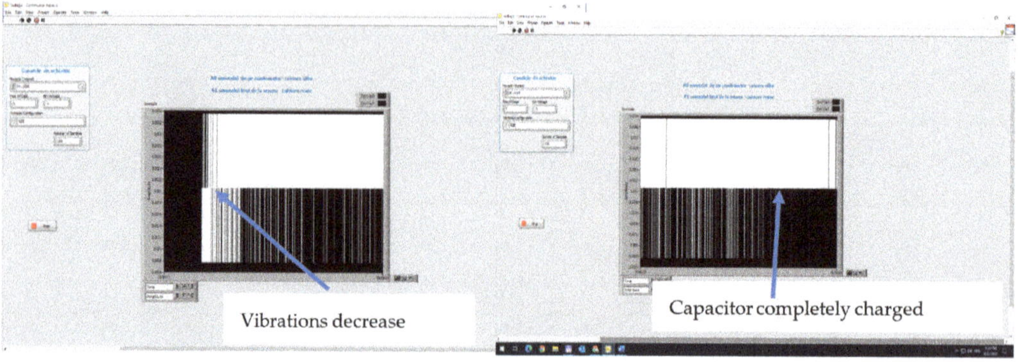

(a) (b)

Figure 20. The charging of the capacitor when vibrations decrease: (**a**) capacitor on charge; (**b**) capacitor fully charged.

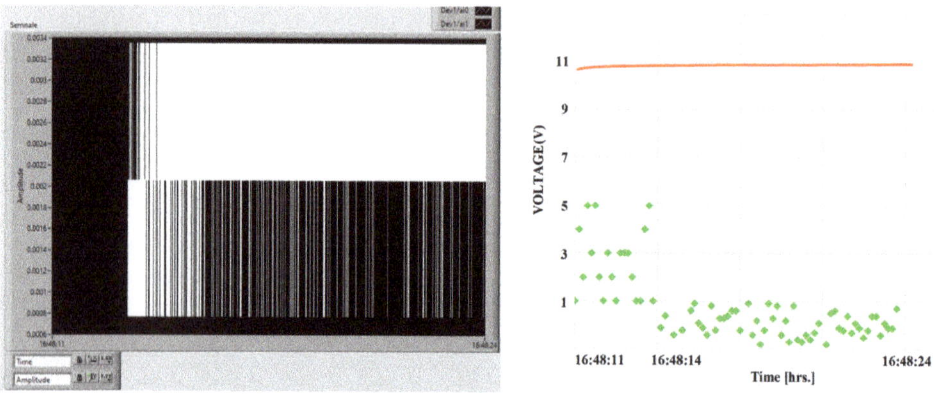

(a) (b)

Figure 21. Digital signal on capacitor (**a**) and charging characteristic (**b**), when vibrations decrease for the activated sensor.

6. Conclusions

The overall picture based on the actual literature shows that many researchers are carrying out research with the aim of testing piezoelectricity as the main source for energy harvesting, and the applications are very innovative, especially in the case of flexible piezoelectric systems.

Composite materials from polymer powders (TPU and TPO—herein polypropylene) with the addition of $BaTiO_3$ until 35% were manufactured and interdisciplinary tested, and it was proved that the filler is well homogenized within the polymer structure with a uniform dispersion. For such composite material, in addition to glass transition processes and chemical oxidation processes, there is also a first-order phase transformation process—melting, due to the TPO (PP) melting process.

The addition of $BaTiO_3$ dramatically increases the relative dielectric permittivity, from about 3 for pure polymers to about 10 for an addition of 12% $BaTiO_3$, and finally to about 30 for an addition of 35% $BaTiO_3$. In regard to the dielectric loss, at lower frequencies the effect of interfacial polarization is detrimental, with a very significant increase for an addition of 35% $BaTiO_3$. The piezoelectric characteristics of the composite with 35% $BaTiO_3$, without preliminary activation under a tailored electric field, are remarkable, especially at lower frequencies.

The filament extruded from the composite material with 35% $BaTiO_3$ was very elastic, close to the elasticity of rubber, but it provided good features to be used for 3D printing applications and presented a variation in a maximum diameter of 10%.

The importance and degree of novelty of our study consist in the demonstration of the operation in energy harvesting mode of the new concept of sensors made by 3D printing technology, made in the form of thermoplastic composites deposited in various forms; here, the testing was performed on a structure of rectangular network type, acting as a piezoelectric sensor in a dedicated signal processing circuit.

It was technically demonstrated that the 3D thermal deposition of composite filament with 35% $BaTiO_3$ was a convenient process with no clogging of the nozzle, no problems with multilayer deposition, and no 3D printing defects.

For the development and integration of the signal processing circuits and the demonstration of the functionality of the energy harvesting features of 3D printed devices, an innovative test stand was developed and is assisted by a virtual instrument to produce and analyze the tailored vibrations to activate devices with functionality as piezoelectric sensors. As brief conclusions upon the electronic tests performed with non-activated and electrically activated devices:

- The electrically activated device has superior performance in terms of the generated electrical signal; the device with 35% $BaTiO_3$ has an immediate and proportional response to periodic mechanical excitation.
- The device, with a higher concentration of $BaTiO_3$ and a quasi-4D effect with frequency, responds better at lower frequencies to the variation of mechanical oscillations and achieves a quasi-uniform charging of the capacitor of the energy harvesting circuit, even if the mechanical oscillations become reduced in intensity.
- The higher the concentration of $BaTiO_3$, the lower the minimum response threshold at low levels of mechanical oscillations, with an induced mechanical hysteresis effect when the mechanical oscillations decrease.

Such flexible piezoelectric systems can be used in various biomedical devices (as wearable electronics or intelligent prosthesis), generating enough energy to make such devices completely autonomous only by exploiting body movements at variable low frequencies.

Author Contributions: Conceptualization, R.C.C. and A.R.C.; methodology, M.A., R.C.C., C.S., A.R.C. and M.P.; software, M.P. and J.P.; validation, R.C.C., C.S., A.R.C., M.P., J.P. and M.B.; formal analysis, M.A., R.C.C. and M.B.; investigation, M.A., R.C.C., C.S., A.R.C., J.P. and M.B.; data curation, R.C.C.; writing—original draft preparation, M.A. and R.C.C.; writing—review and editing, M.A. and R.C.C.; supervision, R.C.C. All authors have read and agreed to the published version of the manuscript.

Funding: This work was supported by a publication grant from Technical Univ. Iasi, and it was based on the joint experimental research of participating organizations, partially financed by a grant (4DPrintEN) under COFUND-ERANET MANUNET III, including the contribution of the Spanish company IZERTIS, which technically modified and updated a specialized 3D printer for the project purpose.

Institutional Review Board Statement: Not applicable.

Data Availability Statement: The data presented in this study are available on request from the corresponding author.

Conflicts of Interest: The authors declare no conflict of interest.

References

1. Kozioł, M.; Szperlich, P.; Toro, B.; Olesik, P.; Jesionek, M. Assessment of the Piezoelectric Response of an Epoxy Resin/SbSINanowires Composite Filling FDM Printed Grid. *Materials* **2020**, *13*, 5281. [CrossRef] [PubMed]
2. Lee, J.; Choi, W.; Yoo, Y.K.; Hwang, K.S.; Lee, S.-M.; Kang, S.; Lee, J. A micro-fabricated force sensor using an all thin film piezoelectric active sensor. *Sensors* **2014**, *14*, 22199–22207. [CrossRef] [PubMed]
3. Tadaki, D.; Ma, T.; Yamamiya, S.; Matsumoto, S.; Imai, Y.; Hirano-Iwata, A.; Niwano, M. Piezoelectric PVDF-based sensors with high pressure sensitivity induced by chemical modification of electrode surfaces. *Sens. Actuators A Phys.* **2020**, *316*, 112424. [CrossRef]
4. Wan, C.; Bowe, C.R. Multiscale-structuring of polyvinylidene fluoride for energy harvesting: The impact of molecular-, micro- and macro-structure. *J. Mater. Chem. A Mater. Energy Sustain.* **2017**, *5*, 3091–3128. [CrossRef]
5. Jia, H.; Li, H.; Lin, B.; Hu, Y.; Peng, L.; Xu, D.; Cheng, X. Fine scale 2-2 connectivity PZT/epoxy piezoelectric fiber composite for high frequency ultrasonic application. *Sens. Actuators A Phys.* **2021**, *324*, 112672. [CrossRef]
6. Meshkinzar, A.; Al-Jumaily, A. Cylindrical Piezoelectric PZT Transducers for Sensing and Actuation. *Sensors* **2023**, *23*, 3042. [CrossRef]
7. Jian, G.; Jiao, Y.; Feng, L.; Meng, Q.; Yang, N.; Zhu, S.; Lü, M.; Wong, C.P. High energy density of $BaTiO_3$-TiO_2 nanosheet/polymer composites via ping-pong-like electron area scattering and interface engineering. *NPG Asia Mater.* **2022**, *14*, 4. [CrossRef]
8. Su, J.; Zhang, J. Recent development on modification of synthesized barium titanate ($BaTiO_3$) and polymer/$BaTiO_3$ dielectric composites. *J. Mater.Sci. Mater. Electron.* **2019**, *30*, 1957–1975. [CrossRef]
9. Jian, G.; Jiao, Y.; Meng, Q.; Wei, Z.; Zhang, J.; Yan, C.; Moon, K.S.; Wong, C.P. Enhanced dielectric constant and energy density in a BaTiO3/polymer-matrix composite sponge. *Commun. Mater.* **2020**, *1*, 91. [CrossRef]
10. Pratap, A. Dielectric behavior of nano barium titanate filled polymeric composites. *Int. J. Mod.Phys. Conf. Ser.* **2013**, *22*, 1–10. [CrossRef]
11. Charif, A.R.C.; Diorio, N.; Fodor-Csorba, K.; Jákli, A.; Puskas, J. A piezoelectric thermoplastic elastomer containing a bent-core liquid crystal. *RSC Adv.* **2013**, *3*, 7446–17452. [CrossRef]
12. Qiu, L.; Coskun, M.B.; Tang, Y.; Liu, J.Z.; Alan, T.; Ding, J.; Truong, V.; Li, D. Ultrafast Dynamic Piezoresistive Response of Graphene-Based Cellular Elastomers. *Adv. Mater.* **2016**, *28*, 194–200. [CrossRef] [PubMed]
13. Xie, S.H.; Zhu, B.K.; Wei, X.Z.; Xu, Z.K.; Xu, Y.Y. Polyimide/$BaTiO_3$ composites with controllable dielectric properties. *Compos. Part A Appl. Sci. Manuf.* **2005**, *36*, 1152–1157. [CrossRef]
14. Stojchevska, E.; Popeski-Dimovski, R.; Kokolanski, Ž.; Gualandi, C.; Bužarovska, A. Effect of Particle Functionalization on Structural and Dielectric Properties of Flexible TPU/$BaTiO_3$/MWCNTs Composite Films. *Macromol. Chem. Phys.* **2023**, *224*, 2200401. [CrossRef]
15. Wondu, E.; Lule, Z.C.; Kim, J. Fabrication of thermoplastic polyurethane composites with a high dielectric constant and thermal conductivity using a hybrid filler of CNT-$BaTiO_3$. *Mater. Today Chem.* **2023**, *27*, 101287. [CrossRef]
16. Luo, Z.; Zhang, L.; Liang, Y.; Wen, S.; Liu, L. Improved the dielectric properties of thermoplastic polyurethane elastomer filled with MXene nanosheets and $BaTiO_3$ nanofibers. *Polym.Test.* **2022**, *111*, 107592. [CrossRef]
17. Gupta, R.; Badel, B.; Gupta, P.; Bucknall, D.G.; Flynn, D.; Pancholi, K. Flexible low-density polyethylene–$BaTiO_3$ nanoparticle composites for monitoring leakage current in high-tension equipment. *ACS Appl. Nano Mater.* **2021**, *4*, 2413–2422. [CrossRef]
18. Ma, Z.; Xie, Y.; Mao, J.; Yang, X.; Li, T.; Luo, Y. Thermoplastic Dielectric Elastomer of Triblock Copolymer with High Electromechanical Performance. *Macromol. Rapid Commun.* **2017**, *3*, 1700268. [CrossRef]
19. Ko, Y.S.; Nüesch, F.A.; Opris, D.M. Charge generation by ultra-stretchable elastomeric electrets. *J. Mater. Chem. C* **2017**, *5*, 1826–1835. [CrossRef]
20. Yaqoob, U.; Uddin, A.S.M.I.; Chung, G. A novel tri-layer flexible piezoelectric nanogenerator based on surface- modified graphene and PVDF-$BaTiO_3$ nanocomposites. *Appl. Surf. Sci.* **2017**, *405*, 420–426. [CrossRef]
21. Dagdeviren, C.; Joe, P.; Tuzman, O.L.; Park, K.I.; Lee, K.J.; Shi, Y.; Huang, Y.; Rogers, J.A. Recent progress in flexible and stretchable piezoelectric devices for mechanical energy harvesting, sensing and actuation. *Extrem. Mech. Lett.* **2016**, *9*, 269–281. [CrossRef]
22. Göksenin, H.; Sümer, Ç.B. A Flexible Piezoelectric Energy Harvesting System for Broadband and Low-frequency Vibrations. *Procedia Eng.* **2015**, *120*, 345–348.
23. Park, S.; Lee, H.; Yeon, S.; Park, J.; Lee, N. Flexible and Stretchable Piezoelectric Sensor with Thickness-Tunable Configuration of Electrospun Nanofiber Mat and Elastomeric Substrates. *ACS Appl. Mater. Interfaces* **2016**, *8*, 24773–24781. [CrossRef] [PubMed]
24. Wei, Y.; Chen, S.; Dong, X.; Lin, Y.; Liu, L. Flexible piezoresistive sensors based on "dynamic bridging effect" of silver nanowires toward grapheme. *Carbon* **2017**, *113*, 395–403. [CrossRef]
25. Niu, D.; Jiang, W.; Ye, G.; Wang, K.; Yin, L.; Shi, Y.; Chen, B.; Luo, F.; Liu, H. Graphene-elastomer nanocomposites based flexible piezoresistive sensors for strain and pressure detection. *Mater. Res. Bull.* **2018**, *102*, 92–99. [CrossRef]
26. Yan, Y.; Potts, M.; Jiang, Z.; Sencadas, V. Synthesis of highly-stretchable graphene—Poly(glycerol sebacate) elastomeric nanocomposites piezoresistive sensors for human motion detection applications. *Compos. Sci. Technol.* **2018**, *162*, 14–22. [CrossRef]
27. Shen, Y.; Wang, S.; Zhi, Y.; Han, B. Properties of novel 0–3 PZT/silicone resin flexible piezoelectric composites for ultrasonic guided wave sensor applications. *Front. Mater. Sec. Smart Mater.* **2022**, *9*, 958775.
28. Arul, K.T.; Rao, M.R. Ferroelectric properties of flexible PZT composite films. *J. Phys. Chem. Solids* **2020**, *146*, 109371. [CrossRef]

29. Khoo, Z.X.; Teoh, J.M.; Liu, Y.; Chua, C.K.; Yang, S.; An, J. 3D printing of smart materials: A review on recent progresses in 4D printing. *Virtual Phys. Prototyp.* **2015**, *10*, 103–122. [CrossRef]
30. Lee, J.; Kim, H.; Choi, J.; Lee, I.H. A review on 3D printed smart devices for 4D printing. *Int. J. Precis. Eng. Manuf. -Green Technol.* **2017**, *4*, 373–383. [CrossRef]
31. Li, X.; Shang, J.; Wang, Z. Intelligent materials: A review of applications in 4D printing. *Assem. Autom.* **2017**, *37*, 170–185. [CrossRef]
32. Baibarac, M.; Nila, A.; Smaranda, I.; Stroe, M.; Singescu, L.; Cristea, M.; Cercel, R.C.; Lorinczi, A.; Ganea, P.; Mercioniu, I.; et al. Optical, structural and Dielectric properties of composites based on thermoplastic polymers of the polyolefin and polyurethane type and $BaTiO_3$ nanoparticles. *Materials* **2021**, *14*, 753. [CrossRef] [PubMed]
33. Smaranda, I.; Nila, A.; Ganea, P.; Daescu, M.; Zgura, I.; Ciobanu, R.C.; Trandabat, A.; Baibarac, M. The influence of the ceramic nanoparticles on the thermoplastic polymers matrix: Their structural, optical and conductive properties. *Polymers* **2021**, *13*, 2773. [CrossRef] [PubMed]
34. Ciobanu, R.; Schreiner, C.; Aradoaei, M.; Hitruc, G.E.; Rusu, B.G.; Aflori, M. Characteristics of composite materials of the type: TPU/PP/$BaTiO_3$ Powder for 3D printing applications. *Polymers* **2023**, *15*, 73. [CrossRef] [PubMed]
35. Baibarac, M.; Arzumanyan, G.; Daescu, M.; Udrescu, A. Anisotropic photoluminescence of poly(3-hexyl thiophene) and their composites with single-walled carbon nanotubes highly separated in metallic and semiconducting tubes. *Molecules* **2021**, *26*, 294. [CrossRef]
36. *ISO 10109-7:2001*; Optics and Optical Instruments. Environmental Requirements—Part 7: Test Requirements for Optical Measuring Instruments. International Organization for Standardization: Geneva, Switzerland, 2001.
37. Available online: https://www.novocontrol.de/php/turn_key_10_90.php (accessed on 20 April 2023).
38. Operating Instructions-Simultaneous TG-DTA/DSC Apparatus (STA 449F3) Jupiter), Selb, October 2008, NE-TZSCH Gerätebau GmbH. Available online: https://www.cif.iastate.edu/sites/default/files/uploads/Other_Inst/TGA/STA449%20F1%20Operating%20Instructions.pdf (accessed on 20 April 2023).
39. Tao, R.; Granier, F.; Therriault, D. Multi-material freeform 3D printing of flexible piezoelectric composite sensors using a supporting fluid. *Addit. Manuf.* **2022**, *60*, 103243.
40. Available online: https://www.ni.com/ro-ro/shop/labview.html (accessed on 20 April 2023).

polymers MDPI

Article

Experimental Study of In-Process Heat Treatment on the Mechanical Properties of 3D Printed Thermoplastic Polymer PLA

Ioan Tamașag *, Irina Beșliu-Băncescu *, Traian-Lucian Severin, Constantin Dulucheanu and Delia-Aurora Cerlincă

Faculty of Mechanical Engineering, Automotive and Robotics, Stefan cel Mare University, 720229 Suceava, Romania; severin.traian@usm.ro (T.-L.S.); dulu@usm.ro (C.D.); delia@usm.ro (D.-A.C.)
* Correspondence: ioan.tamasag@usm.ro (I.T.); irina.besliu@usm.ro (I.B.-B.); Tel.: +40-741-564-971 (I.T.); +40-744-322-449 (I.B.-B.)

Abstract: The scientific literature regarding additive manufacturing, mainly the material extrusion method, suggests that the mechanical characteristics of the parts obtained by this technology depend on a number of the input factors specific to the printing process, such as printing temperature, printing trajectory, layer height, etc., and also on the post-process operations for parts, which, unfortunately, requires supplementary setups, equipment, and multiple steps that raise the overall costs. Therefore, this paper aims to investigate the influence of the printing direction, the thickness of the deposited material layer, and the temperature of the previously deposited material layer on the part tensile strength, hardness by means of Shore D and Martens hardness, and surface finish by using an in-process annealing method. A Taguchi L9 DOE plan was developed for this purpose, where the test specimens, with dimensions according to ISO 527-2 type B, were analysed. The results showed that the presented in-process treatment method is possible and could lead to sustainable and cost-effective manufacturing processes. The varied input factors influenced all the studied parameters. Tensile strength tended to increase, up to 12.5%, when the in-process heat treatment was applied, showed a positive linear variation with nozzle diameter, and presented considerable variations with the printing direction. Shore D and Martens hardness had similar variations, and it could be observed that by applying the mentioned in-process heat treatment, the overall values tended to decrease. Printing direction had a negligible impact on the additively manufactured parts' hardness. At the same time, the nozzle diameter presented considerable variations, up to 36% for Martens hardness and 4% for Shore D, when higher diameter nozzles were used. The ANOVA analysis highlighted that the statistically significant factors were the nozzle diameter for the part's hardness and the printing direction for the tensile strength.

Keywords: additive manufacturing; PLA; annealing; surface quality; mechanical proprieties

Citation: Tamașag, I.; Beșliu-Băncescu, I.; Severin, T.-L.; Dulucheanu, C.; Cerlincă, D.-A. Experimental Study of In-Process Heat Treatment on the Mechanical Properties of 3D Printed Thermoplastic Polymer PLA. *Polymers* **2023**, *15*, 2367. https://doi.org/10.3390/polym15102367

Academic Editors: Cristina-Elisabeta Pelin and Anton Ficai

Received: 29 April 2023
Revised: 15 May 2023
Accepted: 17 May 2023
Published: 18 May 2023

1. Introduction

Recently, additive manufacturing technology has been gaining momentum both in industry, with applications in the automotive, medical, aerospace, etc. industries [1], and in academia, due to the possibility of manufacturing parts with complex surfaces [2] at low costs, especially when using FDM (fused deposition modelling) [3].

In the FDM process, also known as the MEX (material extrusion) process [4], one of the most widely used materials, at the expense of petroleum-based polymers [5,6], is polylactic acid (PLA), a polymer that has a semi-crystalline structure, is biodegradable, and has good physical properties. Since the manufacturing process results in parts with low mechanical and thermal properties [6–9], the literature also presents different methods to improve these properties by reinforcing the polymer with other materials such as carbon fibre [10], glass fibre [11], and natural fibre [12], and post-processing methods, such as heat treatment [13–15] or chemical treatment [16,17], which involve additional steps and costs, thus also increasing the manufacturing time. Regarding the application of heat treatment on parts made of PLA by the FDM method, several research studies have been carried out [14,18–22], and it has been proven that the mechanical properties of the tested specimens improved drastically when the obtained parts were subjected to heat treatment post-processing. The results obtained by the authors showed that the mechanical properties are enhanced by applying a heat treatment at an optimal temperature between the polymer's glass transition temperature and the cold crystallisation temperature [20], in the range of 80–120 °C. The literature published on these issues [14] highlights that by applying heat treatment, the material's crystallinity level increases, which leads to an increase in mechanical properties, by up to 49%, and thermal resistance, consequently raising the glass transition point. In addition, in the case of parts made of PLA polymers, the influence of heat treatment at a temperature in the range of 65–95 °C leads to an improvement in the mechanical properties by 11–17% and in the crystallinity level by about 25% [19].

Other research [23] addresses the problem of interlayer adhesion. When using a pre-deposition heating method such as near-layer laser heating to increase the interlayer interface temperature above the critical interpenetrating diffusion temperature, the interlayer bond strength is enhanced. Results obtained by the authors showed an increase of up to 50% in the mechanical properties; the fracture behaviour of these interlayer interfaces becomes ductile, and plastic deformation is observed.

Regarding the surface hardness of additively manufactured parts, the literature presents several investigations for different types of materials used in the MEX process [24–27], including Martens-type micro-hardness [28–30]. Several research studies have been carried out on the influence of heat treatment on the hardness of PLA additively manufactured parts [31,32]. Following the experimental determinations in [31], the results obtained by the authors showed that by applying heat treatment, the tensile strength increased considerably, by up to 35%, while the hardness of the parts subjected to heat treatment decreased, with the values fluctuating depending on the printing direction of the specimens. Furthermore, the parts did not show plastic deformation when applying a heat treatment at a temperature in the range of 55–80 °C. Still, at a temperature of 95 °C, the parts showed a degree of dimensional deviations and were not analysed further. However, regardless of the printing direction, applying a heat treatment with temperature values in the above-mentioned range increases the ultimate tensile strength (UTS). The same trend was also observed in [32], where, by applying heat treatment in the range 70 °C−90 °C, authors observed a considerable improvement in UTS values. However, the values for surface hardness showed an increased variability, which requires further studies on the influence of heat treatment on hardness, as present in this paper.

Furthermore, in the research domain, the tendency to replace post-processing with in-processing technologies was observed. The speciality literature offers a series of research studies based on in-process applications meant to improve the mechanical properties, surface quality, or even colourisation of the parts [33–35]. However, related to the influence of in-process heat treatment, the literature provides limited information, especially on Martens hardness, which represents a research opportunity and provides this study the novelty of this research.

In this paper, the possibility of applying heat treatment with a hot air jet, at temperatures of 80 °C, locally, throughout the actual printing process is explored, reducing working time, manufacturing space, and the effects on the adhesion between the deposited layers. In order to solve these objectives, a Taguchi DOE L9 design was developed, which allowed the study of the influence of temperature, printing direction, and nozzle size on the tensile strength, roughness, and surface micro-hardness of the printed parts. The tests were carried out in accordance with the ISO 527 standard: Plastics—Determination of tensile properties, Parts 1 and 2 [36,37].

In Section 2, the experimental design and equipment used to achieve the proposed objectives are presented in detail, and in Section 3, the results are analysed and discussed.

The experimental values obtained and included in the experimental design were analysed using Minitab software, which allowed for the plotting of variation graphs of the mean effects and Pareto graphs representing the degree of influence of each factor. At the same time, by applying the Taguchi experimental design, it was possible to apply ANOVA-type (analysis of variance) analyses to determine the statistically significant factors. The experimental results revealed that when using the in-process local heat treatment method, the values for UTS increased, and the values for hardness and surface roughness decreased. However, as observed in the literature [26], the values obtained vary depending on the printing direction and nozzle size.

2. Experimental Setup

To achieve the proposed objectives, a research plan shown in Figure 1 was created, which graphically illustrates the steps of this article. The ISO 527 type 1B specimens were printed according to the description in Section 2.1. They were then tested in terms of tensile strength, hardness, micro-hardness, and surface roughness with the equipment described in Section 2.2. The values obtained after the testing made in accordance with the Taguchi L9 DOE plan are analysed and discussed in Section 3.

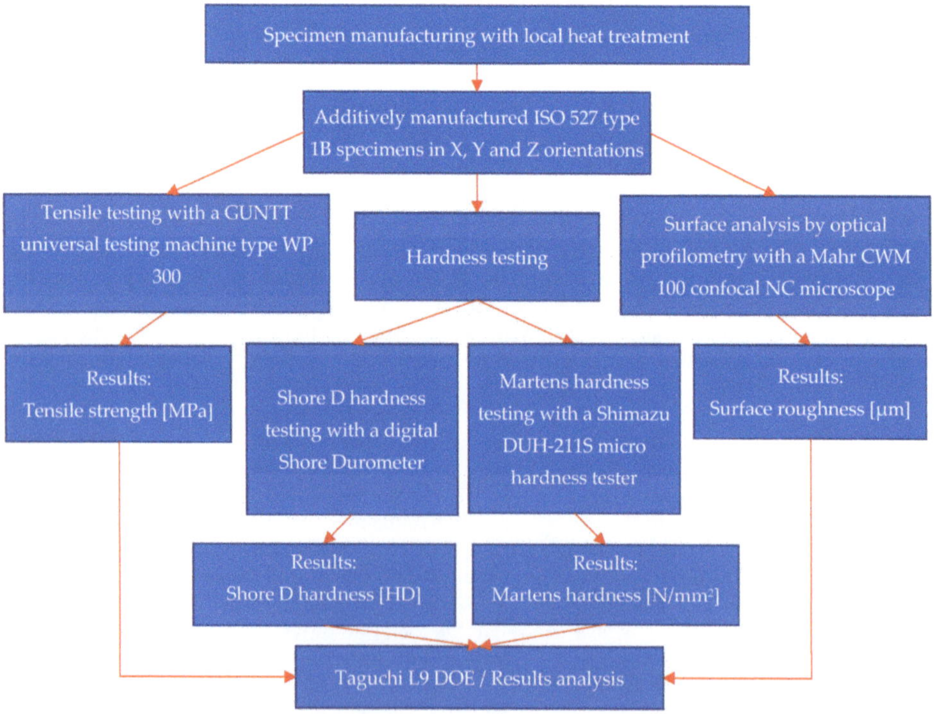

Figure 1. The experimental program.

The proposed Taguchi design of experiments (DOE) is shown in Table 1. In order to study the influence of the above-mentioned factors, three levels of variation were considered for the three factors (nozzle diameter, part temperature, and printing direction), noted with 1, 2 and 3 in Table 1 and corresponding with the values presented in Table 2.

Table 1. Taguchi L27 DOE.

Test Number	Nozzle Size	Heat Temperature	Printing Direction
1	1	1	1
2		2	2
3		3	3
4	2	1	2
5		2	3
6		3	1
7	3	1	3
8		2	1
9		3	2

Table 2. Levels of variation for the factors used in the Taguchi DOE design.

Levels	A Nozzle Size [mm]	B Part Temperature [°C]	C Printing Direction
1	0.5 mm	No heat	X
2	0.6 mm	Heated build platform at 50 °C	Y
3	0.8 mm	Local heating at 80 °C	Z

2.1. Sample Preparation and Materials

The specimens made according to ISO 527-2 type B [36] have the geometry shown in Figure 2a. They were fabricated in all three directions (Figure 2b) using a Creality 3D printer, type CR6-Se; a SUNLU filament dryer, where the filament was dried before and during use at 40 °C; and a Wagner Furno 750 hot air gun, with the possibility of adjusting the temperature by 10 °C, in the range 50–630 °C (Figure 3). Furthermore, in the printing process, the Z seam position was selected to be placed in the corners of the parts to remove the influence of layer height shifting on the UTS of the specimens, as seen in the literature [13,38].

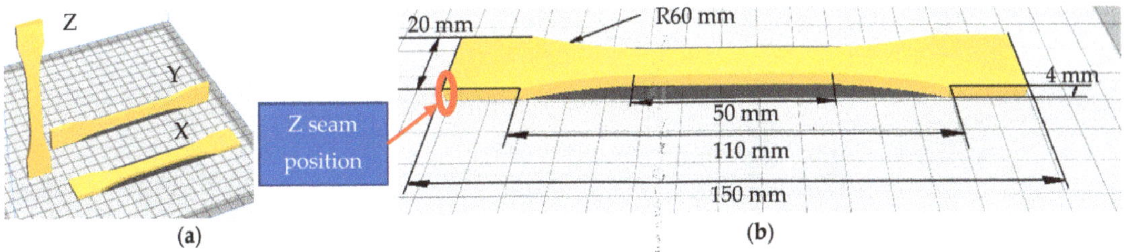

Figure 2. ISO 527 type 1B specimens: (**a**) printing directions; (**b**) specimen dimensions.

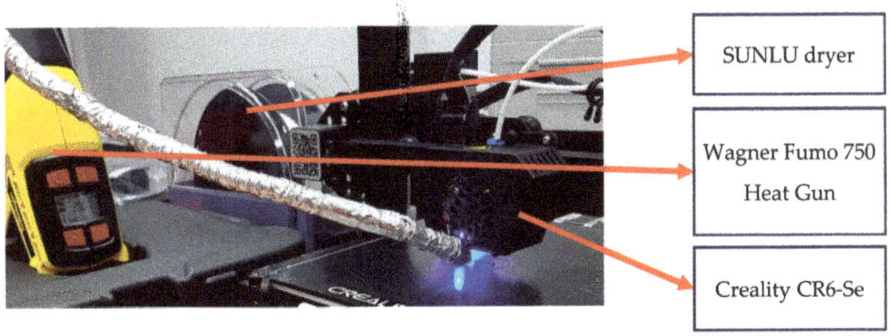

Figure 3. 3D printer assembly.

The tested material is PLA (polylactic acid) (produced by eSUN—Shenzhen Esun Industrial Co., Ltd., based in Shenzhen, China), a biodegradable and bioactive thermoplastic polymer that is commonly used in additive manufacturing due to its ease of use, low toxicity, and availability. The tested material's chemical composition and properties according to the technical and material safety data sheets [39,40] are shown in Table 3. The printing parameters that were kept constant have the values shown in Table 4.

Table 3. Material properties.

Material	Density (g/cm^3)	Heat Distortion Temp (°C, 0.45 MPa)	Melt Flow Index (g/10 min)	Tensile Strength [MPa]	Flexural Modulus [MPa]	Polylactic Acid
PLA	1.2	53	3.5	75	1915	99.8%

Table 4. Additive manufacturing constant parameters.

Parameter	Values	Parameter	Values
Layer height	0.3 [mm]	Print speed	50 [mm/s]
Infill	20%	Travel speed	150 [mm/s]
Infill type	Cubic	Retraction distance	6.5 [mm]
Printing temperature	215 [°C]	Cooling	ON

A FLIR X6540sc thermal imaging camera was used to study the temperature of the printing area. It was found that the temperature of the printing area shows significant variations during the printing process, with considerable decreases when new material is deposited (Figure 4).

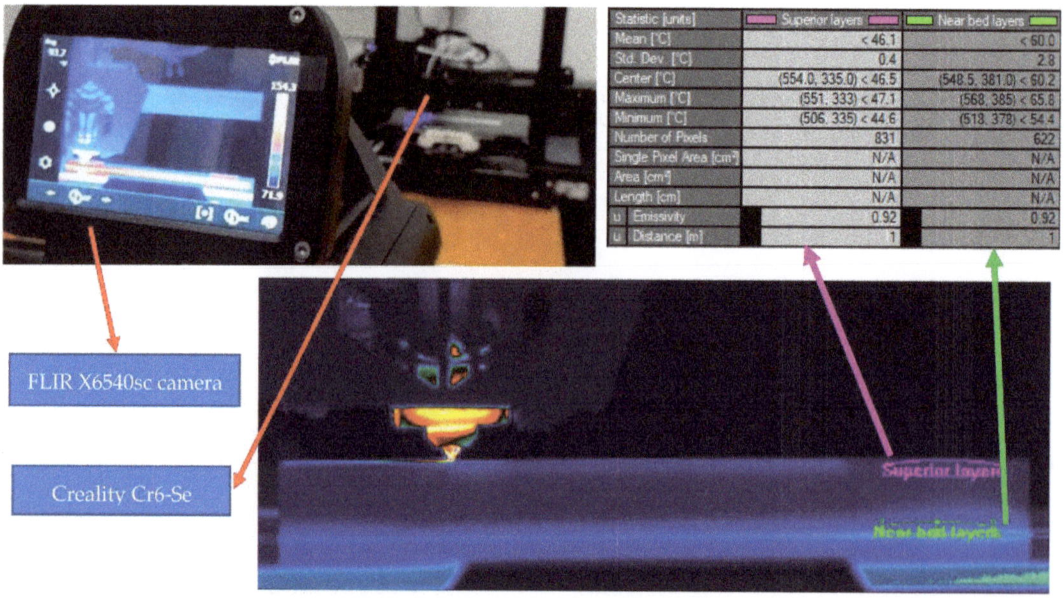

Figure 4. Temperature difference between printed layers.

In order to eliminate this problem, the deposition process was assisted by a hot air heating system using a Wagner blower, which maintains a constant temperature (Figure 5) throughout the deposition process (80 °C).

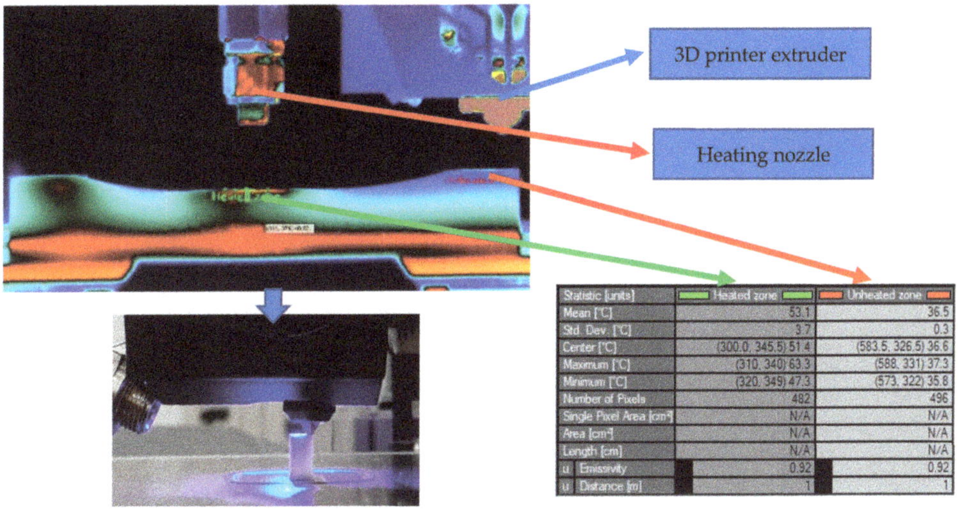

Statistic [units]	Heated zone	Unheated zone
Mean [°C]	53.1	36.5
Std. Dev. [°C]	3.7	0.3
Center [°C]	(300.0, 345.5) 51.4	(583.5, 326.5) 36.6
Maximum [°C]	(310, 340) 63.3	(588, 331) 37.3
Minimum [°C]	(320, 349) 47.3	(573, 322) 35.8
Number of Pixels	482	496
Single Pixel Area [cm²]	N/A	N/A
Area [cm²]	N/A	N/A
Length [cm]	N/A	N/A
u Emissivity	0.92	0.92
u Distance [m]	1	1

Figure 5. Local heating of the part.

2.2. Testing Equipment

To determine the tensile strength, a GUNT testing machine was used (Figure 6a), assisted by a computer equipped with software for data processing. The values obtained experimentally were analysed, and variation graphs were made.

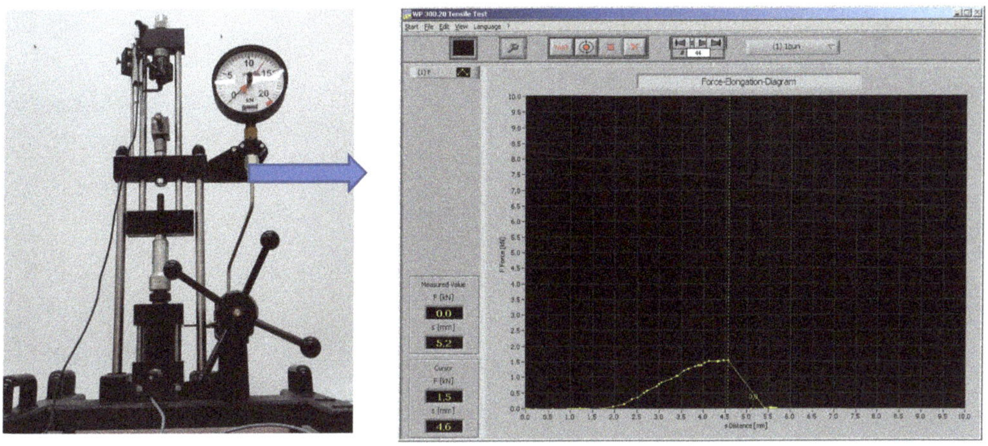

(a) (b)

Figure 6. GUNT universal testing unit: (**a**) testing machine; (**b**) obtained tensile strength result image.

The surface roughness study was carried out with a Mahr CWM 100 confocal microscope and interferometer, without contact, and with NC positioning of the samples in the central region of the specimen (Figure 7). The obtained surface topography was analysed using the related MountainsLab 8.1 software, where it was possible to obtain the surface texture parameter (Sq). This parameter is an (ISO 25178) equivalent to the standard deviation of heights, which uses a predefined 0.8 mm robust Gaussian L-filter. Furthermore, the surface roughness measurements were made in a 2 × 1.5 mm^2 area, and both directions of the roughness were taken into consideration.

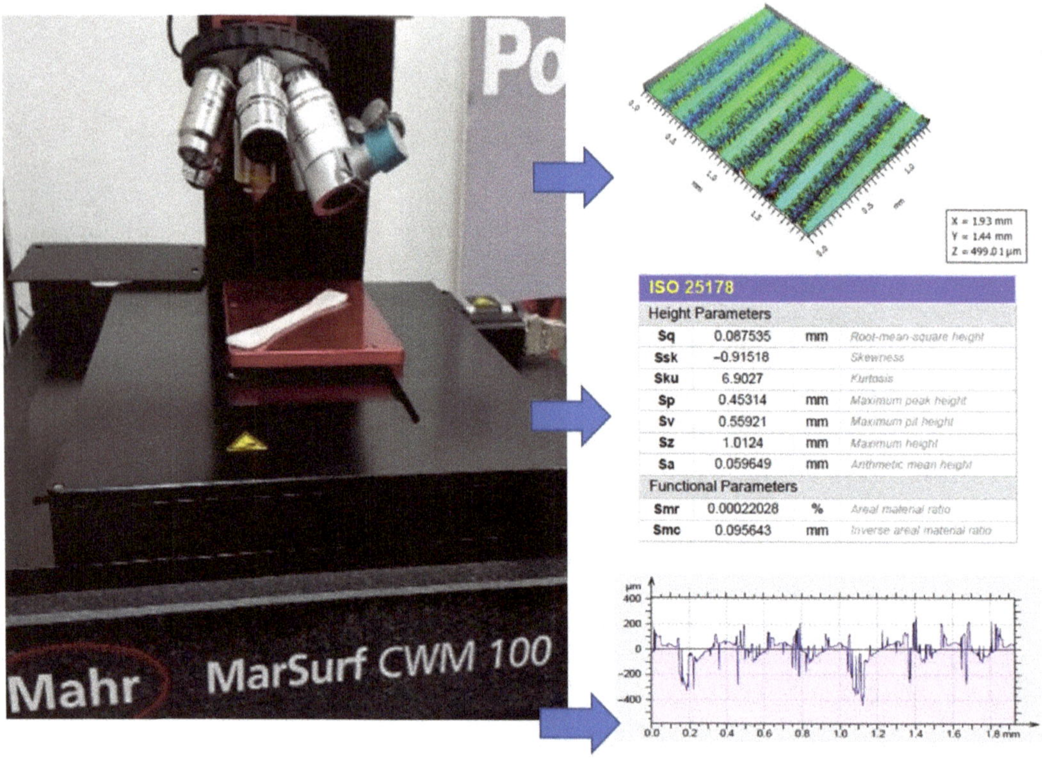

Figure 7. Mahr CWM 100 confocal microscope.

The hardness of the tested parts was considered from multiple points of view. Firstly, the microhardness of test specimens was investigated and evaluated using the Shimadzu DUH-211S microhardness tester (Figure 8 to obtain the Martens hardness (Figure 9), as well as using a digital Shore D durometer to obtain the Shore hardness.

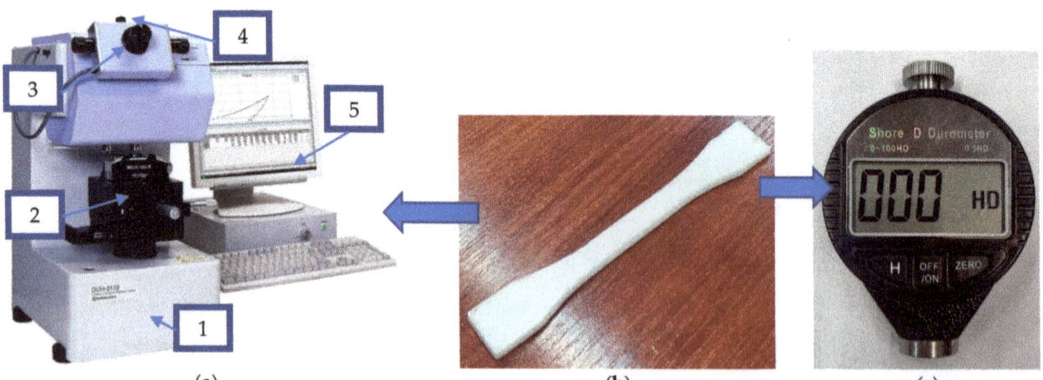

Figure 8. Hardness testing: (**a**) Shimadzu DUH-211S microhardness tester (1), sample manual positioning system (2), footprint optical viewing system (3), image pickup video system CCD (4), hardness measurement and footprint inspection software (5); (**b**) test specimen; (**c**) Shore D durometer.

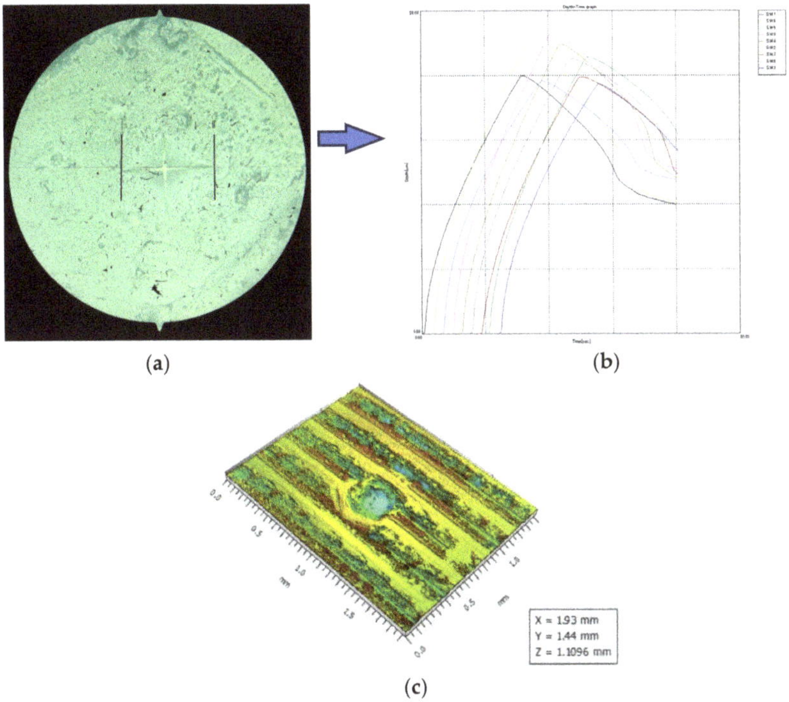

Figure 9. Hardness testing results: (**a**) optical image; (**b**) resulted graph; (**c**) indentation point.

Secondly, the tests were carried out according to ISO 868:2003 [41], where the indentation position and specimen size parameters were repeated (Figure 10a) and the Shore D tests were also repeated at the edge of the specimens (Figure 10b), in the area where the part wall is created and can be considered 100% infill. The measuring tests were repeated five times, and the average value was considered.

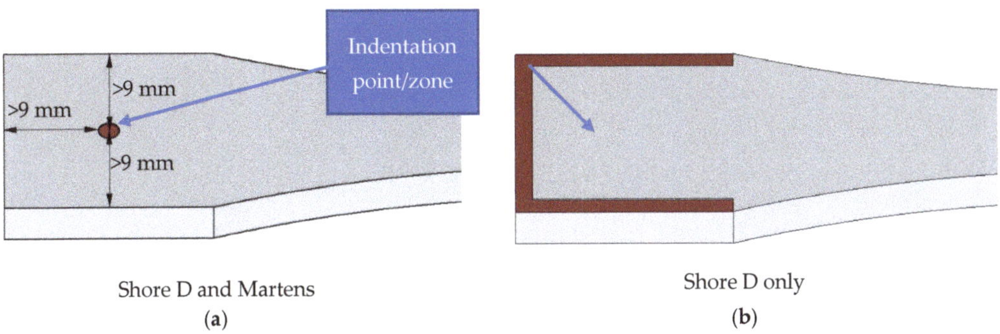

Figure 10. Indentation position: (**a**) in accordance with ISO 969:2003; (**b**) on walls.

In this study, the surface on the build platform was not taken into consideration, given that the part was printed on a tempered glass build platform and, therefore, the surface roughness is much lower than that of the rest of the part surfaces. Even more, on that specific surface, the hardness is higher. Usually, the first layers that coincide with that surface have different printing conditions, of which the most relevant are lower layer height and lower printing speeds. Therefore, the testing was made on the surfaces that differed from the build platform, and the indentation point was at least 9 mm farther from that specific surface.

3. Results and Discussions

Following the experimental determinations, the values obtained have been included in Table 5, which coincides with the Taguchi L9 plan presented in Section 2, Table 1.

Table 5. Experimental values.

Test Number	Tensile Strength [MPa]	Shore D Hardness [HD]	Martens Hardness, HMV [N/mm^2]	Printing Duration [min]	Surface Roughness, Sq [μm]
1	35.0	84.0	160.765	32	49.237
2	37.5	83.0	156.843	38	57.613
3	12.5	81.7	153.672	61	37.431
4	45.0	81.5	150.406	33	56.215
5	15.0	81.2	133.415	64	49.923
6	42.5	81.0	132.185	28	38.683
7	20.0	80.3	119.970	59	65.826
8	45.0	79.2	113.623	24	53.660
9	57.5	79.7	111.014	29	28.857

3.1. Tensile Strength

Obtained values for tensile strength ranged between 12.5 MPa and 57.5 MPa, which is also observed in other research studies [20,24,42]. Compared with the values offered by the manufacturer (Table 3), it can be seen that the results for the tensile strength from this study are lower. However, this phenomenon can be explained by the manufacturer providing the tensile strength for parts with 100% infill.

The experimental results obtained for the tensile strength allowed for drawing of dependence curves for the mean effects of the three factors considered (Figure 11). It can be seen that the tensile strength of the specimen increases substantially with increasing nozzle diameter (20.56% for 0.6 mm nozzle and 44.13% for 0.8 mm nozzle), decreases by 2.5% with increasing build platform temperature and increases by 12.5% if the temperature is kept constant at 80 °C during the printing process.

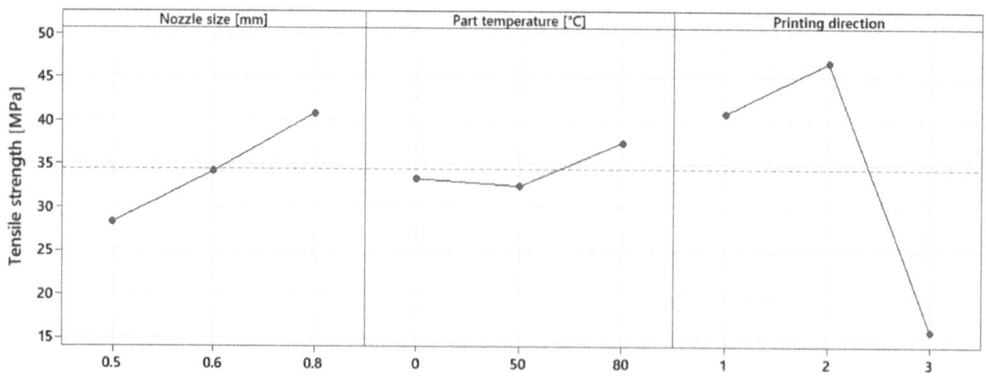

Figure 11. Variation in mean effects for tensile strength.

By analysing the values obtained and the graphical representation in the above graph, it was possible to draw up Table 6, where the symbols ↑ and ↓ represent growth and decline, respectively.

Table 6. Obtained results' percentage variation for tensile strength.

Level	Nozzle Size		Part Temperature		Printing Direction	
	Value	Variation	Value	Variation	Value	Variation
1	0.5	-	0	-	X	-
2	0.6	↑ 20.56%	50	↓ 2.5%	Y	↑ 14.11%
3	0.8	↑ 44.13%	80	↑ 12.5%	Z	↓ 61.35%

The results obtained confirm the conclusions presented in the works [19,43] concerning the tensile strength and the influence of the nozzle size. An increase of 20.5% in the tensile strength can be observed when using the 0.6 mm nozzle and 19.5% when using the 0.8 mm nozzle. These values are explained by the reduction in the number of layers used for printing the walls, but also in the interlayer contact dimension. Other explanations for this phenomenon may be due to the increased adhesion between the layers due to the density changes of the test specimens, proven in the literature [43,44], where the authors concluded that by increasing the nozzle size, the density of the specimens also increases. Another explanation may be the fact that the wider the nozzle diameter, the longer the layers stay at a maintenance temperature that allows the material to settle better due to thermal convection [34].

Regarding the influence of the printing direction on the tensile strength, the experimental tests showed an increase of 14.11% in the case of the Y-direction and 61.35% in the case of the Z-direction, variations similar to those found in the literature [18].

However, compared to the X-direction, higher values for the tensile strength were obtained when parts were printed in the Y-direction. This may also be due to changes in the cross-section of the specimens. Due to the positioning and orientation of the lines of the layers, the tensile strength varies considerably. Figure 12 shows the settings used for printing in the X and Y directions.

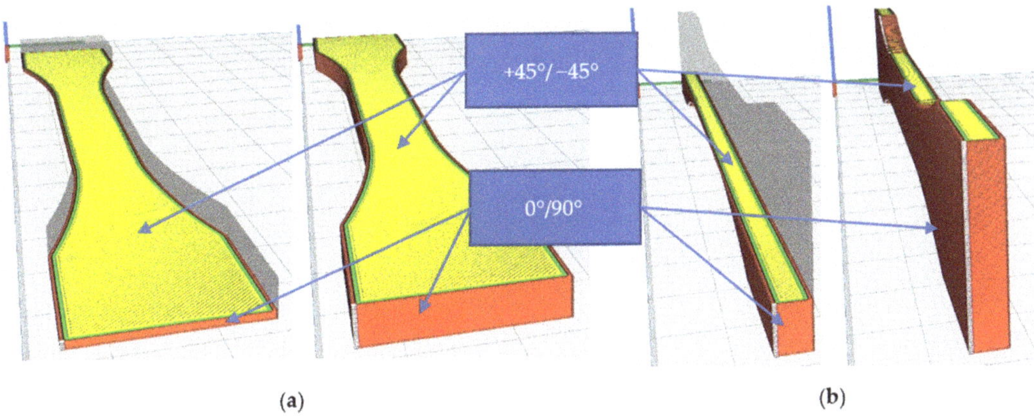

Figure 12. Test specimens: (**a**) flat specimen, printed in the X-direction; (**b**) specimen printed in Y-direction.

The lines of the outer walls are different from one printing direction to another, which influences the area of occurrence of the bottleneck where the material breaks and the morphology of the apparent section of the specimen, and, therefore, the tensile strength. In the present case, the high results of tensile strength for parts printed in the Y-direction are due to the fact that the total volume of the upper and lower surfaces of the specimen is kept constant by depositing material in the form of continuous, straight lines at 0–90°.

By applying the experimental design presented in subchapter 2.1, it was possible to employ an ANOVA analysis (Table 7), from which resulted the level of statistical significance (Figure 13).

Table 7. ANOVA analysis for tensile strength.

Source	DF	Adj SS	Adj MS	F-Value	*p*-Value
Regression	6	1887.50	314.583	64.71	0.015
Nozzle size	2	234.72	117.361	24.14	0.040
Part temperature	2	43.06	21.528	4.43	0.184
Printing direction	2	1609.72	804.861	165.57	0.006
Error	2	9.72	4.861		
Total	8	1897.22			

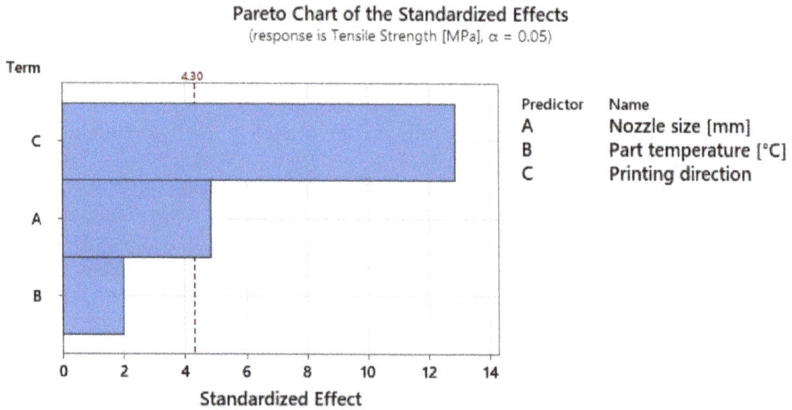

Figure 13. Graphical illustration of the degree of influence of studied factors on tensile strength.

The results obtained from the ANOVA analysis showed an almost insignificant variation in the tensile strength when only the temperature of the printing platform was used, whereas when the heating method described in Section 2.1 was used, the values showed an improvement of up to 12.5% in tensile strength. Given this, it can be concluded that by applying local heating, the tensile strength increases, confirming other research studies from the literature related to the application of heat treatment [31,32,45]. However, the values obtained for the tensile strength have a smaller variation than those obtained by other researchers between parts without annealing and parts with annealing. This may be due to the duration of the treatment application. In most of the research studies, the experiments involved increased treatment application times, whereas in this case, the exposure time of the parts to local heat treatment depends on the part size and printing speed. In this case, the duration of local heat treatment on the parts coincides with the printing duration, calculated in the slicing software Cura 5.2.2, shown in Table 5. Of the factors studied, the ANOVA analysis revealed that the statistically significant factors influencing the tensile strength variation were printing direction with 65.2% influence, followed by nozzle diameter with 24.68% influence, and finally, temperature with 10.12% influence.

3.2. Hardness

The surface hardness results gave similar values regardless of indentation position. For this reason, only the results of tests performed according to ISO 969:2003 were considered. Similarly, the variations in the type of test show the same trends. The graphs in Figures 14 and 15 represent the variation in the mean effects of Shore D hardness and Martens, respectively.

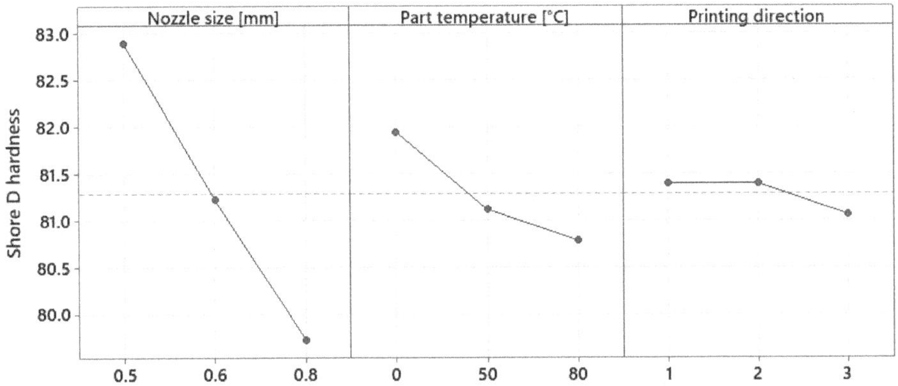

Figure 14. Variation in mean effects for Shore D hardness.

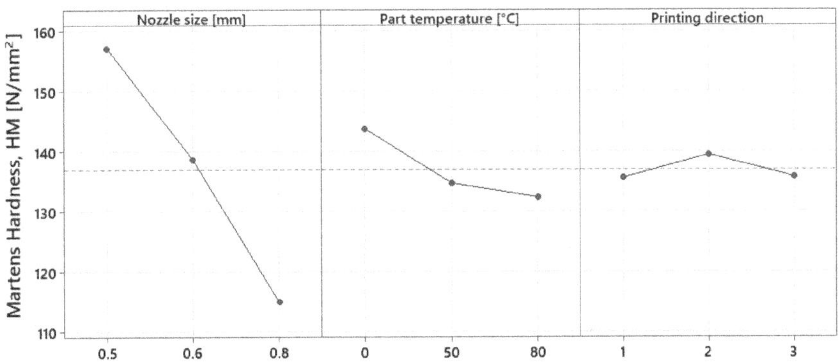

Figure 15. Variation in mean effects for Martens hardness.

Following the variation graphs shown above, it can be considered that the values obtained for hardness are opposite to those for tensile strength. Comparing the results obtained with those found in the literature, it is found that they fall within the range of 79–84 HD, values also obtained by other researchers [26,31], raising the degree of confidence in the accuracy of the values obtained in this study.

The level of variation and direction of hardness trends for both types of tests are presented in Table 8.

Table 8. The level of variation and direction of hardness trends for both types of tests.

	Level	Nozzle Size		Part Temperature		Printing Direction	
		Value	**Variation**	**Value**	**Variation**	**Value**	**Variation**
	1	0.5	-	0	-	X	-
Shore D	2	0.6	↓ 2.05%	50	↓ 1.03%	Y	- 0%
	3	0.8	↓ 3.97%	80	↓ 1.44%	Z	↓ 0.41%
		Value	Variation	Value	Variation	Value	Variation
	1	0.5	-	0	-	X	-
Martens	2	0.6	↓ 13.29%	50	↓ 6.75%	Y	↑ 2.88%
	3	0.8	↓ 36.76%	80	↓ 8.64%	Z	↑ 0.08%

Contrary to the trends observed for the tensile tests, it can be seen that surface hardness decreases with increasing nozzle diameter by up to 3.97% for Shore D hardness and 36.76% for Martens hardness when nozzles larger than 0.5 mm in diameter were used. Research in the field [43] has shown that in MEX fabrication, even when 100% infill was used, air voids occur between the deposited layers due to improper adhesion. It was observed that by increasing the nozzle diameter, the apparent density of the specimens changes, and significant voids appear in the internal structure of the specimens with sizes up to 254 μm when using 0.8 mm diameter nozzles. Moreover, in some cases, the material shows air pores due to nozzle wear or gas evacuation produced by filament moisture [34], leading to morphological changes in the material and, consequently, to the size and homogeneity of the specimen's apparent section (Figure 16).

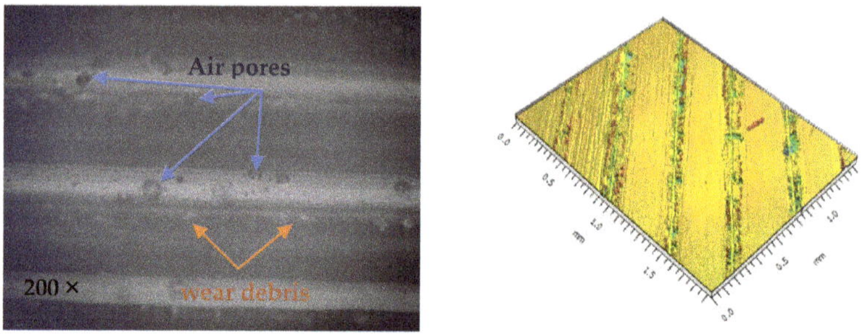

(**a**) (**b**)

Figure 16. Surface analysis: (**a**) microscopical view; (**b**) surface topography [34].

In this sense, it can be concluded that the variation for hardness obtained with the nozzle diameter is due to the thickness of the deposited material layer's lines, which implicitly leads to a larger volume of air pores between the layers, thus resulting in lower values for the hardness of the material studied.

By using the local heating method, the hardness trend is decreasing, an aspect also observed by other researchers [31] who used conventional post-processing annealing methods. The recorded values showed up to 1.44% variation for Shore hardness and 8.64% for Martens hardness when 80 °C temperature was used. This trend, in addition to the morphological changes of the semi-crystalline structure material, may also be due to the reduction in the internal stresses of the heat-treated material, as concluded by other researchers [15,18].

The hardness values as a function of printing direction have slight variations. However, their trend has also been observed in the literature [45], where it was deduced that the highest hardness was obtained when the horizontal printing direction was used, and the lowest values were recorded when the parts were printed in the Z-direction.

The ANOVA analysis revealed that, of all the factors studied for both types of hardness, the statistically significant factor (p-value < 0.05) was nozzle diameter (Table 9 for Shore hardness and Table 10 for Martens hardness).

Table 9. ANOVA analysis for Shore D hardness.

Source	DF	Adj SS	Adj MS	F-Value	*p*-Value
Regression	3	16.6474	5.5491	14.13	0.007
Nozzle size	1	14.3353	14.3353	36.50	0.002
Part temperature	1	2.1454	2.1454	5.46	0.067
Printing direction	1	0.1667	0.1667	0.42	0.544
Error	5	1.9637	0.3927		
Total	8	18.6111			

Table 10. ANOVA analysis for Martens hardness.

Source	DF	Adj SS	Adj MS	F-Value	*p*-Value
Regression	6	2936.47	489.41	35.72	0.027
Nozzle size	2	2688.79	1344.39	98.12	0.010
Part temperature	2	218.52	109.26	7.97	0.111
Printing direction	2	29.16	14.58	1.06	0.484
Error	2	27.40	13.70		
Total	8	2963.87			

However, considering the Pareto charts in Figures 17 and 18, it can be seen that the local heating application also has a high degree of influence. The significance levels for both types of hardness have been listed in Table 10.

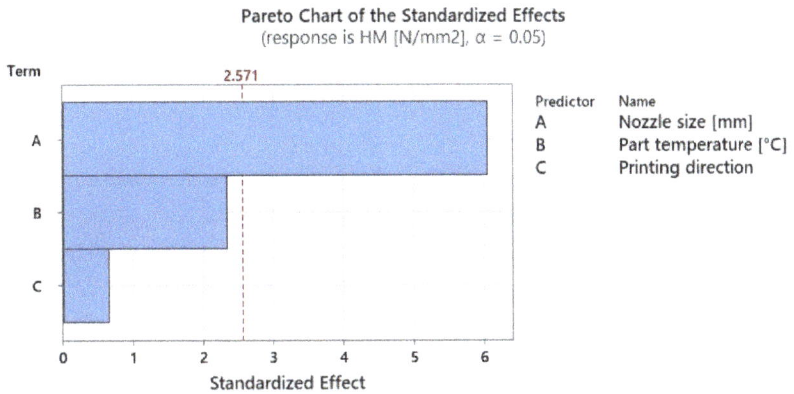

Figure 17. Graphical illustration of the degree of influence of studied factors on Shore D hardness.

From Table 11, it can be seen that temperature influences Martens hardness by about 20% and Shore hardness by 20%, showing that the application of this type of local heating has an influence on the mechanical characteristics of the resulting parts. Moreover, the variations presented in this paper in both the hardness of the parts and the tensile strength have the same trends as those found in the literature, raising the confidence in the application of the in-process local heat treatment method.

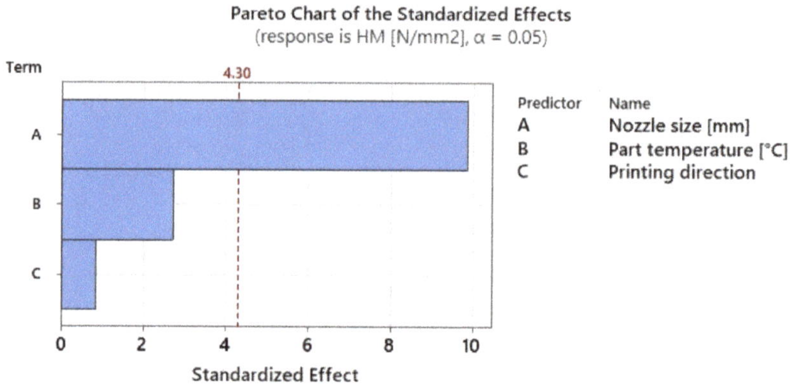

Figure 18. Graphical illustration of the degree of influence of studied factors on Martens hardness.

Table 11. Obtained results' percentage variation for hardness.

Factor	Shore D	Martens
Nozzle size	66.90%	73.35%
Temperature	25.88%	20.34%
Printing direction	7.210%	6.320%

3.3. *Surface Roughness*

Surface roughness measurements were made on the surfaces coinciding with the layer height, in this case, as shown in Figure 19, on the red surfaces. However, the values for the surfaces parallel to the build platform were also compared as a function of nozzle diameter.

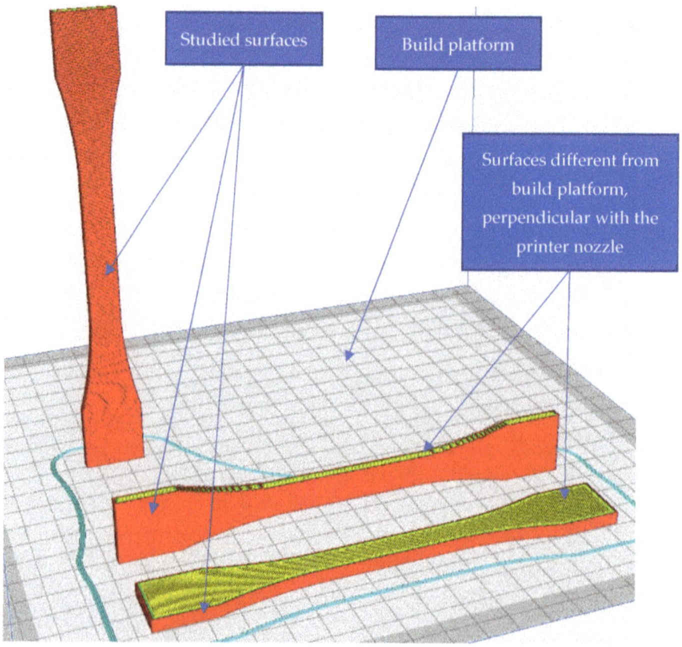

Figure 19. Graphical illustration of the measured zones for surface roughness.

As specified in Section 2 for hardness, the surface roughness tests also did not take into account the surface on the printer platform. However, Table 12 shows comparative results between a surface on the build platform and an opposing surface. Moreover, the table also shows a comparison between a simple tempered glass build platform and a build platform covered by microporous coating. It can be seen that due to the different manufacturing conditions for the first layers and the use of tempered glass as the build platform, the values showed a big difference, with the lowest roughness being obtained for the surface on the build platform.

Table 12. Roughness values for surfaces on the build platform and opposite.

	Surface on Simple Tempered Glass Build Platform	**Surface on Coated Tempered Glass Build Platform**	**Surface Opposite to Build Platform**
3D image	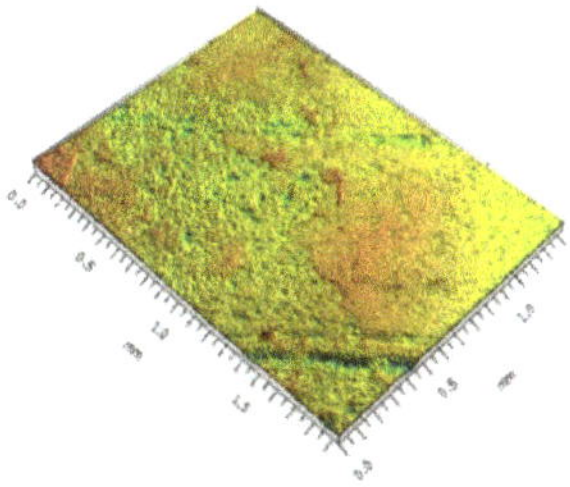	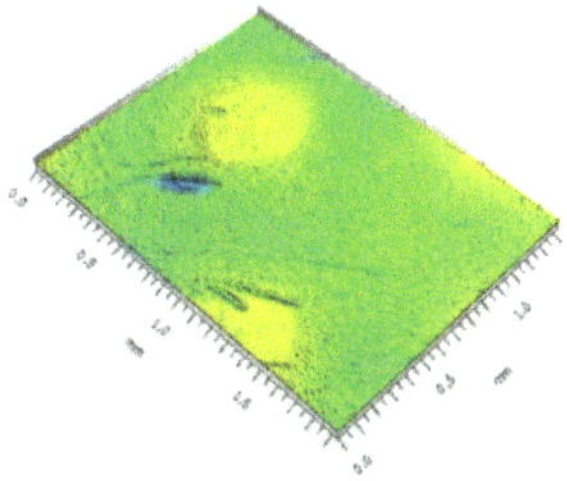	
2D image	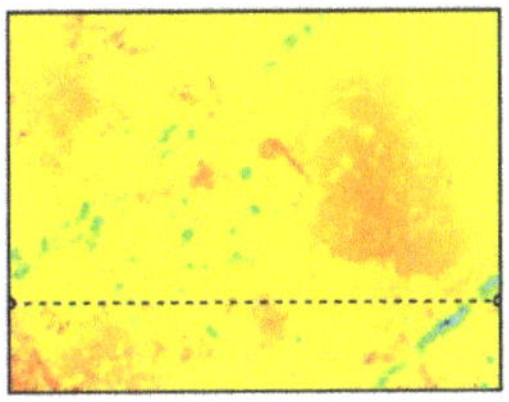	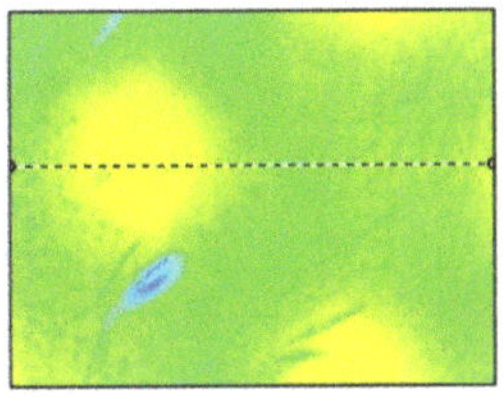	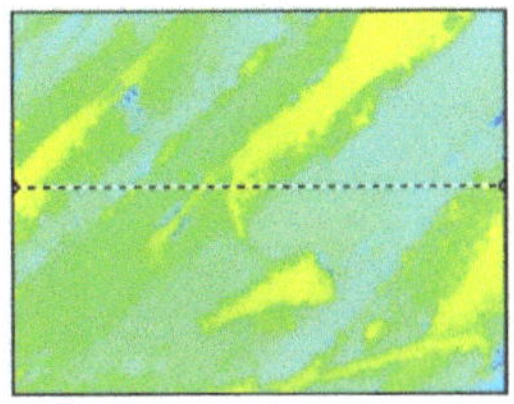
Profile	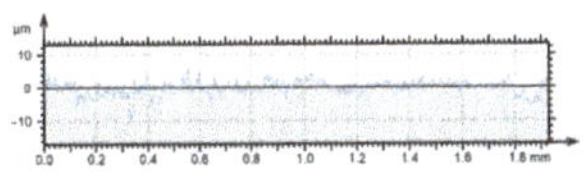	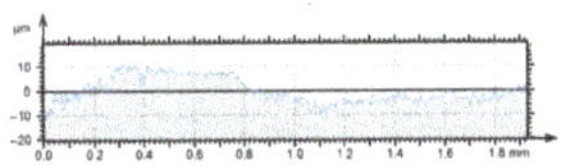	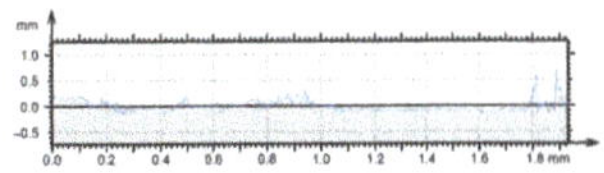
Sq	2.63 μm	5.10 μm	99.3 μm

The results from Table 12 show a major difference between the surface roughness on the build platform and the surface roughness of the surfaces opposite to the build platform. In this case, the simple tempered glass build platform corresponded to the best surface roughness values (Sq = 2.63 μm, compared to Sq = 5.10 μm obtained for the coated build platform). The microporous coating used for the build platform was a carborundum (SiC) produced by Creality 3D, which is specific for the 3D printer used in the 3D experiments. However, the surface opposite from the build platform presented up to 38× higher surface roughness values compared to those obtained for the two types of build platforms.

On the basis of the experimental results for surface roughness, it was possible to generate the mean effect plots (Figure 20) and significance plots of the factors studied, resulting from an ANOVA analysis (Table 13, Figure 21).

Table 13. ANOVA analysis for surface roughness.

Source	DF	Adj SS	Adj MS	F-Value	*p*-Value
Regression	3	639.94	213.313	2.55	0.169
Nozzle size	1	2.80	2.796	0.03	0.862
Part temperature	1	618.42	618.416	7.40	0.042
Printing direction	1	18.73	18.727	0.22	0.656
Error	5	417.70	83.541		
Total	8	1057.64			

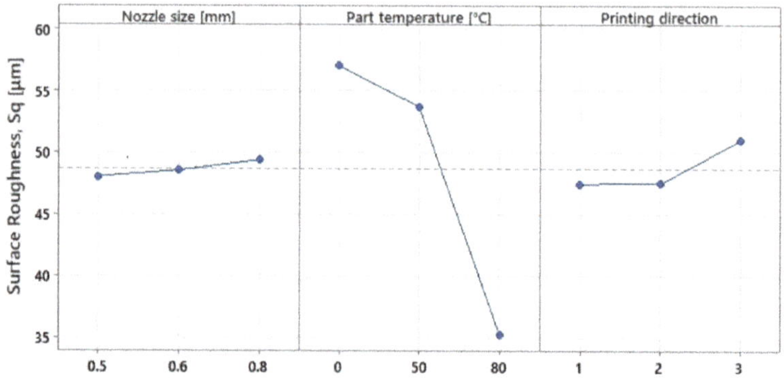

Figure 20. Variation in mean effects for surface roughness.

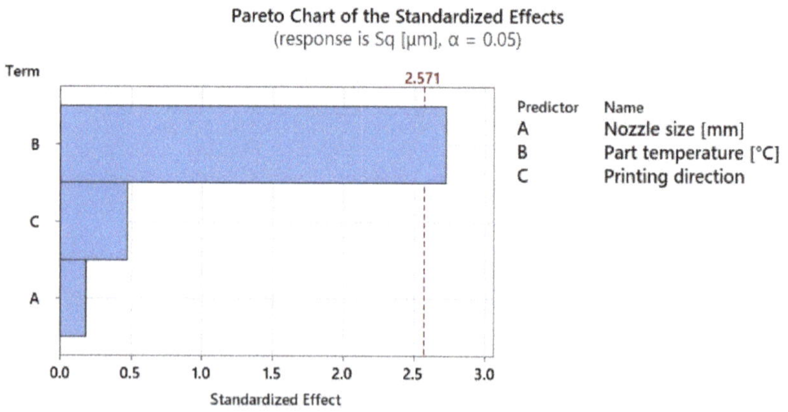

Figure 21. Graphical illustration of the degree of influence of studied factors on surface roughness.

In the case of parts printed in any direction, and regardless of the heat treatment, due to the fact that the layer height was kept constant, the results for surface roughness as a function of nozzle diameter showed very small variations, up to 2.81%, with the values falling within the range of 48–49.5 µm. However, the presented analysis shows an increasing trend for surface roughness values with increasing nozzle diameter, an aspect also observed in other studies [46,47].

The analysis highlighted that the in-process heat treatment had the most impact on the resulting surface roughness. A small variation of 6.25% was obtained when only the build platform temperature was used, followed by a considerable improvement in the surface roughness by 63.17% when the in-process heat treatment was used, compared with no heat. Additionally, an ANOVA analysis showed that the statistically significant factor, on surface roughness, with 79.53% significance, was temperature.

Printing direction shows neglectable variation for the X and Y directions, while an 8.19% increase in surface roughness was observed for parts manufactured in the Z direction. This aspect may be explained by the parts' dimensions and the used 3D printer type. The used 3D printer was a bed slinger type, which uses the build platform to move in the Y-direction, influencing the stability of thin and tall parts.

Table 14 presents the overall variation percentage and the degree of influence resulting from the analysis. An ANOVA analysis shows the degree of influence of all the studied factors. In this study, the statistically significant factor (p-value < 0.5) was the part temperature with 79.53% influence, followed by printing direction with 14.92% influence, and finally, the nozzle size with 5.55% influence.

Table 14. Obtained results' percentage variation and degree of influence for surface roughness.

Level	Nozzle Size		Part Temperature		Printing Direction	
	Value	Variation	Value	Variation	Value	Variation
1	0.5	-	0	-	X	-
2	0.6	↑ 0.37%	50	↓ 6.25%	Y	↑ 0.78%
3	0.8	↑ 2.81%	80	↓ 63.17%	Z	↑ 8.19%
	Degree of variation according to the ANOVA analysis					
Percentage	5.55%		79.53%		14.92%	

Another important aspect was observed when the results for surface roughness were compared with the values obtained for the flat surfaces, perpendicular with the nozzle (Figure 19). Figure 22 shows the results for surface roughness as a function of nozzle diameter, for the surfaces of the part parallel and opposite to the build platform.

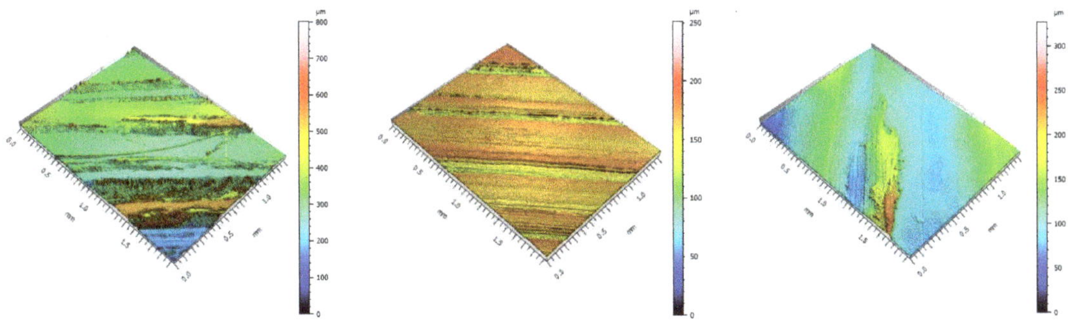

Taguchiprobe number 1
Nozzle size = 0.5 mm
Part temperature = 0
Printing direction = X
Sq value = 65.096 μm

Taguchiprobe number 6
Nozzle size = 0.6 mm
Part temperature = 80 °C
Printing direction = X
Sq value = 38.683 μm

Taguchiprobe number 8
Nozzle size = 0.8 mm
Part temperature = 50 °C
Printing direction = X
Sq value = 22.659 μm

Figure 22. Nozzle diameter influence on the top surface roughness of the tested specimens.

From Figure 22, it can be seen that, contrary to the results for surface roughness for surfaces that coincide with layer height (Figure 20), the values tend to decrease with increasing nozzle size. This aspect is justified by the fact that with increasing nozzle diameter, the interlayer adhesions are fewer and depend on the extrusion width. With increasing nozzle diameter, extrusion width also increases, requiring fewer layers to print a surface.

4. Conclusions

The analysis of the results obtained in this paper, related to the factors of influence considered, shows that the direction of printing and the nozzle diameter have the most significant influence on the tensile strength. In this case, given the input factors used, the temperature of the part does not have a major influence on the tensile strength, but it is not negligible. With the results obtained, it can be considered that the application of heat treatment during the process is possible. Keeping the temperature constant has influences on the tensile strength up to the critical melting temperature (the temperature at which the material plasticises massively), after which the deposition process cannot be continued.

The experimental values and trends are in agreement with the results obtained by other researchers in the literature [18,31,32,47,48]. This convergence of results from multiple sources significantly bolsters the confidence level associated with the outcomes presented in this paper, further confirms the possibility of applying in-process annealing during additive manufacturing, and also strengthens the reliability and significance of the multiple research studies from the literature.

For this study, the presented in-process heat treatment was applied on small parts with flat surfaces, with the heat being blown in one direction. Following the previous conclusion and the results from this study, the in-process heat treatment methodology will be optimised in the future, which offers the opportunity for more studies regarding the in-process heat treatment impact on larger parts with complex shapes.

The in-process heat treatment helps with the adhesion of the layers, resulting in an improvement in tensile strength by about 12.5% and an improved roughness by 12%, which demonstrates that the application of in-process heat treatment is feasible and can provide opportunities for further research by optimising the process. Regarding the hardness of the material, the results obtained confirmed the results of other research [28] and showed that by applying in-process heat treatment, due to morphological changes of the semi-crystalline material, the hardness decreases.

Nozzle diameter positively influences tensile strength, leading to an increase of about 44% when using a 0.8 mm diameter nozzle. At the same time, air voids in the material, as demonstrated in the literature [43], lead to a significant decrease in material hardness, with nozzle diameter being the factor with the highest statistical significance (p-value < 0.05).

Surface texture is an important factor for part functionality. In MEX processes the surface roughness is determined by the specific way in which melted material flows through the extruding nozzle and is deposited in the part as layers. The extruded material's proprieties are also strongly influenced by the printing temperature and cooling speed. The in-process heat treatment technique reheats the deposited material layers and slows the cooling of the new material layers that are being deposited. The experimental data showed that, for parts with higher surface quality demands, if in-process heat treatment is considered, the previous layer temperature should be one of the main parameters that needs careful selection. The surface roughness Sq parameter improves by 63.17% when in-process heat treatment is used and is approximately constant over the entire printed part. For the specific printing conditions considered in this study, the nozzle diameter and the printing direction parameter did not exhibit a significant influence on the surface roughness.

The printing direction greatly impacted the tensile strength, where the lowest values were obtained when the parts were printed in the Z-direction and the highest when printed in the Y-direction. Upon hardness, the printing direction had a low influence.

Author Contributions: Conceptualization, I.T.; methodology, I.T. and I.B.-B.; software, I.T., I.B.-B. and T.-L.S.; validation, T.-L.S., D.-A.C. and C.D.; formal analysis, C.D., I.B.-B. and T.-L.S.; investigation, I.T., I.B.-B. and T.-L.S.; resources, D.-A.C.; data curation, I.T.; writing—original draft preparation, I.T. and I.B.-B.; writing—review and editing, I.B.-B. and T.-L.S.; visualization, I.T. and D.-A.C.; supervision, C.D.; project administration, C.D.; funding acquisition, I.T. All authors have read and agreed to the published version of the manuscript.

Funding: This research was funded by The Human Capital Operational Programme 2014–2020 (POCU), Romania (Contract no. 62487/03.06.2022—POCU/993/6/13—Code 153299).

Institutional Review Board Statement: Not applicable.

Data Availability Statement: Some or all data, models, or code generated or used during the study are available from the corresponding author by request.

Acknowledgments: The work of the author Ioan Tamașag was supported by the project "PROINVENT", Contract no. 62487/03.06.2022—POCU/993/6/13—Code 153299, financed by The Human Capital Operational Programme 2014–2020 (POCU), Romania.

Conflicts of Interest: The authors declare no conflict of interest.

References

1. Ngo, T.D.; Kashani, A.; Imbalzano, G.; Nguyen, K.T.Q.; Hui, D. Additive Manufacturing (3D Printing): A Review of Materials, Methods, Applications and Challenges. *Compos. Part B Eng.* **2018**, *143*, 172–196. [CrossRef]
2. Vaezi, M.; Seitz, H.; Yang, S. A Review on 3D Micro-Additive Manufacturing Technologies. *Int. J. Adv. Manuf. Technol.* **2013**, *67*, 1721–1754. [CrossRef]
3. Zhou, W.D.; Chen, J.S. 3D Printing of Carbon Fiber Reinforced Plastics and Their Applications. *Mater. Sci. Forum* **2018**, *913*, 558–563. [CrossRef]
4. *ISO/ASTM 52900*; International Standard: 2021 Additive Manufacturing—General Principles—Fundamentals and Vocabulary. ISO/ASTM International: Geneva, Switzerland, 2021.
5. *Poly(Lactic Acid): Synthesis, Structures, Properties, Processing, and Applications*; Auras, R.; Lim, L.-T.; Selke, S.E.M.; Tsuji, H. (Eds.) John Wiley & Sons, Inc.: Hoboken, NJ, USA, 2010; ISBN 978-0-470-64984-8.
6. Drumright, R.E.; Gruber, P.R.; Henton, D.E. Polylactic Acid Technology. *Adv. Mater.* **2000**, *12*, 1841–1846. [CrossRef]
7. Rasal, R.M.; Janorkar, A.V.; Hirt, D.E. Poly(Lactic Acid) Modifications. *Prog. Polym. Sci.* **2010**, *35*, 338–356. [CrossRef]
8. Dorgan, J.R.; Janzen, J.; Clayton, M.P.; Hait, S.B.; Knauss, D.M. Melt Rheology of Variable L -Content Poly(Lactic Acid). *J. Rheol.* **2005**, *49*, 607–619. [CrossRef]
9. Palade, L.-I.; Lehermeier, H.J.; Dorgan, J.R. Melt Rheology of High l-Content Poly(Lactic Acid). *Macromolecules* **2001**, *34*, 1384–1390. [CrossRef]
10. Cao, M.; Cui, T.; Yue, Y.; Li, C.; Guo, X.; Jia, X.; Wang, B. Investigation of Carbon Fiber on the Tensile Property of FDM-Produced PLA Specimen. *Polymers* **2022**, *14*, 5230. [CrossRef] [PubMed]
11. Chicos, L.-A.; Pop, M.A.; Zaharia, S.-M.; Lancea, C.; Buican, G.R.; Pascariu, I.S.; Stamate, V.-M. Fused Filament Fabrication of Short Glass Fiber-Reinforced Polylactic Acid Composites: Infill Density Influence on Mechanical and Thermal Properties. *Polymers* **2022**, *14*, 4988. [CrossRef]
12. Müller, M.; Šleger, V.; Kolář, V.; Hromasová, M.; Piš, D.; Mishra, R.K. Low-Cycle Fatigue Behavior of 3D-Printed PLA Reinforced with Natural Filler. *Polymers* **2022**, *14*, 1301. [CrossRef]
13. Tamașag, I.; Ispas, L.; Picus, C.M.; Severin, T.L.; Dulucheanu, C. Annealing influence on 3d printed parts with amorfous structures. *TEHNOMUS* **2021**, *28*, 124–127.
14. Simmons, H.; Tiwary, P.; Colwell, J.E.; Kontopoulou, M. Improvements in the Crystallinity and Mechanical Properties of PLA by Nucleation and Annealing. *Polym. Degrad. Stab.* **2019**, *166*, 248–257. [CrossRef]
15. Singh, S.; Singh, M.; Prakash, C.; Gupta, M.K.; Mia, M.; Singh, R. Optimization and Reliability Analysis to Improve Surface Quality and Mechanical Characteristics of Heat-Treated Fused Filament Fabricated Parts. *Int. J. Adv. Manuf. Technol.* **2019**, *102*, 1521–1536. [CrossRef]
16. Castro-Casado, D. Chemical Treatments to Enhance Surface Quality of FFF Manufactured Parts: A Systematic Review. *Prog. Addit. Manuf.* **2021**, *6*, 307–319. [CrossRef]
17. Tamaşag, I.; Beşliu, I.; Amarandei, D. Application of Reverse Engineering for Automotive Plastic Components–Case Study. *Macromol. Symp.* **2021**, *395*, 2000265. [CrossRef]
18. Slavković, V.; Grujović, N.; Dišić, A.; Radovanović, A. Influence of Annealing and Printing Directions on Mechanical Properties of Pla Shape Memory Polymer Produced by Fused Deposition Modeling. In Proceedings of the 6th International Congress of Serbian Society of Mechanics Mountain Tara, Mountain Tara, Serbia, 19–21 June 2017.
19. Wach, R.A.; Wolszczak, P.; Adamus-Wlodarczyk, A. Enhancement of Mechanical Properties of FDM-PLA Parts via Thermal Annealing. *Macromol. Mater. Eng.* **2018**, *303*, 1800169. [CrossRef]
20. Bhandari, S.; Lopez-Anido, R.A.; Gardner, D.J. Enhancing the interlayer tensile strength of 3D printed short carbon fiber reinforced PETG and PLA composites via annealing. *Addit. Manuf.* **2019**, *30*, 100922. [CrossRef]
21. Park, S.-D.; Todo, M.; Arakawa, K. Effect of Annealing on the Fracture Toughness of Poly(Lactic Acid). *J. Mater. Sci.* **2004**, *39*, 1113–1116. [CrossRef]
22. Szust, A.; Adamski, G. Using Thermal Annealing and Salt Remelting to Increase Tensile Properties of 3D FDM Prints. *Eng. Fail. Anal.* **2022**, *132*, 105932. [CrossRef]

23. Ravi, A.K.; Deshpande, A.; Hsu, K.H. An In-Process Laser Localized Pre-Deposition Heating Approach to Inter-Layer Bond Strengthening in Extrusion Based Polymer Additive Manufacturing. *J. Manuf. Process.* **2016**, *24*, 179–185. [CrossRef]
24. Sikora, P.; Gnatowski, A.; Gołębski, R. Tests of Mechanical Properties of Semicrystalline and Amorphous Polymeric Materials Produced by 3D Printing. *MATEC Web Conf.* **2019**, *254*, 06003. [CrossRef]
25. Ansari, A.A.; Kamil, M. Izod Impact and Hardness Properties of 3D Printed Lightweight CF-Reinforced PLA Composites Using Design of Experiment. *Int. J. Lightweight Mater. Manuf.* **2022**, *5*, 369–383. [CrossRef]
26. Zeng, Y.-S.; Hsueh, M.-H.; Lai, C.-J.; Hsiao, T.-C.; Pan, C.-Y.; Huang, W.-C.; Chang, C.-H.; Wang, S.-H. An Investigation on the Hardness of Polylactic Acid Parts Fabricated via Fused Deposition Modeling. *Polymers* **2022**, *14*, 2789. [CrossRef] [PubMed]
27. Maguluri, N.; Suresh, G.; Guntur, S.R. Effect of Printing Parameters on the Hardness of 3D Printed Poly-Lactic Acid Parts Using DOE Approach. *IOP Conf. Ser. Mater. Sci. Eng.* **2022**, *1248*, 012004. [CrossRef]
28. Prechtel, A.; Reymus, M.; Edelhoff, D.; Hickel, R.; Stawarczyk, B. Comparison of Various 3D Printed and Milled PAEK Materials: Effect of Printing Direction and Artificial Aging on Martens Parameters. *Dent. Mater.* **2020**, *36*, 197–209. [CrossRef]
29. Schönhoff, L.M.; Mayinger, F.; Eichberger, M.; Reznikova, E.; Stawarczyk, B. 3D Printing of Dental Restorations: Mechanical Properties of Thermoplastic Polymer Materials. *J. Mech. Behav. Biomed. Mater.* **2021**, *119*, 104544. [CrossRef]
30. Zinelis, S.; Panayi, N.; Polychronis, G.; Papageorgiou, S.N.; Eliades, T. Comparative Analysis of Mechanical Properties of Orthodontic Aligners Produced by Different Contemporary 3D Printers. *Orthod. Craniofacial Res.* **2022**, *25*, 336–341. [CrossRef]
31. Shbanah, M.; Jordanov, M.; Nyikes, Z.; Tóth, L.; Kovács, T.A. The Effect of Heat Treatment on a 3D-Printed PLA Polymer's Mechanical Properties. *Polymers* **2023**, *15*, 1587. [CrossRef]
32. Butt, J.; Bhaskar, R. Investigating the Effects of Annealing on the Mechanical Properties of FFF-Printed Thermoplastics. *J. Manuf. Mater. Process.* **2020**, *4*, 38. [CrossRef]
33. Gan, X.; Wang, Z.; Xing, Z.; Chong, P.L.; Yazdi, M.H. Analysis of the Carreau Model Mixed Mechanism with a Stir Shaft in Color FDM Printing. *Processes* **2023**, *11*, 559. [CrossRef]
34. Tamașag, I.; Suciu, C.; Beșliu-Băncescu, I.; Dulucheanu, C.; Cerlincă, D.-A. Experimental Study on the Possibilities of FDM Direct Colour Printing and Its Implications on Mechanical Properties and Surface Quality of the Resulting Parts. *Polymers* **2022**, *14*, 5173. [CrossRef] [PubMed]
35. Kim, G.-T.; Go, H.-B.; Yu, J.-H.; Yang, S.-Y.; Kim, K.-M.; Choi, S.-H.; Kwon, J.-S. Cytotoxicity, Colour Stability and Dimensional Accuracy of 3D Printing Resin with Three Different Photoinitiators. *Polymers* **2022**, *14*, 979. [CrossRef] [PubMed]
36. *ISO 527*; International Standard: Plastics—Determination of Tensile Properties—Part 1: General Principles. International Organization for Standardization: Geneva, Switzerland, 2012.
37. *ISO 527*; International Standard: Plastics—Determination of Tensile Properties—Part 2: Test Conditions for Moulding and Extrusion Plastics. International Organization for Standardization: Geneva, Switzerland, 2019.
38. Sardinha, M.; Frutuoso, N.; Vicente, C.M.S.; Ribeiro, R.; Leite, M.; Reis, L. Influence of Seams in the Mechanical Properties of PLA Produced with Multiple Extrusion Modules. *Procedia Struct. Integr.* **2020**, *28*, 358–363. [CrossRef]
39. eSUN PLA Technical Data Sheet. Available online: https://www.esun3d.com/uploads/eSUN_PLA-Filament_TDS_V4.0.pdf (accessed on 26 April 2023).
40. eSUN PLA Material Safety Data Sheet. Available online: https://www.esun3d.com/uploads/PLA-MSDS.pdf (accessed on 26 April 2023).
41. *ISO 868:2003*; International Standard: Plastics and Ebonite—Determination of Indentation Hardness by Means of a Durometer (Shore Hardness). International Organization for Standardization: Geneva, Switzerland, 2003.
42. Singh, I.; Kumar, S.; Koloor, S.S.R.; Kumar, D.; Yahya, M.Y.; Mago, J. On Comparison of Heat Treated and Non-Heat-Treated LOM Manufactured Sample for Poly(Lactic)Acid: Mechanical and Morphological View Point. *Polymers* **2022**, *14*, 5098. [CrossRef]
43. Czyżewski, P.; Marciniak, D.; Nowinka, B.; Borowiak, M.; Bieliński, M. Influence of Extruder's Nozzle Diameter on the Improvement of Functional Properties of 3D-Printed PLA Products. *Polymers* **2022**, *14*, 356. [CrossRef]
44. Triyono, J.; Sukanto, H.; Saputra, R.M.; Smaradhana, D.F. The Effect of Nozzle Hole Diameter of 3D Printing on Porosity and Tensile Strength Parts Using Polylactic Acid Material. *Open Eng.* **2020**, *10*, 762–768. [CrossRef]
45. Beniak, J.; Holdy, M.; Križan, P.; Matúš, M. Research on Parameters Optimization for the Additive Manufacturing Process. *Transp. Res. Procedia* **2019**, *40*, 144–149. [CrossRef]
46. Shirmohammadi, M.; Goushchi, S.J.; Keshtiban, P.M. Optimization of 3D Printing Process Parameters to Minimize Surface Roughness with Hybrid Artificial Neural Network Model and Particle Swarm Algorithm. *Prog. Addit. Manuf.* **2021**, *6*, 199–215. [CrossRef]
47. Alsoufi, M.S.; Elsayed, A.E. How Surface Roughness Performance of Printed Parts Manufactured by Desktop FDM 3D Printer with PLA+ Is Influenced by Measuring Direction. *Am. J. Mech. Eng.* **2017**, *5*, 211–222.
48. Hervan, S.Z.; Altınkaynak, A.; Parlar, Z. Hardness, Friction and Wear Characteristics of 3D-Printed PLA Polymer. *Proc. Inst. Mech. Eng. Part J J. Eng. Tribol.* **2021**, *235*, 1590–1598. [CrossRef]

polymers

MDPI

Article

Effect of Short Carbon Fiber Reinforcement on Mechanical Properties of 3D-Printed Acrylonitrile Butadiene Styrene

Evgeniy Lobov, Anastasia Dobrydneva, Ilia Vindokurov and Mikhail Tashkinov *

Faculty of Applied Mathematics and Mechanics, Perm National Research Polytechnic University, Perm 614990, Russia; eslobov@pstu.ru (E.L.); a.dobrydneva@pstu.ru (A.D.); ivv@pstu.ru (I.V.)
* Correspondence: m.tashkinov@pstu.ru

Abstract: The effect of short carbon fiber (SCF) filler on the mechanical properties of 3D-printed acrylonitrile butadiene styrene (ABS) was investigated. The fused filament fabrication (FFF) method was used for the manufacturing of samples. Elastic properties and strength characteristics of samples made of conventional ABS and SCF-reinforced ABS were compared in tensile and bending tests. Fracture toughness and critical strain energy release rate were also determined. In addition, 3D-printed monofilament SCF-reinforced samples were fabricated, the internal structure of which was analyzed using microcomputed tomography (micro-CT). Based on the tomography data, finite-element (FE) models of representative volume elements (RVEs) of the reinforced material were created and used for the numerical calculation of effective characteristics. Numerical and experimental results for the effective elastic properties were compared with the Mori-Tanaka homogenization technique. The ABS samples filled with SCF showed considerably higher mechanical characteristics than those of the conventional ABS. Finally, the dependence between the strength characteristics and elastic properties of the samples on the diameter of the nozzle used for 3D printing was established. 3D-printed ABS reinforced with SCF demonstrated a gain in tensile strength and fracture toughness by 30% and 20%, respectively. Interlayer adhesion strength in flexure tests showed an increase of 28% compared to pure ABS samples.

Keywords: 3D printing; ABS; short carbon fiber; mechanical tests; FE model; elastic characteristics; strength; fracture toughness

Citation: Lobov, E.; Dobrydneva, A.; Vindokurov, I.; Tashkinov, M. Effect of Short Carbon Fiber Reinforcement on Mechanical Properties of 3D-Printed Acrylonitrile Butadiene Styrene. *Polymers* **2023**, *15*, 2011. https://doi.org/10.3390/polym15092011

Academic Editors: Cristina-Elisabeta Pelin and Anton Ficai

Received: 29 March 2023
Revised: 19 April 2023
Accepted: 21 April 2023
Published: 24 April 2023

1. Introduction

Additive manufacturing in the form of 3D printing based on layer-by-layer material deposition is currently used for manufacturing parts with complex geometry [1]. One of the most accessible methods of 3D printing is fused filament fabrication (FFF), which works with polymeric materials in the form of a filament [2,3]. A large variety of thermoplastic polymers were adapted to be used with this method. In addition to the initial material properties, the effective mechanical characteristics of the manufactured products are influenced by printing parameters, such as nozzle diameter, infill pattern and angle, infill density, printing speed and others [4], as well as different types of postprocessing treatment [5,6]. Additive composites with a polymer matrix and natural or synthetic microscale additives were developed to improve stiffness, elastic modulus and strength of 3D-printed polymers [7–9]. In particular, short carbon fibers [10–15], as well as continuous carbon fibers Field [16,17], are widely used as the most promising reinforcing agents.

Various issues related to thermoplastic polymers with randomly distributed SCF were addressed in many papers. For example, according to the latest studies, short carbon fibers were introduced into the designed 3D orthogonal preforms, which compressive behavior was characterized in [18]. The tensile strength, elastic modulus and fracture toughness of 3D-printed samples of SCF-reinforced ABS were studied in [19–21]. The interlayer mode-I fracture toughness of FFF-printed SCF-reinforced ABS materials was examined in [20] using a modified double cantilever beam (DCB) test. Paper [22] presents results of the interfacial bonding strength between printed wires of ABS, carbon nanotube-reinforced ABS and SCF-reinforced ABS specimens. The overheat FFF printing was used in [23] to fabricate ABS composite specimens with enhanced mechanical performance. The effect of carbon fiber concentration and type, infill pattern and environmental temperatures on the mechanical properties of the printed polyamide samples were investigated in [24]. The results of [25] indicated that the gradual change in fiber reinforcement reduced the stress concentrations at the interface zone and increased the strength of ABS. The numerical methods were also applied to predict mechanical properties and analyze the effect of various compositions and manufacturing parameters. Mechanical properties of SCF-reinforced PLA specimens were predicted using three-scale asymptotic homogenization for both random and aligned fiber distributions [26]. In [27], attention was given to the impact of process parameters from the micro- to macro-level by integrating image-based statistical analysis with physics-based modeling.

Although additive composites with SCF have been successfully fabricated and are commercially available, understanding their deformation and fracture mechanisms is a relevant subject of research because of the complexities of the microstructural morphology. In particular, the presence of microscale defects in the form of matrix voids, fiber damage and irregular fiber orientation can alter the targeted properties of these composites.

The aim of this work is to investigate the effect of manufacturing parameters and resulting microstructural characteristics on the elastic and fracture properties of 3D-printed SCF-reinforced ABS samples by comparing the results of experimental and numerical studies. The mechanical properties of samples were evaluated through a series of tensile and bending tests. The internal microstructure was examined using micro-CT scans and images from scanning electronic microscopy (SEM). Finite element simulations were performed to assess the effective response of RVEs based on the internal structure of the real monofilament samples. Some novel testing procedures and methods were suggested and implemented. New results on stiffness and strength properties of SCF-reinforced printed samples were obtained and compared with those made of standard ABS material.

2. Materials and Methods

2.1. Samples Manufacturing

A commercial ABS filament, as well as a filament with ABS matrix and SCF filler from REC (Russia) with a diameter of 1.75 mm, was used for the experiments. A Raise3D Pro2 Plus 3D printer was used for sample manufacturing. Standard samples for tensile and three-point bending tests, as well as monofilament samples for tensile tests, were produced with nozzle diameters of 0.4 mm and 0.8 mm. The tensile samples (Figure 1a) were made according to ISO 527-2:2012, with two infill angles: 0 and 90 degrees. The three-point flexural testing method was used to investigate the interlayer adhesion. The samples for bending tests were rods (length 100 mm, width 20 mm, height 10 mm) with an inner hole (length 100 mm, width 15.2 mm, height 5.2 mm). The size of the hole was chosen so that the wall thickness was divided by integer parts of the line widths when printing with different nozzles (Figure 1b). Single-layer monofilament samples were produced in such a way that the thickness of the working area of the sample corresponded to the thickness of the nozzle (Figure 1c).

The samples were printed with the infill angle of 0° and 90°, as shown in Figure 2a,b. For SCF-reinforced samples, the following printing parameters were taken: layer thickness 0.2 mm, table temperature 110 °C, nozzle temperature 290 °C, infill/drawing ratio 100% straight, printing speed 30 mm/s. In the case of pure ABS, the table temperature (100 °C) and nozzle temperature (255 °C) were different. All samples were printed in a closed chamber and cooled to room temperature naturally.

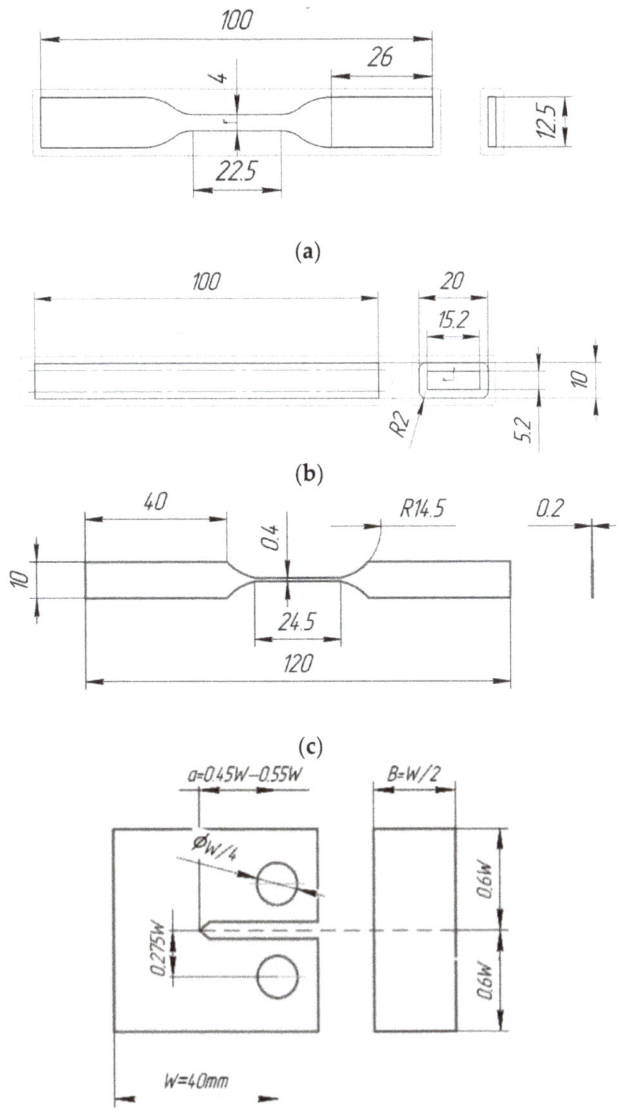

Figure 1. Dimensions of samples: (**a**) for tensile tests; (**b**) for bending tests; (**c**) monofilament tensile tests, (**d**) Geometry of CT sample.

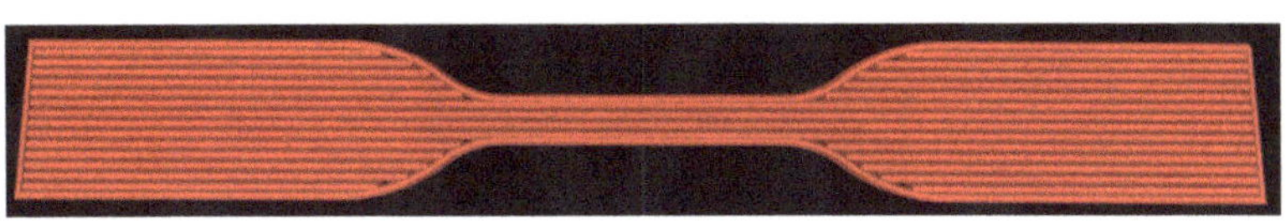

(**a**)

Figure 2. *Cont.*

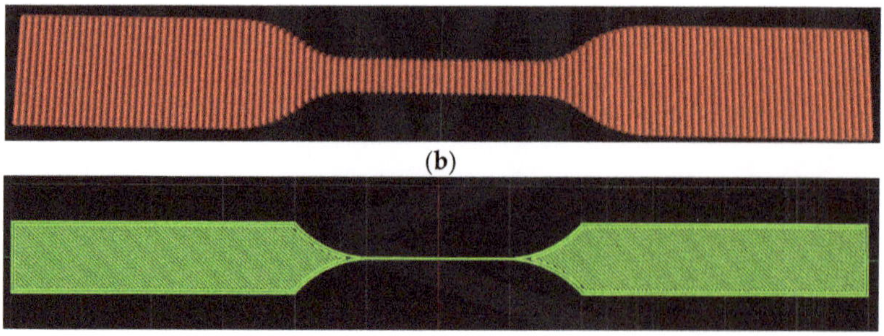

(**b**)

(**c**)

Figure 2. Tensile sample models and filling: (**a**) infill angle 0°; (**b**) infill angle 90°; (**c**) monofilament.

2.2. Mechanical Characterization

Tensile tests were performed at a constant displacement rate of 0.5 mm/min and ambient temperature (~20 °C) on an Instron 68SC-5 universal testing machine with a 5 kN load cell. The accuracy of load measurement is 0.5% of the measured value in the range from 5 N to 5 kN, and the resolution of the movement is 0.0095 μm. An AVE2 video extensometer was used to measure displacements. Two points were marked on each sample surface at an equal distance from the sample middle plane, and the video extensometer was used to track the movement of these points to measure the elongation of the sample. The elastic tensile properties of the samples were obtained from the analysis of the elastic section of the stress-strain curve. In order to collect statistics, each of the sample configurations was tested at least 15 times to assess repeatability.

Bending tests were performed at a constant displacement rate of 0.1 mm/min with a 500 N load cell. The ratio used was:

$$\sigma = \frac{6Flh}{4\left(bh^3 - b_1 h_1^3\right)} \tag{1}$$

where h is sample height, b is sample width, h_1 is the height of the interior part of a sample and b_1 is interior sample width.

The tensile fracture properties were evaluated according to the ASTM D5045-14 standard for measuring the fracture toughness of polymers under plane strain state. Compact tension (CT) samples (Figure 3) were used. Their infill pattern and slicing parameters were chosen in a way that the crack propagated perpendicular to the printing layers stacking. The such configuration ensures the highest achievable fracture toughness value.

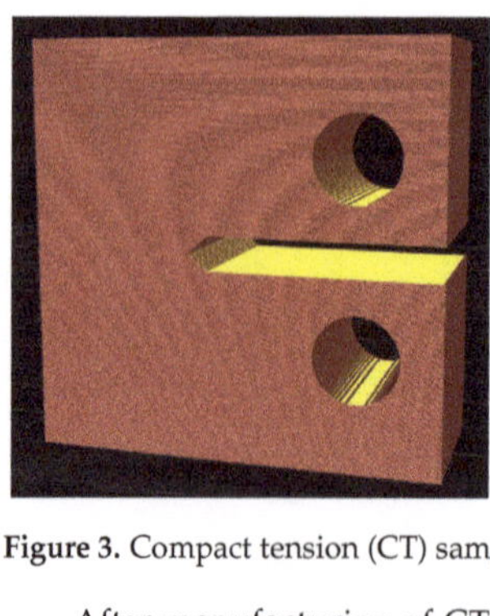

Figure 3. Compact tension (CT) sample model.

After manufacturing of CT samples, the supports in the crack area were removed using a table-top drilling and milling machine. A razor blade with a thickness of 0.08 mm was used to form a sharp notch marking the initial position of the crack. The crosshead speed of an Instron 68SC-5 universal testing machine during the tests was 1 mm/min. A VIC-3D Micro-DIC system (Correlated Solutions, Irmo, SC, USA) was used to measure the displacement field near the opening crack. Before the test, a white background and a black pattern were applied to the samples using an airbrush. Sequential speckle images were obtained using two 5.0-megapixel cameras to evaluate the displacement field. The average strain in the crack opening zone was estimated using Vic-3D 9 software (Correlated Solutions, Irmo, SC, USA). The fracture properties of the samples were obtained from the stress-strain curve.

2.3. Analysis of Microstructure

The surface of the samples was analyzed using an EM-30+ scanning electron microscope (COXEM, Daejeon, Korea) at magnifications of 100×, 200×, 500×, and 1000×. Before the morphological analysis, a thin layer of gold was disposed of on the samples. An accelerating voltage of 15 kV was used, and the analysis was performed using a secondary electron detector.

To study the internal microstructural features of additive composites, tomographic imaging of samples of the monofilament 3D printed samples was performed on a SkyScan 1272 Bruker micro-CT with the following parameters: X-ray tube voltage 37 kV, current 57 μA, resolution (voxel edge size) 1 μm, exposure time 3000 μs, sample rotation step 0.1° with 360° imaging with averaging over 4 frames. The duration of the sample scanning was 16 h and 45 min.

2.4. Numerical Methods

For numerical analysis of the elastic properties of an SCF-reinforced material, it is necessary to create a representative volume element (RVE)—a small material sample containing a sufficient statistical description number of microstructural elements of each component. These RVEs, together with information on geometry, volume fraction, fiber orientation, and fiber length distribution, were obtained from the micro-CT data. Each RVE was discretized using tetrahedral elements and subjected to uniaxial tensile loading. Calculations were performed in SIMULIA Abaqus 2022 (Dassault Systemes, Montréal, QC, Canada) with the C3D4 element type.

In addition to numerical calculations, analytical estimation of the effective elastic properties was performed with the Mori-Tanaka homogenization scheme [28], which gives estimates based only on the volume fraction and elastic properties of individual constituents.

3. Results

3.1. Strength and Stiffness

3.1.1. Tensile Properties

The tensile strength and Young's modulus values grouped by material, nozzle diameter and sample type are shown in Table 1. The stress-strain curves obtained from the tensile tests are shown in Figure 4, and the corresponding block diagram is shown in Figure 5. The results are grouped according to the nozzle diameter and infill angle parameter. Hereinafter in tables and figures, ABS + CF, stands for SCF-reinforced ABS material. The accepted short notation also includes the used nozzle diameter and infill angle. E.g., "ABS + CF_04_90" means a sample of SCF-reinforced ABS, printed using a nozzle with a diameter of 0.4 mm and with a 90° infill angle.

The highest tensile strength (79.12 ± 1.6 MPa) was observed for ABS + CF samples printed with a 0.8 mm nozzle and 0° infill angle. The use of a 0.4 mm nozzle resulted in the highest tensile modulus value (9242.48 MPa) for the ABS + CF samples with a 0° infill angle (Table 1, Figure 5). The strength of pure ABS samples printed with a 0.4 mm nozzle and 0° infill angle was 67% higher than that of samples with a 90° infill angle. For pure ABS samples printed with a 0.8 mm nozzle, the difference between the strength of samples with different infill angles is 52%.

Table 1. Results of tensile tests.

	Nozzle Diameter, mm	Infill Angle	Tensile Strength (MPa)	Tensile Elastic Modulus (MPa)
ABS	0.4	0°	49.22 ± 1.66	5260.62 ± 101.45
		90°	29.43 ± 2.14	3819.44 ± 171.18
		Monofilament	26.94 ± 8.58	7648.32 ± 1077.95
	0.8	0°	58.14 ± 3.38	4978.21 ± 296.14
		90°	38.32 ± 4.41	4459.64 ± 427.69
		Monofilament	9.59 ± 7.41	6042.31 ± 119.20
ABS + CF	0.4	0°	70.14 ± 2.61	9242.44 ± 155.99
		90°	27.36 ± 2.87	4229.15 ± 248.56
		Monofilament	28.35 ± 4.32	11,709.16 ± 889.50
	0.8	0°	79.12 ± 1.65	9137.99 ± 197.26
		90°	44.62 ± 3.12	6023.59 ± 403.70
		Monofilament	12.51 ± 7.25	8704.72 ± 510.52

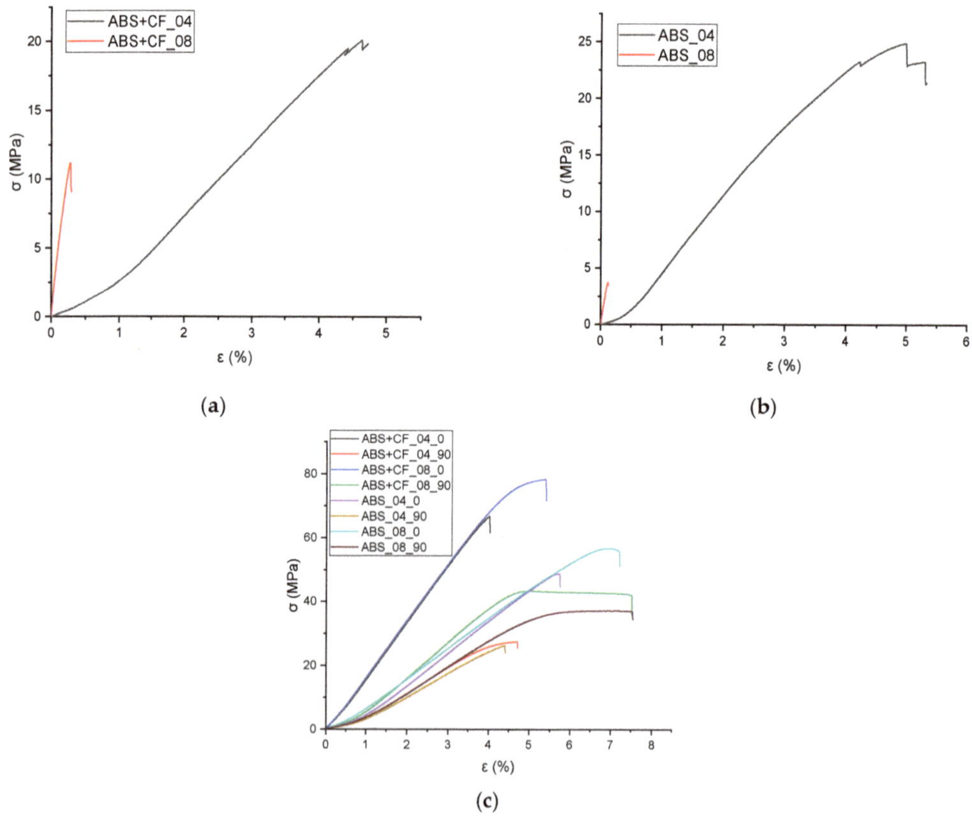

Figure 4. Tensile tests stress-strain diagrams for a series of samples: (**a**) ABS + CF monofilament printed with different nozzles; (**b**) ABS monofilament printed with different nozzles, (**c**) ABS + CF and ABS dog-bone samples printed with different nozzles and infill angles.

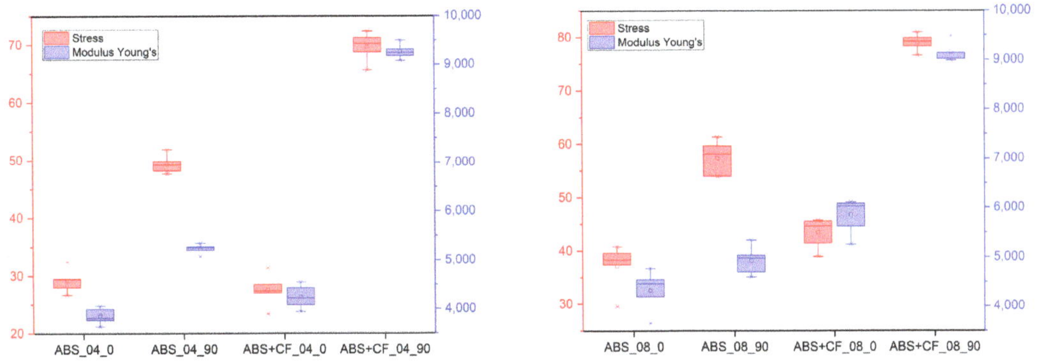

Figure 5. Tensile strength values and elastic moduli of samples printed with different nozzle diameters: (**a**) 0.4 mm; (**b**) 0.8 mm.

For ABS + CF, the strength upon change of an infill angle differs by 150% for samples manufactured with a 0.4 mm nozzle and by 77% for samples manufactured with a 0.8 mm nozzle. The maximum strength of the ABS + CF samples is 42.5% higher than that of pure ABS samples when printed with a 0.4 mm nozzle and 36% higher when printed with a 0.8 mm nozzle.

Similar trends were observed for the modulus of elasticity: samples with 0° infill angle exhibited stiffer behavior in all cases. However, in contrast to the strength characteristics, the samples printed using a nozzle with a smaller diameter have a higher (by 1–5%) stiffness at a 0° infill angle. For 90° infill samples, the elastic properties are significantly lower than those of 0° infill samples for both pure ABS and reinforced ABS.

Tensile tests of the monofilament samples showed ambiguous results—the strength characteristics of the samples printed with the 0.8 mm nozzle are significantly lower than those of the samples printed with the 0.4 mm nozzle (Figure 4). This is observed for both pure and reinforced materials. At the same time, on average, the elastic modulus of the monofilament exceeds that of the tested standard samples but has a large scatter of values.

Figure 6 shows SEM images of a surface of a fractured 3D printed ABS + CF monofilament sample after a tensile test at a magnification of 100×, 200×, 500× and 1000× times. Gaps are observed between the matrix and short fibers at the location of the fracture, indicating possible imperfect adhesion of the short fiber in the filament. The presence of defects in the form of free cavities can be related to the pullout of short fibers. Pulled short fibers in the field of vision have a smooth surfaces without any residual polymer material on them.

3.1.2. Flexural Properties

Bending strength and bending modulus values grouped by material type and nozzle diameter are shown in Table 2. The stress-strain curves obtained from the flexural tests are shown in Figure 7, and the corresponding block diagram is presented in Figure 8. The highest bending strength was observed for samples printed using a 0.8 mm nozzle with a printing speed of 30 mm/s and a nozzle temperature of 290 °C. In the three-point bending test, all printed samples first showed linear elastic deformation and then reached the maximum bending stress. Samples printed with the 0.8 mm nozzle showed higher deformation than samples printed with the 0.4 mm nozzle while maintaining a linear behavior. SCF reinforcement does not solve the problem of the influence of nozzle diameter on the interlayer sintering and strength of the samples: in the case of pure ABS, the difference between the 0.4 mm and 0.8 mm nozzle samples was 32%, and 42% in the case of ABS + CF. The flexural modulus is 12% higher for ABS samples made with a 0.8 mm nozzle and 4% higher for ABS + CF samples.

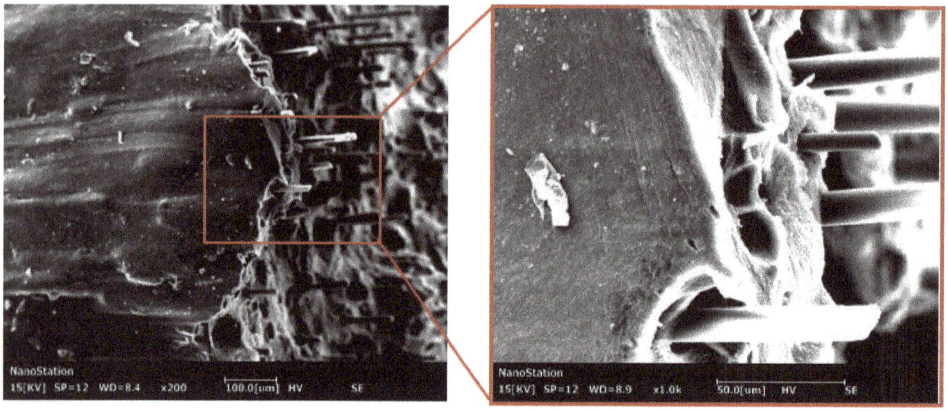

(a)

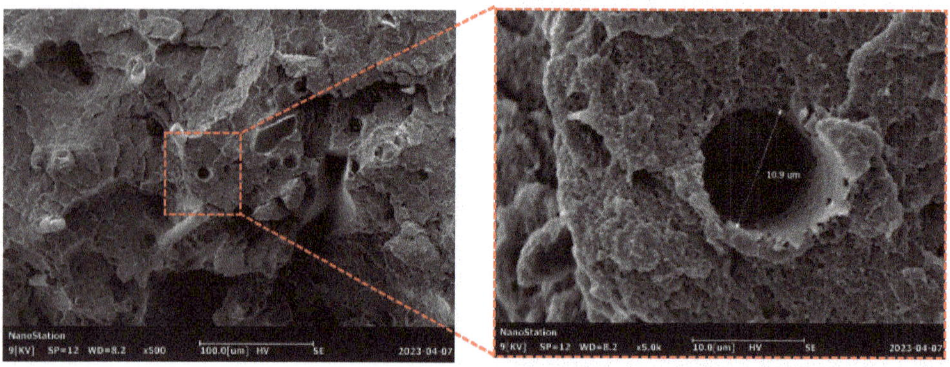

(b)

Figure 6. SEM images of the surface: (**a**) side view on the fracture plane; (**b**) front view on the fracture plane.

Table 2. Results of bending tests.

	Nozzle Diameter, mm	Bending Strength (MPa)	Bending Modulus (MPa)
ABS	0.4	15.30 ± 1.77	1321.25 ± 27.95
	0.8	20.23 ± 1.47	1481.46 ± 12.24
ABS + CF	0.4	20.80 ± 1.72	1556.99 ± 34.49
	0.8	29.47 ± 1.16	1621.77 ± 45.67

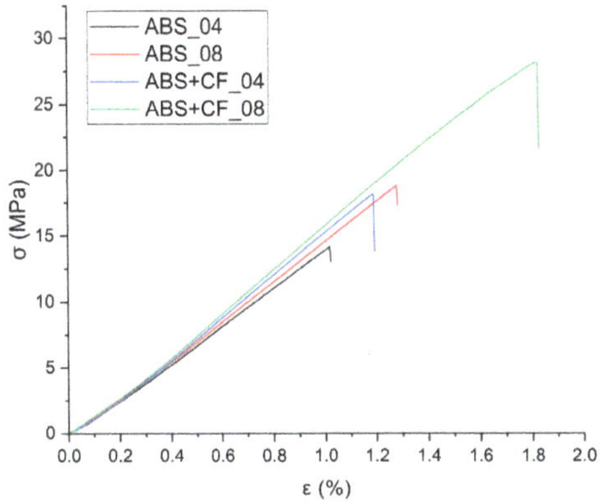

Figure 7. Bending tests stress-strain dependences for a series of samples printed with 0.4 mm and 0.8 mm nozzles.

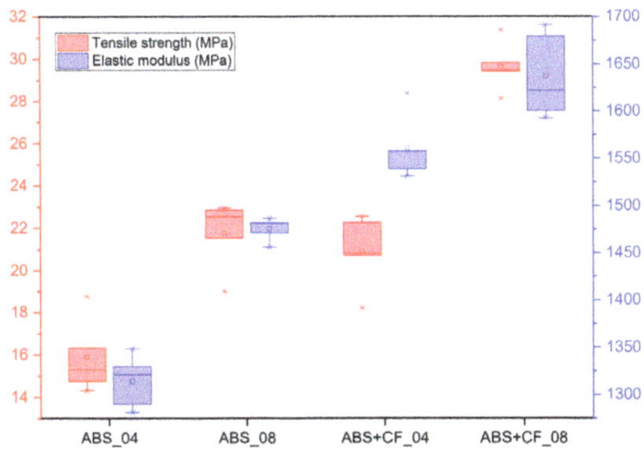

Figure 8. Bending strength and bending modulus of samples printed with different nozzle diameters.

3.1.3. Fracture Toughness and Critical Strain Energy Release Rate

Fracture toughness (critical stress intensity factor) K_{Ic} of the first fracture mode was determined by formula according to the method given in ASTM D5045 [29,30]:

$$K_{Ic} = \frac{P_Q}{BW^{\frac{1}{2}}} f(x) \tag{2}$$

where P_Q is load, B is sample thickness, W is sample width, a is crack length, $x = a/W$,

$$f(x) = \frac{(2+x)\left(0.886 + 4.64x - 13.32x^2 + 14.72x^3 - 5.6x^4\right)}{(1-x)^{\frac{3}{2}}} \tag{3}$$

The critical strain energy release rate then can be determined as [29,30]:

$$G_{Ic} = \frac{(1-\nu^2){K_{Ic}}^2}{E} \tag{4}$$

where E is elastic modulus, obtained during fracture toughness tests, ν is Poisson's ratio, K_{Ic} is critical stress intensity factor. The adopted scheme assumes linear elastic fracture.

The criterion described in Section 9.1.1 of the ASTM D5045 [29] was used to ensure the reliability of K_{Ic}. In addition, the sample size was chosen according to this criterion, which ensured a flat deformed state at the apex of the crack.

The ABS + CF sample made with a 0.4 mm nozzle showed better fracture toughness (by 28%) and critical strain energy release rate (by 48%) characteristics compared to conventional ABS (see Table 3). Crack propagation in the samples was sequential, with delamination along the layers. The crack tip in samples with compact tension has a higher probability of advancing across the layer-layer interface, where the interfacial zone between the layers acted as the weakest link, contributing to softening. This effect was stronger for pure ABS samples.

Table 3. Fracture toughness and critical strain energy release rate.

	K_{Ic} MPa∗m$^{1/2}$	G_{Ic} kJ/m^2	Strength (MPa)
ABS	1.28	0.031	0.62
ABS + CF	1.65	0.046	0.77

Figure 9a demonstrates SEM images of the fractured ABS + CF CT-sample surface. A number of randomly oriented short fibers imprinted in the matrix, as well as fibers pulled out from the matrix, are clearly visible on the surface. Although the crack plane is perpendicular to the printing layers stacking, there were no abrupt changes in the direction of crack propagation. Force-displacement diagrams in Figure 9b show differences in the mechanical behavior of 3D-printed CT samples made of pure ABS and SCF-reinforced ABS. The latter is able to withstand a tensile load of about 20% higher than the former.

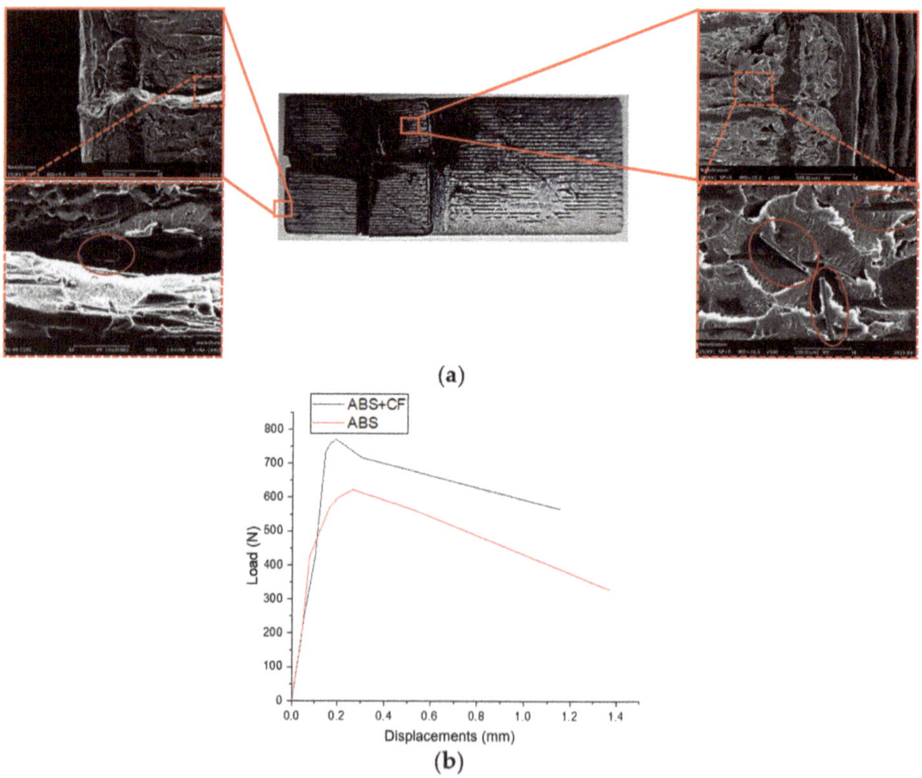

Figure 9. (**a**) SEM images of fractured CT-sample printed with ABS + CF, two sections at the start and at the end of crack formation are zoomed in; (**b**) force-displacement diagrams for CT samples tension tests averaged over a series of samples.

3.2. Internal Structure Characterization

Computed tomography data was used for the generation of finite-element models of the material's RVEs, which can be used for the investigation of mechanical properties and microscale mechanical behavior. This paper compares the effective elastic characteristics of these RVEs with the results of experimental studies of the samples, as well as with analytical estimates based on the mean-field method with a Mori-Tanaka homogenization technique.

The obtained stacks of micro-CT X-ray images were used for stereological reconstructions and visualization of 3D models of the two-phase structure of 3D printed ABS + CF monofilament samples. Figure 10 shows a reconstructed model of a monofilament sample printed with a 0.4 mm nozzle. The average volumetric content of the short fiber in the sample, estimated from the 3D model, is 2.92%. The average diameter of the carbon fibers is 12 μm. The quantitative analysis of individual fibers was implemented to evaluate their parameters. The predominant number of fibers in the sample is less than 100 μm in length, which agrees with the results from [18], while there are single fibers longer than 400 μm also present. The fibers are predominantly oriented along with the printing direction, with single fibers located transversely. Deviation from the axis aligned with the printing direction in most cases does not exceed 15–20 degrees. Spatial distribution of fibers in the volume of the sample after the 3D-printing process remains uniform.

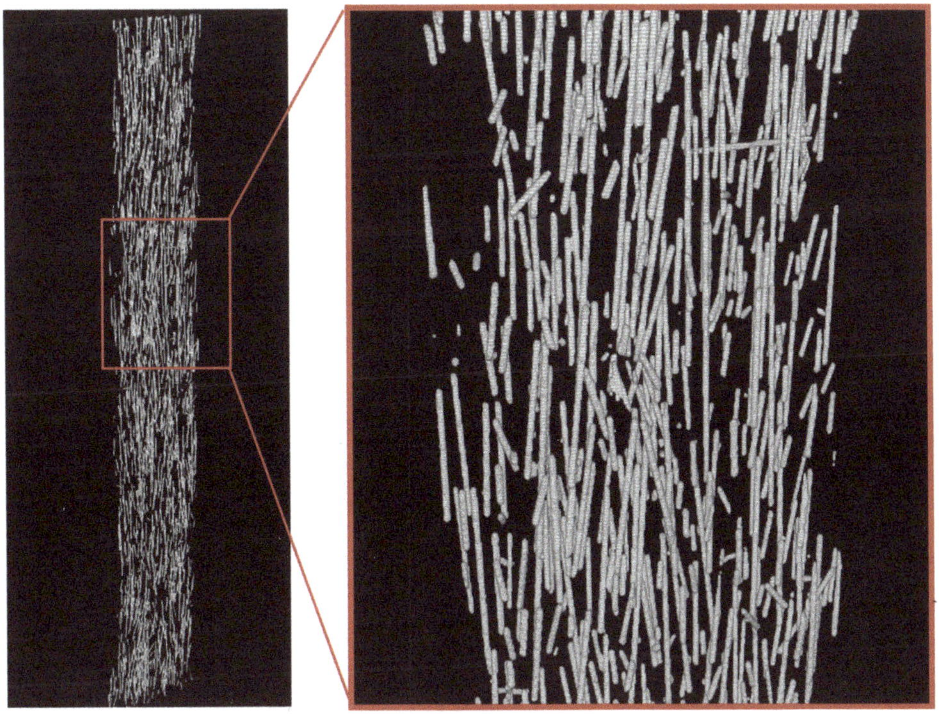

Figure 10. 3D model of the fiber structure of a single-layer monofilament sample (general view and enlarged fragment) obtained from micro-CT data.

Figure 11 shows histograms of fiber distribution by length, as well as by the angle of their deviation from the axis coinciding with the printing direction in 3D-printed monofilament SCF-reinforced samples. Most of the fibers have lengths in a range between 0.01 mm and 0.2 mm, while their orientation angle is predominantly between 0 and 20 degrees.

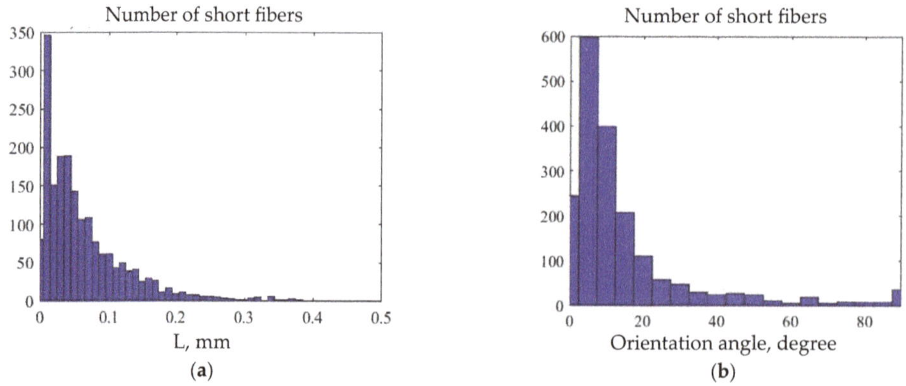

Figure 11. Histogram of fibers distribution in 3D-printed SCF-reinforced monofilament sample: (**a**) by fiber length; (**b**) by the angle of deviation from the vertical axis.

3.3. Numerical Results

Based on the micro-CT images, RVEs with dimensions X = 300 μm, Y = 160 μm, and Z = 600 μm were created. The dimensions were chosen in relation to the average fiber length: three fibers with maximal length fit in the Z dimension, which coincides with the printing direction, and, accordingly, is axial to the primary orientation of short fibers. According to the fiber length distribution analysis (Figure 6a), any RVE with the selected size will accommodate even the longest fiber from the distribution. The X and Y RVE dimensions were equal to 80% of the full monofilament sample size. Three areas were selected from the three-dimensional micro-CT model as RVEs, as shown in Figure 12.

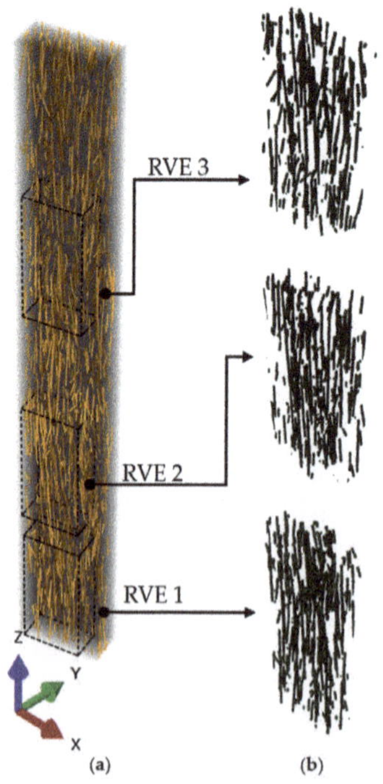

Figure 12. Three-dimensional model of the internal structure of monofilament sample printed with a 0.4 mm nozzle (**a**) and three RVEs selected from it (**b**).

The initial elastic characteristics of the matrix and SCF are presented in Table 4. The properties of the ABS matrix were obtained experimentally for samples printed with 0.4 mm and 0.8 mm nozzle and, in both cases, with 0° infill angle. The elastic properties and density of the carbon fiber were determined from other authors' works [31,32].

Table 4. Elastic mechanical properties of components.

Component	Density kg/m^3	Young's Modulus, MPa	Poisson's Ratio
ABS matrix—sample printed with a 0.4 mm nozzle	1.05	5235.75	0.392
ABS matrix—sample printed with a 0.8 mm nozzle	1.05	5841.26	0.392
Carbon fiber	1.72	220,000	0.15

To obtain the effective characteristics, RVEs were subjected to tensile loading with lower face fixed. Displacements of the upper face were restricted in the XY plane, and the load directed along Z axis was applied to the upper face. The effective Young's modulus was determined by Hooke's law:

$$E = \frac{Fl}{S\Delta l} = \frac{F}{\varepsilon \times S}, \tag{5}$$

where ε is the percentage of elongation, F is force, S is surface area.

Table 5 presents the results of the FE analysis and their comparison with the experimental data and with the results of the Mori-Tanaka homogenization.

Table 5. Effective elastic characteristics of SCF-reinforced ABS monofilament sample obtained with different methods.

			Nozzle Diameter 0.4 mm	Nozzle Diameter 0.8 mm
FE analysis	Young's modulus, MPa	RVE 1	8290	9035
		RVE 2	7955	8675
		RVE 3	7816	8540
Tensile test			11,709 ± 890	8704 ± 511
Mori-Tanaka estimation			7730	8414

The volume fraction of SCF in three investigated representative volumes of 3D-printed monofilament samples varied within the range from 2.81% to 2.96%. The Young's moduli values obtained on the basis of FE-models for different RVEs from the same sample have shown insignificant deviation (up to 1.26%). The difference between Young's moduli of the monofilament samples printed with different nozzle diameters is about 9%. Numerically calculated values of Young's modulus differ from the experimental tension test values for monofilament printed samples on average by 46% in the case of a 0.4 mm nozzle and practically coincide in the case of a 0.8 mm nozzle. When comparing the numerically calculated values of the modulus with the properties of the standard tensile samples with 0° infill angle (see Table 1), 0.4 mm nozzle gives a 15.23% difference, while for 0.8 mm nozzle, it decreases to 4.43%. The Young's modulus values obtained with the Mori-Tanaka homogenization technique differ downwards from the experimental values by 12% and from the results of the FE analysis by 3.5%. With different matrix properties defined by the samples printed with different nozzle diameters, the Mori-Tanaka model gives estimations with an 8% difference.

4. Discussion

An experimental study of the elastic moduli, tensile strength, and fracture toughness of samples of 3D printed SCF-reinforced was performed and compared to pure ABS. The dependence of the results on the nozzle diameter and infill angle was established. The obtained characteristics were compared to analytical predictions and the numerical results of the FE modeling of RVEs created using micro-CT three-dimensional data.

According to the tensile and flexural tests, the nozzle diameter was the most important parameter affecting the final mechanical properties of the printed samples. SCF reinforcement significantly increased the uniaxial tensile strength in samples with 0° infill angle (lines printed along load application). In the cases of transversal printing (90° infill angle, interlayer interaction between extrusion lines), SCF reinforcement does not give a clear increase in the tensile strength compared to pure ABS.

The numerical results from RVE models with morphology based on the micro-CT data predicts the effective elastic modulus better than the Mori-Tanaka homogenization technique, which, by definition, depends only on volume fraction, form, and properties of inclusions. This confirms that the parameters of short fibers, such as their length and alignment, have an effect on the elastic mechanical response of the additively manufactured composite material. However, in this particular case, results for Young's modulus based on FE simulation and Mori-Tanaka estimations were quite close. On the contrary, experimental results for the monofilament samples were much higher than theoretically predicted. This means that not only microstructural parameters and distribution of SCF play a significant role, but also manufacturing process parameters and quality of the printed samples.

The diameter of the nozzle affects the printing process of the samples and, consequently, their effective mechanical properties. The width of the extruded filaments is indirectly determined by a nozzle diameter, but slicers allow some adjustments to these configurations [33]. Depending on the nozzle diameter, the number of printed lines required to fill the same area changes. Hence, the number of contact interfaces between printing lines changes as well. The surface area of these interfaces, as well as adhesion quality, play an important role and affect the resulting mechanical properties. Besides, in the case of short fiber reinforced polymers, some minor effects could be added by changing fiber orientation during the transition of the thermoplastic polymer through the nozzle. The less the nozzle diameter, the more pronounced this dependence.

The layer height of all samples was set to 50% of the nozzle diameter. The width of the contour lines and infill lines in the cross-sectional plane was set to 100% of the nozzle diameter. The tensile strength of samples printed with the 0.8 mm nozzle was generally higher than those of the 0.4 mm nozzle samples. This may be due to the larger sintering area of the layers printed with the 0.8 mm nozzle along the Y axis, which lead to an even distribution of stress along each line of the structure and prevents stress concentration. However, Young's moduli were higher for samples printed with a 0.4 mm nozzle. This can be explained by the fact that samples printed with the smaller nozzle accommodate a greater number of material lines in the same volume. Oriented in the direction of the load, they redistribute this load better but have a lesser area of contact between printing lines which results in lower strength. In addition, other factors, such as temperature fluctuations of the layers due to the heater, fan, etc., during the experiment were not considered in this study but could also potentially have an effect on the resulting mechanical response.

5. Conclusions

The mechanical properties of 3D-printed samples made of pure ABS and SCF-reinforced ABS were compared. Tensile, flexural, and compact tension tests were performed. The standard, monofilament, and compact tension samples that were investigated in these tests were manufactured using nozzles of different diameters, as well as with two different infill angle patterns. The experimental results on the effective elastic properties were compared to the results of finite-element simulations and Mori-Tanaka analytical estimations, which showed sufficient agreement. FE models were created using the data from micro-CT characterization of the monofilament samples.

According to the results, adding short carbon fibers to the ABS matrix increased the tensile strength of the printed samples by 29.8%, that, combined with high-performance characteristics, such as surface quality and post-processability, opens up a wide range of possibilities for the solution of various engineering tasks using such material. Fracture and bending tests showed that the SCF-reinforced ABS performed well. However, the interlayer adhesion, as with the standard material, could be further improved. By selecting numerical simulation parameters, it is possible to create models that exploit the positive properties of short fiber reinforced polymer to the greatest extent possible.

Author Contributions: Conceptualization, M.T. and E.L.; methodology, M.T., E.L. and I.V.; validation, E.L., A.D. and I.V.; formal analysis, E.L., A.D. and I.V.; investigation, E.L., A.D. and I.V.; writing—original draft preparation, E.L., A.D. and I.V.; writing—review and editing, M.T.; visualization, E.L., A.D. and I.V.; supervision, M.T.; project administration, M.T.; funding acquisition, M.T. All authors have read and agreed to the published version of the manuscript.

Funding: The research was performed at Perm National Research Polytechnic University with the support of the Russian Science Foundation (project No. 22-79-10350).

Institutional Review Board Statement: Not applicable.

Informed Consent Statement: Not applicable.

Data Availability Statement: Data available on request due to restrictions eg privacy or ethical.

Conflicts of Interest: The authors declare no conflict of interest.

References

1. Ngo, T.D.; Kashani, A.; Imbalzano, G.; Nguyen, K.T.Q.; Hui, D. Additive Manufacturing (3D Printing): A Review of Materials, Methods, Applications and Challenges. *Compos. Part B Eng.* **2018**, *143*, 172–196. [CrossRef]
2. Brenken, B.; Barocio, E.; Favaloro, A.; Kunc, V.; Pipes, R.B. Fused Filament Fabrication of Fiber-Reinforced Polymers: A Review. *Addit. Manuf.* **2018**, *21*, 1–16. [CrossRef]
3. Pandelidi, C.; Bateman, S.; Piegert, S.; Hoehner, R.; Kelbassa, I.; Brandt, M. The Technology of Continuous Fibre-Reinforced Polymers: A Review on Extrusion Additive Manufacturing Methods. *Int. J. Adv. Manuf. Technol.* **2021**, *113*, 3057–3077. [CrossRef]
4. Khan, S.; Joshi, K.; Deshmukh, S. A Comprehensive Review on Effect of Printing Parameters on Mechanical Properties of FDM Printed Parts. *Mater. Today Proc.* **2021**, *50*, 2119–2127. [CrossRef]
5. Vindokurov, I.; Pirogova, Y.; Tashkinov, M.; Silberschmidt, V.V. Effect of Heat Treatment on Elastic Properties and Fracture Toughness of Fused Filament Fabricated PEEK for Biomedical Applications. *Polymers* **2022**, *14*, 5521. [CrossRef]
6. Dizon, J.R.C.; Gache, C.C.L.; Cascolan, H.M.S.; Cancino, L.T.; Advincula, R.C. Post-Processing of 3D-Printed Polymers. *Technologies* **2021**, *9*, 61. [CrossRef]
7. Agarwal, K.; Kuchipudi, S.K.; Girard, B.; Houser, M. Mechanical Properties of Fiber Reinforced Polymer Composites: A Comparative Study of Conventional and Additive Manufacturing Methods. *J. Compos. Mater.* **2018**, *52*, 3173–3181. [CrossRef]
8. Ghabezi, P.; Flanagan, T.; Harrison, N. Short Basalt Fibre Reinforced Recycled Polypropylene Filaments for 3D Printing. *Mater. Lett.* **2022**, *326*, 132942. [CrossRef]
9. Sodeifian, G.; Ghaseminejad, S.; Yousefi, A.A. Preparation of Polypropylene/Short Glass Fiber Composite as Fused Deposition Modeling (FDM) Filament. *Results Phys.* **2019**, *12*, 205–222. [CrossRef]
10. Nawafleh, N.; Celik, E. Additive Manufacturing of Short Fiber Reinforced Thermoset Composites with Unprecedented Mechanical Performance. *Addit. Manuf.* **2020**, *33*, 101109. [CrossRef]
11. Hofstätter, T.; Pedersen, D.B.; Tosello, G.; Hansen, H.N. State-of-the-Art of Fiber-Reinforced Polymers in Additive Manufacturing Technologies. *J. Reinf. Plast. Compos.* **2017**, *36*, 1061–1073. [CrossRef]
12. Wang, X.; Jiang, M.; Zhou, Z.; Gou, J.; Hui, D. 3D Printing of Polymer Matrix Composites: A Review and Prospective. *Compos. Part B Eng.* **2017**, *110*, 442–458. [CrossRef]
13. Parandoush, P.; Lin, D. A Review on Additive Manufacturing of Polymer-Fiber Composites. *Compos. Struct.* **2017**, *182*, 36–53. [CrossRef]
14. R., M.K.H.; Benal, M.G.M.; G.S., P.K.; Tambrallimath, V.; H.R., G.; Khan, T.M.Y.; Rajhi, A.A.; Baig, M.A.A. Influence of Short Glass Fibre Reinforcement on Mechanical Properties of 3D Printed ABS-Based Polymer Composites. *Polymers* **2022**, *14*, 1182. [CrossRef]
15. HR, M.; Benal, M.G.M.; G.S., P.; Tambrallimath, V.; Ramaiah, K.; Khan, T.M.Y.; Bhutto, J.K.; Ali, M.A. Effect of Short Glass Fiber Addition on Flexural and Impact Behavior of 3D Printed Polymer Composites. *ACS Omega* **2023**, *8*, 9212–9220. [CrossRef]

16. Zhang, H.; Huang, T.; Jiang, Q.; He, L.; Bismarck, A.; Hu, Q. Recent Progress of 3D Printed Continuous Fiber Reinforced Polymer Composites Based on Fused Deposition Modeling: A Review. *J. Mater. Sci.* **2021**, *56*, 12999–13022. [CrossRef]
17. Chacón, J.M.; Caminero, M.A.; Núñez, P.J.; García-Plaza, E.; García-Moreno, I.; Reverte, J.M. Additive Manufacturing of Continuous Fibre Reinforced Thermoplastic Composites Using Fused Deposition Modelling: Effect of Process Parameters on Mechanical Properties. *Compos. Sci. Technol.* **2019**, *181*, 107688. [CrossRef]
18. Quan, Z.; Larimore, Z.; Wu, A.; Yu, J.; Qin, X.; Mirotznik, M.; Suhr, J.; Byun, J.-H.; Oh, Y.; Chou, T.-W. Microstructural Design and Additive Manufacturing and Characterization of 3D Orthogonal Short Carbon Fiber/Acrylonitrile-Butadiene-Styrene Preform and Composite. *Compos. Sci. Technol.* **2016**, *126*, 139–148. [CrossRef]
19. Srinivasan Ganesh Iyer, S.; Keles, O. Effect of Raster Angle on Mechanical Properties of 3D Printed Short Carbon Fiber Reinforced Acrylonitrile Butadiene Styrene. *Compos. Commun.* **2022**, *32*, 101163. [CrossRef]
20. Young, D.; Wetmore, N.; Czabaj, M. Interlayer Fracture Toughness of Additively Manufactured Unreinforced and Carbon-Fiber-Reinforced Acrylonitrile Butadiene Styrene. *Addit. Manuf.* **2018**, *22*, 883–890. [CrossRef]
21. Goh, G.D.; Dikshit, V.; Nagalingam, A.P.; Goh, G.L.; Agarwala, S.; Sing, S.L.; Wei, J.; Yeong, W.Y. Characterization of Mechanical Properties and Fracture Mode of Additively Manufactured Carbon Fiber and Glass Fiber Reinforced Thermoplastics. *Mater. Des.* **2018**, *137*, 79–89. [CrossRef]
22. Zhang, W.; Cotton, C.; Sun, J.; Heider, D.; Gu, B.; Sun, B.; Chou, T.-W. Interfacial Bonding Strength of Short Carbon Fiber/Acrylonitrile-Butadiene-Styrene Composites Fabricated by Fused Deposition Modeling. *Compos. Part B Eng.* **2018**, *137*, 51–59. [CrossRef]
23. Tran, T.Q.; Canturri, C.; Deng, X.; Tham, C.L.; Ng, F.L. Enhanced Tensile Strength of Acrylonitrile Butadiene Styrene Composite Specimens Fabricated by Overheat Fused Filament Fabrication Printing. *Compos. Part B Eng.* **2022**, *235*, 109783. [CrossRef]
24. Abderrafai, Y.; Hadi Mahdavi, M.; Sosa-Rey, F.; Hérard, C.; Otero Navas, I.; Piccirelli, N.; Lévesque, M.; Therriault, D. Additive Manufacturing of Short Carbon Fiber-Reinforced Polyamide Composites by Fused Filament Fabrication: Formulation, Manufacturing and Characterization. *Mater. Des.* **2022**, *214*, 110358. [CrossRef]
25. Hasanov, S.; Gupta, A.; Alifui-Segbaya, F.; Fidan, I. Hierarchical Homogenization and Experimental Evaluation of Functionally Graded Materials Manufactured by the Fused Filament Fabrication Process. *Compos. Struct.* **2021**, *275*, 114488. [CrossRef]
26. Nasirov, A.; Gupta, A.; Hasanov, S.; Fidan, I. Three-Scale Asymptotic Homogenization of Short Fiber Reinforced Additively Manufactured Polymer Composites. *Compos. Part B Eng.* **2020**, *202*, 108269. [CrossRef]
27. Pei, S.; Wang, K.; Chen, C.B.; Li, J.; Li, Y.; Zeng, D.; Su, X.; Yang, H. Process-Structure-Property Analysis of Short Carbon Fiber Reinforced Polymer Composite via Fused Filament Fabrication. *J. Manuf. Process.* **2021**, *64*, 544–556. [CrossRef]
28. Mori, T.; Tanaka, K. Average Stress in Matrix and Average Elastic Energy of Materials with Misfitting Inclusions. *Acta Metall.* **1973**, *21*, 571–574. [CrossRef]
29. *ASTM D 5045-99*; Standard Test Methods for Plane-Strain Fracture Toughness and Strain Energy Release Rate of Plastic Materials 1. ASTM Inernational: West Conshohocken, PA, USA, 2007; pp. 1–10. [CrossRef]
30. *Fracture Toughness Testing and Its Applications*; ASTM International: West Conshohocken, PA, USA, 1965; ISBN 978-0-8031-0105-0.
31. Cai, H.; Ye, J.; Wang, Y.; Saafi, M.; Huang, B.; Yang, D.; Ye, J. An Effective Microscale Approach for Determining the Anisotropy of Polymer Composites Reinforced with Randomly Distributed Short Fibers. *Compos. Struct.* **2020**, *240*, 112087. [CrossRef]
32. Sommacal, S.; Matschinski, A.; Drechsler, K.; Compston, P. Characterisation of Void and Fiber Distribution in 3D Printed Carbon-Fiber/PEEK Using X-Ray Computed Tomography. *Compos. Part A Appl. Sci. Manuf.* **2021**, *149*, 106487. [CrossRef]
33. Torrado, A.R.; Shemelya, C.M.; English, J.D.; Lin, Y.; Wicker, R.B.; Roberson, D.A. Characterizing the Effect of Additives to ABS on the Mechanical Property Anisotropy of Specimens Fabricated by Material Extrusion 3D Printing. *Addit. Manuf.* **2015**, *6*, 16–29. [CrossRef]

polymers

MDPI

Article

Effect of UV-C Radiation on 3D Printed ABS-PC Polymers

Catalin Gheorghe Amza *, Aurelian Zapciu, Florin Baciu and Constantin Radu

Faculty of Industrial Engineering and Robotics, University Politehnica of Bucharest, 060042 Bucharest, Romania; aurelianzapciu@yahoo.com (A.Z.); florin.baciu@upb.ro (F.B.); rcdnd@yahoo.com (C.R.)
* Correspondence: acatal@camis.pub.ro; Tel.: +40-744-500-803

Abstract: During the initial stages of the COVID-19 pandemic, healthcare facilities experienced severe shortages of personal protective equipment (PPE) and other medical supplies. Employing 3D printing to rapidly fabricate functional parts and equipment was one of the emergency solutions used to tackle these shortages. Using ultraviolet light in the UV-C band (wavelengths of 200 nm to 280 nm) might prove useful in sterilizing 3D printed parts, enabling their reusability. Most polymers, however, degrade under UV-C radiation, so it becomes necessary to determine what 3D printing materials can withstand the conditions found during medical equipment sterilization with UV-C. This paper analyzes the effect of accelerated aging through prolonged exposure to UV-C on the mechanical properties of parts 3D printed from a polycarbonate and acrylonitrile butadiene styrene polymer (ABS-PC). Samples 3D printed using a material extrusion process (MEX) went through a 24-h UV-C exposure aging cycle and then were tested versus a control group for changes in tensile strength, compressive strength and some selected material creep characteristics. Testing showed minimal mechanical property degradation following the irradiation procedure, with tensile strength being statistically the same for irradiated parts as those in the control group. Irradiated parts showed small losses in stiffness (5.2%) and compressive strength (6.5%). Scanning electron microscopy (SEM) was employed in order to assess if any changes occurred in the material structure.

Keywords: 3D-printing polymers; ABS-PC; PCABS; accelerated aging; ultraviolet; UV-C

check for updates

Citation: Amza, C.G.; Zapciu, A.; Baciu, F.; Radu, C. Effect of UV-C Radiation on 3D Printed ABS-PC Polymers. *Polymers* **2023**, *15*, 1966. https://doi.org/10.3390/polym15081966

Academic Editors: Cristina-Elisabeta Pelin and Anton Ficai

Received: 30 March 2023
Revised: 13 April 2023
Accepted: 19 April 2023
Published: 21 April 2023

1. Introduction

In the early weeks of the COVID-19 pandemic, healthcare facilities were overwhelmed by the influx of people requiring medical assistance, leading to problems with personal protective equipment supply and availability, as stated by the World Health Organization [1]. These shortages were widespread and happened at all levels of medical care, from nursing homes, local hospitals and clinics to emergency rooms at some of the world's biggest hospitals [2–4]. The distribution of locations that needed equipment and the diversity of items needed made resupplying extremely difficult, both from a manufacturing perspective and from a logistical one [5–8]. The necessary equipment had to be manufactured quickly and had to be distributed widely, which led to even more problems occurring because of resource misallocation [9,10]. Additive manufacturing (AM), also known as 3D printing, has a decentralized aspect and a unique ability to adapt to producing new part designs without additional capital investment [11–14]. With AM, parts can be produced on-site and on-demand, lowering the delay between necessity and availability. For this reason, 3D printing saw increased use during the initial responses to the medical equipment shortages [15–17]. Hospital technicians, small businesses and even private citizens who owned hobby-grade 3D printers began creating, distributing and manufacturing designs of protective equipment [18–20].

While the majority of the manufactured products meant to replace medical equipment in case of critical and urgent supply issues were designed for single use (e.g., swabs for nasopharyngeal testing [21–23]), other products could be used multiple times if sterilized under inadequate conditions. Examples of such products are face shields [24,25].

Material Extrusion 3D Printing (MEX) is an additive manufacturing process [26] that uses thermoplastic material feedstock. MEX 3D printing enables several advantages over more traditional manufacturing.

Studies performed on N95 respirators during the COVID-19 pandemic showed that filtering efficiency and infection prevention increase when the respirator is properly fitted [27–29]. The fact that 3D printing can be used to rapidly manufacture bespoke products for each healthcare worker can be leveraged to provide better and tighter seals for respirators. Ballard et al. produced respirators using real 3D data obtained using computer tomography that successfully passed OSHA-certified testing [30], while McAvoy et al. produced polymer frames that improve the fit of available N95 respirators [31].

In the medical field, the sterilization of medical equipment is performed through various methods, such as autoclaving, dry heating, ultrasonic sterilization [32–35], or chemical sterilization using various chemicals, such as hydrogen peroxide or ethylene oxide [36]. However, some of these methods are not applicable to certain polymers. For example, acrylonitrile butadiene styrene (ABS) parts deform when going through repeated autoclaving treatments [37] and suffer from a degradation of mechanical properties when sterilized using alcohol or other disinfectants [38,39].

In recent years, researchers have paid increasing attention to the sterilizing action of ultraviolet radiation, specifically UV-C radiation [40–43]. The International Commission on Illumination (in French: Commission Internationale de l'Eclairage, CIE) uses a wavelength of 280 nm to separate UV-C and UV-B bands, 315 nm to separate UV-B and UV-A bands, and 400 nm to separate UV-A and photosynthetically active radiation. The effects of UV-B/UV-A radiation have been studied extensively, as radiation of these wavelengths reaches Earth's troposphere naturally. However, UV-C radiation is entirely filtered by the Earth's atmosphere and has to be produced artificially. For this reason, there is no accelerated aging testing standard for the action of UV-C on materials, unlike that for radiation of longer wavelengths [44]. A workshop organized by the United States National Institute of Standards and Technology (NIST) and the International UV Association (IUVA) [45] aimed to gather information about applications, new certifications and new guidelines regarding UV-C disinfection and sterilization [46]. This workshop also highlighted the need to study the effects of UV-C radiation on materials. Standards for UV-C applications and testing are currently under development as a cooperation between NIST and the Illuminating Engineering Society (IES.org).

To get ahead of this trend, materials testing is important in determining their applicability and compatibility with novel sterilization methods. In the case of UV-C sterilization, materials testing is also crucial to the standardization efforts.

Acrylonitrile Butadiene Styrene (ABS) is a 3D printing material with widespread use [47] due to its good mechanical characteristics [48,49] and low cost. The material is known for its weak ultraviolet resistance [50]. Through the addition of polycarbonate fibers (PC) to an ABS base, manufacturers have successfully created a copolymer blend with enhanced mechanical properties and ultraviolet resistance [51] while still maintaining its ease of processing through MEX 3D printing [52]. This copolymer has high strength and stiffness, high heat resistance [53,54], and good impact resistance [55]. The newly created acrylonitrile butadiene styrene and polycarbonate polymer blends (ABS-PC) are being marketed for use in industrial applications, prototyping, toolmaking or end-use parts manufacturing [56].

Given this context, this work investigates the changes in the mechanical behavior of 3D-printed parts made from ABS-PC after exposure to artificial UV-C radiation. This work pursues one of the research directions raised by the NIST-IUVA workshop highlighted previously related to identifying and studying suitable materials. Unlike more common polymers, such as ABS, polylactic acid (PLA), or modified polyethylene terephthalate-glycol (PETG), literature on 3D printed ABS-PC blends is scarce and does not include mechanical behavior after exposure to artificial radiation in the UV-C spectrum. Additionally, the study of 3D printed materials is necessary for the implementation and approval of 3D printing techniques and 3D printed devices for medical applications, including those produced at the point of care (3DPOC) [57,58]. By using a controlled irradiation treatment in an irradiation chamber, an accelerated effect that simulates many cycles of ultraviolet sterilization can be observed. The mechanical properties assessed in this work are tensile and compression strength, material stiffness, as well as changes in creep behavior of ABS-PC samples subjected to prolonged tensile loads.

2. Materials and Methods

Accelerated aging of materials under UV-C radiation does not have well-defined testing standards. Testing parameters and procedures should be determined based on existing literature. An appendix of the ISO 4892-2 standard [59] mentions modifying testing conditions to use a mercury lamp that generates 10 W/m^2 of 254 nm wavelength radiation, a wavelength that falls in the UV-C band. A testing protocol for the assessment of material behavior under UV-C radiation is defined by the Business and Institutional Furniture Manufacturers Association (BIFMA, Grand Rapids, MI, USA) [60]. These guidelines were created for healthcare furniture design and specify that materials should be tested with 291 kJ/m^2 of UV-C radiation for a period of 12 to 24 h.

The material analyzed in this study is a copolymer made from ABS (Acrylonitrile Butadiene Styrene) and PC (PolyCarbonate). The tested material is commercially available under the brand name Z-PCABS, manufactured by Zortrax (Olsztyn, Poland). According to the manufacturer, this Z-PCABS blend contains 55–60% ABS, 30–35% PC and up to 10% additives and colorants. The material has a glass transition temperature of 104 °C and melts at a temperature of 260 °C (manufacturer specifications). The rated tensile strength is 36.89 MPa (ISO 527:1998 [61]), and the material's specific density is 1.14 g/cm^3. The material was sourced as a 1 kg spool of 1.75 mm diameter filament, opaque, ivory color. The spool of material was provided in a sealed reflective bag that included silica desiccant and was unsealed only prior to printing. Decisive in material selection was the fact that many filament manufacturers have settled on this composition based on process-specific requirements. When used for 3D printing, ABS should be extruded at temperatures between 230 °C and 260 °C while PC requires a higher temperature, 260 °C to 300 °C. As PC content in the blend increases, so does the temperature needed to process the material. The selected ABS/PC ratio places this material at the upper end of what common 3D printers can process in terms of extrusion temperature. Above this temperature, MEX 3D printers need to be equipped with high-temperature resistant components, higher-power ceramic heaters, temperature sensors for high-temperature applications and heated enclosures.

All specimens were 3D printed on a Zortrax M200 3D printer equipped with a 0.4 mm diameter nozzle. Z-Suite software, produced by the same company, was used to section the virtual model into layers. Printing parameters were selected based on what is known to provide consistent printing results [62–65] in order to minimize potential errors related to the manufacturing process. Layer settings used for slicing are layer height = 0.19 mm and layer width = 0.4 mm, with 2 outside perimeters and solid infill using a grid pattern (45° raster angle, alternating each layer [66]). The extrusion temperature was set at 265 °C, and the build platform temperature was set at 85 °C. Parts were printed horizontally [67]. Table 1 shows some of the process parameters used for 3D printing test samples.

Table 1. Main parameters used in the 3D-printing process for sample manufacturing.

Material	Layer Height	Perimeters	Infill	Infill Pattern	Extrusion Temperature	Bed Temperature
Z-PCABS	0.19 mm	2	100%	"Grid" 45°/−45°	265 °C	85 °C

Considering the aspects mentioned above, the 3D-printed test samples (Figure 1a) were subjected to ultraviolet radiation in an Opsytec Dr. Grobel BS-02 irradiation chamber (Ettlingen, Germany) (Figure 1b). The irradiation chamber comes equipped with two groups of fluorescent lamps, one group emitting UV-C, λ = 254 nm, and the second group emitting UV-B λ = 315 nm. The radiation dose is measured with calibrated sensors and controlled using a UV-MAT controller produced by the same manufacturer [68].

(**a**)

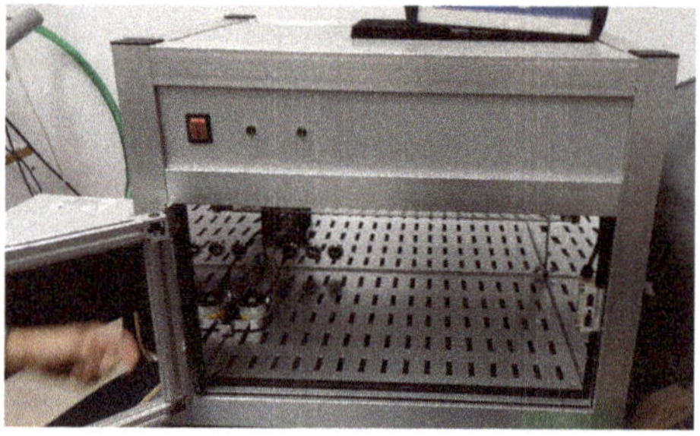

(**b**)

Figure 1. Accelerated aging in irradiation chamber (**a**) Test parts positioned for irradiation cycle; (**b**) Irradiation chamber.

A group of 3D-printed samples was exposed to 254 nm radiation for 24 h. During the irradiation cycle, a back panel temperature of 50 °C was maintained in the chamber, and the samples were hit with 10 W/m^2 radiating power. After 24 h, the lamps were turned off, and the test samples were left in the chamber for an additional 4 h to cool down to room temperature. In order to have another reference point for determining radiation effects on the material, a second set of sample parts were exposed to UV-B radiation in the same irradiation chamber.

The effects of UV-B and UV-C radiation were analyzed according to ISO 4892-1:2016, which indicates how data from accelerated aging using light radiation exposure [69] should be analyzed. Tensile and compressive strength tests were designed according to ISO 4582:2017 [70].

For tensile strength determination, 15 specimens were 3D printed from Z-PCABS according to ASTM type I dimensions [71].

For compressive strength tests, 15 test samples measuring 15 mm × 15 mm × 15 mm were 3D printed from the same material.

For each mechanical property test, the samples were split into 3 groups, forming groups of 5 randomly selected samples.

For strength testing (tensile strength, compressive strength), accelerated aging using UV-B radiation using the parameters highlighted previously was performed on the first group of samples. The second group was exposed to UV-C radiation. The third group did not undergo any radiation exposure and represented the control group. Before and after the accelerated aging treatment, all samples were measured using electronic calipers and were visually inspected.

In total, 15 3D-printed samples (dog bone-shaped) were tested for tensile strength on an Instron 8872 machine (Norwood, MA, USA) (Figure 2a). Tests were performed starting with a preload force of 5 N with 1 mm/minute loading speed. Elongation of the test part under tensile load was measured using an electronic extensometer and was used to determine part stiffness. Fifteen 3D-printed cubic samples were tested for compressive strength on an Instron 8801 machine (Norwood, MA, USA) (Figure 2b). The compressive force is applied along the sample Z-axis, perpendicular to the horizontal 3D-printed layers. Preload for this test was 5 N. All strength tests were in controlled conditions of 50% RH at 24 °C.

(a) (b)

Figure 2. Mechanical strength tests: (**a**) Tensile strength testing of 3D printed dog-bone ABS-PC samples; (**b**) Compression strength testing of 3D printed cubic ABS-PC samples.

Section 3.1 details the results of tensile tests (strength, stiffness), while Section 3.2 discusses the results of compressive tests (strength).

Polymers and fibers have a known tendency to deform plastically when subjected to loads for long periods of time, a characteristic commonly referred to as cold flow or creep. For certain polymers, creep-inducing stress can occur even at a small fraction of the material's ultimate strength. ABS-PC blends experience creep, with creep resistance increasing with increased PC content [72,73]. Thus, the effects of radiation on the creep behavior of ABS-PC parts was also investigated. Tensile creep testing was performed on 10 3D-printed specimens with a narrow section of 5.2 mm × 3 mm (Figure 3a). Testing was performed in compliance with standard ASTM D2990-17 [74]. The creep behavior of parts exposed to UV-C radiation was compared to that of parts in a control group.

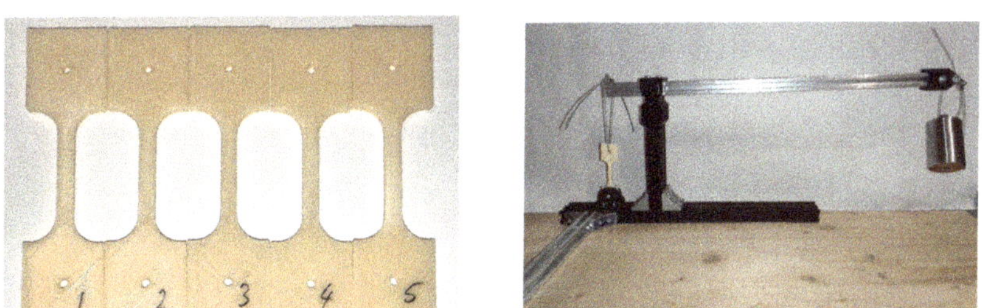

(a) (b)

Figure 3. Creep properties testing: (**a**) 3D printed parts made for tension creep testing; (**b**) Testing rig for determining tensile creep.

For finding creep under tensile load, parts were loaded in tension in a purposefully designed rig (Figure 3b). Creep was determined by applying a load that produces mechanical stress of 25% of the ultimate tensile strength measured previously for parts in the control group. Sample elongation under this tensile load was measured after 2 h, 6 h, 21 h, 24 h, then once every 24 h for a total of 168 h (7 days). The distance between the ends of the samples was measured using a micrometer on each side of the narrow section (left and right sides). The average change in these two distances was considered as elongation. The results of creep testing are discussed in Section 3.3.

The exterior surfaces and internal structure of the samples were analyzed using Scanning Electron Microscopy analysis (SEM). One sample each was selected from the group exposed to 254 nm radiation and from the control group. The analysis was performed using a Quanta Inspect F50 scanning electron microscope from Thermo Fisher Scientific (Eindhoven, The Netherlands) at a resolution of 1.2 nm. A layer of Au was sputter-coated on the samples (coverage time 90 s) using a Quorum Technologies Q150 PlusSeries coater (Lewes, UK). Section 3.4 discusses SEM results.

3. Results

Dimensional and visual checks were performed on the parts before and after accelerated aging under 254 nm radiation. Dimensions of the UV-C exposed parts were measured using electronic calipers before and after irradiation, revealing no statistically relevant changes in dimensions after UV-C exposure. Table 2 shows the average dimensions of samples in the aged group vs. the control group.

Table 2. Sample dimensions were measured using electronic calipers.

	Tensile Strength Sample		Compression Strength Sample	
	Width	Height	X	Z
3D model [mm]	13.00	4.20	15.00	15.00
Control samples [mm]	13.07 ± 0.012	4.15 ± 0.016	15.00 ± 0.016	15.16 ± 0.010
UV-C samples [mm]	13.09 ± 0.010	4.14 ± 0.010	15.01 ± 0.010	15.18 ± 0.012

3.1. *Tensile Strength and Young's Modulus*

Visual analysis of the failure mechanism found that the ABS-PC samples fractured along the deposited polymer filaments, creating a zig-zag failure pattern. The same pattern was observed in the control group and both groups exposed to radiation (UV-B/UV-C) (Figure 4). This fracture pattern can be explained by considering the anisotropic mechanical properties of parts produced through MEX 3D printing [75]. The anisotropy is caused by higher tensile strength along the filaments than the tensile strength of the adhesion between adjacent filaments [76] and is influenced by print orientation, infill type and amount, raster angle, and layer height [77–79]. The similar failure modes for all tested groups also indicate that changes generated by the irradiation treatments were uniform throughout the parts. The parts displayed a slight browning of the ivory-colored material following irradiation. A similar amount of browning is found in both UV-B and UV-C groups.

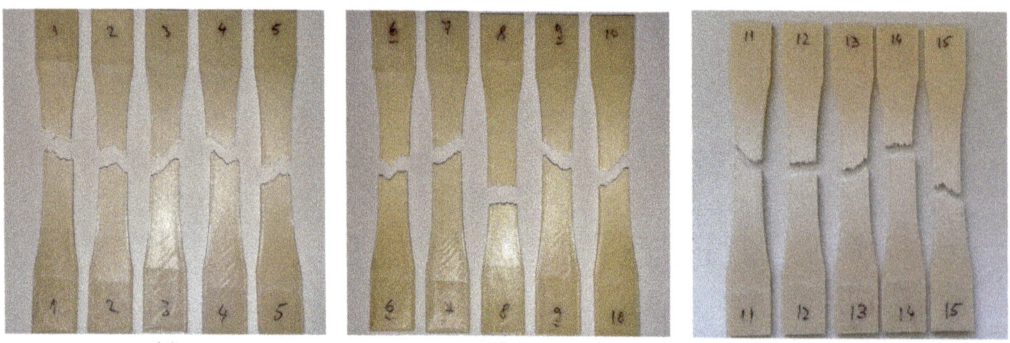

(a) (b) (c)

Figure 4. Fracture modes of tensile-tested parts: (**a**) Fractured ABS-PC parts from the control group; (**b**) Fractured ABS-PC parts exposed to UV-B; (**c**) Fractured ABS-PC parts exposed to UV-C.

Experimental data from tensile testing of PC-ABS are shown in Figure 5 as stress-strain graphs.

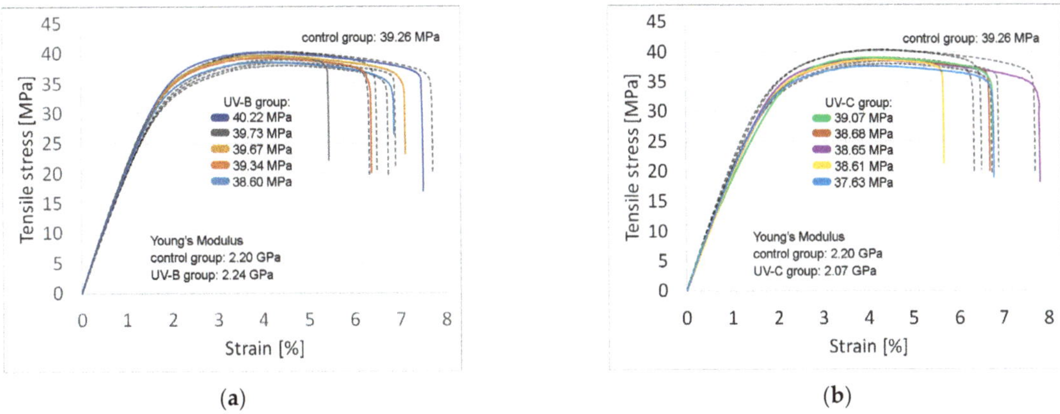

Figure 5. Stress-strain diagrams (tensile tests): (**a**) Samples exposed to UV-B λ = 315 nm (solid line) and samples from the control group (dashed line); (**b**) Samples exposed to UV-C λ = 254 nm (solid line) and samples from the control group (dashed line).

Dog bone-shaped samples made from ABS-PC subjected to the previously described dose of UV-B radiation had no statistically significant difference in tensile strength (39.51 MPa vs. 39.26 MPa; $F = 0.23$, $p = 0.65$) and stiffness (2238.06 MPa vs. 2196 MPa; $F = 1.57$, $p = 0.24$) compared to the parts in the control group.

Parts made from ABS-PC aged under UV-C exhibited 1.86% lower tensile strength compared to control samples (38.53 MPa vs. 39.26 MPa), a result that is not statistically significant ($F = 1.91$, $p = 0.20$). On average, the Young's Modulus of aged samples decreased by 5.51% vs. control samples (2075 MPa vs. 2196 MPa), a result that was determined to be statistically significant ($F = 5.70$, $p = 0.04$).

Results for average tensile strength after calculating the standard error can be found in Table 3. The same table shows the average Young's Modulus.

Table 3. Experimental results—average tensile strength and Young's Modulus.

Property	ABS-PC (no UV)	ABS-PC (UV-B)	ABS-PC (UV-C)
Tensile strength [MPa]	39.26 ± 0.47	39.51 ± 0.27	38.53 ± 0.24
Young's Modulus [MPa]	2196.4 ± 31.7	2238.1 ± 9.78	2074.9 ± 39.8

3.2. Compressive Strength

Compressive loads applied during compressive strength testing deformed the parts plastically without visible material rupture at their surface. Compared to parts in the control group (no UV exposure), all samples in the accelerated aging groups performed worse. Overall, ABS-PC samples have 5.2% lower compressive strength after UV-B irradiation (58.60 MPa vs. 61.81 MPa; $F = 33.8$, $p = 4 \times 10^{-4}$) while samples subjected to UV-C radiation have 6.5% lower compressive strength than those in the control group (57.82 MPa vs. 61.81 MPa; $F = 36.3$, $p = 3 \times 10^{-4}$). Stress-strain diagrams for ABS-PC are shown in Figure 6.

Results for average compressive strength after calculating standard error are shown in Table 4.

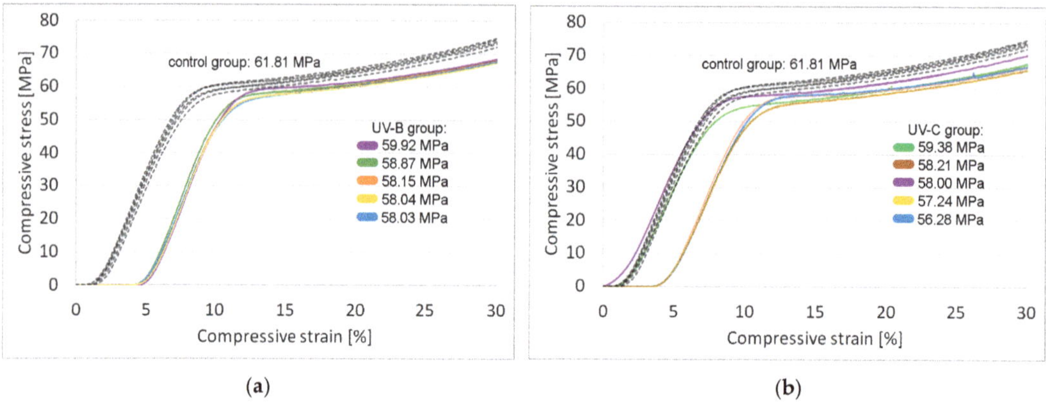

Figure 6. Stress-strain diagrams (compressive tests): (**a**) ABS-PC samples exposed to UV-B λ = 315 nm (solid line) and control samples (dashed line); (**b**) ABS-PC samples exposed to UV-C λ = 254 nm (solid line) and control samples (dashed line).

Table 4. Experimental results—average compressive strength.

Property	ABS-PC (no UV)	ABS-PC (UV-B)	ABS-PC (UV-C)
Compressive str. [MPa]	61.81 ± 0.41	58.60 ± 0.37	57.82 ± 0.52

3.3. *Creep Characteristics of the Analyzed Material*

Tensile creep testing was performed with loads appropriate to the tested material based on the tensile strength value identified previously ($\sigma_{Z\text{-}PCABS}$ = 39.26 MPa). The parts are placed in a testing rig that uses mechanical advantage to amplify the load applied to one of its ends. The cross-section area of the tested parts is 15.6 mm^2 (5.2 mm × 3 mm). Table 5 shows the main parameters (loads) used in this experiment.

Table 5. Creep testing parameters.

Material	Test Parameter			
	Strength [MPa]	Creep Test Stress [MPa]	Targeted Tensile Force on Sample [N]	Test Rig Load Configuration
ABS-PC	39.26	9.82	154	22 N × 7

Figure 7 shows the tensile creep of ABS-PC samples while being loaded (9.82 MPa) over a period of 168 h. The majority of the elongation took place right after the load was applied, and the rate of creep continued to slow down with the rise in strain. Despite the fact that the parts from the UV-C exposed group experienced more deformation under prolonged tensile stress, they exhibited very similar creep curves.

3.4. *Scanning Electron Microscopy (SEM)*

To gain insight into how the samples failed under tensile stress and to spot any potential changes in the internal structure following exposure to UV-C radiation, a fractographic analysis was performed using SEM. In Figure 8a, an SEM image is presented, depicting an ABS-PC sample from the control group. Figure 8b displays an SEM image of a sample from the irradiated group (UV-C).

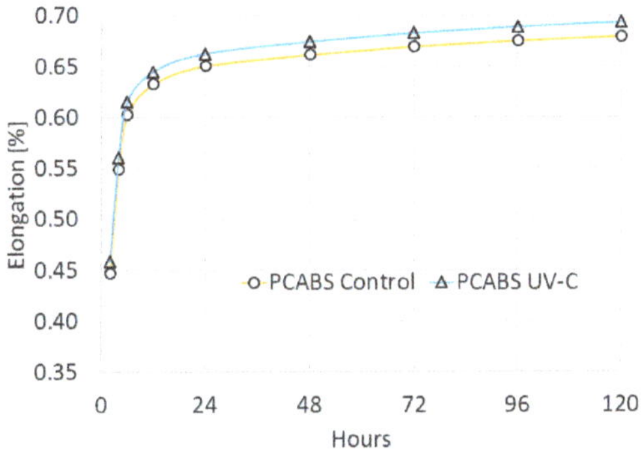

Figure 7. Tensile creep of Z-PCABS; graph showing an increase in elongation over time.

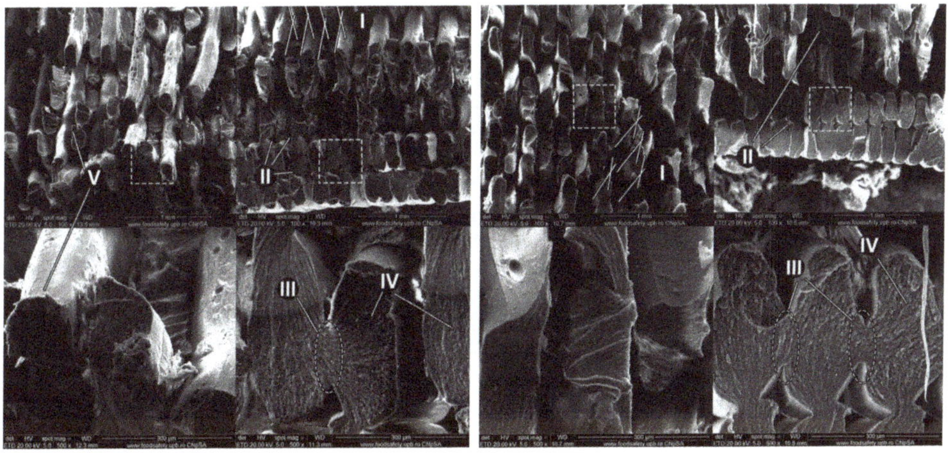

(**a**) (**b**)

Figure 8. SEM imaging of fractured ABS-PC sample with the left column showing rupture surfaces in part infill; right column showing rupture surface in part contour: (**a**) Z-PCABS sample from the control group; (**b**) Z-PCABS sample exposed to UV-C λ = 254 nm; I—fracture parallel with filaments; II—microvoids; III—interlayer fusion; IV—extensive flaking at the fractured surface; V—reduced necking of the ruptured filaments.

The ABS-PC samples fractured along the deposited filament lines (Figure 8a,b-I), as previously described in Section 3.1. Overlap of the deposited filaments, along with flattening following deposition, can be observed, giving the filaments a characteristic oblong shape. The MEX process introduces in the manufactured parts specific gaps usually referred to as "microvoids". These voids are triangular in shape due to the angled raster used with infill deposition and the oblong shape of deposited filaments and form weak points in the material structure [80]. The formation of microvoids is generally associated with the lack of pressure applied to the molten filament while it is being deposited, unlike other manufacturing processes, such as injection molding [81]. Microvoids can be observed in both SEM-analyzed samples, with or without radiation exposure (Figure 8a,b-II).

Interlayer adhesion is distinctly noticeable in both the control sample (Figure 8a-III) and the radiation-aged sample (Figure 8b-III). Samples from both groups present flaking at the rupture surface with small areas of smooth fracture surface (Figure 8a,b-IV), with the control part also showing reduced necking of ruptured filaments (Figure 8a-V). This indicates the parts do not display increased brittleness after 24 h of UV-C exposure.

4. Discussion

This work investigated how sterilizing UV-C radiation λ = 254 nm affects the mechanical properties of ABS-PC components after 24 h of exposure. This is particularly significant when considering the growing use of 3D printing for functional parts. The study evaluated the tensile and compressive strength of samples that were 3D printed using the same process parameters and then exposed to 2 different radiation wavelengths (UV-B λ = 315 nm, UV-C λ = 254 nm). The group of parts exposed to a UV-C accelerated aging cycle was compared to a control group.

Following exposure to UV-C radiation ABS-PC parts produced a change in color at their surface, consistent with the degradation of the ABS component [82]. The changes were only visual, with no dimensional differences between parts from both groups. This is true for both bulky parts like the compressive strength tested specimens (15 mm × 15 mm × 15 mm cubes) and for high length-to-width ratio parts like the samples used for tensile strength testing (115 mm × 13 mm dog bone shape).

Laureto et al. [83] found that a difference exists between ASTM D638-14 Type I and Type IV dimensions when testing 3D printed parts, with the type I test specimens producing slightly better results tensile strength results. However, the difference in specimen performance is not expected to influence the conclusions drawn in this paper.

Following tensile strength testing, there was no statistically significant difference between unexposed samples and samples exposed to 24 h of 10 W/m^2. UV-C radiation. A decrease in stiffness was observed after irradiation which can be attributed to the scission of molecular bonds.

Compressive strength testing showed a weakening of ABS-PC following radiation exposure, with aged parts (UV-C) having 6.5% less compressive strength.

Tensile creep testing resulted in similar elongation-time curves for the control group and the irradiated group, mirroring the findings regarding material stiffness. Mohamed et al. found that different printing parameters influence the creep resistance of PC-ABS 3D printed parts [84], a fact that highlights the need for further testing with other parameter sets. The amount of creep and the creep rate should only be associated with the particular use of a $45°/-45°$ infill raster angle. As Zhang et al. have found, the creep resistance of MEX 3D-printed parts is anisotropic and is influenced by infill orientation, with parts printed using 90° infill having the highest creep resistance [85]. However, infill orientation is not expected to produce different relative results (unexposed parts vs UV-C exposed).

SEM imaging was used to investigate whether any changes in the material occurred. Samples from both groups presented features common to the manufacturing process, such as microvoids and interlayer fusion regions. Key aspects commonly used to determine material behavior, such as fracture appearance and fracture surface smoothness, were also similar, mirroring the small differences observed following destructive testing.

UV radiation can cause polymers to age by breaking down their molecular chains, weakening their physical properties. This is because UV radiation contains high-energy photons that are capable of breaking chemical bonds in the polymer chains. When this happens, the polymer molecules become shorter, and the polymer chains become weaker, which can lead to the degradation of the material over time. This process is often referred to as photo-oxidation. One of the primary ways photo-oxidation degrades the material is by initiating the formation of free radicals—highly reactive molecules that can cause chain reactions in the polymer—leading to the formation of new chemical bonds and altering the polymer's structure. In the presence of oxygen, free radicals can also react with oxygen molecules to form peroxides. Peroxides can then react with other polymer molecules to form more free radicals, perpetuating the chain reaction and causing further degradation of the polymer. With short wavelength radiation, such as UV-C, another important effect occurs, namely Fries rearrangement. When high-energy photons are absorbed by the polymer, primarily free radicals can form by breaking carbonate bonds in a process that does not need the presence of oxygen [86].

It is known that PC has better UV resistance compared to non-oxygen polymers, such as ABS [87,88]. We can hypothesize that the good behavior of ABS-PC is not only due to the PC component being more UV resistant, but its presence could aid the ABS component as well. It is known that at low wavelengths, ultraviolet radiation consists of photons with sufficient energy to break molecular bonds, creating cross-linking and scission effects on polymers [89,90]. It is, thus, possible for the radicals formed in the PC component to have a scavenging effect on other free radicals formed during the cleavage of molecules in the ABS component. This hypothesis requires further investigation.

In addition to causing physical changes in the polymer, UV radiation can also affect its optical properties. Polymers that are exposed to UV radiation can become discolored or yellowed (as was observed in these experiments), as photons can break down chromophores, which are chemical groups responsible for a polymer's color. This is an important aspect to consider when designing parts where functionality depends on the material's optical properties.

A large compilation of 431 studies on the UV dose (fluence) needed to inactivate common pathogens reveals that most pathogens see a 1-log reduction in numbers (90% pathogen inactivation) at UV doses lower than 20 mJ/cm^2 [91,92]. This is equivalent to 20 s of exposure to the UV-C radiation source used in this experiment. These factors indicate that the investigated aging treatment is relevant to the common practices of sterilizing materials and surfaces. Given the good response of ABS-PC 3D-printed parts in terms of tensile and compressive strength, stiffness and creep behavior before and after UV exposure, this material should be considered for applications where UV sterilizing is employed. This study looked at the performance of a common commercially available ABS-PC blend. As discussed previously, a higher ratio of PC would increase UV stability while decreasing processability through MEX 3D printing. Further improvement of the material's performance and stability under ultraviolet radiation could be obtained through various methods. Additives, such as organo-modified layered double hydroxides are known to increase the photostability of polymers while providing anti-bacterial properties [93]. Other UV absorbers and UV stabilizers can also be used [94]. Another aspect worth studying is the replacement of ABS with acrylonitrile styrene acrylate (ASA) in a PC/ASA copolymer. ASA has superior UV resistance compared to ABS, but a PC/ASA copolymer requires higher processing temperatures (275 °C extrusion temperature, 110 °C bed temperature). It is important to note that the presence of MEX process-specific features, such as layer lines, microvoids, etc., may play a role in determining the sterilization efficiency of 3D-printed parts, and future studies need to address this fact.

5. Conclusions

The objective of this paper was to examine the impact of UV-C exposure-induced accelerated aging on the mechanical characteristics of ABS-PC copolymer samples manufactured through MEX 3D printing. Mechanical testing found minor tensile and compressive strength decreases following irradiation. The same tests found that UV-C radiation has slightly decreased part stiffness.

After conducting a tensile creep test, it was observed that the creep behavior remained unchanged even after exposure to UV-C radiation, and both exposed and unexposed materials exhibited comparable creep curves.

The results overall suggest a minor reduction in mechanical properties of 3D printed ABS-PC parts following an irradiation treatment. These minor changes in mechanical properties, coupled with the proven long-term stability of ABS-PC blends, suggest that these materials are suitable for use in the presented scenarios. This is also true for parts subjected to continuous tensile stress and vulnerable to polymer creep. SEM investigation mirrored the findings of destructive testing,

Testing is needed to assess the medical efficiency of ultraviolet UV-C sterilization on such parts, given the inherent characteristics of MEX 3D printed materials. Elements, such as surface porosity, hygroscopy, etc., may represent a hurdle in the adoption of these materials.

It is important to mention that variations in feedstock material compositions, 3D printing equipment and software, and different printing parameters may yield slightly different outcomes for MEX 3D printed parts, as emphasized by Popescu et al. in a review of mechanical property testing [95]. In-house assessment of material properties, 3D-printing machine characteristics and parameter sets is recommended before producing end-use polymer parts.

Author Contributions: Conceived the experiments, A.Z. and C.G.A.; performed the experiments, A.Z., F.B. and C.R.; analyzed the data obtained from the 3D prints, A.Z., F.B. and C.G.A. All authors have read and agreed to the published version of the manuscript.

Funding: This research received no external funding.

Institutional Review Board Statement: Not applicable.

Informed Consent Statement: Not applicable.

Data Availability Statement: Data will be made available on request.

Conflicts of Interest: The authors declare no conflict of interest. The funders had no role in the design of the study; in the collection, analyses, or interpretation of data; in the writing of the manuscript, or in the decision to publish the results.

References

1. World Health Organization. Shortage of Personal Protective Equipment Endangering Health Workers Worldwide. 2020. Available online: https://www.who.int/news/item/03-03-2020-shortage-of-personal-protective-equipment-endangering-health-workers-worldwide. (accessed on 15 March 2023).
2. Blanco-Donoso, L.M.; Moreno-Jiménez, J.; Amutio, A.; Gallego-Alberto, L.; Moreno-Jiménez, B.; Garrosa, E. Stressors, job re-sources, fear of contagion, and secondary traumatic stress among nursing home workers in face of the COVID-19: The case of Spain. *J. Appl. Gerontol.* **2021**, *40*, 244–256. [CrossRef]
3. Ranney, M.L.; Griffeth, V.; Jha, A.K. Critical Supply Shortages—The Need for Ventilators and Personal Protective Equipment during the COVID-19 Pandemic. *N. Engl. J. Med.* **2020**, *382*, e41. [CrossRef] [PubMed]
4. Webber, L.; Jewett, C. Testing Swabs Run in Short Supply as Makers Try to Speed up Production. Available online: https://www.npr.org/sections/health-shots/2020/03/18/817801222/testing-swabs-run-in-short-supply-as-makers-try-to-speed-up-production (accessed on 18 March 2023).
5. Hufford, A. 3M CEO on N95 Masks: 'Demand Exceeds Our Production Capacity'. 2020. Available online: https://www.wsj.com/articles/3m-ceo-on-n95-masks-demand-exceeds-our-production-capacity-11585842928 (accessed on 20 March 2023).
6. Chen, P.G.; Chan, E.W.; Qureshi, N.; Shelton, S.; Mulcahy, A.W. RAND Health Care–Project Report. Medical Device Supply Chains An Overview and Description of Challenges During the COVID-19 Pandemic. 2021. Available online: https://aspe.hhs.gov/sites/default/files/documents/e48047020834c0c34cf6baf08a9428d0/PR-A328-2-medicaldevices.pdf (accessed on 20 March 2023).
7. Park, C.-Y.; Kim, K.; Roth, S.; Beck, S.; Kang, J.W.; Tayag, M.C.; Griffin, M. Global Shortage of Personal Protective Equipment amid COVID-19: Supply Chains, Bottlenecks, and Policy Implications. *ADB Briefs* **2020**, 130. [CrossRef]
8. Bown, C.P. How COVID-19 Medical Supply Shortages Led to Extraordinary Trade and Industrial Policy. *Asian Econ. Policy Rev.* **2022**, *17*, 114–135. [CrossRef]
9. Ibn-Mohammed, T.; Mustapha, K.B.; Godsell, J.; Adamu, Z.; Babatunde, K.A.; Akintade, D.D.; Acquaye, A.; Fujii, H.; Ndiaye, M.M.; Yamoah, F.A.; et al. A critical analysis of the impacts of COVID-19 on the global economy and ecosystems and opportunities for circular economy strategies. *Resour. Conserv. Recycl.* **2021**, *164*, 105169. [CrossRef] [PubMed]
10. Cohen, J.; Rodgers, Y. Contributing factors to personal protective equipment shortages during the COVID-19 pandemic. *Prev. Med.* **2020**, *141*, 106263. [CrossRef] [PubMed]
11. Jordan, J.M. Additive manufacturing ("3D printing") and the future of organizational design: Some early notes from the field. *J. Org. Des.* **2019**, *8*, 5. [CrossRef]
12. Mueller, T.; Elkaseer, A.; Charles, A.; Fauth, J.; Rabsch, D.; Scholz, A.; Marquardt, C.; Nau, K.; Scholz, S.G. Eight Weeks Later—The Unprecedented Rise of 3D Printing during the COVID-19 Pandemic—A Case Study, Lessons Learned, and Implications on the Future of Global Decentralized Manufacturing. *Appl. Sci.* **2020**, *10*, 4135. [CrossRef]
13. Ben-Ner, A.; Siemsen, E. Decentralization and Localization of Production: The Organizational and Economic Consequences of Additive Manufacturing (3D Printing). *Calif. Manag. Rev.* **2017**, *59*, 000812561769528. [CrossRef]
14. Malik, A.; Ul Haq, M.I.; Raina, A.; Gupta, K. 3D printing towards implementing Industry 4.0: Sustainability aspects, barriers and challenges. *Ind. Rob.* **2022**, *49*, 491–511. [CrossRef]

15. Choong, Y.Y.C.; Tan, H.W.; Patel, D.C.; Choong, W.T.N.; Chen, C.-H.; Low, H.Y.; Tan, M.J.; Patel, C.D.; Chua, C.K. The global rise of 3D printing during the COVID-19 pandemic. *Nat. Rev. Mater.* **2020**, *5*, 637–639. [CrossRef]
16. Maracaja, L.; Blitz, D.; Maracaja, D.; Walker, C. How 3D Printing Can Prevent Spread of COVID-19 Among Healthcare Professionals During Times of Critical Shortage of Protective Personal Equipment. *J. Cardiothorac. Vasc. Anesth.* **2020**, *34*, 2847–2849. [CrossRef]
17. Manero, A.; Smith, P.; Koontz, A.; Dombrowski, M.; Sparkman, J.; Courbin, D.; Chi, A. Leveraging 3D Printing Capacity in Times of Crisis: Recommendations for COVID-19 Distributed Manufacturing for Medical Equipment Rapid Response. *Int. J. Environ. Res. Public Health* **2020**, *17*, 4634. [CrossRef] [PubMed]
18. Hagen, A.; Chisling, M.; House, K.; Katz, T.; Abelseth, L.; Fraser, I.; Bradley, S.; Kirsch, R.; Morris, J.; Giles, J.W.; et al. 3D Printing for Medical Applications: Current State of the Art and Perspectives during the COVID-19 Crisis. *Surgeries* **2021**, *2*, 244–259. [CrossRef]
19. Longhitano, G.A.; Nunes, G.B.; Candido, G.; da Silva, J.V.L. The role of 3D printing during COVID-19 pandemic: A review. *Prog. Addit. Manuf.* **2021**, *6*, 19–37. [CrossRef]
20. Radfar, P.; Bazaz, S.R.; Mirakhorli, F.; Warkiani, M.E. The role of 3D printing in the fight against COVID-19 outbreak. *J. 3D Print. Med.* **2021**, *5*, 51–60. [CrossRef]
21. Ford, J.; Goldstein, T.; Trahan, S.; Neuwirth, A.; Tatoris, K.; Decker, S. A 3D-printed nasopharyngeal swab for COVID-19 diagnostic testing. *3D Print. Med.* **2020**, *6*, 21. [CrossRef]
22. Alyouha, S.; Almazeedi, S.; Alghounaim, M.; Al-Mutawa, Y.; Alsabah, S. Polyester tipped 3-dimensionally printed swab that costs less than US$0.05 and can easily and rapidly be mass produced. *BMJ Innov.* **2020**, *6*, 262–264. [CrossRef]
23. Manoj, A.; Bhuyan, M.; Banik, S.R.; Sankard, M.R. 3D printing of nasopharyngeal swabs for COVID-19 diagnose: Past and current trends. *Mater. Today Proc.* **2021**, *44*, 1361–1368. [CrossRef]
24. Wesemann, C.; Pieralli, S.; Fretwurst, T.; Nold, J.; Nelson, K.; Schmelzeisen, R.; Hellwig, E.; Spies, B.C. 3-D Printed Protective Equipment during COVID-19 Pandemic. *Materials* **2020**, *13*, 1997. [CrossRef]
25. Novak, J.I.; Loy, J. A quantitative analysis of 3D printed face shields and masks during COVID-19. *Emerald Open Research* **2020**, *2*, 42. [CrossRef]
26. Wong, K.V.; Hernandez, A. A Review of Additive Manufacturing. *Int. Sch. Res. Not.* **2012**, *2012*, 208760. [CrossRef]
27. Ju, J.; Boisvert, L.N.; Zuo, Y.Y. Face masks against COVID-19: Standards, efficacy, testing and decontamination methods. *Adv. Colloid Interface Sci.* **2021**, *292*, 102435. [CrossRef]
28. O'Kelly, E.; Arora, A.; Pirog, S.; Ward, J.; Clarkson, P.J. Comparing the fit of N95, KN95, surgical, and cloth face masks and assessing the accuracy of fit checking. *PLoS ONE* **2021**, *16*, e0245688. [CrossRef]
29. Dugdale, C.M.; Walensky, R.P. Filtration Efficiency, Effectiveness, and Availability of N95 Face Masks for COVID-19 Prevention. *JAMA Intern. Med.* **2020**, *180*, 1612–1613. [CrossRef]
30. Ballard, D.H.; Jammalamadaka, U.; Meacham, K.W.; Hoegger, M.J.; Burke, B.A.; Morris, J.A.; Scott, A.R.; O'Connor, Z.; Gan, C.; Hu, J.; et al. Quantitative Fit Tested N95 Respirator-Alternatives Generated with CT Imaging and 3D Printing: A Response to Potential Shortages During the COVID-19 Pandemic. *Acad. Radiol.* **2021**, *28*, 158–165. [CrossRef] [PubMed]
31. McAvoy, M.; Bui, A.T.N.; Hansen, C.; Plana, D.; Said, J.T.; Yu, Z.; Yang, H.; Freake, J.; Van, C.; Krikorian, D.; et al. 3D Printed frames to enable reuse and improve the fit of N95 and KN95 respirators. *BMC Biomed. Eng.* **2021**, *3*, 10. [CrossRef] [PubMed]
32. Tipnis, N.P.; Burgess, D.J. Sterilization of implantable polymer-based medical devices: A review. *Int. J. Pharm.* **2018**, *544*, 455–460. [CrossRef] [PubMed]
33. Ghobeira, R.; Philips, C.; Declercq, H.; Cools, P.; De Geyter, N.; Cornelissen, R.; Morent, R. Effects of different sterilization methods on the physico-chemical and bioresponsive properties of plasma-treated polycaprolactone films. *Biomed. Mater.* **2017**, *12*, 015017. [CrossRef] [PubMed]
34. Holmes, S. An overview of current surgical instrument and other medical device decontamination practices. In *Woodhead Publishing Series in Biomaterials, Decontamination in Hospitals and Healthcare*, 2nd ed.; Walker, J., Ed.; Woodhead Publishing: Sawston, UK, 2020; pp. 443–482. [CrossRef]
35. Sadeque, M.; Balachandran, S.K. Overview of medical device processing. In *Trends in Development of Medical Devices*; Shanmugam, P.S.T., Chokkalingam, L., Bakthavachalam, P., Eds.; Academic Press: Cambridge, MA, USA, 2020; pp. 177–188. [CrossRef]
36. McEvoy, B.; Rowan, N. Terminal sterilization of medical devices using vaporized hydrogen peroxide: A review of current methods and emerging opportunities. *J. Appl. Microbiol.* **2019**, *127*, 1403–1420. [CrossRef] [PubMed]
37. Popescu, D.; Baciu, F.; Vlasceanu, D.; Cotrut, M.C.; Marinescu, R. Effects of multiple sterilizations and natural aging on the mechanical behavior of 3D-printed ABS. *Mech. Mater.* **2020**, *148*, 103423. [CrossRef]
38. Grzelak, K.; Łaszcz, J.; Polkowski, J.; Mastalski, P.; Kluczyński, J.; Łuszczek, J.; Torzewski, J.; Szachogłuchowicz, I.; Szymaniuk, R. Additive Manufacturing of Plastics Used for Protection against COVID19—The Influence of Chemical Disinfection by Alcohol on the Properties of ABS and PETG Polymers. *Materials* **2021**, *14*, 4823. [CrossRef] [PubMed]
39. Popescu, D.; Baciu, F.; Amza, C.G.; Cotrut, C.M.; Marinescu, R. The Effect of Disinfectants Absorption and Medical Decontamina-
:al Performance of 3D-Printed ABS Parts. *Polymers* **2021**, *13*, 4249. [CrossRef] [PubMed]
in, S.B., Jr.; Thewlis, R.E.; Sarkisian, K.; Nwoko, J.O.; Mead, K.R.; Noti, J.D. Effects of Ultraviolet Germicidal
n N95 Respirator Filtration Performance and Structural Integrity. *J. Occup. Environ. Hyg.* **2015**, *12*, 509–517.

41. van Doremalen, N.; Bushmaker, T.; Morris, D.H. Letter to the Editor: Aerosol and Surface Stability of HCoV-19 (SARS-CoV-2) Compared to SARS-CoV-1. *N. Engl. J. Med.* **2020**, *382*, 1564–1567. [CrossRef]
42. Stojalowski, P.S.; Fairfoull, J. Comparison of Reflective Properties of Materials Exposed to Ultraviolet-C Radiation. *J. Res. Natl. Inst. Stand. Technol.* **2021**, *126*, 126017. [CrossRef]
43. Spicer, D.B. Methods and Mechanisms of Photonic Disinfection. *J. Res. Natl. Inst. Stand. Technol.* **2021**, *126*, 126016. [CrossRef]
44. *Standard ISO 4892-3:2016*; Plastics–Methods of Exposure to Laboratory Light Sources–Part 3: Fluorescent UV Lamps. Available online: https://www.iso.org/standard/67793.html (accessed on 25 March 2023).
45. The National Institute of Standards and Technology. Summary of Event and Post-Workshop Activities. 2020. Available online: https://www.nist.gov/news-events/events/2020/01/workshop-ultraviolet-disinfection-technologies-healthcare-associated-4 (accessed on 25 March 2023).
46. IUVA Fact Sheet on UV Disinfection for COVID-19. Available online: https://www.iuva.org/IUVA-Fact-Sheet-on-UV-Disinfection-for-COVID-19 (accessed on 25 March 2023).
47. Ngo, T.D.; Kashani, A.; Imbalanzo, G.; Nguyen, K.; Hui, D. Additive manufacturing (3D printing): A review of materials, methods, applications and challenges. *Comp. Part B Eng.* **2018**, *143*, 172–196. [CrossRef]
48. Cantrell, J.; Rohde, S.; Damiani, D.; Gurnani, R.; DiSandro, L.; Anton, J.; Young, A.; Jerez, A.; Steinbach, D.; Kroese, C.; et al. Experimental Characterization of the Mechanical Properties of 3D-Printed ABS and Polycarbonate Parts. In *Advancement of Optical Methods in Experimental Mechanics*; Yoshida, S., Lamberti, L., Sciammarella, C., Eds.; Conference Proceedings of the Society for Experimental Mechanics Series; Springer: Cham, Switzerland, 2016; Volume 3. [CrossRef]
49. Arivazhagan, A.; Masood, S.H. Dynamic mechanical properties of ABS material processed by fused deposition modelling. *Int. J. Eng. Res. Appl. (IJERA)* **2012**, *2*, 2009–2014.
50. Pérez, J.; Vilas, J.; Laza, J.; Arnaiz, S.; Mijangos, F.; Bilbao, E.; León, L. Effect of Reprocessing and Accelerated Weathering on ABS Properties. *J. Environ. Polym. Degrad.* **2010**, *18*, 71–78. [CrossRef]
51. Li, J.; Chen, F.; Yang, L.; Jiang, L.; Dan, Y. FTIR analysis on aging characteristics of ABS/PC blend under UV-irradiation in air. *Spectrochim. Acta A Mol. Biomol. Spectrosc.* **2017**, *184*, 361–367. [CrossRef]
52. Greco, R.; Astarita, M.F.; Dong, L.; Sorrentino, A. Polycarbonate/ABS blends: Processability, thermal properties, and mechanical and impact behavior. *Adv. Polym. Technol.* **1994**, *13*, 259–274. [CrossRef]
53. Suarez, H.; Barlow, W.; Paul, D.R. Mechanical properties of ABS/PC blends. *J. Appl. Polym. Sci.* **1984**, *29*, 3253–3259. [CrossRef]
54. Krache, R.; Debbah, I. Some Mechanical and Thermal Properties of PC/ABS Blends. *Mater. Sci. Appl.* **2011**, *2*, 404–410. [CrossRef]
55. Wildes, G.; Keskkula, H.; Paul, D.R. Fracture characterization of PC/ABS blends: Effect of reactive compatibilization, ABS type and rubber concentration. *Polymer* **1999**, *40*, 7089–7107. [CrossRef]
56. Zortrax Product Page–Z-PCABS. Available online: https://zortrax.com/filaments/z-pcabs/ (accessed on 25 March 2023).
57. U.S. Food & Drug Administration–Discussion Paper: 3D Printing Medical Devices at the Point of Care. 2021. Available online: https://www.fda.gov/medical-devices/3d-printing-medical-devices/3d-printing-medical-devices-point-care-discussion-paper (accessed on 13 April 2023).
58. The Pew Charitable Trusts–Issue Brief: FDA's Regulatory Framework for 3D Printing of Medical Devices at the Point of Care Needs More Clarity. 2022. Available online: https://www.pewtrusts.org/en/research-and-analysis/issue-briefs/2022/07/fdas-regulatory-framework-for-3d-printing-of-medical-devices-needs-more-clarity (accessed on 13 April 2023).
59. *Standard ISO 4892-2:2013*; Plastics—Methods of Exposure to Laboratory Light Sources—Part 2: Xenon-Arc Lamps. Available online: https://www.iso.org/standard/55481.html (accessed on 20 April 2023).
60. *Standard BIFMA HCF 8.1-2014*; Health Care Furniture Design–Guidelines for Cleanability. Business and Institutional Furniture Manufacturers Association: Grand Rapids, MI, USA, 2014.
61. *Standard ISO 527-2*; Plastics—Determination of Tensile Properties—Part 2: Test Conditions for Moulding and Extrusion Plastics. Available online: https://www.iso.org/standard/4593.html (accessed on 20 April 2023).
62. Samykano, M.; Selvamani, S.K.; Kadirgama, K.; Ngui, W.K.; Kanagaraj, G.; Sudhakar, K. Mechanical property of FDM printed ABS: Influence of printing parameters. *Int. J. Adv. Manuf. Technol.* **2019**, *12*, 2779–2796. [CrossRef]
63. Vicente, C.M.; Martins, T.S.; Leite, M.; Ribeiro, A.; Reis, L. Influence of fused deposition modeling parameters on the mechanical properties of ABS parts. *Polym. Adv. Technol.* **2020**, *31*, 501–507. [CrossRef]
64. Giannatsis, J.; Sofos, K.; Canellidis, V.; Karalekas, D.; Dedoussis, V. Investigating the influence of build parameters on the mechanical properties of FDM parts. In *Innovative Developments in Virtual and Physical Prototyping, Proceedings of the 5th International Conference on Advanced Research and Rapid Prototyping, Leiria, Portugal, 28 September—1 October 2011*; Taylor & Francis: Abingdon, UK, 2012; pp. 525–529.
65. Tymrak, B.M.; Kreiger, M.; Pearce, J.M. Mechanical properties of components fabricated with open-source 3-D printers under realistic environmental conditions. *Mater. Des.* **2014**, *58*, 242–246. [CrossRef]
66. Galeja, M.; Hejna, A.; Kosmela, P.; Kulawik, A. Static and Dynamic Mechanical Properties of 3D Printed ABS as a Function of Raster Angle. *Materials* **2020**, *13*, 297. [CrossRef] [PubMed]
67. Górski, F.; Wichniarek, R.; Kuczko, W.; Zawadzki, P.; Buń, P. Strength of ABS parts produced by Fused Deposition Modelling technology–a critical orientation problem. *Adv. Sci. Technol.* **2015**, *9*, 12–19. [CrossRef]
68. Opsytec Dr. Grobel. Product Datasheet-Irradiation Chamber BS-02. Available online: https://www.opsytec.com/fileadmin/user_upload/products/downloads/e_bs02.pdf. (accessed on 11 April 2023).

69. *Standard ISO 4892-1:2016*; Plastics—Methods of Exposure to Laboratory Light Sources—Part 1: General Guidance. Available online: https://www.iso.org/standard/60048.html (accessed on 20 March 2023).
70. *Standard ISO 4582:2017*; Plastics—Determination of Changes in Colour and Variations in Properties after Exposure to Glass-Filtered Solar Radiation, Natural Weathering or Laboratory Radiation Sources. Available online: https://www.iso.org/standard/67791.html (accessed on 20 March 2023).
71. *Standard ASTM D638–22*; Standard Test Method for Tensile Properties of Plastics. Available online: https://www.astm.org/Standards/D638 (accessed on 25 March 2023).
72. Hassan, A.; Jwu, W.Y. Mechanical properties of high impact ABS/PC blends–effect of blend ratio. In Proceedings of the Polymer Symposium, Kebangsaan Ke-V, Selangor, Malaysia, 23–24 August 2005; pp. 65–76.
73. Bano, S.; Iqbal, T.; Ramzan, N.; Farooq, U. Study of Surface Mechanical Characteristics of ABS/PC Blends Using Nanoindentation. *Processes* **2021**, *9*, 637. [CrossRef]
74. *Standard ASTM D2990–17*; Standard Test Methods for Tensile, Compressive, and Flexural Creep and Creep-Rupture of Plastics. Available online: https://www.astm.org/Standards/D2990.htm (accessed on 18 March 2023).
75. Zohdi, N.; Yang, R. Material Anisotropy in Additively Manufactured Polymers and Polymer Composites: A Review. *Polymers* **2021**, *13*, 3368. [CrossRef]
76. Zou, R.; Xia, Y.; Liu, S.; Ping Hu, P.; Hou, W.; Hu, Q.; Shan, C. Isotropic and anisotropic elasticity and yielding of 3D printed material. *Compos. B Eng.* **2016**, *99*, 506–513. [CrossRef]
77. Gordelier, T.J.; Thies, P.R.; Turner, L.; Johanning, L. Optimising the FDM additive manufacturing process to achieve maximum tensile strength: A state-of-the-art review. *Rapid Prototyp. J.* **2019**, *25*, 953–971. [CrossRef]
78. Rajpurohit, S.R.; Dave, H.K. Effect of process parameters on tensile strength of FDM printed PLA parts. *Rapid Prototyp. J.* **2018**, *24*, 1317–1324. [CrossRef]
79. Zhang, J.; Yang, B.; Fu, F.; You, F.; Dong, X.; Dai, M. Resistivity and its anisotropy characterization of 3D-printed acrylonitrile butadiene styrene copolymer (ABS)/carbon black (CB) composites. *Appl. Sci.* **2017**, *7*, 20. [CrossRef]
80. Wickramasinghe, S.; Do, T.; Tran, P. FDM-Based 3D Printing of Polymer and Associated Composite: A Review on Mechanical Properties, Defects and Treatments. *Polymers* **2020**, *12*, 1529. [CrossRef]
81. Dawoud, M.; Taha, I.; Ebeid, S.J. Mechanical behaviour of ABS: An experimental study using FDM and injection moulding techniques. *J. Manuf. Process.* **2016**, *21*, 39–45. [CrossRef]
82. Iannuzzi, G.; Mattsson, B.; Rigdahl, M. Color changes due to thermal ageing and artificial weathering of pigmented and textured ABS. *Polym Eng Sci* **2013**, *53*, 1687–1695. [CrossRef]
83. Laureto, J.; Pearce, J. Anisotropic mechanical property variance between ASTM D638-14 type I and type IV fused filament fabricated specimens. *Polym. Test.* **2018**, *68*, 294–301. [CrossRef]
84. Mohamed, O.A.; Masood, S.H.; Bhowmik, J. Creep Deformation Behaviour of PC-ABS Parts Processed by Fused Deposition Additive Manufacturing under Different Extrusion Parameters. In Proceedings of the Society of Plastics Engineers Conference, SPE ANTECA, Anaheim, CA, USA, 8–10 May 2017; pp. 13–16.
85. Zhang, H.; Cai, L.; Golub, M.; Zhang, Y.; Yang, X.; Schlarman, K.; Zhang, J. Tensile, Creep, and Fatigue Behaviors of 3D-Printed Acrylonitrile Butadiene Styrene. *J. Materi. Eng. Perform.* **2018**, *27*, 57–62. [CrossRef]
86. Rivaton, A. Recent advances in bisphenol-A polycarbonate photodegradation. *Polym. Degrad. Stab.* **1995**, *49*, 163–179. [CrossRef]
87. Torikai, A.; Mitsuoka, T.; Fueki, K. Wavelength sensitivity of the photoinduced reaction in polycarbonate. *J. Polym. Sci. Part A Polym. Chem.* **1993**, *31*, 2785–2788. [CrossRef]
88. Blakey, I.; Yu, A.; Blinco, J.; Jack, K.S.; Liu, H.; Leeson, M.; Yueh, W.; Younkin, T.; Whittaker, A.K. Polycarbonate Based Nonchemically Amplified Photoresists for Extreme Ultraviolet Lithography. *Extrem. Ultrav. (EUV) Lithogr.* **2010**, *7636*, 952–959. [CrossRef]
89. Ramani, R.; Ranganathaiah, C. Degradation of acrylonitrile-butadiene-styrene and polycarbonate by UV irradiation. *Polym. Degrad. Stab.* **2000**, *69*, 347–354. [CrossRef]
90. Yousif, E.; Haddad, A. Photodegradation and photostabilization of polymers, especially polystyrene: Review. *Springerplus* **2013**, *2*, 398. [CrossRef]
91. Malayeri, A.H.; Mohseni, M.; Cairns, B.; Bolton, J.; Chevrefils, G.; Caron, E. Fluence (UV Dose) Required to Achieve Incremental Log Inactivation of Bacteria, Protozoa, Viruses and Algae, IUVA News. Available online: https://iuva.org/resources/Resource%20Documents/Malayeri-Fluence%20Required%20to%20Achieve%20Incremental%20Log%20Inactivation%20of%20Bacteria,%20Protozoa,%20Viruses%20and%20Algae.pdf (accessed on 25 March 2023).
92. Ledrise Led Professional. Achieving Effective Germicidal Action with UVC Radiation: A Comprehensive Guide. 2021. Available online: https://www.ledrise.eu/blog/uv-fluence-for-disinfection/ (accessed on 25 March 2023).
93. Marek, A.; Verney, V.; Totaro, G.; Sisti, L.; Celli, A.; Bozzi Cionci, N.; Di Gioia, D.; Massacrier, L.; Leroux, F. Organo-modified LDH fillers endowing multi-functionality to bio-based poly(butylene succinate): An extended study from the laboratory to possible market. *Appl. Clay Sci.* **2020**, *188*, 105502. [CrossRef]

94. Rajan, V.; Waber, R.; Wieser, J. Influence of different types of UV absorber/UV stabilizer combination on the photodegradation of PC/ABS blend. *J. Appl. Polym. Sci.* **2011**, *124*, 4007–4015. [CrossRef]
95. Popescu, D.; Zapciu, A.; Amza, C.G.; Baciu, F.; Marinescu, R. FDM process parameters influence over the mechanical properties of polymer specimens: A review. *Polym. Test.* **2018**, *69*, 157–166. [CrossRef]

polymers

MDPI

Article

Selection and Optimization of Carbon-Reinforced Polyether Ether Ketone Process Parameters in 3D Printing—A Rotating Component Application

Raja Subramani [1], Praveenkumar Vijayakumar [2], Maher Ali Rusho [3], Anil Kumar [4], Karthik Venkitaraman Shankar [2,5,*] and Arun Kumar Thirugnanasambandam [1]

1 Center for Sustainable Materials and Surface Metamorphosis, Chennai Institute of Technology, Chennai 600069, India; sraja@citchennai.net (R.S.); arunkumar.t@citchennai.net (A.K.T.)
2 Department of Mechanical Engineering, Amrita Vishwa Vidyapeetham, Amirtapuri 690525, India; praveenkv@am.amrita.edu
3 Lockheed Martin Engineering Management, University of Colorado, Boulder, CO 80308, USA; maru4732@colorado.edu
4 Department of Mechanical Engineering, Kamala Nehru Institute of Technology, Sultanpur 228118, India; anilk@knit.ac.in
5 Centre for Flexible Electronics and Advanced Materials, Amrita Vishwa Vidyapeetham, Amritapuri 690525, India
* Correspondence: karthikvs@am.amrita.edu

Abstract: The selection of process parameters is crucial in 3D printing for product manufacturing. These parameters govern the operation of production machinery and influence the mechanical properties, production time, and other aspects of the final product. The optimal process parameter settings vary depending on the product and printing application. This study identifies the most suitable cluster of process parameters for producing rotating components, specifically impellers, using carbon-reinforced Polyether Ether Ketone (CF-PEEK) thermoplastic filament. A mathematical programming technique using a rating method was employed to select the appropriate process parameters. The research concludes that an infill density of 70%, a layer height of 0.15 mm, a printing speed of 60 mm/s, a platform temperature of 195 °C, an extruder temperature of 445 °C, and an extruder travel speed of 95 mm/s are optimal process parameters for manufacturing rotating components using carbon-reinforced PEEK material.

Keywords: additive manufacturing; Material Extrusion; carbon-reinforced Polyether Ether Ketone; MCDM; Fuzzy-AHP-TOPSIS; process parameter selection; optimization

Citation: Subramani, R.; Vijayakumar, P.; Rusho, M.A.; Kumar, A.; Shankar, K.V.; Thirugnanasambandam, A.K. Selection and Optimization of Carbon-Reinforced Polyether Ether Ketone Process Parameters in 3D Printing—A Rotating Component Application. *Polymers* **2024**, *16*, 1443. https://doi.org/10.3390/polym16101443

Academic Editors: Francesco Mollica and Annalisa Chiappone

Received: 7 April 2024
Revised: 26 April 2024
Accepted: 17 May 2024
Published: 20 May 2024

1. Introduction

Additive manufacturing (AM) technology, as an alternative to subtractive techniques within the manufacturing industry, constructs objects layer by layer and garners substantial interest across various domains. Within this paradigm of 3D printing, objects are generated by directly segmenting their geometries from digital formats and feeding this data into the 3D printing machine. The literature has delineated seven distinct processes within additive manufacturing, categorized based on the raw materials employed and the printing methodologies utilized. The Material Extrusion (MEx) process is the simplest and most prevalent 3D printing technique [1–3]. The Material Extrusion (MEx) process involves using thermoplastic polymer filaments to manufacture the end product. Optimization of MEx process parameters is deemed crucial for product printing, as it directly influences the control of printing machine operations, printing duration, and mechanical properties of the final product. This optimization factor fine-tunes input data for the 3D printing machine following the feed design. Key parameters include printing speed, travel speed, extruder temperature, and printing pattern [4–6]. Previous studies have extensively investigated the selection of process parameters for carbon reinforcement. This section highlights some of the notable research papers in this area. A comprehensive explanation is also provided regarding the reasoning behind the meticulous selection and optimization process of carbon-reinforced polymers, coupled with using mathematical programming techniques that previous researchers have explored or employed to select the final product of rotating components.

Lu et al. [7] conducted a study in which they produced carbon-reinforced PEEK composites in various ratios and examined their mechanical properties. Their findings revealed that blending carbon fibre with PEEK at a ratio of 20 weight % significantly impacts the mechanical properties. Therefore, in the present research, carbon was incorporated into the PEEK material at a lower ratio of 10 weight %. Chidambaram et al. [8] studied the process parameters of layer thickness and printing speed to optimize PEEK material's hardness and wear characteristics. The investigation determined that a printing speed of 20 mm/s and a layer thickness of 0.15 mm resulted in superior hardness and wear behaviour. Sivakumar et al. [9] explored the application of machine learning algorithms to optimize the selection of process parameters for manufacturing final fusion cages using PEEK material. Parameters such as layer height, printing temperature, printing speed, infill density, built orientation, and line width were carefully considered in the investigation. The research determined that an optimal range for layer thickness falls between 0.1 mm and 0.3 mm, printing temperatures range from 370 to 410 degrees Celsius, printing speeds vary from 10 mm/s to 50 mm/s, infill densities range from 40 percent to 100 percent, the optimal built orientation spans from 0 degrees to 90 degrees, and line widths range from 0.1 mm to 0.3 mm.

Kechagias et al. [10] have investigated to ascertain the porosity levels of components produced through 3D printing process parameters. This exploration involved considering various FDM process parameters, including infill rate, infill density, built orientation, part orientation, printing speed, bed temperature, environmental conditions, and nozzle temperature. Ultimately, the study revealed a significant impact of printing parameters on mechanical loading and porosity. Jiang et al. [11] have explored process parameters employing the Taguchi method to attain high-strength products utilizing PEEK material. The Taguchi L9 experiment encompassed crucial process variables such as printing speed, layer thickness, printing temperature, and extruder strand width. The culmination of the investigation revealed that for the production of high-strength final products using PEEK material, optimal parameters include a printing speed of 5 mm/s, a layer thickness of 0.1 mm, an extrusion strand width of 0.4 mm, and a printing temperature of 395 °C.

Liu et al. [12] have investigated the optimization of process parameters for bone manufacturing with PEEK material. Parameters such as layer thickness and infill density were carefully considered in this study. The findings revealed that the optimal process parameters for PEEK material are a layer thickness of 0.43 mm and an infill density of 55 percent. While various studies have explored process parameter selection for PEEK material, there is limited literature on the selection of process parameters for carbon-reinforced PEEK materials. Particularly in applications involving rotating components, no research has yet determined the process parameters for carbon-reinforced PEEK material.

Thermoplastic Polymers in Rotating Component Application

Polymers have been increasingly utilized to manufacture rotors in various systems, enhancing the efficiency of blowers and siphons and improving cost-effectiveness [13–15]. Industry favours these materials for their affordability, ease of manufacturing, water resistance, and adaptability [16,17]. The extrusion process for polymer materials can involve a variety of raw materials, including powders, granules, filaments, and resins, depending on the manufacturing method and raw material used. Thermoplastic polymers such as Acrylonitrile butadiene styrene (ABS) [18], Polylactic Acid (PLA) [19], Polyethylene Terephthalate (PETG) [20], Polyether ether Ketone (PEEK) [21], and Polyphenylene sulfide (PPS) [22] have been utilized in the production of rotating parts. For instance, in a miniature organic rankine cycle (mORC), Hernandez-Carrillo et al. [23] investigated using an ABS impeller. In their study, the traditional load of the impeller was reduced to enhance the mORC's efficiency. Operating conditions for the impeller included an input temperature of 55 °C, a pressure of 4 Bar, an exit temperature of 44.9 °C, and a pressure of 2.5 Bar, with a rotational speed of 36,000 rpm. After considering the operating conditions and the factor of safety (FoS), which is the ratio of yield strength to the maximum equivalent stress, the ABS impeller met the required performance criteria. One significant advantage of using this polymer was its ability to enable mass production of mORCs by reducing the cost of impeller manufacturing using ABS. The impeller's isentropic efficiency was also estimated to be between 76% and 86% based on their simulations for the penta-fluor propane working fluid. However, one limitation noted was the restricted operation of ABS at temperatures above 89.9 °C.

Pavlovic et al. [24] assessed the mechanical properties of ABS for use in impeller pump construction, demonstrating its reliability for impeller manufacturing. Similarly, Polak [25] confirmed the hydraulic qualities of an ABS impeller for a spiral diffusive siphon, noting an efficiency increase at an alternating velocity of 2950 rpm. The remarkably smooth surface of the ABS impeller was identified as a key factor contributing to this efficiency improvement. A thermoplastic material called PLA, derived from renewable resources such as sugar beet or maize starch, offers biodegradability and composability as some of its key attributes. Compared to other polymers, PLA's low cost, environmental friendliness, biocompatibility, and suitable physical and mechanical properties make it a preferable choice. PLA has been used to manufacture impellers for pumps [26,27], compressors [28], and maritime applications [29]. Despite the availability of both PLA and ABS, these two thermoplastic types have been extensively studied as materials for pump impellers. ABS is often chosen due to its higher resistance to stress compared to PLA, which exhibits high fragility in comparison [30].

Birosz et al. [31] examined PLA impellers for blowers, focusing on PLA's tensile strength, creep behaviour, and bending characteristics. Creep testing results indicated that PLA behaved more like a poorly cross-linked elastomer under low loads, with the creep curve maintaining stability over time. This study contributed to the advancement of PLA as a potential material for long-term use. PLA is a thermoplastic that is semi-crystalline. Pump and mORC applications have utilized PLA impellers. For the creation of pump sharp edges, this polymer was chosen on the grounds of its brilliant water resistance and biodegradability. The PLA impeller was concentrated by Odetti et al. [32] for a Pump Jetting Module (PJM) application. A PLA impeller tried for this application showed satisfactory qualities with a speed of turn of 1200 rpm and a push of 14 N.

Zywica et al. [33] conducted a study on plastics and identified them as potential materials for ORC framework impellers. This investigation utilized two thermoplastic polymers: PPS and PEEK. The impeller was subjected to 120,000 cycles per minute. The study concluded that PEEK polymer is well-suited for impeller production based on simulation results pertaining to heat resistance, chemical resistance, tensile strength, and thermal expansion. Rotational components, such as gears, shafts, and impellers, are crucial parts of various mechanical systems where durability, wear resistance, and chemical resistance are paramount. Carbon-reinforced PEEK is a high-performance thermoplastic compared to neat PEEK and offers exceptional mechanical properties, including high strength, stiffness, and fatigue resistance, making it an ideal material for such applications. However, the literature gap regarding CF-PEEK in impeller applications remains unexplored.

Mathematical Programming Technique

The three primary categories for optimization techniques in operations research are statistical methods, stochastic techniques, and mathematical programming techniques. Using a numerical value, the statistical technique selects the best option from the experiment data. Response surface methodology (RSM), Taguchi, and design of experiment (DOE) are a few instances that may be used conventionally. Optimization via approximation theories is known as a stochastic approach. Selecting and optimizing using mathematical programming involves building quantitative numerical models based on qualitative data and solving them. This study employs a mathematical programming technique called multicriteria decision-making (MCDM). Multicriteria decision-making has the structure of selecting the best choice from multiple options based on multiple criteria for a single objective. Also, the commonly used statistical optimization tools (Taguchi method, Design of Experiment, RSM method etc.) and optimal solutions are found after multi-stage algorithms. Optimization, maximization, or minimization is the final result, but as far as the multicriteria decision-making technique is concerned, selection and optimization will be available as the final result. In this, multiple choices will be analyzed using a pair-wise matrix, depending on the objective based on each criterion. Finally, the choices are sorted, and the ranking is given. This helps prioritize and choose choices according to the situation, and a higher ranking indicates maximization or minimization. The accompanying previous research utilized MCDM approaches to select problems in additive manufacturing. In order to choose the best additive manufacturing technology for a customized product in Industry 4.0, Malaga et al. [34] used the hybrid MCDM method. Raja et al. [4] applied the MCDM strategy to choose the appropriate additive manufacturing machine from the 3D printer client's perspective, utilizing the Fuzzy TOPSIS technique. The neutrosophic best-worst MCDM technique was utilized by Ghuge et al. [35] to pick the provider of additive manufacturing services. The hybrid MCDM technique was used by Chandra et al. to choose the sustainable additive manufacturing concept for solving environmental challenges [36].

Raja et al. [5] applied the analytical hierarchy process (AHP) MCDM technique to choose the best Material Extrusion machine in view of the additive manufacturing machine clients' opinions. Subramani et al. [6] explored the different MCDM strategies and accentuated the choice examination issues in additive manufacturing. For that reason, numerous MCDM systems are utilized to address choice challenges in additive manufacturing. The Fuzzy-AHP-TOPSIS strategy again utilizes a five-guide rating framework to assess the algorithm and choices. Moreover, five similar choices were utilized in the examination level headed, and the reason for the simple linguistic terms is that the TOPSIS approach was applied in this research. Comprehensive details will be presented in the forthcoming section.

This study aims to identify the optimal cluster of Material Extrusion (MEx) process parameters for producing high mechanical properties in end products using carbon-reinforced Polyether Ether Ketone (CF-PEEK) filament. This filament is not typically employed in the fabrication of rotating components, specifically impellers. The objective is to achieve superior mechanical properties in a shorter production time. To address these research gaps, this research outlines Part two describes the tests necessary for rotating component applications, along with the material and method. Detailed results and discussions are presented in the final section.

2. Materials and Methods

In this study, we investigate the application of carbon-reinforced polyetheretherketone (PEEK) in additive manufacturing processes to enhance the mechanical properties of manufactured components. The selection of the optimal material and process parameters is crucial to achieving high-quality prints with the desired properties. To address this, we employ the fuzzy analytical hierarchy process (AHP) in conjunction with the technique for order of preference by similarity to the ideal solution (TOPSIS). These methods allow for a comprehensive evaluation and selection of materials and process parameters, considering multiple criteria such as mechanical properties, cost, and environmental impact. This combined approach provides a systematic and effective way to optimize the additive manufacturing process for carbon-reinforced PEEK, contributing to the advancement of polymer-based additive manufacturing technology.

2.1. Assumptions of the Research

This study aims to identify the optimal cluster of printing parameters for carbon-reinforced PEEK based on the analysis of tensile, flexural, and morphology data results. To achieve this objective, the following assumption is made:

- This study considers multiple process parameters that can affect the mechanical properties of the final product. For instance, if the extrusion temperature decreases while the printing speed is high during the fabrication of a product, the final result may be compromised if the raw material does not solidify as expected. Similarly, variations in mechanical characteristics can occur if the printing speed is high, the extruder travel speed is slow, and the infill amount fluctuates.
- The evaluation framework is based on the results of the available data. Criteria 1 (C1) corresponds to the observation of ultimate tensile strength (UTS), Criteria 2 (C2) corresponds to the observation of Young's modulus, Criteria 3 (C3) corresponds to the observation of ultimate flexural test, and Criteria 4 (C4) corresponds to the observation of surface defects. Notably, the options mentioned are labelled explicitly as Sample 1 to Sample 5, rather than Alternative (A1, A2, A3, A4, and A5).
- As reported in previous literature, printing parameter values range from the minimum to maximum values for both PEEK and carbon-reinforced PEEK, refer Table 1.
- Furthermore, the infill printing parameter was set to a normal "line" as per standard practice, although it can be adjusted to accommodate other infill designs such as hexagonal or triangular patterns. Each of the alternatives mentioned represents a distinct set of parameters (referred to as a cluster of process parameters), as outlined in Table 1.

2.2. Material Extrusion Printing

The MEx process transforms a product's design from CAD models or other design software into its physical form. Modern slicing technology (Flash Forge 6.0 Zhejiang Flashforege 3D Technology Ltd., Zhejiang, China) is utilised. STL (Standard Triangle Language) files are divided into layers with varying printing parameters before slicing. This sliced data is then inputted into the MEx machine (Zortrax Endureal 3D printer, Olsztyn, Poland) for printing. The extruder moves the carbon-reinforced PEEK filament from the spool to the melting zone, where it is melted and deposited layer by layer on the build platform according to the input data. Previous studies have investigated polymer composite filaments to analyze the effects of carbon reinforcement on their properties.

Material Extrusion Process Parameter

The fused deposition method (FDM) process parameters have been significantly enhanced, exerting a notable influence on the durability of the final products. Numerous researchers have explored a wide array of controllable characteristics to achieve desired part features, with many convergent on a set of critical factors. The process parameters under scrutiny in this study are delineated below. The internal structure of the printed component is formed using the infill pattern printing technique, offering a variety of filling patterns, including cubic, diamond, triangle, hexagonal, honeycomb, linear, and line. Honeycomb is the optimal inner structure material for superior mechanical qualities, albeit with longer production times [37]. Consequently, employing the appropriate infill pattern without altering the build time or other print quality settings is strongly recommended.

The infill density indicates how much material is printed on a specific component and directly impacts its quality. A fully infilled density, when build delays are inconsequential, and an optimal infill density, when material customization and build times are feasible, yield the best mechanical strength [38]. At lower infill densities, a solid cross-section has a minimal effect on material failure. The volume of material deposited along the machine's vertical axis in a single pass determines the layer height of a Material Extrusion machine.

Before commencing AM, the 3D CAD model must be sliced, as shallow slice heights prolong the construction process, while high slice heights result in a pronounced staircase effect and poor surface quality. Mechanical strength increases with layer height due to decreased void density [39]. Print speed, which influences build time and print quality, along with material deposition quality, plays a crucial role [40]. A faster print speed increases tensile strength due to the speed at which successive layer surfaces bond [41].

The temperature condition denotes the temperature at which the model material is heated by the system, controlling both the amount of molten material extruded from the nozzle and the platform temperature. Extrusion temperature refers to the temperature maintained in the heating head nozzle of the FDM prior to filament extrusion, enhancing fusing within and between layers as the temperature rises [42].

The speed at which an object is built also depends on the extruder's velocity and the parts' orientation in the build bed relative to the X, Y, and Z axes. The Z axis moves toward a part's thickness or height while parallel to the build platform. Part travel speed impacts both mechanical and surface characteristics [43]. Figure 1 presents a schematic diagram illustrating the process parameters of the Material Extrusion method in additive manufacturing.

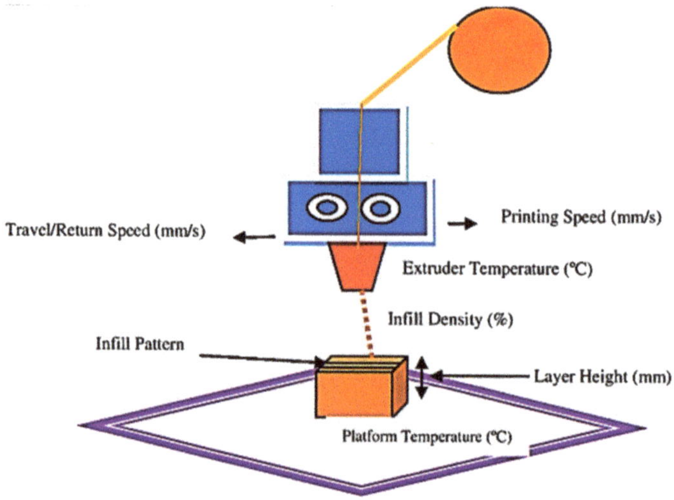

Figure 1. Schematic of MEx Process Parameters.

Table 1. Process parameters are taken for printing CF-PEEK [7–12,21,33].

Process Parameter	Infill Pattern	Layer Height (mm)	Print Speed (mm/s)	Platform Temperature (°C)	Extruder Temperature (°C)	Travel Speed (mm/s)	Infill Density (%)
Alternative 1	Line	0.30	30	180	430	80	55
Alternative 2		0.25	40	185	435	85	60
Alternative 3		0.20	50	190	440	90	65
Alternative 4		0.15	60	195	445	95	70
Alternative 5		0.10	70	200	450	100	75

Figure 2 illustrates the geometric standard for the ASTM D638 type V tensile test [44], while Figure 3 portrays the geometric standard for the ISO-178 flexural test of printed specimens produced on an FDM machine, with parameters outlined in Table 1.

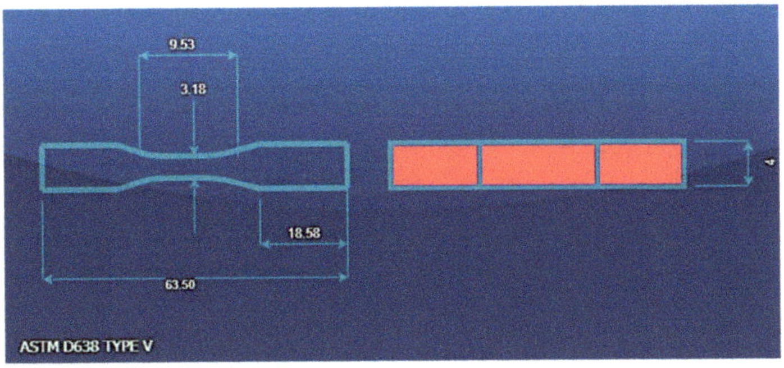

Figure 2. Experimental specimens.

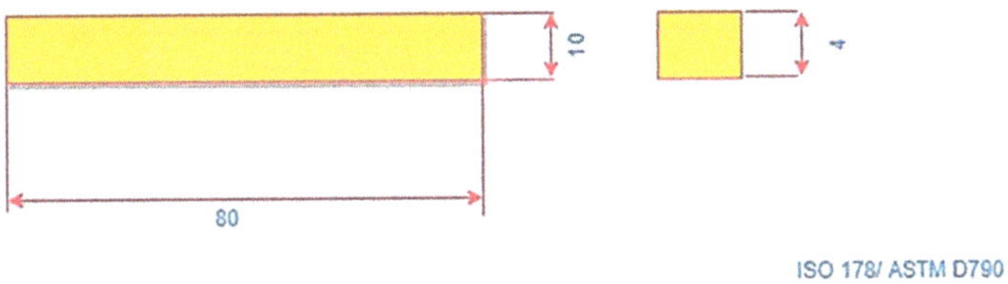

Figure 3. Experimental specimens [45].

2.3. Fuzzy AHP-TOPSIS

Fuzzy-AHP-TOPSIS depends on the rule of tracking down the best option, in contrast to the biggest mathematical separation from the negative ideal solution and the briefest mathematical separation from the positive ideal solution. By assessing how intently the choices look like the best response, Fuzzy-AHP-TOPSIS additionally decides the most ideal choice. The Fuzzy-AHP-TOPSIS approach is applied as continues in this study article:

The Analytic Hierarchy Process (AHP), created by Thomas Saaty in the 1970s, is expanded upon by the Fuzzy Analytic Hierarchy Process (AHP). The Analytic Hierarchy Process (AHP) is a systematic approach to handling complex decisions using a hierarchical structure to represent the decision criteria at various levels of abstraction. It is extensively used in many business, engineering, and healthcare domains for decision-making. Pair-wise comparisons of the alternatives and decision criteria are performed using the Analytic Hierarchy Process (AHP) to create priority scales determining the ultimate choice. Nonetheless, decisions made by decision-makers may not always be exact or free from uncertainty in many real-world scenarios. The Fuzzy AHP, which uses fuzzy logic to handle imprecise and uncertain information, was created to overcome this limitation. By employing linguistic variables such as "very low," "low," "average," "high," and "very high" to characterize the relationships between criteria and alternatives, fuzzy logic enables the representation of ambiguous and subjective judgments [5,42,46]. Instead of using exact numerical values for pair-wise comparisons, fuzzy numbers are used in the Fuzzy AHP. A membership function that gives each linguistic variable a degree of membership defines these fuzzy numbers. The priority scales for the criteria and alternatives are then obtained by combining the comparisons using fuzzy arithmetic operations. Figure 4 illustrates the fundamental concept and hierarchical structure of the research framework's Fuzzy Analytic Hierarchy Process (AHP) Technique for Order of Preference by Similarity to the Ideal Solution (TOPSIS) method.

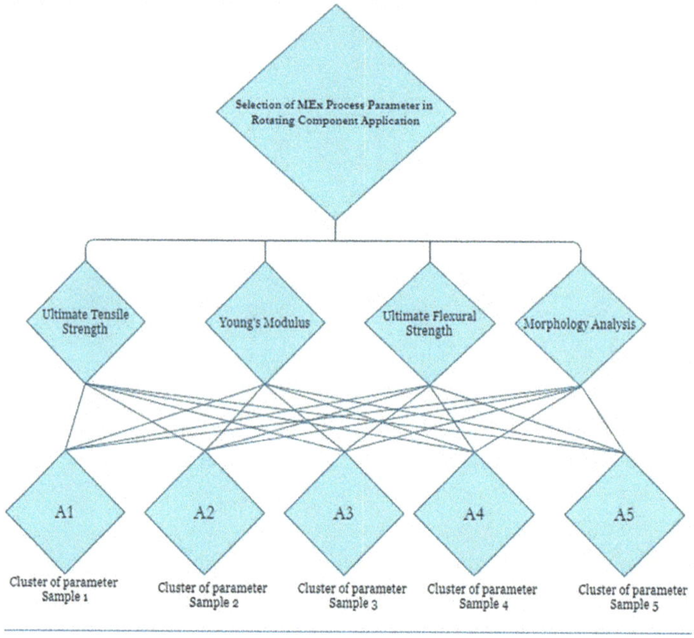

Figure 4. Concept of this research.

In this paper, the triangle membership function is utilized. The accompanying image (μå) is ordinarily used to address a Fuzzy value. Figure 5 depicts the triangle membership function in this exploration.

$$\mu å\ (X) = Å = (1, 2, 3) \tag{1}$$

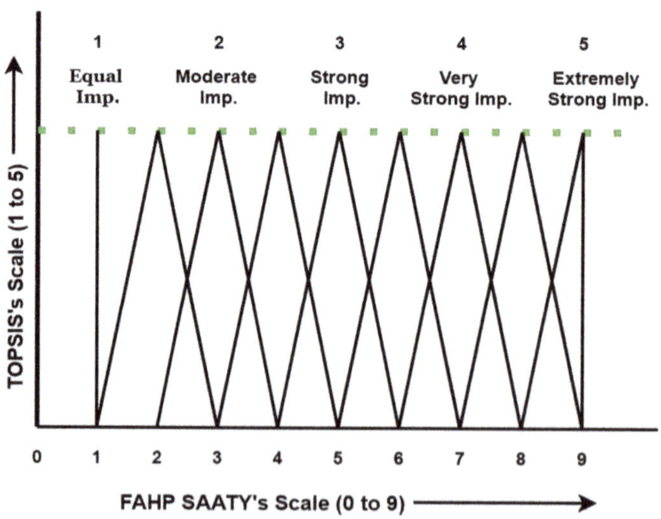

Figure 5. Scale conversion using the triangular membership function.

As per Equation (1), the lowest, middle, and upper fuzzy numbers in triangle membership functions are 1, 2, and 3 separately [4].

This study employs the linguistic terms outlined in Table 2. These terms are consistent with the previously described fuzzy numbers derived from the same triangle membership function.

Table 2. Scale conversion using Triangular Membership Function [4,5].

Saaty Parameters	Saaty Scale	Fuzzified Using Triangle Membership Function
Equal consideration	1	1,1,1
Moderate consideration	3	2,3,4
Strong consideration	5	4,5,6
Very strong consideration	7	6,7,8
Extremely strong consideration	9	9,9,9
Intermediate consideration	2	1,2,3
	4	3,4,5
	6	5,6,7
	8	7,8,9

This study employs various integrated decision-making methods alongside systematic group decision-making. Specifically, linguistic terms are utilized to construct a collaborative decision matrix encompassing alternatives and criteria. Subsequently, the corresponding fuzzy values are adjusted accordingly, as depicted in Table 3.

Table 3. Terms related to linguistics and fuzzy numbers.

Linguistics Terms	Fuzzy Numbers Based on Triangular Membership Function	Linguistics Scales
Very Low (VL)	1,1,3	1
Low (L)	1,3,5	2
Average (A)	3,5,7	3
High (H)	5,7,9	4
Very High (VH)	7,9,9	5

In the initial stage of implementing the Fuzzy Analytic Hierarchy Process (AHP) and Technique for Order of Preference by Similarity to the Ideal Solution (TOPSIS) method, the first step involves constructing a Pair-wise matrix using the provided data. Subsequently, the process normalises the decision matrix, constituting the second step. Following this, the third step involves distinguishing between beneficial and non-beneficial criteria. Using these identified criteria, the subsequent step involves determining the fuzzy positive ideal solution and the fuzzy negative ideal solution. Once these solutions are established, the differences between them are computed, leading to the calculation of the closeness coefficient. This detailed procedure is exclusively elaborated upon in the discussion section, which focuses on the results.

2.4. Mechanical Testing

Tensile testing was conducted using the Tinius Olsen H10KL machine, employing each set of parameters for three specimens, totalling 15 specimens across five parameter sets. Flexural tests were employed to evaluate both the composites' flexural strength and the failure surface's characteristics. Through these tests, the flexural stress of each specimen was measured across three trials. Criteria 3 was established to evaluate the different samples based on the average flexural test results and the flexural modulus findings. To ensure consistent testing conditions, the span length of the specimen was multiplied by four times its thickness, maintaining an equal distance from the bottom solid support.

2.5. Morphology Analysis

The sample prepared with revised set of printing parameter was analyzed using FESEM (Field Emission Scanning Electron Microscopy). This analysis revealed microstructural imperfections in the samples. Subsequently, morphology analysis was conducted to evaluate criteria 4, assigning points on a linguistic scale ranging from 1 to 5.

3. Results and Discussion

As per prior research, ultimate tensile strength (UTS) significantly influences tensile outcomes. The values for Young's modulus and ultimate tensile strength (UTS) are presented in Tables 4 and 5, respectively. Trials A, B, and C were performed on each specimen, and their average results for assessing the criteria using linguistic terms were calculated for both Young's modulus and UTS.

Table 4. Observation of Young's Modulus (in GPa).

	A1	A2	A3	A4	A5
A	3.61	3.25	3.71	3.05	3.17
B	3.54	3.21	3.64	2.99	3.08
C	3.49	3.11	3.69	3.01	3.01
Average	3.54	3.19	3.68	3.01	3.08
Importance	H	A	VH	VL	L

Table 5. Observation of UTS (in MPa).

	A1	A2	A3	A 4	A5
A	82.1	83.7	80.2	80.3	79.8
B	82.6	84.2	80.6	79.6	78.6
C	83.4	84.1	79.8	79.2	78.2
Average	82.7	84	80.2	79.7	78.8
Importance	H	VH	A	L	VL

According to the observations in Table 4, A4 exhibits the lowest average Young's modulus value, whereas Sample 3 demonstrates the highest average value.

The observation of Table 5 indicates that A2 exhibits the highest average UTS value, whereas A5 demonstrates the lowest value. As per the linguistic term scale, A2 was rated very high (VH) and A5 was rated Very Low (VL).

According to Young's modulus, based on the available data, Alternative 3 exhibits the highest strength at 3.68 GPa, securing the top position, followed by Alternative 1 at 3.54 GPa, ranking second, Alternative 2 at 3.19 GPa, ranking third, Alternative 5 at 3.08 GPa, ranking fourth, and Alternative 4 at 3.01 GPa, ranking fifth. Consequently, the TOPSIS Linguistics scale attributes a significance score of 5 to high-strength Alternative 3, indicating its utmost importance, while assigning a score of 1 to low-strength Alternative 4, reflecting its comparatively lesser significance.

Consistent with this, Alternative 2 exhibited the highest strength (84 MPa), followed by Alternative 1 in second place (82.7 MPa), Alternative 3 in third place (80.2 MPa), Alternative 4 in fourth place (79.7 MPa), and Alternative 5 in fifth place (78.8 MPa). This indicates that for the high-strength Alternative 2, the TOPSIS Linguistic scale assigned very high importance (5 points), while for the low-strength Alternative 2, it assigned very low importance (1 point).

The assessment of flexural performance heavily relies on the ultimate flexural strength. Table 6 displays the average ultimate flexural strength values for each specimen. Additionally, Table 6 includes linguistic terms corresponding to these average values. The analysis indicates that Sample 2 exhibits superior performance in terms of ultimate flexural strength.

Table 6. Observation of Ultimate Flexure Strength (in MPa).

	A1	A2	A3	A4	A5
A	148	149.6	144.5	140.3	138.6
B	147.3	148.2	143.6	141.2	137.9
C	146.9	148.9	144.1	139.7	138.1
Average	147.4	148.9	144.06	140.4	138.2
Importance	H	VH	A	L	VL

In the tested alternatives, alternatives 2 demonstrated the highest strength at 148.9 MPa, followed by alternatives 1 at 147.4 MPa, alternatives 3 at 144.06 MPa, alternatives 4 at 140.4 MPa, and alternatives 5 at 138.2 MPa. Consequently, the Fuzzy-AHP-TOPSIS linguistic scale assigned a high importance rating of 5 points to Alternative 2 for its high strength, while Alternative 5 received a low importance rating of 1 point due to its lower strength.

In terms of morphological outcomes, this study focuses on surface roughness and defects. Based on the morphological analysis, it was found that A5 exhibits fewer defects and less pronounced line pattern dents, attributed to the extruder melting temperature. Conversely, the surface of A1 shows defects and improper infill, also due to the extruder melting temperature. Consequently, A5 has been assigned a very high (VH) priority. In Figure 6a, the microstructure of Alternative 1 is presented, followed by the microstructures of Alternative 2 in Figure 6b, Alternative 3 in Figure 6c, Alternative 4 in Figure 6d, and Alternative 5 in Figure 6e. During FESEM analysis, the printed carbon-reinforced PEEK samples were coated with gold sputter, as depicted in Figure 6f. This process enhances the conductivity of non-conductive materials.

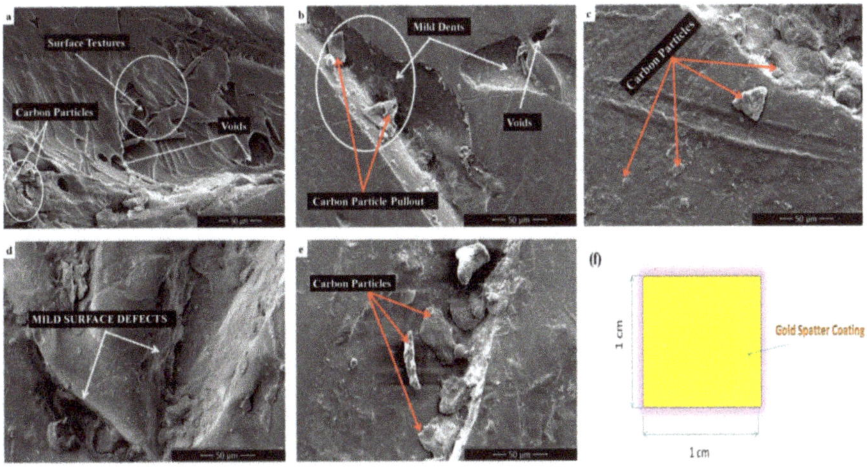

Figure 6. Morphology analysis of printed alternatives (**a**) A 1 (**b**) A 2 (**c**) A 3 (**d**) A 4 (**e**) A 5 (**f**) Geometric of gold-sputter-coated samples.

Alternative 5 exhibited a smooth surface finish with fewer defects compared to Alternative 3. Alternative 2 showed a slightly smoother surface finish with fewer lines than Alternative 4. However, Alternative 1 presented the roughest surface among the specimens, ranking fifth due to numerous surface flaws such as pores, gaps, and other imperfections. These rankings are consistent with the TOPSIS Linguistic Scale, which assigns very high importance (5 points) to Alternative 5 for its high strength and very low importance (1 point) to Alternative 1 for its lower strength. Table 7 illustrates the pair-wise matrix derived from observational data.

Table 7. Linguistic Scale of Evaluation matrix.

	C1	C2	C3	C4
Alternative 1	H	H	H	VL
Alternative 2	VH	A	VH	A
Alternative 3	A	VH	A	H
Alternative 4	L	VL	L	L
Alternative 5	VL	L	VL	VH

The fuzzified values of 5, 7, and 9 in the first column and first row of Table 8 represent Alternative 1 (Sample 1), which is deemed highly important (H) relative to Criteria 1.

Table 8. Fuzzified Evaluation Matrix.

		C1			C2			C3			C4	
Alternative 1	5	7	9	5	7	9	5	7	9	1	1	3
Alternative 2	7	9	9	3	5	7	7	9	9	3	5	7
Alternative 3	3	5	7	7	9	9	3	5	7	5	7	9
Alternative 4	1	3	5	1	1	3	1	3	5	1	3	5
Alternative 5	1	1	3	1	3	5	1	1	3	7	9	9

Subsequently, the total number of individual values is normalized by dividing it by the maximum value of each column. Table 9 illustrates the FPIS (A^+), which is derived by selecting the maximum value from each column representing beneficial criteria. This study employed useful criteria and morphological criteria to enhance mechanical properties. The subsequent step involves minimizing each column's cost criterion (A^-) to ascertain the fuzzy negative ideal solution.

Table 9. Compute A^+ and A^-.

		C1			C2			C3			C4	
A1	0.555	0.777	1	0.555	0.777	1	0.555	0.777	1	0.111	0.111	0.333
A2	0.777	1	1	0.333	0.555	0.777	0.777	1	1	0.333	0.555	0.777
A3	0.333	0.555	0.777	0.777	1	1	0.333	0.555	0.777	0.555	0.777	1
A 4	0.111	0.333	0.555	0.111	0.111	0.333	0.111	0.333	0.555	0.111	0.333	0.555
A 5	0.111	0.111	0.333	0.111	0.333	0.555	0.111	0.111	0.333	0.777	1	1
A^+	0.777	1	1	0.777	1	1	0.777	1	1	0.777	1	1
A^-	0.111	0.111	0.333	0.111	0.111	0.333	0.111	0.111	0.333	0.111	0.111	0.333

Table 10 illustrates the distance of each alternative from the fuzzy positive ideal solution (FPIS), while Table 11 displays the distance of each alternative from the fuzzy ideal solution. Equation (2) below can be utilized to compute this distance [4].

$$d\,(\dot{z},) = sq\sqrt{((a_1 - a_2)^2 + (b_1 - b_2)^2 + (c_1 - c_2)^2)} \quad (2)$$

Table 10. Compute the FPIS (A^+).

	C1	C2	C3	C4	di^+
Alternative 1	0	0.181	0	0.748	0.929
Alternative 2	0	0.384	0	0.384	0.769
Alternative 3	0.384	0	0.384	0.181	0.951
Alternative 4	0.379	0.748	0.601	0.601	2.331
Alternative 5	0.748	0	0.748	0	1.496

Table 11. Compute the FNIS (A^-).

	C1	C2	C3	C4
Alternative 1	0.601	0	0.601	0
Alternative 2	0	0.384	0.748	0.384
Alternative 3	0.384	0.748	0.384	0.601
Alternative 4	0.181	0	0	0
Alternative 5	0	0.181	0	0.748

The FPIS is determined by selecting the value of a1 for each column if a_1 represents the A^+ value of each column and a_2 represents the individual value of each column. Similarly, to calculate the fuzzy negative ideal solution (FNIS), the values of each column are substituted by A1 and A2. The resulting totals are presented in Tables 11 and 12.

Table 12. Decision Matrix.

	Closeness Co-Efficient (Cci)	Rank
Alternative 1	0.564	III
Alternative 2	0.663	II
Alternative 3	0.060	V
Alternative 4	0.722	I
Alternative 5	0.383	IV

The decision matrix below was constructed after thoroughly analysing the criteria and alternatives. Table 12 displays the ranking of the alternatives based on the coefficient of closeness, as calculated by the formula employed.

Alternative 4 exhibited the highest coefficient of closeness compared to other samples, as indicated by the observations. The parameters of Alternative 4 demonstrated the highest mechanical properties among all samples for manufacturing impeller applications. The fourth set of parameters, comprising an infill density of 70%, a layer height of 0.15 mm, a printing speed of 60 mm/s, a platform temperature of 195 °C, an extruder temperature of 445 °C, and an extruder travel speed of 95 mm/s, were found to be the most suitable for CF-PEEK filament used in the production of rotating components.

Sensitivity analysis involves adjusting the weights of criteria to observe how the ranking of alternatives changes. This process helps ensure the reliability of the solutions and the ranking of alternatives by testing their robustness. Accordingly, altering the weight given to tensile and flexural properties does not change the ranking obtained. Similarly, increasing the weighting of tensile and surface defects also does not change the ranking. These sensitivity analyses confirm the correctness of the ranking, as it depicted in Table 13. Table 14 compares rankings between the PSI MCDM and fuzzy AHP-TOPSIS rating techniques. Maniya and Bhatt [47] used the preference selection index method to calculate the final preference score for each alternative by determining the preference value between each alternative and its variant. Overall, the comparative analysis has helped to elucidate the differences in rankings obtained from various MCDM techniques. In this case, the rankings from different MCDM techniques are similar to those from the fuzzy AHP-TOPSIS rating technique.

Table 13. Sensitivity Assessment.

	Set 1—High Weightage to C1, C2	Set 2—High Weightage to C1, C2, C3	Set 3—High Weightage to C1, C2, C4	Set 4 (Original)—Equal Weightage to All Criteria	Rank
Alternative 1	0.589	0.498	0.509	0.564	III
Alternative 2	0.699	0.572	0.589	0.663	II
Alternative 3	0.092	0.049	0.054	0.060	V
Alternative 4	0.813	0.633	0.676	0.722	I
Alternative 5	0.412	0.282	0.302	0.383	IV

Table 14. Comparative Analysis Using the Preference Selection Index Method.

	Preference Selection Index (PSI) Method	Fuzzy-AHP TOPSIS (This Method)	Rank
Alternative 1	0.522	0.564	III
Alternative 2	0.654	0.663	II
Alternative 3	0.085	0.060	V
Alternative 4	0.746	0.722	I
Alternative 5	0.392	0.383	IV

In this study, the validation of the data is based on the specific criteria applied, which aligns with previous research where tensile and flexural properties have been prioritized as key mechanical properties for impeller production. This approach ensures the validity of the selected criteria. The chosen process parameters are also validated, as previous researchers commonly employ them to determine suitable process parameters in FDM printing, which directly impact the mechanical properties of the printed parts. The selected optimization tool for decision-making is highly suitable, offering a slightly simpler procedure compared to other MCDM tools without compromising reliability. Furthermore, the research model has been validated by experts in the MCDM field, adding an additional layer of credibility to the study.

4. Conclusions

Various studies have explored the selection of process parameters in additive manufacturing. However, this research specifically focuses on determining the suitable process parameters for manufacturing rotating components using carbon-reinforced PEEK polymer, incorporating new and innovative assumptions. Unlike conventional approaches that rely on maximization or minimization after multi-stage optimization, this study employs a multicriteria decision-making method. This method effectively selects optimal process parameters by considering multiple criteria and alternative parameter clusters. The mechanical properties are used as criteria, while different clusters of parameters serve as alternatives. Based on previous research findings regarding mechanical properties, the study identifies the best process parameters from five different samples, ranging from the minimum to the maximum values for the selected material. Moreover, the criteria are selected based on the application of rotating components, ensuring the importance of these criteria and alternatives in the final product production, as validated by previous studies. Furthermore, the model developed in this research is tested and validated on another material with the same process parameters. Sensitivity analysis confirms the accuracy and consistency of the decision-making and optimization models. For instance, when assigning equal importance to all criteria, Alternative 4 is identified as suitable for carbon fiber PEEK. Even when different weight ages are assigned to criteria, this result remains unchanged, demonstrating the model's reliability. Future researchers can use similar approaches to conduct selection and optimization studies with different materials or alternative multicriteria decision-making methods.

Author Contributions: Conceptualization, R.S.; Methodology, R.S.; Software, R.S.; Validation, R.S and A.K.; Investigation, R.S., A.K. and A.K.T.; Resources, R.S.; Data curation, P.V., A.K. and K.V.S.; Writing—original draft, R.S.; Writing—review & editing, M.A.R., A.K., K.V.S. and A.K.T.; Visualization, P.V.; Supervision, A.K. and K.V.S.; Funding acquisition, A.K., A.K.T. and K.V.S. All authors have read and agreed to the published version of the manuscript.

Funding: This work is partially funded by the Center for Sustainable Materials and Surface Metamorphosis. Chennai Institute of Technology, India, vide funding number CIT/CSMSM/2024/RP/002.

Institutional Review Board Statement: Not applicable.

Data Availability Statement: Data are contained within the article.

Conflicts of Interest: The authors declare no conflict of interest.

References

1. Marino, S.G.; Košťáková, E.K.; Czél, G. Development of pseudo-ductile interlayer hybrid composites of standard thickness plies by interleaving polyamide 6 nanofibrous layers. *Compos. Sci. Technol.* **2023**, *234*, 109924. [CrossRef]
2. Adapa, S.K. Jagadish Prospects of Natural Fiber-Reinforced Polymer Composites for Additive Manufacturing Applications: A Review. *JOM* **2023**, *75*, 920–940. [CrossRef]
3. Subramani, R.; Kaliappan, S.; Sekar, S.; Patil, P.P.; Usha, R.; Manasa, N.; Esakkiraj, E.S. Polymer Filament Process Parameter Optimization with Mechanical Test and Morphology Analysis. *Adv. Mater. Sci. Eng.* **2022**, *2022*, 8259804. [CrossRef]
4. Raja, S.; Rajan, A.J. A Decision-Making Model for Selection of the Suitable FDM Machine Using Fuzzy TOPSIS. *Math. Probl. Eng.* **2022**, *2022*, 7653292. [CrossRef]
5. Raja, S.; Rajan, A.J.; Kumar, V.P.; Rajeswari, N.; Girija, M.; Modak, S.; Kumar, R.V.; Mammo, W.D. Selection of Additive Manufacturing Machine Using Analytical Hierarchy Process. *Sci. Program.* **2022**, *2022*, 1596590. [CrossRef]
6. Subramani, R.; Kaliappan, S.; Arul, P.V.; Sekar, S.; Poures, M.V.; De Patil, P.P.; Esakki, E.S. A Recent Trend on Additive Manufacturing Sustainability with Supply Chain Management Concept, Multicriteria Decision Making Techniques. *Adv. Mater. Sci. Eng.* **2022**, *2022*, 9151839. [CrossRef]
7. Lu, S.; Zhang, B.; Niu, J.; Yang, C.; Sun, C.; Wang, L.; Li, D. Effect of fiber content on mechanical properties of carbon fiber-reinforced polyether-ether-ketone composites prepared using screw extrusion-based online mixing 3D printing. *Addit. Manuf.* **2024**, *80*, 103976. [CrossRef]
8. Chithambaram, K.; Senthilnathan, N. Effects of printing parameters on hardness and wear characteristics of 3D printed polyetheretherketone (PEEK) polymer. *Mater. Lett.* **2024**, *356*, 135588. [CrossRef]
9. Sivakumar, N.K.; Palaniyappan, S.; Bodaghi, M.; Azeem, P.; Nandhakumar, G.; Basavarajappa, S.; Pandiaraj, S.; Hashem, M.I. Predictive modeling of compressive strength for additively manufactured PEEK spinal fusion cages using machine learning techniques. *Mater. Today Commun.* **2024**, *38*, 108307. [CrossRef]
10. Kechagias, J.; Zaoutsos, S. Effects of 3D-printing processing parameters on FFF parts' porosity: Outlook and trends. *Mater. Manuf. Process.* **2024**, *39*, 804–814. [CrossRef]
11. Jiang, C.-P.; Cheng, Y.-C.; Lin, H.-W.; Chang, Y.-L.; Pasang, T.; Lee, S.-Y. Optimization of FDM 3D printing parameters for high strength PEEK using the Taguchi method and experimental validation. *Rapid Prototyp. J.* **2022**, *28*, 1260–1271. [CrossRef]
12. Liu, H.; Cheng, X.; Yang, X.H.; Zheng, G.M.; Guo, Q.J. Experimental study on parameters of 3D printing process for PEEK materials. In Proceedings of the IOP Conference Series: Materials Science and Engineering, the 2nd International Workshop on Materials Science and Mechanical Engineering (IWMSME2018), Qingdao, China, 26–28 October 2018; IOP Publishing: Bristol, UK, 2018; Volume 504, p. 012001. [CrossRef]
13. Olaiya, N.G.; Maraveas, C.; Salem, M.A.; Raja, S.; Rashedi, A.; Alzahrani, A.Y.; El-Bahy, Z.M.; Olaiya, F.G. Viscoelastic and Properties of Amphiphilic Chitin in Plasticised Polylactic Acid/Starch Biocomposite. *Polymers* **2022**, *14*, 2268. [CrossRef] [PubMed]
14. Raja, S.; Logeshwaran, J.; Venkatasubramanian, S.; Jayalakshmi, M.; Rajeswari, N.; Olaiya, N.G.; Mammo, W.D. OCHSA: Designing Energy-Efficient Lifetime-Aware Leisure Degree Adaptive Routing Protocol with Optimal Cluster Head Selection for 5G Communication Network Disaster Management. *Sci. Program.* **2022**, *2022*, 5424356. [CrossRef]
15. Praveenkumar, V.; Raja, S.; Jamadon, N.H.; Yishak, S. Role of laser power and scan speed combination on the surface quality of additive manufactured nickel-based superalloy. *Inst. Mech. Eng. Part L J. Mater. Des. Appl.* **2023**, *in press*. [CrossRef]
16. Raja, S.; Rajan, J. Selection of Polymer Extrusion Parameters By Factorial Experimental Design—A Decision Making Model. *Sci. Iran.* **2023**, *in press*.
17. Subramani, R.; Kalidass, A.K.; Muneeswaran, M.D.; Lakshmipathi, B.G. Effect of fused deposition modeling process parameter in influence of mechanical property of acrylonitrile butadiene styrene polymer. *Appl. Chem. Eng.* **2024**, *7*, 3576. [CrossRef]
18. Andrés, M.S.; Chércoles, R.; Navarro, E.; de la Roja, J.M.; Gorostiza, J.; Higueras, M.; Blanch, E. Use of 3D printing PLA and ABS materials for fine art. Analysis of composition and long-term behaviour of raw filament and printed parts. *J. Cult. Herit.* **2023**, *59*, 181–189. [CrossRef]
19. Raja, S.; Agrawal, A.P.; Patil, P.P.; Thimothy, P.; Capangpangan, R.Y.; Singhal, P.; Wotango, M.T. Optimization of 3D Printing Process Parameters of Polylactic Acid Filament Based on the Mechanical Test. *Int. J. Chem. Eng.* **2022**, *2022*, 5830869. [CrossRef]

20. Joseph, T.M.; Kallingal, A.; Suresh, A.M.; Mahapatra, D.K.; Hasanin, M.S.; Haponiuk, J.; Thomas, S. 3D printing of polylactic acid: Recent advances and opportunities. *Int. J. Adv. Manuf. Technol.* **2023**, *125*, 1015–1035. [CrossRef]
21. Pulipaka, A.; Gide, K.M.; Beheshti, A.; Bagheri, Z.S. Effect of 3D printing process parameters on surface and mechanical properties of FFF-printed PEEK. *J. Manuf. Process.* **2023**, *85*, 368–386. [CrossRef]
22. French, A.; Anguiano, S.; Bliss, M.; Christ, J.; di Vacri, M.; Erikson, R.; Harouaka, K.; Hoppe, E.; Grate, J.; Arnquist, I. Mass spectrometric investigations into 3D printed parts to assess radiopurity as ultralow background materials for rare event physics detectors. *Nucl. Instrum. Methods Phys. Res. Sect. A Accel. Spectrometers Detect. Assoc. Equip.* **2023**, *1047*, 167830. [CrossRef]
23. Hernandez-Carrillo, I.; Wood, C.J.; Liu, H. Advanced materials for the impeller in an ORC radial microturbine. *Energy Procedia* **2017**, *129*, 1047–1054. [CrossRef]
24. Pavlović, A.; Šljivić, M.; Kraisnik, M.; Ilić, J.; Anić, J. Polymers in additive manufacturing: The case of a water pump impeller. *FME Trans.* 2017; 45, 354–359.
25. Polák, M. Behaviour of 3D printed impellers in performance tests of hydrodynamic pump. In Proceedings of the 7th International Conference on Trends in Agricultural Engineering, Prague, Czech Republic, 17–20 September 2019; pp. 17–20.
26. Kyzyrov, U.; Turgali, D. Performance Enhancement of a Centrifugal Pump by Impeller Retrofitting; Nazarbayev University School of Engineering and Digital Sciences: 2019. Available online: https://nur.nu.edu.kz/bitstream/handle/123456789/4476/Performance%20Enhancement%20of%20a%20Centrifugal%20Pump%20by%20Impeller%20Retrofitting.pdf;jsessionid=3A3B30B32BD593170B53A63F8FFC63C9?sequence=5 (accessed on 3 December 2023).
27. Kopparapu, R.; Mathew, S.; Siciliano, E.; Stasick, G.; Dias, M. Designing a Centrifugal Pump System for High Altitude Water Crises. 2017. Available online: https://engineering.rutgers.edu/sites/default/files/imce/gov2017/Designing%20a%20Centrifugal%20Pump%20System%20for%20High%20Altitude%20Water%20Crises.pdf.pdfing.rutgers.edu (accessed on 10 December 2023).
28. Warner, J.; Celli, D.; Scott-Emuakpor, O.; George, T.; Tomlin, T. Fused Deposition Modeling Fabrication Evaluation of a Ti-6Al-4V Centrifugal Compressor. *J. Eng. Gas Turbines Power* **2022**, *145*, 031008. [CrossRef]
29. Matos, T.; Pinto, V.; Sousa, P.; Martins, M.; Fernández, E.; Henriques, R.; Gonçalves, L.M. Design and In Situ Validation of Low-Cost and Easy to Apply Anti-Biofouling Techniques for Oceanographic Continuous Monitoring with Optical Instruments. *Sensors* **2023**, *23*, 605. [CrossRef] [PubMed]
30. Mishra, V.; Negi, S.; Kar, S. FDM-based additive manufacturing of recycled thermoplastics and associated composites. *J. Mater. Cycles Waste Manag.* **2023**, *25*, 758–784. [CrossRef] [PubMed]
31. Birosz, M.T.; Andó, M.; Jeganmohan, S. Finite Element Method modeling of Additive Manufactured Compressor Wheel. *J. Inst. Eng. Ser. D* **2021**, *102*, 79–85. [CrossRef]
32. Odetti, A.; Altosole, M.; Bruzzone, G.; Caccia, M.; Viviani, M. Design and construction of a modular pump-jet thruster for autonomous surface vehicle operations in extremely shallow water. *J. Mar. Sci. Eng.* **2019**, *7*, 222. [CrossRef]
33. Zywica, G.; Kaczmarczyk, T.Z.; Ihnatowicz, E.; Baginski, P.; Andrearczyk, A. Application Of a heat resistant plastic IN a high-speed microturbine designed for the domestic ORC system. In Proceedings of the 5th International Seminar on ORC Power Systems, Athens, Greece, 9–11 September 2019; pp. 1–8.
34. Malaga, A.; Vinodh, S. Technology Selection for Additive Manufacturing in Industry 4.0 Scenario Using Hybrid MCDM Approach. In *Industry 4.0 and Advanced Manufacturing*; Springer: Singapore, 2023; pp. 207–217.
35. Ghuge, S.; Parhi, S. Additive Manufacturing Service Provider Selection Using a Neutrosophic Best Worst Method. *Procedia Comput. Sci.* **2023**, *217*, 1550–1559. [CrossRef]
36. Chandra, M.; Shahab, F.; Kek, V.; Rajak, S. Selection for additive manufacturing using hybrid MCDM technique considering sustainable concepts. *Rapid Prototyp. J.* **2022**, *28*, 1297–1311. [CrossRef]
37. Rinaldi, M.; Ghidini, T.; Cecchini, F.; Brandao, A.; Nanni, F. Additive layer manufacturing of poly (ether ether ketone) via FDM. *Compos. Part B Eng.* **2018**, *145*, 162–172. [CrossRef]
38. Algarni, M.; Ghazali, S. Comparative Study of the Sensitivity of PLA, ABS, PEEK, and PETG's mechanical properties to FDM printing process parameters. *Crystals* **2021**, *11*, 995. [CrossRef]
39. Fountasa, N.; Kechagiasb, J.; Vaxevanidisa, N. Statistical Modeling and Optimization of Surface Roughness for PLA and PLA/Wood FDM Fabricated Items. *J. Mater. Eng.* **2023**, *1*, 38–44. [CrossRef]
40. Mustafa, M.A.; Raja, S.; Asadi, L.A.A.L.; Jamadon, N.H.; Rajeswari, N.; Kumar, A.P. A Decision-Making Carbon Reinforced Material Selection Model for Composite Polymers in Pipeline Applications. *Adv. Polym. Technol.* **2023**, *2023*, 6344193. [CrossRef]
41. Shekar, A.C.; Djilani, A.H.; Zitoune, R.; Toubal, L.; Hof, L.A. Effect of input variables on the mechanical properties of additively manufactured PEEK thermoplastics. *Mater. Today Proc.* **2023**, *in press*. [CrossRef]
42. Raja, S.; Al-Tmimi, H.M.; Ghadir, G.K.; Mustafa, M.A.; Alani, Z.K.; Rusho, M.A.; Rajeswari, N. An analysis of polymer material selection and design optimization to improve Structural Integrity in 3D printed aerospace components. *Appl. Chem. Eng.* **2024**, *7*, 1875. [CrossRef]
43. Patel, A.; Taufik, M. Extrusion-Based Technology in Additive Manufacturing: A Comprehensive Review. *Arab. J. Sci. Eng.* **2024**, *49*, 1309–1342. [CrossRef]
44. *ASTM D638-14*; Standard Practice for Preparation of Metallographic Specimens. ASTM International: West Conshohocken, PA, USA, 2016; pp. 1–15.
45. *ASTM D790*; Standard Test Methods for Flexural Properties of Unreinforced and Reinforced Plastics and Electrical Insulating Materials. ASTM International: West Conshohocken, PA, USA, 2002; pp. 1–12.

46. Raja, S.; Rajan, A.J. Challenges and Opportunities in Additive Manufacturing Polymer Technology: A Review Based on Optimization Perspective. *Adv. Polym. Technol.* **2023**, *2023*, 8639185. [CrossRef]
47. Maniya, K.; Bhatt, M. An alternative multiple attribute decision making methodology for solving optimal facility layout design selection problems. *Comput. Ind. Eng.* **2011**, *61*, 542–549. [CrossRef]

MDPI AG
Grosspeteranlage 5
4052 Basel
Switzerland
Tel.: +41 61 683 77 34

Polymers Editorial Office
E-mail: polymers@mdpi.com
www.mdpi.com/journal/polymers